MECHANICS OF OPTIMAL STRUCTURAL DESIGN

MECHANICS OF OPTIMAL STRUCTURAL DESIGN

Minimum Weight Structures

David W. A. Rees

School of Engineering and Design, Brunel University, Uxbridge, UK

WILEY

A John Wiley and Sons, Ltd., Publication

This edition first published 2009

Registered office
John Wiley & Sons Ltd, The Atrium, Southern Gate, Chichester, West Sussex, PO19 8SQ, United Kingdom

For details of our global editorial offices, for customer services and for information about how to apply for permission to reuse the copyright material in this book please see our website at www.wiley.com.

Library of Congress Cataloguing-in-Publication Data

Rees, D. W. A. (David W. A.), 1947-
Mechanics of optimal structural design : minimum weight structures / David Rees.
p. cm.
Includes bibliographical references and index.
ISBN 978-0-470-74623-3 (cloth)
1. Lightweight construction. I. Title.
TA663.R43 2009
624.1′7–dc22

2009025913

A catalogue record for this book is available from the British Library.
ISBN: 978-0-470-74623-3

Typeset in 10/12 Times by Laserwords Private Limited, Chennai, India.
Printed and bound in Singapore by Markono

Contents

Preface

Design is the process of finding a solution to a problem, with all the constraints and requirements that it presents. Accordingly, there would normally be more than one design solution, depending upon the requirements perceived from the application. Specifically, an optimum design will narrow the solutions to one or two depending upon the criteria adopted. For example, the criteria may relate to structural form, the material characteristics, ease of manufacture and assembly, a demand for standard parts, or a specified mechanical performance involving aerodynamics, heat transfer, wear resistance, etc. In principle, the designer would wish to optimise some merit factor common to all of these. In practice, an optimisation can be achieved when related items are considered together but more usually specific requirements are optimised in isolation, as with attaining minimal waste in machining and maximal heat transfer in material selection. In this book we are concerned with minimising the weight of load bearing structures through making informed choices upon their structural forms. In fact, we find that this problem can be treated in terms of its building blocks. This amounts to optimising all the key elements of a structure – struts, ties, beams, shear panels, plates, etc. – in terms of their strength and weight contributions.

An almost infinite number of structural forms can bear load. Whilst a strut, a beam, a braced frame, a tripod would usually be adequate, the choice rests with which is best for the particular application. In the fields of aerospace, transport and structural engineering the reduction in weight has become a key factor with their associated environmental issues. Quite apart from the obvious reduction in material cost, the improvement in performance from the reduction in weight becomes an overriding factor when designing to reduce emissions in road vehicles and aircraft. In this context the usual reference is made to a strength-to-weight ratio, which implies a maximising of strength for a given weight. In fact, it is the weight that is to be minimised with respect to a required strength and so the term weight-to-strength is a more precise description of the approach to be adopted here.

Though the structures considered in this book are many and varied, they share the common theme of optimising their weight to strength. The question as to whether an optimised design exists for thin-walled tubes, Ts, Is and channel sections is answered by applying additional design criteria involving both global and local buckling. Generally, the solution is based upon similar loading stresses from all sources being reached simultaneously within each design criterion adopted. For example, in a thin-walled open strut, with an I or T cross-section, Euler's global stress for flexural buckling in the length and the local buckling stress for the sections' flanges are made to coincide with the limiting design stress. Struts as compression members, both wide and narrow, appear typically as columns, pillars and walls (Chapters 1 and 2). Beams and bar sections under uniform bending moments and torques are treated in Chapters 3 and 4. Various types of strut and beam sections are covered. Solid sections in standard shapes are usually more amenable to an optimum design, but tubular and thin-walled sections obviously present less weight at the outset. Thin sections present wall thickness as a further variable to an optimum design and an additional, local buckling failure mode within their walls.

Transverse shear and torsion in thin-walled tubes are given special consideration (Chapters 5 and 6) because of their ability to sustain a constant shear flow irrespective of the wall thickness, unlike an open section. However, a pure flexure is only possible when shear forces act at the shear centre of a tube. With shear forces displaced from the shear centre, an example of combined loading upon a structure arises through bending, torsion and shear. The effect is similar to when a torque is not applied at the centre of twist, this giving rise to a similar loading combination (Chapter 6). Here we might simplify the problem by idealising a section in which webs carry shear force and booms carry bending moments. The interested reader can find the detail of these more subtle aspects of optimum design within Chapter 6 and 7. Methods of reducing weight while preserving the shear resistance of webs are considered in Chapter 8. Corrugations, Z and top-hat stiffeners are effective in offsetting the risk of buckling in thin webs over long lengths and depths. A method of optimising the design of pin-jointed frame assemblies is given in Chapter 9. Whilst the structures considered consist of relatively few bars, they are sufficient to show how the approach may be extended to plane and three-dimensional frames with many bars.

Beams which feature as lateral carriers for vertical loading are allowed in Chapters 10 and 11 to vary in cross-section. In particular, the beam supports are altered from resting upon simple knife edges or rollers to having one or both ends built in. The influence of both shear force and bending moment upon an optimum beam design has a significant influence upon weight reductions compared to standard designs in uniform section. Savings of over 50% are possible by allowing the section to vary whilst maintaining the load bearing integrity. In the case of combined bending and shear in beams, their effects can be separated because at the surface, where the bending stress is at its maximum, the shear stress is zero. Conversely, at the neutral axis, where the bending stress is zero, the shear stress is at its maximum. So in a given cross-section we need only alternate from their respective maximum stress positions as either bending or shear governs the design of the section depth. However, we might seek an ideal optimum design in which all points in the depth of a section are stressed equally, say at a given equivalent (von Mises) stress. Chapter 11 shows that all the common solid beam sections are inefficient, even with the aforementioned weight savings. The ideal section is found numerically, so that when its depth is allowed to change with the varying force and moment, the greatest possible weight saving is achieved.

In Chapter 12 further examples of combined loading in structures present themselves in various ways: tension-torsion in shafts, bending and torsion of beams, compression and bending of struts and combinations of concentrated and distributed loadings in beams. Limbs are separated within free-body diagrams within which the design stress is limited by a yield criterion before the appropriate loading criteria are applied. Here the von Mises yield criterion is used to account for combined loading and also in converting material yield stress from the shear to the tension mode.

The fully encastré beam (Chapter 13), being a statically indeterminate structure, is optimised for minimum weight as it is known that these provide a stiffer, lighter structure than simply-supported beams bearing similar loading. It is shown that encastré beams can operate at minimum weight with a variable profile in circular and square cross-sections for central point loading and uniformly distributed loading acting separately and in combination. Both bending and shear effects are combined within a true optimum design. In applying the limiting stresses from bending and shear, regions of the length appear over which the beam section is optimised, depending upon which design criterion dominates.

In all that has been mentioned these designs are essentially elastic since the limiting stress is normally set at the yield or at a proof stress involving less than 1% plastic strain. Some allowance for a greater degree of plasticity is allowed in the buckling of plates by adopting a tangent modulus in the critical stress formula. Gross plasticity is broached in the final two Chapters 14 and 15 where an optimum design at the ultimate strength of a material is given. This is made possible by the application of energy balance equations to a collapsing structure. A collapsing structure is taken to behave as a rigid mechanism following the development of plastic hinges at the ultimate stress level of its elements. Given that there may be many ways for a combination of these hinges to form possible collapse mechanisms, the optimum plastic design is based upon collapse under the least loading. In achieving an efficient design, we can alter the section geometry to make possible the coincidence of a number of mechanisms under the applied loading. Graphical and dynamic programming techniques are shown to assist with this procedure in Chapter 15.

There are a number of books on the subject of optimum structural design (see the Further Reading section of Chapter 9) but none present the material in quite the same way as the reader will find here. While the basic method of weight to strength optimisation remains common to all structures, it has been embellished with all the necessary mechanics to allow for this. The following topics feature among the supporting material which require an understanding through their optimum design applications: Euler strut buckling, torsion of non-circular sections, buckling of plates in compression and shear, shear flow in thin sections, the construction of shear force and bending moment diagrams, properties of areas, the principle of virtual work and plastic collapse.

The author would like to express his gratitude to J. J. Richardson for his invitation some years ago to attend his inspiring short course, 'An Introduction to Optimum Structural Design', within the Department of Mechanical, Aeronautical and Production Engineering at Kingston Polytechnic (now Kingston University). The course content is reflected within earlier chapters of this book and underpins the author's developments of all integrated new material. As a student receiving new material there appeared, within this carefully structured and thought-provoking course, a clear need for further work to be done in certain areas. Faced with that challenge, the author has produced this book to inform optimum design practice and perhaps inspire the reader in turn to advance the subject further. The challenge is extended to all readers to make a contribution to weight saving in engineering applications, especially to those environmentally sensitive designs involving the planet's dwindling resources and fragile ecological system.

Glossary of Terms

The following are among the most common terms that appear on the subject of mechanics in the optimum design of structures. The terms are centred upon the common theme of optimising the weight to strength of a structure. This involves applying design criteria to all failure modes, both global and local. The optimum solution is based upon similar loading stresses from each design criterion adopted being reached simultaneously.

Design constraints: Many factors impose limits on the design variables. For example, restrictions may be placed upon material, geometry, available space, use of non standard parts, and fabrication methods.

Design variables: These refer to the control of the shape and proportion of a structure as imposed by the designer.

Failure criteria: These refer to the limits placed upon the structural response to the applied loading. Commonly, this amounts to limiting stress to that which will produce the onset of plastic yielding, local necking, unstable buckling and brittle fracture. Optimisation requires that, where such modes co-exist within a structure, their critical stresses are attained simultaneously.

Loading: This refers to all types of external actions, including tension, compression, bending, shear and torsion, all of which must be specified in magnitude and direction in terms of the internal stress that they induce.

Material factor (or material efficiency) M or m: An independent variable appearing as an argument of the objective function expression. The analyses that follow reveal that the material efficiency appears consistently as a ratio between particular material properties such as density, elastic modulus and yield stress, typically $M = \sqrt{E}/\rho$ for struts, $m = \rho/\sigma_y^{2/3}$ for beams, thereby revealing how material choice influences the objective function.

Objective function: Generally, a mathematical function of a number of independent structural variables arises from imposing any criterion that is to be maximised or minimised. Here the objective function g in equation (1) refers to an expression that will serve to minimise the weight for a given strength. Often this is more easily achieved by expressing the weight/strength ratio as an equivalent ratio R between two properties. This ratio may either be minimised or maximised, depending upon whether it is directly proportional to or inversely proportional to the weight/strength ratio. The objective function itself may be split into the product of three independent variables – the shape factor F, the material factor M and the structural index S – to allow their separate influences upon weight to be examined. Thus, we may express these dependencies in the objective function equation

$$R = g(F, M, S) \tag{1}$$

Hence the essential pre-requisite for an optimisation procedure is to derive the arguments F, M and S in equation (1). Despite there being a large number of sections available as struts, beams and shear webs, the procedure for the optimisation of R is constant for each type of loading. For example, in the struts considered in Chapters 1 and 2, the independent variables in equation (1) are found to be separable as

$$R = F \times M \times S^n \tag{2}$$

where the exponent n is fractional. Thus, for a given M and S, a maximum in F gives a maximum in R, this being a common route to minimising the weight of struts.

Problem definition: This refers to a clear analysis of all factors that impose upon the eventual design, including the loading, geometry and material. An optimum design will depend entirely upon the criteria set within its definition.

Shape efficiency factor F: This is used with struts, where it is inversely proportional to weight. For standard sections, e.g. square, circular and tubular, F appears as a numerical coefficient in which a maximum is sought. Here F-values usually lie in the range $0 < F < 1$ (see Table 1.4, p. 25), indicating that the effect of F upon R is less influential than M and S in equation (1). For non-uniform, thin-walled sections, F depends upon the section geometry, so allowing their dimensions to be optimised.

Shape factor f: This is used with beams and shear webs, where it is proportional to weight. For standard sections, e.g. square, circular and tubular, f appears as a numerical coefficient in which a minimum is sought. Here $f > 1$ (see Table 3.1, p. 88). For open and closed sections, with non-uniform thin walls, f depends upon the section geometry, so allowing their dimensions to be optimised.

Structural index S or s: This refers to the loading expressed in a form to facilitate the minimum weight to strength analysis of a structure. The unit of the index S corresponds to that of stress, i.e MPa or N/m^2. Thus, for struts and plates of length L under a compression (P or w), the index is P/L^2 and w/L, while for beams in bending, the index M/D^3 refers to a circular section of diameter D. Here the loading symbols are: a concentrated force P, a force/unit width w and an applied moment M. In equation (1) the index S, as an argument of the objective function g, is found to bear the non-linear relation to R for struts and beams given in equation (2). The index S that optimises R is then sought by mathematical means, i.e. differentiation for the function's turning points or by alternative graphical methods. For many structures a non-dimensional structural index $s = S/\sigma_y$, where σ_y is the yield stress, may be used across five decades (i.e. $10^{-3} \leq s \leq 10^2$) to show the optimum weight dependence (below).

Transitions in s between beam section designs based upon the bending moment M and shear force F are denoted by s_{FM} and s_{MF}.

Weight parameter W or n: Weight W is the quantity to be minimised in all structures. Weight arises from the mass of their structural elements: beams, bars plates, tubes, etc. A non-dimensional weight parameter n is used with s (above) to show graphically the optimum weight range of most structural geometries for a practical range of working loads. The normalised weight parameter is defined as $n = (1/\rho)W/L^3$, where ρ is the material density (see equation (3) below). In a structure with varying section the s versus n relation is found from the integration of an optimum section profile (see below).

Key Symbols

In addition to the symbols reserved for the glossary terms above, other general symbols are employed to denote the following. The list is not intended to be exhaustive as various specialist terms and their symbols are defined more meaningfully where they first appear in the text.

Applied loading: This is a generic term denoting the relatively few external actions that induce stress within a structure. These are: tension and compression P, bending moment M, shear force F, torque T, pressure and distributed loading w. The problem is complicated when two or more loads act in combination.

Buckling coefficient: Appearing within both global and local buckling formulae, K connects the critical stress to an aspect ratio, e.g. the thickness/width ratio for a thin plate. Moreover, the dependence of K upon different edge fixings – hinged, fixed, etc. – is represented typically in design data sheets (derived from the listed published sources). Buckling is an especially important criterion of failure to consider under shear and compression of thin-walled sections. In slender struts, for example, a local buckling of webs and flanges within the section is taken concurrently with flexural buckling of the length.

Dimensionless parameters: Non-dimensional load, weight and length parameters s, n, q are especially useful for the analysis of beams (see Glossary). Thus the optimum weight is modified to $n = (1/\rho)W/L^3$ and the structural index is written as $s = (1/\sigma_y)F/L^2$. Here, when finding the beam contour, it is found convenient to normalise with length both the half depth ($d/2L$) and the length position ($q = z/L$). Correspondingly, these might appear within a non-dimensional form of the objective function (2), as follows:

$$\frac{W}{L^3} = f\left(\frac{\rho}{\sigma_y{}^n}\right)\left(\frac{F}{L^2}\right)^n, \quad \Rightarrow \quad \frac{1}{\rho}\frac{W}{L^3} = f\left(\frac{1}{\sigma_y}\frac{F}{L^2}\right)^n, \quad \Rightarrow \quad n = fs^n \qquad (3)$$

Plots of n versus s appear for four standard cases of beam loadings in which their sections are circular, square and rectangular in their optimised forms. By allowing s to scan 5 decades within these plots all manner of applications are contained, both within short studs and rivets placed under shear and in the extended lengths required of transversely loaded beams with simple and encastré supports.

Elastic modulus: The elastic modulus in tension E and the shear modulus G appear with various subscripts: S and T to denote secant and tangent moduli. These are employed where some measure of plasticity is admitted to the design. The tangent modulus $E_T = d\sigma/d\varepsilon$ is the gradient to the stress–strain curve at a point in the plastic region corresponding to a given offset plastic strain. Alternatively, the ratio between the coordinates for this point define a secant modulus $E_S = \sigma/\varepsilon$. For either definition, a suitable empirical description of the stress–strain curve is required. For this the Ramberg–Osgood equation is used in which elastic and plastic strains are additive (see Appendix B, p. 531).

Properties of areas: Any section area A has four important properties that appear throughout this book. They are $(\overline{x}, \overline{y})$ denoting the coordinates of the centroid position, i the first moment of area, I the second moment of area and $k(= \sqrt{I/A})$ the radius of gyration of the section. Both i and I refer to axes passing through the centroid. Subscripts (x, y) are used for general centroidal axes and (u, v) for principal axes.

Section geometry: This appears within symbols b, d, t and p, denoting breadth, depth, thickness and pitch, respectively. These may appear with subscripts, say, t_w and t_f, denoting the thickness of the web and the flange of a T-section. Having minimised weight, the section geometry is said to be optimised. That is, a unique combination of dimensions will appear to ensure that all the design criteria are met. The preferred design is one which ensures that failures from all potential sources occur together rather than a design based upon one or other failure mode.

Strain and displacement: Corresponding to direct and shear stress (below), the direct and shear strains are denoted as ε and γ. Direct strain arises from a dimensional change, i.e. a lengthening or a shortening (say, x) arising from tension and compression, respectively. Shear strain refers to the angular distortion (in radians), arising from shear force and torsion, between two initially perpendicular directions. Here a shear distortion or a twist (θ) occurs between the unstrained reference directions. Beam curvature R is involved where surface tension and compression occur together from bending. In-line displacements (δ or Δ) appear beneath the loads applied to pin-jointed structures.

Stress: This is the measure of a material's resistance to the applied loading, whether it be a direct stress σ or a shear stress τ. A safe design is ensured by placing a limit upon stress, and an optimum structural design is achieved where this limit (the design stress) is reached simultaneously from its various sources: flexure, compression, buckling, torsion, etc. The stress measure known as *shear flow* q refers to the product τt reserved for the stress measure in thin-walled sections under torsion and flexural shear.

Weight: The two weight measures, W and n, depend upon section geometry (see Glossary above), length L and material density ρ. For varying cross-sections, the weight integral depends upon position z in the length. Where, for example, an optimum design proposes that diameter d varies with length z, we have the weight integral

$$W = \frac{\rho\pi}{4}\int_{z=0}^{z=L} d^2\,\mathrm{d}z \quad \Rightarrow \quad n = \frac{1}{\rho}\frac{W}{L^3} = \rho\pi\int_{q=0}^{q=1}\left(\frac{d}{2L}\right)^2 \mathrm{d}q \qquad (4)$$

The corresponding normalised weight parameter n in equation (4) employs $q = z/L$ and $d/2L$. The integration of equation (4) leads to the form of equation (3), in which it appears that n depends upon both s and q.

1

Compression of Slender Struts

1.1 Introduction

The stress σ within a long slender strut of uniform cross-section is affected by the magnitude of the load applied P and its length L. It will be shown that the weight of the strut is a minimum when the stress is a maximum. It is therefore necessary to investigate how σ varies with both P and L and the shape of the cross-section, each being under the control of the designer. The general approach is to seek an objective function in which the strut's weight W is expressed as the product of the strut's volume and density ρ. This gives

$$W = \rho AL = \rho\left(\frac{P}{\sigma}\right)L = \frac{PL}{\sigma/\rho} \tag{1.1}$$

where the elastic stress σ in the strut prior to its buckling is equated to the axial load per unit area, i.e. $\sigma = P/A$, and L is the 'pinned' strut length. Hence, to minimise W it follows from equation (1.1) that σ/ρ, the equivalent objective function, is to be maximised. The following failure criteria provide the various limiting stress measures upon which the strut's minimum weight is to be based.

1.2 Failure Criteria

The failure criteria for a strut would need to be expressed in terms of the section's shape, whether this be solid, hollow, thin-walled, tubular, etc.

1.2.1 Flexural Buckling

Euler's theory [1] expresses the critical elastic buckling load P_c for a pinned-end strut as

$$P_c = \frac{\pi^2 EI}{L^2} \tag{1.2a}$$

Mechanics of Optimal Structural Design: Minimum Weight Structures D. W. A. Rees

where $I = Ak^2$ is the least second moment of area, which depends upon the radius of gyration of the section. Hence, the critical flexural buckling stress σ_F may be expressed as

$$\sigma_F = \frac{\pi^2 E}{(L/k)^2} \tag{1.2b}$$

In shorter, stockier struts, where buckling is elastic-plastic, the tangent modulus E_T may replace the elastic modulus E in equations (1.2a,b).

1.2.2 Local Buckling

A local buckling failure refers specifically to struts with thin walls in their cross-sections. Typically, this mode of failure appears as an indentation of diamond shape upon the surface or in a bowing of the section walls [2]. Local buckling does not arise in struts with solid sections. For buckling of the flat plates (i.e. the walls) within thin-walled tubular sections, the local buckling stress takes the form [3]

$$\sigma_L = K_L\ E_T \left(\frac{t}{d}\right) \tag{1.3}$$

where K_L is a buckling coefficient that depends upon the plates aspect ratio and the support provided to its edges (see equation (D.1a,b) in Appendix D).

1.2.3 Working Stress

In the absence of buckling, the axial, compressive, working stress σ_W is found simply by dividing the applied load P by the section area A:

$$\sigma_W = \frac{P}{A} \tag{1.4}$$

1.2.4 Limiting Stress

The stress in equations (1.2b), (1.3) and (1.4) would normally be limited to the yield stress where buckling is elastic. In the case of plastic buckling the limiting stress is raised to correspond to a given offset (plastic) strain, i.e. the 0.1% proof stress. Let σ_y be the limiting yield or proof stress of the strut material appropriate to its buckling behaviour. Then, its relation to the applied stress (σ_F, σ_L and σ_W above) from those sources in equations (1.2b)–(1.4), is simply

$$\sigma \leq \sigma_y \tag{1.5}$$

1.2.5 Objective Function

An optimum section size is found by equating (1.2b), (1.3) (where appropriate) and (1.4), where they all have been limited by equation (1.5). Finally, all appropriate failure criteria in § 1.2.1–1.2.4 are combined within the objective function to minimise the weight. We

Figure 1.1 Standard, solid strut cross-sections

shall demonstrate this design procedure, firstly, with the more common solid cross-sections in Figure 1.1.

1.3 Solid Cross-Sections

The four solid cross-section shown in Figure 1.1 are the most likely contenders for strut cross-sections as these are available in long bars of extruded stock.

1.3.1 Circular Section, Diameter d (see Figure 1.1a)

With $I = \pi d^4/64$, then $k^2 = d^2/16$ and the buckling failure criterion (1.2a), becomes

$$\sigma_F = \frac{\pi^2 E d^2}{16L^2} \tag{1.6}$$

When equation (1.6) is combined with the axial stress formula (1.4) with $A = \pi d^2/4$, so that $\sigma_F = \sigma_W$, this gives the optimum diameter:

$$d_{\text{opt}} = \frac{(64P)^{1/4} L^{1/2}}{\pi^{3/4} E^{1/4}} = 1.199 \left(\frac{PL^2}{E}\right)^{1/4} \tag{1.7a}$$

It will be seen that all solid sections will conform to an equation of similar form for an optimum section dimension (here the diameter) as:

$$d_{\text{opt}} = C \left(\frac{PL^2}{E}\right)^{1/4} \tag{1.7b}$$

where C is a *shape coefficient*. Substituting equation (1.7a) into equation (1.6) and dividing by ρ leads to the equivalent objective function for a solid, circular-section strut:

$$\left(\frac{\sigma}{\rho}\right)_{\text{opt}} = 0.886 \left(\frac{E^{1/2}}{\rho}\right) \left(\frac{P}{L^2}\right)^{1/2} \tag{1.8a}$$

Here, a *shape efficiency factor* $F = 0.886$ appears. The *material efficiency factor* is $M = \sqrt{E}/\rho$ and the *structural index* $S = P/L^2$ is raised to the fractional power ½. Hence, we may write the quantity to be maximised, the *objective function* R, more generally as

$$R = \left(\frac{\sigma}{\rho}\right)_{\text{opt}} = F \times M \times S^n \tag{1.8b}$$

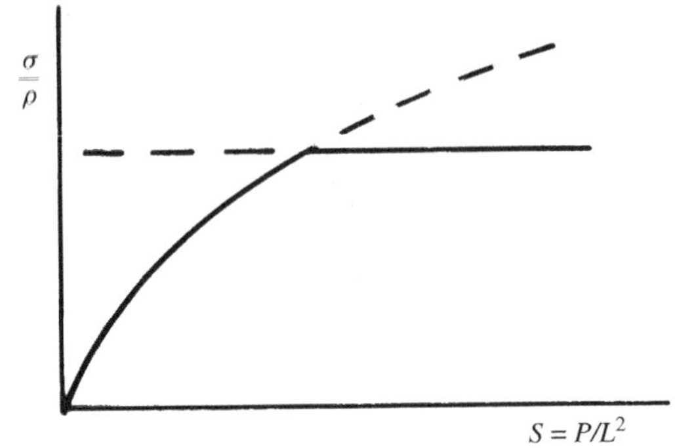

Figure 1.2 Objective function plot from equation (1.8a) showing limiting stress cut-off

where, for a circular cross-section, $n = ½$. If we wish to employ a tangent modulus E_T this will reduce M by the ratio $\sqrt{(E_T/E)}$. We can derive from equation (1.8a) the plot given in Figure 1.2 with limiting stress cut-offs at an appropriate yield, proof or ultimate stress level.

1.3.2 Solid Square Bar $a \times a$ (see Figure 1.1b)

With $I = a^4/12$ and $A = a^2$, then $k^2 = a^2/12$ and the buckling failure criterion, equation (1.2b), becomes

$$\sigma_F = \frac{\pi^2 E a^2}{12L^2} \tag{1.9}$$

Equate (1.9) to the axial stress formula (1.4), i.e. $\sigma_W = \sigma_F$, and on setting $A = a^2$, this gives the square side length as

$$a_{\text{opt}} = \frac{(12P)^{1/4} L^{1/2}}{\pi^{1/2} E^{1/4}} = 1.050 \left(\frac{PL^2}{E}\right)^{1/4} \tag{1.10}$$

Substituting equation (1.10) into equation (1.9) and dividing by ρ leads to the objective function required:

$$\left(\frac{\sigma}{\rho}\right)_{\text{opt}} = 0.907 \left(\frac{E^{1/2}}{\rho}\right) \left(\frac{P}{L^2}\right)^{1/2} \tag{1.11}$$

Equation (1.11) is similar in form to the circular section's objective function, equation (1.8a). Note here that the greater value of the shape efficiency factor $F = 0.907$ indicates that more of the material in the square section is fully stressed. Figure 1.3 presents equation (1.11) graphically for four materials whose properties and relevant ratios appear in Table 1.1 (see also Appendix A). The figure shows working stress ranges cut off by the limiting stress. The latter is taken to be the 0.1% proof stress for the metallic materials and the ultimate stress for Douglas fir and the glass fibre-reinforced composite (GFRC).

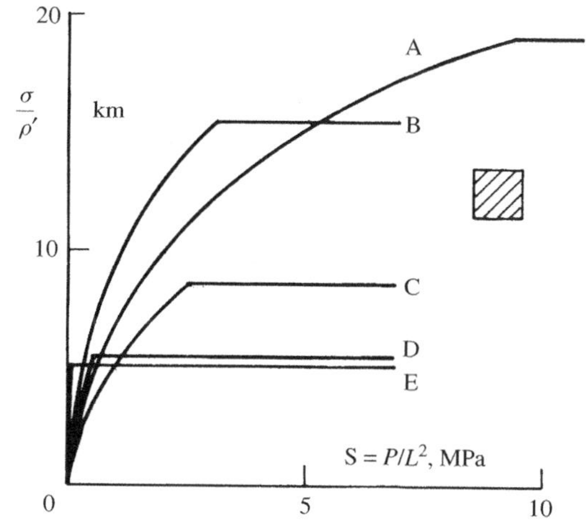

Figure 1.3 Objective function versus structural index for struts with solid square sections (key: A, Ti alloy; B, Al alloy; C, steel; D, GFRC; E, Douglas fir (see Table 1.1))

Within the range of the index $S = P/L^2$ shown, the aluminium alloy (L65) appears to optimise the stress most efficiently at a given S. Notably, in extending the range of S threefold, titanium alloy (DTD 5053) allows higher objective functions to be reached. The high grade bolt steel (S96) is a poor performer on a weight/strength basis. GFRC lies between the aluminium and titanium alloys but is cut off at a much lower value, $\sigma_{\text{ult}}/\rho = 5.7\,\text{km}$. Douglas fir has a similar cut-off at 5.1 km and, within a very restricted range of structural indices ($P/L^2 < 0.2$), provides the greatest objective function of all the materials considered within this figure. In this case both E and σ_{ult} are properties of fir taken parallel to the grain.

A similar analysis may be applied to a strut of any solid cross-section. Table 1.2 summarises the results obtained here for the circular and square sections together with those that apply to semi-circular and equilateral triangular sections (see Figure 1.1c,d). The comparison between the four sections may be made simply through the shape coefficient C and the shape factor F, appearing in equations (1.7b) and (1.8b). Of these four sections an equilateral triangle appears to bear the greatest stress. The implication of this is that when $(\sigma/\rho)_{\text{opt}}$ is to be maximised, in order to minimise the strut's weight, the triangular

Table 1.1 Properties of common structural materials

Material\Property → ↓ Unit →	E GPa	ρ kg/m^3	$E^{1/2}/\rho$ m^2/N$^{1/2}$	$E^{2/3}/\rho$ (m^5/N)$^{1/3}$	$E^{3/5}/\rho$ (m^9/N^2)$^{1/5}$
A Ti alloy (DTD 5053)	118	4540	7.22	494.6	9.2
B Al alloy (L65)	75	2790	9.83	634.6	119.8
C Steel (S96)	207	7800	5.79	441.7	78
D GFRC	20	1800	8	41.73	85.84
E Douglas fir	11	497	21.51	1014.47	217.18

Table 1.2 Shape coefficients for slender struts of solid section

Cross-section	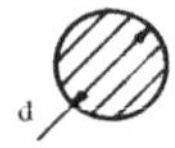	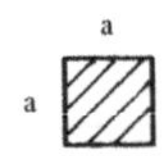	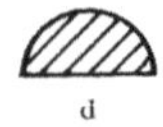	
C	1.199	1.050	1.960	1.539
F	0.886	0.907	0.663	0.975

section strut would provide the lowest weight for supporting a predetermined compressive load in a given material.

1.4 Thin-Walled, Tubular Sections

The objective function is again $R = \sigma/\rho$, but two cross-section variables arise in tubular sections: a mean section dimension d and the wall thickness t (see Figure 1.4a). Appendix D shows that failure criteria must now include local buckling in addition to flexural buckling. In deriving the objective function R, the usual procedure is to establish the failure criteria first. Then, by ensuring that the critical stresses by these criteria are attained simultaneously, the geometry of the tubular section is optimised, from which the usual form for R will follow.

1.4.1 Thin-Walled Circular Tube

Local, inelastic buckling in thin-walled circular tubes (Figure 1.4a) of moderate length under compression has been reported in [4]. Mostly, buckling appeared in the two-lobe failure mode (Figure 1.4b), even though four lobes are generally assumed for an elastic failure [2].

(i) Flexural Buckling
With $I = \pi d^3 t/8$ and $A \approx \pi dt$, then $k^2 = I/A \approx d^2/8$. Hence the buckling stress in equation (1.2b) becomes

$$\sigma_F = \frac{\pi^2 E_T d^2}{8L^2} \tag{1.12}$$

(ii) Limiting Stress
The working, compressive stress σ_W in the strut is given as

$$\sigma_W = \frac{P}{A} = \frac{P}{\pi dt} \tag{1.13}$$

which is limited to σ_y, i.e. $\sigma_W \le \sigma_y$. (Here we take: $\sigma_W = \sigma_y$)

(iii) Local Buckling
A collapse of the wall surface occurs (see Figure 1.4b) when local depressions appear under an axial stress:

$$\sigma_L = K_L E_T \left(\frac{t}{d}\right) \tag{1.14}$$

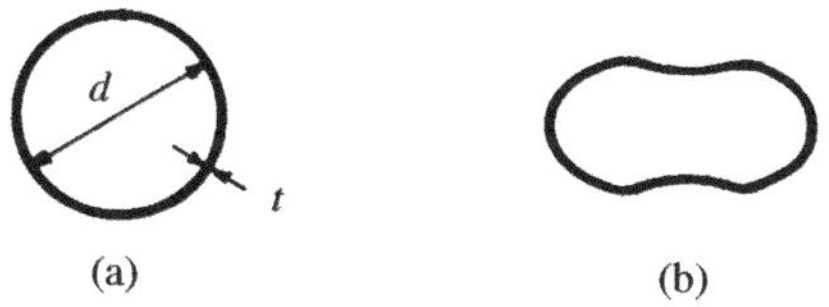

Figure 1.4 Circular tube section showing local wall buckling

The theoretical value of the buckling coefficient is $K_L = 1.212$ but here the value $K_L = 0.4$ is used as it has been found to match experimental data more closely [1].

(iv) Optimisation
A strut with optimum geometry is found by combining equations (1.12)–(1.14), so that the stresses by the three criteria are made equal, $\sigma_F = \sigma_W = \sigma_L$, which gives

$$\frac{\pi^2 E_T d^2}{8L^2} = \frac{P}{\pi dt} = 0.40\ E_T \left(\frac{t}{d}\right) \tag{1.15}$$

Equation (1.15) allows optimum dimensions d and t to be found as

$$t_{\text{opt}} = \left(\frac{P}{\pi K_L E_T}\right)^{1/2} = 0.892 \left(\frac{P}{E_T}\right)^{1/2} \tag{1.16a}$$

$$d_{\text{opt}} = \left(\frac{8\sqrt{K_L}}{\sqrt{\pi^5}}\right)^{1/3} \left(\frac{PL^4}{E_T}\right)^{1/6} = 0.661 \left(\frac{PL^4}{E_T}\right)^{1/6} \tag{1.16b}$$

The optimum stress follows from substituting d_{opt} and t_{opt} into any of the three expressions in equation (1.15). Dividing by ρ gives the objective function

$$\left(\frac{\sigma}{\rho}\right)_{\text{opt}} = 0.540 \left(\frac{E_T^{2/3}}{\rho}\right) \left(\frac{P}{L^2}\right)^{1/3} \tag{1.17}$$

As the exponent of E_T, within the material factor M, differs from that of a solid circular section (see equation (1.8a)) a comparison between the contribution to R from the separate factors F, M and S for the two sections would be inappropriate. However, the result of the product between F, M and S, which gives R, provides the basis for the required weight to strength comparison. Figure 1.5 presents equation (1.17) graphically for four materials whose properties and relevant ratios appear in Table 1.1. Again, the aluminium alloy (L65) is the most efficient when the structural index $S < 100$, but beyond this the titanium alloy (DTD 5053) takes over, allowing S to reach 400 at its greatest limiting stress level.

1.4.2 *Thin-Walled Square Tube*

The uniform square tube (Figure 1.6a) has a mean side length a and wall thickness t. Local buckling (Figure 1.6b) appears in a bowing of the flat plates that comprise the walls.

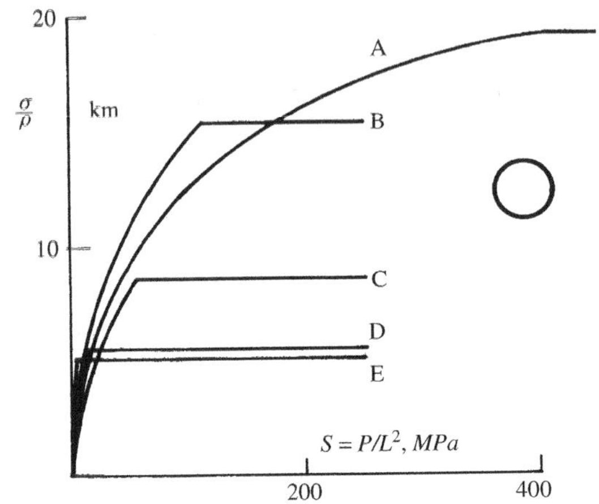

Figure 1.5 Objective function versus S for struts with thin-walled, circular tube sections (key: A, Ti alloy; B, Al alloy; C, steel; D, GFRC; E, Douglas fir (see Table 1.1))

As this form distortion of the section occurs without translation or rotation it is said to be local, i.e. it may exist together with the usual flexural buckling behaviour of a strut [2].

(i) Flexural Buckling
With $I = 2a^3t/3$ and $A \approx 4at$, then $k^2 = I/A \approx a^2/6$. Hence, the buckling stress in equation (1.2b) becomes

$$\sigma_F = \frac{\pi^2 E_T a^2}{6L^2} \tag{1.18}$$

(ii) Limiting Stress
The working, compressive stress in the strut is given as

$$\sigma_W = \frac{P}{A} = \frac{P}{4at} \tag{1.19}$$

where $\sigma_W \leq \sigma_y$, for a proof stress σ_y defined by an offset strain (typically 0.1% or 0.2%). Here the gradient to the stress–strain curve, corresponding to the stress level σ_y, is used to define the tangent modulus E_T in equation (1.18).

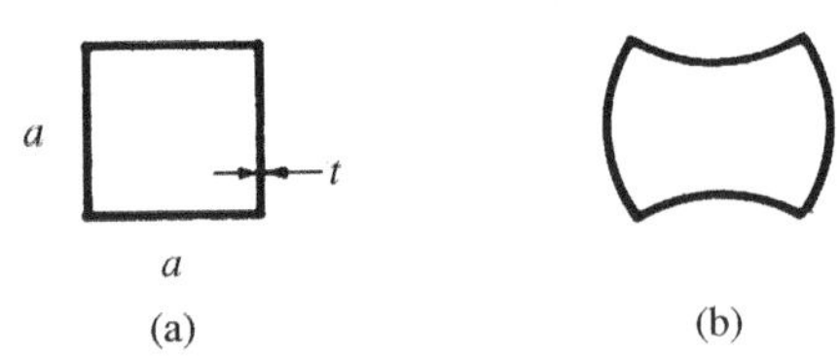

Figure 1.6 Square tube section showing local wall buckling

(iii) Local Buckling

When each wall is taken as a long flat plate with simply supported edges (see Figure 1.6b), a lateral collapse occurs under an axial stress:

$$\sigma_L = K_L\ E_T \left(\frac{t}{a}\right)^2 \tag{1.20}$$

where the buckling coefficient $K_L = 3.62$ for $\nu = 0.3$ [3] (see Appendix D).

(iv) Optimisation

An optimum geometry is found by combining equations (1.18)–(1.20) so that the stresses by the three criteria (i), (ii) and (iii) above are made equal, $\sigma_F = \sigma_W = \sigma_L$, which gives

$$\frac{\pi^2 E_T a^2}{6L^2} = \frac{P}{4at} = 3.62\ E_T \left(\frac{t}{a}\right)^2 \tag{1.21}$$

Equation (1.21) allows the optimum dimensions a and t to be written as

$$a_{\text{opt}} = 0.743 \left(\frac{PL^3}{E_T}\right)^{1/5}$$

$$t_{\text{opt}} = 0.371 \left(\frac{P^2 L}{E_T^2}\right)^{1/5} \tag{1.22a,b}$$

The optimum stress follows from substituting a_{opt} and t_{opt} into any of the three parts within equation (1.21). Dividing by ρ gives the required objective function:

$$\left(\frac{\sigma}{\rho}\right)_{\text{opt}} = 0.907 \left(\frac{E_T^{3/5}}{\rho}\right)\left(\frac{P}{L^2}\right)^{2/5} \tag{1.23}$$

Again, the material factor M differs from that of solid section, equation (1.11). Hence, a comparison between the separate contributions of the factors F, M and S to $R = (\sigma/\rho)_{\text{opt}}$ for the two sections cannot be made. However, the result of their products which define R in equation (1.23) provides the basis for the required weight to strength comparison (see Table 1.3).

1.4.3 Thin-Walled Hexagonal Tube

Following the analyses of the circular and square tubes, we may consider closed tubes with any number of sides. Similar analyses [5] to those given here reveal that regular polygons all have identical material efficiency factors and structural indices to that for the square tube. We give one example of a thin-walled, regular hexagon in Figure 1.7a.

Table 1.3 Shape efficiency factors and buckling coefficients for thin-walled, tubular polygon strut sections

No of sides	F	K_L
3	0.789	4.31
4	0.907	3.62
5	1.031	3.87
6	1.112	3.62
7	1.203	3.76
8	1.268	3.62
9	1.340	3.69
10	1.396	3.62

(i) Flexural Buckling

The second moment of area about axis x has contributions from the two vertical sides and the four sloping sides (Figure 1.7b) as follows:

$$I_x = 2\frac{tb^3}{12} + 4\int_0^b \left(\frac{b}{2} + s\sin\theta\right)^2 t\,ds = \frac{5tb^3}{2} \tag{1.24a}$$

The second moment of area about axis y has contributions from the same two vertical sides and four sloping sides but in a different disposition:

$$I_y = 2\left[\frac{bt^3}{12} + bt(b\sin\theta)\right] + 4\int_0^b (s\sin\theta)^2 t ds \tag{1.24b}$$

$$= \frac{bt^3}{6} + \frac{5tb^3}{2} \approx \frac{5tb^3}{2}$$

For practical purposes the two second moments may be assumed equal. Hence, buckling will occur about each axis simultaneously under a critical compressive load. With $I = 5tb^3/2$ and $A \approx 6bt$, then $k^2 = I/A = 5b^2/12$. Hence the critical buckling stress in

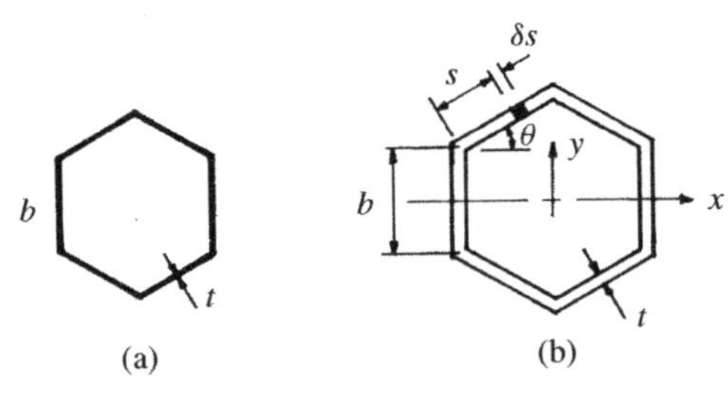

Figure 1.7 Hexagonal tube section showing the I_x calculation for a sloping wall

equation (1.2b) becomes

$$\sigma_F = \frac{5\pi^2 E_T b^2}{12L^2} \tag{1.25}$$

(ii) Limiting Stress

The working, compressive stress in the strut is given as

$$\sigma_W = \frac{P}{A} = \frac{P}{6bt} \tag{1.26}$$

where $\sigma_W \leq \sigma_y$, for a proof stress σ_y defined by an offset strain (typically 0.1% or 0.2%). The gradient to the stress–strain curve, corresponding to σ_y, defines the tangent modulus E_T in equation (1.25).

(iii) Local Buckling

When each wall surface is modelled as a long flat plate with simply supported corners, a critical, lateral bowing occurs in the manner of Figure 1.6b, under an axial stress:

$$\sigma_L = K_L E_T \left(\frac{t}{b}\right)^2 \tag{1.27}$$

where $K_L = 3.62$ applies consistently to polygons with even numbers of sides (see Table 1.3).

(iv) Optimisation

Method 1

An optimum geometry is found by the usual method of combining (i), (ii) and (iii) above so that the stresses by the three criteria are made equal, $\sigma_F = \sigma_W = \sigma_L$, between equations (1.25)–(1.27), which gives

$$\frac{5\pi^2 E_T b^2}{12L^2} = \frac{P}{6bt} = 3.62 E_T \left(\frac{t}{b}\right)^2 \tag{1.28}$$

Equation (1.28) provides the optimum dimensions d and t as

$$t_{\text{opt}} = 0.289 \left(\frac{P^2 L}{E_T^2}\right)^{1/5} \tag{1.29a}$$

$$b_{\text{opt}} = 0.519 \left(\frac{P L^3}{E_T}\right)^{1/5} \tag{1.29b}$$

Substituting d_{opt} and t_{opt} from equations (1.29a,b) into any one of the three parts of equation (1.28) gives the optimum stress. Dividing by ρ gives the objective function:

$$\left(\frac{\sigma}{\rho}\right)_{\text{opt}} = 1.112 \left(\frac{E_T^{3/5}}{\rho}\right) \left(\frac{P}{L^2}\right)^{2/5} \tag{1.30}$$

Method 2

A check on equation (1.30) is provided by an alternative non-dimensional analysis in which the optimum stress is taken as the product of fractional powers in σ_F, σ_W and σ_L. We may write this from equations (1.25)–(1.27) in two ways:

$$\sigma_{\text{opt}} = {\sigma_F}^{\alpha}{\sigma_W}^{\beta}{\sigma_L}^{\gamma} = \left(\frac{5\pi^2 E b^2}{12L^2}\right)^{\alpha}\left(\frac{P}{6bt}\right)^{\beta}\left[3.62E_T\left(\frac{t}{b}\right)^2\right]^{\gamma}$$

from which equalities between the indices in stress and the dimensions b and t follow:

$$[\sigma] : \alpha + \beta + \gamma = 1$$

$$[t] : -\beta + 2\gamma = 0$$

$$[b] : 2\alpha - \beta - 2\gamma = 0$$

giving: $\alpha = 2/5$, $\beta = 2/5$ and $\gamma = 1/5$. Consequently,

$$\sigma_{\text{opt}} = \left(\frac{5\pi^2 E_T b^2}{12L^2}\right)^{2/5}\left(\frac{P}{6bt}\right)^{2/5}\left[3.62E_T\left(\frac{t}{b}\right)^2\right]^{1/5}$$

$$= \left(\frac{5\sqrt{3.62}\pi^2}{12 \times 6}\right)^{2/5} E_T^{3/5}\left(\frac{P}{L^2}\right)^{2/5}$$

in which the shape coefficient value of 1.112 in equation (1.30) is confirmed.

The exponents within the material efficiency factor M and the structural index S differ from those for the circular tube but are identical to those found for the square tube. In fact, all regular polygons conform to a similar, optimum objective function expression. This fact is a consequence of the stress in any closed, regular polygon arising from similar causes: flexure and local buckling of the plate within each side wall. This function appears consistently as

$$R = \left(\frac{\sigma}{\rho}\right)_{\text{opt}} = F \times M \times S^n \qquad (1.31)$$

where $n = 2/5$, $M = E_T^{3/5}/\rho$, $S = P/L^2$ and F varies as in Table 1.3. The differences in their shape efficiency factors F are attributed in part to the different local buckling coefficients K_L, when the polygon has an odd number of sides (see Table 1.3). The explanation given for this is that while each side of an even-sided polygon is simply supported, an odd number of sides has a partial restraint that depends upon the wavelength of buckling [5]. Despite this, F is seen to increase with the number of sides, though this is limited when the sides of a multiple-sided polygon do not remain straight [6]. We will see, however, that closed tubes are always more efficient than thin-walled open sections. Attempts to improve the structural efficiency of a strut further using closed cellular sections have only been partially successful. The triangular and square tubes with matching interior honeycomb show no improvement irrespective of the number of cells, i.e. F is unchanged as the gains in local buckling and flexural buckling stresses are cancelled by the weight increase. While interior hexagonal cells provide a marginal benefit, a better alternative appears to be where a cellular wall in each geometry retains its hollow interior. For example, when a square tube has 10 square cells within each of its 4 walls F is raised to 1.42, compared to 0.907 for an 'uninterrupted' wall in Table 1.3.

It is left as an exercise for the reader to confirm the shape factors given in Table 1.3 for regular polygons with seven sides and more.

1.5 Thin-Walled, Open Sections

The objective function is again $R = \sigma/\rho$ but more cross-sectional dimensions will require optimisation within the straight limbs of an open section. Moreover, given the particular shape (I, T, L, U, etc.), it is necessary to: (i) ensure that flexural buckling occurs about its two principal axes simultaneously and (ii) establish whether the flange or web is less resistant to local buckling. There follows the usual procedure of ensuring that critical stresses by the governing criteria are attained simultaneously. This condition will serve to optimise the section and thereafter allow the objective function R to be determined.

1.5.1 I-Section of Uniform Thickness

With $2b$ and d as the section's mean outer dimensions (Figure 1.8a), the modes of local buckling for a uniform I-section are shown in Figure 1.8b. For the optimum design condition both the flange and the web are taken as plates that buckle in flexure under a critical compressive stress. The web has both its long edges simply supported at the flange connection. Each flange half-length has one edge simply-supported and the other edge free.

(i) Flexural Buckling

Firstly, it is noted that k in equation (1.2b) is identified with the least second moment of area for the strut section. However, by allowing b and d to vary we can ensure that centroidal axes x and y are equally resistant to buckling, as was the case with our two earlier tube examples. Of course, in practice, flexural buckling would only occur about one axis only due to a slight eccentricity in the load axis and material imperfections. That k has equal values for the section about axes x and y in Figure 1.8a, follows from setting

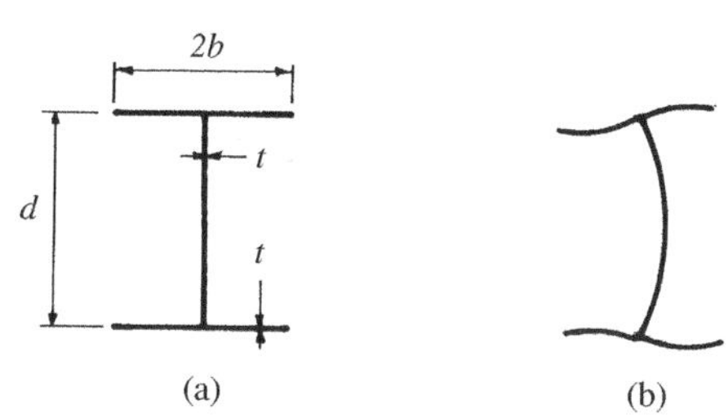

Figure 1.8 I-section showing local web and flange buckling

$I_x = I_y$. This gives

$$2\left[2bt\left(\frac{d}{2}\right)^2\right] + \frac{t(d-t)^3}{12} = 2\left[\frac{t(2b)^3}{12}\right] + \frac{(d-t)t^3}{12} \tag{1.32a}$$

Provided $t \ll d$ and $t \ll b$ the final term on the right-hand side of equation (1.32a) may be ignored. This also allows us to set $d - t \simeq d$ in the second term, leading to a cubic equation in b/d:

$$16(b/d)^3 - 12(b/d) - 1 = 0 \tag{1.32b}$$

Of the three roots of equation (1.32b), $b/d = 0.905$, -0.821 and -0.084, the one viable solution is $b/d = 0.905$, which corresponds to the I values

$$I_x = I_y = 2\left[\frac{(2 \times 0.905\,d)^3 t}{12}\right] = 0.988\,d^3 t \tag{1.33a}$$

The section area is

$$\begin{aligned} A &= 4bt + (d-t)t \approx 4bt + dt \\ &= 4(0.905\,d)t + dt = 4.62\,dt \end{aligned} \tag{1.33b}$$

when from equations (1.33a,b):

$$k^2 = \frac{I_x}{A} = \frac{0.988\,d^3 t}{4.62\,dt} = 0.214\,d^2 \tag{1.33c}$$

Hence, from equations (1.33c) and (1.2b) the buckling stress becomes

$$\sigma_F = \frac{\pi^2 E_T (0.214 d^2)}{L^2} = \frac{2.112 E_T d^2}{L^2} \tag{1.34}$$

(ii) Limiting Stress

The working, compressive stress in this strut becomes

$$\sigma_W = \frac{P}{A} = \frac{P}{4bt + dt} = \frac{P}{4.62dt} \tag{1.35}$$

where $\sigma_W \le \sigma_y$, for a proof stress σ_y defined by an offset strain (typically 0.1% or 0.2%). The gradient to the stress–strain curve, corresponding to σ_y, defines the tangent modulus E_T in equation (1.34).

(iii) Local Buckling

Referring to Figure 1.8b, when the ends of the web are assumed to be simply supported, the stress condition required for it to bow, as shown in Figure 1.8b, is given by equation (1.20),

$$\sigma_L = 3.62 E_T \left(\frac{t}{d}\right)^2 \tag{1.36a}$$

for $v = 0.3$. Equation (1.20) will also apply to the stress for local flange buckling, when it is written as [3]:

$$\sigma_L = K_L E_T \left(\frac{t}{b}\right)^2 \tag{1.36b}$$

where, because one end is free, K_L is lowered to 0.385 [7]. Moreover, with a flange/web ratio $b/d = 0.905$ we have, from equation (1.36b), the stress condition for local buckling in each of the four half-flanges:

$$\sigma_L = 0.385\ E_T \left(\frac{t}{0.905\,d}\right)^2 = 0.470\ E_T \left(\frac{t}{d}\right)^2 \tag{1.36c}$$

Comparing equations (1.36a) and (1.36c) shows that flange will buckle well before the web. Hence, equation (1.36c) provides the local buckling criterion for an I-section with uniform thickness.

(iv) Optimisation

An optimum geometry is found by combining the criteria (i), (ii) and (iii) above to make the stresses by equations (1.34), (1.35) and (1.36c) equal, $\sigma_F = \sigma_W = \sigma_L$, which gives

$$\frac{2.112\ E_T d^2}{L^2} = \frac{P}{4.62dt} = 0.470\ E_T \left(\frac{t}{d}\right)^2 \tag{1.37}$$

Taking the components of equation (1.37) in pairs allows optimum dimensions d and t to be found:

$$t_{\text{opt}} = 0.630 \left(\frac{P^2 L}{E_T^2}\right)^{1/5}, \quad d_{\text{opt}} = 0.546 \left(\frac{PL^3}{E_T}\right)^{1/5} \tag{1.38a,b}$$

The optimum stress follows from substituting d_{opt} and t_{opt} from equation (1.38a,b) into any one of the three (now equal) stress expressions σ_F, σ_W, or σ_L respectively in equation (1.37). Then, dividing by ρ gives the optimum, equivalent objective function:

$$\left(\frac{\sigma}{\rho}\right)_{\text{opt}} = 0.629 \left(\frac{E_T^{3/5}}{\rho}\right)\left(\frac{P}{L^2}\right)^{2/5} \tag{1.39}$$

Figure 1.9 presents equation (1.39) graphically for four materials whose properties and relevant ratios appear in Table 1.1.

Here the structural index S is reduced twofold compared to closed, thin-walled tubes. Within this restricted range, aluminium alloy (L65) is the most efficient material in raising the objective function to 15 km when $S \approx 0.5$, but titanium alloy extends S to beyond 1.5, raising the objective function to $\approx$ 20 km. Figure 1.10 compares four sections in the titanium alloy, including a solid square, two closed tubes and an I-section.

This shows that both the shape and material efficiency factors influence these plots, along with the appropriate exponent on the structural index S. Clearly, for a given $S < 0.5$,

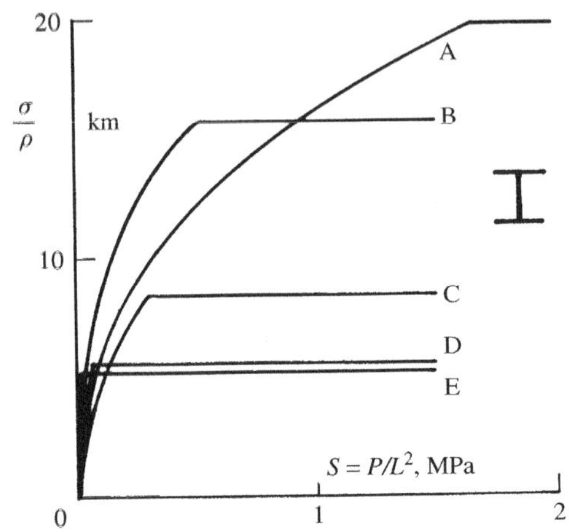

Figure 1.9 Objective function versus S for I-section struts with uniform thickness (key: A, Ti alloy; B, Al alloy; C, steel; D, GFRC; E, Douglas fir (see Table 1.1))

the circular tube is most efficient in supporting a compressive load. Note that an extended range for S appears in the earlier figures. For example, Figure 1.3 shows a maximum value of $S \approx 10$ at the stress limit for a solid square section in titanium.

1.5.2 Non-Uniform I-Section

The dimensions of an I-section having different web and flange thicknesses are shown in Figure 1.11a.

(i) Flexural Buckling

With $2b$ and d as the section's mean outer dimensions and web and flange thicknesses t_w and t_f respectively, the axes x and y are made equally least resistant to buckling when

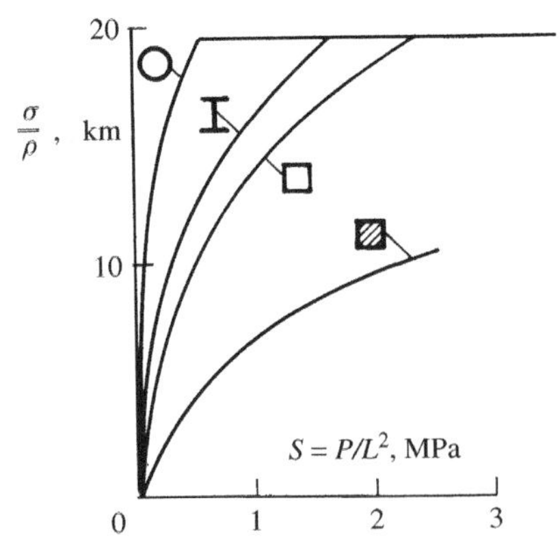

Figure 1.10 Objective function versus S for Ti alloy struts in various cross-sections

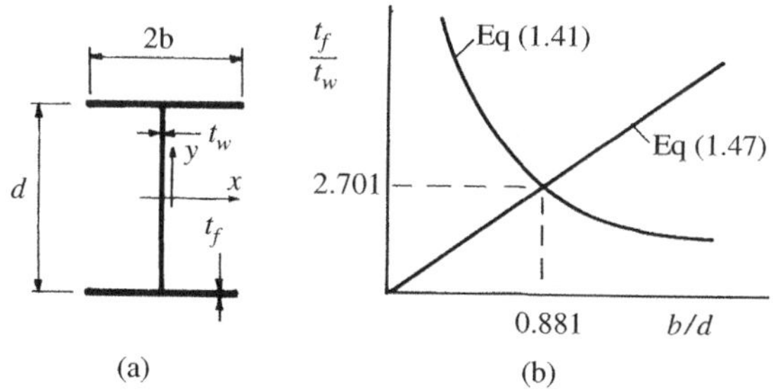

Figure 1.11 I-section showing web and flange geometry

$I_x = I_y$. This gives

$$2\left[2bt_f\left(\frac{d}{2}\right)^2\right] + \frac{t_w(d-t_f)^3}{12} = 2\left[\frac{t_f(2b)^3}{12}\right] + \frac{(d-t_f)t_w^3}{12} \tag{1.40a}$$

Provided b and d are each more than an order of magnitude greater than t_f and t_w, a good approximation to equation (1.40a) is

$$2\left[2bt_f\left(\frac{d}{2}\right)^2\right] + \frac{t_w d^3}{12} = 2\left[\frac{t_f(2b)^3}{12}\right] \tag{1.40b}$$

Equation (1.40b) gives the required thickness ratio

$$\frac{t_f}{t_w} = \left[16\left(\frac{b}{d}\right)^3 - 12\frac{b}{d}\right]^{-1} \tag{1.41}$$

The right-hand side of equation (1.40b) gives

$$I_x = I_y = \frac{4}{3}b^3 t_f \tag{1.42a}$$

and the section area is

$$A = 4bt_f + (d-t_f)t_w \approx 4bt_f + dt_w \tag{1.42b}$$

Hence, from equations (1.42a,b),

$$k^2 = \frac{I_x}{A} = \frac{4b^3 t_f}{3(4bt_f + dt_w)} \tag{1.43c}$$

Substituting equation (1.43c) into equation (1.2b), the buckling stress becomes

$$\sigma_F = \frac{\pi^2 E_T}{L^2}\left[\frac{4b^3 t_f}{3(4bt_f + dt_w)}\right] = \frac{4}{3}\frac{E_T}{}\left(\frac{\pi d}{L}\right)^2\left[\frac{(t_f/t_w)(b/d)^3}{1 + 4(t_f/t_w)b/d}\right] \tag{1.44}$$

(ii) Limiting Stress
The working, compressive stress within the strut is to be

$$\sigma_W = \frac{P}{A} = \frac{P}{4bt_f + dt_w} = \frac{P}{dt_w[1 + 4(t_f/t_w)b/d]} \tag{1.45}$$

where $\sigma_W \leq \sigma_y$, for a proof stress σ_y defined by an offset strain (typically 0.1% or 0.2%). Again, the gradient to the stress–strain curve at the stress level σ_y defines the tangent modulus E_T in equation (1.44).

(iii) Local Buckling
Local buckling expressions are similar to those discussed previously for a uniform I-section [6]. Allowing for the different thicknesses, local web buckling occurs under a critical stress:

$$\sigma_L = 3.62\ E_T \left(\frac{t_w}{d}\right)^2 \tag{1.46a}$$

and local flange buckling occurs under the stress:

$$\sigma_L = 0.385\ E_T \left(\frac{t_f}{b}\right)^2 \tag{1.46b}$$

Here we can arrange for web and flange buckling to occur simultaneously by equating the stresses in equations (1.46a,b). This gives

$$\frac{t_f}{t_w} = 3.066\,\frac{b}{d} \tag{1.47}$$

(iv) Optimisation
Firstly, we equate the thickness ratio t_f/t_w from each of equations (1.41) and (1.47):

$$3.066\,\frac{b}{d}\left[16\left(\frac{b}{d}\right)^3 - 12\frac{b}{d}\right] = 1 \tag{1.48}$$

Equation (1.48) may be reduced to a quadratic equation whose positive root leads to optimised geometric ratios:

$$\left(\frac{b}{d}\right)_{\text{opt}} = 0.881, \quad \left(\frac{t_f}{t_w}\right)_{\text{opt}} = 2.701 \tag{1.49a,b}$$

Alternatively, the graphical solution to equating (1.41) and (1.47) is shown in Figure 1.11b. Now, we equate σ_F to σ_W, from equations (1.44) and (1.45), to give

$$\frac{4\ E_T}{3}\left(\frac{\pi\, d}{L}\right)^2\left[\frac{(t_f/t_w)(b/d)^3}{1 + 4(t_f/t_w)b/d}\right] = \frac{P}{dt_w[1 + 4(t_f/t_w)b/d]}$$

where, from equations (1.49a,b), we find

$$d^3 t_w = \frac{0.0411PL^2}{E_T}$$

which leads to the optimum outer dimensions and thicknesses:

$$b_{\text{opt}} = 0.486\left(\frac{PL^3}{E_T}\right)^{1/5}, \quad d_{\text{opt}} = 0.552\left(\frac{PL^3}{E_T}\right)^{1/5} \tag{1.50a,b}$$

$$t_{w_{\text{opt}}} = 0.244\left(\frac{P^2L}{E_T^2}\right)^{1/5}, \quad t_{f_{\text{opt}}} = 0.659\left(\frac{P^2L}{E_T^2}\right)^{1/5} \tag{1.51a,b}$$

The optimum stress follows from substituting the appropriate optimum dimensions into any one of the three (now equal) stress expressions σ_F, σ_W, or σ_L in equations (1.44)–(1.46). Dividing by ρ gives the optimum objective function

$$\left(\frac{\sigma}{\rho}\right)_{\text{opt}} = 0.705\left(\frac{E_T^{3/5}}{\rho}\right)\left(\frac{P}{L^2}\right)^{2/5} \tag{1.52}$$

Equation (1.52) is similar in form to equation (1.39). Between equations (1.50)–(1.52) it can be seen how this optimised design increases the structural efficiency compared to an I-section with uniform thickness, where $F = 0.629$.

1.5.3 Equal-Angle

Though the equal-angle is the simplest thin-walled, open strut section (see Figure 1.12a), it is complicated by having principal axes of flexure at 45° to its long sides. As there is no web, a single local flange buckling criterion need only be combined with the criterion for flexural buckling about the principal axis with the lesser I-value.

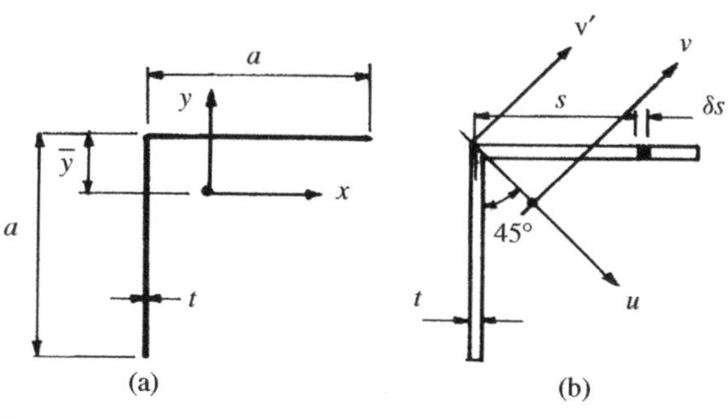

Figure 1.12 Equal-angle showing principal axes u and v

(i) Flexural Buckling

Taking a as the mean side length, it is required to identify k in equation (1.2b) with the least second moment of area. With the side lengths being equal, it is not possible for the section to be equally resistant to buckling about its principal axes u and v in Figure 1.12b. That is, $I_u \neq I_v$ and the lesser I-value follows, firstly, from locating the centroid position $\overline{y} = (\overline{x})$ in Figure 1.12a:

$$(at)t/2 + at(a/2) = 2at\overline{y}, \quad \Rightarrow \quad \overline{y} = (a+t)/4 \simeq a/4 \tag{1.53}$$

With each limb being equally inclined to the u-axis (see Figure 1.12b), $I_u = \int_A v^2 \mathrm{d}A$ is applied with $v = s \sin 45^\circ$ and $\mathrm{d}A = t\,\mathrm{d}s$. This gives:

$$I_u = 2\int_0^a (s \sin 45^\circ)^2 \times t\,\mathrm{d}s = \frac{a^3 t}{3} \tag{1.54a}$$

To find I_v we first find the second moment of the area about an axis v' passing through the corner (see Figure 1.11b) and then transfer it to the parallel centroidal axis v by using equation (1.53) and the parallel axis theorem:

$$I_v = I_{v'} - Ah^2 = 2\int_0^a (s \cos 45^\circ)^2 t ds - 2at(\surd 2\overline{y})^2$$

$$= \frac{a^3 t}{3} - \frac{a^3 t}{4} = \frac{a^3 t}{12} \tag{1.54b}$$

Clearly, from equations (1.54a,b), I_v is the lesser of the two I-values, from which

$$k^2 = \frac{I_v}{A} = \frac{a^3 t/12}{2at} = \frac{a^2}{24} \tag{1.55a}$$

Substituting equation (1.55a) into the buckling stress equation (1.2b) gives

$$\sigma_F = \frac{\pi^2 E_T a^2}{24 L^2} \tag{1.55b}$$

(ii) Limiting Stress

The working, compressive stress for this strut is simply

$$\sigma_W = \frac{P}{A} = \frac{P}{2at} \tag{1.56}$$

where $\sigma_W \leq \sigma_y$, for a proof stress σ_y defined by an offset strain (typically 0.1% or 0.2%). The gradient to the stress–strain curve corresponding to σ_y defines the tangent modulus E_T in equation (1.55b). If σ_y lies at a the limit of proportionality, then E replaces E_T.

(iii) Local Buckling
Referring to Figure 1.12a, a local buckling of both limbs occurs simultaneously when the stress attains it limiting value:

$$\sigma_L = K_L\ E_T \left(\frac{t}{a}\right)^2 \tag{1.57}$$

where, as for the previous section considered, $K_L = 0.385$ for $v = 0.3$ [3].

(iv) Optimisation
An optimum geometry is found by combining criteria (i), (ii) and (iii) above so that the stresses by equations (1.55)–(1.57) are made equal, $\sigma_F = \sigma_W = \sigma_L$, which gives

$$\frac{\pi E_T a^2}{24L^2} = \frac{P}{2at} = 0.385\ E_T \left(\frac{t}{a}\right)^2 \tag{1.58}$$

Equation (1.58) allows optimum dimensions a and t to be found:

$$t_{\text{opt}} = 1.103 \left(\frac{P^2 L}{E_T^2}\right)^{1/5}, \quad a_{\text{opt}} = 1.033 \left(\frac{PL^3}{E_T}\right)^{1/5} \tag{1.59a,b}$$

The optimum stress follows from substituting a_{opt} and t_{opt} into any one of the three (now equal) stress expressions σ_F, σ_W, or σ_L in equation (1.58). Dividing by ρ gives this section's optimum objective function

$$\left(\frac{\sigma}{\rho}\right)_{\text{opt}} = 0.439 \left(\frac{E_T^{3/5}}{\rho}\right)\left(\frac{P}{L^2}\right)^{2/5} \tag{1.60}$$

1.5.4 Unequal-Angle

The previous analysis of an equal-angle section suggests that a greater degree of optimisation is possible when the lengths a, b and thicknesses t_a, t_b of the two limbs are different. The principal axes of flexure will no longer lie at 45°. Again, the axis having the minimum second moment of area will control flexural buckling while, simultaneously, a local buckling in each limb can be arranged to occur for the optimum design. Here, we shall firstly equalise the local flange buckling stress for each limb. The constraint that this imposes upon the angle's geometry is then applied to the flexural buckling of an angle section with any limb length ratio $L = b/a$, where $b > a$, in Figure 1.13a.

(i) Local Buckling
If the limb lengths a and b and their respective thicknesses t_a and t_b are different, we can ensure that their local buckling occurs simultaneously. From equation (1.57), this condition gives

$$\sigma_L = K_L\ E_T \left(\frac{t_a}{a}\right)^2 = K_L E_T \left(\frac{t_b}{b}\right)^2 \tag{1.61a}$$

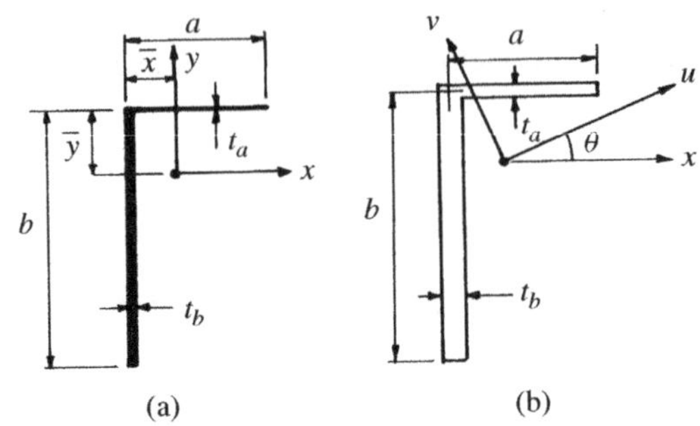

Figure 1.13 Unequal angle showing principal axes

where $K_L = 0.385$. Hence, the simple geometrical relationship

$$r = b/a = t_b/t_a \tag{1.61b}$$

follows from equation (1.61a).

(ii) Flexural Buckling

The centroidal coordinates, $(\overline{x}, \overline{y})$ in Figure 1.13a, are found from

$$bt_b^2/2 + a^2t_a/2 = (at_a + bt_b)\overline{x}, \Rightarrow \overline{x} \approx \frac{a}{2[1 + (b/a)(t_b/t_a)]} = \frac{a}{2(1 + r^2)}$$

$$at_a^2/2 + b^2t_b/2 = (at_a + bt_b)\overline{y}, \Rightarrow \overline{y} \approx \frac{b}{2[1 + (a/b)(t_a/t_b)]} = \frac{br^2}{2(1 + r^2)} \tag{1.62a,b}$$

The second moments of area, referred to centroidal axes x and y, are

$$I_x = at_a\overline{y}^2 + t_b b^3/12 + bt_b(b/2 - \overline{y})^2 = \frac{a^3t_ar^4(r^2 + 4)}{12(1 + r^2)}$$

$$I_y = bt_b\overline{x}^2 + t_a a^3/12 + at_a(a/2 - \overline{x})^2 = \frac{a^3t_a(4r^2 + 1)}{12(1 + r^2)} \tag{1.63a,b,c}$$

$$I_{xy} = at_a(a/2 - \overline{x})\overline{y} + bt_b(b/2 - \overline{y})\overline{x} = \frac{a^3t_ar^3}{4(1 + r^2)}$$

The principal second moments of area $I_{u,v}$ and their orientation θ with respect to axis x (see Figure 1.13b) are found from

$$I_u, I_v = \frac{1}{2}(I_x + I_y) \pm \frac{1}{2}\sqrt{(I_x - I_y)^2 + 4I_{xy}^2}$$

$$\tan 2\theta = 2I_{xy}/(I_x - I_y) \tag{1.64a,b,c}$$

where $I_u > I_v$ when θ refers to the inclination of u with respect to x (see Figure 1.13b). Substituting equations (1.63a–c) into equation (1.64b), the lesser I is seen to be

$$I_v = \left[\frac{(r^6 + 4r^4 + 4r^2 + 1) - \sqrt{(r^6 + 4r^4 - 4r^2 - 1)^2 + 36r^6}}{24(1 + r^2)}\right] a^3 t_a = \frac{a^3 t_a n(r)}{24(1 + r^2)} \tag{1.65a}$$

where the function $n(r)$ is the expression in square brackets. The section area is

$$A = at_a + bt_b = at_a(1 + r^2) \tag{1.65b}$$

Dividing equations (1.65a,b), the least radius of gyration $k^2 = I_v/A$ is written as

$$k^2 = \frac{a^2 n(r)}{24(1 + r^2)^2} \tag{1.65c}$$

It follows from equation (1.65c) that the buckling stress in equation (1.2b) becomes

$$\sigma_F = \frac{\pi^2 E_T a^2 n(r)}{24(1 + r^2)^2 L^2} \tag{1.66}$$

(iii) Limiting Stress

The working, compressive stress in the strut is given as

$$\sigma_W = \frac{P}{A} = \frac{P}{at_a + bt_b} = \frac{P}{at_a(1 + r^2)} \tag{1.67}$$

where $\sigma_W \leq \sigma_y$, for a proof stress σ_y defined by an offset strain (typically 0.1% or 0.2%). The gradient to the stress–strain curve, corresponding to σ_y, defines the tangent modulus E_T in equation (1.66).

(iv) Optimisation

An optimum geometry is found by combining criteria (i), (ii) and (iii) above so that the stresses by equations (1.61), (1.66) and (1.67) are made equal, $\sigma_F = \sigma_W = \sigma_L$, which gives

$$\frac{\pi^2 E_T a^2 n(r)}{24(1 + r^2)^2 L^2} = \frac{P}{at_a(1 + r^2)} = 0.385\, E_T \left(\frac{t_a}{a}\right)^2 \tag{1.68}$$

Equation (1.68) allows the optimum dimensions a_{opt} and $(t_a)_{opt}$ to be found for an angle-section with a particular $R = b/a$ value:

$$(t_a)_{opt} = \frac{1.455}{\{(1 + r^2)^2\,[n(r)]\}^{1/10}} \left(\frac{P^2 L}{E_T^2}\right)^{1/5} \tag{1.69a,b}$$

$$a_{opt} = 1.187 \left\{\frac{(1 + r^2)^4}{[n(r)]^3}\right\}^{1/10} \left(\frac{PL^3}{E_T}\right)^{1/5}$$

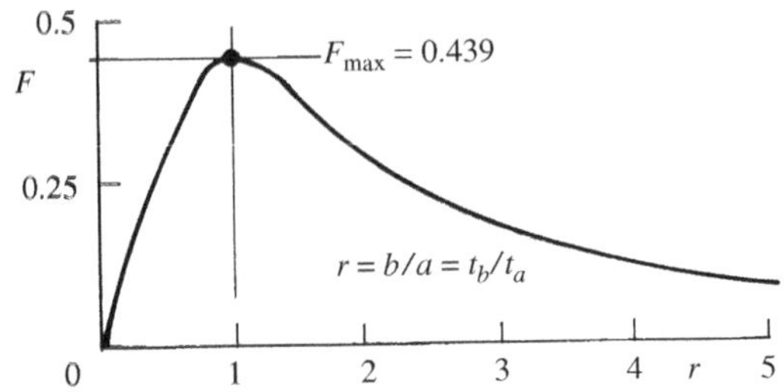

Figure 1.14 Dependence of shape function F upon limb ratio r

Correspondingly, the optimum stress follows from substituting a_{opt} and $(t_a)_{opt}$ into any one of the three (now equal) stress expressions σ_F, σ_W, or σ_L in equation (1.68). Dividing by ρ gives the objective function:

$$\left(\frac{\sigma}{\rho}\right)_{opt} = 0.579\left[\frac{n(r)}{(1+r^2)^3}\right]^{2/5}\left(\frac{E_T^{3/5}}{\rho}\right)\left(\frac{P}{L^2}\right)^{2/5} = f\left(\frac{E_T^{3/5}}{\rho}\right)\left(\frac{P}{L^2}\right)^{2/5} \quad (1.70)$$

Now it becomes clear that if the strut weight is to be minimised, how this optimum stress should be maximised. Thus, for a strut of given material, length and loading, a maximum for the coefficient term in r is required. Identifying this coefficient term within the shape efficiency factor, $F = 0.579[n(r)/(1+r^2)^3]^{2/5}$, the value of r that maximises F is shown in Figure 1.14.

Having used $r = b/a = t_b/t_a$ for the above analysis, the graph shows that it is the equal angle, i.e. $r = 1$, which has the minimum weight possible for this condition. Hence the shape efficiency factor expression in equation (1.70) reduces to $F = 0.439$, agreeing with that found previously for an equal-angle. The optimum values of F, for $1 < r \leq 5$ may be read from the graph. It is left as an exercise for the reader to investigate the effect upon the optimum condition of any alternative geometric constraint to that in equation (1.61b). For example, BS sections [8] use $R = 1.16$, 1.2, 1.33, 1.5, 1.6, 2, 2.25, with thicknesses having an inner taper.

1.6 Summary of Results

Table 1.4 summarises the results for the objective functions of the 11 strut sections considered here. It is seen that only the structural index $S = P/L^2$ is common to all sections. However, its exponent n differs along with the material efficiency factor within the three categories of section shown in Table 1.4. These categories refer to flexural buckling only and flexural buckling combined with local buckling in curved and flat plates. Where more than one section lies within a given category it becomes possible to compare their shape efficiency factors. Given the inverse relationship between the equivalent objective function and the weight, we may conclude that those strut sections which minimise weight in solid, closed and open sections most effectively are: (i) a solid

Table 1.4 Equivalent objective functions for various struts: $R = F \times M \times S^n$ [$R = (\sigma/\rho)_{opt}$, $S = P/L^2$]

Cross-section	Shape efficiency factor F	Material efficiency factor M	Exponent n
	0.886	$E_T^{1/2}/\rho$	1/2
	0.907	$E_T^{1/2}/\rho$	1/2
	0.663	$E_T^{1/2}/\rho$	1/2
	0.975	$E_T^{1/2}/\rho$	1/2
	0.540	$E_T^{2/3}/\rho$	1/3
	0.907	$E_T^{3/5}/\rho$	2/5
	1.112	$E_T^{3/5}/\rho$	2/5
	0.630	$E_T^{3/5}/\rho$	2/5
	0.705	$E_T^{3/5}/\rho$	2/5
	0.439	$E_T^{3/5}/\rho$	2/5

square, (ii) a regular, thin-walled, hexagonal tube and (iii) a non-uniform I-section.

References

[1] Timoshenko, S. J. and Gere, J. M. *Theory of Elastic Stability*, McGraw-Hill, 1961.
[2] ESDU 78021, Guide to items on the strength and stability of struts, October 1978.
[3] ESDU 72019, Buckling of flat isotropic plates under uniaxial and biaxial loading, August 1972.
[4] Gerard, G. Compressive and torsional buckling of thin-walled cylinders in the yield region, NACA Tech. Note 3726, August 1956.
[5] Cox, H. L. *The Design of Structures of Least Weight*, Pergamon, 1965.
[6] ESDU 01.01.08, Local instability of struts with flat sides, November 1990.
[7] ESDU 78020, Local buckling and crippling of I, Z and channel section struts, July 1978.
[8] BS 4848 Part 2: 1991, Specifications for hot rolled structural steel sections: unequal angles.

Exercises

1.1 Confirm the shape factors C and F, given in Table 1.2, that appear within the optimum dimension and the objective function for pinned-end struts, having the following cross-sections: (i) a solid semi-circle, diameter d, and (ii) an equilateral triangle, side length a.

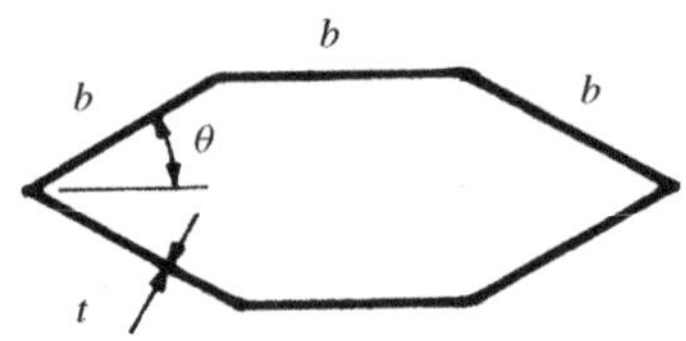

Figure 1.15 Flattened hexagonal tube

1.2 Using the local buckling coefficients K_L, listed in Table 1.3, confirm those shape efficiency factors F given for regular polygons with seven sides and more.

1.3 When the regular hexagonal tube in Section 1.4.3 is flattened so that the included angle θ is not 60°, the second moments of area about the x- and y-axes are no longer equal. Examine the effect upon the shape efficiency factor F when: (i) $0° < \theta < 45°$ and (ii) $45° < \theta < 90°$ (see Figure 1.15). Confirm the known result in Table 1.3 for when $\theta = 60°$ in each case. [Answer: (i) $F = 1.248(\sin\theta)^{4/5}$]

1.4 A long, aluminium-alloy strut ($E = 70$ GPa), with a uniformly square, tubular cross-section and pinned ends, is required to bear a compressive load of 10 kN in a 2 m length. Obtain the values of the mean side length a and wall thickness t that will minimise the weight (i.e. minimise the section area). Show this graphically in a plot of the objective function against cross-section areas ranging from 100 mm^2 to 350 mm^2 in steps of 50 mm^2. Construct a 2D design space with boundaries fixed by the following constraints:

(i) flexural buckling must not occur;
(ii) local bucking of the walls is to be avoided;
(iii) a must lie in the range $40 \leq a \leq 100$ mm;
(iv) t must lie in the range $0.9 \leq t \leq 2.5$ mm.

1.5 Investigate the optimum design condition for a strut with an unequal angle section where $t_b/b < t_a/a$, such that local buckling occurs only in the longer limb.

1.6 Investigate the optimum design condition for a strut having an unequal angle section with $b/a = 3$ and constant thickness such that local buckling occurs only in the longer limb.

1.7 A long, thin-walled, regular hexagonal tube in steel is to support a compressive load of 20 kN over its 5 m pinned length. Construct a 2D design space bounded by the following requirements:

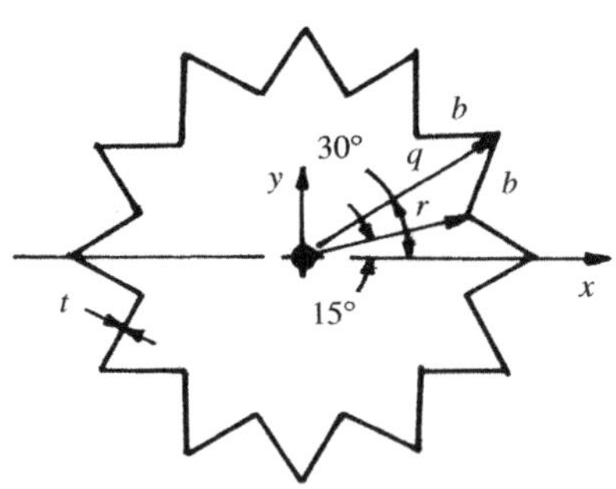

Figure 1.16 Tubular strut having 12 vee-corrugations

(i) flexural buckling must not occur
(ii) local buckling of the four walls must not occur
(iii) the mean side length is to lie between 150 mm and 200 mm
(iv) the wall thickness is to lie between 2.5 mm and 5 mm.

Find the side wall dimensions a and t which minimise the area and so the weight. Take $E = 210$ GPa.

1.8 A strut is made in the form of a closed tube with 12 vee-corrugations, each having flats $b \times t$, as shown in Figure 1.16. Derive an expression for the second moment of area in terms of the geometry given about the axis x. Is this the least value for the section? Derive the objective function when flexural and local buckling occur together under a limiting compressive stress σ_y. Hence, find the shape efficiency factor and compare with that for a thin, circular tube.

2

Compression of Wide Struts

2.1 Introduction

A similar approach is adopted to optimise both wide and slender struts. This ensures that the weight of the strut is a minimum when the stress is at a maximum. Again, it is necessary to investigate how the stress σ, which may increase to a limiting value, is affected by the magnitude of the load P and the strut length L, in consideration of its flexural and lateral buckling behaviour. As with slender struts (Chapter 1) the general approach for wide struts is to maximise an objective function, chosen to contain stress and density. We let A and W be the area and weight respectively of a repeating unit in the width that does not vary in the length, as for a uniform corrugated or cellular section. When a compressive load P is distributed uniformly over the full width, such that w is the load distributed across the pitch length p for a single unit, its axial stress becomes $\sigma = P/A = wp/A$. This gives the required weight function as:

$$W = \rho AL = \rho \frac{wp}{\sigma} L = \frac{wpL}{\sigma/\rho} \tag{2.1}$$

where ρ is the density and L here refers to the pinned strut length. Hence, to minimise the weight of a strut when carrying a given load, it follows that σ/ρ, the equivalent objective function, is to be maximised.

2.2 Failure Criteria

The stress within the denominator of equation (2.1) can be expressed in terms of the section's shape according to the following failure criteria.

2.2.1 *Flexural buckling*

Euler's theory expresses the critical elastic buckling load P_c, for a pinned-end strut, as

$$P_c = \frac{\pi^2 EI}{L^2} \tag{2.2a}$$

Mechanics of Optimal Structural Design: Minimum Weight Structures D. W. A. Rees

where $I = Ak^2$ in which k is the radius of gyration of the section. As a wide strut can only bow in a direction perpendicular to its plane, this predetermines the least second moment in equation (2.2a). Hence, a critical flexural buckling stress may be expressed as

$$\sigma_F = \frac{\pi^2 E}{(L/k)^2} \tag{2.2b}$$

If the flexural buckling involves plasticity, the tangent modulus E_T may replace the elastic modulus E in equations (2.2a,b) [1].

2.2.2 *Local Buckling*

Buckling within the thin walls of a wide strut occurs under the local buckling stress

$$\sigma_L = K_L E_T \left(\frac{t}{d}\right) \tag{2.3}$$

where K_L is a coefficient, usually taken as 3.62 for long plates with simply supported edges [2]. It is seen in Appendix D that other values of the coefficient K_L apply to sections with increased rotational restraint as, for example, in corrugations with unequal lengths of horizontal and vertical flats.

2.2.3 *Working Stress*

A uniform, axial compressive working stress σ_W can also be found from the load w, distributed across the pitch length p and area A, as

$$\sigma_W = \frac{P}{A} = \frac{wp}{A} \tag{2.4}$$

2.2.4 *Limiting Stress*

The stress in equations (2.2)–(2.4) would normally be limited to the yield stress where buckling is elastic. In the case of plastic buckling the limiting stress is raised to that corresponding to a given offset (plastic) strain. i.e. the 0.1% proof stress. The most efficient designs ensure that the material is worked to its maximum capacity, i.e. the stresses that arise from local and flexural buckling occur simultaneously at this limiting stress value. Let σ_y be the limiting yield or proof stress for the strut material. To ensure that the strut bears an applied load safely its working stress must be made less than the material's limiting stress:

$$\sigma \le \sigma_y \tag{2.5}$$

2.2.5 *Objective Function*

It follows from the three failure criteria above that an optimum section size is found by equating the stresses in equation (2.2)–(2.4), to the limit set by equation (2.5).

This ensures that all appropriate failure criteria appear within an objective function for minimum weight. We shall now derive this function for wide cellular and corrugated strut sections. Writing the denominator of equation (2.1) as an objective function ratio $R = \sigma/\rho$, its optimisation will appear as the product of separable independent variables in the form

$$R = F \times M \times S^n \tag{2.6}$$

where, as for slender struts, F is the shape efficiency factor, M is the material efficiency factor and S is the structural index whose exponent n is fractional.

2.3 Cellular Sections

Cells within a wide strut are normally square or hexagonal to provide for the required connection between their adjacent vertical walls. Unlike an open corrugated design, there appears to be little weight saving from selecting other regular polygons for closed cells or from varying their wall thicknesses [3]. However, it will be shown that the connection point between regular cells is influential.

2.3.1 Square Cells $b \times b$, Thickness t (see Figure 2.1)

(i) Flexural Buckling

Three sides per pitch, two horizontal and one vertical (see Figure 2.1), give the area properties

$$A = 3bt,\ I = 7b^3t/12 \text{ and } k^2 = I/A = 7b^2/36$$

The failure criterion (2.2b) gives the flexural buckling stress

$$\sigma_F = \frac{7\pi^2 E_T b^2}{36L^2} \tag{2.7}$$

(ii) Limiting Stress

The working, compressive stress σ in the strut appears from equation (2.4) as

$$\sigma_W = \frac{P}{A} = \frac{wb}{3bt} = \frac{w}{3t} \tag{2.8}$$

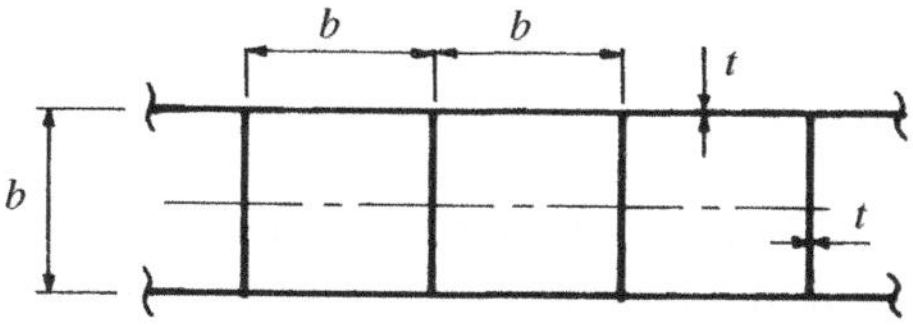

Figure 2.1 Wide strut with square cells

Here $\sigma_W \leq \sigma_y$, where σ_y is a proof stress defined by an offset strain (typically 0.1% or 0.2%). The tangent modulus E_T in equation (2.7) is the gradient to the stress – strain curve at the offset strain corresponding to σ_y.

(iii) Local Buckling

When each wall is taken to be a long flat plate with simply supported corners (see Figure 1.6b), a local, lateral collapse occurs under the axial compressive stress:

$$\sigma_L = K_L E_T \left(\frac{t}{b}\right)^2 \tag{2.9}$$

where $K_L = 3.62$ for each side plate, taking $\nu = 0.3$ and assuming simply supported edges [2].

(iv) Optimisation

An optimum geometry is found by combining (i), (ii) and (iii) above so that the stresses by the three criteria are made equal, $\sigma_F = \sigma_W = \sigma_L$, which gives

$$\frac{7\pi^2 E_T b^2}{36L^2} = \frac{w}{3t} = 3.62\, E_T \left(\frac{t}{b}\right)^2 \tag{2.10}$$

Equation (2.10) allows the optimum dimensions b and t to be found as

$$t_{\text{opt}} = 0.356\left(\frac{wL}{E_T}\right)^{1/2}, \quad b_{\text{opt}} = 0.699\left(\frac{wL^3}{E_T}\right)^{1/4} \tag{2.11a,b}$$

The optimum stress follows from substituting d_{opt} and t_{opt} into any of the three stress expressions in equation (2.10). Dividing by ρ gives the objective function in the form of equation (2.6):

$$\left(\frac{\sigma}{\rho}\right)_{\text{opt}} = 0.937\left(\frac{E_T^{1/2}}{\rho}\right)\left(\frac{w}{L}\right)^{1/2} \tag{2.12}$$

Equation (2.12) shows that the shape efficiency factor $F = 0.937$, the material efficiency factor $M = \sqrt{E_T}/\rho$ and the structural index $S = w/L$, whose exponent is $n = 1/2$.

2.3.2 *Regular Hexagonal Cells – Corner Connection*

In Figure 2.2 each regular hexagonal cell has side lengths b, pitch $2b$ and thickness t. The corner connection ensures straight, 60° sloping sides between the cells.

(i) Flexural Buckling

For a single pitch having six sides – two horizontal and four sloping – with included angles of 120° between them, the second moment of area about the horizontal axis is

$$I = 4\left(\frac{b^3 t}{3}\sin^2 60^\circ\right) + 2\left[(bt)(b\sin 60^\circ)^2\right] = \frac{5}{2}b^3 t \tag{2.13a}$$

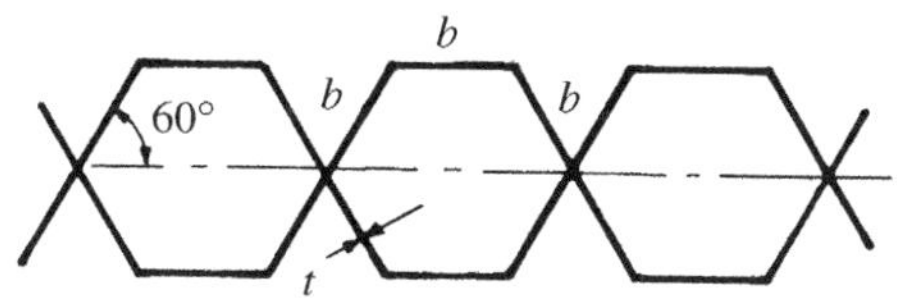

Figure 2.2 Wide strut with hexagonal cells

The cell wall area and the radius of gyration follow as

$$A = 6bt, \quad k^2 = I/A = 5b^2/12 \tag{2.13b,c}$$

The failure criterion (2.2b) becomes

$$\sigma_F = \frac{5\pi^2 E_T b^2}{12L^2} \tag{2.14}$$

(ii) Limiting Stress
The working, compressive stress for the strut is given as

$$\sigma_W = \frac{P}{A} = \frac{w(2b)}{6bt} = \frac{w}{3t} \tag{2.15}$$

where $\sigma_W \leq \sigma_y$, for σ_y a proof stress defined by an offset strain (typically 0.1% or 0.2%).

(iii) Local Buckling
Again, when each wall surface is modelled as a long flat plate with simply supported corners, buckling occurs under the axial compressive stress:

$$\sigma_L = K_L E_T \left(\frac{t}{b}\right)^2 \tag{2.16}$$

where $K_L = 3.62$ applies to each side wall of the hexagonal section.

(iv) Optimisation
An optimum geometry is found by combining criteria (i), (ii) and (iii) above, so that the stresses by equations (2.14)–(2.16) become equal, $\sigma_F = \sigma_W = \sigma_L$, which gives

$$\frac{5\pi^2 E_T b^2}{12L^2} = \frac{w}{3t} = 3.62\, E_T \left(\frac{t}{b}\right)^2 \tag{2.17}$$

Equation(2.17) provides the optimum dimensions b and t as

$$b_{\text{opt}} = 0.525 \left(\frac{wL^3}{E_T}\right)^{1/4}, \quad t_{\text{opt}} = 0.294 \left(\frac{wL}{E_T}\right)^{1/2} \tag{2.18a,b}$$

The optimum stress follows from substituting d_{opt} and t_{opt} into any of the three expressions in equation (2.17). Dividing by ρ gives the objective function in the form

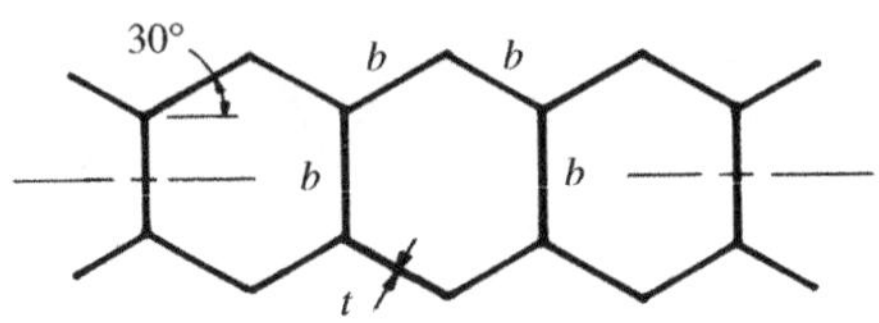

Figure 2.3 Hexagonal, cellular tube section with common walls

of equation (2.6):

$$\left(\frac{\sigma}{\rho}\right)_{\text{opt}} = 1.134\left(\frac{E_T^{1/2}}{\rho}\right)\left(\frac{w}{L}\right)^{1/2} \tag{2.19}$$

2.3.3 *Regular Hexagonal Cells – Wall Connection*

When the hexagons are rearranged as shown in Figure 2.3, the pitch is altered to $\sqrt{3}b$ and five of its walls contribute to I about its horizontal flexural axis.

(i) Flexural Buckling
Using the result $I_x = 7tb^3/12$ for one sloping side found previously (see § 1.4.3), the five sides within a single pitch give

$$I = \left(4 \times \frac{7tb^3}{12}\right) + \frac{tb^3}{12} = \frac{29tb^3}{12} \tag{2.20a}$$

The wall area and radius of gyration become

$$A = 5bt, \quad k^2 = I/A = 29b^2/60 \tag{2.20b,c}$$

The buckling failure criterion, equation (2.2b), becomes

$$\sigma_F = \frac{29\pi^2 E_T b^2}{60L^2} \tag{2.21}$$

(ii) Limiting Stress
The working, compressive stress σ in the strut is given as:

$$\sigma_W = \frac{P}{A} = \frac{w(\sqrt{3}b)}{5bt} = \frac{\sqrt{3}w}{5t} \tag{2.22}$$

where $\sigma_W \leq \sigma_y$, for σ_y a proof stress defined by an offset strain (typically 0.1% or 0.2%).

(iii) Local Buckling
A lateral collapse of each wall surface occurs under an axial stress:

$$\sigma_L = K_L E_T \left(\frac{t}{b}\right)^2 \tag{2.23}$$

where $K_L = 3.62$ for each side plate, taking $\nu = 0.3$ and assuming simply supported edges.

(iv) Optimisation
An optimum geometry is found by combining criteria (i), (ii) and (iii) so that the stresses in equations (2.21)–(2.23) become equal. That is, $\sigma_F = \sigma_W = \sigma_L$, which gives

$$\frac{29\pi^2 E_T b^2}{60L^2} = \frac{\sqrt{3}w}{5t} = 3.62E_T\left(\frac{t}{b}\right)^2 \tag{2.24}$$

This allows optimum dimensions b and t to be found:

$$t_{\text{opt}} = 0.289\left(\frac{wL}{E_T}\right)^{1/2}, \quad b_{\text{opt}} = 0.502\left(\frac{wL^3}{E_T}\right)^{1/4} \tag{2.25a,b}$$

The optimum stress follows from substituting d_{opt} and t_{opt} into any of the three expressions in equation (2.24). Dividing by ρ gives the equivalent objective function, conforming to equation (2.6):

$$\left(\frac{\sigma}{\rho}\right)_{\text{opt}} = 1.199\left(\frac{E_T^{1/2}}{\rho}\right)\left(\frac{w}{L}\right)^{1/2} \tag{2.26}$$

These results show that the common vertical side design is marginally heavier than one with the cells meeting at a corner. However, having a common vertical side is preferable where cells are to be used in-line and would assist with cell manufacture.

2.3.4 *Double-Vee Corrugated Cells*

In Figure 2.4 each 'cell' is formed when two identical corrugated sheets are abutted in the manner shown. With b, t and θ as variables, the pitch becomes $2b\cos\theta$.

(i) Flexural Buckling
The second moment of area about the horizontal axis is

$$I = 4\int_0^b (t\,\mathrm{d}s)(s\sin\theta)^2 = \frac{4tb^3}{3}\sin^2\theta \tag{2.27a}$$

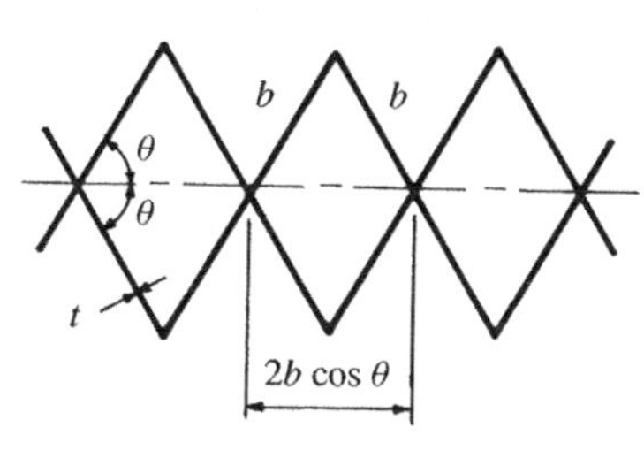

Figure 2.4 Corrugated, cellular construction

The wall area and radius of gyration are

$$A = 4bt, \quad k^2 = I/A = {}^1\!/\!_3 b^2 \sin^2\theta \tag{2.27b,c}$$

Substituting equation (2.27c) into equation (2.2b), the failure criterion becomes

$$\sigma_F = \frac{\pi^2 \sin^2\theta\, E_T b^2}{3L^2} \tag{2.28}$$

(ii) Limiting Stress
The working, compressive stress for this strut is given as

$$\sigma_W = \frac{P}{A} = \frac{w(2b\cos\theta)}{4bt} = \frac{w\cos\theta}{2t} \tag{2.29}$$

where $\sigma_W \leq \sigma_y$, in which σ_y is a proof stress defined by offset strain (typically 0.1% or 0.2%).

(iii) Local Buckling
Taking each side plate to be a long flat surface with simply supported corners, a lateral collapse is expected under an axial stress:

$$\sigma_L = K_L E_T \left(\frac{t}{b}\right)^2 \tag{2.30}$$

where $K_L = 3.62$ when $\nu = 0.3$.

(iv) Optimisation
An optimum geometry is found by combining criteria (i), (ii) and (iii) above so that the stresses by equations (2.28)–(2.30) are made equal, $\sigma_F = \sigma_W = \sigma_L$, which gives

$$\frac{\pi^2 E_T b^2 \sin^2\theta}{3L^2} = \frac{w\cos\theta}{2t} = 3.62\, E_T \left(\frac{t}{b}\right)^2 \tag{2.31}$$

Equation (2.31) allows the optimum dimensions b and t to be found as

$$t_{\text{opt}} = \frac{0.381}{\sqrt{\tan\theta}} \left(\frac{wL}{E_T}\right)^{1/2}, \quad b_{\text{opt}} = 0.632 \left(\frac{\cos\theta}{\sin^3\theta}\right)^{1/4} \left(\frac{wL^3}{E_T}\right)^{1/4} \tag{2.32a,b}$$

The optimum stress follows from substituting d_{opt} and t_{opt} into any of the three expressions in equation (2.31). Dividing σ by ρ gives an objective function conforming to equation (2.6):

$$\left(\frac{\sigma}{\rho}\right)_{\text{opt}} = 0.929\sqrt{\sin 2\theta} \left(\frac{E_T^{1/2}}{\rho}\right) \left(\frac{w}{L}\right)^{1/2} \tag{2.33}$$

Clearly, the shape efficiency factor $F = 0.929\sqrt{(\sin 2\theta)}$ attains a maximum when $\theta = 45^\circ$, giving the constant coefficients for t_{opt}, b_{opt} and $(\sigma/\rho)_{\text{opt}}$ in equations (2.32a,b) and (2.33) as 0.381, 0.751 and 0.929, respectively.

2.4 Open Sections

2.4.1 Corrugated Vee Section

A single sheet of V-corrugations has its flexural axis at half depth, where the pitch is $b\cos\theta$, as shown in Figure 2.5.

(i) Flexural Buckling
The second moment of area about the flexural axis is

$$I = 2\int_0^{b/2} (t\mathrm{d}s)(s\sin\theta)^2 = \frac{tb^3}{12}\sin^2\theta \tag{2.34a}$$

corresponding to an area and radius of gyration given by

$$A = bt, \quad k^2 = I/A = (b^2/12)\sin^2\theta \tag{2.34b,c}$$

The failure criterion (2.2b) becomes

$$\sigma_F = \frac{\pi^2\sin^2\theta E_T b^2}{12L^2} \tag{2.35}$$

(ii) Limiting Stress
The working, compressive stress in the strut is given as

$$\sigma_W = \frac{P}{A} = \frac{w(b\cos\theta)}{bt} = \frac{w\cos\theta}{t} \tag{2.36}$$

where $\sigma_W \le \sigma_y$, for σ_y is a proof stress defined by an offset strain (typically 0.1% or 0.2%).

(iii) Local Buckling
As before, when the end connection of each leg of the V-section is taken to be simply supported, a lateral collapse (bowing) of the sloping side plate is expected under a critical axial, compressive stress:

$$\sigma_L = K_L E_T\left(\frac{t}{b}\right)^2 \tag{2.37}$$

where $K_L = 3.62$, assuming $\nu = 0.3$.

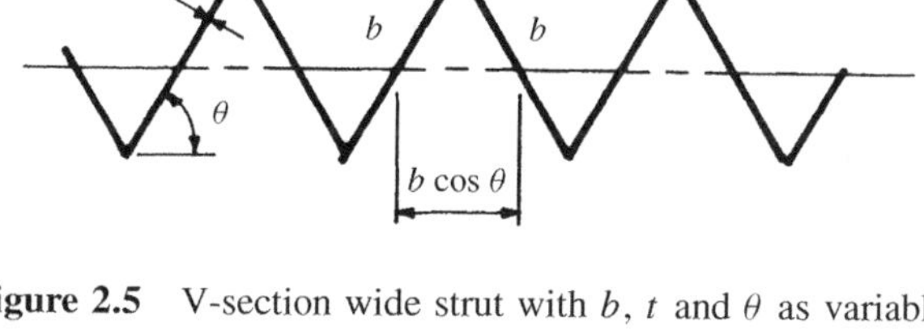

Figure 2.5 V-section wide strut with b, t and θ as variables

(iv) Optimisation
An optimum geometry is found by combining criteria (i), (ii) and (iii) so that the stresses by equations (2.35)–(2.37) become equal, $\sigma_F = \sigma_W = \sigma_L$, which gives

$$\frac{\pi^2 E_T b^2 \sin^2\theta}{12L^2} = \frac{w\cos\theta}{t} = 3.62\, E_T \left(\frac{t}{b}\right)^2 \tag{2.38}$$

This allows the optimum dimensions b and t to be found:

$$t_{opt} = \frac{0.761}{\sqrt{\tan\theta}}\left(\frac{wL}{E_T}\right)^{1/2}, \quad b_{opt} = 1.265\left(\frac{\cos\theta}{\sin^3\theta}\right)^{1/4}\left(\frac{wL^3}{E_T}\right)^{1/4} \tag{2.39a,b}$$

The optimum stress follows from substituting d_{opt} and t_{opt} into any of the three expressions in equation (2.38). Dividing by ρ gives the optimum, equivalent objective function:

$$\left(\frac{\sigma}{\rho}\right)_{opt} = 0.929\sqrt{\sin 2\theta}\left(\frac{E_T^{1/2}}{\rho}\right)\left(\frac{w}{L}\right)^{1/2} \tag{2.40}$$

Figure 2.6 shows the influence of θ upon the shape efficiency factor $F = 0.929\sqrt{(\sin 2\theta)}$. Clearly, this objective function attains a maximum when $\theta = 45°$, giving $F_{max} = 0.929$.

Correspondingly, the constant coefficients for t_{opt}, b_{opt} and $(\sigma/\rho)_{opt}$ in equations (2.39a,b) and (2.40) become 0.761, 1.503 and 0.929, respectively. The fact that the shape efficiency factor remains at 0.929 indicates that no further resistance to the flexural buckling mode is provided through adding a second corrugated sheet in the manner of the double-V design in § 2.3.4. However, a double-V will provide an enormous increase in torsional stiffness to offset torsional buckling when required. The torsion analysis is simplified when the path of each corrugation is taken as a closed tube (see Chapter 4).

2.4.2 Corrugated Section – Equal Flats at 90°

The sheet of square corrugations shown in Figure 2.7 has equal vertical and horizontal flats $b \times t$, with its flexural axis at half depth and a pitch b.

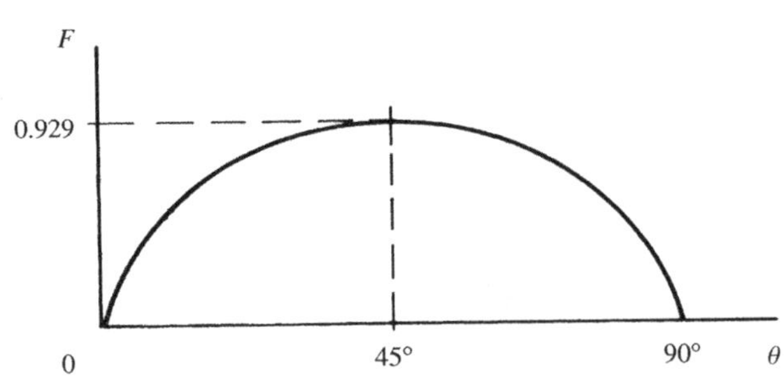

Figure 2.6 Effect of θ upon the shape factor F for a V-corrugation

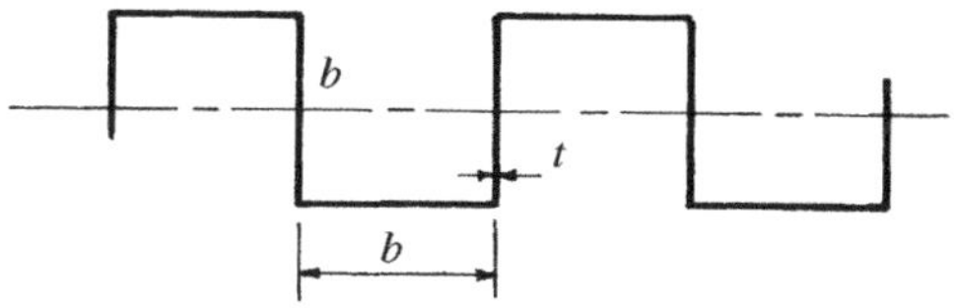

Figure 2.7 Square corrugations with b and t as variables

(i) Flexural Buckling
The second moment of area about the flexural axis is

$$I = (bt)(b/2)^2 + tb^3/12 = tb^3/3 \tag{2.41a}$$

The section area and radius of gyration are

$$A = 2bt, \quad k^2 = I/A = b^2/6 \tag{2.41b,c}$$

The failure criterion in equation (2.2b) becomes

$$\sigma_F = \frac{\pi^2 E_T b^2}{6L^2} \tag{2.42}$$

(ii) Limiting Stress
The working, compressive stress in the strut is given as

$$\sigma_W = \frac{P}{A} = \frac{wb}{2bt} = \frac{w}{2t} \tag{2.43}$$

where $\sigma_W \leq \sigma_y$, in which σ_y is a proof stress defined by an offset strain (typically 0.1% or 0.2%).

(iii) Local Buckling
A lateral bowing collapse of the sloping side plates (flats) occurs under an axial stress:

$$\sigma_L = K_L E_T \left(\frac{t}{b}\right)^2 \tag{2.44}$$

where $K_L = 3.62$, for $\nu = 0.3$.

(iv) Optimisation
An optimum geometry is found by combining criteria (i), (ii) and (iii) above so that the stresses by equations (2.42)–(2.44) become equal, $\sigma_F = \sigma_W = \sigma_L$, which gives

$$\frac{\pi^2 E_T b^2}{6L^2} = \frac{w}{2t} = 3.62\, E_T \left(\frac{t}{b}\right)^2 \tag{2.45}$$

Equation (2.45) allows optimum dimensions b and t to be found:

$$t_{opt} = 0.453\left(\frac{wL}{E_T}\right)^{1/2}, \quad b_{opt} = 0.820\left(\frac{wL^3}{E_T}\right)^{1/4} \tag{2.46a,b}$$

The optimum stress follows from substituting d_{opt} and t_{opt} into any of the three expressions in equation (2.42)–(2.44). Dividing by ρ gives the optimum objective function

$$\left(\frac{\sigma}{\rho}\right)_{opt} = 1.105\,\frac{E_T^{1/2}}{\rho}\left(\frac{w}{L}\right)^{1/2} \tag{2.47}$$

The high shape efficiency factor, $F = 1.105$, found for this section is a consequence of its flat areas lying at a maximum distance from the flexural axis.

2.4.3 *Corrugated Section – Equal Flats at θ*

The corrugated sheet shown in Figure 2.8 has the variables as b, t and θ, giving a pitch length $b(1+\cos\theta)$.

(i) Flexural Buckling
The second moment of area about the flexural axis is

$$I = (bt)\left(\frac{b}{2}\sin\theta\right)^2 + \int_0^{b/2}(t\,\mathrm{d}s)(s\sin\theta)^2 = \frac{tb^3}{3}\sin^2\theta \tag{2.48a}$$

The section area and radius of gyration are

$$A = 2bt,\ k^2 = I/A = (b^2/6)\sin^2\theta \tag{2.48b,c}$$

The failure criterion in equation (2.2b) becomes

$$\sigma_F = \frac{\pi^2\sin^2\theta\, E_T b^2}{6L^2} \tag{2.49}$$

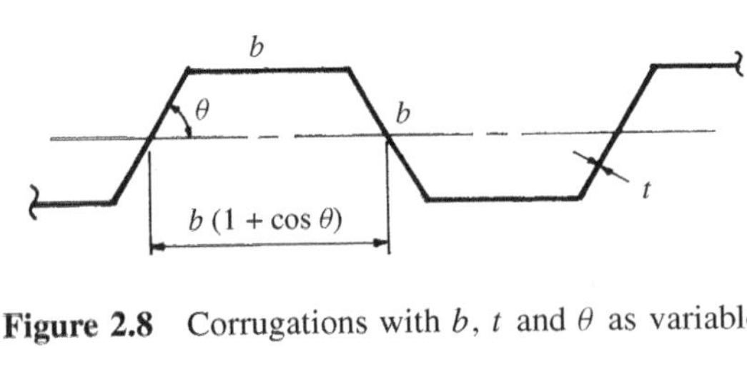

Figure 2.8 Corrugations with b, t and θ as variables

(ii) Limiting Stress
The working compressive stress in the strut is given as

$$\sigma_W = \frac{P}{A} = \frac{wb(1+\cos\theta)}{2bt} = \frac{w(1+\cos\theta)}{2t} \tag{2.50}$$

where $\sigma_W \leq \sigma_y$, where σ_y is, say, a 0.1% proof stress.

(iii) Local Buckling
Lateral collapse of each simply supported side plate (flat) occurs under an axial stress:

$$\sigma_L = K_L E_T \left(\frac{t}{b}\right)^2 \tag{2.51}$$

where $K_L = 3.62$, for $\nu = 0.3$.

(iv) Optimisation
An optimum geometry is found by combining criteria (i), (ii) and (iii) above, so that the stresses by equations (2.49)–(2.51) become equal, $\sigma_F = \sigma_W = \sigma_L$, which gives

$$\frac{\pi^2 E_T b^2 \sin^2\theta}{6L^2} = \frac{w(1+\cos\theta)}{2t} = 3.62\, E_T \left(\frac{t}{b}\right)^2 \tag{2.52}$$

Equation (2.52) allows optimum dimensions b and t to be expressed as

$$t_{\text{opt}} = 0.452\left(\frac{1+\cos\theta}{\sin\theta}\right)^{1/4}\left(\frac{wL}{E_T}\right)^{1/2}, \quad b_{\text{opt}} = 0.820\left(\frac{1+\cos\theta}{\sin^3\theta}\right)^{1/4}\left(\frac{wL^3}{E_T}\right)^{1/4} \tag{2.53a,b}$$

The optimum stress follows from substituting d_{opt} and t_{opt} into any of the three stress expressions in equation (2.52). Dividing σ by ρ gives the objective function conforming to equation (2.6):

$$\left(\frac{\sigma}{\rho}\right)_{\text{opt}} = 1.105\sqrt{\sin\theta(1+\cos\theta)}\left(\frac{E_T^{1/2}}{\rho}\right)\left(\frac{w}{L}\right)^{1/2} \tag{2.54}$$

The shape efficiency factor $F = 1.105\sqrt{\sin\theta(1+\cos\theta)}$ in equation (2.54) is now maximised:

$$\frac{\mathrm{d}F}{\mathrm{d}\theta} = -\sin^2\theta + \cos\theta(1+\cos\theta) = 0 \quad \Rightarrow \quad (2\cos\theta - 1)(\cos\theta + 1) = 0$$

for which $\theta = 60^\circ$ and $F = 1.259$. Correspondingly, the constant coefficients for t_{opt} and b_{opt} in equations (2.53a,b) become 0.595 and 1.011, respectively. The plot of F versus θ in Figure 2.9 shows this maximum and reveals that two orientations apply to lower F-values. For example, two orientations 43° and 80° apply where, from equation (2.1), there is a 5% loss in $F(= 1.196)$, resulting in a 5.3% increase in weight.

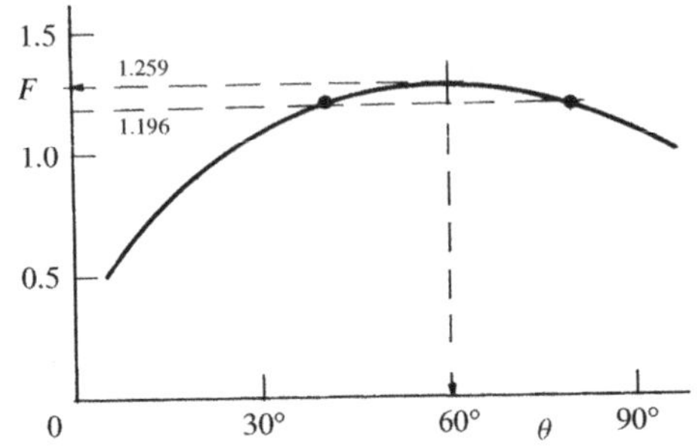

Figure 2.9 Dependence of F upon θ

2.4.4 *Corrugated Section – Unequal Flats at 90°*

Figure 2.10 shows a sheet of 90° corrugations, with unequal horizontal and vertical flats $b \times t$ and $d \times t$, respectively. The pitch is b and the flexural axis lies at half depth $d/2$.

The difference in flat lengths alters their degree of restraint to local buckling, so invalidating the usual assumption of simple, wall-to-wall support for regular polygon tubes, where $K_L = 3.62$. The increased restraint depends upon the ratio b/d in a manner that may be estimated from the known local buckling behaviour of the walls within a uniformly thin, rectangular tube with section dimensions $q \times h \times t$ [4], as shown in Figure 2.11 (see also Figure D.2 in Appendix D).

Though the corrugation is open, the ends of each flat are connected in a similar manner to the walls within a closed tube. The local buckling stress expression equation (2.51) is modified to

$$\sigma_L = K_L E_T \left(\frac{t}{q}\right)^2 \tag{2.55}$$

where q is the greater flat length, which we may identify with b when $b \geq d$ and with d when $d \geq b$. Two analyses are required for $h/q \leq 1$ in Figure 2.11 with q equated, in turn, to the larger critical flat length (b in Figure 2.10a and d in Figure 2.10b).

(a) Condition $b \geq d (b/d \geq 1)$
Here b is the longer flat (see Figure 2.10a) and so h/q becomes d/b in Figure 2.11.

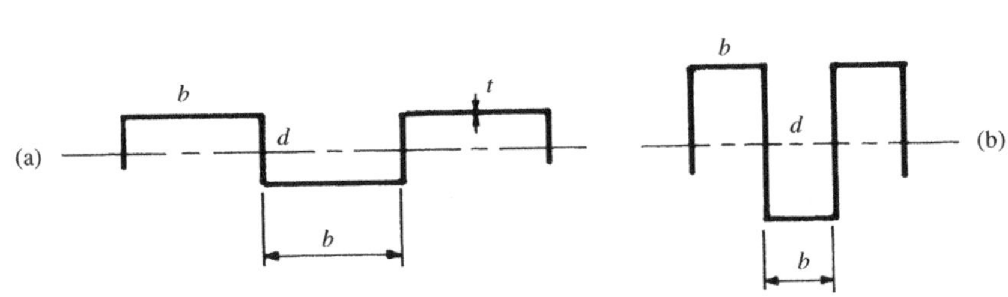

Figure 2.10 Sheet with 90° corrugations having unequal flats, pitch b and depth d

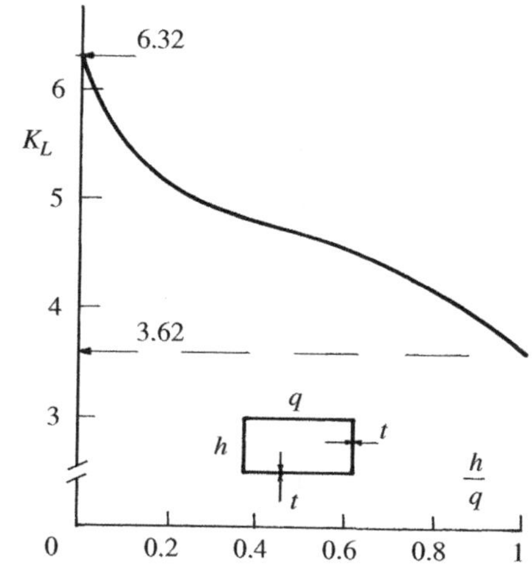

Figure 2.11 Local buckling coefficient K_L for a thin rectangular tube

(i) Flexural Buckling

The second moment of area about the flexural axis is

$$I = (bt)(d/2)^2 + td^3/12 = tbd^2/4 + td^3/12 \tag{2.56a}$$

The wall area and radius of gyration are:

$$A = (b + d)t, \tag{2.56b}$$

$$k^2 = \frac{I}{A} = \frac{b^2}{12} \times \frac{(3 + d/b)(d/b)^2}{1 + d/b} \tag{2.56c}$$

The failure criterion (2.2b) becomes

$$\sigma_F = \frac{\pi^2 E_T b^2}{12L^2} \times \frac{(3 + d/b)(d/b)^2}{1 + d/b} = \frac{\pi^2 E_T b^2 r_1}{12L^2} \tag{2.57}$$

where r_1 is the multiplying factor in d/b.

(ii) Limiting Stress

The working, compressive stress in the strut is given as:

$$\sigma_W = \frac{P}{A} = \frac{wb}{(b + d)t} = \frac{w}{(1 + d/b)t} \tag{2.58}$$

where $\sigma_W \leq \sigma_y$, where σ_y is a proof stress defined typically by a 0.1% or 0.2% offset strain.

(iii) Local Buckling
Lateral collapse of each flat, having simply supported corners, occurs under an axial stress:

$$\sigma_L = K_L E_T \left(\frac{t}{b}\right)^2 \tag{2.59}$$

where K_L is now read from Figure 2.11 at the given $d/b(= h/q)$ ratio [4].

(iv) Optimisation
An optimum geometry is found by combining criteria (i), (ii) and (iii) above so that the stresses by equations (2.57)–(2.59) are made equal. Setting $\sigma_F = \sigma_W = \sigma_L$ allows b and t to be eliminated in turn, this leading to their optimum dimensions

$$t_{\text{opt}} = \left[\frac{1.216}{r_1 K_L(1 + d/b)^2}\right]^{1/4} \left(\frac{wL}{E_T}\right)^{1/2}, \quad b_{\text{opt}} = \left[\frac{1.797 K_L}{r_1^3(1 + d/b)^2}\right]^{1/8} \left(\frac{wL^3}{E_T}\right)^{1/4} \tag{2.60a,b}$$

Back-substituting t_{opt} and b_{opt} into any one of the stress expressions in equations (2.57)–(2.59) and then dividing by ρ gives the required equivalent objective function:

$$\left(\frac{\sigma}{\rho}\right)_{\text{opt}} = \left[\frac{\pi^2 K_L(3 + d/b)(d/b)^2}{12(1 + d/b)^3}\right]^{1/4} \frac{E_T^{1/2}}{\rho} \left(\frac{w}{L}\right)^{1/2} = F_a \left(\frac{E_T^{1/2}}{\rho}\right) \left(\frac{w}{L}\right)^{1/2} \tag{2.61}$$

where the shape efficiency factor F_a depends upon b/d as shown in Figure 2.12. The variation arises from extracting appropriate values of K_L from Figure 2.11, with $d/b(= h/q)$ lying in the range $0 \le d/b \le 1$.

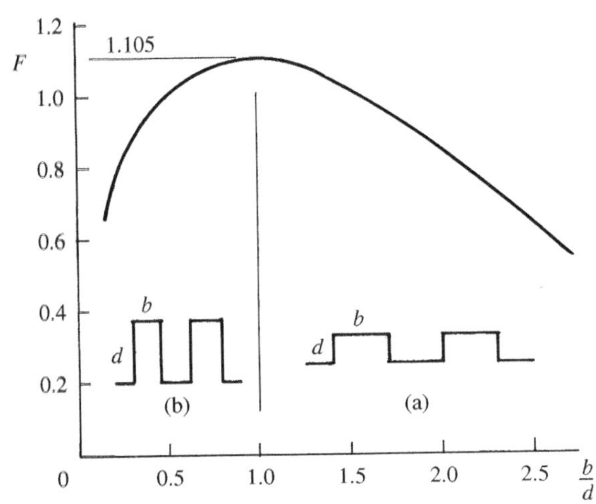

Figure 2.12 Dependence of $F_{a,b}$ upon flat aspect ratio when: (a) $b/d \ge 1$ and (b) $b/d \le 1$

(b) Condition $b \le d(b/d \le 1)$
Here d is the longer flat (see Figure 2.10b) and so b/d identifies with h/q in Figure 2.11.

(i) Flexural Buckling
The second moment of area about the flexural axis is

$$I = (bt)(d/2)^2 + td^3/12 = tbd^2/4 + td^3/12 \tag{2.62a}$$

The area of the material within this section and its radius of gyration are

$$A = (b + d)t \tag{2.62b}$$

$$k^2 = \frac{I}{A} = \frac{d^2}{12} \times \frac{3b/d + 1}{b/d + 1} \tag{2.62c}$$

The failure criterion (2.2b) becomes

$$\sigma_F = \frac{\pi^2 E_T d^2}{12L^2} \times \frac{3b/d + 1}{b/d + 1} = \frac{\pi^2 E_T d^2 r_2}{12L^2} \tag{2.63}$$

where r_2 is the multiplying factor in the ratio b/d.

(ii) Limiting Stress
The working, compressive stress in the strut is written as

$$\sigma_W = \frac{P}{A} = \frac{wb}{(b + d)t} = \frac{w(b/d)}{(b/d + 1)t} \tag{2.64}$$

where $\sigma_W \le \sigma_y$, for σ_y as a 0.1% or 0.2% proof stress.

(iii) Local Buckling
Assuming simply supported corners, a lateral collapse of each long, flat wall surface occurs under an axial stress:

$$\sigma_L = K_L E_T \left(\frac{t}{d}\right)^2 \tag{2.65}$$

where K_L is read from Figure 2.11 for the given $b/d(= h/q)$ ratio [4].

(iv) Optimisation
An optimum geometry is found by combining criteria (i), (ii) and (iii) so that the stresses by equations (2.63)–(2.65) are made equal. Setting $\sigma_F = \sigma_W = \sigma_L$ allows b and t to be eliminated in the usual way giving the optimum dimensions t and d:

$$t_{opt} = \left[\frac{1.216(b/d)^2}{r_2 K_L (1 + b/d)^2}\right]^{1/4} \left(\frac{wL}{E_T}\right)^{1/2}, \quad d_{opt} = \left[\frac{1.797 K_L (b/d)^2}{r_2^3 (1 + b/d)^2}\right]^{1/8} \left(\frac{wL^3}{E_T}\right)^{1/4} \tag{2.66a,b}$$

Back-substituting t_{opt} and b_{opt} into any one of the stress equations (2.63)–(2.65), then dividing by ρ, gives the required objective function

$$\left(\frac{\sigma}{\rho}\right)_{opt} = \left[\frac{\pi^2 K_L(3 + b/d)(b/d)^2}{12(1 + b/d)^3}\right]^{1/4} \left(\frac{E_T^{1/2}}{\rho}\right)\left(\frac{w}{L}\right)^{1/2} = F_b\left(\frac{E_T^{1/2}}{\rho}\right)\left(\frac{w}{L}\right)^{1/2} \quad (2.67)$$

Here the shape efficiency factor F_b depends upon $b/d < 1$ in a slightly different manner to F_a's dependence upon $b/d > 1$ (see Figure 2.12). This figure shows that both F_a and F_b are less than their optimum value of 1.105 when their respective flat ratios b/d are greater and less than unity.

2.4.5 *Corrugated Section – Unequal Flats at* θ

Figure 2.13 shows a wide strut with corrugations, having unequal flat lengths with an inclination θ. Here the influence of varying θ, in addition to varying b, d and t, upon the shape efficiency factor is to be established.

Again, the differing flat lengths b and d alter the degree of restraint to local buckling. The restraint K_L is found from Figure 2.11 as h/q identifies with a flat ratio less than unity, d/b and b/d in Figure 2.13a and Figure 2.13b, respectively. Correspondingly, the following two regimes will fit the requirement $h/q \leq 1$ in Figure 2.11 when q is taken as the larger critical flat length, b in Figure 2.13a and then d in Figure 2.13b.

(a) Condition $b \geq d$ $(d/b \leq 1)$
Figure 2.13a shows the corrugations with b as the longer, horizontal flat. The range of inclinations θ for the shorter, sloping flat d is now to be established.

(i) Flexural Buckling
The second moment of area about the flexural axis for one pitch length: $b + d\cos\theta$, is

$$I = \frac{bt^3}{12} + bt\left(\frac{d}{2}\sin\theta\right)^2 + 2\int_0^{d/2}(t\,\mathrm{d}s)(s\sin\theta)^2 \quad (2.68a)$$

$$\approx \frac{d^2bt}{4}\sin^2\theta + \frac{d^3t}{12}\sin^2\theta = \left(\frac{d}{b}\right)^2\left(3 + \frac{d}{b}\right)\frac{b^3t}{12}\sin^2\theta$$

in which the area and radius of gyration are

$$A = (b + d)t = (1 + d/b)bt \quad (2.68b)$$

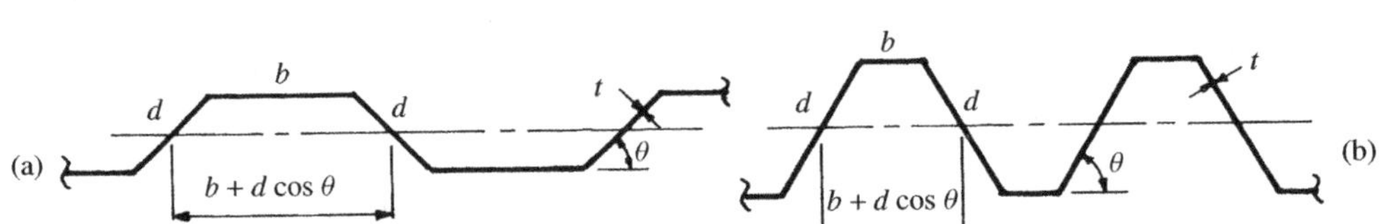

Figure 2.13 Corrugations at θ with unequal b and d

$$k^2 = \frac{I}{A} = \frac{(3 + d/b)(d/b)^2 b^2 \sin^2\theta}{12(1 + d/b)} \tag{2.68c}$$

The flexural buckling criterion (2.2b) becomes

$$\sigma_F = \frac{\pi^2 E_T b^2 \sin^2\theta}{12L^2} \times \frac{(3 + d/b)(d/b)^2}{1 + d/b} = \frac{\pi^2 E_T b^2 r_1 \sin^2\theta}{12L^2} \tag{2.69}$$

where r_1 is the multiplying factor in d/b.

(ii) Limiting Stress

The working, compressive stress in the strut is given as

$$\sigma_W = \frac{P}{A} = \frac{w(b + d\cos\theta)}{(b + d)t} = \frac{w[1 + (d/b)\cos\theta]}{t(1 + d/b)} \tag{2.70}$$

where $\sigma_W \leq \sigma_y$, and σ_y is a 0.1% or 0.2% proof stress.

(iii) Local Buckling

Lateral bowing of each simply supported flat occurs under an axial compressive stress:

$$\sigma_L = K_L E_T (t/b)^2 \tag{2.71}$$

With b as the longer flat, h/q becomes d/b when reading the constraint coefficient K_L from Figure 2.11 [5].

(iv) Optimisation

An optimum geometry is found by combining criteria (i), (ii) and (iii) so that the stresses by equations (2.69)–(2.71) are made equal. Setting $\sigma_F = \sigma_W = \sigma_L$ allows b and t to be eliminated in the usual way, giving their optimum dimensions as

$$t_{\text{opt}} = \left[\frac{1.216\,[1 + (d/b)\cos\theta]^2}{r_1 K_L \sin^2\theta (1 + d/b)^2}\right]^{1/4} \left(\frac{wL}{E_T}\right)^{1/2},$$

$$b_{\text{opt}} = \left[\frac{1.797 K_L [1 + (d/b)\cos\theta]^2}{r_1^3 \sin^6\theta (1 + d/b)^2}\right]^{1/8} \left(\frac{wL^3}{E_T}\right)^{1/4} \tag{2.72a,b}$$

which reduce to equations (2.66a,b) when $\theta = 90^\circ$. Back-substituting t_{opt} and b_{opt} into any one of the stress expressions above, then dividing by ρ, gives the required objective function:

$$\left(\frac{\sigma}{\rho}\right)_{\text{opt}} = F_a \left(\frac{E_T^{1/2}}{\rho}\right) \left(\frac{w}{L}\right)^{1/2} \tag{2.73a}$$

where the shape efficiency factor F_a depends upon the ratio d/b and θ as

$$F_a = \left[\frac{\pi^2 K_L \sin^2\theta[1 + (d/b)\cos\theta]^2(3 + d/b)(d/b)^2}{12(1 + d/b)^3}\right]^{1/4} \tag{2.73b}$$

To optimise F_a, with respect to θ, we set within equation (2.73b),

$$m_a = [1 + (d/b)\cos\theta]\sin\theta \tag{2.74a}$$

and apply the condition $\mathrm{d}m_a/\mathrm{d}\theta = 0$ to equation (2.74a), giving a maximum in m_a when

$$\frac{d}{b} = \frac{\cos\theta}{\sin^2\theta - \cos^2\theta} \tag{2.74b}$$

for which $60^\circ < \theta < 90^\circ$ yields a valid ratio $d/b < 1$ (i.e. positive and less than unity). Over this range of θ the ratio d/b is calculated from equation (2.74b) and, correspondingly, K_L is read from Figure 2.11. Then F_a can then be found from equation (2.73b) and plotted against θ, as shown in Figure 2.14.
Figure 2.14 shows that F_{max} will lie in the second regime which follows.

(b) Condition $b \le d$ $(b/d \le 1)$
Figure 2.13b shows corrugations with b as the shorter flat in which the valid range of inclinations will be seen to be $45^\circ < \theta < 60^\circ$.

(i) Flexural Buckling
The second moment of area, equation (2.68a) is written as

$$I = \left(1 + \frac{3b}{d}\right)\frac{d^3 t}{12}\sin^2\theta \tag{2.75a}$$

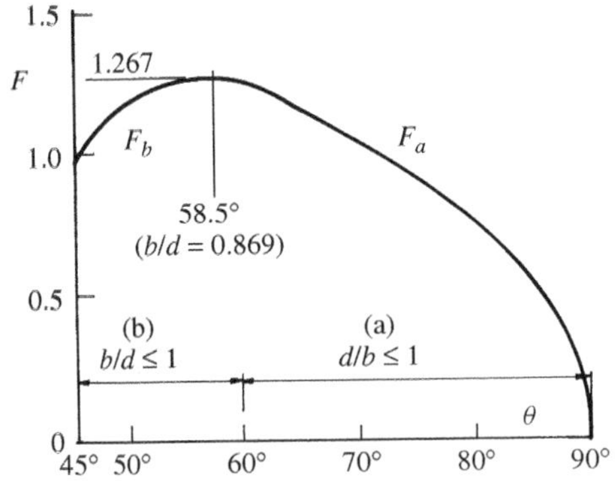

Figure 2.14 Variation in F with θ for flat ratios $b/d \le 1$ and $d/b \le 1$

in which the wall area and radius of gyration for one pitch length are

$$A = (1 + b/d)dt \tag{2.75b}$$

$$k^2 = \frac{I}{A} = \frac{d^2}{12} \times \frac{3b/d + 1}{b/d + 1} \tag{2.75c}$$

The failure criterion (2.2b) becomes

$$\sigma_F = \frac{\pi^2 E_T d^2 \sin^2\theta}{12L^2} \times \frac{1 + 3b/d}{1 + b/d} = \frac{\pi^2 E_T d^2 r_2 \sin^2\theta}{12L^2} \tag{2.76}$$

where r_2 is the multiplying factor in b/d.

(ii) Limiting Stress

The working, compressive stress in the strut is written as

$$\sigma_W = \frac{P}{A} = \frac{w(b/d + \cos\theta)}{t(1 + b/d)} \tag{2.77}$$

where $\sigma_W \le \sigma_y$, in which σ_y may be taken as a 0.1% or a 0.2% proof stress.

(iii) Local Buckling

Taking each wall as a long flat plate with simply supported corners, a lateral collapse occurs under an axial stress:

$$\sigma_L = K_L E_T \left(\frac{t}{d}\right)^2 \tag{2.78}$$

With d as the longer sloping flat, K_L is read from Figure 2.11 for h/q identified with b/d.

(iv) Optimisation

An optimum geometry is found by combining criteria (i), (ii) and (iii) so that the stresses by equations (2.76)–(2.78) are made equal. Setting $\sigma_F = \sigma_W = \sigma_L$ allows b and t to be eliminated in turn to provide their optimum dimensions as

$$t_{\text{opt}} = \left[\frac{1.216\,[b/d + \cos\theta]^2}{r_2 K_L \sin^2\theta (1 + b/d)^2}\right]^{1/4} \left(\frac{wL}{E_T}\right)^{1/2},$$

$$d_{\text{opt}} = \left[\frac{1.797 K_L [b/d + \cos\theta]^2}{r_2^3 \sin^6\theta (1 + b/d)^2}\right]^{1/8} \left(\frac{wL^3}{E_T}\right)^{1/4} \tag{2.79a,b}$$

which reduce to equations (2.66a,b) for $\theta = 90^\circ$. Back-substituting t_{opt} and b_{opt} into any one of the stress expressions above, then dividing by ρ, gives the required objective function:

$$\left(\frac{\sigma}{\rho}\right)_{\text{opt}} = F_b \left(\frac{E_T^{1/2}}{\rho}\right) \left(\frac{w}{L}\right)^{1/2} \tag{2.80a}$$

where the shape efficiency factor F_b depends upon the ratio b/d and θ as

$$F_b = \left[\frac{\pi^2 K_L \sin^2\theta (b/d + \cos\theta)^2 (1 + 3b/d)}{(1 + b/d)^3} \right]^{1/4} \tag{2.80b}$$

To optimise F_b with respect to θ, we set

$$m_b = (b/d + \cos\theta)\sin\theta \tag{2.81a}$$

and applying the condition for a maximum, $dm_b/d\theta = 0$, gives

$$b/d = (\sin^2\theta - \cos^2\theta)/\cos\theta \tag{2.81b}$$

for which $45^\circ < \theta < 60^\circ$ yields a valid ratio $b/d < 1$ (i.e. positive and less than unity). Over this range of θ the ratio b/d is calculated from equation (2.81b), this allowing K_L to be read from Figure 2.11. Then, F_b can then be found from equation (2.80b) and plotted (with F_a) over the full range of $\theta(45^\circ < \theta < 90^\circ)$, as shown in Figure 2.14. This figure shows that weight is minimised with $F_{\max} = 1.267$ at $\theta = 58.5^\circ$ for which $(b/d)_{\text{opt}} = 0.869$.

2.4.6 Z-stiffened Plate

The classical case of Z-stiffening a thin-aerofoil section, whilst minimising overall weight, was examined by Farrar [5]. The arrangement in Figure 2.15 shows stiffeners with pitch spacing b along a plate of thickness t. The stiffener thickness is t_s, the height is h, and to offset local buckling the flange lengths are set at $0.3h$.

(i) Flexural Buckling

With four rectangles – three horizontal and one vertical – per pitch b (see Figure 2.15), the area of the pitch section is

$$A = 1.6ht_s + bt \tag{2.82a}$$

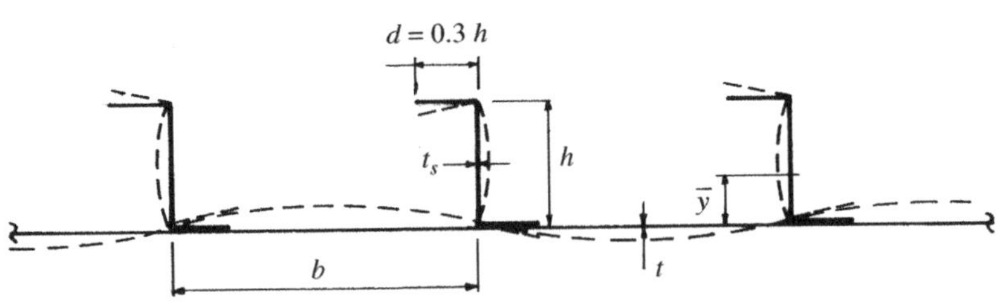

Figure 2.15 Z-stiffened plate showing local buckling modes

The flexural axis lies nearer the base where there is a greater weight of area. This axis passes through the centroid, whose distance from the base $\overline{y}$ is found by equating the sum of the first moments of these four rectangles to that of the whole pitch area:

$$\overline{y} = \frac{1.6ht_s t + 0.8h^2 t_s + bt^2/2}{1.6ht_s + bt} \tag{2.82b}$$

This centroid position reappears in the expression for the second moment of area:

$$I = 0.3ht_s(h - \overline{y})^2 + 0.3ht_s\overline{y}^2 + bt\overline{y}^2 + t_s h^3/12 + ht_s(h/2 - \overline{y})^2 \tag{2.83a}$$

and substituting equation (2.82b) into equation (2.83a) gives a good approximation to I:

$$I = \frac{h^3 t_s(0.633bt + 0.373ht_s)}{1.6ht_s + bt} \tag{2.83b}$$

Now, from equations (2.82a) and (2.83b)

$$k^2 = \frac{I}{A} = \frac{h^3 t_s(0.633bt + 0.373ht_s)}{(1.6ht_s + bt)^2} \tag{2.83c}$$

Combining equation (2.83c) with equation (2.2b), the buckling failure criterion becomes

$$\sigma_F = \frac{\pi^2 E_T h^3 t_s(0.633bt + 0.373ht_s)}{(1.6ht_s + bt)^2 L^2} \tag{2.84}$$

(ii) Limiting Stress
The compressive stress in the strut is given by

$$\sigma_W = \frac{P}{A} = \frac{wb}{1.6ht_s + bt} = \frac{w}{t[1 + A_s/(bt)]} \tag{2.85}$$

where A_s is the area of the Z-stiffener. Again $\sigma_W \le \sigma_y$, where σ_y is a limiting proof stress defined by an offset strain (typically 0.1% or 0.2%). Correspondingly, the tangent modulus E_T in equation (2.84) is taken as the gradient to the stress–strain curve at the chosen offset strain.

(iii) Local Buckling
Under compression, local panel buckling (see Figure 2.15) is attributed to a complex mixture of three possible modes:

(a) flexural buckling of the stiffener plates;
(b) local buckling of the plate between stiffeners;
(c) torsional buckling of the stiffeners.

One mode may dominate, depending upon the section geometry, as shown within the regimes of Figure 2.16.

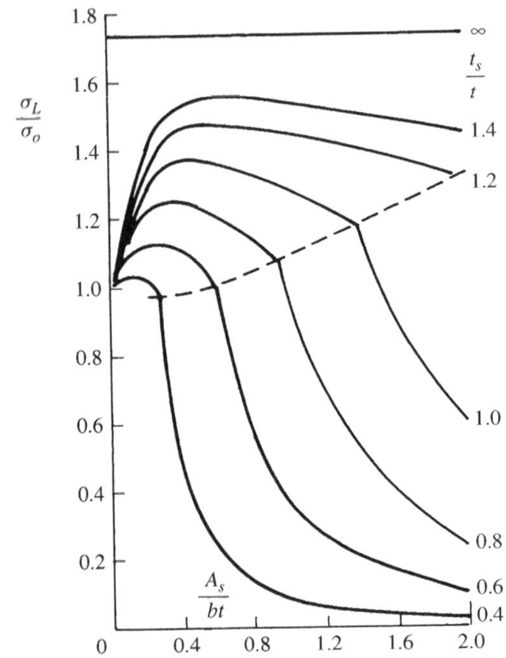

Figure 2.16 Local panel buckling with Z-section stringers

The ordinate expresses the ratio between the stress required for local buckling of the whole panel and that of the connecting plate, when, specifically, $d/h = 0.3$ and $h/t_s = 25$ in a long strut ($L/b > 3$). It is likely [6] that the figure may be applied to local buckling over a broader range, $15 < h/t_s < 35$ and $d/h \leq 0.5$, when, for buckling of the plate between the stiffeners,

$$\sigma_o = K_L E_T \left(\frac{t}{b}\right)^2 \tag{2.86a}$$

where $K_L = 3.62$ for each simply supported side plate in which $\nu = 0.3$ is implied [2]. The condition for buckling of the complete panel to occur is:

$$\sigma_L = \frac{\sigma_L}{\sigma_o} K_L E_T \left(\frac{t}{b}\right)^2 \tag{2.86b}$$

where σ_L/σ_o appears as a function of $A_S/(bt)$ and t_s/t in Figure 2.16. Alternatively, the buckling coefficient K_L for the whole panel may be found directly according to the ratios h/b and t_s/t [6] (see Figure 2.28 in Exercise 2.6).

(iv) Optimisation

An optimum geometry is found by combining criteria (i), (ii) and (iii) above so that the stresses by equations (2.84)–(2.86) become equal, $\sigma_F = \sigma_W = \sigma_L$, which leads to optimum dimensions b, h, t and t_s:

$$\frac{t_{\text{opt}}}{L} = 0.412\left(\frac{w}{LE_T}\right)^{1/2}, \quad \frac{(t_s)_{\text{opt}}}{L} = 0.437\left(\frac{w}{LE_T}\right)^{1/2} \tag{2.87a,b}$$

$$\frac{b_{\text{opt}}}{L} = 0.923\left(\frac{w}{LE_T}\right)^{1/4}, \quad \frac{h_{\text{opt}}}{L} = 0.817\left(\frac{w}{LE_T}\right)^{1/2} \tag{2.87c,d}$$

Eliminating b and t between equations (2.84)–(2.86b), the optimum stress is found by substituting b_{opt}, t_{opt}, etc. into any one of these three equations. Dividing by ρ gives the equivalent objective function, conforming to equation (2.6):

$$\left(\frac{\sigma}{\rho}\right)_{\text{opt}} = F\left(\frac{E_T^{1/2}}{\rho}\right)\left(\frac{w}{L}\right)^{1/2} \tag{2.88a}$$

where the shape efficiency factor F is given by

$$F = \frac{1.719}{1 + A_s/(bt)}\left[\frac{(\sigma_L/\sigma_o)(A_s/bt)^3[0.633 + 0.233(A_s/(bt)]}{(t_s/t)^2}\right]^{1/4} \tag{2.88b}$$

which has a maximum value $F = 0.95$ for $t_s/t = 1.05$ and $A_s/bt \approx 1.5$ [5]. With non-optimum designs, contours of constant F-values appear in the axes of t_s/t versus

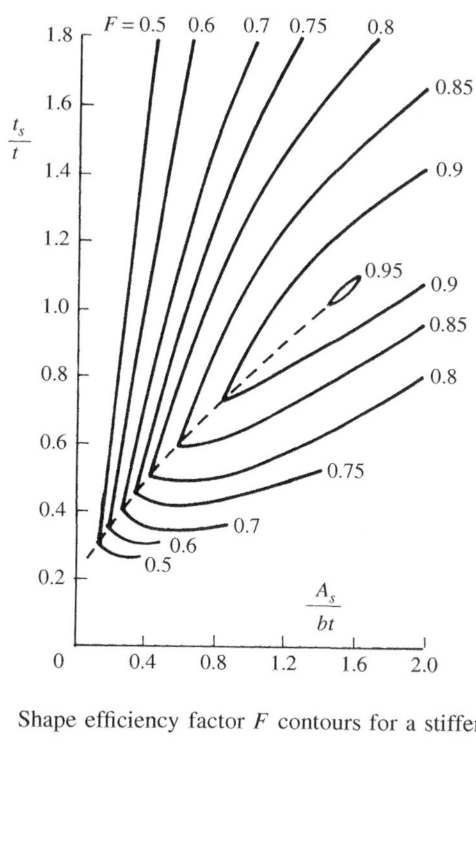

Figure 2.17 Shape efficiency factor F contours for a stiffened stringer

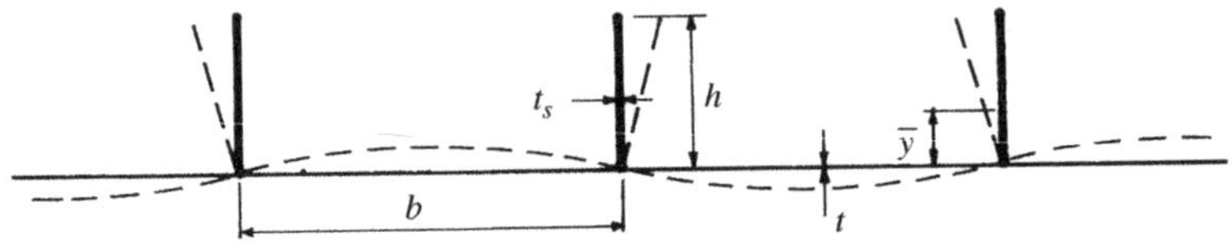

Figure 2.18 Integrally machined, unflanged stiffeners showing local plate buckling

A_s/bt in the manner of Figure 2.17. The contours reveal that only when significant departures from the optimum values t_s/t and A_s/bt arise are losses in the shape efficiency factor appreciable.

2.4.7 Machined Stiffeners

Machined vertical webs, integral with a horizontal plate, form a common wide strut design. The arrangement shown in Figure 2.18 has web thickness t_s, height h, pitch spacing b along a plate with thickness t. All dimensions are chosen to minimise weight as follows.

(i) Flexural Buckling

With two rectangles per pitch length b (see Figure 2.18) the pitch section area is

$$A = ht_s + bt \tag{2.89a}$$

The flexural axis lies nearer the base where there is a greater weight of area. This axis passes through the centroid whose distance from the base is found by equating the sum of the first moments of the two rectangles to that of the whole:

$$\bar{y} = \frac{bt^2/2 + h^2t_s/2}{ht_s + bt} \tag{2.89b}$$

The centroid position reappears in the second moment of area expression:

$$I = t_s h^3/12 + ht_s(h/2 - \bar{y})^2 + bt\bar{y}^2 \tag{2.90a}$$

and substituting equation (2.89b) into equation (2.90a) gives a good approximation to k^2 as

$$k^2 = \frac{I}{A} = \frac{b^2}{12} \times \frac{[4 + (t_s/t)(h/b)](t_s/t)(h/b)^3}{[1 + (t_s/t)(h/b)]^2} \tag{2.90b}$$

so that the flexural failure stress equation (2.2b) becomes

$$\sigma_F = \frac{\pi^2 E_T b^2}{12L^2} \times \frac{[4 + (t_s/t)(h/b)](t_s/t)(h/b)^3}{[1 + (t_s/t)(h/b)]^2} \tag{2.91}$$

(ii) Limiting Stress

The compressive stress in the strut is given as

$$\sigma_W = \frac{P}{A} = \frac{wb}{ht_s + bt} = \frac{w}{t[1 + (t_s/t)(h/b)]} \tag{2.92}$$

where A_s is the Z-stiffener area and again $\sigma_W \leq \sigma_y$, in which σ_y is a limiting proof stress defined by an offset strain (typically 0.1% or 0.2%). The tangent modulus E_T in equation (2.91) is the gradient to the stress–strain curve at an offset strain corresponding to σ_y.

(iii) Local Buckling

The local buckling modes are similar to those for Z-stringers. Thus, for the buckling of the plate between the stiffeners (see Figure 2.18) the datum stress is

$$\sigma_o = K_L E_T \left(\frac{t}{b}\right)^2 \tag{2.93a}$$

where $K_L = 3.62$ for a simply supported base plate in which $\nu = 0.3$ is assumed [2]. For local buckling of the complete panel,

$$\sigma_L = \frac{\sigma_L}{\sigma_o} K_L E_T \left(\frac{t}{b}\right)^2 \tag{2.93b}$$

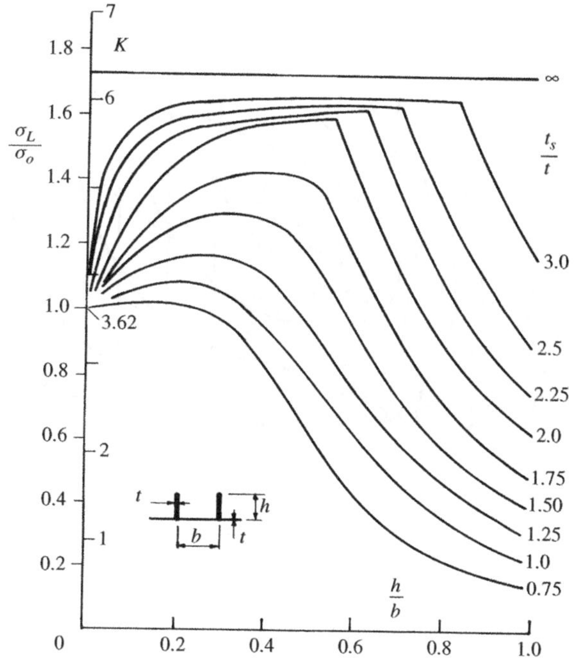

Figure 2.19 Initial buckling stress for a panel with an integral stiffener

where σ_L/σ_o appears in Figure 2.19 as a function of the thickness ratio t_S/t and the length ratio h/b [7].
Alternatively, the local buckling coefficient K for the whole panel may be read from the inner scale [8]. This provides the buckling stress directly as

$$\sigma_L = KE_T\left(\frac{t}{b}\right)^2 \tag{2.93c}$$

(iv) Optimisation
The stresses by the three criteria (2.91)–(2.93b) are made equal, $\sigma_F = \sigma_W = \sigma_L$, and the optimum stress follows from eliminating b and t between these equations. Dividing by ρ gives an objective function conforming to equation (2.6):

$$\left(\frac{\sigma}{\rho}\right)_{\text{opt}} = F\left(\frac{E_T^{1/2}}{\rho}\right)\left(\frac{w}{L}\right)^{1/2} \tag{2.94a}$$

where the shape efficiency factor F depends upon the ratios t_s/t and h/b as

$$F = \frac{1.314}{1+(t_s/t)(h/b)}\left\{\frac{\sigma_L}{\sigma_o}\frac{t_s}{t}\left(\frac{h}{b}\right)^3\left(4+\frac{t_s}{t}\frac{h}{b}\right)\right\}^{1/4} \tag{2.94b}$$

Catchpole [7] showed that equation (2.94b) has its maximum value $F = 0.81$ for $t_s/t = 2.25$ and $h/b \approx 0.65$. This maximum lies at the centre of the design chart in Figure 2.20.

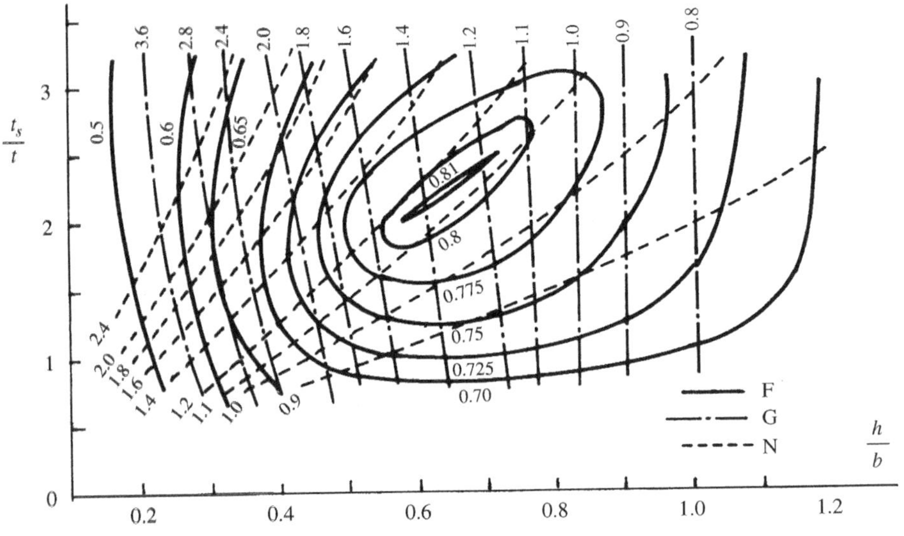

Figure 2.20 Shape efficiency factor contours F for integrally stiffened stringers

The chart employs the optimum geometry, found by combining equations (2.91)–(2.93):

$$(t_s)_{\text{opt}} = N\left(\frac{wL}{E_T}\right)^{1/2}, \quad t_{\text{opt}} = P\left(\frac{wL}{E_T}\right)^{1/2}$$

$$b_{\text{opt}} = G\left(\frac{wL^3}{E_T}\right)^{1/4}, \quad h_{\text{opt}} = Q\left(\frac{wL^3}{E_T}\right)^{1/4} \qquad (2.95\text{a–d})$$

where the coefficients G and P depend upon the ratios t_s/t and h/b:

$$G = \frac{1.103\,F^{1/2}[1 + (t_s/t)(h/b)]}{\{(t_s/t)(h/b)^3[4 + (t_s/t)(h/b)]\}^{1/2}}, \quad P = \frac{1}{F[1 + (t_s/t)(h/b)]} \qquad (2.95\text{e,f})$$

The remaining coefficients follow as $N = P(t_s/t)_{\text{opt}}$ and $Q = G(h/b)_{\text{opt}}$. For non-optimum designs, contours of constant values of F, G and N appear in the axes of t_s/t versus h/b, as shown in Figure 2.20.

2.5 Corrugated Sandwich Panel

The corrugated sections considered in §§ 2.4.5 and 2.4.6, not being of an aerodynamic shape, would normally be used to support a smooth external skin. Let us consider the general case of a sandwich panel in which the corrugations have skins on both sides. Figure 2.21 shows the arrangement in which the flat lengths of the corrugation are b' and d, with an inclination θ. Also, the thicknesses of the skin and the corrugation, t_b and t_d respectively, are allowed to differ.

The method of skin attachment is important to restraining a local buckling failure. A corner rivet attachment offers less restraint than when the skin is adhesively bonded across the whole of the horizontal flat area $b' \times L$. The latter will be considered here in which the effective increase in thickness $t_b + t_d$ in length b', offsets local buckling in this horizontal flat. Instead, the local buckling is confined to the sloping flat and to the unsupported skin length $b = b' + 2d\cos\theta$. Note that, with a corner fixing, a bowing between the two plates could occur and so again the two regimes $b' > d$ and $b' < d$ would need additional consideration (as in Figures 2.13a,b). A similar approach would

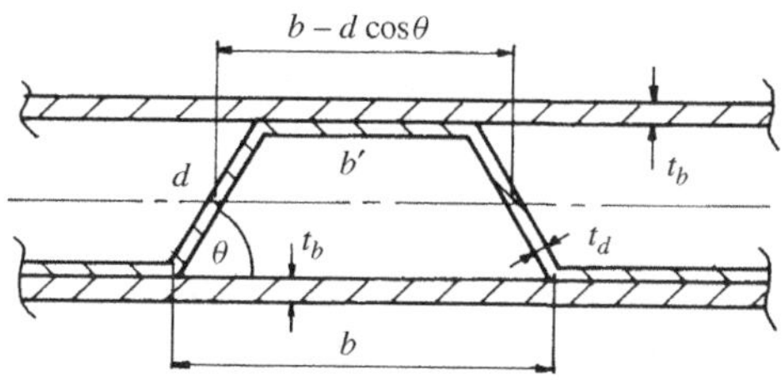

Figure 2.21 Sandwich panel corrugations at θ with b, b', d, t_b and t_d as variables

apply to a rivet fixing central to length b'. Analyses of optimum designs with these alternative methods of attachment are left as exercises for the reader. With an adhesive bond, we need only arrange for the two unsupported flats of lengths b and d to buckle simultaneously to provide the optimum design.

(i) Flexural Buckling

The second moment of area about the flexural axis for one pitch length $b - d\cos\theta$ is

$$I = 2(b' - d\cos\theta)t_b\left(\frac{d}{2}\sin\theta\right)^2 + b't_d\left(\frac{d}{2}\sin\theta\right)^2 + 2\int_0^{d/2}(t_d\,\mathrm{d}s)(s\sin\theta)^2$$

$$= \frac{b^3 t_b \sin^2\theta}{12}\left[6\left(1 - \frac{d}{b}\cos\theta\right)\left(\frac{d}{b}\right)^2 + 3\left(1 - \frac{2d}{b}\cos\theta\right)\frac{t_d}{t_b}\left(\frac{d}{b}\right)^2 + \left(\frac{d}{b}\right)^3\frac{t_d}{t_b}\right] \tag{2.96a}$$

The cross-sectional area contained within this pitch is

$$A = 2(b - d\cos\theta)t_b + (b - 2d\cos\theta)t_d + dt_d$$

$$= bt_b\left[2\left(1 - \frac{d}{b}\cos\theta\right) + \left(1 - \frac{2d}{b}\cos\theta\right)\frac{t_d}{t_b} + \frac{d}{b}\frac{t_d}{t_b}\right] \tag{2.96b}$$

Abbreviating the square bracketed terms in equations (2.96a,b) as [.] and [..] respectively, the radius of gyration becomes

$$k^2 = \frac{I}{A} = \frac{b^2}{12}\sin^2\theta\,\frac{[.]}{[..]} \tag{2.96c}$$

The flexural buckling criterion (2.2b) becomes

$$\sigma_F = \frac{\pi^2 E_T k^2}{L^2} = \frac{\pi^2 E_T b^2 \sin^2\theta}{12L^2}\frac{[.]}{[..]} \tag{2.97}$$

(ii) Limiting Stress

The working compressive stress in the strut is found from the pitch length p as

$$\sigma_W = \frac{wp}{A} = \frac{w(1 - d/b\cos\theta)}{t_b[..]} \tag{2.98}$$

where $\sigma_W \le \sigma_y$, for σ_y a proof stress, defined by an offset strain (typically 0.1% or 0.2%).

(iii) Local Buckling

Each flat has a cross-section $b \times t_b$ and $d \times t_d$. As long, flat plates with simply supported edges, their lateral collapse would occur under the respective compressive stresses

$$\sigma_{Lb} = K_L E_T\left(\frac{t_b}{b}\right)^2, \quad \sigma_{Ld} = K_L E_T\left(\frac{t_d}{d}\right)^2 \tag{2.99a,b}$$

For simultaneous buckling of each flat length under a common stress, it follows from equations (2.99a,b) that $d/b = t_d/t_b$, when $K_L = 3.72$ is estimated from Figure 2.11.

(iv) Optimisation

An optimum geometry is found by combining criteria (i), (ii) and (iii) so that the stresses by the three equations (2.97)–(2.99) are made equal. That is, setting $\sigma_F = \sigma_W = \sigma_L$ allows the optimum dimensions for t_b and b to be solved. Writing $x = d/b = t_d/t_b$, this gives

$$(t_b)_{\text{opt}} = \left(\frac{12}{\pi^2 K_L}\right)^{1/4} [A(x,\theta)]^{1/4} \left(\frac{wL}{E_T}\right)^{1/2} \tag{2.100a}$$

where

$$A(x,\theta) = \left\{ \frac{(1 - x\cos\theta)^2}{[2(1 - x\cos\theta) + x(1 - 2x\cos\theta) + x^2][6x^2(1 - x\cos\theta) + 3x^3(1 - 2x\cos\theta) + x^4]\sin^2\theta} \right\}$$

and

$$b_{\text{opt}} = \left(\frac{12^3 K_L}{\pi^6}\right)^{1/8} \times \left\{ \frac{[2(1 - x\cos\theta) + x(1 - 2x\cos\theta) + x^2]^2 (1 - x\cos\theta)^2}{[6x^2(1 - x\cos\theta) + 3x^3(1 - 2x\cos\theta) + x^4]^3 (\sin^2\theta)^3} \right\}^{1/8} \left(\frac{wL^3}{E_T}\right)^{1/4} \tag{2.100b}$$

Equations (2.100a,b) are simplified for rectangular corrugations when $\theta = 90^\circ$. Back-substituting t_{opt} and b_{opt} into any one of the stress expressions in (i), (ii) or (iii) above and dividing by ρ gives the required objective function:

$$\left(\frac{\sigma}{\rho}\right)_{\text{opt}} = F\left(\frac{E_T^{1/2}}{\rho}\right)\left(\frac{w}{L}\right)^{1/2} \tag{2.101a}$$

where the shape efficiency factor F depends upon the ratio $x = d/b$ and θ as follows:

$$F = \left(\frac{\pi^2 K_L}{12}\right)^{1/4} \left\{ \frac{x^2(1 - x\cos\theta)^2[6(1 - x\cos\theta) + 3x(1 - 2x\cos\theta) + x^2]\sin^2\theta}{[2(1 - x\cos\theta) + x(1 - 2x\cos\theta) + x^2]^3} \right\}^{1/4} \tag{2.101b}$$

In this case x and θ within equations (2.100a,b) and (2.101b) are dependent. From the geometry in Figure 2.21, $b = b' + 2d\cos\theta$. Setting the flat length ratio $r = b'/d$ gives the relationship

$$x = \frac{1}{r + 2\cos\theta} \tag{2.102}$$

Hence if we choose the ratio $r = 0.5$, 1 and 2 (say) then x and F may be evaluated from equations (2.102) and (2.101b) over the range $0^\circ < \theta \le 90^\circ$, as shown in Figure 2.22.

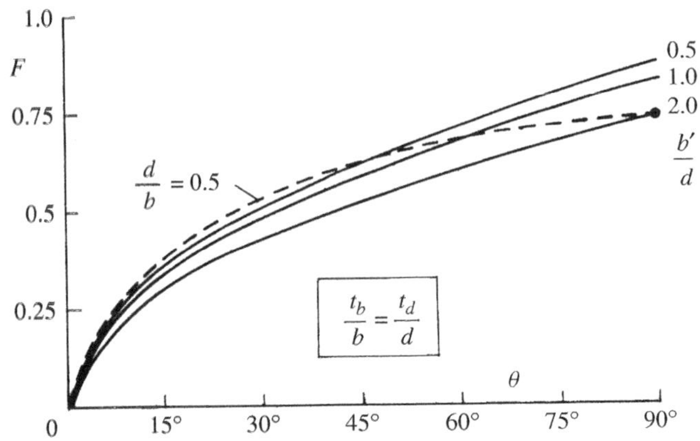

Figure 2.22 Effect of the geometry of sandwich corrugations upon F

None of these plots reveal an optimum θ, but the largest F occurs for $\theta = 90^\circ$ with $r = 0.5$. This suggests that a sandwich corrugation with bonded skin can be made more structurally efficient by shortening the horizontal flat length b' and allowing θ to approach 90°.

A similar conclusion is drawn when $x = d/b$ is set to a constant within equation (2.101b). If, for example, $x = 0.5$ then from equation (2.102), $r = 2(1 - \cos\theta)$. Thus, as F increases with θ, when $d/b = 0.5$ as shown, so too $r = b'/d$ increases with θ. Here a maximum ratio $r = 2$ applies to $\theta = 90^\circ$, this being in agreement with the previous plot in which $r = 2$ is constant.

2.6 Summary of Results

Table 2.1 summarises the objective functions found for the 11 wide-strut sections considered here. It is seen that both the material efficiency factor $M = \sqrt{E_T}/\rho$ and the structural index $S = w/L$, along with its exponent value $n = 1/2$, are common to all sections. Thus, the equivalent objective function $R = (\sigma/\rho)_{\text{opt}}$, that has now been maximised for each strut, becomes

$$R = F \times M \times S^n = F \times \sqrt{E_T}/\rho \times \sqrt{(w/L)} \tag{2.103}$$

We may compare wide struts quite simply on the basis of their shape efficiency factors F alone. Given the inverse relationship between the objective function R in equation (2.103) and the strut's weight, we may conclude that corrugations with unequal flats, having an inclination 58.5°, will provide a wide strut with the least weight. Corrugations having equal flats with a 60° inclination would increase the weight by only 0.64%. The cellular hexagonal section minimises weight most effectively when adjacent cells share a common side The section machined with plate and integral stiffener is least efficient, increasing the weight by 56% compared to the most efficient wide-strut section. Note that these weight comparisons have used the reciprocals of the F-values given in Table 2.1. (see equation (2.1)).

Table 2.1 Shape efficiency factor ranking for wide-strut sections

Cross-Section	Shape Efficiency Factor F
	1.267
	1.259
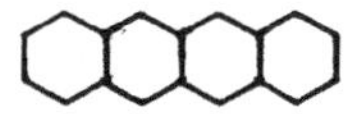	1.199
	1.134
	>1.105
	1.105
	0.977
	0.950
	0.929
	0.929
	0.810

References

[1] ESDU 78021, Guide to items on the strength and stability of struts, October 1978.
[2] ESDU 72019, Buckling of flat isotropic plates under uniaxial and biaxial loading, August 1972.
[3] Cox, H. L. *The Design of Structures of Least Weight*, Chapter 2, Pergamon, 1965.
[4] ESDU 01.01.19, Local buckling and crippling of rectangular tube section struts, September 1976.
[5] Farrar, D. J. The design of compression surfaces for minimum weight, *Aeronautical Journal*, 1949.
[6] ESDU 71014, Local buckling of compression panels with flanged stiffeners, August 1971.
[7] Catchpole, E. J. The optimum design of compression surfaces having un-flanged integral stiffeners, *Journal of the Royal Aeronautical Society*, November 1954.
[8] ESDU 70003, Local buckling of compression panels with un-flanged integral stiffeners, January 1970.

Exercise

2.1 Derive the objective function for a corner connection between adjacent square cells. Does this provide a more efficient wide square-cell strut than one with a common wall between adjacent square cells? Find the optimum side length and thickness in the case of a corner connection.

2.2 An external skin is added to a wide, 60° corrugated strut with equal flats of length b for the design of an aircraft wing (see Figure 2.23). What is the effect of the skin upon the shape efficiency factor F when the attachment is made: (i) at the centre of the horizontal flats; (ii) at the corners of the corrugations. Express F for (i) and (ii) as a percentage change to the $F = 1.259$ value for the corrugation alone (see Table 2.1). Using $1/F$ values, establish the corresponding percentage weight change. Assume a common thickness throughout.

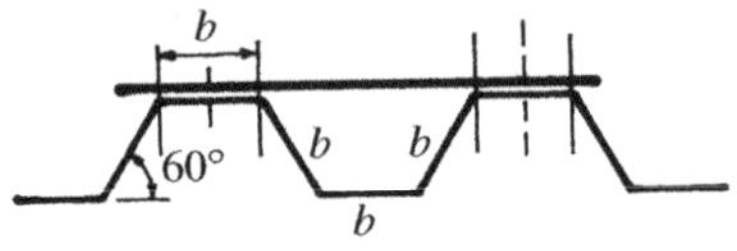

Figure 2.23 Corrugations at 60° with single skin

2.3 What is the effect of adding a skin to both sides of the corrugation in Exercise 2.2 when the attachments are made at the corners of the corrugation in the manner of Figure 2.24?

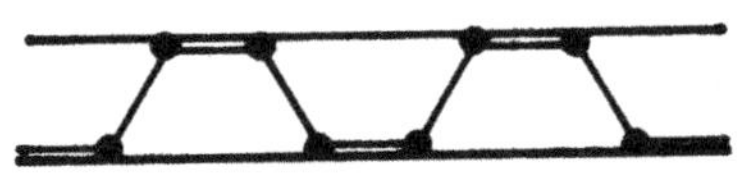

Figure 2.24 Corrugation with double skin

2.4 The panel with skin and Z-section stringers shown in Figure 2.25 applies to $d/h = 0.3$. The flange and web have equal thicknesses, but the thickness of the stiffener differs from the skin. A bonded attachment is used between the panel and the stringer's flange. Derive the optimum dimensions b and t and the shape factor for the panel when $h/b = 0.5$ and $t_s/t = 1.25$.

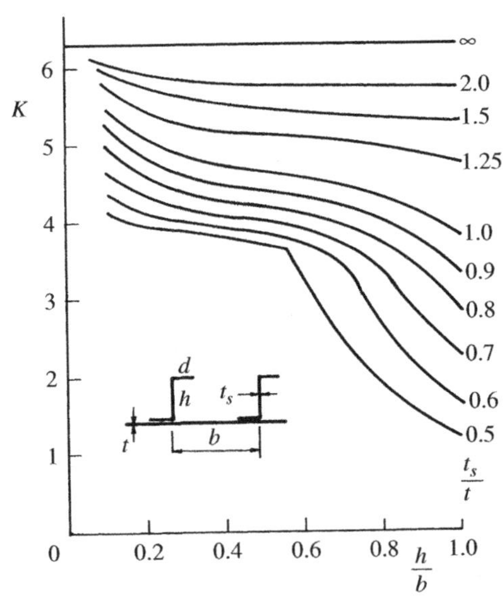

Figure 2.25 Local buckling coefficient for Z-section stringers [6]

2.5 Derive the optimum dimensions and the shape efficiency factor for the I-stiffened plate given in Figure 2.26. Assume that the local buckling coefficients are similar to those given for the Z-stiffened plate in Figure 2.25.

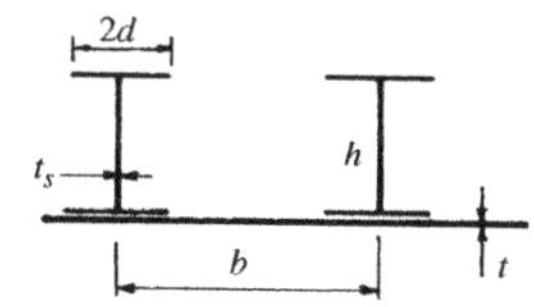

Figure 2.26 I-stiffened plate

2.6 A skin with the top hat section stringers is shown in Figure 2.27. The thickness t_s of the hat differs from the skin thickness t. The ratio $d/b = 1.5$ applies and a flanged-base attachment is used. Determine the shape efficiency factor for edge compression of the panel when the local bucking coefficients are provided by Figure 2.11.

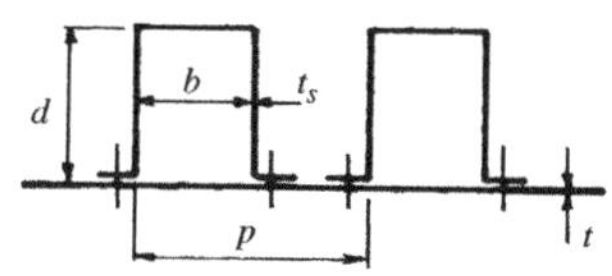

Figure 2.27 Top-hat stringers

2.7 Consider how the analysis of the corrugated sandwich panel given in Figure 2.21 would be altered when an adhesive bond over flat length b' on both sides is replaced by (i) a corner rivet fixing at the ends of b' and (ii) a single rivet fixing at its centre of b'.

2.8 The panels with integrally manufactured stringers shown in Figure 2.28 apply to $d/h = 0.3$ in which the flange and web have equal thicknesses t_s. Derive the optimum dimensions and the shape efficiency factors for the stringer when: (i) $h/b = 0.8$ with $t_s/t = 0.9$; (ii) $h/b = 0.4$ with $t_s/t = 1.25$.

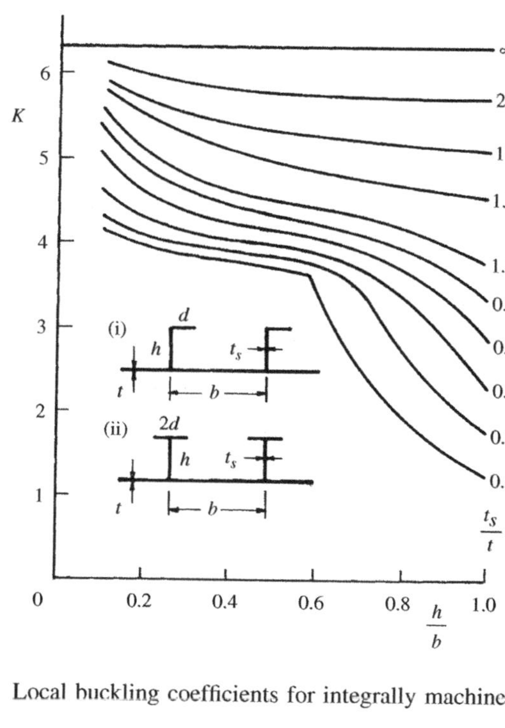

Figure 2.28 Local buckling coefficients for integrally machined stiffeners [6]

3

Bending of Slender Beams

3.1 Introduction

The procedure for minimising the weight of a beam is similar to that adopted for a slender strut, except that the strut's objective function σ/ρ does not apply to a beam. More directly, for a beam with a uniform cross-sectional area A and length L, its weight $W = \rho AL$ is the objective function, i.e. the quantity to be minimised. In a beam the stress is not uniform but distributed throughout the depth depending upon the section's area properties. In contrast, the stress in a slender strut is uniformly distributed to a level of loading at which the strut becomes unstable. In this chapter a hogging bending moment M is taken to be constant over the length. Therefore, no shear force exists and M stresses the beam in tension and in compression above and below the neutral axis (NA) of bending, respectively (see Figure 3.1).

A slender beam is one whose cross-sectional dimensions are small in comparison to its length. The maximum stresses in tension and in compression lie at the outer edges. Between these maxima the stress is distributed through the depth linearly as shown. A similar linear stress distribution applies to every section in the length [1]. In Chapters 10 and 11 the variable moments and shear forces that arise in laterally loaded beams are considered for optimum designs over their lengths and cross-sections. The properties of the cross-section, i.e. the position of its centroid C, through which the neutral axis passes, and the second moment of its area, determine the peak stress values. Coupled to this, the known strengths of the beam material in tension and in compression place a limit upon the level of stress that can be allowed in bending. The following minimal weight analyses apply to solid, hollow and open beam cross-sections. The minimum weight of a beam will be seen to appear within the objective function:

$$(W/L^3)_{\min} = fmS^n \tag{3.1}$$

where the structural index $S = M/L^3$ is common to all beam sections but the shape factor f, the material factor m and the exponent n will depend upon the cross-section.

Mechanics of Optimal Structural Design: Minimum Weight Structures D. W. A. Rees

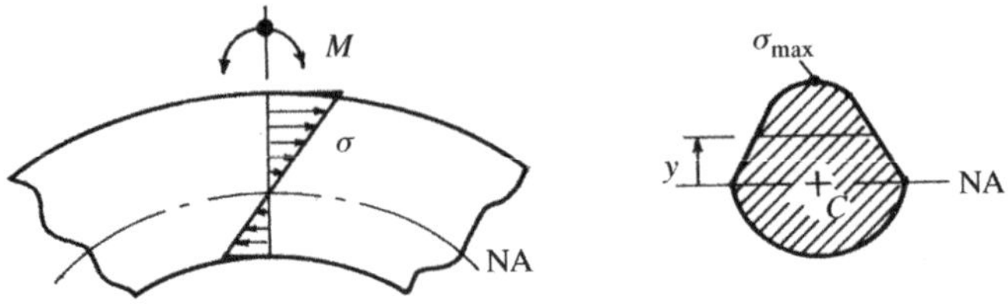

Figure 3.1 Stress distribution in a beam

3.2 Solid Cross-Sections

3.2.1 Circular

As the maximum bending stress applies to single points at the top and bottom of the vertical diameter (see Figure 3.2), the circular section cannot rank highly as a structurally efficient section. However, as circular beams appear in many structures, typically as rods, bars, pins and bolts, a knowledge of their minimal weight conditions is useful.

(i) Objective Function

$$W = \rho AL = \rho \pi d^2 L/4 \tag{3.2}$$

(ii) Failure Criterion
When the peak bending stress σ_{max} is limited to the yield (or proof) stress σ_y value for the beam material, we write $\sigma_{max} \leq \sigma_y$ (see § 3.5), where

$$\sigma_{max} = \frac{My_{max}}{I} = \frac{M(d/2)}{(\pi d^4/64)} = \frac{32M}{\pi d^3} \tag{3.3a}$$

When $\sigma_{max} = \sigma_y$, the diameter of a fully stressed beam becomes

$$d_{opt} = \left(\frac{32M}{\pi \sigma_y}\right)^{1/3} \tag{3.3b}$$

Substituting d_{opt} for d into equation (3.2),

$$W_{min} = \frac{\pi \rho L}{4}\left(\frac{32M}{\pi \sigma_y}\right)^{2/3} = 3.691 \frac{\rho}{\sigma_y^{2/3}} M^{2/3} L \tag{3.4a}$$

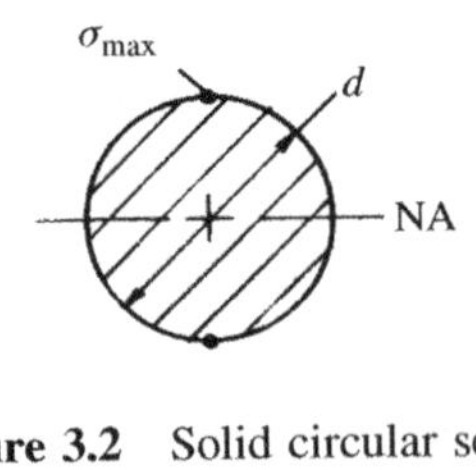

Figure 3.2 Solid circular section

(iii) Optimisation
To provide the structural index in stress units, divide equation (3.4a) by L^3 to give

$$\left(\frac{W}{L^3}\right)_{\min} = 3.691\left(\frac{\rho}{\sigma_y^{2/3}}\right)\left(\frac{M}{L^3}\right)^{2/3} \tag{3.4b}$$

3.2.2 Solid Square Section

In Figure 3.3 the maximum bending stress is reached at all points across the top and bottom edges, which facilitates a more structurally efficient section.

(i) Objective Function

$$W = \rho A L = \rho a^2 L \tag{3.5}$$

(ii) Failure Criterion
When the peak bending stress $\sigma_{\max}$ is limited to the yield (or proof) stress σ_y for the beam material, we write $\sigma_{\max} \le \sigma_y$, where

$$\sigma_{\max} = \frac{M y_{\max}}{I} = \frac{Ma/2}{a^4/12} = \frac{6M}{a^3} \tag{3.6a}$$

When $\sigma_{\max} = \sigma_y$, the section size of a fully stressed beam becomes

$$a_{\text{opt}} = \left(\frac{6M}{\sigma_y}\right)^{1/3} \tag{3.6b}$$

(iii) Optimisation
Substituting a_{opt} for a into equation (3.6a),

$$W_{\min} = \rho L\left(\frac{6M}{\sigma_y}\right)^{2/3} = 3.302\,\frac{\rho}{\sigma_y^{2/3}} M^{2/3} L \tag{3.7a}$$

To provide the structural index in stress units, divide equation (3.7a) by L^3 to give

$$\left(\frac{W}{L^3}\right)_{\min} = 3.302\left(\frac{\rho}{\sigma_y^{2/3}}\right)\left(\frac{M}{L^3}\right)^{2/3} \tag{3.7b}$$

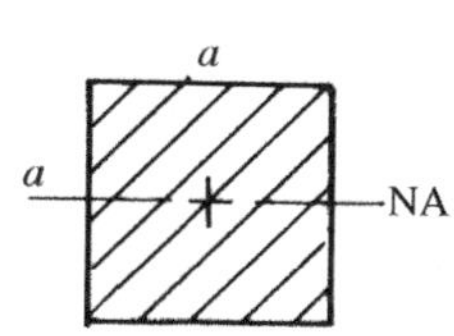

Figure 3.3 Solid square section

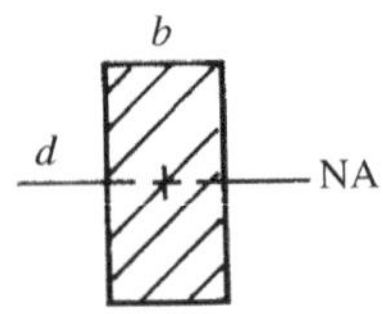

Figure 3.4 Solid, rectangular section

The lower shape factor in equation (3.7b), compared to equation (3.4b), confirms that a square section beam with lesser weight can carry a similar moment to the circular section.

3.2.3 Solid Rectangular Section

When a rectangular section $b \times d$ (Figure 3.4) is mounted as shown, the maximum stress is reached at all points along the top and bottom edges ($y_{max} = \pm d/2$) from the neutral axis. When $b \ll d$ the possibility of lateral buckling (side-bowing) presents a further failure criterion.

(i) Objective Function

$$W = \rho AL = \rho bdL \tag{3.8}$$

(ii) Bending Failure Criterion
When the peak bending stress σ_{max} is limited to the yield (or proof) stress σ_y for the beam material we write $\sigma_{max} \le \sigma_y$, where

$$\sigma_{max} = \frac{My_{max}}{I} = \frac{Md/2}{bd^3/12} = \frac{6M}{bd^2} \tag{3.9}$$

(iii) Lateral Buckling Criterion
When the breadth b is small the lateral stiffness may not be sufficient to offset a bowing from occurring across the depth d. The bending stress level at which this form of lateral buckling occurs is given by [1]

$$\sigma_L = 2.61E_T\left(\frac{b}{L}\right) \tag{3.10}$$

where E_T is the tangent modulus at σ_y. Equating (3.9) and (3.10) gives an optimum product

$$(bd)_{opt} = \left(\frac{6ML}{2.61E_T}\right)^{1/2} \tag{3.11}$$

(iv) Optimisation
Substituting $(bd)_{opt}$ into equation (3.8),

$$W_{min} = \rho L\left(\frac{6ML}{2.61E_T}\right)^{1/2} = 3.302\frac{\rho}{E_T^{1/2}}(ML)^{1/2} \tag{3.12a}$$

To provide the structural index in stress units, divide equation (3.12a) by L^3 to give

$$\left(\frac{W}{L^3}\right)_{\min} = 1.516\left(\frac{\rho}{E_T^{1/2}}\right)\left(\frac{M}{L^3}\right)^{1/2} \tag{3.12b}$$

Now b and d may be separated when the maximum stress in equation (3.9) attains the yield stress, i.e. $\sigma_{\max} = \sigma_y$. This gives $bd = 6M/d\sigma_y$ which, when equated to (3.11), gives

$$d_{\text{opt}} = 3.957\,\frac{E^{1/2}}{\sigma_y}\left(\frac{M}{L}\right)^{1/2}, \quad b_{\text{opt}} = 0.383\,\frac{\sigma_y}{E}L \tag{3.13a,b}$$

Equations (3.12b) and (3.13a,b) provide the optimum dimensions for a rectangular section.

3.3 Thin-Walled, Tubular Sections

Lateral buckling is a common failure mode for all thin-walled, closed tubular sections. The buckling stress expression depends upon the tubular section but, generally, is similar to the local buckling stress for slender struts. Recall the derivation of this expression (see Appendix D) in which plate buckling theory is applied to each simply supported flat. An obvious exception to this is the circular tube section with which this analysis begins.

3.3.1 Circular Tube

When a circular tube, with mean diameter d and wall thickness t (Figure 3.5), is used as a beam section, the maximum bending stress $\sigma_{\max}$ is reached at the top and bottom positions, i.e. at distances $y_{\max} = \pm d/2$ from the neutral axis. It follows that a more efficient, non-circular tubular design would arrange for more material to be fully stressed at each location.

As the wall is thin, the area and its second moment may be simplified to

$$A = \pi\,dt, \quad I = 4\int_0^{\pi/2} t(r\,\mathrm{d}\theta)(r\sin\theta)^2 = \pi d^3 t/8 \tag{3.14a,b}$$

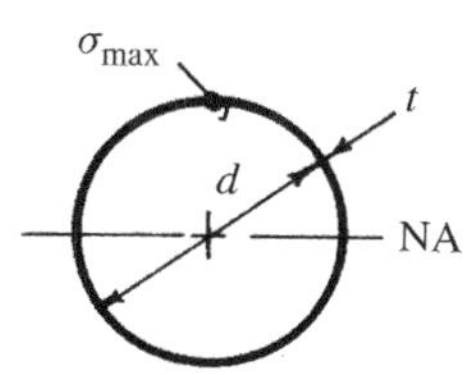

Figure 3.5 Thin-walled, circular tube section

(i) Objective Function

$$W = \rho AL = \rho\pi dtL \tag{3.15}$$

(ii) Bending Failure Criterion
When the peak bending stress σ_{max} is limited to the yield (or proof) stress σ_y for the beam material, we write $\sigma_{max} \le \sigma_y$, where

$$\sigma_{max} = \frac{My_{max}}{I} = \frac{Md/2}{\pi d^3 t/8} = \frac{4M}{\pi d^2 t} \tag{3.16}$$

(iii) Local Buckling Criterion
Wrinkles arise over the compressive side of the length under a local buckling stress:

$$\sigma_L = K_L E_T \left(\frac{t}{d}\right) \tag{3.17}$$

where $K_L = 0.798$ [2] and E_T is the tangent modulus at the chosen yield stress. The latter would normally correspond to a given offset strain where a yield point is not defined sharply.

(iv) Optimisation
Eliminating d and t between equations (3.16) and (3.17) and then dividing equation (3.15) by L^3 gives the objective function

$$\left(\frac{W}{L^3}\right)_{min} = 3.975\left(\frac{\rho}{E_T^{1/3}\sigma_y^{1/3}}\right)\left(\frac{M}{L^3}\right)^{2/3} \tag{3.18}$$

In equations (3.16) and (3.17) both σ_{max} and σ_L have attained the yield stress σ_y simultaneously. Correspondingly, the optimum d and t may be separated from within (3.16) and (3.17):

$$d_{opt} = 1.005\left(\frac{ME_T}{\sigma_y^2}\right)^{1/3}, \quad t_{opt} = 1.259\left(\frac{M\sigma_y}{E_T^2}\right)^{1/3} \tag{3.19a,b}$$

Equations (3.18) and (3.19a,b) provide the optimum solution for this tubular section.

3.3.2 Thin-Walled Square Tube

When a square tube, with mean side length a and wall thickness t (Figure 3.6), is used for a beam section, the maximum bending stress is reached at the top and bottom positions (i.e. $y_{max} = \pm a/2$) from the neutral axis. All the material across the full width at the top and bottom becomes fully stressed, which results in a more efficient tubular design.

As the wall is thin, the area and its second moment may be simplified to

$$A = 4at, \quad I = 2ta^3/12 + 2at(a/2)^2 = 2ta^3/3 \tag{3.20a,b}$$

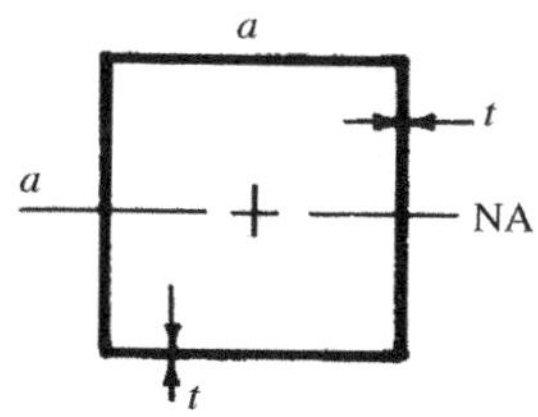

Figure 3.6 Thin-walled, tubular square section

(i) Objective Function

$$W = \rho AL = \rho(4at)L \tag{3.21}$$

(ii) Bending Failure Criterion
When the peak bending stress σ_{max} is limited to the yield (or proof) stress σ_y for the beam materiall, we write $\sigma_{max} \leq \sigma_y$, where

$$\sigma_{max} = \frac{My_{max}}{I} = \frac{Ma/2}{2ta^3/3} = \frac{3M}{4a^2t} \tag{3.22}$$

(iii) Local Buckling Criterion
Taking the side walls to be simply supported plates, the local buckling stress appears as [1]

$$\sigma_L = K_L E_T \left(\frac{t}{a}\right)^2 \tag{3.23}$$

where $K_L = 3.62$ for the horizontal sides and $K_L = 21.7$ for the vertical sides. Clearly, the lower K_L-value, which applies to uniform compression, is the critical coefficient here.

(iv) Optimisation
When both σ_{max} and σ_L attain the yield stress σ_y simultaneously, we may eliminate d and t between equations (3.21)–(3.23). Dividing the weight in equation (3.21) by L^3 gives the objective function :

$$\left(\frac{W}{L^3}\right)_{min} = 2.666\left(\frac{\rho}{E_T^{1/6}\sigma_y^{1/2}}\right)\left(\frac{M}{L^3}\right)^{2/3} \tag{3.24}$$

Different indices appear compared to those appearing in the objective function for a circular tube in equation (3.18). Correspondingly, from equations (3.22) and (3.23), the optimum a and t become:

$$a_{opt} = 1.126\,\frac{M^{1/3}E_T^{1/6}}{\sigma_y^{1/2}}, \quad t_{opt} = 0.592\left(\frac{M}{E_T}\right)^{1/3} \tag{3.25a,b}$$

Equations (3.24) and (3.25a,b) provide the optimum solution for this tubular section.

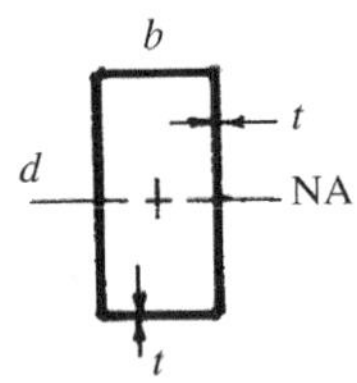

Figure 3.7 Thin-walled, rectangular tube

3.3.3 Thin-Walled, Rectangular Tube

When a rectangular tube, with breadth b, depth d and wall thickness t (Figure 3.7), is used for a beam section, the peak stress is reached across the full breadth at the top and bottom positions, i.e. $y_{\max} = \pm d/2$, from the neutral axis (NA).

As the wall is uniformly thin, the area and its second moment may be simplified to

$$A = 2dt\left(1 + \frac{b}{d}\right) \tag{3.26a}$$

$$I = \frac{2td^3}{12} + 2bt\left(\frac{d}{2}\right)^2 = \frac{td^3}{6}\left[1 + 3\left(\frac{b}{d}\right)\right] \tag{3.26b}$$

(i) Objective Function

$$W = \rho AL = \rho\left[2\,dt\left(1 + \frac{b}{d}\right)\right]L \tag{3.27}$$

(ii) Bending Failure Criterion
When the peak bending stress $\sigma_{\max}$ is limited to the yield (or proof) stress σ_y for the beam material, we may write $\sigma_{\max} \le \sigma_y$, where

$$\sigma_{\max} = \frac{My_{\max}}{I} = \frac{3M}{td^2[1 + 3(b/d)]} \tag{3.28}$$

(iii) Local Buckling Criteria
The possibility of local buckling (bowing) within both the horizontal and vertical walls of this section must be considered. As with equation (3.23), the local buckling stresses for these 'plates' are respectively

$$\sigma_{Lb} = 3.62\ E_T(t/b)^2, \quad \sigma_{Ld} = 21.7\ E_T(t/d)^2 \tag{3.29a,b}$$

(iv) Optimisation
Three strategies apply to the optimisation of this section depending upon which pair of equations (3.28) and (3.29a,b) are met simultaneously.

A. Combining equations (3.28) and (3.29a)

When by each equation both $\sigma_{\max}$ and σ_{Lb} attain the yield stress σ_y simultaneously, we may eliminate d and t. Dividing by L^3 gives the objective function

$$\left(\frac{W}{L^3}\right)_{\min} = 3.357\,\frac{(1+b/d)(b/d)^{1/3}}{(1+3b/d)^{2/3}}\left(\frac{\rho}{E_T^{1/6}\sigma_y^{1/2}}\right)\left(\frac{M}{L^3}\right)^{2/3} \tag{3.30}$$

There appears to be no minimum to this shape factor coefficient other than to confirm the value $f = 2.666$ when $b = d$, as was found previously for the square section.

B. Combining equations (3.28) and (3.29b)

Similarly, when the stresses $\sigma_{\max}$ and σ_{Ld} by these two equations are each equated to the yield stress, this allows for the elimination of d an t within an alternative objective function,

$$\left(\frac{W}{L^3}\right)_{\min} = 2.491\,\frac{1+b/d}{(1+3b/d)^{2/3}}\left(\frac{\rho}{E_T^{1/6}\sigma_y^{1/2}}\right)\left(\frac{M}{L^3}\right)^{2/3} \tag{3.31}$$

In this case the shape factor has a minimum value $f = 1.977$ for $b/d = 1$.

C. Combining equations (3.29a) and (3.29b)

When local buckling of both horizontal and vertical plates occurs simultaneously we may equate equations (3.29a,b) to give $b/d = 0.409$. This relation allows b and d to be eliminated from equations (3.27) and (3.29a,b), resulting in an objective function with a constant shape factor:

$$\left(\frac{W}{L^3}\right)_{\min} = 2.058\left(\frac{\rho}{E_T^{1/6}\sigma_y^{1/2}}\right)\left(\frac{M}{L^3}\right)^{2/3} \tag{3.32}$$

Figure 3.8 shows the shape factors as a function of b/d established by the three strategies A–C. It is evident from this figure that the minimum in curve B does not lie at its intersection with A, indicating that this section cannot be optimised fully.

Taking the optimum design at the intersection between curves A and B (i.e. strategy C), the minimum weight corresponds to $f = 2.058$ with $b/d = 0.409$. This compromise solution provides the corresponding dimensions for the tube:

$$d_{\text{opt}} = 1.845\,\frac{M^{1/3}\,E_T^{1/6}}{\sigma_y^{1/2}}, \qquad t_{\text{opt}} = 0.396\left(\frac{M}{E_T}\right)^{1/3} \tag{3.33a,b}$$

Equations (3.32) and (3.33a,b) provide the best optimum solution for this tubular section.

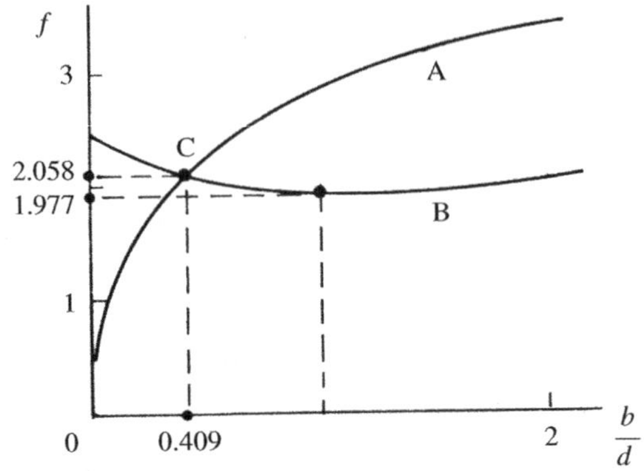

Figure 3.8 Shape factors from strategies A, B and C

3.3.4 *Non-Uniform, Thin-Walled Rectangular Tube*

Here the rectangular tube, of breadth b and depth d, has different wall thickness t_b and t_d (see Figure 3.9). The tube's area and its second moment may be simplified to

$$A = 2(dt_d + bt_b) = 2t_b\, d\left(1 + \frac{b}{d}\frac{t_b}{t_d}\right) \tag{3.34a}$$

$$I = \frac{2t_d\, d^3}{12} + 2bt_b\left(\frac{d}{2}\right)^2 = \frac{t_d\, d^3}{6}\left(1 + 3\frac{b}{d}\frac{t_b}{t_d}\right) \tag{3.34b}$$

(i) Objective Function

$$W = \rho AL = \rho 2t_b\, d\left(1 + \frac{b}{d}\frac{t_b}{t_d}\right)L \tag{3.35}$$

(ii) Bending Failure Criterion
When the peak bending stress $\sigma_{\max}$ is limited to the yield (or proof) stress σ_y for the beam material, we write $\sigma_{\max} \le \sigma_y$ where

$$\sigma_{\max} = \frac{M y_{\max}}{I} = \frac{3M}{t_d\, d^2\left[1 + 3(b/d)(t_b/t_d)\right]} \tag{3.36}$$

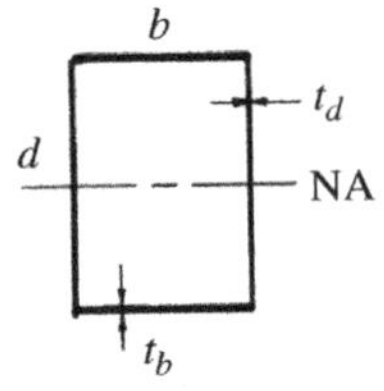

Figure 3.9 Thin-walled, rectangular tube

(iii) Local Buckling Criteria

The possibility of local buckling within both horizontal and vertical 'plates' must be considered. With differing thicknesses, the local buckling stresses for these plates are respectively

$$\sigma_{Lb} = 3.62\ E_T(t_b/b)^2, \quad \sigma_{Ld} = 21.7\ E_T(t_d/d)^2 \tag{3.37a,b}$$

(iv) Optimisation

Here we use the experience gained from optimising the uniform rectangular tube in § 3.3.3 to simplify this process. Firstly, the local buckling stress is allowed to be reached in each side simultaneously. Thus, equating (3.37a,b) gives

$$\frac{t_b/t_d}{b/d} = 2.448 \tag{3.38}$$

Next, equations (3.35), (3.36) and (3.37b) are combined by eliminating d and t_d. This gives an objective function whose shape factor can be minimised:

$$\left(\frac{W}{L^3}\right)_{\min} = 2.491\left\{\frac{(1+(b/d)(t_b/t_d))}{[(1+3(b/d)(t_b/t_d)]^{2/3}}\right\}\left(\frac{\rho}{E_T^{1/6}\sigma_y^{1/2}}\right)\left(\frac{M}{L^3}\right)^{2/3} \tag{3.39}$$

Correspondingly, d and t_d follow from equations (3.36) and (3.37b) as:

$$d = \frac{2.409}{[1+3(b/d)(t_b/t_d)]^{1/3}}\frac{M^{1/3}\,E_T^{1/6}}{\sigma_y^{1/2}}, \quad t_d = \frac{0.517}{[1+3(b/d)(t_b/t_d)]^{1/3}}\left(\frac{M}{E_T}\right)^{1/3} \tag{3.40a,b}$$

The product of the geometric ratios becomes unity for a minimum value of the shape factor, given by the expression in braces in equation (3.39):

$$(b/d)(t_b/t_d) = 1 \tag{3.41a}$$

Simultaneously solving equations (3.38) and (3.41a) separates each ratio to its optimum value:

$$(b/d)_{\text{opt}} = 0.639, \quad (t_b/t_d)_{\text{opt}} = 1.565 \tag{3.41b,c}$$

The solution (3.41b,c) is shown graphically in Figure 3.10.

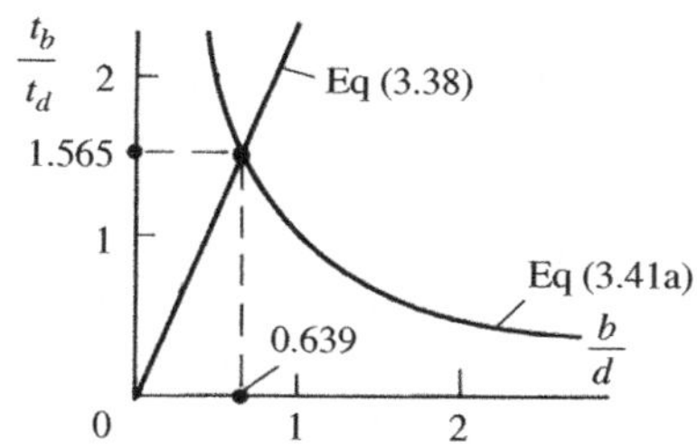

Figure 3.10 Length and thickness ratios from matched strategies

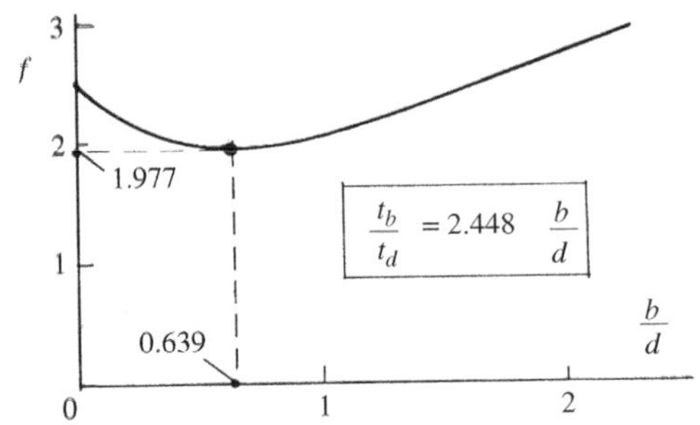

Figure 3.11 Dependence of shape factor upon b/d when equation (3.38) applies

Substituting the optimum ratios from equations (3.41b,c) into equations (3.39) and (3.40a,b) provides the minimum weight solutions for this cross-section:

$$\left(\frac{W}{L^3}\right)_{\min} = 1.977 \frac{\rho}{E_T^{1/6}\sigma_y^{1/2}} \left(\frac{M}{L^3}\right)^{2/3} \tag{3.42}$$

Correspondingly, the optimum dimensions for the horizontal sides become

$$b_{\text{opt}} = 0.969 \frac{M^{1/3} E_T^{1/6}}{\sigma_y^{1/2}}, \quad t_{b_{\text{opt}}} = 0.510 \left(\frac{M}{E_T}\right)^{1/3} \tag{3.43a,b}$$

and the optimum dimensions for the vertical sides become

$$d_{\text{opt}} = 1.517 \frac{M^{1/3} E_T^{1/6}}{\sigma_y^{1/2}}, \quad t_{d_{\text{opt}}} = 0.326 \left(\frac{M}{E_T}\right)^{1/3} \tag{3.44a,b}$$

Figure 3.11 interprets the minimum weight result graphically at the stationary point in the plot of f verus b/d. The f curve is described by the expression in braces in equation (3.39), for which the dimension ratios are controlled by equation (3.38). We see that, by allowing for different thicknesses, the minimum in the curve corresponds to equation (3.41b).

3.4 Open Sections

Thin-walled, open sections are common among beams used in the construction of bridges, high-rise buildings and vehicles. With the selection of an appropriate material, it is understood that weight will be reduced considerably for a section having webs and flanges, as in I-, T-, U- and L-sections, compared to solid sections [3–9]. We shall examine how to reduce the weight of an I-section to a minimum, firstly with uniform thickness and secondly where the web and flange thicknesses differ. Here the neutral axis bisects the depth, which is not the case for the T- and U-sections that follow. However, their principal axes of bending remain vertical and horizontal. For the flexure of an L-section, the principal axes are inclined, as was shown when a similar section was employed as a strut (see § 1.5.3).

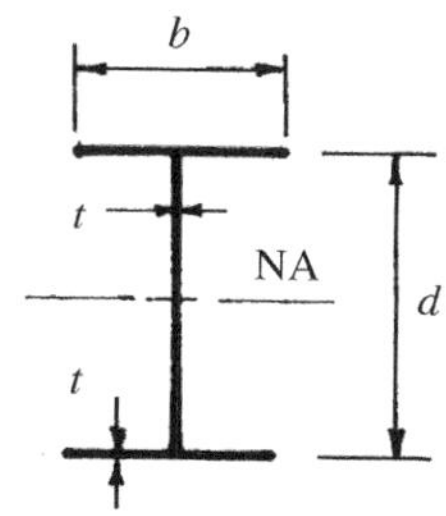

Figure 3.12 Thin-walled, I-section

3.4.1 *I-Section, Uniform Thickness*

When a uniform I-section, with breadth b, depth d and wall thickness t (see Figure 3.12), is used for a beam section, the peak stress is reached at the flange top and bottom across the full breadth, distance $y_{\max} = \pm d/2$ from the neutral axis (NA).

As the wall is uniformly thin, the area and its second moment may be simplified to

$$A = 2bt + dt = dt(1 + 2b/d) \tag{3.45a}$$

$$I = \frac{td^3}{12} + 2bt\left(\frac{d}{2}\right)^2 = \frac{td^3}{12}\left(1 + 6\frac{b}{d}\right) \tag{3.45b}$$

(i) Objective Function

$$W = \rho AL = \rho\, dt(1 + 2b/d)L \tag{3.46}$$

(ii) Bending Failure Criterion

When the peak bending stress $\sigma_{\max}$ at position $y_{\max} = \pm d/2$, is limited to the yield (or proof) stress σ_y for the beam material, we write $\sigma_{\max} \le \sigma_y$, where

$$\sigma_{\max} = \frac{My_{\max}}{I} = \frac{6M}{td^2[1 + 6(b/d)]} \tag{3.47}$$

(iii) Local Buckling Criteria

The possibility of local buckling within the flanges and the web must be considered. The local buckling stresses for these 'plates' are respectively

$$\sigma_{Lf} = 0.385\ E_T\left(\frac{t}{b/2}\right)^2 = 1.54\ E_T\left(\frac{t}{b}\right)^2 \tag{3.48a}$$

$$\sigma_{Lw} = 21.7\ E_T\left(\frac{t}{d}\right)^2 \tag{3.48b}$$

(iv) Optimisation

There are three strategies for an optimisation of this section, depending upon which pair of stresses within equations (3.47), (3.48a) and (3.48b) are met simultaneously.

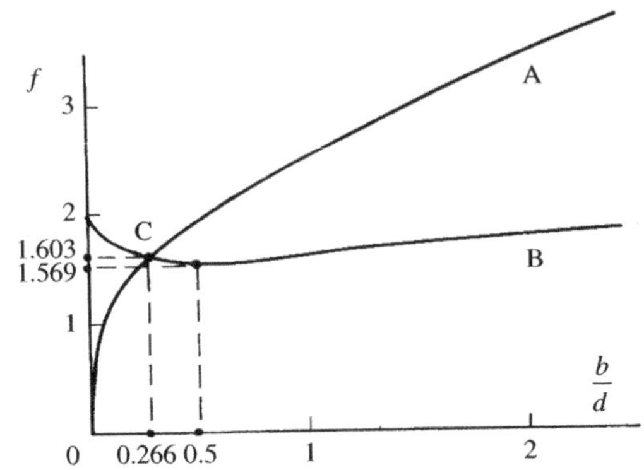

Figure 3.13 Shape factors from strategies A, B and C

A. Combining equations (3.47) and (3.48a)
When, by each equation, both $\sigma_{\max}$ and σ_{Lf} attain the yield stress σ_y simultaneously, we may eliminate d and t. Dividing by L^3 gives the objective function:

$$\left(\frac{W}{L^3}\right)_{\min} = 3.073\left[\frac{(1+2b/d)(b/d)^{1/3}}{(1+6b/d)^{2/3}}\right]\left(\frac{\rho}{E_T^{1/6}\sigma_y^{1/2}}\right)\left(\frac{M}{L^3}\right)^{2/3} \tag{3.49}$$

There is no value of b/d which minimises the shape factor expression (in square brackets) within equation (3.49) and hence the weight will increase with increasing b/d (see Figure 3.13).

B. Combining equations (3.47) and (3.48b)
Similarly, when the stresses $\sigma_{\max}$ and σ_{Lw} in equations (3.47) and (3.48b) are each equated to σ_y, this allows for the elimination of d and t, which leads to an alternative objective function,

$$\left(\frac{W}{L^3}\right)_{\min} = \left[\frac{1.977(1+2b/d)}{(1+6b/d)^{2/3}}\right]\left(\frac{\rho}{E_T^{1/6}\sigma_y^{1/2}}\right)\left(\frac{M}{L^3}\right)^{2/3} \tag{3.50}$$

In this case the shape factor expression (in square brackets) attains a minimum value $f = 1.569$ for $b/d = 0.5$ but, as will be seen, this is not the ratio chosen to optimise the weight.

C. Combining equations (3.48a) and (3.48b)
When local buckling of both horizontal and vertical plates occurs simultaneously we may equate (3.48a,b) to give $b/d = 0.266$. This particular ratio allows b and d to be eliminated from equations (3.46), (3.47) and (3.48b), so providing a third objective function with a constant shape factor:

$$\left(\frac{W}{L^3}\right)_{\min} = 1.603\left(\frac{\rho}{E_T^{1/6}\sigma_y^{1/2}}\right)\left(\frac{M}{L^3}\right)^{2/3} \tag{3.51}$$

Figure 3.13 shows the shape factor as a function of b/d established by these three strategies. It is evident that strategy C will lie at the intersection between the strategy A and B curves, but this does not coincide with the minimum in curve B. Here we take the weight to be optimised under C, where $b/d = 0.266$, to give $f = 1.603$.

Correspondingly, the optimum dimensions become

$$d_{\text{opt}} = 2.208\,\frac{M^{1/3}\,E_T^{1/6}}{\sigma_y^{1/2}}, \qquad t_{\text{opt}} = 0.474\left(\frac{M}{E_T}\right)^{1/3} \tag{3.52a,b}$$

We shall now show that only when the thicknesses of the web differs from that of the flanges can the minimum for the B curve be made to intersect the A curve, thereby truly minimising the weight.

3.4.2 I-Section, Non-Uniform Thickness

Let a non-uniform I-section, of breadth b and depth d, with flange and web thickness, t_f and t_w, be used for a beam section (see Figure 3.14). The peak stress $\sigma_{\max}$ in bending, which is reached at the flange top and bottom across the full breadth, is limited by the yield stress of the beam material.

The area and its second moment may be simplified to

$$A = dt_w + 2bt_f = t_w\,d\left(1 + 2\frac{b}{d}\frac{t_f}{t_w}\right) \tag{3.53a}$$

$$I = \frac{t_w d^3}{12} + 2bt_f\left(\frac{d}{2}\right)^2 = \frac{t_w d^3}{12}\left(1 + 6\frac{b}{d}\frac{t_f}{t_w}\right) \tag{3.53b}$$

(i) Objective Function

$$W = \rho AL = \rho t_w\,d\left(1 + 2\frac{b}{d}\frac{t_f}{t_w}\right)L \tag{3.54}$$

(ii) Bending Failure Criterion

The peak bending stress $\sigma_{\max}$ in bending, which is reached at $y_{\max} \pm d/2$, becomes

$$\sigma_{\max} = \frac{M y_{\max}}{I} = \frac{6M}{t_w\,d^2\left[1 + 6(b/d)(t_f/t_w)\right]} \tag{3.55}$$

where $\sigma_{\max}$ is limited to the yield, or proof, stress σ_y for the beam material, i.e. $\sigma_{\max} \le \sigma_y$.

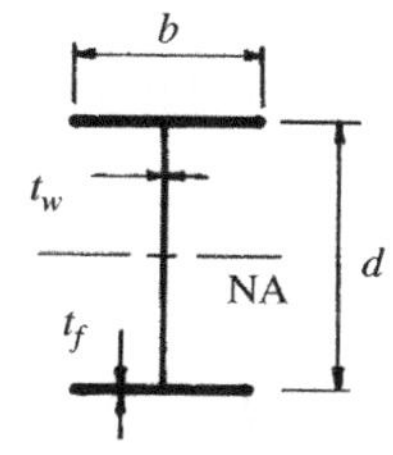

Figure 3.14 Thin-walled, non-uniform I-section

(iii) Local Buckling Criteria

The possibility of local buckling within both horizontal and vertical flats must be considered. Given their differing thicknesses and support positions, the local buckling stresses for the flange and web are respectively

$$\sigma_{Lf} = 0.385\ E_T \left(\frac{t_f}{b/2}\right)^2 = 1.54\ E_T \left(\frac{t_f}{b}\right)^2 \tag{3.56a}$$

$$\sigma_{Lw} = 21.7\ E_T (t_w/d)^2 \tag{3.56b}$$

(iv) Optimisation

To optimise the non-uniform I-section we use the experience gained from optimising the uniform I-section in § 3.4.1 to simplify this process. Firstly, the local buckling stress is allowed to be reached in each side simultaneously. Equating (3.56a) and (3.56b) gives

$$\frac{t_f/t_w}{b/d} = 3.754 \tag{3.57}$$

Next, equations (3.54), (3.55) and (3.56b) are combined by eliminating d and t_w. This gives an objective function whose shape factor can be minimised:

$$\left(\frac{W}{L^3}\right)_{\min} = \left\{\frac{1.977\left[(1 + 2(b/d)(t_f/t_w)\right]}{\left[(1 + 6(b/d)(t_f/t_w)\right]^{2/3}}\right\} \left(\frac{\rho}{E_T^{1/6}\sigma_y^{1/2}}\right) \left(\frac{M}{L^3}\right)^{2/3} \tag{3.58}$$

Correspondingly, d and t_w follow from equations (3.55) and (3.56b) as

$$d = \frac{3.305}{\left[1 + 6(b/d)(t_f/t_w)\right]^{1/3}} \frac{M^{1/3}\ E_T^{1/6}}{\sigma_y^{1/2}}, \quad t_w = \frac{0.651}{\left[1 + 6(b/d)(t_f/t_w)\right]^{1/3}} \left(\frac{M}{E_T}\right)^{1/3} \tag{3.59a,b}$$

The minimum shape factor in braces in equation (3.58) applies to a geometrical product of 0.5:

$$(b/d)(t_f/t_w) = 1/2 \tag{3.60}$$

Solving equations (3.57) and (3.60) separates the optimum ratios

$$(b/d)_{\text{opt}} = 0.365, \quad (t_f/t_w)_{\text{opt}} = 1.370 \tag{3.61a,b}$$

The graphical interpretation of this solution appears in Figure 3.15.

Substituting the optimum ratios from equations (3.61a,b) into equations (3.58) and (3.59a,b) provides the minimum weight solution for this section:

$$\left(\frac{W}{L^3}\right)_{\min} = 1.569 \left(\frac{\rho}{E_T^{1/6}\sigma_y^{1/2}}\right) \left(\frac{M}{L^3}\right)^{2/3} \tag{3.62}$$

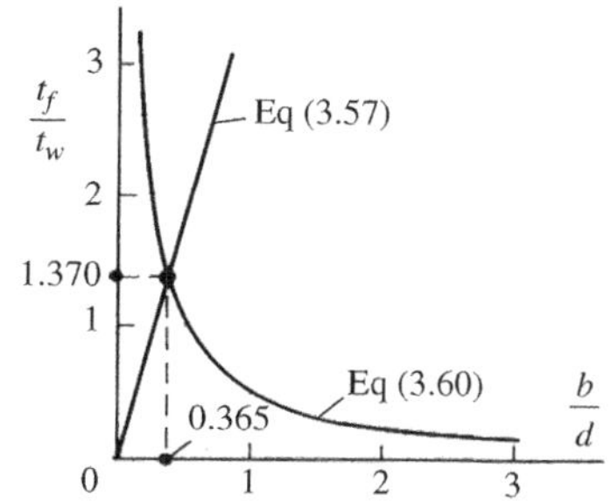

Figure 3.15 Optimum geometry for minimum shape factor

The optimum dimensions for each flange become

$$b_{\text{opt}} = 0.698 \frac{M^{1/3}\, E_T^{1/6}}{\sigma_y^{1/2}}, \quad t_{f_{\text{opt}}} = 0.562 \left(\frac{M}{E_T}\right)^{1/3} \qquad (3.63\text{a,b})$$

and the optimum dimensions of the web become

$$d_{\text{opt}} = 1.912 \frac{M^{1/3}\, E_T^{1/6}}{\sigma_y^{1/2}}, \quad t_{w_{\text{opt}}} = 0.410 \left(\frac{M}{E_T}\right)^{1/3} \qquad (3.63\text{c,d})$$

Figure 3.16 interprets the unique minimum weight solution for this section graphically. The curve is described by shape factor f, i.e. the quantity in braces in equation (3.58). At the stationary point in the plot of f versus b/d the weight is minimised when the dimensions conform exactly to the coincidence between equations (3.57) and (3.60). This is in contrast to the uniform section, where a similar coincidence was not possible.

3.4.3 T-Section, Uniform Thickness

Figure 3.17 shows a T-section, with flange breadth b and web depth d each with uniform wall thickness t. When used as a beam section, the neutral axis (NA) does not bisect the

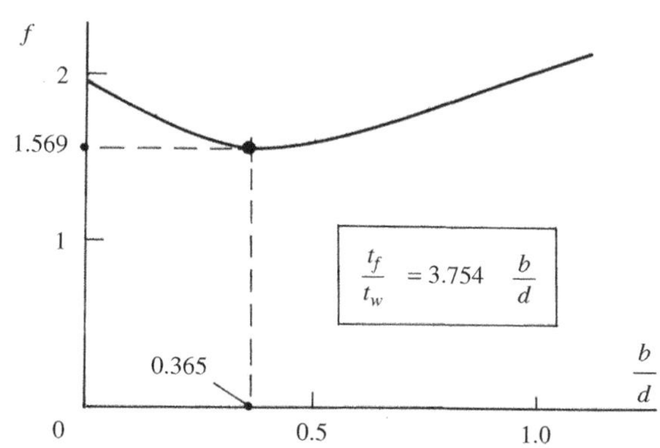

Figure 3.16 Minimum in the shape factor curve satisfying equations (3.57) and (3.60)

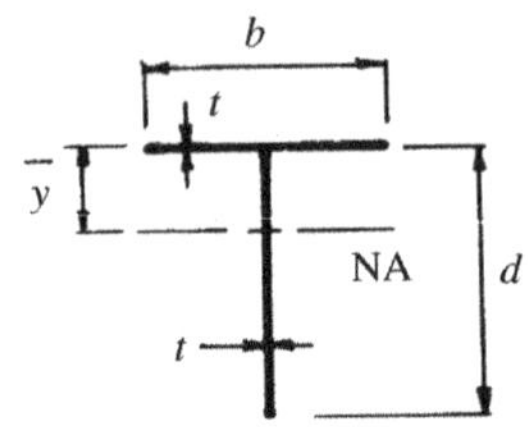

Figure 3.17 Thin-walled, T-section

depth but lies at the closer distance ($\bar{y}$) to the flange top. Consequently, the maximum bending stress is reached at the web bottom.

As the wall is uniformly thin, the area and its second moment may be simplified to

$$A = bt + dt = dt(1 + b/d) \tag{3.64a}$$

The $\bar{y}$ position of the neutral axis is found from taking moments of area about the flange top:

$$A\bar{y} = \frac{bt^2}{2} + \frac{d^2t}{2} \quad \Rightarrow \quad \bar{y} = \frac{d\left[1 + (b/d)(t/d)\right]}{2(1 + b/d)} \tag{3.64b}$$

The second moment of area of the T-section, referred to this axis, is given by

$$I = bt\bar{y}^2 + \frac{td^3}{12} + dt(d/2 - \bar{y})^2 \tag{3.65a}$$

$$= \frac{td^3}{12}\left\{\frac{3(b/d)[1 + (b/d)(t/d)]^2 + [1 + (b/d)]^2 + 3(b/d)^2[1 - (t/d)]^2}{[1 + (b/d)]^2}\right\}$$

A good approximation to I in equation (3.65a) holds when $d/t > 20$:

$$I = \frac{td^3}{12}\left[\frac{1 + 4(b/d)}{1 + (b/d)}\right] \tag{3.65b}$$

(i) Objective Function

$$W = \rho AL = \rho\, dt(1 + b/d)L \tag{3.66}$$

(ii) Bending Failure Criterion

The maximum bending stress σ_{max} occurs at the web bottom:

$$y_{max} = d - \bar{y} = \frac{d}{2}\left[\frac{1 + 2(b/d) - (b/d)(t/d)}{1 + (b/d)}\right] \approx \frac{d}{2}\left[\frac{1 + 2(b/d)}{1 + (b/d)}\right] \tag{3.67a}$$

such that

$$\sigma_{max} = \frac{My_{max}}{I} = \frac{6M[1 + 2(b/d)]}{td^2[1 + 4(b/d)]} \tag{3.67b}$$

in which $\sigma_{\max}$ is limited to the yield (or proof) stress σ_y for the beam material, i.e. $\sigma_{\max} \leq \sigma_y$

(iii) Local Buckling Criteria

The possibility of local buckling within the flange and the web must be considered. The local buckling stresses for these 'plates' are respectively

$$\sigma_{L_f} = 0.385E\left(\frac{t}{b/2}\right)^2 = 1.54E\left(\frac{t}{b}\right)^2 \tag{3.68a}$$

$$\sigma_{L_w} = 21.7E\left(\frac{t}{2d}\right)^2 = 5.425E\left(\frac{t}{d}\right)^2 \tag{3.68b}$$

Because of the asymmetric stress distribution (as in Figure 3.1) it is assumed that the local buckling of the flange and web is elastic. With the absence of a lower flange the effective length of the web is doubled, to offer less resistance to bowing within a half wavelength.

(iv) Optimisation

The following three strategies A, B and C are considered for optimising this cross-section, depending upon which pair of equations (3.67a) and (3.68a,b) are met simultaneously.

A. Combining equations (3.67b) and (3.68a)

When by each equation both $\sigma_{\max}$ and σ_{Lf} reach the yield stress σ_y simultaneously, we may eliminate d and t. Dividing equation (3.66) by L^3 leads to the objective function

$$\left(\frac{W}{L^3}\right)_{\min} = 3.073\left[\frac{(1+b/d)(1+2b/d)^{2/3}(b/d)^{1/3}}{(1+4b/d)^{2/3}}\right]\left(\frac{\rho}{E_T^{1/6}\sigma_y^{1/2}}\right)\left(\frac{M}{L^3}\right)^{2/3} \tag{3.69}$$

There is no value of b/d which minimises the shape factor (in square brackets), i.e. the weight increases with the ratio b/d. Equation (3.69) must be rejected because the yield stress is first reached at the web bottom. This means that the flange can only attain the yield stress when plastic penetration into the web is allowed (see Chapter 14).

B. Combining equations (3.67b) and (3.68b)

When the stresses $\sigma_{\max}$ and σ_{Lw} by equations (3.67b) and (3.68b) are each set to σ_y, the elimination of d an t leads to an acceptable objective function:

$$\left(\frac{W}{L^3}\right)_{\min} = 2.491\left[\frac{(1+b/d)(1+2b/d)^{2/3}}{(1+4b/d)^{2/3}}\right]\left(\frac{\rho}{E_T^{1/6}\sigma_y^{1/2}}\right)\left(\frac{M}{L^3}\right)^{2/3} \tag{3.70}$$

In this case the expression in square brackets attains a minimum value of 0.9904 for $b/d = 0.0643$, which gives a minimum $f = 2.491 \times 0.9904 = 2.467$. However, the following strategy is a more likely choice for optimising the beam's weight.

C. Combining equations (3.68a) and (3.68b)
When local buckling of both horizontal and vertical plates occurs simultaneously, we may equate (3.68a,b) to give $b/d = 0.533$. Setting $b/d = 0.533$ within equations (3.67b) and (3.68a) and equalising them at σ_y optimises the objective function:

$$\left(\frac{W}{L^3}\right)_{\min} = 2.893\left(\frac{\rho}{E_T^{1/6}\sigma_y^{1/2}}\right)\left(\frac{M}{L^3}\right)^{2/3} \tag{3.71}$$

Correspondingly, the optimum dimensions follow as:

$$d_{\text{opt}} = 1.876 b_{\text{opt}} = 2.097\frac{M^{1/3}E_T^{1/6}}{\sigma_y^{1/2}}, \quad t_{\text{opt}} = 0.90\left(\frac{M}{E_T}\right)^{1/3} \tag{3.72a,b}$$

Figure 3.18 shows the shape factor as a function of b/d, according to the two strategies B and C. It is evident from this that strategy C is a point on the strategy B curve at the position shown.

At C it is evident that the weight is more than that which would be found from the minimum in the strategy B curve. That is, $f = 2.467$, but this corresponds to a restricted flange width $b = 0.0643 \times d$. That neither design is particularly efficient is a consequence of a section that is largely understressed in bending, i.e. only the web bottom attains the design stress. Some improvement is found when the thicknesses of the web differs from the flange thickness. Then the minimum in the B curve can be made to coincide with the strategy C point, as we shall now show.

3.4.4 T-Section, Non-Uniform Thickness

Let a non-uniform T-section, of breadth b and depth d, with flange and web thickness t_f and t_w respectively, be used for a beam section (see Figure 3.19). The peak stress $\sigma_{\max}$ in bending is again first reached at the web bottom.

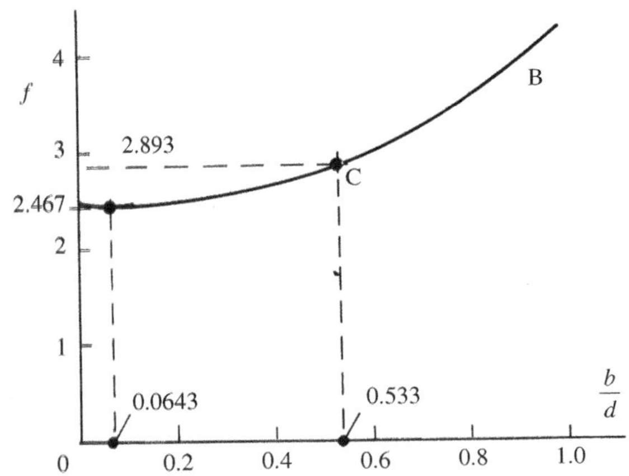

Figure 3.18 Shape factors from strategies B and C

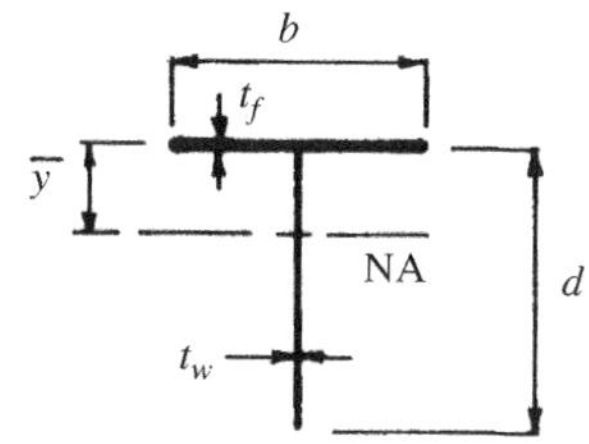

Figure 3.19 Thin-walled, non-uniform T-section

The section area A, centroid $\overline{y}$ and its second moment I may be simplified as follows

$$A = dt_w + bt_f = t_w d\left(1 + \frac{b}{d}\frac{t_f}{t_w}\right) \tag{3.73a}$$

The centroid is located at the position $\overline{y}$,

$$\overline{y} = \frac{d\left[1 + (b/d)(t_f/t_w)(t_f/d)\right]}{2\left[1 + (b/d)(t_f/t_w)\right]} \tag{3.73b}$$

and with $d/t_f > 20$, the second moment of area approximates to

$$I = \frac{t_w d^3}{12}\left[\frac{1 + 4(b/d)(t_f/t_w)}{1 + (b/d)(t_f/t_w)}\right] \tag{3.73c}$$

(i) Objective Function

$$W = \rho AL = \rho t_w d\left(1 + \frac{b}{d}\frac{t_f}{t_w}\right)L \tag{3.74}$$

(ii) Bending Failure Criterion
The maximum bending stress σ_{max} in bending is reached at the web bottom:

$$y_{max} = d + t_f - \overline{y} \approx \frac{d}{2}\left[\frac{1 + 2(b/d)(t_f/t_w)}{1 + (b/d)(t_f/t_w)}\right] \tag{3.75a}$$

which, when taken with the bending stress expression, gives

$$\sigma_{max} = \frac{My_{max}}{I} = \frac{6M\left[1 + 2(b/d)(t_f/t_w)\right]}{t_w d^2\left[1 + 4(b/d)(t_f/t_w)\right]} \tag{3.75b}$$

where $\sigma_{max} \leq \sigma_y$ is limited to the yield (or proof) stress σ_y for the beam material.

(iii) Local Buckling Criteria
The possibility of local buckling occurring within both flange and web must be considered. With differing lengths and thicknesses, the local elastic buckling stresses for these plates are respectively

$$\sigma_{L_f} = 0.385E\left(\frac{t_f}{b/2}\right)^2 = 1.54E\left(\frac{t_f}{b}\right)^2 \tag{3.76a}$$

$$\sigma_{L_w} = 21.7E\left(\frac{t_w}{2d}\right)^2 = 5.43E\left(\frac{t_w}{d}\right)^2 \tag{3.76b}$$

(iv) Optimisation

Here we use the experience gained from optimising the uniform T-section in § 3.4.3 to direct this process. Firstly, the local buckling stress is allowed to be reached in each plate simultaneously. Thus, equating (3.76a) to (3.76b) gives

$$\frac{b/d}{t_f/t_w} = 0.533 \tag{3.77}$$

Next, equations (3.75b) and (3.76b) are each equated to σ_y. This gives d and t_w as

$$d = 2.409\left[\frac{1+2(b/d)(t_f/t_w)}{1+4(b/d)(t_f/t_w)}\right]^{1/3}\frac{M^{1/3}\,E_T^{1/6}}{\sigma_y^{1/2}},$$

$$t_w = 1.034\left[\frac{1+2(b/d)(t_f/t_w)}{1+4(b/d)(t_f/t_w)}\right]^{1/3}\left(\frac{M}{E_T}\right)^{1/3} \tag{3.78a,b}$$

Substituting d and t_w from equations (3.78a,b) into equation (3.74) and then dividing by L^3 gives

$$\left(\frac{W}{L^3}\right)_{\min} = 2.491\left\{\frac{[(1+(b/d)(t_f/t_w)]\,[(1+2(b/d)(t_f/t_w)]^{2/3}}{[(1+4(b/d)(t_f/t_w)]^{2/3}}\right\}\left(\frac{\rho}{E_T^{1/6}\sigma_y^{1/2}}\right)\left(\frac{M}{L^3}\right)^{2/3} \tag{3.79}$$

The shape factor in braces in equation (3.79) is a minimum for the product ratio:

$$(b/d)(t_f/t_w) = 0.0643 \tag{3.80}$$

Note that a similar minimum was found previously for $b/d = 0.0643$ in equation (3.70), i.e. where $t_f/t_w = 1$. Solving equations (3.77) and (3.80) simultaneously separates the optimum ratios:

$$(b/d)_{\text{opt}} = 0.185, \quad (t_f/t_w)_{\text{opt}} = 0.347 \tag{3.81a,b}$$

Figure 3.20 shows the solutions (3.81a,b) graphically.

Substituting equations (3.81a,b) into (3.79) provides the minimum shape factor $f = 2.467$. Clearly, this is an improvement upon the uniform T-section, where $f = 2.893$, but a non-uniform T still remains a relatively inefficient section, whose minimum weight solution is

$$\left(\frac{W}{L^3}\right)_{\min} = 2.467\left(\frac{\rho}{E_T^{1/6}\sigma_y^{1/2}}\right)\left(\frac{M}{L^3}\right)^{2/3} \tag{3.82}$$

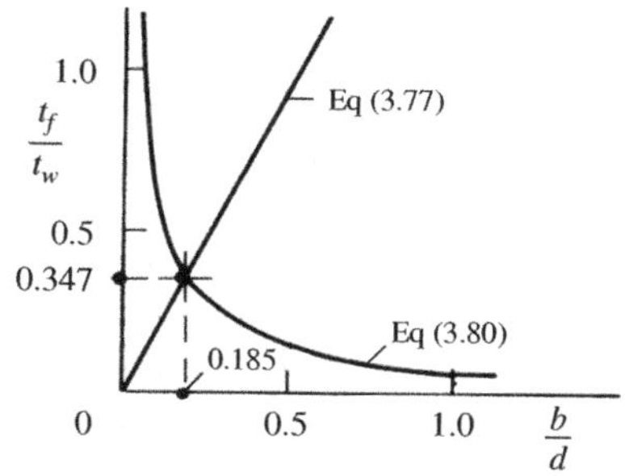

Figure 3.20 Optimum section dimension ratios

Equations (3.75) and (3.76) give the optimum dimensions for the web:

$$d_{\text{opt}} = 2.324\,\frac{M^{1/3}\,E_T^{1/6}}{\sigma_y^{1/2}},\qquad t_{w_{\text{opt}}} = 0.998\left(\frac{M}{E_T}\right)^{1/3} \tag{3.83a,b}$$

Correspondingly, from equations (3.81a,b), the optimum dimensions for the flange become:

$$b_{\text{opt}} = 0.185\,d_{\text{opt}} = 0.430\,\frac{M^{1/3}\,E_T^{1/6}}{\sigma_y^{1/2}},\qquad t_{f_{\text{opt}}} = 0.347 t_{w_{\text{opt}}} = 0.346\left(\frac{M}{E_T}\right)^{1/3} \tag{3.83c,d}$$

Figure 3.21 shows the dependence of the shape factor f within equation (3.79) upon b/d. The minimum weight solution corresponds to the minimum f for the single ratio $b/d = 0.185$.

Figure 3.21 also provides other non-optimum shape factors f for when equation (3.79) is constrained only by the restrictive condition $t_f/t_w = 1.876b/d$, i.e. when $b/d \neq 0.185$.

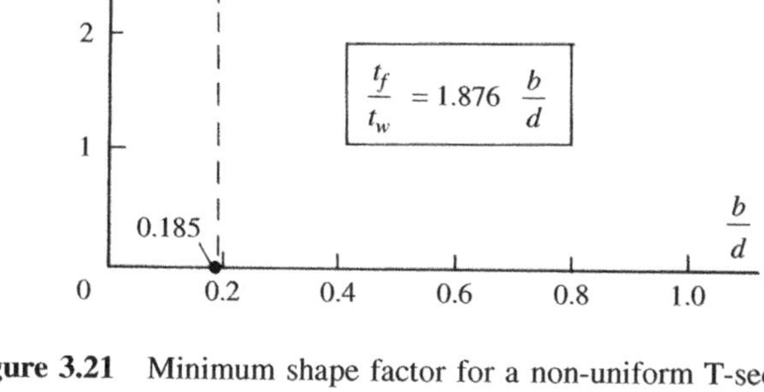

Figure 3.21 Minimum shape factor for a non-uniform T-section

3.5 Summary of Results

Table 3.1 summarises the results for the objective functions found for the 11 beam sections – solid, tubular and thin-walled – considered here.

While the structural index $S = M/L^3$ is common to all 11 sections, the exponent n on this index differs for two of the four categories of the material factor identified within this table. Where more than one section lies within a given category it becomes possible to compare their shape factors, e.g. a solid square section is more efficient than a solid circular section and the non-uniform I-section is the most efficient thin-walled section. Hence, we may conclude that those beams which minimise weight most effectively are:

(i) a rectangular solid square section;
(ii) a non-uniform rectangular tube;
(iii) a non-uniform I-section.

The last of these, which ranks the most highly among thin open sections, accounts for the popularity of I-sections when used as beams in structural engineering applications.

One final comment should be made upon the failure criterion $\sigma_{max} \le \sigma_y$ adopted here. Suppose that we have to choose a square section timber beam with limiting stress $\sigma_y = 5$ MPa to support a uniform bending moment of $M = 2.5$ kNm over a length of $L = 4$ m.

Table 3.1 Objective function for beam sections: $(W/L^3)_{min} = f \times m \times S^n$, where $S = M/L^3$

Cross-Section	Shape Factor, f	Material Factor, m	Exponent n
	3.691	$\rho/\sigma_y^{1/3}$	2/3
	3.302	$\rho/\sigma_y^{1/3}$	2/3
	1.516	$\rho/E^{1/2}$	1/2
	3.975	$\rho/(E\sigma_y)^{1/3}$	1/3
	2.666	$\rho/(E^{1/6}\sigma_y{}^{1/2})$	2/3
	2.058	$\rho/(E^{1/6}\sigma_y{}^{1/2})$	2/3
	1.978	$\rho/(E^{1/6}\sigma_y{}^{1/2})$	2/3
	1.603	$\rho/(E^{1/6}\sigma_y{}^{1/2})$	2/3
	1.569	$\rho/(E^{1/6}\sigma_y{}^{1/2})$	2/3
	2.893	$\rho/(E^{1/6}\sigma_y{}^{1/2})$	2/3
	2.467	$\rho/(E^{1/6}\sigma_y{}^{1/2})$	2/3

Firstly, taking $\sigma_{\max} = \sigma_y$, equation 3.6b provides the optimum, square section size (in units of N and m):

$$a_{\text{opt}} = \left(\frac{6M}{\sigma_y}\right)^{1/3} = \left(\frac{6 \times 2.5 \times 10^3}{5 \times 10^6}\right)^{1/3} = 0.1442 \text{ m}$$

Given that the density of wood is $\rho = 800$ kg/m^3, the weight of this beam is

$$W = \rho a_{\text{opt}}^2 L = 800 \times (0.1442)^2 \times 4 = 66.56 \text{ kg}$$

When the beam is allowed to rest upon simple supports this self-weight adds to the applied moment from external loading. Assuming that both moments act in the same sense the greatest moment occurs at the beam centre (in units of kN and m):

$$M_{\max} = M + \frac{wL^2}{8} = M + \frac{(\rho a_{\text{opt}}^2)L^2}{8}$$
$$= 2.5 + \frac{(800 \times 9.81 \times 0.1442^2 \times 4^2)}{(8 \times 10^3)} = 2.826 \text{ kN m}$$

which modifies the initial, optimum square dimension to 150.24 mm. A second iteration then gives $W = 72$ kg and $M_{\max} = 2.854$ kN m. The third iteration settles the size, weight and maximum moment at $a_{\text{opt}} = 150.7$ mm, $W = 72.71$ kg and $M_{\max} = 2.856$ kN m, respectively. Note that we could have allowed for this self-weight effect by using $\sigma_{\max} = 4.38$ MPa for our initial a_{opt} calculation. This example illustrates how to interpret the inequailty meaningfully within the limiting stress failure criterion $\sigma_{\max} < \sigma_y$. More generally, the application of a safety factor to σ_y will offset all increases in stress from unforseen sources.

References

[1] Timoshenko, S. and Gere, J. M. *Theory of Elastic Stability*, McGraw-Hill, 1961.
[2] Roarke, R. J. and Young, W. C. *Formulas for Stress and Strain*, McGraw-Hill, 1989.
[3] Cox, H. L. *The Design of Structures of Least Weight*, Pergamon, 1965.
[4] Farkas, J. *Optimum Design of Metal Structures*, Ellis-Horwood, 1984.
[5] Hemp, W. S. *Optimum Structures*, Clarendon Press, 1973.
[6] Majid, K. I. (ed.) *Optimum Design of Structures*, Newnes-Butterworth, 1974.
[7] Kirsch, U. *Optimum Structural Design*, McGraw-Hill, 1981.
[8] Spunt, L. *Optimum Structural Design*, Prentice-Hall, 1972.
[9] Save, M. and Prager, W. *Structural Optimization*, Plenum Press, 1990.

Exercises

3.1 Derive the shape factors for (i) a regular, solid hexagonal and (ii) a regular, solid octagonal sections, with side lengths a, where the neutral axis of bending passes through opposite corners, such that one flat face lies at the top and another at the bottom.

3.2 Investigate the efficiency of a thin-walled, equilateral triangular tube (Figure 3.22) for use as a beam. The failure criteria are that the maximum bending stress must attain the yield strength

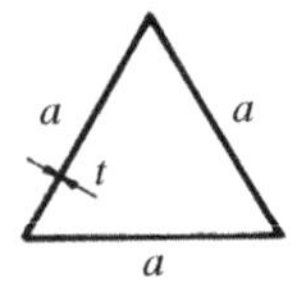

Figure 3.22 Equilateral, triangular tube beam section

simultaneously with the attainment of the critical local buckling stress in each wall. Assume pure bending throughout.

3.3 What is the effect on the shape factor of partitioning the thin-walled square tube beam section into four square cells as shown in Figure 3.23? Assume a common thickness throughout and apply similar failure criteria to those which established a shape factor of 2.67 for the plain tube.

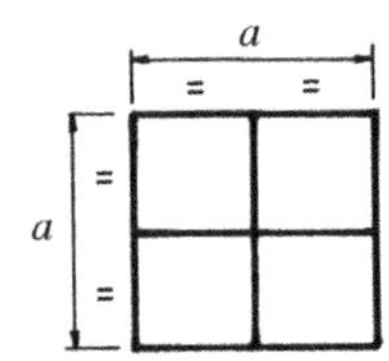

Figure 3.23 Partitioned square tube beam section

3.4 Derive the shape factor for a hexagonal tubular beam section with side length a where the neutral axis of bending passes through the corners such that flat faces lie at the top and bottom surfaces.

3.5 Show that the structural index S varies with the length L between the supports of a rectangular section beam $b \times d$, under four-point loading, as follows:

$$S = \frac{\sigma_y b d^2}{6L^3}$$

where σ_y is the limiting (design) stress. Hence, show graphically how the optimum weight varies with the support length.

3.6 Derive the objective function for a uniform channel section having constant thickness t, with flange length b and web depth d. Take the flange top to support maximum tension.

3.7 Derive the objective function for a non-uniform channel section, with flange length b and web depths d with respective wall thicknesses t_b and t_d. Take the flange top to support maximum tension.

3.8 Calculate the minimum possible weight of a non-uniform, T-section, steel beam with flange/web length in the ratio $b/d = 0.5$ (see Figure 3.19) when it is to carry a uniform moment of 50 kN m over a length of 3 m with the flange top surface placed under maximum tension. Hence determine the full section dimensions. Take $E = 200$ GPa and $\rho = 7860$ kg/m^3.

3.9 Two identical, rectangular, aluminium alloy box sections, each with breadth b and depth d, have respective wall thicknesses t_b and t_d (see Figure 3.9). The beams are mounted vertically 6 m apart in support of a boat hull of weight 50 kN suspended centrally between them at a distance of 3 m from the shorter centroidal axis of each tube (NA in Figure 3.9). Assuming pure bending in each tube, determine the optimum size for each dimension b, d, t_b and t_d. Take, for aluminum, $E = 70$ GPa, $\rho = 2800$ kg/m^3 and $\sigma_y = 160$ MPa. What is the weight per metre length of each tube?

4

Torsion of Bars and Tubes

4.1 Introduction

The procedure for minimising the weight of a torsion bar is similar to that adopted for beams except that fewer closed solutions are available for solid, non-circular sections. Referring to Figure 4.1, three torsion theories are available, depending upon the section shape [1]. We shall restrict our use of the classical torsion theory to bars and tubes with circular sections (Figure 4.1a,b). Solid rectangular sections and other open sections composed of thin rectangular limbs (see Figure 4.1c) are dealt with using St Venant's theory. The Bredt–Batho torsion theory applies to closed, thin-walled tubes of any shape, as in Figure 4.1d.

In each case the axial torque T is taken to be applied uniformly about a length axis passing through the centre of twist, i.e. the one point within the cross-section that does not rotate. For axisymmetric cross-sections (Figure 4.1a,b) this point coincides with the centroid. For non-axisymmetric cross-sections (Figure 4.1c,d) this point coincides with the shear centre. The effect of applying torque along respective axes passing through the centroid or the shear centre is to place the section under pure shear (i.e. with no bending). Peak shear stresses apply to the outer surfaces of Figure 4.1 and fall linearly to zero at the relevant torque axis.

As with beam bending, the weight $W = \rho AL$ of a bar or a tube under torsion, having a uniform sectional area A and length L, becomes the objective function, i.e. W is the quantity to be minimised directly. The properties of the cross-section – the position of its centroid or shear centre and its polar second moment of area – will set the peak shear stress value. Coupled to this is the known strength of a material in shear, this placing a limit upon the level of shear stress that can be allowed under torsion. The following minimum weight analyses apply to solid, hollow and open sections. The minimum weight will be shown to appear consistently for each cross-section within the objective function

$$(W/L^3)_{\min} = fmS^n \tag{4.1}$$

In equation (4.1) it will be seen that the structural index $S = T/L^3$ is common to all cross-sections but the shape factor f, the material factor m and the exponent n will depend upon the particular bar section.

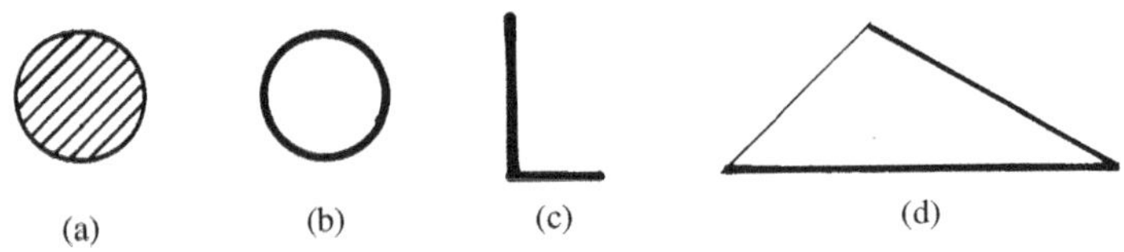

Figure 4.1 Typical torsion sections for optimisation

4.2 Solid Cross-Sections

4.2.1 Circular (Figure 4.2a)

As the peak shear stress τ_{max} applies to the whole of the outer surface (see Figure 4.2b), the circular section ranks highly in its structurally efficiency. Circular sections are common among torsion bars, appearing in spindles, axles, bolts and drive shafts. Thus, a knowledge of the minimum weight condition for a circular bar is an obvious starting place for the optimisation analyses to follow.

(i) Objective Function

$$W = \rho A L = \frac{\rho \pi d^2 L}{4} \tag{4.2}$$

(ii) Failure Criterion
When the peak shear stress τ_{max} is limited to the yield (or proof) stress τ_y for the bar material, we write $\tau_{max} \leq \tau_y$, where

$$\tau_{max} = \frac{Tr}{J} = \frac{T(d/2)}{\pi d^4/32} = \frac{16T}{\pi d^3} \tag{4.3a}$$

From equation (4.3a), the diameter of the fully stressed bar becomes

$$d_{opt} = \left(\frac{16T}{\pi \tau_y}\right)^{1/3} \tag{4.3b}$$

Substituting d_{opt} for d into equation (4.2),

$$W_{min} = \frac{\pi \rho L}{4}\left(\frac{16T}{\pi \tau_y}\right)^{2/3} = 2.325\frac{\rho}{\tau_y^{2/3}}T^{2/3}L \tag{4.4a}$$

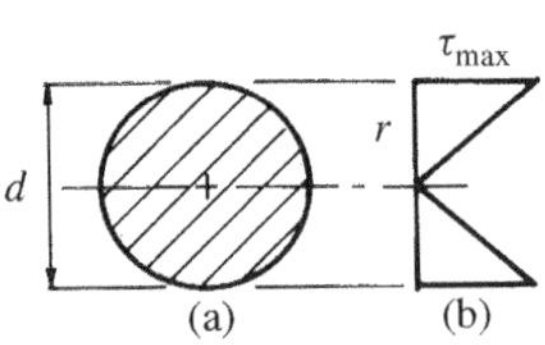

Figure 4.2 Solid circular section

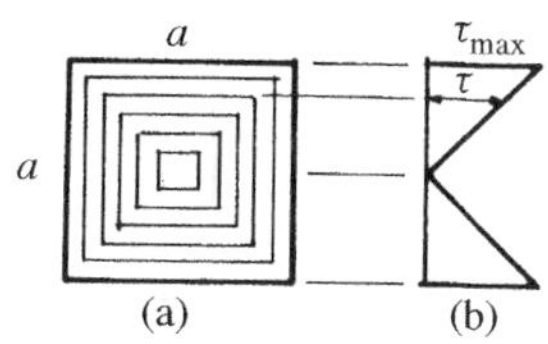

Figure 4.3 Solid square section showing constant shear stress contours

To provide the structural index in stress units, divide equation (4.4a) by L^3 to give

$$\left(\frac{W}{L^3}\right)_{min} = 2.325\left(\frac{\rho}{\tau_y^{2/3}}\right)\left(\frac{T}{L^3}\right)^{2/3} \tag{4.4b}$$

4.2.2 Solid Square and Rectangular Sections

In solid, non-circular sections the shear stress is constant along contours that contract the outer shape. However, it is only for the square section that such contours expand from within to contain the outer surface (Figure 4.3a). Thus, the peak shear stress τ_{max} applies to the whole of an outer, square surface (see Figure 4.3b). Therefore, we should expect a square-section torsion bar to rank most highly in minimising weight among non-circular sections.

A rectangular section $b \times t$, in which $b > t$, interrupts the iso-stress contours such that the peak shear stress τ_{max} is reached along the longer horizontal sides only. The maximum shear stress τ'_{max} along the shorter vertical side is proportionately lower at $(t/b)\tau_{max}$. A linear distribution applies to both the vertical and horizontal shear stresses, as shown in Figure 4.4.

(i) Objective Function

$$W = \rho AL = \rho(bt)L = \rho b^2 L/\lambda \tag{4.5}$$

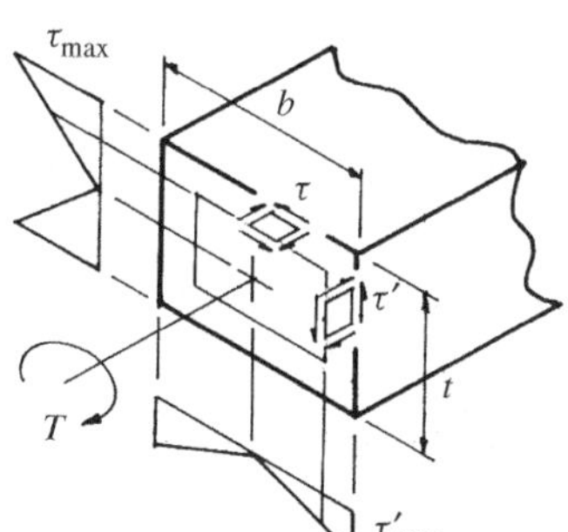

Figure 4.4 Maximum shear stresses in a rectangular section

Table 4.1 Dependence of torsion coefficients α and β upon strip aspect ratio $\lambda = b/t$

$\lambda = b/t$	1	2	4	6	8	10	∞
α	0.208	0.246	0.282	0.299	0.307	0.313	0.333
β	0.141	0.229	0.281	0.299	0.307	0.313	0.333

where $\lambda = b/t$ is the aspect ratio, taken within the range $1 \leq \lambda \leq 10$. Thin strips, with $\lambda > 10$, require an additional shear buckling failure criterion but, by limiting the λ range, the optimum analyses of square and rectangular sections are similar to that for a circular bar.

(ii) Failure Criterion

When the peak shear stress τ_{max} is limited to the yield (or proof) stress τ_y for the bar material we write $\tau_{max} \leq \tau_y$, where

$$\tau_{max} = \frac{T}{\alpha b t^2} = \frac{T\lambda^2}{\alpha b^3} \tag{4.6a}$$

from which the optimum breadth becomes

$$b_{opt} = \left(\frac{T\lambda^2}{\alpha\tau_y}\right)^{1/3} \tag{4.6b}$$

where α is a function of λ. Values of α and λ are tabulated for the appropriate range of solid rectangular sections in Table 4.1.

Note: The β-coefficient is employed later in § 4.3.5. Substituting b_{opt} for b in equation (4.5),

$$W_{min} = \frac{\rho L}{\lambda}\left(\frac{T\lambda^2}{\alpha\tau_y}\right)^{2/3} = \rho L\left(\frac{\lambda}{\alpha^2}\right)^{1/3}\left(\frac{T}{\tau_y}\right)^{2/3} \tag{4.7a}$$

Dividing equation (4.7a) by L^3 gives the structural index $S = T/L^3$ in stress units:

$$\left(\frac{W}{L^3}\right)_{min} = \left(\frac{\lambda}{\alpha^2}\right)^{1/3}\left(\frac{\rho}{\tau_y^{2/3}}\right)\left(\frac{T}{L^3}\right)^{2/3} \tag{4.7b}$$

Hence the shape factor $f = (\lambda/\alpha^2)^{1/3}$ is derived from the α and λ values given in Table 4.1. The plot of f versus λ, given in Figure 4.5, shows a minimum shape factor $f = 2.85$ for the square section ($\lambda = 1$), increasing to $f = 4.67$ for a rectangular section with $\lambda = 10$.

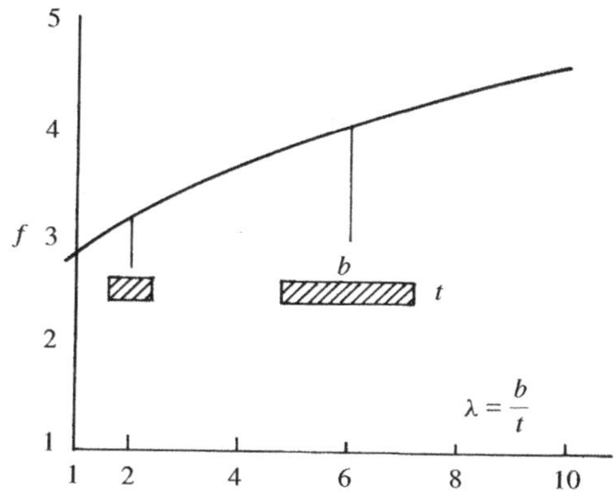

Figure 4.5 Shape factor $f = (\lambda/\alpha^2)^{1/3}$ for rectangular sections with increasing $\lambda = b/t$

4.2.3 Thin Rectangular Strip

When $\lambda = b/t > 10$ the coefficient $\alpha \rightarrow 1/3$ in equation (4.6a), so defining the maximum shear stress in a thin strip. The vertical shear stress τ' (see Figure 4.4) then becomes negligible. Shear buckling of thin plates depends upon the plate section $b \times t$, its length L, together with any restraints present along the sides b and L (see Figure 4.6). The most likely side restraint imposed upon a flat strip is that which arises from gripping one or both ends $(b \times t)$ during torque application while the sides $(L \times t)$ remain free (i.e. simply supported). It will be shown later that gently curved thin strips may also be treated with this flat, thin-strip theory.

(i) Objective Function

$$W = \rho AL = \rho(bt)L \tag{4.8}$$

where $b/t > 10$.

(ii) Failure Criterion

The maximum shear stress $\tau_{\max}$ in a thin strip [1] is

$$\tau_{\max} = \frac{3T}{bt^2} \tag{4.9}$$

which we may limit to a predetermined stress, usually at the elastic limit or corresponding to a given offset strain (stress symbols k and τ_y respectively). The stress k at the elastic limit (the limit of proportionality) involves a purely elastic analysis. On the other hand, the use of a proof stress τ_y will admit a limited amount of plasticity. The choice between k and τ_y will affect the shear bucking stress as the following analyses show.

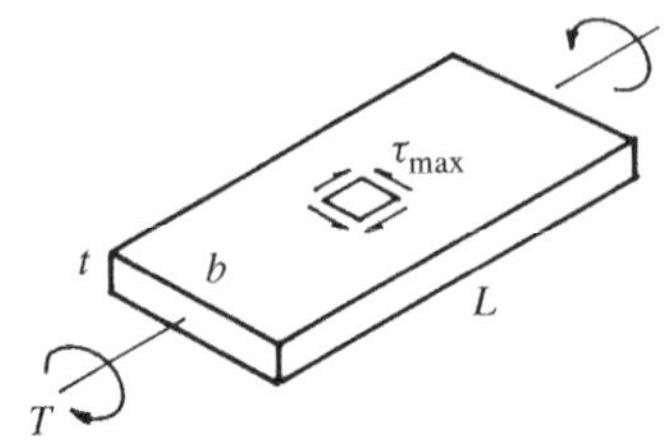

Figure 4.6 Thin rectangular strip under torsion

(iii) Elastic Shear Buckling
The thin, flat rectangular plate in Figure 4.6 buckles under an elastic shear stress [2]:

$$\tau_e = K_e E \left(\frac{t}{b}\right)^2 \tag{4.10}$$

In Appendix C, it is shown that the appropriate edge restraints all appear within the elastic coefficient $K_e = K_e(b/L)$. The function $K_e(b/L)$ is interpreted graphically in Figure 4.7 for plates under torsion with various combinations of simply supported and clamped sides. Similar plots (see Figure C.3) apply to plates under direct shear [2]. Equating (4.9) and (4.10) and limiting the shear stress to k:

$$k = \frac{3T}{bt^2} = K_e E \left(\frac{t}{b}\right)^2 \tag{4.11}$$

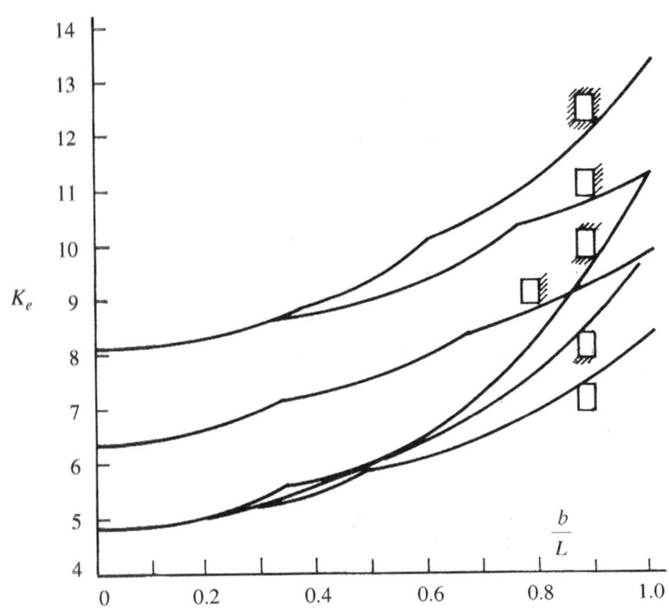

Figure 4.7 Dependence of K_e upon *b/L* for plates in shear with side restraints

Equation (4.11) allows optimum dimensions t and b to be found:

$$t_{\text{opt}} = \left[\frac{3T}{(K_e E k)^{1/2}}\right]^{1/3} \quad \text{and} \quad b_{\text{opt}} = \left(\frac{3T K_e E}{k^2}\right)^{1/3} \tag{4.12a,b}$$

Substituting t_{opt} and b_{opt} from equation (4.12a,b) into equation (4.8) and dividing by L^3 minimises the weight within the objective function:

$$\left(\frac{W}{L^3}\right)_{\min} = (3^4 K_e)^{1/6}\left(\frac{\rho E^{1/6}}{k^{5/6}}\right)\left(\frac{T}{L^3}\right)^{2/3} \tag{4.13}$$

(iv) Plastic Shear Buckling

Equation (4.10) can be made to admit plasticity when the elastic modulus is replaced with the tangent modulus E_T and used with an accompanying plasticity reduction factor $\eta = \tau_p/\tau_e$. The latter gives $\tau_p = \eta\tau_e$, as the buckling stress where the effect of plasticity has reduced the elastic modulus by 50%, i.e. $E_T/E = 1/2$. A graphical interpretation (see Figure 4.8) of the effect of plasticity upon strip buckling accompanies Figure 4.7, for when $\nu = 0.3$ [2].

In Figure 4.8 it is assumed that a uni-axial (tensile) stress–strain curve is represented with a hardening exponent m in the Ramberg–Osgood relation (see Appendix B) and that σ_n is the tensile stress corresponding to $E_T/E = 1/2$. Thus, we may modify equation (4.10)

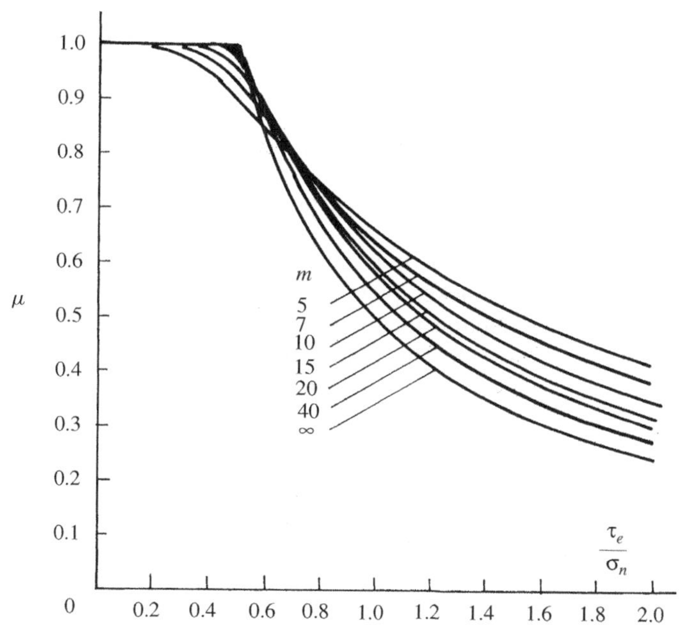

Figure 4.8 Plasticity reduction factor for thin, flat plates in shear

to give the shear stress for plastic buckling as follows:

$$\tau_p = \eta\tau_e = \eta K_e E\left(\frac{t}{b}\right)^2 = K_p E_T\left(\frac{t}{b}\right)^2 \tag{4.14a}$$

where $K_p = 2\eta K_e$, in which this effect of plasticity upon the elastic buckling coefficient K_e applies to a 50% reduction to the elastic modulus. A far lesser plastic reduction applies when the elastic modulus is retained:

$$\tau_p = \eta\tau_e = \eta K_e E\left(\frac{t}{b}\right)^2 = K_p E\left(\frac{t}{b}\right)^2 \tag{4.14b}$$

where $K_p = \eta K_e$. It appears that the use of tensile material properties E, ν, σ_y and m to describe shear buckling behaviour is largely a matter of convenience given that engineering handbooks list tensile data mostly (as in Appendix A). The following example shows how to apply these tensile data to shear buckling.

Example 4.1 Find the shear buckling stress for an aluminium alloy plate (see Figure 4.6) of width $b = 200$ mm and length $L = 400$ mm in two thicknesses $t = 2.5$ mm and 5 mm. The tensile properties for the plate material are: $m = 15$, $E = 72$ GPa, $\nu = 0.33$ and $\sigma_n =$ 350 MPa. The shear stress at the limit of proportionality is $k = 150$ MPa.

Here we have $b/L = 200/400 = 0.5$, when, from Figure 4.7, $K_e' = 6.1$. Firstly, a slight correction is applied for a Poisson's ratio that differs from the 0.3 value upon which Figure 4.7 is based. The true elastic stress buckling coefficient is:

$$K_e = K_e' \times (1 - 0.3^2)/(1 - \nu^2)$$

which gives $K_e = 6.23$ for $\nu = 0.33$. The elastic buckling stress for each plate thickness is found from equation (4.10):

$$\tau_e = 6.23 \times 72 \times 10^3 \times (2.5/200)^2 = 70.09 \text{ MPa}$$

$$\tau_e = 6.23 \times 72 \times 10^3 \times (5/200)^2 = 280.35 \text{ MPa}$$

Comparing these with $k = 150$ MPa, it is clear that buckling is elastic in the thinner plate and elastic-plastic in the thicker plate. Therefore, only the thicker plate requires a plasticity reduction. The reduction factor is read from Figure 4.8 as $\eta = 0.72$ for an abscissa $\tau_e/\sigma_n = 280.35/350 = 0.8$. This reduces the higher τ_e, which still remains within the plastic range:

$$\tau_p = \eta\tau_e = 0.72 \times 280.35 = 201.85 \text{ MPa}$$

This correction is tantamount to setting either $K_p = 2\eta K_e = 8.97$ in equation (4.14a), or $K_p = \eta K_e = 4.49$ in equation (4.14b). The plastic limiting stress is required for the following optimisation.

(v) Optimisation

An optimum geometry follows when the maximum stress from strip torsion and its plastic buckling stress under shear are made equal. Taking $\tau_p = \tau_y$ at the plastic stress limit, we may equate (4.9) to either one of equations (4.14a,b) as follows:

$$\tau_y = \frac{3T}{bt^2} = K_p E_T \left(\frac{t}{b}\right)^2 \tag{4.15}$$

Here τ_y is the shear equivalent of σ_y, provided by the von Mises yield criterion as $\tau_y = \sigma_y/\sqrt{3}$. Equation (4.15) allows optimum dimensions t and b to be found as:

$$t_{opt} = \left[\frac{3T}{(K_p E_T \tau_y)^{1/2}}\right]^{1/3} \quad \text{and} \quad b_{opt} = \left(\frac{3T K_p E_T}{\tau_y^2}\right)^{1/3} \tag{4.16a,b}$$

Substituting t_{opt} and b_{opt} from equations (4.16a,b) into equation (4.8) and dividing by L^3 minimises the weight within the objective function:

$$\left(\frac{W}{L^3}\right)_{min} = (3^4 K_p)^{1/6} \left(\frac{\rho E_T^{1/6}}{\tau_y^{5/6}}\right) \left(\frac{T}{L^3}\right)^{2/3} \tag{4.17}$$

Conveniently, the limiting elastic and plastic optimum designs require only an appropriate interchange between K_e and K_p and between k and τ_y within equations (4.13) and (4.17). As the linear elastic analysis is more fundamental to buckling theory we shall work to the lower, shear stress k at the limit of proportionality to optimise the cross-sections that follow.

4.3 Thin-Walled, Open Sections

Again, when $b/t > 10$, the coefficient $\alpha \to 1/3$ in equation (4.6a), this providing the maximum shear stress in a thin strip. We have seen that shear buckling of thin plates depends upon the plate section geometry $b \times t$, its length L, together with any restraints present along the sides b and L (see Figures 4.6 and 4.7). The most likely restraint imposed upon a flat strip is that which arises from gripping one or both of its ends ($b \times t$) during the torque application while the sides ($L \times t$) remain free, i.e. simply supported. Thin-walled sections with gentle curvature along the perimeter length b (see Figure 4.9) may be treated as flat plates, but additional restraint will exist in the length along the sharp right-angled junction of an equal-angle section. A pure torque is applied so that these sections are not subjected to simultaneous bending and torsion. This is only possible when the torque

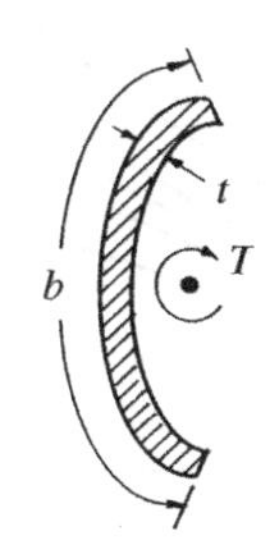

Figure 4.9 Plate section with gentle curvature along the perimeter length b

acts about an axis passing through the centre of twist. This centre lies at the intersection between axes of symmetry in axisymmetric sections, thereby coinciding with the centroid. In non-symmetric sections, the centre of twist coincides with the shear centre and not the centroid. The shear centre refers to a point in the section where a shear force must be applied if is it to bend and not twist the section [3]. This point lies at the intersection between the sides, e.g. at the corner in a L-angle section or at the intersection between web and flange in a T-section.

4.3.1 Curved Plates and Panels

Consider a gently curved plate, of dimensions $b \times t$, as shown in Figure 4.9. The torque is applied about an axis passing through the shear centre, which in this case lies outside the wall. A similar location also applies to the position of the shear centre for semi-circular and channel sections. Here the position of the shear centre may be found by applying a vertical transverse shear force at this position. Then the torque produced by this force will be equilibrated by the internal shear stress along the walls (see Chapter 5). Being applied at the shear centre means that the shear force bends but does not twist the section.

The optimum dimensions and minimum weight expressions for a curved plate remain as those found previously in § 4.2.3 for a thin strip. Hence, from equation (4.13), the minimum weight for a curved plate when it is to operate at its elastic limit is

$$\left(\frac{W}{L^3}\right)_{\min} = (3^4 K_e)^{1/6}\left(\frac{\rho E^{1/6}}{k^{5/6}}\right)\left(\frac{T}{L^3}\right)^{2/3} \tag{4.18}$$

In particular, for a circular arc of radius R, where the length $L > b$, the elastic buckling coefficient K_e in equation (4.18) must be read from Figure 4.10 at the appropriate values of the abscissa $b/\sqrt{(Rt)}$ and the L/b ratio [4].

Figure 4.11 applies to a shorter, curved panel in which $L < b$ [4]. Here L replaces b in the plate buckling stress equation (4.10). Equating this modified shear buckling stress

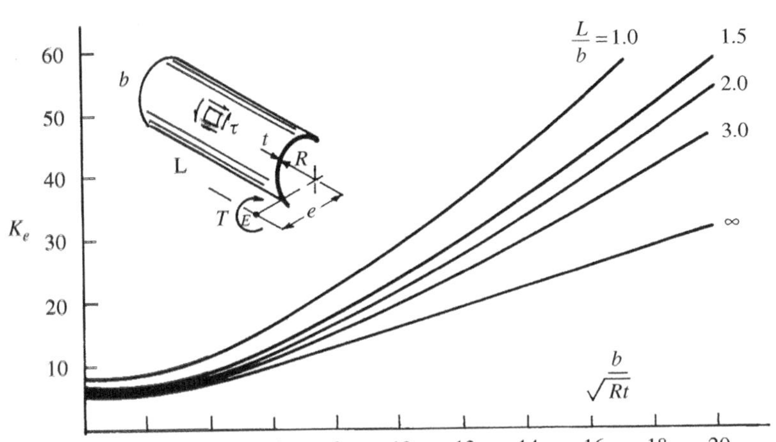

Figure 4.10 K_e for a long, thin, curved panel

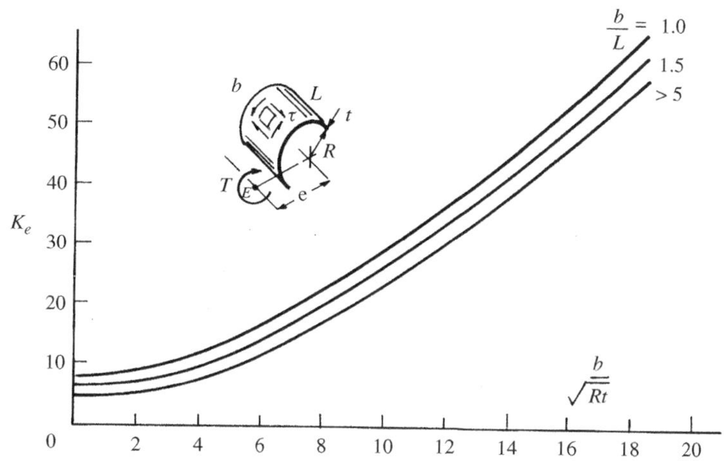

Figure 4.11 Buckling coefficient K_e for a short, thin, curved panel

expression to equation (4.9), with the shear stress at its elastic limit k, gives

$$k = \frac{3T}{bt^2} = K_e E \left(\frac{t}{L}\right)^2 \tag{4.19a}$$

The three variables b, L and t within equation (4.19a) can be optimised at each b/L ratio plotted in Figure 4.11. Taking $C = b/L$ alters the shape factor coefficient in equation (4.18) to

$$f = (3^4 C^2 K_e)^{1/6} \tag{4.19b}$$

For both long and short curved panels, the torque is taken to be applied about an axis through the centre of twist (the shear centre) that will depend upon the geometry. For example, it can be shown [1] that the shear centre E for a large semi-circular panel lies on a horizontal axis of symmetry, at a distance $e = 4R/\pi$ left of centre (see Figures 4.10 and 4.11).

4.3.2 Equal-Angle Section

The two perpendicular limbs, each of identical dimension $b \times t$, meet at the shear centre E, which also coincides with the centre of twist (see Figure 4.12). Hence, a torque T applied about a longitudinal axis through E will not bend the bar.

When $b/t \geq 10$, the limbs may be treated as separate thin strips using the theory outlined above. The maximum shear stress is now found from the applied torque as

$$\tau_{\max} = \frac{T}{\sum_{i=1}^{N} (\alpha bt^2)_i} = \frac{3T}{2bt^2} \tag{4.20}$$

where $\alpha = 1/3$ and $N = 2$. The critical shear stress for elastic buckling in each limb is again given by equation (4.10). If $\tau_{\max}$ is limited to the shear stress k at the proportional

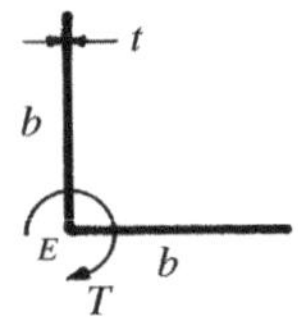

Figure 4.12 Equal-angle section

limit, we can ensure that k also becomes the buckling stress when

$$k = \frac{3T}{2bt^2} = K_e E \left(\frac{t}{b}\right)^2 \tag{4.21}$$

Equation (4.21) provides the optimum dimensions as

$$t_{\text{opt}} = \left[\frac{3T}{2(K_e Ek)^{1/2}}\right]^{1/3} \quad \text{and} \quad b_{\text{opt}} = \left(\frac{3TK_e E}{2k^2}\right)^{1/3} \tag{4.22a,b}$$

The weight $W = 2btL\rho$, when minimised from equations (4.22a,b), follows as

$$\left(\frac{W}{L^3}\right)_{\min} = (2^2 3^4 K_e)^{1/6} \frac{\rho E^{1/6}}{k^{5/6}} \left(\frac{T}{L^3}\right)^{2/3} \tag{4.23}$$

4.3.3 *Unequal Angle Sections*

We may extend the previous analysis to a section with perpendicular limbs of unequal lengths. In Figure 4.13a, the thickness t is uniform but the limb lengths differ so that $a > b$.

Using equation (4.20), equation (4.21) becomes:

$$k = \frac{3T}{at^2(1 + b/a)} = K_e E \left(\frac{t}{a}\right)^2 \tag{4.24}$$

in which the lesser, local buckling stress for the greater flat length is the critical one for an optimisation. Equation (4.24) leads to the optimum dimensions for this section:

$$t_{\text{opt}} = \left[\frac{3T}{(K_e Ek)^{1/2}(1 + b/a)}\right]^{1/3} \quad \text{and} \quad a_{\text{opt}} = \left[\frac{3TK_e E}{k^2(1 + b/a)}\right]^{1/3} \tag{4.25a,b}$$

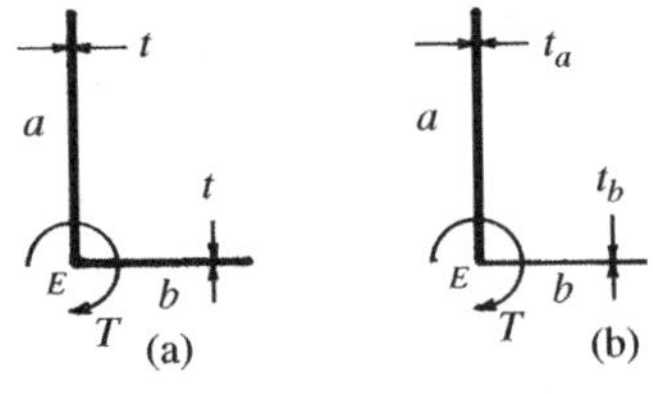

Figure 4.13 Unequal angle sections with and without uniform thickness

The minimum weight, $W = (a + b)tL\rho$, provides the objective function

$$\left(\frac{W}{L^3}\right)_{\min} = \left[3^4(1 + b/a)^2 K_e\right]^{1/6}\left(\frac{\rho E^{1/6}}{k^{5/6}}\right)\left(\frac{T}{L^3}\right)^{2/3} \tag{4.26}$$

When both thicknesses and flat lengths for the two limbs differ such that $a > b$ and $t_a > t_b$, (as in Figure 4.13b) the thin strip theory employs the St Venant torsion constant $J = {}^1\!/_3 \sum bt^3$. This provides the maximum shear stress $\tau_{\max} = Tt_a/J$ as

$$\tau_{\max} = \frac{3Tt_a}{at_a^3 + bt_b^3} = \frac{3T}{at_a^2\left[1 + (b/a)(t_b/t_a)^3\right]} \tag{4.27}$$

It is now possible to arrange for a common, critical buckling stress within each limb when

$$\tau_{\text{cr}} = K_a E\left(\frac{t_a}{a}\right)^2 = K_b E\left(\frac{t_b}{b}\right)^2 \tag{4.28a}$$

where K_a and K_b are elastic buckling coefficients to be read from Figure 4.7 for plates having aspect ratios a/L and b/L (the abscissa), with one long edge being fixed and the other edge being free. Equation (4.28a) gives the limb length ratio

$$\frac{b}{a} = \left(\frac{K_b}{K_a}\right)^{1/2}\frac{t_b}{t_a} \tag{4.28b}$$

The optimisation lies within an equalisation of the shear stress by equations (4.27) and (4.28a) at the elastic stress limit k. This gives

$$k = \frac{3T}{at_a^2\left[1 + (b/a)(t_b/t_a)^3\right]} = K_a E\left(\frac{t_a}{a}\right)^2 \tag{4.29}$$

Equations (4.28a) and (4.29) lead to the optimum section dimensions

$$(t_a)_{\text{opt}} = \left\{\frac{3T}{(K_a E k)^{1/2}\left[1 + (b/a)^4(K_a/K_b)^{3/2}\right]}\right\}^{1/3} \quad \text{and}$$

$$a_{\text{opt}} = \left\{\frac{3TK_a E}{k^2\left[1 + (b/a)^4(K_a/K_b)^{3/2}\right]}\right\}^{1/3} \tag{4.30a,b}$$

The weight is written as

$$W = at_a\left[1 + (b/a)(t_b/t_a)\right]\rho L \tag{4.31a}$$

Substituting equations (4.28b) and (4.30a,b) into equation (4.31a), gives an optimum objective function:

$$\left(\frac{W}{L^3}\right)_{\text{opt}} = \frac{(3^4 K_a)^{1/6}\left[1 + (K_a/K_b)^{1/2}(b/a)^2\right]}{\left[1 + (K_a/K_b)^{3/2}(b/a)^4\right]^{2/3}}\left(\frac{\rho E^{1/6}}{k^{5/6}}\right)\left(\frac{T}{L^3}\right)^{2/3} \tag{4.31b}$$

where the bucking coefficients K_a and K_b depend upon a/L and b/L in a manner provided by the central plot in Figure 4.7 In seeking to minimise the weight, we might assume, as a first approximation, that the K_a/K_b ratio is independent of the b/a ratio. Differentiating the shape factor in equation (4.31b) and equating to zero gives the corresponding b/a ratio:

$$\left(\frac{b}{a}\right)^2 = \left[4\frac{K_b}{K_a} + 3\left(\frac{K_b}{K_a}\right)^{3/2}\right]^{1/2} - 2\left(\frac{K_b}{K_a}\right)^{1/2} \tag{4.32a}$$

in which $K_b/K_a < 1$. It follows from equations (4.31b) and (4.32a) that the minimum shape factor becomes

$$f_{\min} = \frac{(3^4 K_a)^{1/6}\left[\sqrt{4 + 3(K_b/K_a)^{1/2}} - 1\right]}{\left[4 + 8(K_a/K_b)^{1/2} - 4\sqrt{4(K_a/K_b) + 3(K_a/K_b)^{1/2}}\right]^{2/3}} \tag{4.32b}$$

The stationary values (4.32a,b) could be employed, say, as the initial values within an iteration procedure where the dependency of K_a/K_b upon the b/a ratio is to be admitted.

4.3.4 Symmetric Sections

The channel, T- and Z-sections shown in Figure 4.14 have a symmetry in which the shape is reflected about one or both centroidal axes x and y. Each section has an additional flange area $b \times t_b$ compared to the angle section considered previously in Figure 4.14b. Therefore, by adding in this area to the previous analysis we may deduce from equation (4.30a,b) that each section has similar optimal dimensions:

$$(t_a)_{\text{opt}} = \left\{\frac{3T}{(K_a E k)^{1/2}\left[1 + 2(b/a)^4 (K_a/K_b)^{3/2}\right]}\right\}^{1/3} \quad \text{and}$$

$$a_{\text{opt}} = \left\{\frac{3T K_a E}{k^2\left[1 + 2(b/a)^4 (K_a/K_b)^{3/2}\right]}\right\}^{1/3} \tag{4.33a,b}$$

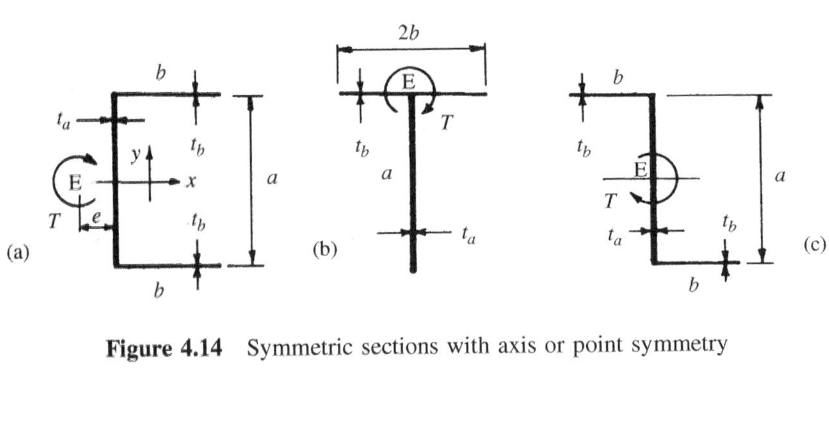

Figure 4.14 Symmetric sections with axis or point symmetry

The maximum stress in the section becomes

$$\tau_{\max} = \frac{3Tt_a}{at_a^3 + 2bt_b^3} = \frac{3T}{at_a^2\left[1 + 2(b/a)(t_b/t_a)^3\right]} \tag{4.34a}$$

and the objective function for minimising weight modifies equation (4.31b) to

$$\left(\frac{W}{L^3}\right)_{\text{opt}} = \frac{(3^4K_a)^{1/6}\left[1 + 2(K_a/K_b)^{1/2}(b/a)^2\right]}{\left[1 + 2(K_a/K_b)^{3/2}(b/a)^4\right]^{2/3}} \frac{\rho E^{1/6}}{k^{5/6}} \left(\frac{T}{L^3}\right)^{2/3} \tag{4.34b}$$

Identical shape factors apply to these sections within which the stationary values are

$$\left(\frac{b}{a}\right)^2 = \sqrt{\frac{K_b}{K_a} + \frac{3}{2}\left(\frac{K_b}{K_a}\right)^{3/2}} - \left(\frac{K_b}{K_a}\right)^{1/2} \tag{4.35a}$$

and

$$f_{\min} = \frac{(3^4K_a)^{1/6}\left[2\sqrt{1 + (3/2)(K_b/K_a)^{1/2}} - 1\right]}{4^{2/3}\left[1 + (K_a/K_b)^{1/2} - \sqrt{(K_a/K_b) + (3/2)(K_a/K_b)^{1/2}}\right]^{2/3}} \tag{4.35b}$$

The equivalence between these sections will only apply when the axial torque T is applied at the shear (twist) centre E for each section, indicated in Figure 4.14. In § 5.3 it is shown that the shear centre for a channel section lies at a distance $e = 3b^2/(a + 6b)$ to the left of the vertical web in Figure 4.14a. The shear centre lies at the intersection between web and flange in Figure 4.14b and coincides with the centroid in Figure 4.14c. Recall that the shear centre is the position at which a vertical force must be applied if that force is to bend and not twist the section. More appropriately here, the converse is true. That is, when a torque is applied at the shear (twist) centre it will twist and not bend the section. Thus the shear centre may also be regarded as the centre of twist for identifying that singular point in the cross-section which does not rotate under torsion. It is left to the reader to establish the shape factor f for a uniform, thin-walled, symmetrical I-section (see Exercise 4.2).

4.3.5 Unsymmetrical Sections

So far we have considered thin-walled open sections where $\alpha = 1/3$ in equation (4.6a). If the ratio $b/t < 10$ then it becomes necessary to allow for $\alpha < 1/3$ and $\beta < 1/3$ according to Table 4.1.

(a) Channel Section

Consider, for example, an unsymmetrical channel in Figure 4.15 whose geometrical ratios conform to $a/t_a > b/t_b > c/t_c$, in which each ratio is less than 10/1.

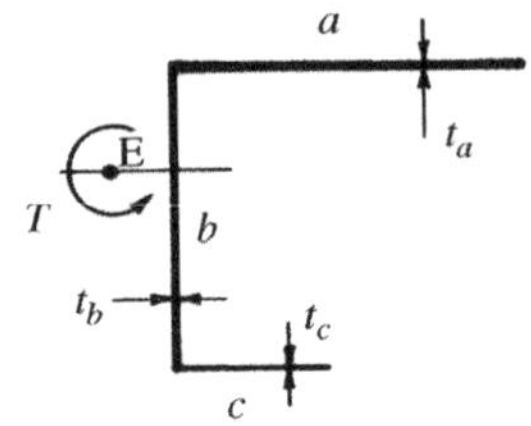

Figure 4.15 Unsymmetrical thick-walled channel under torsion

When the torque T is applied at the centre of twist E, the angular twist θ is found from

$$\frac{G\theta}{L} = \frac{T}{\sum(\beta bt^3)} = \frac{T}{\beta_a a t_a^3 + \beta_b b t_b^3 + \beta_c c t_c^3} \tag{4.36a}$$

in which G is the shear modulus and L is the length. Since the same angular twist θ must apply to each rectangle,

$$\frac{G\theta}{L} = \frac{T_a}{\beta_a a t_a^3} = \frac{T_b}{\beta_b b t_b^3} = \frac{T_c}{\beta_c c t_c^3} \tag{4.36b}$$

where T_a, T_b and T_c are the torques carried by the rectangles $a \times t_a$, $b \times t_b$ and $c \times t_c$, such that

$$T = T_a + T_b + T_c \tag{4.36c}$$

Hence, from equations (4.36a,b), the applied torque T may be partitioned as:

$$T_a = \frac{T\beta_a a t_a^3}{\beta_a a t_a^3 + \beta_b b t_b^3 + \beta_c c t_c^3}, \quad T_b = \frac{\beta_b b t_b^3}{\beta_a a t_a^3} T_a, \quad T_c = \frac{\beta_c c t_c^3}{\beta_a a t_a^3} T_a \tag{4.37a,b,c}$$

The maximum shear stress in each rectangle is then found to be

$$\tau_a = \frac{T_a}{\alpha_a a t_a^2}, \quad \tau_b = \frac{T_b}{\alpha_b b t_b^2}, \quad \tau_c = \frac{T_c}{\alpha_c c t_c^2} \tag{4.38a,b,c}$$

The greatest of these stress maxima applies to rectangle $a \times t_a$, since it carries the greatest proportion of the applied torque. Thus, when we limit τ_a to the shear stress at elastic limit, i.e. $\tau_a \leq k$, both τ_b and τ_c will remain elastic. Hence, from equations (4.37a) and (4.38a), the limiting stress expression becomes

$$k = \frac{T}{\alpha_a a t_a^2 \left[1 + (\beta_b/\beta_a)(b/a)(t_b/t_a)^3 + (\beta_c/\beta_a)(c/a)(t_c/t_a)^3\right]} \tag{4.39}$$

Local elastic buckling of the web and the two flanges may be arranged to occur simultaneously when the buckling coefficients K_a, K_b and K_c are taken to vary with each plate's width/length ratio, together with its long edge supports, as in Figure 4.7. This gives

$$\tau_{\mathrm{cr}} = K_a E_T \left(\frac{t_a}{a}\right)^2 = K_b E_T \left(\frac{t_b}{b}\right)^2 = K_c E_T \left(\frac{t_c}{c}\right)^2 \tag{4.40a}$$

from which

$$\left(\frac{b}{a}\right)^2 = \frac{K_b}{K_a}\left(\frac{t_b}{t_a}\right)^2, \quad \left(\frac{c}{a}\right)^2 = \frac{K_c}{K_a}\left(\frac{t_c}{t_a}\right)^2, \quad \left(\frac{c}{b}\right)^2 = \frac{K_c}{K_b}\left(\frac{t_c}{t_b}\right)^2 \tag{4.40b,c,d}$$

The optimisation lies within the equalisation of the shear stress from equations (4.39) and (4.40a), each having attained the stress k at the elastic limit:

$$k = \frac{T}{\alpha_a a t_a^2\left[1 + (\beta_b/\beta_a)(b/a)(t_b/t_a)^3 + (\beta_c/\beta_a)(c/a)(t_c/t_a)^3\right]} = K_a E \left(\frac{t_a}{a}\right)^2 \tag{4.41}$$

Equation (4.41) optimises the thickness of rectangle $a \times t_a$ as:

$$(t_a)_{\mathrm{opt}} = \left\{\frac{T}{(K_a E k)^{1/2}\alpha_a\left[1+(\beta_b/\beta_a)(K_a/K_b)^{3/2}(b/a)^4+(\beta_c/\beta_a)(K_c/K_a)^{3/2}(c/a)^4\right]}\right\}^{1/3} \tag{4.42a}$$

where its optimum width a is

$$a_{\mathrm{opt}} = \left\{\frac{T K_a E}{k^2\alpha_a\left[1 + (\beta_b/\beta_a)(K_a/K_b)^{3/2}(b/a)^4 + (\beta_c/\beta_a)(K_c/K_a)^{3/2}(c/a)^4\right]}\right\}^{1/3} \tag{4.42b}$$

Using equations (4.40b,c), the weight is written as

$$W = a t_a\left(1 + \frac{b}{a}\frac{t_b}{t_a} + \frac{c}{a}\frac{t_c}{t_a}\right)\rho L = a t_a\left[1 + \left(\frac{K_a}{K_b}\right)^{1/2}\left(\frac{b}{a}\right)^2 + \left(\frac{K_a}{K_c}\right)^{1/2}\left(\frac{c}{a}\right)^2\right]\rho L \tag{4.43a}$$

Substituting the optimum t_a and a from equations (4.42a,b) into equation (4.43a) gives the optimal weight within the objective function:

$$\left(\frac{W}{L^3}\right)_{\mathrm{opt}} = \frac{K_a^{1/6}\left[1 + (K_a/K_b)^{1/2}(b/a)^2 + (K_a/K_c)^{1/2}(c/a)^2\right]}{\alpha_a^{2/3}\left[1 + (\beta_b/\beta_a)(K_a/K_b)^{3/2}(b/a)^4 + (\beta_c/\beta_a)(K_a/K_c)^{3/2}(c/a)^4\right]^{2/3}} \times \left(\frac{\rho E^{1/6}}{k^{5/6}}\right)\left(\frac{T}{L^3}\right)^{2/3} \tag{4.43b}$$

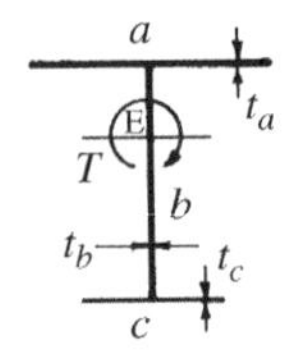

Figure 4.16 Thick-walled, unsymmetrical I-section

The following notes are drawn from the shape factor expression within equation (4.43b):

(i) The buckling coefficients K_a, K_b and K_c depend upon a/L, b/L and c/L, but if we assume that each ratio is independent of b/a and c/a then we can seek to minimise f as before. The simplest method is to minimise b/a for a fixed c/a, as might be typical of extruded stock. Differentiation for a minimum, $\mathrm{d}f/\mathrm{d}(b/a) = 0$, provides the b/a ratio that minimises f as

$$\left(\frac{b}{a}\right)^2 = \frac{2}{p}(1+r)\left[\sqrt{1+\frac{3p^2(1+s)}{4q(1+r)^2}} - 1\right] \tag{4.44a}$$

where $p = (K_a/K_b)^{1/2}$, $q = (\beta_b/\beta_a)(K_a/K_b)^{3/2}$, $r = (K_a/K_c)^{1/2}(c/a)^2$ and $s = (\beta_c/\beta_a)(K_a/K_c)^{1/2}(c/a)^4$.

(ii) The corresponding, shape factor for a thin section follows from setting $\alpha = \beta = 1/3$ in equation (4.43b):

$$f = \frac{(3^4 K_a)^{1/6}\left[1 + (K_a/K_b)^{1/2}(b/a)^2 + (K_a/K_c)^{1/2}(c/a)^2\right]}{\left[1 + (K_a/K_b)^{3/2}(b/a)^4 + (K_a/K_c)^{3/2}(c/a)^4\right]^{2/3}} \tag{4.44b}$$

(iii) In what follows the shape factor f in equation (4.43b) is modified for one other thick open-section composed of three rectangles.

(b) I-Section

Figure 4.16 shows an unsymmetrical I-section in which the limb lengths a, b and c and their respective thicknesses t_a, t_b and t_c conform to $(a/2)/t_a > b/t_b > (c/2)/t_c$ where each of these ratios is less than 10/1. With the limiting torque applied at the shear centre E, elastic buckling is taken to occur over one-half of each flange length, i.e. $a/2$ and $c/2$. Following a similar analysis to that given above for the channel section, the following shape factor replaces that appearing in equation (4.43b):

$$f = \frac{(2^2 K_a)^{1/6}\left[1 + 2(K_a/K_b)^{1/2}(b/a)^2 + (K_a/K_c)^{1/2}(c/a)^2\right]}{\alpha_a^{2/3}\left[1 + (\beta_b/\beta_a)(4K_a/K_b)^{3/2}(b/a)^4 + (\beta_c/\beta_a)(K_a/K_c)^{3/2}(c/a)^4\right]^{2/3}} \tag{4.44c}$$

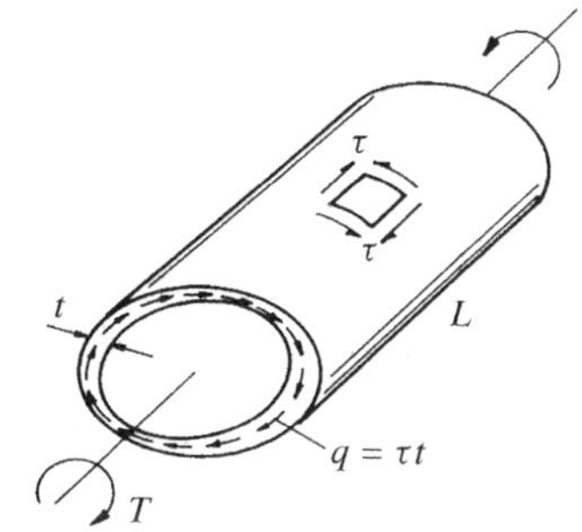

Figure 4.17 Shear flow in a thin-walled tube with varying thickness

4.4 Thin-Walled, Closed Tubes

The Bredt–Batho torsion theory provides the torque $T = 2A_o q$ in terms of the wall shear flow q and the cross-sectional area A_o *enclosed* by the tube wall's mean centre line. The shear flow $q = \tau t$ is always constant in a closed tube even though τ and t may vary within irregular shaped tubes with varying thickness, an example of which is shown in Figure 4.17.

Hence, in a tube with varying wall thickness we may write the torque as

$$T = 2A_o \tau t \tag{4.45}$$

where τ and t refer to a single point in the wall. Here, an optimum design ensures that the limiting shear stress from torsion matches that required to cause shear buckling in the wall. We shall see firstly that the axial length and the curvature of a circular tube wall play an important role in its shear buckling behaviour. Thereafter, for simplicity, we restrict our attention to long, non-circular tubes with flat wall segments that behave as rectangular plates under shear. The weight to be minimised is common to all thin-walled tubes:

$$W = \rho A L \tag{4.46}$$

where A now refers to the wall area within the tube cross-section (not the enclosed area A_o).

4.4.1 Circular Tubes

The symbols used for circular tubes – mean radius r_m, wall thickness t and length L – are shown in Figure 4.18. Torque T is applied about the central, longitudinal axis, and this produces shear flow $q = \tau t$ around the wall centre-line as shown. The following two failure criteria apply:

(i) Limiting Stress

This failure criterion follows from knowing the maximum shear stress under the applied torque. Taking $A_o = \pi r_m^2$ in equation (4.45), this stress is

$$\tau_{\max} = \frac{T}{2A_o t} = \frac{T}{2\pi r_m^2 t} \tag{4.47a}$$

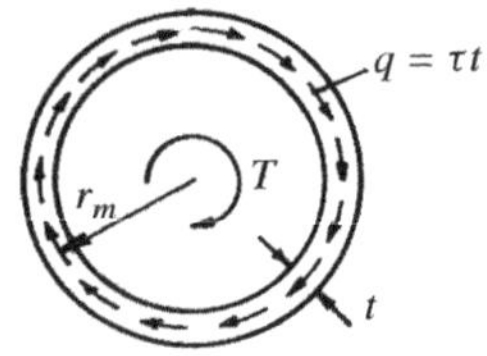

Figure 4.18 Circular tube geometry

where $\tau_{\max}$ is limited by

$$\tau_{\max} \leq \tau_y \tag{4.47b}$$

in which τ_y is a limiting yield (or proof) stress for the tube material. In earlier sections this limiting stress had been set at k, the stress at the proportional limit, but from now on we shall admit an amount of offset plastic strain (typically 0.1%) and so define τ_y as a proof stress.

(ii) Torsional Buckling
Whilst the shear stress limit from equation (4.47b) applies to any length of tube, the buckling criterion needs to be divided between three length regimes. The development of buckling formulae to be employed within these regimes is largely due to Batdorf et al. [5, 6] and Gerrard et al. [7–10] who reinterpreted Donnell's original work [11] on the elastic stability of thin-walled tubes under torsion.

I Tubes with Short Lengths
A more practical interpretation of the condition $r_m/L \to \infty$ for a short length tube is to apply $L < \sqrt{(r_m t)}$, in order to identify shear buckling with a known flat plate solution. Thus, when the cylinder perimeter $2\pi r_m$ is rolled out flat this length becomes infinitely long compared to the cylinder's short length L. The 'flat plate' buckles under a critical shear stress [7]:

$$\tau_{cr} = \frac{\pi^2 K_s \eta_s E}{12(1 - \nu_e^2)} \left(\frac{t}{L}\right)^2 \tag{4.48a}$$

where $K_s = 5.35$ for a tube with simply supported ends and $K_s = 8.98$ for clamped ends. For the former K_s-value the ends are free to rotate and for the latter the ends are held perpendicular to the axis of torsion. The plasticity reduction factor η_s is required when the critical shear stress exceeds that at the elastic limit [8]:

$$\eta_s = \left(\frac{1 - \nu_e^2}{1 - \nu_p^2}\right)^{3/4} \frac{E_s}{E} \tag{4.48b}$$

where the elastic and plastic Poisson's ratios are taken to be $\nu_e = 0.3$ and $\nu_p = 0.5$ respectively and E_s is the secant modulus at the plastic buckling stress level. Here we may combine equations (4.48a,b) in a more conservative form,

$$\tau_{cr} = KE_T\left(\frac{t}{L}\right)^2 \tag{4.48c}$$

in which the tangent modulus E_T is the gradient to the flow curve at the plastic buckling stress. If E is preferred then, depending upon the end condition, K takes one of two values, these differing from those stated above only by the multiplying factor $\pi^2\eta_s/\left[12(1-\nu_e^2)\right]$.

II Tubes with Medium Lengths

The manner in which torque is applied to the tube ends becomes less important when the tube geometry falls within a range defined by

$$50\frac{t}{r} < \left(\frac{L}{r}\right)^2 < 10\frac{r}{t} \tag{4.49}$$

For a simply supported tube, the critical buckling stress is expressed as [9, 10]

$$\tau_{cr} = \frac{0.7}{(1-\nu^2)^{5/8}}E\left(\frac{t}{r}\right)^{5/4}\left(\frac{r}{L}\right)^{1/2} \tag{4.50a}$$

To allow for plasticity and other end constraints we shall modify equation (4.50a) to

$$\tau_{cr} = KE_T\left(\frac{t}{r}\right)^{5/4}\left(\frac{r}{L}\right)^{1/2} \tag{4.50b}$$

where K follows from a plasticity reduction factor η by the usual method (see Example 4.1).

III Very Long Tubes

In the elastic case the cylinder distorts into an elliptical shape, with two-lobe buckling, when the geometry conforms to

$$\left(\frac{L}{r}\right)^2\left(\frac{t}{r}\right) > a \tag{4.51}$$

where $a = 42$ for simple supports and $a = 60$ for clamped ends, though these and other intermediate end-fixings were found to have a negligible effect on the elastic buckling stress:

$$\tau_{cr} = \frac{0.272}{(1-\nu^2)^{3/4}}E\left(\frac{t}{r}\right)^{3/2} \tag{4.52a}$$

With Gerrard's allowance for plasticity [8], equation (4.48b) modifies equation (4.52a) to

$$\tau_{cr} = KE_T\left(\frac{t}{r}\right)^{3/2} \tag{4.53b}$$

in which K becomes a 'plastic' buckling coefficient as with equation (4.50b).

(iii) Optimum Geometry and Minimum Weight

The three length regimes given above have optimal cross-sectional dimensions when the Batho shear stress is equated to the buckling stress at the material's limiting stress, $\tau_y = \tau_{\max} = \tau_{cr}$. Now, as $\tau_{\max}$ is common to all lengths, equation (4.47a) gives

$$\tau_y = \frac{T}{2\pi r^2 t} = \tau_{cr} \tag{4.54}$$

where τ_{cr} within regimes I, II and III above apply to short, medium and long tubes. In what follows, the respective solutions to r_{opt} and t_{opt} allow the weight to be minimised for each length regime as

$$W_{\min} = 2\pi r_{opt} t_{opt} L\rho \tag{4.55}$$

I Short Tubes

From equations (4.48b) and (4.54),

$$\tau_y = \frac{T}{2\pi r^2 t} = KE_T\left(\frac{t}{L}\right)^2 \tag{4.56}$$

Solving equation (4.56) simultaneously for r and t,

$$\left(\frac{t}{L}\right)_{opt} = \left(\frac{\tau_y}{KE_T}\right)^{1/2}, \quad r_{opt} = \left(\frac{\sqrt{K}}{2\pi}\right)^{1/2}\left(\frac{E_T}{\tau_y^3}\right)^{1/4}\left(\frac{T}{L}\right)^{1/2} \tag{4.57a,b}$$

Substituting equations (4.57a,b) into equation (4.55) and dividing $W_{\min}$ by L^3 gives the objective function

$$\left(\frac{W}{L^3}\right)_{\min} = \left(\frac{2\pi}{\sqrt{K}}\right)^{1/2}\left[\frac{\rho}{(E_T\tau_y)^{1/4}}\right]\left(\frac{T}{L^3}\right)^{1/2} \tag{4.58}$$

II Medium Length Tubes

From equations (4.50b) and (4.54),

$$\tau_y = \frac{T}{2\pi r^2 t} = KE_T\left(\frac{t}{r}\right)^{5/4}\left(\frac{r}{L}\right)^{1/2} \tag{4.59}$$

Solving simultaneously for r and t,

$$t_{opt} = \left[\frac{1}{(2\pi)^3 K^8}\right]^{1/13}\left(\frac{\tau_y^3}{E_T^8}\right)^{1/13}(T^3L^4)^{1/13}, \quad r_{opt} = \left[\frac{K^4}{(2\pi)^5}\right]^{1/13}\left(\frac{E_T^4}{\tau_y^9}\right)^{1/13}\left(\frac{T^5}{L^2}\right)^{1/13} \tag{4.60a,b}$$

Substituting equations (4.60a,b) into equation (4.55) and dividing $W_{\min}$ by L^3,

$$\left(\frac{W}{L^3}\right)_{\min} = \left[\frac{(2\pi)^5}{K^4}\right]^{1/13}\left[\frac{\rho}{(E_T\tau_y)^{4/13}}\right]\left(\frac{T}{L^3}\right)^{8/13} \tag{4.61}$$

III Long Tubes
From equations (4.53b) and (4.54),

$$\tau_y = \frac{T}{2\pi r^2 t} = KE_T \left(\frac{t}{r}\right)^{\frac{3}{2}} \tag{4.62}$$

Solving simultaneously for r and t,

$$t_{\text{opt}} = \left[\frac{1}{(2\pi)^3 K^4}\right]^{\frac{1}{9}} \left(\frac{T^3 \tau_y}{E_T^4}\right)^{\frac{1}{9}}, \quad r_{\text{opt}} = \left[\frac{K^2}{(2\pi)^3}\right]^{\frac{1}{9}} \left(\frac{E_T^2 T^3}{\tau_y^5}\right)^{\frac{1}{9}} \tag{4.63a,b}$$

Substituting equations (4.63a,b) into equation (4.55) and dividing $W_{\min}$ by L^3,

$$\left(\frac{W}{L^3}\right)_{\min} = \left[\frac{(2\pi)^3}{K^2}\right]^{\frac{1}{9}} \left(\frac{\rho}{E_T^{2/9} \tau_y^{4/9}}\right) \left(\frac{T}{L^3}\right)^{\frac{2}{3}} \tag{4.64}$$

Similar length regimes for other non-circular, thin tubes could be identified, but we will continue with these only as long tubes where the end constraint is less important. Hence, the optimum solutions which follow are independent of the manner in which torque is applied to their ends.

4.4.2 Square Tube

In a long, uniformly thin square tube (see Figure 4.19) the side plates are allowed to buckle under a shear stress made equal to the limiting stress from Batho shear.

(i) Limiting Batho Shear
The constant shear stress in the uniform section follows from the applied torque as

$$T = 2A_o \tau t = 2a^2 \tau t, \quad \Rightarrow \quad \tau = \frac{T}{2a^2 t} \le \tau_y \tag{4.65a,b}$$

where τ_y is a predetermined limiting stress, chosen, say, at a point along a shear stress–strain curve where the gradient is one-half the elastic shear modulus, i.e. $G_T = G/2$. This enables a good estimate, $E_T \approx 2.6G_T$, within the shear buckling formulae that follow.

(ii) Shear Buckling
The buckling of the rectangular tube wall depends upon its width a, length L and thickness t and the manner of its side supports. These and plasticity effects are subsumed within

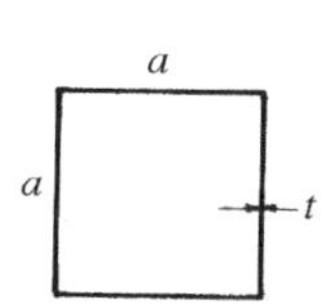

Figure 4.19 Thin-walled square tube geometry

the buckling stress formula [2]

$$\tau_{cr} = KE_T\left(\frac{t}{a}\right)^2 \tag{4.66}$$

where the buckling coefficient function $K = K(a/L)$ is interpreted graphically in Figure 4.7 assuming simply supported edges. Traditionally, equation (4.66) appears with the tensile modulus but, equivalently, the shear modulus could be used. Thus, given the relationship between the elastic constants $E = 2G(1+\nu)$, we may assume $\nu = 0.3$ and then $E = 2.6G$. The use of the tangent modulus E_T instead of E within equation (4.66) provides a more conservative estimate of the buckling stress in anticipation of some plastic strain (see § 4.4.5).

(iii) Optimisation
From equations (4.65b) and (4.66),

$$\tau_y = \frac{T}{2a^2t} = KE_T\left(\frac{t}{a}\right)^2 \tag{4.67}$$

Solving equation (4.67) simultaneously for a and t,

$$t_{opt} = \left(\frac{1}{2K}\right)^{1/3}\left(\frac{T}{E_T}\right)^{1/3}, \quad a_{opt} = \left(\frac{K}{4}\right)^{1/6}\left(\frac{E_T T^2}{\tau_y^3}\right)^{1/6} \tag{4.68a,b}$$

The minimum weight is found from $W_{min} = 4a_{opt}t_{opt}L\rho$ when, on dividing by L^3,

$$\left(\frac{W}{L^3}\right)_{min} = \left(\frac{2^8}{K}\right)^{1/6}\left(\frac{\rho}{E_T^{1/6}\tau_y^{1/2}}\right)\left(\frac{T}{L^3}\right)^{2/3} \tag{4.69}$$

4.4.3 *Triangular Tube*

In a long tube with an equilateral triangular section (see Figure 4.20) the side plates $a \times t$ are allowed to buckle under a shear stress that is to equal to the limiting stress from Batho shear.

(i) Limiting Batho Shear
The constant shear stress in the uniform section follows from the applied torque as

$$T = 2A_o\tau t = \frac{\sqrt{3}}{2}a^2\tau t, \quad \Rightarrow \quad \tau = \frac{2T}{\sqrt{3}a^2t} \le \tau_y \tag{4.70a,b}$$

where τ_y is a predetermined limiting stress, chosen, say, at a point along a shear stress–strain curve where the gradient is one-half the elastic shear modulus, i.e. $G_T = G/2$.

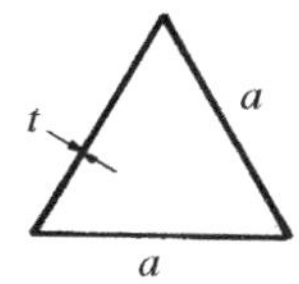

Figure 4.20 Triangular tube geometry

(ii) Shear Buckling

The buckling of each rectangular side wall depends upon its width a, length L, thickness t and the manner in which the adjacent walls support its sides [2]. Again, each influence lies within the buckling stress formula (4.66). The elastic buckling coefficient $K = K(a/L)$ is interpreted graphically in Figure 4.7, where the connection of each plate to its neighbour, i.e. along the bend, assumes a simply supported edge condition.

(iii) Optimisation

From equations (4.66) and (4.70b),

$$\tau_y = \frac{2T}{\sqrt{3}a^2 t} = KE_T\left(\frac{t}{a}\right)^2 \tag{4.71}$$

Solving equation (4.71) simultaneously for a and t,

$$t_{\text{opt}} = \left(\frac{2}{\sqrt{3}K}\right)^{1/3}\left(\frac{T}{E_T}\right)^{1/3}, \quad a_{\text{opt}} = \left(\frac{4K}{3}\right)^{1/6}\left(\frac{E_T T^2}{\tau_y^3}\right)^{1/6} \tag{4.72a,b}$$

The minimum weight is found from $W_{\min} = 3a_{\text{opt}}t_{\text{opt}}L\rho$ when, on dividing by L^3,

$$\left(\frac{W}{L^3}\right)_{\min} = \left(\frac{6^4}{K}\right)^{1/6}\left(\frac{\rho}{E_T^{1/6}\tau_y^{1/2}}\right)\left(\frac{T}{L^3}\right)^{2/3} \tag{4.73}$$

4.4.4 Hexagonal Tube

In a long, thin-walled, hexagonal tube section (see Figure 4.21) the side plates $a \times t$ are allowed to buckle under a shear stress made equal to the limiting stress from Batho shear.

(i) Limiting Batho Shear

The constant shear stress in the uniformly thin wall follows from the applied torque as

$$T = 2A_o\tau t = 3\sqrt{3}a^2\tau t, \quad \Rightarrow \quad \tau = \frac{T}{3\sqrt{3}a^2 t} \le \tau_y \tag{4.74a,b}$$

where τ_y is a limiting yield stress, here chosen to correspond with a tangent modulus in shear that is one-half the elastic shear modulus, i.e. $G_T = G/2$.

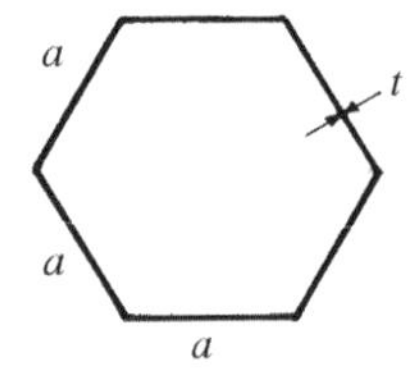

Figure 4.21 Hexagonal tube geometry

(ii) Shear Buckling
The buckling of each rectangular side plate depends upon its width a, length L, thickness t and the manner of its side supports. The buckling stress formula (4.66) again applies, where the function $K = K(a/L)$ is interpreted graphically in Figure 4.7 for simply supported sides.

(iii) Optimisation
From equations (4.66) and (4.74b),

$$\tau_y = \frac{T}{3\sqrt{3}a^2t} = KE_T\left(\frac{t}{a}\right)^2 \tag{4.75}$$

Solving the three parts to equation (4.75) simultaneously, provides the optimum t and a,

$$t_{\text{opt}} = \left(\frac{2}{\sqrt{3}K}\right)^{1/3}\left(\frac{T}{E_T}\right)^{1/3}, \quad a_{\text{opt}} = \left(\frac{K}{3^3}\right)^{1/6}\left(\frac{E_T T^2}{\tau_y^3}\right)^{1/6} \tag{4.76a,b}$$

The minimum weight is found from $W_{\min} = 6a_{\text{opt}}t_{\text{opt}}L\rho$ when, on dividing by L^3,

$$\left(\frac{W}{L^3}\right)_{\min} = \frac{2}{K^{1/6}}\left(\frac{\rho}{E_T^{1/6}\tau_y^{1/2}}\right)\left(\frac{T}{L^3}\right)^{2/3} \tag{4.77}$$

Comparing the three non-circular tubes, it appears that the weight is least for this section. Tubes with greater number of sides reduce the weight further, but only by small amounts. In the limit, as the tube of least weight becomes circular, it is seen from equations (4.64) and (4.77) that there is a further contribution from the different material factors.

4.4.5 *Octagonal Tube*

In a long, thin-walled octagonal tube (see Figure 4.22) the side plates are allowed to buckle under a shear stress made equal to the limiting stress from Batho shear.

(i) Limiting Batho Shear
Since the sides are inclined at 135°, the area enclosed by the octagon is given as:

$$A_o = a^2 + 4\frac{a}{2\sqrt{2}}\frac{a}{\sqrt{2}} + 4a\frac{a}{\sqrt{2}} = 4.828a^2 \tag{4.78}$$

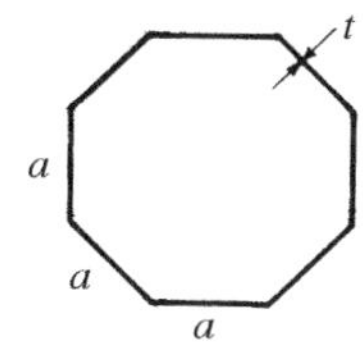

Figure 4.22 Octagonal tube geometry

The constant shear stress in the uniformly thin wall then follows from equation (4.78) and the Batho torque T as

$$T = 2A_o\tau t = 9.657a^2\tau t, \quad \Rightarrow \quad \tau = \frac{T}{9.657a^2 t} \leq \tau_y \tag{4.79a,b}$$

where, again, τ is limited to a plastic shear stress τ_y at which the tangent modulus is one-half the elastic shear modulus, i.e. $G_T = G/2$.

(ii) Shear Buckling

The elastic buckling of a rectangular plate, supported along its long edges by the adjacent plates, occurs under a shear stress given by equation (4.66). The bucking coefficient $K = K(a/L)$ may be read from Figure 4.7 to give an elastic buckling stress and then subsequently modified with a plasticity reduction factor η. We can write the plastic buckling stress as

$$\tau_{cr} = K_p E_T \left(\frac{t}{a}\right)^2 = 3\eta K G_T \left(\frac{t}{a}\right)^2 \tag{4.80a,b}$$

where $K_p = \eta K$. Note that the tangent modulus in shear G_T has replaced E_T in equation (4.80b). Here it has been assumed that $E_T = 2G_T(1+\nu)$ and, in taking $\nu = 0.5$ for incompressible plasticity, this relationship gives $E_T = 3G_T$.

(iii) Optimisation

From equations (4.79b) and (4.80a),

$$\tau_y = \frac{T}{9.657a^2 t} = K_p E_T \left(\frac{t}{a}\right)^2 \tag{4.81}$$

Solving simultaneously for the optimum t and a,

$$t_{opt} = 0.47\left(\frac{T}{K_p E_T}\right)^{1/3}, \quad a_{opt} = 0.47K^{1/6}\left(\frac{E_T T^2}{\tau_y^3}\right)^{1/6} \tag{4.82a,b}$$

The minimum weight is found from $W_{min} = 8a_{opt}t_{opt}L\rho$, when, on dividing by L^3,

$$\left(\frac{W}{L^3}\right)_{min} = \frac{1.767}{K^{1/6}}\left(\frac{\rho}{E_T^{1/6}\tau_y^{1/2}}\right)\left(\frac{T}{L^3}\right)^{2/3} \tag{4.83}$$

Table 4.2 Comparison of shape factors for various tubes with uniform thickness

Section				
$f \times K^{1/6}$	3.302	2.52	2	1.767

Table 4.2 summarises the numerical values for shape factors found for the regular polygon tubular sections given here. This shows that the weight of a torque-carrying tube is reduced as the number of its sides increases. As it is not very practical to manufacture tubes with more than eight sides, we see that the greatest weight saving, which lies between the octagonal and triangular tubes, is a significant 46%.

Though it might appear that the solution to a long, circular tube would provide a lower, limiting f-value, we see that equation (4.64) contains different exponents on the two properties E_T and τ_y within its material factor. This reflects the differing buckling behaviour between curved and flat surfaces in shear. Their weight comparison becomes possible when E_T and τ_y are known. For example, taking the ratio between equations (4.64) and (4.73) gives a weight saving between circular and triangular tubes,

$$\frac{W_\Delta - W_o}{W_\Delta} = 1 - 0.559\left(\frac{K_\Delta^3 \tau_y}{K_o^4 E_T}\right)^{1/18}$$

where K_Δ and K_o can be read from Figures 4.7 and 4.10 respectively at the appropriate aspect ratio for each tube. We show next how it is possible to lessen the weight of triangular and rectangular tubes further by allowing their wall thicknesses to differ.

4.4.6 Variable Thickness Rectangular Tube

The tube's wall dimensions are $a \times t_a$ and $b \times t_b$, as shown in Figure 4.23. Buckling will occur at a lower shear stress in the longer walls when $t_a/a < t_b/b$. If this stress is to be our limiting plastic buckling stress τ_y, then the optimisation equation becomes

$$\tau_y = \frac{T}{2abt_a} = K_a E_T \left(\frac{t_a}{a}\right)^2 \qquad (4.84)$$

The optimum dimensions follow from equation (4.84) as

$$(t_a)_{\text{opt}} = \left(\frac{1}{2K_a}\right)^{1/3}\left(\frac{a}{b}\right)^{1/3}\left(\frac{T}{E_T}\right)^{1/3}, \quad a_{\text{opt}} = \left(\frac{K_a}{4}\right)^{1/6}\left(\frac{a}{b}\right)^{1/3}\left(\frac{E_T T^2}{\tau_y^3}\right)^{1/6} \qquad (4.85\text{a,b})$$

The weight is expressed as

$$W = 2(at_a + bt_b)\rho L = 2at_a\left(1 + \frac{b\,t_b}{a\,t_a}\right)\rho L \qquad (4.86)$$

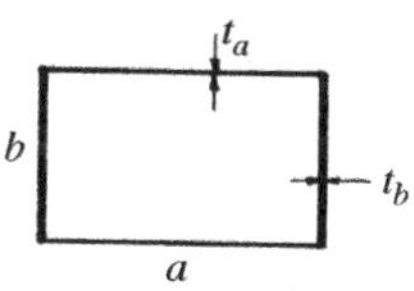

Figure 4.23 Rectangular tube with different wall thicknesses

in which the thickness ratio may be expressed in terms of the side length ratio from the fact that the shear flow q is constant in a closed tube. That is,

$$q = \tau_a t_a = \tau_b t_b \tag{4.87}$$

It follows from equations (4.66) and (4.87) that shear buckling within the section's perpendicular sides will occur simultaneously when

$$q = K_a E_a \left(\frac{t_a}{a}\right)^2 t_a = K_b E_b \left(\frac{t_b}{b}\right)^2 t_b \tag{4.88a}$$

A rearrangement of equation (4.88a) gives this condition as

$$\left(\frac{t_b}{t_a}\right)^3 = \frac{K_a}{K_b}\left(\frac{b}{a}\right)^2 \tag{4.88b}$$

Substituting equation (4.88b) into equation (4.86), the weight becomes

$$W = 2at_a\left[1 + \left(\frac{K_a}{K_b}\right)^{\frac{1}{3}}\left(\frac{b}{a}\right)^{\frac{5}{3}}\right]\rho L \tag{4.89a}$$

Substituting the optimum a and t_a from equations (4.85a,b) into equation (4.89a) and dividing by L^3 gives the optimum weight function:

$$\left(\frac{W}{L^3}\right)_{\text{opt}} = \left(\frac{4}{K_a}\right)^{1/6} \frac{1 + (K_a/K_b)^{1/3}(b/a)^{5/3}}{(b/a)^{2/3}} \left(\frac{\rho}{E_T^{1/6}\tau_y^{1/2}}\right)\left(\frac{T}{L^3}\right)^{2/3} \tag{4.89b}$$

in which the least weight is found from minimising the shape function. That is,

$$\left(\frac{4}{K_a}\right)^{1/6} \frac{\mathrm{d}}{\mathrm{d}(b/a)} \frac{1 + (K_a/K_b)^{1/3}(b/a)^{5/3}}{(b/a)^{2/3}} = 0 \tag{4.90a}$$

Equations (4.89b) and (4.90a) lead to the stationary values

$$\frac{b}{a} = \left(\frac{2}{3}\right)^{3/5}\left(\frac{K_b}{K_a}\right)^{1/5} \quad \text{and} \quad f_{\min} = \frac{5}{3}\left(\frac{3}{2}\right)^{2/5}\left(\frac{4}{K_a}\right)^{1/6}\left(\frac{K_a}{K_b}\right)^{2/15} = \frac{2.4696}{K_a^{1/6}}\left(\frac{K_a}{K_b}\right)^{2/15} \tag{4.90b,c}$$

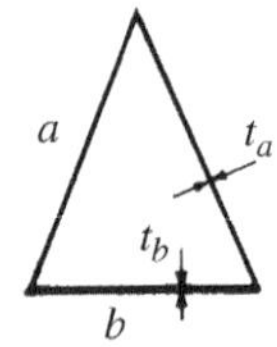

Figure 4.24 Isosceles triangular tube with different wall thicknesses

Hence, to minimise the weight we must first find b/a from equation (4.90b), and then find the optimum a and t_a from equations (4.85a,b). The procedure is assisted when one leading dimension, say a, is known along with the length L. Taking $a = a_{opt}$, we may read K_a from Figure 4.7 at the given a/L ratio and then iterate from an assumed b/a value until both equations (4.85b) and (4.90b) are satisfied.

4.4.7 Variable Thickness Triangular Tube

The sloping walls of the isosceles triangle are $a \times t_a$ and the shorter, thicker base is $b \times t_b$, as shown in Figure 4.24. The maximum stress due to the torque is found from $T = 2A_o q$ in which the Batho shear flow $q = \tau_a t_a = \tau_b t_b$. Thus, as $t_a < t_b$, the maximum stress is found from the enclosed area A_o as

$$\tau_{\max} = \frac{T}{2A_o t_a} = \frac{T}{ab\left[1 - {}^1\!/\!_4(b/a)^2\right]^{1/2} t_a} \tag{4.91a}$$

The critical buckling stress occurs under τ_y in the longer walls, where $t_a/a < t_b/b$. Hence, from equation (4.91a), the equation governing the optimisation becomes

$$\tau_y = \frac{T}{ab\left[1 - {}^1\!/\!_4(b/a)^2\right]^{1/2} t_a} = K_a E_T \left(\frac{t_a}{a}\right)^2 \tag{4.91b}$$

from which the optimum dimensions appear as

$$(t_a)_{opt} = \frac{T^{1/3}(a/b)^{1/3}}{(K_a E_T)^{1/3}\left[1 - {}^1\!/\!_4(b/a)^2\right]^{1/6}}, \quad a_{opt} = \frac{(K_a E_T)^{1/6} T^{1/3} (a/b)^{1/3}}{\tau_y^{1/2}\left[1 - {}^1\!/\!_4(b/a)^2\right]^{1/6}} \tag{4.92a,b}$$

The weight is expressed as

$$W = (2at_a + bt_b)\rho L = at_a\left(2 + \frac{b}{a}\frac{t_b}{t_a}\right)\rho L \tag{4.93}$$

in which the thickness ratio may be expressed in terms of the length ratio from the constant shear flow,

$$q = \tau_a t_a = \tau_b t_b \tag{4.94a}$$

It follows from equation (4.7) and (4.94a) that, with a simultaneous buckling of the tube's walls,

$$q = K_a E_a \left(\frac{t_a}{a}\right)^2 t_a = K_b E_b \left(\frac{t_b}{b}\right)^2 t_b \tag{4.94b}$$

from which

$$\left(\frac{t_b}{t_a}\right)^3 = \frac{K_a}{K_b}\left(\frac{b}{a}\right)^2 \tag{4.94c}$$

Substituting equation (4.94c) into (4.93), the weight becomes

$$W = at_a\left[2 + \left(\frac{K_a}{K_b}\right)^{1/3}\left(\frac{b}{a}\right)^{5/3}\right]\rho L \tag{4.95a}$$

Substituting the optimum t_a and a from equations (4.92a,b) into equation (4.93) and dividing by L^3 gives the objective function

$$\left(\frac{W}{L^3}\right)_{\text{opt}} = \left(\frac{1}{K_a}\right)^{1/6} \frac{2 + (K_a/K_b)^{1/3}(b/a)^{5/3}}{(b/a)^{2/3}\left[1 - \frac{1}{4}(b/a)^2\right]^{1/3}} \left(\frac{\rho}{E_T^{1/6}\tau_y^{1/2}}\right)\left(\frac{T}{L^3}\right)^{2/3} \tag{4.95b}$$

in which the minimum weight is found from minimising the shape function f:

$$\frac{\mathrm{d}f}{\mathrm{d}(b/a)} = \left(\frac{1}{K_a}\right)^{1/6} \frac{\mathrm{d}}{\mathrm{d}(b/a)} \frac{2 + (K_a/K_b)^{1/3}(b/a)^{5/3}}{\left[1 - \frac{1}{4}(b/a)^2\right]^{1/3}(b/a)^{2/3}} = 0 \tag{4.96a}$$

Equation (4.96a) provides the wall width ratio b/a which minimises f as the solution to

$$12\left(\frac{b}{a}\right)^{5/3} - \left(\frac{b}{a}\right)^{11/3} - 8\left(\frac{K_b}{K_a}\right)^{1/3}\left[2 - \left(\frac{b}{a}\right)^2\right] = 0 \tag{4.96b}$$

An initial iteration is required to find the b/a ratio satisfying equation (4.96b) and remain consistent with the K_b/K_a ratio. A pre-selection of one wall width, a or b, assists with this procedure. Once determined, this stationary b/a ratio may be substituted into equations (4.92a,b) to find the remaining optimum dimensions that minimise weight.

4.5 Multi-Cell Tubes

The addition of internal walls or partitions can serve to increase torsional stiffness at the expense of weight. However, the benefit to stiffness can only be gained when the internal wall is stressed. Three of the more common cell shapes are square, hexagonal and rectangular. Cells with uniform or variable wall thicknesses may be designed to give certain features to multi-cell tubes, as will now be discussed.

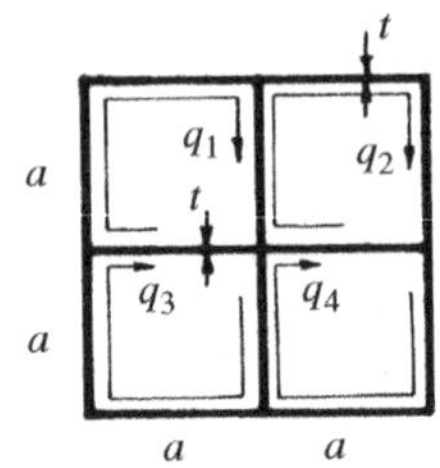

Figure 4.25 Four, square cell-tube showing shear flows

4.5.1 *Square Multi-Cell Tube*

Consider, for example, the tube with four identical square cells, each $a \times a$ with uniform thickness t, in Figure 4.25. The total torque carried is $T = 4 \times 2Aq = 8a^2\tau t$, from which the wall shear stress is found to be $\tau = T/(8a^2 t)$. When this stress is limited to τ_y and made to correspond with the critical buckling stress for one cell wall, the optimisation lies within

$$\tau_y = \frac{T}{8a^2 t} = K_a E_T \left(\frac{t}{a}\right)^2 \tag{4.97}$$

Equation (4.97) optimises the square cell thickness and width to

$$t_{\text{opt}} = \frac{1}{2}\left(\frac{T}{K_a E_T}\right)^{1/3} \quad \text{and} \quad a_{\text{opt}} = \frac{1}{2}\left(\frac{K_a E_T T^2}{\tau_y^3}\right)^{1/6} \tag{4.98a,b}$$

The weight expression $W = 12at\rho L$ is minimised by identifying t and a with equations (4.98a,b) respectively. Dividing by L^3 gives

$$\left(\frac{W}{L^3}\right)_{\min} = \frac{3}{K_a^{1/6}}\left(\frac{\rho}{E_T^{1/6}\tau_y^{1/2}}\right)\left(\frac{T}{L^3}\right)^{2/3} \tag{4.99}$$

Now remove the inner partitions to produce a square tube of width $2a$ and thickness t. The previous single-cell analysis applies where, from equation (4.69), the shape factor was found to be $f = (2^8/K_a)^{1/6}$. Comparing with $f = 3/K_a^{1/6}$ from equation (4.99), we see that partitioning increases the optimum weight by $\approx 19\%$.

Each cell twists at the same rate (θ rad/m), which must be the same as that for the whole tube. Hence, for cell 1 the rate of twist $\delta\theta/\delta z$ (i.e. with respect to length z) is [1]

$$2A_1 G\left(\frac{\delta\theta}{\delta z}\right)_1 = q_1 \oint_1 \frac{\delta s}{t} - q_2 \int_2 \frac{\delta s}{t} - q_3 \int_3 \frac{\delta s}{t} \tag{4.100}$$

where G is the shear modulus, q_1 is the shear flow in cell 1, q_2 and q_3 are the opposing shear flows in its adjacent cells 2 and 3 (see Figure 4.25). The path integral sums the width thickness ratios around cell 1, from which are subtracted similar ratios within those widths where q_2 and q_3 oppose q_1. This is simplified from knowing that shear flows are the same for identical cells. Thus, with $q_1 = q_2 = q_3$, equation (4.100) gives the twist rate

$$2a^2G\left(\frac{\delta\theta}{\delta z}\right) = q_1\left(\frac{4a}{t} - \frac{a}{t} - \frac{a}{t}\right), \quad \Rightarrow \quad \frac{\delta\theta}{\delta z} = \frac{q_1}{atG} \qquad (4.101a,b)$$

Hence, the torsional stiffness for the multi-cell tube is given by

$$\frac{T}{\delta\theta/\delta z} = \frac{8a^2q_1}{q_1/(atG)} = 8a^3tG \qquad (4.101c)$$

With the interior partitions removed, the twist rate for the single closed tube, of width $2a$, is found from the reduction to equation (4.100):

$$2AG\left(\frac{\delta\theta}{\delta z}\right) = q\oint\frac{\delta s}{t}, \quad \frac{\delta\theta}{\delta z} = \frac{q \times 4(2a/t)}{2(2a)^2 \times G} = \frac{q}{atG} \qquad (4.102a)$$

and, from equations (4.65a) and (4.102a), the torsional stiffness becomes

$$\frac{T}{\delta\theta/\delta z} = \frac{2(2a)^2q}{q/(atG)} = 8a^3tG \qquad (4.102b)$$

which is the same as for the partitioned tube. Similar stiffness will always apply when shear flows, having the same magnitude in neighbouring cells, oppose, to relieve their common walls of stress.

4.5.2 Hexagonal Multi-Cells

Figure 4.26 shows two arrangements of four hexagonal cells with uniform thickness. In Figure 4.26a each cell shares two walls with its neighbouring cells. For this to be possible the cells walls must meet the centre at 90° as shown. This connection produces irregular cells, but, despite this, we may draw similar conclusions to the square cell arrangement given in § 4.5.1 above. They were: (i) the shear flows $q_1 \ldots q_4$ are similar, (ii) the walls remain unstressed, (iii) the partitions increase the weight and (iv) the interior walls do not increase the stiffness. In short, there is more to lose than to gain with this arrangement of cells. The reader is asked to derive the specific properties for this tube in Exercises 4.4 and 4.5.

However, with the rearrangement given in Figure 4.26b, the walls can play a more useful role. Here the shear flows in cells 1 and 3 are similar since they share three walls. The shear flows in cells 2 and 4 are also similar with their sharing two neighbouring walls. Thus, $q_1 = q_3$ and $q_2 = q_4$ but $q_1 \neq q_2$. Firstly, let us examine the *stiffness* implications.

The twist rate for cell 1 is

$$2A_1G\left(\frac{\delta\theta}{\delta z}\right)_1 = q_1\oint_1\frac{\delta s}{t} - q_2\int_2\frac{\delta s}{t} - q_3\int_3\frac{\delta s}{t} - q_4\int_4\frac{\delta s}{t} \qquad (4.103a)$$

$$= q_1\left(\frac{6a}{t}\right) - (q_2 + q_3 + q_4)\left(\frac{a}{t}\right) = \left(\frac{5a}{t}\right)q_1 - \left(\frac{2a}{t}\right)q_2$$

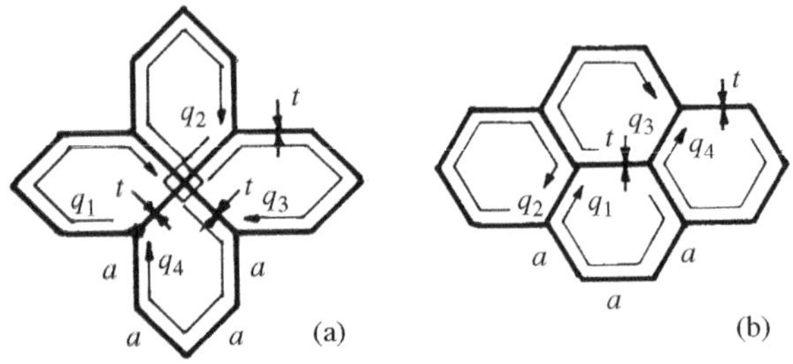

Figure 4.26 Hexagonal multi-cell arrangements

The twist rate for cell 2 is

$$2A_2G\left(\frac{\delta\theta}{\delta z}\right)_2 = q_2\oint_2\frac{\delta s}{t} - q_1\int_1\frac{\delta s}{t} - q_3\int_3\frac{\delta s}{t} \tag{4.103b}$$

$$= q_2\left(\frac{6a}{t}\right) - (q_1+q_3)\left(\frac{a}{t}\right) = -\left(\frac{2a}{t}\right)q_1 + \left(\frac{6a}{t}\right)q_2$$

Since $A_1 = A_2$ we may equate (4.103a) and (4.103b) to give $q_2 = 7/8 \times q_1$. Substituting this into equation (4.103a) gives the twist rate

$$(3\sqrt{3}a^2)G\left(\frac{\delta\theta}{\delta z}\right)_1 = \frac{13aq_1}{4t}, \Rightarrow \left(\frac{\delta\theta}{\delta z}\right)_1 = \frac{13aq_1}{4t} \times \frac{1}{3\sqrt{3}a^2G} = \frac{13q_1}{12\sqrt{3}atG} \tag{4.103c}$$

The torque capacity is given by

$$T = 2(2A_1q_1 + 2A_2q_2) = 4A_1(q_1+q_2) \tag{4.104a}$$

$$= 4 \times \frac{3\sqrt{3}}{2}a^2q_1(1+7/8) = \frac{45\sqrt{3}}{4}a^2q_1$$

Dividing equations (4.104a) and (4.103c) gives the torsional stiffness

$$\frac{T}{(\delta\theta/\delta z)_1} = \frac{45\sqrt{3}a^2q_1}{4} \times \frac{12\sqrt{3}atG}{13q_1} = 31.15a^3tG \tag{4.104b}$$

With regard to the optimum weight implications, the torque expression (4.104a) shows that the maximum wall shear stress is $\tau = 4T/(45\sqrt{3}a^2t)$. In limiting this stress to τ_y, it can be made to correspond with the critical buckling stress within each wall for cells 1 and 3 when

$$\tau_y = \frac{4T}{45\sqrt{3}a^2t} = K_aE_T\left(\frac{t}{a}\right)^2 \tag{4.105}$$

from which the optimum dimensions are derived:

$$t_{\text{opt}} = \frac{1}{3^{1/2}3.75^{1/3}}\left(\frac{T}{K_aE_T}\right)^{1/3} \quad \text{and} \quad a_{\text{opt}} = \frac{1}{3^{1/2}3.75^{1/3}}\left(\frac{K_aE_TT^2}{\tau_y^3}\right)^{1/6} \tag{4.106a,b}$$

The weight expression is $W = 19at\rho L$. Substituting a and t from equations (4.106a,b) and then dividing by L^3, this weight is optimised to

$$\left(\frac{W}{L^3}\right)_{\text{opt}} = \frac{19}{3 \times 3.75^{2/3} K_a^{1/6}} \left(\frac{\rho}{E_T^{1/6}\tau_y^{1/2}}\right)\left(\frac{T}{L^3}\right)^{2/3} \qquad (4.107)$$

When the internal walls are removed, the twist rate for the outer, closed tube is found:

$$2AG\left(\frac{\delta\theta}{\delta z}\right) = q\oint \frac{\delta s}{t}, \quad \Rightarrow \quad \frac{\delta\theta}{\delta z} = \frac{q \times 14(a/t)}{2(4 \times 3\sqrt{3}a^2/2) \times G} = \frac{14q}{12\sqrt{3}atG} \qquad (4.108\text{a,b})$$

With a constant shear flow q in the wall, the torque capacity $T = 2Aq$ is increased slightly compared to equation (4.104a). With A as the area enclosed by the outer wall, the torsional stiffness is found from equation (4.108b) to be

$$\frac{T}{\delta\theta/\delta z} = \frac{2(4 \times 3\sqrt{3}a^2)q}{2} \times \frac{12\sqrt{3}atG}{14q} = 30.86a^3tG \qquad (4.109)$$

Though the stiffness equation (4.109) is only marginally less than that of the multi-cell tube (see equation (4.104b)), the example serves to illustrate how partitions should be employed. With regard to the optimum weight implications, equation (4.74a) shows that the total torque carried by this tube is $T = 12\sqrt{3}a^2\tau t$, from which the wall shear stress is $\tau = T/(12\sqrt{3}a^2 t)$. Limiting this stress to τ_y, it will correspond to the critical buckling stress for wall widths a in cells 1 and 3, when

$$\tau_y = \frac{T}{12\sqrt{3}a^2 t} = K_a E_T \left(\frac{t}{a}\right)^2 \qquad (4.110)$$

in which the usual allowance for plasticity is made. From equation (4.110),

$$t_{\text{opt}} = \frac{1}{3^{1/2}4^{1/3}}\left(\frac{T}{K_p E_T}\right)^{1/3} \quad \text{and} \quad a_{\text{opt}} = \frac{1}{3^{1/2}4^{1/3}}\left(\frac{K_a E_T T^2}{\tau_y^3}\right)^{1/6} \qquad (4.111\text{a,b})$$

The weight expression is $W = 14at\rho L$. Substituting equations (4.111a,b) for t and a respectively into the W expression then dividing by L^3, this weight is optimised to

$$\left(\frac{W}{L^3}\right)_{\text{opt}} = \frac{14}{3 \times 4^{2/3} K_a^{1/6}} \left(\frac{\rho}{E_T^{1/6}\tau_y^{1/2}}\right)\left(\frac{T}{L^3}\right)^{2/3} \qquad (4.112)$$

Comparing equations (4.107) and (4.112) shows that the additional partitions increase the optimum weight by approximately 42%! Hence, there would have to be good reason to raise stiffness by about 1% with this weight penalty. Compare this with the direct removal of the interior walls from within Figure 4.26b. Counting sides of the two tubes, the non-optimum, multi-cell design has increased its weight by $(19 - 14)/14 = 35.8\%$, in allowing q to differ between the cells.

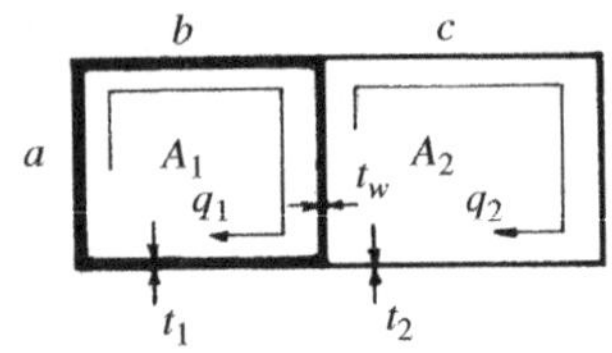

Figure 4.27 Tube with two, dissimilar rectangular cells

4.5.3 Two Rectangular Cells

Allowing both the thicknesses and the widths of the walls in multi-cell tubes to vary can provide more control over a weight optimisation. The rectangular cells in Figure 4.27 have three of their walls differing in width and thickness, say with $c > b > a$ and $t_1 > t_w > t_2$, as shown. Thus, all the walls are stressed by amounts that follow from the shear flows q_1 and q_2, which in turn are provided by the twist rates in each cell:

$$\left(\frac{\delta\theta}{\delta z}\right)_1 = \frac{q_1}{2A_1G}\left(\frac{a}{t_1} + \frac{2b}{t_1} + \frac{a}{t_w}\right) - \frac{q_2}{2A_1G}\left(\frac{a}{t_w}\right) \tag{4.113a}$$

$$\left(\frac{\delta\theta}{\delta z}\right)_2 = \frac{q_2}{2A_2G}\left(\frac{a}{t_2} + \frac{2c}{t_2} + \frac{a}{t_w}\right) - \frac{q_1}{2A_2G}\left(\frac{a}{t_w}\right) \tag{4.113b}$$

Setting $A_1 = ab$ and $A_2 = ac$ with $(\delta\theta/\delta z)_1 = (\delta\theta/\delta z)_2$ in equation (4.113a,b) leads to the ratio between the shear flows q_1 and q_2:

$$\frac{q_1}{q_2} = \frac{(1 + 2c/a)(t_w/t_2) + (1 + c/b)}{(1 + 2b/a)(c/b)(t_w/t_1) + (1 + c/b)} \tag{4.114}$$

The torque capacity is written as

$$T = 2(abq_1 + acq_2) = 2c^2\left(\frac{a}{c}\right)\left(1 + \frac{b}{c}\frac{q_1}{q_2}\right)q_2 \tag{4.115}$$

and setting $q_2 = \tau_2 t_2$, equation (4.115) may be rearranged to give the wall stress in cell 2. Allowing the greater wall width c within cell 2 to buckle under τ_y, its optimisation follows from

$$\tau_y = \frac{T}{2c^2(a/c)t_2\left[1 + (b/c)(q_1/q_2)\right]} = K_c E_T\left(\frac{t_2}{c}\right)^2 \tag{4.116}$$

Equation (4.116) gives the optimum dimensions for cell 2 as

$$(t_2)_{\text{opt}} = \frac{1}{\left\{2(a/c)\left[1 + (b/c)(q_1/q_2)\right]\right\}^{1/3}}\left(\frac{T}{K_cE_T}\right)^{1/3} \tag{4.177a}$$

$$c_{\text{opt}} = \frac{1}{\left\{2(a/c)\left[1 + (b/c)(q_1/q_2)\right]\right\}^{1/3}}\left(\frac{K_cE_TT^2}{\tau_y^3}\right)^{1/6} \tag{4.117b}$$

The weight is written as

$$\frac{W}{\rho L} = (a + 2b)t_1 + (a + 2c)t_2 + at_w = ct_2\left[2 + \frac{a}{c}\left(1 + \frac{t_w}{t_2} + \frac{t_1}{t_2}\right) + 2\frac{t_1}{t_2}\frac{b}{c}\right] \quad (4.118)$$

Substituting for the optimum c and t_2 from equations (4.117a,b) into equation (4.118) leads to the usual form for W/L^3. That is, the material factor and structural index are as they appear in equation (4.112), but where the shape factor is now

$$f = \frac{2 + (a/c)(1 + t_w/t_2 + t_1/t_2) + 2(b/c)(t_1/t_2)}{2^{2/3}K_c^{1/6}(a/c)^{2/3}\left[1 + (b/c)(q_1/q_2)\right]^{2/3}} \quad (4.119)$$

in which q_1/q_2 is given by equation (4.114). Equation (4.119) may be simplified further when we arrange for the wall width b and the web depth a to buckle simultaneously with c. This condition is realised from the simultaneous solution to the following:

$$\frac{q_1}{t_1} = K_bE_T\left(\frac{t_1}{b}\right)^2, \frac{q_2}{t_2} = K_cE_T\left(\frac{t_2}{c}\right)^2 \quad \text{and} \quad \frac{q_1 - q_2}{t_w} = K_wE_T\left(\frac{t_w}{a}\right)^2 \quad (4.120\text{a,b,c})$$

From the division of equations (4.120a–c), thickness ratios may be expressed in terms of the width ratios as follows:

$$\left(\frac{t_w}{t_1}\right)^3 = \frac{q_1 - q_2}{q_1}\frac{K_b}{K_w}\left(\frac{a}{b}\right)^2, \left(\frac{t_w}{t_2}\right)^3 = \frac{(q_1 - q_2)}{q_2}\frac{K_c}{K_w}\left(\frac{a}{c}\right)^2, \left(\frac{t_1}{t_2}\right)^3 = \frac{q_1}{q_2}\frac{K_c}{K_b}\left(\frac{b}{c}\right)^2 \quad (4.121\text{a,b,c})$$

Thus, it appears from equations (4.114) and (4.121a–c) how the shape factor expression (4.119) depends upon the two width ratios a/c and b/c. Naturally, here we should seek each of these ratios to minimise f and hence W. One way of doing this is to pre-select b/c, say, and then seek a stationary a/c by differentiation. Putting $x = a/c$ and $y = b/c$, we may then differentiate f partially with respect to x and equate to zero:

$$\begin{aligned}\frac{\partial f}{\partial x} = &\left\{x^{2/3}\left[1 + y\left(\frac{q_1}{q_2}\right)\right]^{2/3}\right\}\left\{x\left[\frac{\partial}{\partial x}\left(\frac{t_w}{t_2}\right) + \frac{\partial}{\partial x}\left(\frac{t_1}{t_2}\right)\right] + \left(1 + \frac{t_w}{t_2} + \frac{t_1}{t_2}\right)\right.\\ &\left. + 2y\frac{\partial}{\partial x}\left(\frac{t_1}{t_2}\right)\right\} - \left[2 + x\left(1 + \frac{t_w}{t_2} + \frac{t_1}{t_2}\right) + 2y\left(\frac{t_1}{t_2}\right)\right]\\ &\left\{\frac{2}{3}yx^{2/3}\left[1 + y\left(\frac{q_1}{q_2}\right)\right]^{-1/3}\frac{\partial}{\partial x}\left(\frac{q_1}{q_2}\right) + \frac{2}{3}x^{-1/3}\left[1 + y\left(\frac{q_1}{q_2}\right)\right]^{2/3}\right\} = 0\end{aligned} \quad (4.122\text{a})$$

where the following partial derivatives apply to equations (4.114) and (4.121a–c):

$$\frac{\partial}{\partial x}\left(\frac{t_w}{t_2}\right) = \frac{1}{3}\left[\frac{K_c}{K_w}\left(\frac{q_1}{q_2} - 1\right)x^2\right]^{-2/3}\frac{K_c}{K_w}\left[2\left(\frac{q_1}{q_2} - 1\right)x + x^2\frac{\partial}{\partial x}\left(\frac{q_1}{q_2}\right)\right]$$

$$\frac{\partial}{\partial x}\left(\frac{t_1}{t_2}\right) = \frac{y^2}{3}\frac{K_c}{K_b}\left[\frac{K_c}{K_b}\left(\frac{q_1}{q_2}\right)y^2\right]^{-2/3}\frac{\partial}{\partial x}\left(\frac{q_1}{q_2}\right)$$

$$\frac{\partial}{\partial x}\left(\frac{t_w}{t_1}\right)=\frac{1}{3}\left[\frac{K_b}{K_w}\left(1-\frac{q_2}{q_1}\right)\left(\frac{x}{y}\right)^2\right]^{-2/3}\frac{K_b}{K_w}\left[\left(1-\frac{q_2}{q_1}\right)\frac{2x}{y^2}-\left(\frac{x^2}{y^2}\right)^2\frac{\partial}{\partial x}\left(\frac{q_1}{q_2}\right)\right]$$

$$\frac{\partial}{\partial x}\left(\frac{q_1}{q_2}\right)=\left\{\left[(x+2y)\frac{t_w}{t_2}+x(y+1)\right]\left[y(x+2)\frac{\partial}{\partial x}\left(\frac{t_w}{t_1}\right)+y\left(\frac{t_w}{t_1}\right)+(y+1)\right]\right.$$
$$\left.-\left[y(x+2)\left(\frac{t_w}{t_2}\right)+x(y+1)\right]\left[(x+2y)\frac{\partial}{\partial x}\left(\frac{t_w}{t_1}\right)+\frac{t_w}{t_1}+(1+y)\right]\right\}$$
$$\div\left\{\left[(x+2y)\left(\frac{t_w}{t_1}\right)+x(y+1)\right]^2\right\}$$

$$\frac{\partial}{\partial x}\left(\frac{q_2}{q_1}\right)=\frac{\partial}{\partial x}\left(\frac{q_1}{q_2}\right)^{-1}=-\frac{\partial}{\partial x}\left(\frac{q_1}{q_2}\right)\Big/\left(\frac{q_1}{q_2}\right)^2 \qquad (4.122\text{b-f})$$

The system of equations (4.122a–f) may be programmed to find a stationary x-value for a given y. As with the open sections considered in § 4.3, an iteration is required within this to match the buckling coefficients K_b, K_c and K_w. The latter are read from Figure 4.7, corresponding to the ratios a/L, b/L and c/L that serve to minimise weight.

4.5.4 *Two Triangular Cells*

Here the wall thicknesses t_a, t_b and t_w and the widths a, b and c of two isosceles triangular cells are as shown in Figure 4.28. It is assumed that three of their walls differ in width and thickness, say with $a > c > b$ and $t_b > t_c > t_a$, as shown. The final geometrical ratios are determined by applying the following failure criteria. The walls are stressed within their shear flows q_1 and q_2, which in turn are provided by the twist rates in each cell:

$$\left(\frac{\delta\theta}{\delta z}\right)_1=\frac{q_1}{2A_1G}\left(\frac{2a}{t_a}+\frac{c}{t_c}\right)-\frac{q_2}{2A_1G}\left(\frac{c}{t_c}\right) \qquad (4.123\text{a,b})$$
$$\left(\frac{\delta\theta}{\delta z}\right)_2=\frac{q_2}{2A_2G}\left(\frac{2b}{t_b}+\frac{c}{t_c}\right)-\frac{q_1}{2A_2G}\left(\frac{c}{t_c}\right)$$

Setting $(\delta\theta/\delta z)_1 = (\delta\theta/\delta z)_2$ leads to the ratio between q_1 and q_2,

$$\frac{q_1}{q_2}=\frac{A_1\left[1+A_2/A_1+2(t_c/t_b)(b/c)\right]}{A_2\left[1+A_1/A_2+2(t_c/t_a)(a/c)\right]} \qquad (4.124)$$

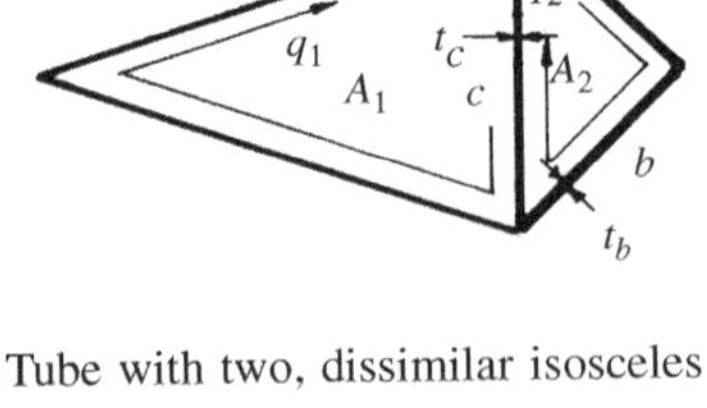

Figure 4.28 Tube with two, dissimilar isosceles triangular cells

in which the areas enclosed by the triangles are in the ratio

$$A_1 = \frac{c}{2}\left(a^2 - \frac{c^2}{4}\right)^{1/2}, A_2 = \frac{c}{2}\left(b^2 - \frac{c^2}{4}\right)^{1/2}, \quad \Rightarrow \quad \frac{A_1}{A_2} = \left[\frac{4(a/c)^2 - 1}{4(b/c)^2 - 1}\right]^{1/2} \tag{4.125a,b,c}$$

The torque capacity follows from equation (4.125c) as

$$T = 2A_1q_1 + 2A_2q_2 = 2A_1q_1\left(1 + \frac{A_2}{A_1}\frac{q_2}{q_1}\right) \tag{4.126}$$

Setting $q_1 = \tau_1 t_a$ in equation (4.126) gives the wall stress in cell 1. When we allow the greater wall width a to buckle by limiting this stress to $\tau_1 \leq \tau_y$, the optimisation of this cell follows from

$$\tau_y = \frac{T}{2A_1t_a\left[1 + (A_2/A_1)(q_2/q_1)\right]} = K_aE_T\left(\frac{t_a}{a}\right)^2 \tag{4.127}$$

Equations (4.125b) and (4.127) lead to optimum dimensions

$$t_{a_{\text{opt}}} = \frac{2^{1/3}(a/c)^{2/3}}{\left\{\left[4(a/c)^2 - 1\right]^{1/2}\left[1 + (A_2/A_1)(q_2/q_1)\right]\right\}^{1/3}}\left(\frac{T}{K_aE_T}\right)^{1/3} \tag{4.128a}$$

$$c_{\text{opt}} = \frac{2^{1/3}(a/c)^{2/3}}{\left\{\left[4(a/c)^2 - 1\right]^{1/2}\left[1 + (A_2/A_1)(q_2/q_1)\right]\right\}^{1/3}}\left(\frac{K_aE_TT^2}{\tau_y^3}\right)^{1/6} \tag{4.128b}$$

The weight is written as

$$\frac{W}{\rho L} = 2(at_a + bt_b) + ct_c = at_a\left\{2 + (c/a)(t_c/t_a)\left[1 + 2(b/c)(t_b/t_c)\right]\right\} \tag{4.129}$$

Substituting the optimum c and t_a from equations (4.128a,b) into equation (4.129) leads to the usual form for an optimum W/L^3 (see, for example, equation (4.112)), where the shape factor f is now

$$f = \frac{\left[4(a/c)\right]^{1/3}\left\{2(a/c) + (t_c/t_a)\left[1 + 2(b/c)(t_b/t_c)\right]\right\}}{K_a^{1/6}\left[4(a/c)^{1/3} - 1\right]^{1/3}\left[1 + (A_2/A_1)(q_2/q_1)\right]^{2/3}} \tag{4.130}$$

in which the inverse ratios q_1/q_2 and A_1/A_2 are given by equations (4.124) and (4.125c), respectively. The thickness ratios appearing in equations (4.129) and (4.130) may be expressed in terms of the wall width ratios when we arrange for the wall length b and the web length c to buckle simultaneously with a. This requires the simultaneous solution of

$$\frac{q_1}{t_a} = K_aE_T\left(\frac{t_a}{a}\right)^2, \frac{q_2}{t_b} = K_bE_T\left(\frac{t_b}{b}\right)^2 \quad \text{and} \quad \frac{q_1 - q_2}{t_c} = K_cE_T\left(\frac{t_c}{c}\right)^2 \tag{4.131a,b,c}$$

and, from the division between equations (4.131a–c),

$$\left(\frac{t_c}{t_a}\right)^3 = \frac{(q_1 - q_2)}{q_1}\left(\frac{K_a}{K_c}\right)\left(\frac{c}{a}\right)^2 \quad \text{and} \quad \left(\frac{t_c}{t_b}\right)^3 = \frac{(q_1 - q_2)}{q_2}\left(\frac{K_b}{K_c}\right)\left(\frac{c}{b}\right)^2 \qquad (4.132\text{a,b})$$

Thus, between equations (4.124), (4.125c) and (4.132a,b) it can be seen how the shape factor expression (4.130) is made to depend only upon the two width ratios $x = a/c$ and $y = b/c$. At this stage we should seek ratios x and y that minimise f and hence W. This is left for the reader (see Exercise 4.9), who should be guided by the previous procedure adopted when seeking stationary values for the two-cell rectangular tube given in § 4.5.3.

References

[1] Rees, D. W. A. *Mechanics of Solids and Structures*, Imperial College Press, 2000.
[2] ESDU 71005, Buckling of flat plates in shear, September 1976.
[3] ESDU 01.01.10, Torsional instability of stringers and struts of angle section, November 1978.
[4] ESDU 02.03.18, Buckling of curved plates and panels in shear, April 1977.
[5] Batdorf, S. B. A simplified method of elastic-stability analysis for thin cylindrical shells I – Donnell's equations, NACA Tech Note 1341, 1947.
[6] Batdorf, S. B., Stein, M. and Schildercrout, M. Critical stress of thin-walled cylinders in torsion, NACA Tech Note 1345, 1947.
[7] Gerard, G. Critical shear stress of plates above the proportional limit, *Journal of Applied Mechanics*, **15**(1), March 1948, 7– 12.
[8] Gerard, G. Compressive and torsional buckling of thin-walled cylinders in the yield region, NACA Tech Note 3726, 1956.
[9] Gerard, G. and Becker, H. *Handbook of Structural Stability*, Part III: Buckling of curved plates and shells, NACA Tech Note 3783, 1957.
[10] Gerard, G. *Introduction to Structural Stability Theory*, McGraw-Hill, 1962
[11] Donnell, L. H. Stability of thin-walled tubes under torsion, NACA Report 479, 1933.

Exercises

4.1 Show, for a curved panel of short length, where equations (4.19a,b) apply, that the optimum cross-sectional dimensions are

$$t_{\text{opt}} = \left[\frac{3T}{C(K_e E k)^{1/2}}\right]^{1/3} \quad \text{and} \quad b_{\text{opt}} = \left(\frac{3TC^2 K_e E}{k^2}\right)^{1/3} = C L_{\text{opt}}$$

where C is the particular b/L ratio, appearing in Figure 4.11, at which K_e is read. Devise a strategy in which L_{opt} can be found from the abscissa of this figure for a panel of given radius R and thickness t.

4.2 Figure 4.29 shows a symmetrical I-section with flanges (width $2b \times$ thickness t_b) and web (depth $a \times$ thickness t_a), subjected to an axial torque T about its centroidal axis. Show that the optimum dimensions are given by

$$(t_a)_{\text{opt}} = \left\{\frac{3T}{(K_a E k)^{1/2}\left[1 + 4(b/a)^4 (K_a/K_b)^{3/2}\right]}\right\}^{1/3} \quad \text{and} \quad a_{\text{opt}} = \left\{\frac{3T K_a E}{k^2\left[1 + 4(b/a)^4 (K_a/K_b)^{3/2}\right]}\right\}^{1/3}$$

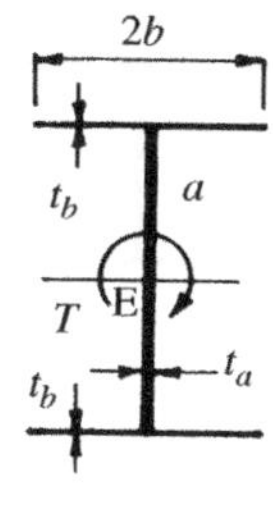

Figure 4.29 Symmetrical I-section

and that the optimum weight is, correspondingly,

$$\left(\frac{W}{L^3}\right)_{\text{opt}} = \frac{(3^4K_a)^{1/6}\left[1+4(K_a/K_b)^{1/2}(b/a)^2\right]}{\left[1+4(K_a/K_b)^{3/2}(b/a)^4\right]^{2/3}}\frac{\rho E^{1/6}}{k^{5/6}}\left(\frac{T}{L^3}\right)^{2/3}$$

[*Hint*: Add two further flange half-lengths ($b \times t_b$) to any of the sections shown in Figures 4.14 as is appropriate to creating an I-section.]

4.3 Show that the minimum shape factor for the I-section in Exercise 4.2 is given by

$$f_{\min} = \frac{(3^4K_a)^{1/6}\left[2\sqrt{1+(3/2)(K_b/K_a)^{1/2}}-1\right]}{4^{2/3}\left[1+(K_a/K_b)^{1/2}-\sqrt{(K_a/K_b)+(3/2)(K_a/K_b)^{1/2}}\right]^{2/3}}$$

corresponding to the condition

$$\left(\frac{b}{a}\right)^2 = \sqrt{\left(\frac{K_b}{K_a}\right)+\frac{3}{2}\left(\frac{K_b}{K_a}\right)^{3/2}} - \left(\frac{K_b}{K_a}\right)^{1/2}$$

4.4 Show, for the regular arrangement of hexagonal cells in Figure 4.26a, that the optimum flat length a and thickness t are

$$t_{\text{opt}} = \frac{1}{2.683K^{1/3}}\left(\frac{T}{E_T}\right)^{1/3} \quad \text{and} \quad a_{\text{opt}} = \frac{K^{1/6}}{2.683}\left(\frac{T^2E_T}{\tau_y^3}\right)^{1/6}$$

and that the minimum weight is, correspondingly,

$$\left(\frac{W}{L^3}\right)_{\text{opt}} = \frac{2.778}{K^{1/6}}\left(\frac{\rho}{E_T^{1/6}\tau_y^{1/2}}\right)\left(\frac{T}{L^3}\right)^{2/3}$$

4.5 Confirm that the torsional stiffness for the hexagonal cells in Figure 4.26a is given by

$$\frac{T}{\delta\theta/\delta z} = 23.31a^3tG$$

Show that when the inner walls are removed there is no change to the torsional stiffness with the consequent 25% saving in weight. Explain why this is so.

4.6 Show that when the double square cell tube in Figure 4.30 is placed under torsion, the shape factor is given by $f = 2.778/K^{1/6}$. What is the shear stress in the walls?

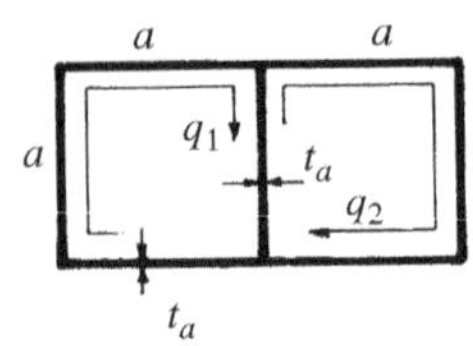

Figure 4.30 Double-square cell tube

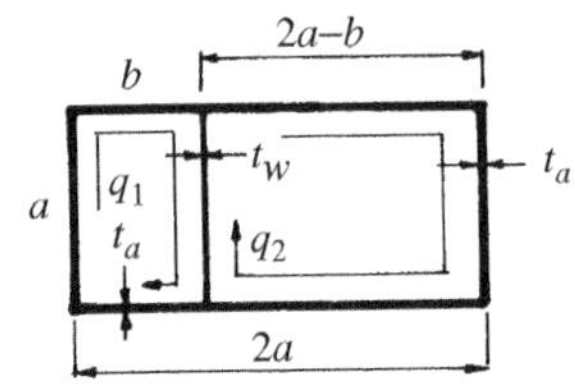

Figure 4.31 Two rectangular cells with thinner partition

4.7 The double-square cell tube in Figure 4.30 is modified with a thinner partition, moved to divide the area into two rectangles in the manner shown in Figure 4.31. Show that when buckling occurs under τ_y in the greater wall width $2a - b$, the optimum dimensions t_a and a are expressed as

$$(t_a)_{\text{opt}} = \frac{[2-(b/a)]^{1/3}}{\{2K_{2a-b}[1+(A_1/A_2)(q_1/q_2)]\}^{1/3}}\left(\frac{T}{E_T}\right)^{1/3}$$

$$a_{\text{opt}} = \frac{K_p^{1/6}}{[2-(b/a)]^{2/3}\{2[1+(A_1/A_2)(q_1/q_2)]\}^{1/3}}\left(\frac{E_T T^2}{\tau_y^3}\right)^{1/6}$$

and that the shape factor is, correspondingly,

$$f = \frac{6 + t_w/t_a}{(2^4 K_{2a-b})^{1/6}[2-(b/a)]^{1/3}[1+(A_1/A_2)(q_1/q_2)]^{2/3}}$$

where

$$\frac{A_1}{A_2} = \frac{(b/a)}{2-(b/a)} \quad \text{and} \quad \frac{q_1}{q_2} = \frac{2+(b/a)[5-2(b/a)](t_w/t_a)}{2+[1+2(b/a)][2-(b/a)](t_w/t_a)}$$

4.8 Show that when the two walls $a \times t_a$ and $a \times t_w$ in Figure 4.31 are allowed to buckle simultaneously with the wall $(2a-b) \times t_a$ the web/wall thickness ratio becomes

$$\left(\frac{t_w}{t_a}\right)^3 = \frac{q_1 - q_2}{q_2}\left(\frac{K_{2a-b}}{K_w}\right)\left(\frac{a}{2a-b}\right)^2$$

Hence, set up the system of equations that will allow the tube weight to be minimised.

4.9 Combine equations (4.128a,b) with (4.129) to show how the shape factor f for a double triangular cell (see Figure 4.28) depends only upon $x = a/c$ and $y = b/c$. Show how a stationary value of x may be found corresponding to minimum weight W, when the ratio y is fixed from the following system of equations:

$$\begin{aligned}\frac{\partial f}{\partial x} = 0 = (4x-1)^{1/3}\left[1+\left(\frac{A_2}{A_1}\right)\left(\frac{q_1}{q_2}\right)\right]^{2/3}\left[x^{1/3}\frac{\partial}{\partial x}\{\ \ \} + \frac{1}{3}\{\ \ \}x^{-2/3}\right] \\ -x^{1/3}\{\ \ \}\left\{(4x^2-1)^{1/3}\frac{\partial}{\partial x}\left[1+\left(\frac{A_2}{A_1}\right)\left(\frac{q_2}{q_1}\right)\right]^{2/3}\right. \\ \left. +\frac{8x}{3}(4x^2-1)^{-2/3}\left[1+\left(\frac{A_2}{A_1}\right)\left(\frac{q_2}{q_1}\right)\right]^{2/3}\right\}\end{aligned}$$

where

$$\begin{aligned}\frac{\partial}{\partial x}\{\ \ \} &= \frac{\partial}{\partial x}\left\{2x+\left(\frac{t_c}{t_a}\right)\left[1+2y\left(\frac{t_b}{t_c}\right)\right]\right\} \\ &= 2\left[1+y\left(\frac{t_c}{t_a}\right)\frac{\partial}{\partial x}\left(\frac{t_b}{t_c}\right)\right]+\left[1+2y\left(\frac{t_b}{t_c}\right)\right]\frac{\partial}{\partial x}\left(\frac{t_c}{t_a}\right)\end{aligned}$$

in which the following partial derivatives apply:

$$\frac{\partial}{\partial x}\left(\frac{t_b}{t_c}\right) = -\frac{1}{3}\left[\frac{K_b}{K_c}\left(\frac{q_1}{q_2}-1\right)\left(\frac{1}{y^2}\right)\right]^{-4/3}\frac{K_b}{K_c}\left(\frac{1}{y^2}\right)\frac{\partial}{\partial x}\left(\frac{q_1}{q_2}\right)$$

$$\frac{\partial}{\partial x}\left(\frac{t_c}{t_a}\right) = -\frac{1}{3}\left[\frac{K_a}{K_c}\left(1-\frac{q_2}{q_1}\right)\left(\frac{1}{x^2}\right)\right]^{-2/3}\frac{K_a}{K_c}\left(\frac{1}{x^2}\right)\left[\frac{\partial}{\partial x}\left(\frac{q_2}{q_1}\right)+\frac{2}{x}\left(1-\frac{q_2}{q_1}\right)\right]$$

$$\begin{aligned}\frac{\partial}{\partial x}\left(\frac{q_2}{q_1}\right) = \left\{\left[1+\frac{A_2}{A_1}+2y\left(\frac{t_c}{t_b}\right)\right]\left\{\frac{A_2}{A_1}\left[\frac{\partial}{\partial x}\left(\frac{A_1}{A_2}\right)+2x\frac{\partial}{\partial x}\left(\frac{t_c}{t_a}\right)+2\left(\frac{t_c}{t_a}\right)\right]\right.\right. \\ \left. +\left[1+\frac{A_1}{A_2}+2x\left(\frac{t_c}{t_a}\right)\right]\frac{\partial}{\partial x}\left(\frac{A_2}{A_1}\right)\right\} - \frac{A_2}{A_1}\left[1+\frac{A_1}{A_2}+2x\left(\frac{t_c}{t_a}\right)\right]\left[\frac{\partial}{\partial x}\left(\frac{A_2}{A_1}\right)\right. \\ \left.\left. +2y\frac{\partial}{\partial x}\left(\frac{t_c}{t_b}\right)\right]\right\} \div \left\{\left[1+\frac{A_2}{A_1}+2y\left(\frac{t_c}{t_b}\right)\right]^2\right\}\end{aligned}$$

$$\frac{\partial}{\partial x}\left(\frac{q_1}{q_2}\right) = \frac{\partial}{\partial x}\left(\frac{q_2}{q_1}\right)^{-1} = -\frac{\partial}{\partial x}\left(\frac{q_2}{q_1}\right)\bigg/\left(\frac{q_2}{q_1}\right)^2$$

$$\frac{\partial}{\partial x}\left(\frac{A_2}{A_1}\right) = \frac{\partial}{\partial x}\left(\frac{4y^2-1}{4x^2-1}\right)^{1/2} = -4x\left(\frac{4y^2-1}{4x^2-1}\right)^{-1/2}\frac{4y^2-1}{(4x^2-1)^2}$$

5

Shear of Solid Bars, Tubes and Thin Sections

5.1 Introduction

Solid bars of standard section, thin-walled open and closed sections are often required to support transverse shear loads. In shorter length beams shear becomes the dominant load as in gear teeth, rivetted and bolted joints, for example. When transverse forces are applied to a long beam its transverse sections will also carry accompanying bending moments. Moreover, unless the load passes through a particular point in the cross-section, torsion will be introduced. In solid sections, with y-axis symmetry, this point is the centroid. Without this symmetry, the transverse force should pass through the shear centre if torsion is to be avoided. In the previous chapter on torsion the torque was required to be applied about an axis passing through the centre of twist if bending was to be avoided. While the centre of twist and the shear centre are coincident, they are not necessarily coincident with the centroid. These positions become important to locate in thin-walled sections where the centroid is displaced from the shear centre. The combination of torsion and shear that arises in sections having this displacement is considered in the following chapter. The minimum weight analysis for bars and tubes given here will follow a similar line to that for torsion bars initially, i.e. with the derivation of a governing equation containing the structural index, the material and shape efficiencies. However, it will be seen in later chapters that the webs of thin-walled open sections carry the greater part of the shear force and the flanges carry the bending moments. Consequently, an optimisation of the weight of these structures is simplified where, for example, it may only be necessary to optimise the weight of the shear web. This is the approach adopted for idealised structures with webbed-boom connections and webs stiffened with corrugations of various shapes in Chapters 7 and 8.

To find the maximum shear stress τ_{max} within the cross-section of a solid bar or beam under a transverse shear force F, we should realise firstly that the manner in which the stress is distributed through the depth is expressed in either one of the equivalent formulae [1]

$$\tau = \frac{F}{Ix_0}\int_{y_0}^{y_1} yx\,\mathrm{d}y = \frac{FA\overline{y}}{Ix_0} \qquad (5.1\mathrm{a,b})$$

Mechanics of Optimal Structural Design: Minimum Weight Structures D. W. A. Rees

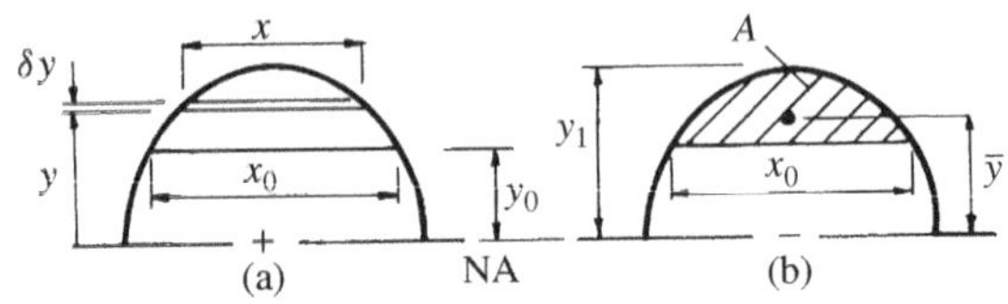

Figure 5.1 Section area under transverse shear

where x_0 and y_0 locate the position of a horizontal fibre in which the shear stress is required, i.e. x_0 is the width and y_0 is its distance from the neutral axis (NA). The second moment of area I refers to the whole section. With symmetry in the section area above and below the neutral axis, equations (5.1a,b) may be referred to its half-sections (see Figure 5.1a). In equation (5.1a) the upper limit of integration y_1 refers to the top of the section where $\tau = 0$. It follows that τ_{max} will lie at the neutral axis. In equation (5.1b) the integral has been identified with the first moment of the area A enclosed between y_0 and y_1 (see Figure 5.1b). This is often a more convenient form to employ where $\bar{y}$ is a recognisable distance from the neutral axis to the centroid of the area A. Where $\bar{y}$ is not known, it is preferable to employ the integral in equation (5.1a) along with two further integrals for finding I and the position $\overline{Y}$ of the neutral axis:

$$I = \int_A y^2 \, \mathrm{d}A, \qquad A\overline{Y} = \int_A y \, \mathrm{d}A \tag{5.2a,b}$$

Equation (5.2b) applies to a non-symmetrical section where the neutral axis does not bisect the depth. In this case $\overline{Y}$ locates the neutral axis from any parallel axis chosen as a datum.

Weight optimisations given for the solid sections that follow will illustrate the use of formulae (5.1a,b) and (5.2a,b).

5.2 Bars of Solid Section

Circular, square and rectangular sections appear in many components that bear shear force, including the pins in hinges and chains, bolts between tensioned plates and key-ways within transmission drives. Elementary texts take the shear stress to be shear force divided by the supporting area. The latter is the section area for single shear but twice the section area for a more efficient double shear joint. This calculation gives only an average shear stress, assumed to be uniform in the depth. It ignores the shear stress distribution and underestimates the maximum stress that equations (5.1a,b) provide. The latter become important to providing more accurate stress estimates where weight restriction is crucial to design.

5.2.1 Rectangular and Square Sections

The rectangle of breadth b and depth d in Figure 5.2a shows the symbols appropriate to equation (5.1a). Applying equation (5.1a) with $x_0 = b$ and $y_1 = d/2$ gives

$$\tau = \frac{F}{Ib} \int_y^{d/2} yb \, \mathrm{d}y = \frac{F}{2I} \left| y^2 \right|_y^{d/2} = \frac{F}{2I}\left(\frac{d^2}{4} - y^2\right) \tag{5.3a}$$

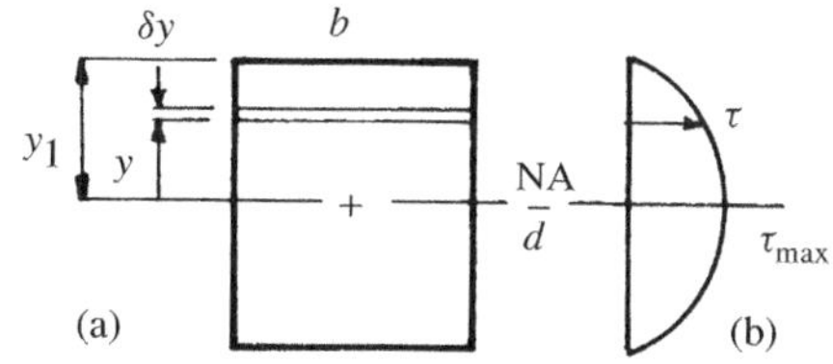

Figure 5.2 Rectangular section

The shear stress is distributed through the depth as shown in Figure 5.2b. With $I = bd^3/12$, equation (5.3a) gives the maximum shear stress at the neutral axis, where $y = 0$, as

$$\tau_{max} = \frac{Fd^2}{8I} = \frac{Fd^2}{8} \times \frac{12}{bd^3} = \frac{3F}{2bd} \tag{5.3b}$$

We may check this result directly from equation (5.1b), where $A = bd/2$ and $\overline{y} = d/4$:

$$\tau_{max} = \frac{FA\overline{y}}{Ib} = \frac{F}{b} \times \frac{bd}{2} \times \frac{d}{4} \times \frac{12}{bd^3} = \frac{3F}{2bd}$$

When τ_{max} is allowed to reach the limiting stress τ_y, the product bd is optimised as

$$(bd)_{opt} = \frac{3F}{2\tau_y} \tag{5.3c}$$

and the minimum weight becomes

$$W = (bd)_{opt}L\rho = \left(\frac{3F}{2\tau_y}\right)L\rho \tag{5.4a}$$

Dividing equation (5.4a) by L^3 reveals the simplified material efficiency factor and structural index within an objective function for shear in a rectangular section:

$$\frac{W}{L^3} = \frac{3}{2}\left(\frac{\rho}{\tau_y}\right)\left(\frac{F}{L^2}\right) \tag{5.4b}$$

For a *square* cross-section we set $b = d = a$ in equation (5.3c), to optimise the side length a:

$$a_{opt} = \left(\frac{3F}{2\tau_y}\right)^{1/2} \tag{5.4c}$$

The high numerical shape factor 3/2 in equation (5.4b) for both the rectangle and square sections indicates that these cannot be weight efficient when supporting shear force. The parabolic shear stress distribution that equation (5.3a) describes (see Figure 5.2b) shows that much material on both sides of the neutral axis is not stressed to its limit. Where $b \ll d$ in a thin rectangular strip, the optimisation requires an additional buckling restriction under $\tau_{max} = \tau_y$:

$$\tau_y = \frac{3F}{2bd} = KE\left(\frac{b}{d}\right)^2 \tag{5.5a}$$

Equation (5.5a) gives the optimum breadth and depth separately as

$$b = \left(\frac{9}{4K}\right)^{1/4}\left(\frac{F^2}{\tau_y E}\right)^{1/4}, \quad d = \left(\frac{9K}{4}\right)^{1/4}\left(\frac{F^2 E}{\tau_y^3}\right)^{1/4} \tag{5.5b,c}$$

A similar, limiting elastic buckling consideration must be made where a thin rectangle describes a shear web or a flange; see, for example, § 5.3.1.

5.2.2 *Oriented Square*

In Figure 5.3a the square section has been rotated 45° beneath the vertical force. As the load bearing material is presented to the force more effectively we should expect a lower shape factor. This orientation ensures that more highly stressed material lies on either side of the neutral axis and any unnecessary material has been removed from the lowly stressed regions at the top and bottom. Applying equation (5.1a) to Figure 5.3a, the maximum shear stress is found:

$$\tau_{\max} = \frac{F}{I(\sqrt{2}a)}\int_0^{a/\sqrt{2}} 2y\left(\frac{a}{\sqrt{2}} - y\right)\,\mathrm{d}y = \frac{\sqrt{2}F}{Ia}\left|\frac{ay^2}{2\sqrt{2}} - \frac{y^3}{3}\right|_0^{a/\sqrt{2}} = \frac{Fa^2}{12I} \tag{5.6a}$$

where, from equation (5.2a),

$$I = 4\int_0^{a/\sqrt{2}} y^2\left(\frac{a}{\sqrt{2}} - y\right)\,\mathrm{d}y = 4\left|\frac{ay^3}{3\sqrt{2}} - \frac{y^4}{4}\right|_0^{a/\sqrt{2}} = \frac{a^4}{12} \tag{5.6b}$$

and/ from equations (5.6a,b),

$$\tau_{\max} = \frac{Fa^2}{12} \times \frac{12}{a^4} = \frac{F}{a^2} \tag{5.6c}$$

Again, restricting $\tau_{\max}$ in equation (5.6c) to the limiting shear stress τ_y, the optimum side length becomes

$$a_{\mathrm{opt}} = \left(\frac{F}{\tau_y}\right)^{1/2} \tag{5.7}$$

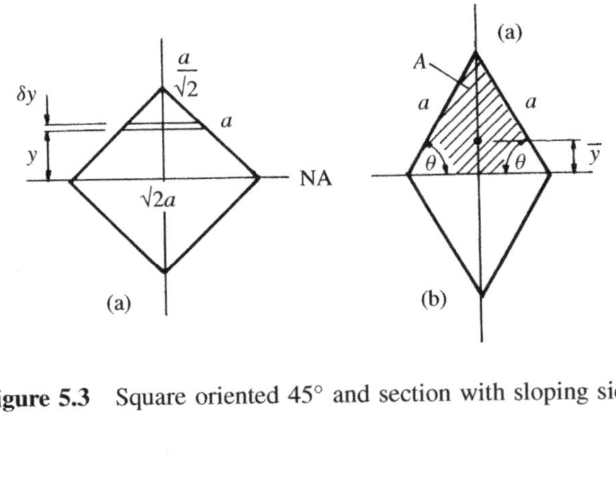

Figure 5.3 Square oriented 45° and section with sloping sides

Comparing equations (5.4c) and (5.7), we see that the side length for a square with 45° orientation may be reduced by a factor of $\surd(3/2)$ for a similar τ_y. The weight $W = a_{opt}^2 L\rho$ provides the objective function

$$\frac{W}{L^3} = \left(\frac{\rho}{\tau_y}\right)\left(\frac{F}{L^2}\right) \tag{5.8}$$

and, with equation (5.8) having a shape factor of unity, the weight saving over a 'right' square is 50%. If the gradient of the sloping sides changes from 45° to θ (see Figure 5.3b), the optimum side length is altered to

$$a_{opt} = \left(\frac{F}{\tau_y \sin 2\theta}\right)^{1/2}$$

but the objective function (5.8) will remain unchanged. Hence, the shape function $f = 1$ is independent of θ for straight sides. This begs the question of whether a symmetrical section having concave curvature within its four sides can provide $f < 1$. This is left for the reader to explore (see Exercise 5.1).

5.2.3 *Equilateral Triangle*

Firstly, the centroid position, through which the neutral axis (NA) passes, is located at $\overline{Y} = a/(2\surd3)$ from the base length a as a datum (see Figure 5.4a). The section width at the centroid position is $x_0 = 2a/3$, from which equation (5.1b) gives

$$\tau_{max} = \frac{3FA\overline{y}}{2aI} = \frac{3F}{2aI} \times \left(\frac{1}{2} \times \frac{2a}{3} \times \frac{a}{\surd3}\right) \times \frac{a}{3\surd3} = \frac{Fa^3}{18I} \tag{5.9a}$$

where, from equation (5.2a),

$$I = \frac{2}{\surd3}\int_{-a/2\surd3}^{a/\surd3} y^2\left(\frac{a}{\surd3} - y\right)\,dy = \frac{2}{\surd3}\left|\frac{ay^3}{3\surd3} - \frac{y^4}{4}\right|_{-a/2\surd3}^{a/\surd3} = \frac{\surd3 a^4}{96} \tag{5.9b}$$

Combining equations (5.9a,b),

$$\tau_{max} = \frac{Fa^2}{18} \times \frac{96}{\surd3 a^4} = \frac{16F}{3\surd3 a^2} \tag{5.9c}$$

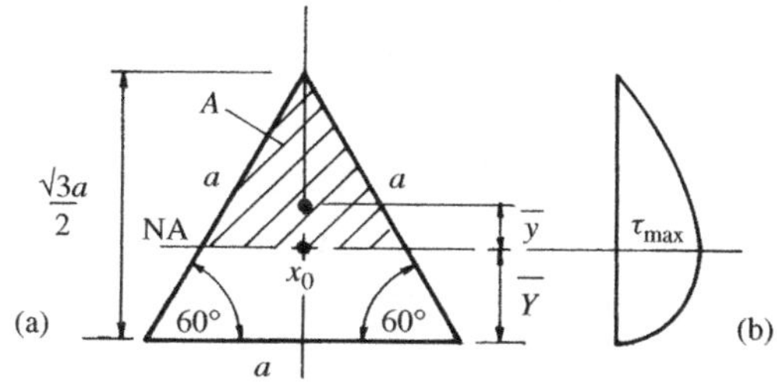

Figure 5.4 Equilateral triangular section

At the maximum shear stress limit τ_y, the optimum side length becomes

$$a_{\mathrm{opt}} = \left(\frac{16F}{3\sqrt{3}\tau_y} \right)^{1/2} \tag{5.10}$$

The objective function follows from equation (5.10) and the weight expression $W = \sqrt{3}a_{\mathrm{opt}}^2 L\rho/4$:

$$\frac{W}{L^3} = \frac{4}{3}\left(\frac{\rho}{\tau_y}\right)\left(\frac{F}{L^2}\right) \tag{5.11}$$

The higher shape factor $f = 4/3$ for this section, compared to the oriented square (Figure 5.3a), is a consequence of an asymmetric stress distribution through the depth (see Figure 5.4b). We show next that the gradient of the section's sloping sides has no influence upon f.

5.2.4 Isosceles Triangle

Let the base length be b and the height be a (see Figure 5.5). The neutral axis (NA) passes horizontally through the centroid, whose position from the base is $\overline{Y} = a/3$. The section width at this position is $x_0 = 2b/3$, from which equation (5.1b) gives

$$\tau_{\max} = \frac{3FA\overline{y}}{2bI} = \frac{3F}{2bI} \times \left(\frac{1}{2} \times \frac{2b}{3} \times \frac{2a}{3}\right) \times \left(\frac{2a}{3 \times 3}\right) = \frac{2Fa^2}{27I} \tag{5.12a}$$

where, from equation (5.2a),

$$I = 2 \times \frac{b}{2a} \int_{-a/3}^{2a/3} y^2 \left(\frac{2a}{3} - y\right) \mathrm{d}y = \frac{b}{a} \left| \frac{2a}{3} \times \frac{y^3}{3} - \frac{y^4}{4} \right|_{-a/3}^{2a/3} = \frac{a^3 b}{36} \tag{5.12b}$$

Combining equations (5.12a,b),

$$\tau_{\max} = \frac{2Fa^2}{27} \times \frac{36}{a^3 b} = \frac{8F}{3ab} \tag{5.12c}$$

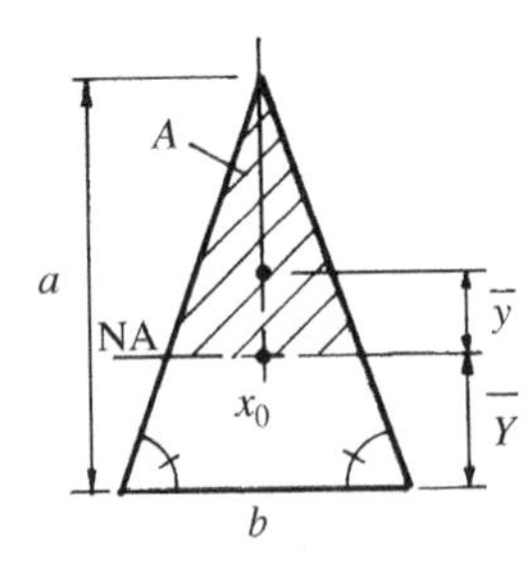

Figure 5.5 Isosceles triangular section

and setting $\tau_{max} = \tau_y$, equation (5.12c) optimises the product ab:

$$(ab)_{opt} = \frac{8F}{3\tau_y} \tag{5.13}$$

With its weight $W = \frac{1}{2}(ab)_{opt}L\rho$, equation (5.13) provides the objective function for this section:

$$\frac{W}{L^3} = \frac{4}{3}\left(\frac{\rho}{\tau_y}\right)\left(\frac{F}{L^2}\right) \tag{5.14}$$

Equation (5.14) is identical to equation (5.11). Hence, there is no weight saving in changing the section from an equilateral to an isosceles triangle under their optimum conditions.

5.2.5 *Circular Section*

Equation (5.1b) requires that $A\overline{y}$ be applied to the area above the neutral axis (see Figure 5.6). Using standard formulae for the centroid position of a semi-circle from its base $\overline{y} = 2d/(3\pi)$ and the second moment of area of a circle about the neutral axis (the horizontal diameter) $I = \pi d^4/64$, equation (5.1b) gives the maximum shear stress as

$$\tau_{max} = \frac{F}{Id} \times \frac{\pi d^2}{8} \times \frac{2d}{3\pi} = \frac{F}{d} \times \frac{64}{\pi d^4} \times \frac{\pi d^2}{8} \times \frac{2d}{3\pi} = \frac{16F}{3\pi d^2} \tag{5.15a}$$

When $\tau_{max} = \tau_y$ in equation (5.15a), the optimum diameter may be set at

$$d_{opt} = \left(\frac{16F}{3\pi\tau_y}\right)^{1/2} \tag{5.15b}$$

The objective function follows from equation (5.15b) and the weight expression $W = \frac{1}{4}\pi d_{opt}^2 L\rho$:

$$\frac{W}{L^3} = \frac{4}{3}\left(\frac{\rho}{\tau_y}\right)\left(\frac{F}{L^2}\right) \tag{5.16}$$

Despite its large shape factor, the circular section is common in supporting transverse shear within pins, rivets etc. Comparing equation (5.16) with equations (5.4b) and (5.11), the circle has a shape factor equal to that of a triangle and less than that of a 'right' square.

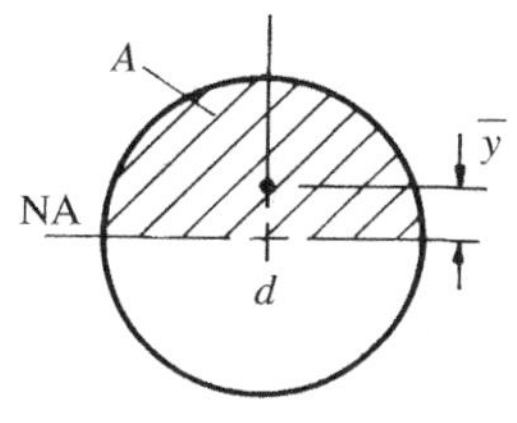

Figure 5.6 Circular Section

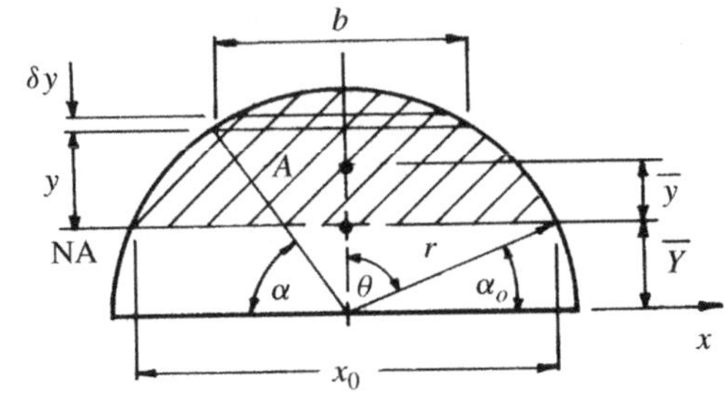

Figure 5.7 Semi-circular section

5.2.6 *Semi-Circular Section*

Circular sections are often machined with flats for location and transmission of shear and torsion. Their optimum design requires that the maximum shear stress be known. In a semicircle, for example (see Figure 5.7), the shaded area A above the neutral axis (NA) position is required, given that its centroid position is $\overline{Y} = 4r/3\pi$.

Firstly, the angle θ is found, giving the semi-base length $\frac{1}{2}x_0$ for that area as

$$\theta = \cos^{-1}\left(\frac{4}{3\pi}\right) = 64.89^\circ = 1.1325 \text{ rad}, \quad \tfrac{1}{2}x_0 = r\sin\theta = 0.9055r \tag{5.17a}$$

Then A follows by subtracting the area of the triangle from that of the segment:

$$A = r^2\theta - \frac{x_0}{2}\frac{4r}{3\pi} = (1.1325 - 0.3843)r^2 = 0.7482r^2 \tag{5.17b}$$

The position $\overline{y}$ of the centroid within this area is found from equation (5.2b) to be

$$A\overline{y} = \int_A by\,\mathrm{d}x = \int_A \sqrt{r^2 - (4r/3\pi + y)^2}\,\mathrm{d}y \tag{5.18a}$$

A conversion to polar coordinates is required for the integration. Using

$$y = r\sin\alpha - \frac{4r}{3\pi} \tag{5.18b}$$

equation (5.18a) becomes

$$A\overline{y} = 2r^3\int_{a_0}^{\pi/2}(\sin\alpha - 4/3\pi)\cos^2\alpha\,\mathrm{d}\alpha \tag{5.18c}$$

Setting $\alpha_0 = \pi/2 - \theta$, the integration gives

$$A\overline{y} = 2r^3\left\{-\frac{1}{3}\left|\cos^3\alpha\right|_{4/3\pi}^{\pi/2} - \frac{2}{3\pi}\left|\alpha + \frac{1}{2}\sin 2\alpha\right|_{4/3\pi}^{\pi/2}\right\} = 0.1774r^3 \tag{5.18d}$$

Now I for the NA is found from the parallel axis theorem:

$$I = I_X - A\overline{Y}^2 = \frac{\pi r^4}{8} - \frac{\pi r^2}{2}\left(\frac{4r}{3\pi}\right)^2 = 0.1098r^4 \tag{5.18e}$$

Finally, substituting x_0, $A\overline{y}$ and I, from equations (5.17a) and (5.18d,e) into equation (5.1b) gives the maximum shear stress:

$$\tau_{\max} = \frac{F \times 0.1774r^3}{2(0.9055r) \times 0.1098r^4} = \frac{0.8921F}{r^2} \tag{5.19a}$$

When $\tau_{\max} = \tau_y$, the optimum radius may be set at

$$r_{\text{opt}} = \left(\frac{0.8921F}{\tau_y}\right)^{1/2} \tag{5.19b}$$

The objective function follows from equation (5.19b) and the weight expression $W = \frac{1}{2}\pi r_{\text{opt}}^2 L\rho$:

$$\frac{W}{L^3} = 1.402\left(\frac{\rho}{\tau_y}\right)\left(\frac{F}{L^2}\right) \tag{5.20}$$

This shows that the shape factor has been raised slightly compared to $f = 4/3$ for a solid circular section (see equation (5.16)).

5.2.7 *Ranking*

Given that the material factors and structural indices for solid sections under transverse shear are similar, we may compare weights from their shape factors alone. Interestingly, Table 5.1 shows that the best that can be obtained from standard extrusions appears to be $f = 1$, simply from rotating the worst performer, the right square, through 45°!

Table 5.1 Shape factors for extruded sections under shear force

Section	◆	▲	●	(semicircle)	■
f	1	4/3	4/3	1.402	3/2

5.3 Thin-Walled Open Sections

Here we shall examine common, thin-walled standard sections having both single and double axial symmetry. We saw in the previous chapter that if we are to avoid the effects of bending in these sections the shear force must act through the shear centre. The latter lie at the position E for the sections shown in Figure 5.8.

Referring to Figure 5.8a, in an I-section the position of the shear centre coincides with the centroid. In L- and T-sections (see Figure 5.8b,c) the shear centre E lies at the intersection between their limbs, i.e. the point where the mid centre-lines of the flange and

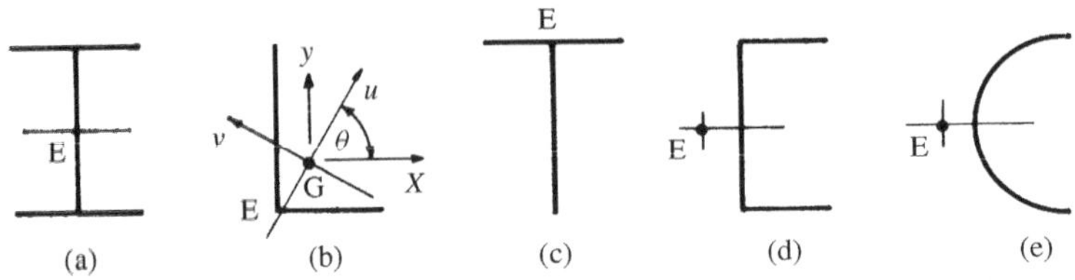

Figure 5.8 Shear centres for thin-walled open sections

web meet. In other sections, such as those in Figure 5.8d,e, the position of the shear centre must be found from an equilibrium equation which eliminates the torque that would arise if the force was not applied through E. Moreover, the asymmetric section (Figure 5.8b) requires that the shear force be resolved along its principal axes u and v, inclined at θ, as shown. Fortunately, this is simplified with the use of a formula to express equivalent shear forces in terms of the three second moments of area I_x, I_y and I_{xy}, whose origin for the reference axes x and y lies at the centroid G.

5.3.1 Symmetrical I-Section, Uniform Thickness

With the dimensions of the I-section shown in Figure 5.9a, we shall apply equation (5.1b) to give the maximum shear stress under a single vertical force F passing through E.

The first moment of the area above the neural axis (NA) follows directly as

$$A\overline{y} = bt\frac{d}{2} + \frac{td}{2}\frac{d}{4} = \frac{d^2t}{8}\left(1 + 4\frac{b}{d}\right) \tag{5.21a}$$

The second moment of area about the x-axis (i.e. the neutral axis) follows from summing the contributions from the three rectangles:

$$I_x = \frac{td^3}{12} + 2\left(\frac{bt^3}{12} + \frac{btd^2}{4}\right) \approx \frac{d^3t}{12}\left(1 + 6\frac{b}{d}\right) \tag{5.21b}$$

Substituting equations (5.21a,b) into equation (5.1b) gives the maximum vertical shear stress in the web at the neutral axis (see Figure 5.9b):

$$\tau_{w,\max} = \frac{F(A\overline{y})}{tI_x} = \frac{12Fd^2t[1 + 4(b/d)]}{8td^3t[1 + 6(b/d)]} = \frac{3F[1 + 4(b/d)]}{2dt[1 + 6(b/d)]} \tag{5.22a}$$

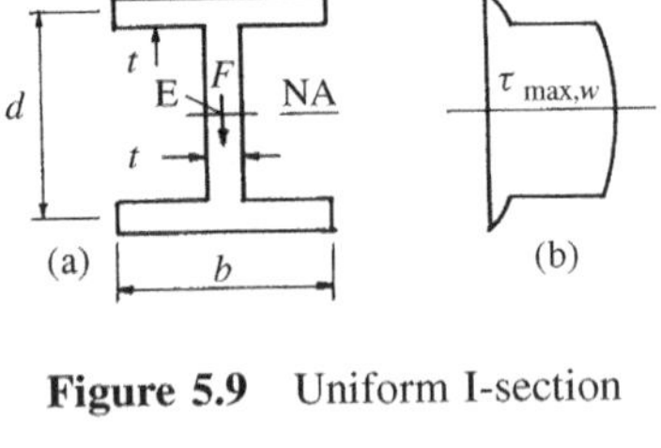

Figure 5.9 Uniform I-section

A horizontal shear stress also exists in the flange rising from zero at the free end to a maximum at its intersection with the web (see Figure 5.9c). Here $A = tb/2$ and $\bar{y} = d/2$ for the flange area. Hence, the maximum flange stress is also given by equation (5.1b) as

$$\tau_{f,\max} = \frac{F(A\bar{y})}{tI_x} = \frac{12F(bt/2)(d/2)}{td^3t[1+6(b/d)]} = \frac{3F(b/d)}{dt[1+6(b/d)]} \quad (5.22b)$$

There is no *b/d* ratio which equalises each maximum in equations (5.22a,b). Thus, the greater shear stress equation (5.22a) is set to the limiting stress τ_y under which buckling in the web also occurs:

$$\tau_y = \frac{3F[1+4(b/d)]}{2dt[1+6(b/d)]} = K_d E_T \left(\frac{t}{d}\right)^2 \quad (5.23a)$$

where K_d is the buckling coefficient for a plate of width d with clamped sides [2]. The optimum dimensions d and t follow from equation (5.23a) as

$$d_{\text{opt}} = \left(\frac{3^2 K_d}{2^2}\right)^{1/4} \frac{[1+4(b/d)]^{1/2}}{[1+6(b/d)]^{1/2}} \left(\frac{E_T F^2}{\tau_y^3}\right)^{1/4} \quad (5.23b)$$

$$t_{\text{opt}} = \left(\frac{3^2}{2^2 K_d}\right)^{1/4} \frac{[1+4(b/d)]^{1/2}}{[1+6(b/d)]^{1/2}} \left(\frac{F^2}{\tau_y E_T}\right)^{1/4} \quad (5.23c)$$

The optimised weight becomes

$$W = d_{\text{opt}} t_{\text{opt}} \left(1 + 2\frac{b}{d}\right) L\rho \quad (5.24a)$$

Substituting equations (5.23b,c) into equation (5.24a) provides the objective function

$$\left(\frac{W}{L^3}\right)_{\text{opt}} = \frac{3[1+2(b/d)][1+4(b/d)]}{2[1+6(b/d)]} \left(\frac{\rho}{\tau_y}\right)\left(\frac{F}{L^2}\right) \quad (5.24b)$$

where, characteristically of shear, neither K_d nor E_T appears directly. Equation (5.24b) shows that the shape factor has no minimum for $b/d > 0$. Hence equation (5.24b) applies to where *b/d* is unrestricted, but with d and t having been optimised from equations (5.23b,c). The lesser restraint that exists in the flange, from having one end unsupported, suggests that buckling in the flange half-length is also a possibility under its maximum shear stress. The limiting stress is proportioned between equations (5.22a,b) to give a second optimising equation:

$$\frac{2(b/d)\tau_y}{1+4(b/d)} = \frac{3F(b/d)]}{dt[1+6(b/d)]} = K_b E_T \left(\frac{2t}{b}\right)^2$$

which limits the optimum dimension b, but not the objective function (5.24b), since the weight product term $d_{\text{opt}} t_{\text{opt}}$ in equation (5.24a) will remain the same. We shall refer to an alternative strategy for the optimising buckling within this I-section as a special case of the following non-uniform section.

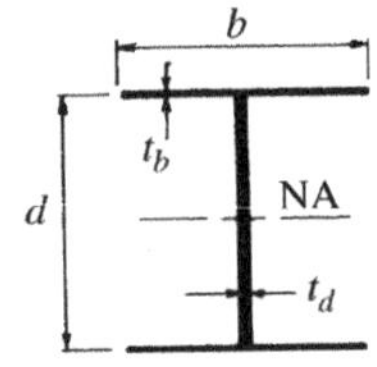

Figure 5.10 Symmetrical I-section with non-uniform thickness

5.3.2 Symmetrical I-Section, Non-Uniform Thickness

The restrictions arising from an I-section with uniform thickness in § 5.3.1 may be partially relaxed when the flange and web thicknesses, t_b and t_d, are allowed to differ (see Figure 5.10).

To find the maximum shear stress, the first moment of the area above the neural axis is required,

$$A\overline{y} = (bt_b)\frac{d}{2} + \frac{t_d d}{2}\frac{d}{4} = \frac{d^2 t_d}{8}\left(1 + 4\frac{b}{d}\frac{t_b}{t_d}\right) \tag{5.25a}$$

along with the second moment of the whole area about the neutral axis,

$$I_x = \frac{t_d d^3}{12} + 2\left(\frac{bt_b^3}{12} + \frac{bt_b d^2}{4}\right) \approx \frac{d^3 t_d}{12}\left(1 + 6\frac{b}{d}\frac{t_b}{t_d}\right) \tag{5.25b}$$

Substituting equations (5.25a,b) into equation (5.1b) gives the maximum vertical shear stress in the web at the neutral axis (NA) position:

$$\tau_{d,\max} = \frac{F(A\overline{y})}{t_d I_x} = \frac{12Fd^2 t_d[1 + 4(b/d)(t_b/t_d)]}{8t_d d^3 t_d[1 + 6(b/d)(t_b/t_d)]} = \frac{3F[1 + 4(b/d)(t_b/t_d)]}{2dt_d[1 + 6(b/d)(t_b/t_d)]} \tag{5.26a}$$

The maximum horizontal shear stress at the flange–web intersection $\tau_{b,\max}$ is found from substituting $A = bt_b/2$ and $\overline{y} = d/2$, into equation (5.1b):

$$\tau_{b,\max} = \frac{FA\overline{y}}{t_b I_x} = \frac{12F(bt_b/2)(d/2)}{t_b d^3 t_d[1 + 6(b/d)(t_b/t_d)]} = \frac{3F(b/d)}{dt_d[1 + 6(b/d)(t_b/t_d)]} \tag{5.26b}$$

Equating (5.26a,b), we may attempt to equalise each maximum shear stress within the relationship

$$\frac{t_b}{t_d} = \frac{2(b/d) - 1}{4(b/d)} \tag{5.27}$$

Equation (5.27) applies to $b/d > 1/2$. Furthermore, if we now impose a simultaneous buckling of the web and flange at their equalised maxima, the condition to be met is

$$\tau_{cr} = K_d E_T\left(\frac{t_d}{d}\right)^2 = K_b E_T\left(\frac{2t_b}{b}\right)^2 \tag{5.28a}$$

According to equations (5.28a) the thickness ratio must satisfy the relation

$$\frac{t_b}{t_d} = \frac{1}{2}\frac{b}{d}\left(\frac{K_d}{K_b}\right)^{1/2} \tag{5.28b}$$

Equating (5.27) and (5.28b), we see from the resulting quadratic equation that the condition for meeting the dual requirements is $K_b/K_d \geq 4$. However, Figure C.3 (Appendix C) shows that a factor of 4 does not exist between the ratio of buckling coefficients for plates with different edge restraints. Hence, the alternative strategies are either (i) buckling of the web only with equalised stress maxima, ($\tau_{b,\max} = \tau_{d,\max}$), where equations (5.26a,b) apply, or (ii) simultaneous buckling of the web and flange under different $\tau_{b,\max}$ and $\tau_{d,\max}$, in which case equations (5.26a,b) are applied within the range of Figure C.3. Setting a limit to the greatest shear stress in the web at τ_y, condition (i) is met when, from equations (5.26a) and (5.28a),

$$\tau_y = \frac{3F[1 + 4(b/d)(t_b/t_d)]}{2dt_d[1 + 6(b/d)(t_b/t_d)]} = K_d E_T \left(\frac{t_d}{d}\right)^2 \tag{5.29}$$

The corresponding optimum dimensions are found from equation (5.29):

$$d_{\text{opt}} = \left(\frac{3^2 K_d}{2^2}\right)^{1/4} \frac{[1 + 4(b/d)(t_b/t_d)]^{1/2}}{[1 + 6(b/d)(t_b/t_d)]^{1/2}} \left(\frac{E_T F^2}{\tau_y^3}\right)^{1/4} \tag{5.30a}$$

$$(t_d)_{\text{opt}} = \left(\frac{3^2}{2^2 K_d}\right)^{1/4} \frac{[1 + 4(b/d)(t_b/t_d)]^{1/2}}{[1 + 6(b/d)(t_b/t_d)]^{1/2}} \left(\frac{F^2}{\tau_y E_T}\right)^{1/4} \tag{5.30b}$$

The optimum weight follows from

$$W = d_{\text{opt}}(t_d)_{\text{opt}}[1 + 2(b/d)]L\rho \tag{5.31a}$$

Substituting equations (5.30a,b) into equation (5.31a) provides the objective function

$$\left(\frac{W}{L^3}\right)_{\text{opt}} = \frac{3[1 + 2(b/d)(t_b/t_d)][1 + 4(b/d)(t_b/t_d)]}{2[1 + 6(b/d)(t_b/t_d)]} \left(\frac{\rho}{\tau_y}\right)\left(\frac{F}{L^2}\right) \tag{5.31b}$$

Great care is needed with this strategy. Suppose we selected $b/d = 1$; then equation (5.27) sets the thickness ratio at $t_b/t_d = 1/4$. However, with $K_b < K_d$ (see Figure C.3) it is likely that the flange would have buckled first. Indeed, this would appear to apply for all *b/d*. Consequently, the second strategy is a more appropriate one. Here the optimised dimension equations (5.30a,b) and the objective function (5.31b) remain the same but another relationship between the breadth and thickness ratios is found by equating stress ratios between equations (5.26a,b) and (5.28a):

$$\left(\frac{\tau_w}{\tau_f}\right)_{\max} = \left(\frac{\tau_w}{\tau_f}\right)_{\text{cr}} \Rightarrow \frac{2(b/d)^3}{1 + 4(b/d)(t_b/t_d)} = 4\frac{K_b}{K_d}\left(\frac{t_b}{t_d}\right)^2 \tag{5.32a,b}$$

Equation (5.32b) leads to a cubic in b/d:

$$\left(\frac{b}{d}\right)^3 - 2\frac{K_b}{K_d}\left(\frac{t_b}{t_d}\right)^2 - 8\frac{K_b}{K_d}\frac{b}{d}\left(\frac{t_b}{t_d}\right)^3 = 0 \quad (5.33a)$$

For example, if we take $b/d = 2$, then the effective buckling lengths for the flange and web are the same. Where b and d are small compared to L, we can read $K_b/K_d = 0.784$ from Figure C.3 for $b/L \to 0$. Equation (5.33a) then gives

$$8 - 1.568(t_b/t_d)^2 - 12.544(t_b/t_d)^3 = 0$$

for which the solution is $t_b/t_d = 0.82$. When the thickness is uniform, as in § 5.3.1, equation (5.33a) reduces to

$$\left(\frac{b}{d}\right)^3 - 2\frac{K_b}{K_d} - 8\frac{K_b}{K_d}\frac{b}{d} = 0 \quad (5.33b)$$

Here it becomes necessary to find b/d by trial using Figure C.3 to satisfy equation (5.33b). The buckling coefficients are read from Figure C.3 for a plate with one long edge fixed (K_b for the flange half-width) and another with both long edges fixed (K_d for the web). Here the abscissa b/a in Figure C.3 is identified, in turn, with $b/2L$ and d/L, respectively. Making similar assumptions, we achieve this when Figure C.3 gives $K_b/K_d = 0.79$ for $b/d = 2.63$. This ratio results in a high shape factor $f_1 = 6.45$ which may not be practical, but if long flanges and uniform thickness are to be a design feature this ratio is the optimum choice. The optimum dimensions follow from equations (5.30a,b) as

$$d_{opt} = 1.719\left(\frac{E_T F^2}{\tau_y^3}\right)^{1/4}, \quad t_{opt} = 0.599\left(\frac{F^2}{\tau_y E_T}\right)^{1/4} \quad (5.34a,b)$$

Allowing the section to have non-uniform thickness provides more control over its geometry. In spite of this, there is not a positive product $(b/d)(t_b/t_d)$ which minimises the shape factor in equation (5.31b). However, when equations (5.28b) and (5.31b) are combined we might seek, numerically, a b/d ratio to minimise this shape factor. Table 5.2 applies to further values of b/d taken within these equations. The shape factors f_1 and f_2 apply to the uniform and non-uniform sections, respectively. A graphical presentation between them is given in Figure 5.11. Whilst there is no minimum in either shape factor for $b/d > 0$, Table 5.2 shows little increase in f_2 to $b/d = 0.5$, compared to f_1 for the uniform section. This useful result lies within the b/d range of standard I-sections, extruded with uniform thickness [3]. An optimum geometry, giving minimum weight for

Table 5.2 Relationship between b/d, t_b/t_d, f_1 and f_2

b/d	0.1	0.25	0.50	1.0	1.5	2.0	2.63
t_b/t_d	0.026	0.095	0.235	0.472	0.66	0.82	1.0
f_2	1.50	1.505	1.597	2.198	3.195	4.477	6.45
f_1	1.575	1.80	2.25	3.21	4.20	5.19	6.45

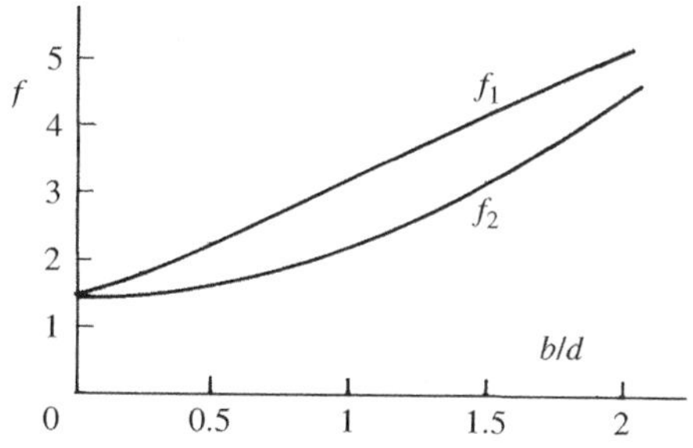

Figure 5.11 Shape factors for uniform and non-uniform I-sections

bearing shear, would require that the flange thickness be reduced to give the corresponding t_b/t_d ratio shown in Table 5.2. A skimming of the flange outer surfaces would suffice, but this should not be done when the section also carries bending moments.

Both forces and moments arise in long beams when they carry transverse loading. Here the shear force is the derivative of a bending moment. Consequently, shear force and bending moment accompany one another and, to accommodate both, the flange is often made thicker than the web. In other, shorter length beams, considered here, shear will govern the design and hence the web thickness should be made greater than the flange thickness.

5.3.3 *Equal-Angle, Uniform Thickness*

Let shear forces F_x and F_y be applied at the shear centre E of the right-angled section with sides of equal length a and thickness t, as shown in Figure 5.12. In this asymmetric section the 'shear flow' $q = \tau t$ in each limb is given by [1]

$$q = \frac{\overline{F}_y D_x}{I_x} + \frac{\overline{F}_x D_y}{I_y} \tag{5.35}$$

in which D_x and D_y are the first moments of the elemental area $t\delta s$ about the centroidal axes x and y in the positions shown:

$$D_x = \int_s yt\,\mathrm{d}s, \qquad D_y = \int_s xt\,\mathrm{d}s \tag{5.36a,b}$$

Equivalent shear forces $\overline{F}_x$ and $\overline{F}_y$ must be used when, as is the case here, x and y are not principal axes. Conventionally, when F_x and F_y are applied in the negative x- and y-directions, as shown, they appear positive within the equivalent force expressions:

$$\overline{F}_x = \frac{F_x - (I_{xy}/I_x)F_y}{1 - I_{xy}^2/(I_x I_y)}, \quad \overline{F}_y = \frac{F_y - (I_{xy}/I_y)F_x}{1 - I_{xy}^2/(I_x I_y)} \tag{5.37a,b}$$

where I_x, I_y and I_{xy} are the three second moment integrals for the area $t \times \delta s$ about the x- and y-axes:

$$I_x = \int_s y^2 t\,\mathrm{d}s, \quad I_y = \int_s x^2 t\,\mathrm{d}s, \quad I_{xy} = \int_s xyt\,\mathrm{d}s \tag{5.38a,b,c}$$

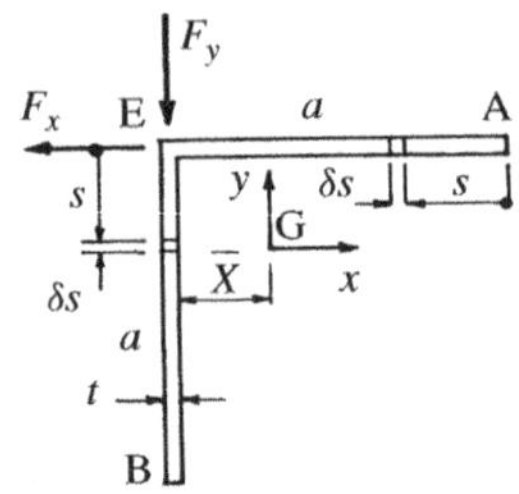

Figure 5.12 Equal-angle with transverse shear forces applied at the shear centre E

Due regard must be paid to the sign of x and y within both first moment expressions, D_x and D_y, in equations (5.36a,b) and to the product moment I_{xy} in equation (5.38c). Equation (5.35) allows reductions in the absence of either $\overline{F}_x$ or $\overline{F}_y$. For axisymmetric sections, i.e. when x and y are principal axes, $I_{xy} = 0$ and $\overline{F}_x$ and $\overline{F}_y$ reduce to F_x and F_y. In fact equation (5.1b), as used previously to find the shear stress in §§ 5.3.1 and 5.3.2, corresponds to the first term in equation (5.35):

$$\tau = \frac{q}{t} = \frac{F_y D_x}{tI_x} \equiv \frac{FA\overline{y}}{tI_x}$$

Returning to our equal-angled section, the position of its centroid $\overline{X}$ is found from taking first moments of its two limb areas about the vertical side:

$$(at)t/2 + at(a/2) = (2at)\overline{X}, \quad \Rightarrow \quad \overline{X}(=\overline{Y}) \approx a/4 \tag{5.39a}$$

in which the approximation holds good when $t \ll a$. With the origin of x and y at the centroid G, the three second moments of area are found from equations (5.38a–c) and (5.39a):

$$I_x = I_y = at^3/12 + at(a/4)^2 + ta^3/12 + at(a/4)^2 = 5a^3t/24 \tag{5.39b,c}$$

$$I_{xy} = at(a/4)(a/4) + at(-a/4)(-a/4) = a^3t/8 \tag{5.39d}$$

Case I

With equal forces acting as shown in Figure 5.12, we put $F_x = F_y = F$. Also, with $I_x = I_y$, the equivalent forces are equal. These forces are, from equations (5.37a,b):

$$\overline{F}_x = \overline{F}_y = F\left[1 - \left(\frac{a^3t}{8}\right) \times \left(\frac{24}{5a^3t}\right)\right] \Big/ \left[1 - \left(\frac{a^3t}{8}\right)^2 \times \left(\frac{24}{5a^3t}\right)^2\right] = \frac{5F}{8} \tag{5.40a}$$

Writing $\overline{F}_x = \overline{F}_y = \overline{F}$ with $I_x = I_y = I$, for simplicity, equation (5.35) becomes

$$q = \frac{\overline{F}}{I}(D_X + D_y) = \frac{\overline{F}}{I}\left(\int ytds + \int xtds\right) + q_0 \tag{5.40b}$$

where q_0 is the shear flow at the starting point of the integration. The multiplication factor in the shear flow expression (5.40b) becomes

$$\frac{\overline{F}}{I} = \frac{5F}{8} \times \frac{24}{5a^3t} = \frac{3F}{a^3t} \tag{5.40c}$$

Both x and y are positive for an origin of s at the free end A of limb AE, i.e. $x = 3a/4 - s$ and $y = a/4$. Also, at A, $q_0 = 0$, when equations (5.40b,c) give

$$q_{AE} = \frac{3F}{a^3t}\left[\int_0^s \frac{at}{4}\mathrm{d}s + \int_0^s \left(\frac{3a}{4} - s\right) t\mathrm{d}s\right] = \frac{3F}{a^3}\left(as - \frac{s^2}{2}\right) \tag{5.41a}$$

Equation (5.41a) describes the parabolic distribution shown in Figure 5.13. The terminal value q_E for $s = a$ follows from equation (5.41a):

$$q_E = \frac{3F}{a^3} \times \frac{a^2}{2} = \frac{3F}{2a} \tag{5.41b}$$

Equation (5.41b) allows the integration to follow the direction s, starting from E within limb EB, where $q_0 = q_E$. With $x = -a/4$ and $y = a/4 - s$, equations (5.40b,c) provide q_{EB}:

$$q_{EB} = \frac{3F}{a^3t}\left[\int_0^s \left(\frac{a}{4} - s\right) t\mathrm{d}s - \int_0^s \left(\frac{a}{4}\right) t\mathrm{d}s\right] + q_E$$

$$= \frac{3F}{a^3}\left(\left|\frac{as}{4} - \frac{s^2}{2}\right|_0^s - \left|\frac{as}{4}\right|_0^s\right) + \frac{3F}{2a} = \frac{3F}{2a^3}(a^2 - s^2) \tag{5.42}$$

Equation (5.42) describes an identical parabolic distribution (see Figure 5.13) with the correct terminal value $q_B = 0$ at the free end ($s = a$). The maximum in each shear flow distribution is $q_E = 3F/2a$, upon which an optimum design will be based. Let us look first at the effect of removing one shear force.

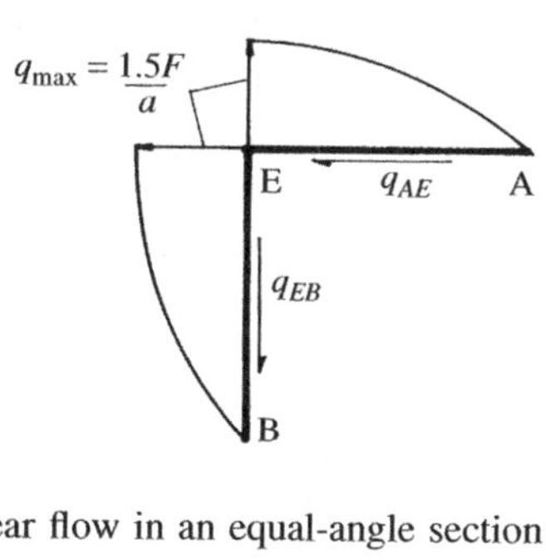

Figure 5.13 Shear flow in an equal-angle section under two forces

Case II

Now put $F_x = 0$ with $F_y = F$. Using the previous expressions (5.39b–d) for I_x, I_y and I_{xy}, the equivalent forces are found from equations (5.37a,b) :

$$\overline{F}_y = F\bigg/\left[1-\left(\frac{a^3t}{8}\right)^2\times\left(\frac{24}{5a^3t}\right)^2\right]=\frac{25F}{16} \quad (5.43a)$$

$$\overline{F}_x = -\left[\left(\frac{a^3t}{8}\right)\times\left(\frac{24}{5a^3t}\right)F\right]\bigg/\left[1-\left(\frac{a^3t}{8}\right)^2\times\left(\frac{24}{5a^3t}\right)^2\right]=-\frac{15F}{16} \quad (5.43b)$$

Hence, the multiplication factors in the shear flow expression (5.35) become

$$\frac{\overline{F}_y}{I_x}=\frac{25F}{16}\times\frac{24}{5a^3t}=\frac{20F}{a^3t},\quad \frac{\overline{F}_x}{I_y}=-\frac{15F}{16}\times\frac{24}{5a^3t}=-\frac{12F}{a^3t} \quad (5.44a,b)$$

Then, working from the free end A of limb AE, where $q_0 = 0$, equations (5.35) and (5.44a,b) give:

$$q_{AE}=\frac{20F}{a^3t}\int_0^s \frac{at}{4}\,\mathrm{d}s-\frac{12F}{a^3t}\int_0^s\left(\frac{3a}{4}-s\right)t\mathrm{d}s=\frac{4F}{a^3}\left(\frac{3s^2}{2}-as\right) \quad (5.45a)$$

Equation (5.45a) describes the distribution shown in Figure 5.14, having the value $-F/2a$ at mid span ($s = a/2$). The terminal value q_E for $s = a$ is:

$$q_E=\frac{4F}{a^3}\times\frac{a^2}{2}=\frac{2F}{a} \quad (5.45b)$$

Next, starting from E within limb EB, where $q_0 = q_E$, we apply equation (5.35) with equations (5.44a,b):

$$q_{EB}=\frac{20F}{a^3t}\int_0^s\left(\frac{a}{4}-s\right)t\mathrm{d}s+\frac{12F}{a^3t}\int_0^s\frac{at}{4}\mathrm{d}s+\frac{2F}{a}=\frac{4F}{a^3}\left(2as-\frac{5s^2}{2}\right)+\frac{2F}{a} \quad (5.46)$$

Equation (5.46) describes the shear flow distribution in Figure 5.14 with a maximum $18F/5a$ for $s = 2a/5$ and returning to zero at the free end B. Comparing Figures 5.13

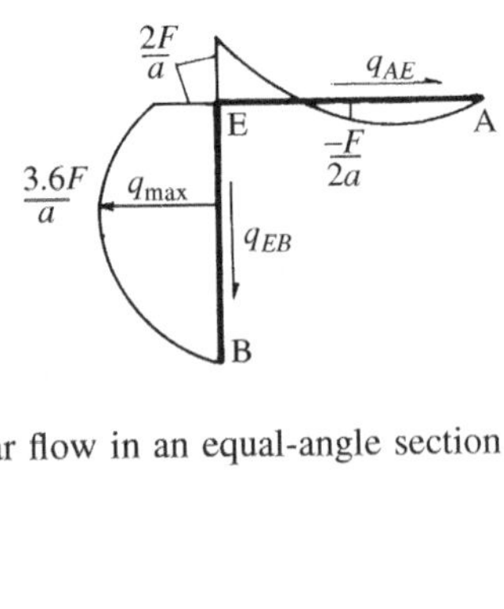

Figure 5.14 Shear flow in an equal-angle section under a single force

and 5.14, we see that greater, maximum q arises with a single vertical force acting. For case I, with the two equal shear forces, acting in the directions shown in Figure 5.12, there is a partial cancellation and reinforcement of their individual shear flows.

Case III

If the direction of F_x is reversed to act inwards with F_y (where $F_x = F_y = F$), its shear flow will mirror that for F_y shown in Figure 5.14 and the two shear flows reinforce. The net shear flow equation may be derived from superposition of equations (5.45a) and (5.46), having reoriented the section to provide the two inward forces. With each equation having a common origin for s at A, this gives

$$q_{\text{net}} = \frac{4F}{a^3}\left[2a(a-s) - \frac{5(a-s)^2}{2}\right] + \frac{2F}{a} + \frac{4F}{a^3}\left(\frac{3s^2}{2} - as\right) \tag{5.47}$$

which has its a maximum value of $4F/a$ at point E as shown in Figure 5.15.

For the optimisation we may summarise cases I, II and II in terms of the maximum shear stresses that equations (5.41a), (5.45a) and (5.47) provide: $\tau_{\max} = q_{\max}/t = 3F/2at$, $18F/5at$ and $4F/at$, respectively. Writing these in a common form $\tau_{\max} = QF/at$, each shear stress may be limited to τ_y under which the limb can be arranged to buckle simultaneously when

$$\tau_y = \frac{QF}{at} = K_a E_T \left(\frac{t}{a}\right)^2 \tag{5.48}$$

The optimised dimensions follow from equation (5.48):

$$t_{\text{opt}} = \left(\frac{Q^2}{K_a}\right)^{1/4}\left(\frac{F^2}{\tau_y E_T}\right)^{1/4}, \quad a_{\text{opt}} = (Q^2 K_a)^{1/4}\left(\frac{F^2 E_T}{\tau_y^3}\right)^{1/4} \tag{5.49a,b}$$

The optimum weight is simply $W_{\text{opt}} = 2a_{\text{opt}}t_{\text{opt}}\rho$, within which equations (5.49a,b) provide the objective function

$$\frac{W}{L^3} = 2Q\left(\frac{\rho}{\tau_y}\right)\left(\frac{F}{L^2}\right) \tag{5.50}$$

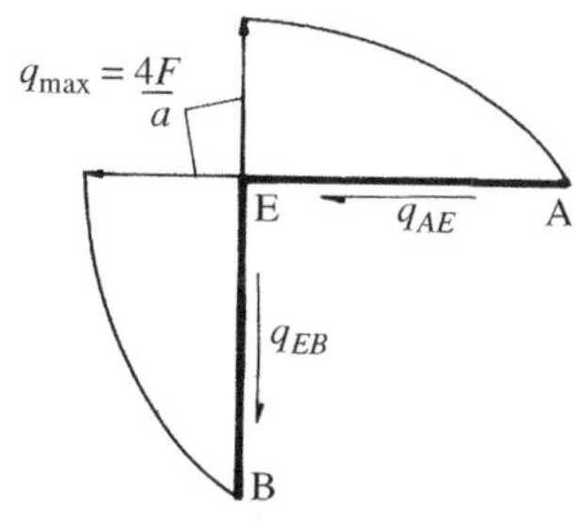

Figure 5.15 Equal-angle with inward corner forces showing reinforcing shear flows

Equation (5.50) shows that the shape factor depends only upon how the loads are applied: case I being the most efficient and case III being the least. In practice, limb thickness of standard angle sections decreases in a taper towards the free end. The effect would be to reduce q_{max} (and Q) at the junction between the two limbs. The shear flow analysis will allow for this when $t = t(s)$ within the integration. Alternatively, the analysis given above could be referred to the thickness at mid length for an approximate solution. A similar average thickness may also be employed for tapered flanges within a channel section.

5.3.4 *Channel Section, Uniform Thickness*

Let a vertical force F_y be applied at the shear centre E of the channel, whose position e in Figure 5.16a is to be found. With channel flanges each of equal length a, web depth d, all of uniform thickness t, and x and y as principal axes, the shear flow is simply

$$q = \frac{F_y D_x}{I_x} \tag{5.51}$$

The position $\overline{X}$ of its centroid G is found from taking first moments of the three rectangular areas about the vertical edge:

$$(dt)t/2 + 2at(a/2) = (2a + d)t\overline{X}, \quad \Rightarrow \quad \frac{\overline{X}}{a} = \frac{a/d}{1 + 2a/d} \tag{5.52a}$$

which is a good approximation when $t \ll a$. With the origin of reference axes x and y at the centroid as shown, I_x is the only second moment of area required:

$$I_x = 2[at^3/12 + at(d/2)^2] + td^3/12 \approx (td^2/2)(a + d/6) \tag{5.52b}$$

Taking the origin for s at A, the shear flow in limb AB is

$$q_{AB} = \frac{F_y}{I_x}\int_0^s yt\mathrm{d}s = \frac{F_y}{I_x}\int_0^s (d/2)t\mathrm{d}s = \frac{F_y s}{d(a + d/6)} \tag{5.53}$$

Equation (5.53) gives a linear distribution having its maximum at B, i.e. q_B for $s = a$ (as shown in Figure 5.16b). This reappears in the shear flow integral for the web BC as

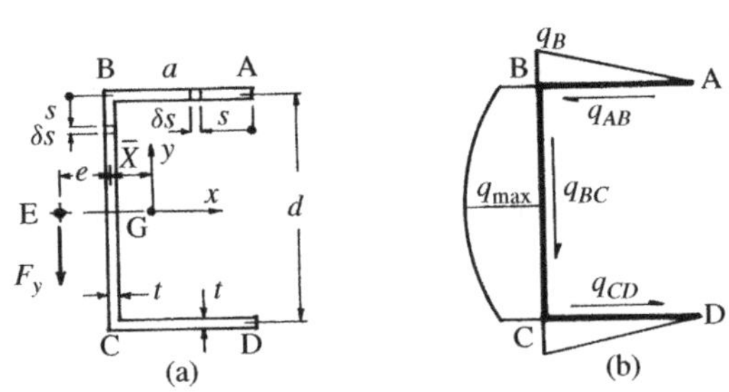

Figure 5.16 Shear flow in a uniformly thick channel with vertical force at E

follows:

$$q_{BC} = \frac{F_y}{I_x}\int_0^s yt\mathrm{d}s + q_B = \frac{F_y}{I_x}\int_0^s (d/2 - s)t\mathrm{d}s + \frac{F_y a}{d(a + d/6)} \tag{5.54a}$$

where the origin for s is at B. The integration of equation (5.54a) and substitution for I_x gives

$$q_{BC} = \frac{F_y t}{I_x}\left|\frac{ds}{2} - \frac{s^2}{2}\right|_0^s + \frac{F_y a}{d(a + d/6)} = \frac{F_y[s(1 - s/d) + a]}{d(a + d/6)} \tag{5.54b}$$

Equation (5.54b) describes the parabolic distribution shown in Figure 5.6b, this revealing that the maximum shear flow occurs at the neutral axis ($s = d/2$):

$$q_{\max} = \frac{F(a + d/4)}{d(a + d/6)} = \frac{3F[1 + 4(a/d)]}{2d[1 + 6(a/d)]} \tag{5.54c}$$

Hence the maximum shear stress $\tau_{\max} = q_{\max}/t$ is identical to that found in the web of a uniform I-beam (see equation (5.22a)). It follows that (with a change of symbol from b to a) the optimised channel dimensions have similar form to equation (5.24b,c):

$$d_{\text{opt}} = \left(\frac{3^2 K_d}{2^2}\right)^{1/4} \frac{[1 + 4(a/d)]^{1/2}}{[1 + 6(a/d)]^{1/2}} \left(\frac{E_T F^2}{\tau_y^3}\right)^{1/4} \tag{5.55a}$$

$$t_{\text{opt}} = \left(\frac{3^2}{2^2 K_d}\right)^{1/4} \frac{[1 + 4(a/d)]^{1/2}}{[1 + 6(a/d)]^{1/2}} \left(\frac{F^2}{\tau_y E_T}\right)^{1/4} \tag{5.55b}$$

Also, the objective function is similar to equation (5.24b):

$$\left(\frac{W}{L^3}\right)_{\text{opt}} = \frac{3[1 + 2(a/d)][1 + 4(a/d)]}{2[1 + 6(a/d)]} \left(\frac{\rho}{\tau_y}\right)\left(\frac{F}{L^2}\right) \tag{5.56}$$

There are however two main differences between these two sections:

(i) Flange buckling occurs over the full length of a channel. If we wish to choose an a/d ratio for the flange and web to buckle under their respective maxima in equations (5.53) and (5.54), the ratio between them is equated to the corresponding ratio between the critical buckling stresses. For the channel:

$$\frac{6F(a/d)}{dt[1 + 6(a/d)]} \times \frac{2dt[1 + 6(a/d)]}{3F[1 + 4(a/d)]} = \frac{K_a(t/a)^2}{K_d(t/d)^2} \tag{5.57a}$$

Equation (5.57a) reduces to

$$\left(\frac{a}{d}\right)^3 - \frac{1}{4}\frac{K_a}{K_d} - \frac{K_a}{K_d}\frac{a}{d} = 0 \tag{5.57b}$$

and, using Figure C.3, we find that equation (5.57b) is satisfied by $a/d \approx 1$ when the length is large, i.e. $d/L \approx 0$. However, in view of what has been said about short lengths being more appropriate to a shear design, Figure C.3 is found to satisfy equation (5.57b) when, typically, $a/d = 1.025$ for $d/L = 0.5$ and when $a/d = 1.055$ for $d/L = 1$. Thus, the optimum design appears to be relatively insensitive to length when the flange width and web depth are approximately the same.

(ii) The shear flow analysis given above refers to when F_y is applied at the shear centre E (see Figure 5.16a). This centre lies outside the section distance e from the web as shown. The requirement that F_y does not twist the section is met when the torque $F_y e$ is equivalent to the net torque due to the q distributions in each limb. Such a static equivalence applies to *any* point that we care to choose, so by taking the corner point C, for example, only the shear flow in AB is influential. That is, taking moments at C,

$$\sum T_C = F_y e - d\int_0^a q_{AB}\,\mathrm{d}s = 0 \tag{5.58a}$$

and substituting q_{AB} from equation (5.53), e is found from equation (5.58a) as

$$F_y e - \frac{F_y d}{d(a + d/6)}\int_0^a s\,\mathrm{d}s = 0, \quad \Rightarrow \quad \frac{e}{a} = \frac{3(a/d)}{1 + 6(a/d)} \tag{5.58b}$$

If F_y is not applied at E then shear stress arises from both shear and torsion. Such a combination arises when the downward vertical force acts in the web or is applied to pass through the centroid G. Both cases are examined in Chapter 6 (see §§ 6.2.1 and 6.2.2).

5.3.5 *Semi-Circular Channel, Uniform Thickness*

Let shear forces F_x and F_y be applied at the shear centre E of a channel, with radius r and uniform thickness t, as shown in Figure 5.17. Again, x and y are principal axes for which the shear flow is

$$q = \frac{F_y D_x}{I_x} + \frac{F_x D_y}{I_y} \tag{5.59}$$

The position $\overline{X}$ of the centroid G is found from summing the first moments of an elemental area $t\delta s$ about the Y-axis. With $\delta s = r\delta\theta$, this gives

$$\left[2rt\int_0^{\pi/2}\mathrm{d}\theta\right]\overline{X} = 2r^2 t\int_0^{\pi/2}\sin\theta\,\mathrm{d}\theta, \quad \Rightarrow \quad \overline{X} = \frac{2r}{\pi} \tag{5.60a}$$

The second moment of the area $t\delta s$ about the x-axis is

$$I_x = 2r^3 t\int_0^{\pi/2}\cos^2\theta\,\mathrm{d}\theta, \quad \Rightarrow \quad I_x = \frac{\pi r^3 t}{2} \tag{5.60b}$$

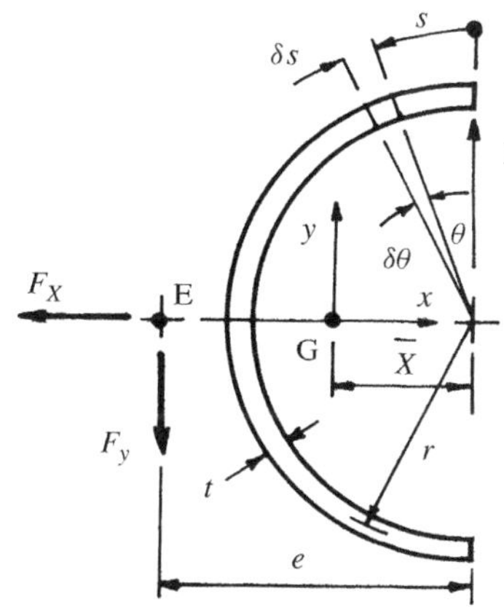

Figure 5.17 Semi-circular channel with shear forces applied at E

and about the Y-axis is

$$I_Y = 2r^3 t \int_0^{\pi/2} \sin^2\theta \mathrm{d}\theta, \quad \Rightarrow \quad I_Y = \frac{\pi r^3 t}{2} \tag{5.61a}$$

Transferring I_Y in equation (5.61a) to the parallel, centroidal axis y,

$$I_y = I_Y - A\overline{X}^2 = \frac{\pi r^3 t}{2} - (\pi r t)(2r/\pi)^2 = r^3 t(\pi/2 - 4/\pi) \tag{5.61b}$$

The first moment of area $t\delta s$ at position s from the top free edge, about the centroidal axis y, becomes

$$D_x = \int_0^s y t \mathrm{d}s = r^2 t \int_0^\theta \cos\theta \mathrm{d}\theta = r^2 t \sin\theta \tag{5.62a}$$

and D_y for axis x, with $x = \overline{X} - r\sin\theta$, becomes

$$D_y = \int_0^s x t \mathrm{d}s = r^2 t \int_0^\theta (2/\pi - \sin\theta)\mathrm{d}\theta = r^2 t(2\theta/\pi + \cos\theta - 1) \tag{5.62b}$$

Substituting the first and second moment equations (5.62a,b) into equation (5.59) gives the shear flow expression

$$q_\theta = \frac{2}{r\pi}\left(F_y \sin\theta + F_x \frac{2\theta/\pi + \cos\theta - 1}{1 - 8/\pi^2}\right) \tag{5.63}$$

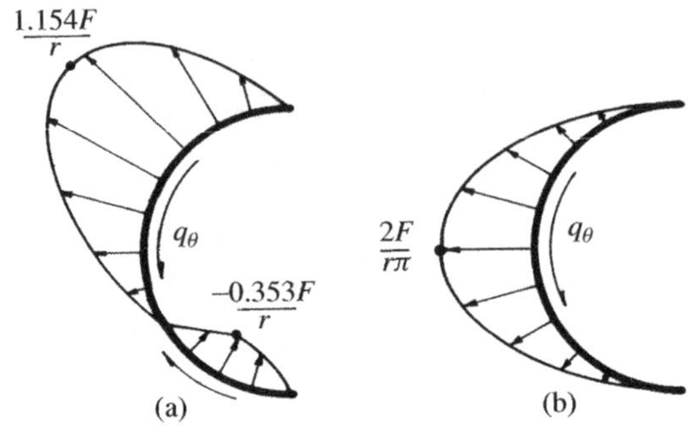

Figure 5.18 Shear flows in a semi-circular tube for cases I and II

Case I

When $F_x = F_y = F$, the q_θ distribution in Figure 5.18a applies. The maximum and minimum q_θ shown are found from applying $\mathrm{d}q_\theta/\mathrm{d}\theta = 0$ to equation (5.63). They are, respectively,

$$\theta = 49.45^\circ, \quad q_\theta = 1.154F/r$$
$$\theta = 152.01^\circ, \quad q_\theta = -0.353F/r$$
$$\theta = 119.3^\circ, \quad q_\theta = 0.$$

Case II

When $F_x = 0$, $F_y = F$, then $q_\theta = (2F/r\pi)\sin\theta$, which has its maximum $q_\theta = 2F/r\pi$ for $\theta = 90^\circ$, as shown in Figure 5.18b. The position e of the shear centre E in Figure 5.17 follows as

$$F_y e = \int_0^\pi q_\theta r^2 \mathrm{d}\theta = \frac{2F_y r^2}{r\pi}\int_0^\pi \sin\theta \mathrm{d}\theta, \quad \Rightarrow \quad e = \frac{4r}{\pi} \tag{5.64}$$

The optimisation of these and other force combinations follows from the maximum shear stress, written generally as $\tau_{\max} = QF/rt$ (case I, $Q = 1.1543$; case II, $Q = 2/\pi$). Setting the limit of $\tau_{\max}$ at τ_y and equating to the critical buckling stress,

$$\tau_y = \frac{QF}{rt} = K_p E_T \left(\frac{t}{\pi r}\right)^2 \tag{5.65a}$$

The optimum dimensions for the channel follow from equations (5.65a):

$$t_{\mathrm{opt}} = \left(\frac{\pi^2 Q^2}{K_p}\right)^{1/4}\left(\frac{F^2}{\tau_y E_T}\right)^{1/4}, \quad r_{\mathrm{opt}} = \left(\frac{Q^2 K_p}{\pi^2}\right)^{1/4}\left(\frac{F^2 E_T}{\tau_y^3}\right)^{1/4} \tag{5.65b,c}$$

The minimum weight follows from equations (5.65b,c) :

$$\left(\frac{W}{L^3}\right)_{\text{opt}} = \pi Q\left(\frac{\rho}{\tau_y}\right)\left(\frac{F}{L^2}\right) \tag{5.66}$$

Note that in this case the buckling coefficient K_p applies to a curved panel [4]. This was also considered in Chapter 4, where the elastic coefficient K_e is found from Figure 4.10a [4] at the given $b/(\sqrt{rt})$. When necessary, K_p is found from applying a plasticity reduction factor, η in Figure 4.8, in the usual manner (see Example 4.1).

5.4 Thin-Walled, Closed Tubes

The net shear flow in a closed, thin-walled tube, under a single vertical force F_y applied through the shear centre E, is written as

$$q = q_b + q_E \tag{5.67a}$$

Here q_b is the flexural shear flow expression, used previously with open sections:

$$q_b = \frac{F_y D_{xb}}{I_x} = \frac{F_y}{I_x}\int_A y\mathrm{d}A \tag{5.67b}$$

However, unlike an open section, we have no free surface to take as an origin for s, so for a closed tube a constant q_E must be added to q_b. Thus, q_E is a constant of integration with an important physical interpretation, namely, that when q_E is added to q_b, it ensures the rate of twist $\delta\theta/\delta z$ is zero at E [1]. That is,

$$\oint q_b\frac{\mathrm{d}s}{t} + q_E\oint\frac{\mathrm{d}s}{t} = 0 \tag{5.68a}$$

which allows q_E to be found. If the position of the shear centre E is not obvious it may be found by combining equation (5.68a) with the counterbalance required between the torque from the net shear flow $q = q_b + q_E$ and the torque with F_y applied at E, each torque being referred to any convenient reference point P. This torque balance is written as:

$$F_y e_x = \oint qR\,\mathrm{d}s = \oint q_b R\,\mathrm{d}s + q_E\oint R\,\mathrm{d}s \tag{5.68b}$$

where e_x is the horizontal distance required, between E and P. Within the first integral R is a perpendicular distance from P to the net incremental force $q\delta s$. Alternatively, the split in this integral recognises q_E as a constant when $\oint R\mathrm{d}s$ becomes twice the area enclosed by the wall's mean centre-line. The following examples, employing relatively simple sections, will demonstrate the use of equations (5.67a,b) and (5.68a,b).

5.4.1 Thin-Walled Circular Tube

Let a vertical force F_y be applied at the centre of a thin, circular tube of mean wall radius r and uniform wall-thickness t, as shown in Figure 5.19a. As F_y lies on the

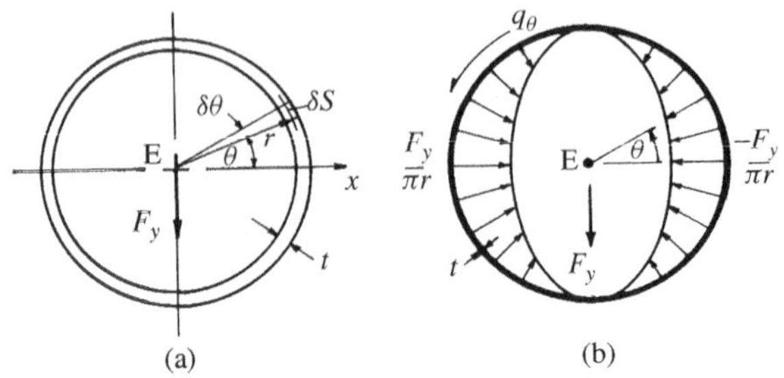

Figure 5.19 Circular tube under a central, vertical shear force

vertical centre-line of the tube cross-section, it passes through the shear centre E. The latter obviously coincides with the circle's centre, which is also the centre of twist, in having no rotation under torsion.

The first moment of the elemental area $t\delta s$ of the tube wall about the x-axis is

$$D_{xb} = \int y(t\,\mathrm{d}s) = 2tr^2 \int_0^{\pi} \sin\theta\,\mathrm{d}\theta = r^2 t(1-\cos\theta) \tag{5.69a}$$

where $\delta s = r\delta\theta$. The second moment of this area about the x-axis is

$$I_x = \int y^2(t\,\mathrm{d}s) = 2tr^3 \int_0^{\pi} \sin^2\theta\,\mathrm{d}\theta = \pi r^3 t \tag{5.69b}$$

Substituting equations (5.69a,b) into equation (5.67b) gives

$$q_b = \frac{F_y r^2 t(1-\cos\theta)}{\pi r^3 t} = \frac{F_y}{\pi r}(1-\cos\theta) \tag{5.70a}$$

Next, we find q_E by applying equation (5.68a) to the q_b distribution in equation (5.70a). This gives

$$\frac{F_y}{\pi t r}\int_0^{2\pi}(1-\cos\theta)r\,\mathrm{d}\theta + \frac{q_E}{t}\int_0^{2\pi} r\,\mathrm{d}\theta = 0 \tag{5.70b}$$

$$\frac{F_y}{\pi t}\left|\theta - \sin\theta\right|_0^{2\pi} + \frac{2\pi r q_E}{t} = 0, \quad \Rightarrow \quad q_E = -\frac{F_y}{\pi r}$$

Hence, the net shear flow distribution $q_\theta = q_b + q_E$, shown in Figure 5.19b, is given by

$$q_\theta = \frac{F_y}{\pi r}(1-\cos\theta) - \frac{F_y}{\pi r} = -\frac{F_y}{\pi r}\cos\theta \tag{5.71}$$

which has its maxima, $q_{max} = \mp(F_y)/(\pi r)$, at $\theta = 0°$ and $180°$ respectively, as shown. Either orientation provides the magnitude of the maximum shear stress as

$$\tau_{max} = \frac{F_y}{\pi r t} \tag{5.72}$$

When τ_{max} is limited to τ_y, equation (5.72) may be combined with the weight expression $W = 2\pi r t L\rho$, from which the objective function follows directly:

$$\frac{W}{L^3} = \frac{2\pi L\rho}{L^3}\frac{F_y}{\pi\tau_y} = 2\left(\frac{\rho}{\tau_y}\right)\left(\frac{F_y}{L^2}\right) \tag{5.73}$$

Equation (5.73) reveals a shape factor of 2, which we may compare with the solid sections given in Table 5.1. This factor is increased to 3 when the force is applied tangentially to the wall (see Chapter 6) and, in general, the weight is increased for forces displaced from the shear centre. It was noted earlier that neither E_T nor K_p appears directly within the objective function for shear. However, E_T and K_p will appear within the optimised dimensions, r and t, as each depends upon the buckling criterion. We know from Chapter 4 that the buckling behaviour of a circular tube in torsion (i.e. under wall shear) is more complicated than that of a flat or curved panel under transverse shear. In particular, § 4.4.1 identified three length regimes for torsion with critical shear stresses given by equations (4.48c), (4.50b) and (4.53b). Let us assume, tentatively, that similar equations apply to transverse shear. Combining each equation in turn with equation (5.72), the respective governing equation will determine their optimum dimensions as follows.

I Short Tube: $L < (rt)^{1/2}$

$$\tau_y = \frac{F_y}{\pi r t} = K_p E_T\left(\frac{t}{L}\right)^2 \tag{5.74a}$$

from which

$$t_{opt} = \frac{1}{\sqrt{K_p}}\left(\frac{\tau_y L^2}{E_T}\right)^{1/2}, \quad r_{opt} = \frac{\sqrt{K_p}}{\pi}\frac{F_y E_T^{1/2}}{\tau_y^{3/2} L} \tag{5.74b,c}$$

II Medium Length Tube: $50(t/r) < (L/r)^2 < 10(r/t)$

$$\tau_y = \frac{F_y}{\pi r t} = K_p E_T\left(\frac{t}{r}\right)^{5/4}\left(\frac{r}{L}\right)^{1/2} \tag{5.75a}$$

from which

$$t_{opt} = \frac{1}{K_p^{1/2}\pi^{3/8}}\left(\frac{F_y^{3/8}\tau_y^{1/8}L^{1/4}}{E_T^{1/2}}\right)^{1/2}, \quad r_{opt} = \frac{K_p^{1/2}}{\pi^{5/8}}\frac{F_y^{5/8}E_T^{1/2}}{\tau_y^{9/8}L^{1/4}} \tag{5.75b,c}$$

III Long Tube: $(L/r)^2(t/r) > 40$

$$\tau_y = \frac{F_y}{\pi r t} = K_p E_T \left(\frac{t}{r}\right)^{3/2} \tag{5.76a}$$

from which

$$t_{\text{opt}} = \frac{1}{K_p^{1/3}\pi^{1/2}}\left(\frac{F_y^{1/2}}{E_T^{1/3}\tau_y^{1/6}}\right), \quad r_{\text{opt}} = \frac{K_p^{1/3}}{\pi^{1/2}}\left(\frac{F_y^{1/2}E_T^{1/3}}{\tau_y^{5/6}}\right) \tag{5.76b,c}$$

Within each region it is seen that the optimum thickness and radius expressions (5.74b,c), (5.75b,c) and (5.76b,c) depend upon the shear force, material and the tube's geometrical instability behaviour. The latter allows for plasticity through the use of E_T and K_p in the usual way (see Example 4.1). Here the plasticity reduction factor η in Figure 4.8 applies to the medium and longer tube categories II and III [5].

5.4.2 *Uniform, Thin-Walled, Equilateral Triangular Tube*

Let a vertical force F_y be applied at the shear centre E within a triangular tube of side length a and uniform wall thickness t, in the manner of Figure 5.20.

The second moment of this area about the x-axis is the sum of the contributions from two sloping sides and one vertical side. From equation (5.2a),

$$I_x = 2\int_0^a (s \sin 30^\circ)^2 (t\, ds) + \frac{ta^3}{12} = \frac{a^3 t}{4} \tag{5.77}$$

The q_b shear flow in each side must be dealt with separately. Applying equation (5.67b) to side AB employs the first moment of the elemental area $D_x = \int t\delta s$ about the x-axis. This gives

$$q_{AB} = \frac{4F_y}{a^3 t}\int_0^a s \sin 30^\circ (t\, ds) = \frac{F_y s^2}{a^3} \tag{5.78a}$$

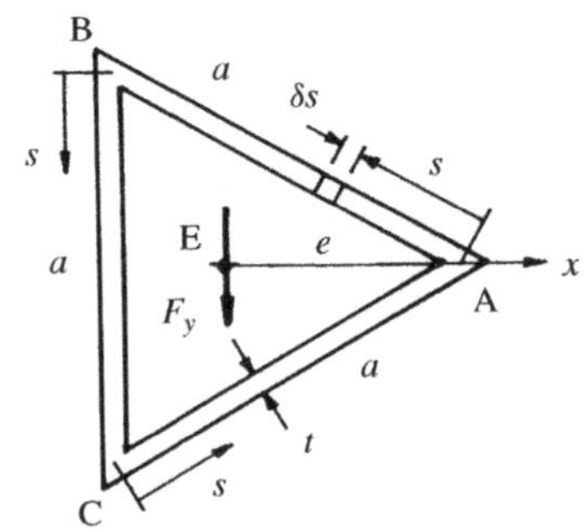

Figure 5.20 Thin-walled equilateral triangular tube

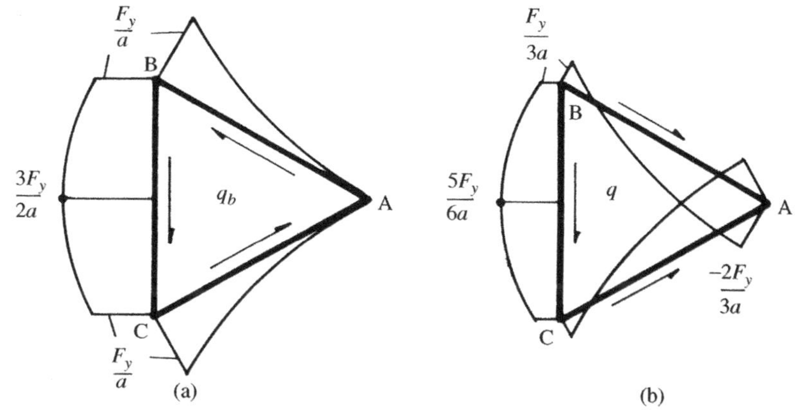

Figure 5.21 Shear flows in a uniform triangular tube

Equation (5.78a) describes the concave, parabolic shear flow distribution given in Figure 5.21a. This has $q_A = 0$ at A and a maximum shear flow $q_B = F_y/a$ at point B. In taking the origin for s at B, where $y = a/2 - s$, we must add q_B when we apply equation (5.67b) to find q_{BC}:

$$q_{BC} = \frac{4F_y}{a^3 t}\int_0^s \left(\frac{a}{2} - s\right) t\,\mathrm{d}s + \frac{F_y}{a} = \frac{2F_y}{a^3}(as - s^2) + \frac{F_y}{a} \tag{5.78b}$$

Equation (5.78b) describes a convex, parabolic variation (see Figure 5.21a) which has its maximum shear flow $q_{b,\max} = 3F_y/2a$ on the x-axis ($s = a/2$). Finally, in taking the origin for s at C, where $y = -(a - s)/2$, we must add $q_C = F_y/a$ (i.e. for $s = 0$) when we apply equation (5.67b) to find q_{CA}:

$$q_{CA} = -\frac{4F_y}{a^3 t}\int_0^s \frac{1}{2}(a - s)t\,\mathrm{d}s + \frac{F_y}{a}$$

$$= -\frac{2F_y}{a^3}\left(as - \frac{s^2}{2}\right) + \frac{F_y}{a} \tag{5.78c}$$

Equation (5.78c) mirrors the distribution found for side AB and, as a check, this returns to $q_A = 0$ correctly when $s = a$. Zero shear flow at the coincident start and finishing points will always apply to q_b irrespective of where that point is taken.

To find q_E for this tube it becomes necessary to apply equation (5.68a) to the q_b shear flows given above. Note that the position of shear centre E along the x-axis is not required for this, only that the force F_y must pass through E. Thus, with E at the centre of twist, the path integral is formed from the three sides:

$$\left[\int_s q_{AB}\frac{\mathrm{d}s}{t} + \int_s q_{BC}\frac{\mathrm{d}s}{t} + \int_s q_{CA}\frac{\mathrm{d}s}{t}\right] + q_E\int_s \frac{\mathrm{d}s}{t} = 0 \tag{5.79a}$$

Substituting into equation (5.79a) the q_b shear flows q_{AB}, q_{BC} and q_{CA} in equations (5.78a–c) lead to

$$\frac{F_y}{a^3 t}\int_0^a s^2\,\mathrm{d}s + \frac{2F_y}{a^3 t}\int_0^a (as - s^2)\,\mathrm{d}s + \frac{F_y}{at}\int_0^a \mathrm{d}s - \frac{2F_y}{a^3 t}\int_0^a (as - s^2/2)\,\mathrm{d}s + \frac{F_y}{at}\int_0^a \mathrm{d}s + \frac{3aq_E}{t} = 0$$

$$\frac{F_y}{a^3 t}\left[\left|\frac{s^3}{3}\right|_0^a + 2\left|\frac{as^2}{2} - \frac{s^3}{3}\right|_0^a - 2\left|\frac{as^2}{2} - \frac{s^3}{6}\right|_0^a\right] + \frac{2F_y}{at}\left|s\right|_0^a + \frac{3aq_E}{t} = 0 \tag{5.79b}$$

Equation (5.79b) gives

$$q_E = -2F_y/3a \tag{5.79c}$$

which must be added to the q_b shear flows (see equation (5.67a)). This results in the net shear flows shown in Figure 5.21b, where the greatest shear flow lies at centre of side BC:

$$q_{\max} = 3F_y/2a - 2F_y/3a = 5F_y/6a \tag{5.80a}$$

Hence, the maximum shear stress at this position is

$$\tau_{\max} = \frac{5F_y}{6at} \tag{5.80b}$$

upon which the optimum design is to be based. We now limit $\tau_{\max}$ to τ_y and use $Q(= 5/6)$ to denote the numerical factor within the greatest shear stress for this tube. Given that the weight is $W = 3at\rho L$, the objective function follows directly from eliminating the product at as supplied by equation (5.80b). This gives the optimum weight for the three sides of this tube as

$$\left(\frac{W}{L^3}\right)_{\text{opt}} = 3Q\left(\frac{\rho}{\tau_y}\right)\left(\frac{F_y}{L^2}\right) \tag{5.81}$$

where the shape factor $3Q = 3 \times 5/6 = 2.5$. The governing equation for optimising the dimensions t and a is similar to one used earlier for the angle section (see equation (5.48)). Hence, equations (5.49a,b), giving t_{opt} and a_{opt}, also apply to this tube:

$$t_{\text{opt}} = \left(\frac{Q^2}{K_a}\right)^{1/4}\left(\frac{F^2}{\tau_y E_T}\right)^{1/4}, \qquad a_{\text{opt}} = (Q^2 K_a)^{1/4}\left(\frac{F^2 E_T}{\tau_y^3}\right)^{1/4} \tag{5.82a,b}$$

Note, however, that K_a in equations (5.82a,b) refers to the plate in shear having both long edges simply supported (Figure C.3).

In Chapter 6 it will be shown that with the force applied down the side BC, the shape factor becomes 3.5. Hence, the present shear centre positioning of the force leads to a more favourable shape factor, i.e. 2.5. Thus there is a weight reduction from applying F_y at the shear centre, so now its position must be found. To apply equation (5.69), let E lie distance e from A (see Figure 5.20). Take moments about A, so that the only contribution to torsion is that from the shear flow in the side BC:

$$F_y e = \left[\frac{\sqrt{3}}{2} a \times \int_0^a (q_b + q_E)\, ds\right]_{BC} \tag{5.83a}$$

Substituting, $q_b = q_{BC}$ from equation (5.78b,c) and q_E from equation (5.79b) into equation (5.83a),

$$F_y e = \frac{\sqrt{3}}{2} a \left[\frac{2F_y}{a^3} \int_0^a (as - s^2)\, ds + \frac{F_y}{a} \int_0^a ds - \frac{2F_y}{3a} \int_0^a ds\right]$$

$$= \frac{\sqrt{3}}{2} a \left[\frac{2F_y}{a^3} \left|\frac{as^2}{2} - \frac{s^3}{3}\right|_0^a + \frac{F_y}{3a} \left|s\right|_0^a\right] \tag{5.83b}$$

from which the solution to equation (5.83b) is $e = a/\sqrt{3}$.

5.4.3 Non-Uniform, Equilateral Triangular Tube

Given that less stress exists in the sloping sides of the uniformly thin tube in Figure 5.20a, a further reduction to the weight should be possible by reducing their thicknesses in the manner of Figure 5.22.

When F_y is applied at E, a simultaneous buckling of AB and BC will occur when

$$\left(\frac{\tau_{AB}}{\tau_{BC}}\right)_{\max} = \left(\frac{\tau_{AB}}{\tau_{BC}}\right)_{cr} \tag{5.84}$$

The same net shear flows apply as were derived for the uniform tube in § 5.4.2. Hence, equation (5.80b) again provides the maximum shear stress at the mid position of side BC:

$$(\tau_{BC})_{\max} = \frac{5F_y}{6at_{BC}} \tag{5.85a}$$

and equation (5.78a) and (5.79c) give the maximum shear stress within side AB for $s = a$ as

$$(\tau_{AB})_{\max} = \frac{F_y a^2}{a^3 t_{AB}} - \frac{2F_y}{3at_{AB}} = \frac{F_y}{3at_{AB}} \tag{5.85b}$$

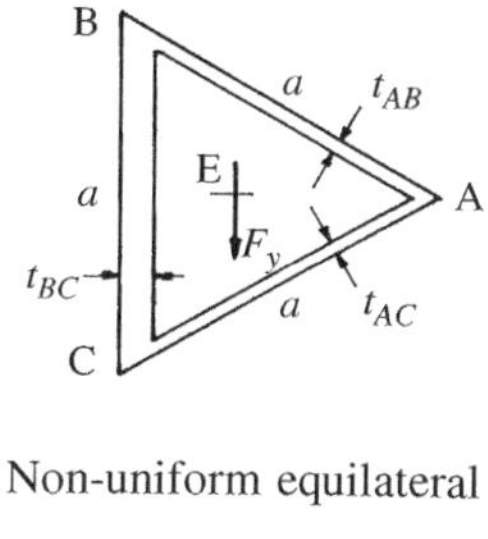

Figure 5.22 Non-uniform equilateral triangular tube

The expressions giving the critical shear stress for buckling are

$$(\tau_{AB})_{\rm cr} = K_a E_T (t_{AB}/a)^2, \quad (\tau_{BC})_{\rm cr} = K_a E_T (t_{BC}/a)^2 \tag{5.86a,b}$$

Substituting equations (5.85a,b) and (5.86a,b) into equation (5.84) leads to

$$\frac{F_y}{3at_{AB}} \times \frac{6at_{BC}}{5F_y} = \frac{t_{AB}^2}{t_{BC}^2}, \quad \Rightarrow \quad \frac{t_{AB}}{t_{BC}} = \left(\frac{2}{5}\right)^{1/3} \tag{5.87a}$$

The objective function now follows from the weight

$$W_{\rm opt} = a(t_{BC} + 2t_{AB})L\rho = at_{BC}[1 + 2t_{AB}/t_{BC})L\rho \tag{5.87b}$$

When equation (5.85a) sets the maximum stress limit at $\tau_y = 5F_y/(6at_{BC})$, we may combine this with equations (5.87a,b) to provide the objective function:

$$\frac{W}{L^3} = at_{BC}\left(1 + 2\frac{t_{AB}}{t_{BC}}\right)\frac{\rho}{L^2} = \frac{5F_y}{6\tau_y}\left[1 + 2\left(\frac{2}{5}\right)^{1/3}\right]\frac{\rho}{L^2} = 2.06\left(\frac{\rho}{\tau_y}\right)\left(\frac{F_y}{L^2}\right) \tag{5.88}$$

i.e. the shape factor is 2.06, compared to a value of 2.5 for the uniformly thin tube when tubes are loaded at their shear centres. Equations (5.82a,b) will again supply the optimum dimensions, with $Q = 5/6$, provided $t_{\rm opt}$ is identified here with t_{BC}.

A similar analysis is made in Chapter 6 (see § 6.3.3) for when F_y is applied down the centre of the vertical side BC. This shows that equations (5.85a,b) are altered to

$$(\tau_{BC})_{\max} = \frac{7F_y}{6at_{BC}}, \quad (\tau_{AB})_{\max} = \frac{2F_y}{3at_{AB}} \tag{5.89a,b}$$

Consequently, equation (5.87a) becomes

$$\frac{t_{AB}}{t_{BC}} = \left(\frac{4}{7}\right)^{1/3} \tag{5.89c}$$

Setting the limiting shear stress to τ_y from equation (5.89a) and combining with equations (5.87b) and (5.89c), we see that the shape factor is increased to 3.103. Chapter 6 shows that f is increased further to 3.5 when the shear force is applied to a uniformly thin tube in a similar manner. Again, when the wall thicknesses are non-uniform the greatest weight reduction that can be found is when the shear force is applied at the shear centre.

5.4.4 *Non-Uniform, Thin-Walled, Rectangular Tube*

Let a vertical force F_y be applied along the vertical centre-line of a rectangular tube, having side lengths b and d with wall-thicknesses t_b and t_d, in the manner of Figure 5.23. The second moment of this area about the x-axis is

$$I_x = 2\left[\frac{bt_b^3}{12} + bt_b\left(\frac{d}{2}\right)^2\right] + 2\frac{t_d d^3}{12} \approx \frac{bt_b d^2}{12} + \frac{t_d d^3}{12} = \frac{t_d d^3}{12}\left(1 + 3\frac{t_b}{t_d}\frac{b}{d}\right) \tag{5.90}$$

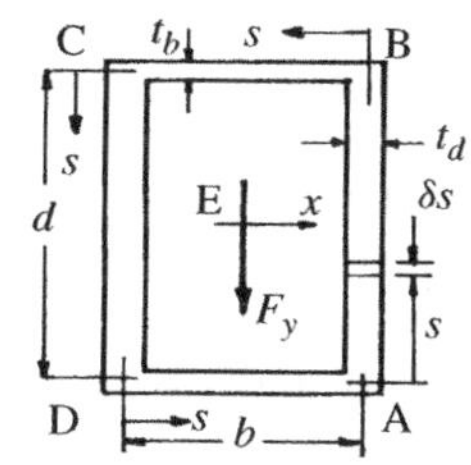

Figure 5.23 Non-uniform, rectangular tube

The q_b shear flow distribution within each side follows from equation (5.67). For side AB, with origin for s at A,

$$q_{AB} = \frac{F_y t_d}{I_x} \int_0^s (s - d/2)\,\mathrm{d}s = \frac{F_y t_d}{2I_x}(s^2 - ds) \tag{5.91a}$$

Equation (5.91a) gives $q_A = q_B = 0$ for $s = 0$ and d, respectively. The maximum in the parabolic distribution occurs for $s = d/2$:

$$q = -\frac{F_y t_d d^2}{8I_x}$$

in which the minus sign implies that q-flow opposes the s-direction (see Figure 5.24a). For side BC, with origin for s at B,

$$q_{BC} = \frac{F_y}{I_x} \int_0^s \frac{d}{2}(t_b\,\mathrm{d}s) + q_B = \frac{dF_y t_b s}{2I_x} \tag{5.91b}$$

Equation (5.91b) matches $q_B = 0$ for $s = 0$. Figure 5.24a shows a maximum in the linear distribution for $s = b$:

$$q_C = \frac{F_y t_b d b}{2I_x} \tag{5.91c}$$

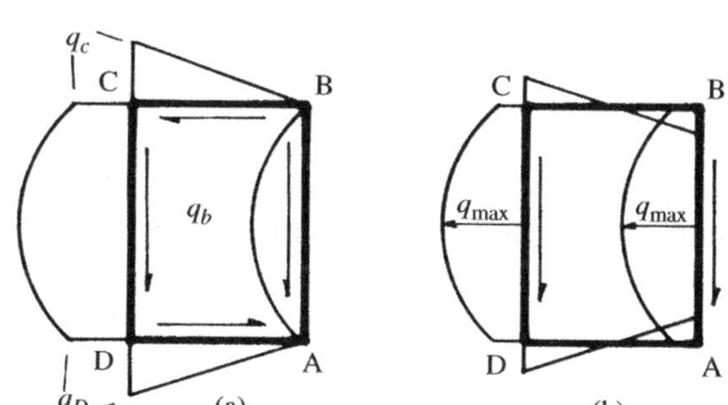

Figure 5.24 Flexural shear flows q_B, and q_{net} with F_y applied at the shear centre

For side CD, with origin for s at C:

$$q_{CD} = \frac{F_y}{I_x}\int_0^s (d/2 - s)\,\mathrm{ds} + q_C = \frac{F_y t_d}{2I_x}(ds - s^2) + \frac{F_y t_b bd}{2I_x} \tag{5.91d}$$

Equation (5.91d) matches q_C for $s = 0$ and gives $q_D = q_C$ for $s = d$. The maximum within the parabolic shear flow distribution for CD occurs at $s = d/2$ (see Figure 5.24a):

$$(q_{CD})_{\max} = \frac{F_y t_d d^2}{8I_x} + \frac{F_y t_b bd}{2I_x} \tag{5.91e}$$

For side DA, with origin for s at D:

$$q_{DA} = \frac{F_y}{I_x}\int_0^s (-d/2)(t_b\,\mathrm{ds}) + q_D = \frac{F_y t_b d}{2I_x}(-s + b) \tag{5.91f}$$

Equation (5.91f) gives $q_A = 0$ for $s = b$ and matches q_D for $s = 0$, with q_D being the maximum in the linear distribution, as shown in Figure 5.24a.

Since F_y acts at the shear centre (the centroid) q_E can be found from equation (5.68a):

$$\left[\int_s q_{AB}\mathrm{ds}/t + \int_s q_{BC}\mathrm{ds}/t + \int_s q_{CD}\mathrm{ds}/t + \int_s q_{DA}\mathrm{ds}/t\right] + q_E \oint \mathrm{ds}/t \tag{5.92a}$$

Substituting the q_b distributions (q_{AB}, q_{BC} etc.) from equations (5.91a–f) into equation (5.92a),

$$\frac{F_y}{2I_x}\int_0^d (s^2 - ds)\,\mathrm{ds} + \frac{F_y d}{2I_x}\int_0^b s\,\mathrm{ds} + \frac{F_y}{2I_x}\int_0^d (ds - s^2)\,\mathrm{ds} + \frac{F_y t_b bd}{2I_x t_d}\int_0^d \mathrm{ds}$$

$$-\frac{F_y d}{2I_x}\int_0^b s\,\mathrm{ds} + \frac{F_y bd}{2I_x}\int_0^b \mathrm{ds} + 2q_E\left(\frac{b}{t_b} + \frac{d}{t_d}\right) = 0$$

from which

$$q_E = -\frac{F_y bdt_b}{4I_x} \tag{5.92b}$$

Equation (5.92b) describes the constant shear flow in the walls of the tube that is to be added to the q_B distribution in Figure 5.24a. This results in the net shear flow shown in Figure 5.24b. In particular, adding q_E from equation (5.92b) to equation (5.91e) leads to the equal maximum, net shear flows at the centres of sides AB and CD:

$$(q_{CD})_{\max} = \frac{F_y t_d d^2}{8I_x}\left(1 + 2\frac{t_b}{t_d}\frac{b}{d}\right) = -(q_{AB})_{\max} \tag{5.93a}$$

Similarly, adding equation (5.92b) to (5.91c) gives the maximum shear flow at C in side BC (and at D in side DA):

$$(q_{BC})_{\max} = \frac{F_y b d t_b}{4I_x} = (q_{DA})_{\max} \tag{5.93b}$$

Net shear flows of similar magnitude to equation (5.93b) apply to points A and B (with a sign change). Substituting for I_x from equation (5.90) into equation (5.93a), the greatest shear stress, which lies at mid-side CD, is

$$(\tau_{CD})_{\max} = \frac{3F_y[1 + 2(t_b/t_d)(b/d)]}{2t_d d[1 + 3(t_b/t_d)(b/d)]} = \frac{QF_y}{t_d d} \tag{5.93c}$$

in which the geometrical coefficient Q may again reappear within the objective function. Equation (5.93b) gives the lesser maximum stress at C in the side BC:

$$(\tau_{BC})_{\max} = \frac{3F_y b}{t_d d^2[1 + 3(t_b/t_d)(b/d)]} \tag{5.93d}$$

The weight is:

$$W = 2(bt_b + dt_d)\rho L = 2t_d d[1 + (b/d)(t_b/t_d)]\rho L \tag{5.94a}$$

Setting $(\tau_{CD})_{\max} = \tau_y$ in equation (5.93c) allows the product $t_d d$ to be eliminated between equations (5.93c) and (5.94a). This leads to the required objective function:

$$\left(\frac{W}{L^3}\right)_{\text{opt}} = 2Q\left(1 + \frac{b}{d}\frac{t_b}{t_d}\right)\frac{\rho}{\tau_y}\frac{F_y}{L^2} = f\left(\frac{\rho}{\tau_y}\right)\left(\frac{F_y}{L^2}\right) \tag{5.94b}$$

in which the shape factor f follows from equations (5.93c) and (5.94b):

$$f = \frac{3[1 + 2(t_b/t_d)(b/d)][1 + (t_b/t_d)(b/d)]}{1 + 3(t_b/t_d)(b/d)} \tag{5.94c}$$

When we allow sides CD (and AB) to buckle under $(\tau_{CD})_{\max} = \tau_y$, equation (5.48) becomes

$$\tau_y = \frac{QF_y}{dt_d} = K_d E_T\left(\frac{t_d}{d}\right)^2 \tag{5.95a}$$

The optimum dimensions follow from equation (5.95a):

$$t_{d_{\text{opt}}} = \left(\frac{Q^2}{K_d}\right)^{1/4}\left(\frac{F_y^2}{\tau_y E_T}\right)^{1/4}, \qquad d_{\text{opt}} = (Q^2 K_d)^{1/4}\left(\frac{F_y^2 E_T}{\tau_y^3}\right)^{1/4} \tag{5.95b,c}$$

Moreover, if we take the ratio between the maximum stresses in the sides CD and BC (or AB and DA) and equate this to the ratio between their critical buckling stresses, as

Table 5.3 Shape factors with flat length and thickness ratios for four-wall buckling in a rectangular tube

b/d	0.25	0.5	1	2	3	4	5
t_b/t_d	0.17	0.420	0.860	1.510	2.025	2.480	2.890
f	3.11	3.16	4.22	8.44	14.52	22.2	31.24

in equation (5.84), all sides will buckle together. Thus, using the ratio between equations (5.93c,d) leads to the following relationship between the side lengths and the thickness ratios:

$$\frac{1 + 2(t_b/t_d)(b/d)}{2b/d} = \frac{K_d}{K_b}\left(\frac{t_d}{t_b}\right)^2\left(\frac{b}{d}\right)^2 \tag{5.96a}$$

or

$$(K_b/K_d)(t_b/t_d)^2[1 + 2(t_b/t_d)(b/d)] = 2(b/d)^3 \tag{5.96b}$$

The solution to equation (5.96b) provides the thickness ratios for the range of b/d ratios shown in Table 5.3. Also given are the shape factors at these ratios. The shape factor, which has no minimum value, increases with b/d to attain highly inefficient values. Clearly, $b/d < 1$ is more desirable for minimising weight as we might expect with the side BC suffering the least net stress.

Note that these solutions apply to a ratio $K_b/K_d = 1$, which may be assumed for long plates (b/L and $d/L < 0.1$) with similar edge restraints (see Figure C.3). The shape factors become less efficient when F_y is displaced from the shear centre (see Chapter 6). When thicknesses vary in this way all sides can be arranged to buckle from equalising their critical shear stresses. This cannot be done in the case of a rectangular tube of uniform thickness. Here only the longer sides buckle under their optimum dimensions given by equations (5.95b,c). Again, Q and f are given by equations (5.93c) and (5.94c) with $t_b = t_d$ (see Exercise 5.5)

5.5 Concluding Remarks

The analysis of shear is complicated by the distribution in shear stress that a section suffers under a transverse force. Nonetheless, whether the section be open or closed, it is possible to optimise its dimensions to provide minimum weight. The technique of limiting the greatest shear stress to a predetermined design stress at which buckling also occurs ensures that the most highly stressed material is used to its full load bearing capacity. In the presence of stress gradients the more lowly stressed material will remain underused. It is possible to use this material more effectively by adjusting dimensions to raise the stress and thereby save weight. It has been seen that the best saving that can be achieved applies to where the force is applied at the shear centre. This allows the material to be used to its full extent where thickness reduction promotes simultaneous buckling, indicating that the various limbs had contributed equally in their support of the load. In practice, of course, such failure is prevented by imposing a margin of safety where the working stress becomes a factored-down design stress, i.e. a fraction of τ_y. Whilst high reduction factors increase the margin of safety, they also increase weight. Where low weight is crucial, the trimming of the section dimensions suggested here would only be effective at low safety margins.

References

[1] Rees, D. W. A. *Mechanics of Solids and Structures*, Imperial College Press, 2000.
[2] ESDU 71005, Buckling of flat plates in shear, September 1976.
[3] BS EN 10034: 1993, Structural steel I and H-sections.
[4] ESDU 02.03.18, Buckling of curved plates and panels in shear, April 1977.
[5] Gerard, G. Compressive and torsional buckling of thin-walled cylinders in the yield region, NACA Tech Note 3726, 1956.

Exercise

5.1 Referring to Figure 5.3(b), examine whether a symmetrical solid section, having a pointed top and bottom and four concave sides, that are quadrants of a circle with concave curvature, can provide a shape factor $f < 1$. The radius of each quadrant is r, with centre positioned to give a length $2r$ for the horizontal neutral axis (NA).

5.2 Using the appropriate formulae developed in this chapter, optimise the equal-angle section given in Figure 5.12 when the forces F_x and F_y, acting in the directions shown, are applied in the ratio $F_x/F_y = 1/2$, 1 and 2. Examine the effect of reversing the direction of F_x at each ratio.

5.3 Find the shape factor for the channel section shown in Figure 5.16 and optimise its dimensions when they are fixed within the ratios $d/a = 2$ and $a/t = 20$. Assuming that no plasticity occurs, take the material properties required from Appendix A (for steel) and the elastic buckling coefficient K_e from Figure C.3 in Appendix C.

5.4 A thin-walled, isosceles triangle, with non-uniform thickness, is to be loaded at its shear centre E with a vertical force F_y parallel to its shortest side, as shown in Figure 5.25. Determine expressions for the side lengths a and b and the respective thicknesses t_a and t_b which minimise the weight. What is the shape factor?

5.5 Determine, from first principles, the optimum dimensions and the shape factor for a uniform rectangular tube with sides $b \times d$ and thickness t, when a vertical force F_y is applied at the shear centre (E in Figure 5.26). Check your answers with the appropriate reductions to the corresponding non-uniform rectangular tube equations given in § 5.4.4.

5.6 Show that there is no minimum shape factor for the rectangular tube in Figure 5.26 when the side ratio b/d is allowed to vary in the range $0 < b/d \leq 1$. What is the shape factor when $b/d = 0.5$ and 1?
[Answers: $f = 3.6, 4.5$]

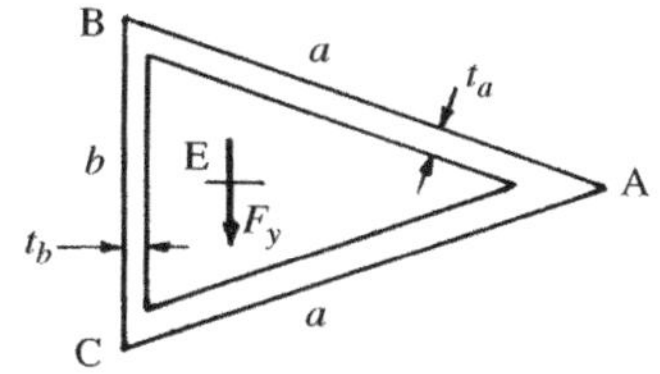

Figure 5.25 Isosceles triangular tube with non-uniform thickness

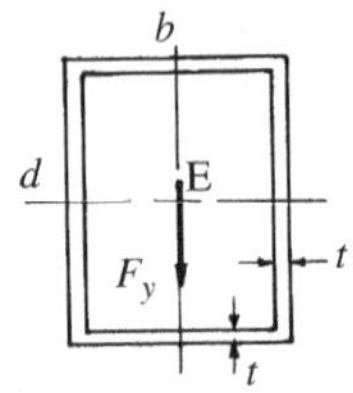

Figure 5.26 Uniformly thin rectangular tube

5.7 A vertical force F_y acts at the shear centre of a uniform, thin-walled, regular hexagonal tube thickness t and side length a. The hexagon is arranged with its two opposite sides to lie parallel to this force, as shown in Figure 5.27. What is the maximum, net shear flow for the section and where is it located? Hence show that the shape factor for this section is 2.1. [Answer: $q_{\max} = 7F_y/20a$, at the centre of each vertical side]

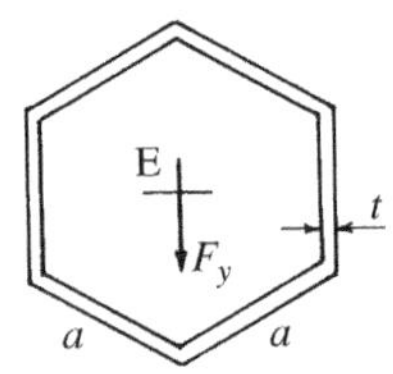

Figure 5.27 Regular hexagonal tube with uniform thickness

5.8 Show that the optimum dimensions of the hexagonal tube in Exercise 5.7 are those given by equations (5.82a,b) in which $Q = 7/20$.

6

Combined Shear and Torsion in Thin-Walled Sections

6.1 Introduction

When a transverse shear force does not act through the shear centre of a given cross-section its net shear flow is a consequence of combined torsion and shear [1]. Thus, in open sections St Venant's torsion is superimposed upon flexural shear, resulting in a combination of their respective shear flows, which were considered separately in Chapters 4 and 5. In closed sections it is more convenient to treat the two effects together, but if they were to be separated, the torsional effect, i.e. the Bredt–Batho shear flow, amounts to taking the difference between the net shear flow given here and that of pure flexure given in Chapter 5. To demonstrate this, we shall restrict our analyses to certain types of thin-walled open and closed cross-sections that can be found in Chapter 5, all of which were subjected to pure flexural shear by virtue of the shear force acting at their shear centres. The optimum designs given here apply to where the shear force is displaced from the shear centre. The maximum shear stress arising from the shear flow in each section corresponds to both the chosen design stress and the shear stress required for buckling. The latter refers to shear buckling of the thin individual limbs within open sections (I, L, T, U, etc.) and within the thin walls of closed polygonal sections (e.g. triangular, rectangular and hexagonal tubes). Their 'flats' are treated as long, flat plates with particular edge supports in a given aspect ratio for which the critical buckling stress is known. Varying the thickness of the flats within these sections allows for a simultaneous buckling of the limbs to occur within the most structurally efficient design, where the weight is reduced to its absolute minimum.

6.2 Thin-Walled, Open Sections

Since the vertical shear force F_y is not applied at the shear centre E for the following cross-sections, the shear stress arises from the combined effects of transverse shear and torsion.

Mechanics of Optimal Structural Design: Minimum Weight Structures D. W. A. Rees

6.2.1 Channel – Vertical Force Aligned with the Web

For the channel section in Figure 6.1a, the force F_y is vertically aligned with the web. Figures 6.1b,c show how its flexural (shear) and torsional effects may be separated.

(a) Greatest Net Stress in the Web

With the force F_y shifted to E (Figure 6.1b), the previous analysis holds (see § 5.3.4) in which equation (5.54c) gave the maximum shear stress at the web's neutral axis position. With a torque $T = F_y e$ applied at E (Figure 6.1c), a St Venant torsion analysis applies (see § 4.3.4). For a channel section with uniform thickness, set $t_a = t_b$ in equation (4.34a) to give the maximum, torsional shear stress around the edge contour as:

$$\tau_{\max} = \frac{3T}{(d+2a)t^2} = \frac{3F_y(e/a)(a/t)}{dt[1+2(a/d)]} \tag{6.1a}$$

where e/a is given by equation (5.58b). When F_y is aligned with the web (Figure 6.1a), equation (6.1a) is added to equation (5.54c) to give the greatest net shear stress in the web at mid-depth:

$$\begin{aligned}\tau_{w_{\text{net}}} &= \frac{3F_y[1+4(a/d)]}{2dt[1+6(a/d)]} + \frac{3F_y}{dt[1+2(a/d)]} \times \frac{3(a/d)(a/t)}{[1+6(a/d)]} \\ &= \frac{3F_y\{[1+4(a/d)][1+2(a/d)]+6(a/d)(a/t)\}}{2dt[1+2(a/d)][1+6(a/d)]} = \frac{QF_y}{dt}\end{aligned} \tag{6.1b}$$

where $Q = Q(a/d)$ reappears within the usual design criterion, in which a limit for equation (6.1b) is set at τ_y, under which the web buckles:

$$\tau_y = \frac{QF_y}{dt} = K_d E_T \left(\frac{t}{d}\right)^2 \tag{6.2a}$$

The optimum dimensions for the section follow from equation (6.2a):

$$t_{\text{opt}} = \left(\frac{Q^2}{K}\right)^{1/4}\left(\frac{F_y^2}{\tau_y E_T}\right)^{1/4}, \qquad d_{\text{opt}} = (Q^2K)^{1/4}\left(\frac{F_y^2 E_T}{\tau_y^3}\right)^{1/4} \tag{6.2b,c}$$

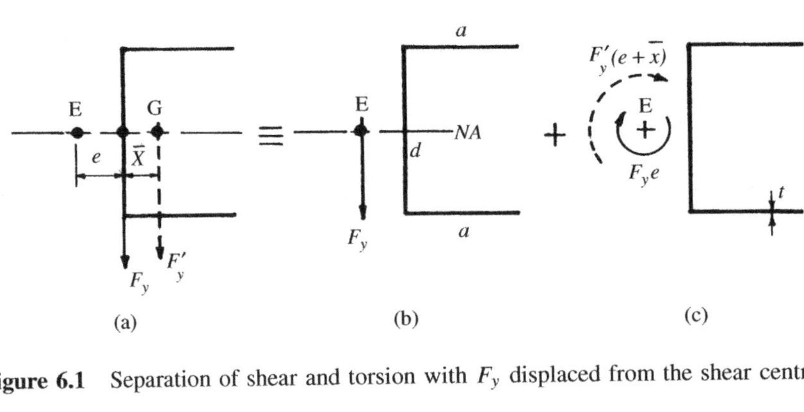

Figure 6.1 Separation of shear and torsion with F_y displaced from the shear centre E

Within the objective function, the shape factor f is identified as follows:

$$\left(\frac{W}{L^3}\right)_{\text{opt}} = f\left(\frac{\rho}{\tau_y}\right)\left(\frac{F}{L^2}\right), \quad \Rightarrow \quad f = \frac{3\{[1+2(a/d)][1+4(a/d)]+6(a/d)(a/t)\}}{2[1+6(a/d)]} \tag{6.3a,b}$$

Comparing equations (5.56) and (6.3b) shows by how much the shape factor is raised when F_y is displaced e from the shear centre.

(b) Greatest Net Stress in the Flanges

Torsion also increases the shear stress in the flange. Adding equation (5.53) to equation (6.1a), the greatest net shear stress in the flange becomes

$$\begin{aligned}\tau_{f\text{net}} &= \frac{6F_y(a/d)}{dt[1+6(a/d)]} + \frac{3F_y}{dt[1+2(a/d)]} \times \frac{3(a/d)(a/t)}{[1+6(a/d)]} \\ &= \frac{6F(a/d)[1+2(a/d)+(3/2)(a/t)]}{dt[1+2(a/d)][1+6(a/d)]}\end{aligned} \tag{6.4}$$

Taking the ratio between the net shear stresses in the flange and the web allows both to buckle simultaneously when

$$\frac{\tau_{f\text{net}}}{\tau_{w\text{net}}} = \frac{K_a}{K_b}\left(\frac{d}{a}\right)^2 \tag{6.5a}$$

Substituting equations (6.1b) and (6.4) into equation (6.5a), the channel's dimension ratios must satisfy

$$\begin{aligned}&4(a/d)^3[1+2(a/d)+(3/2)(a/t)] \\ &\quad = (K_a/K_d)\{[1+2(a/d)][1+4(a/d)]+6(a/d)(a/t)\}\end{aligned} \tag{6.5b}$$

Equation (6.5b) has been solved for a/d by trial for a/t values in the range $5 \le a/t \le 50$. For this, Figure C.3 has been used to find the buckling coefficient ratio K_a/K_d at each a/t. Long lengths are assumed, i.e. $d/L \le 0.1$, when finding this ratio. For example, when $a/t = 10$, we find that equation (6.5b) is satisfied by $a/d = 0.90$ with $K_a/K_d = 0.784$. The difference between K_a and K_d reflects the difference in edge support between this section's web and flanges. Plot (i) in Figure 6.2 shows that a/d is weakly dependent upon a/t in this range, falling from 0.92 to its steady value of 0.89 for $a/t = 50$ and beyond.

Figure 6.2(i) may also be used in combination with equations (6.2b,c). The latter allow the optimum dimensions t, d and a to be separated after having read a/d from this figure at a given a/t.

6.2.2 Channel – Vertical Force Applied at the Centroid

When F_y is applied at positions other than the shear centre, the torque T in equation (6.1a) must be modified. For example, let F_y' be applied vertically downwards at the centroid

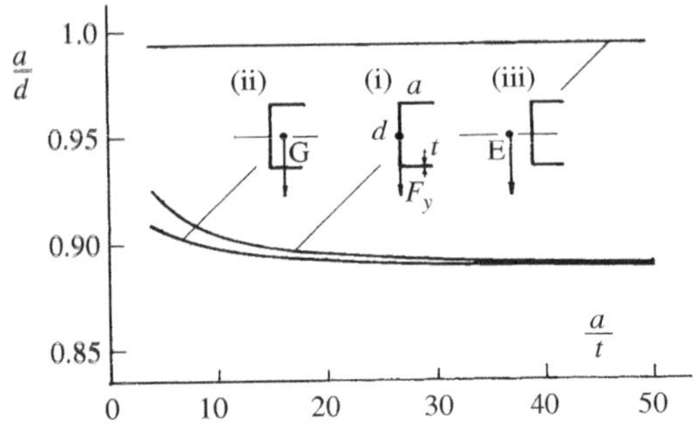

Figure 6.2 Relation between a/d and a/t for simultaneous buckling of web and flange when F_y acts at: (i) the web, (ii) the centroid and (iii) the shear centre

G of the section, as in Figure 6.1a. Equivalently, Figure 6.1b remains unchanged but $T = F'_y(e + \overline{X})$ in Figure 6.1c. Now, from equations (5.52a) and (5.58b),

$$\frac{e}{a} + \frac{\overline{X}}{a} = \frac{4(a/d)[1 + 3(a/d)]}{[1 + 2(a/d)][1 + 6(a/d)]} \tag{6.6a}$$

Equation (6.1a) gives the maximum shear stress due to torsion around the edge contour as

$$\tau_{\max} = \frac{3T}{(d + 2a)t^2} = \frac{3F_y(e/a + \overline{X}/a)(a/t)}{dt[1 + 2(a/d)]} \tag{6.6b}$$

(a) Greatest Net Stress in the Web

Substituting equation (6.6a) into equation (6.6b) and adding to equation (5.54c) gives the greatest net shear stress in the web at its mid-depth (i.e. the neutral axis)

$$\tau_{W\text{net}} = \frac{3F'_y[1 + 4(a/d)]}{2\,dt[1 + 6(a/d)]} + \frac{12F'_y(a/d)(a/t)[1 + 3(a/d)]}{dt[1 + 2(a/d)]^2[1 + 6(a/d)]}$$
$$= \frac{3F'_y}{2dt}\left\{\frac{[1 + 4(a/d)][1 + 2(a/d)]^2 + 8(a/d)(a/t)[1 + 3(a/d)]}{[1 + 2(a/d)]^2[1 + 6(a/d)]}\right\} = \frac{QF'_Y}{dt}$$

With $Q = Q(a/d)$ redefined within equation (6.6c), we may find again the optimum dimensions for web buckling under shear from otherwise similar equations (6.2b,c). Moreover, the shape factor f in equation (6.3a) becomes $f = [1 + 2(a/d)]Q$.

(b) Greatest Net Stress in the Flanges

Adding equations (6.6a,b) to equation (5.53), the increased torque raises the greatest net shear stress in the flange to

$$\tau_{f\text{net}} = \frac{6F'_y(a/d)}{dt[1 + 6(a/d)]}\left\{1 + \frac{2(a/t)[1 + 3(a/d)]}{[1 + 2(a/d)]^2}\right\} \tag{6.7}$$
$$= \frac{6F'_y(a/d)}{dt}\left\{\frac{[1 + 2(a/d)]^2 + 2(a/t)[1 + 3(a/d)]}{[1 + 2(a/d)]^2[1 + 6(a/d)]}\right\}$$

Applying equation (6.5a) to the ratio between the net stresses in equations (6.6c) and (6.7) ensures that, if web and flange buckling are to coincide, their dimensions must satisfy

$$4(a/d)^3\{[1+2(a/d)]^2+2(a/t)[1+3(a/d)]\}$$
$$=(K_a/K_d)\{[1+4(a/d)][1+2(a/d)]^2+8(a/d)(a/t)[1+3(a/d)]\} \qquad (6.8)$$

Equation (6.8) has again been solved for a/d using Figure C.3, taking a/t values in the range $5 \le a/t \le 100$. Plot (ii) in Figure 6.2 shows that the optimum ratio a/d is marginally less than that when F_y is aligned with the web, there being a negligible difference at the highest a/t. These contrast with the constant value $a/d \approx 1$ shown in plot (iii) in Figure 6.2 when F_y acts vertically at the shear centre. Here equation (5.57b) has been applied to the simultaneous buckling condition in which both geometric ratios defining this figure apply to long channels, i.e. where $d/L \le 0.1$. However, we note that as the solution to a/d from equation (5.57b) is relatively insensitive to shorter channels in the range $0.1 \le d/L \le 1$ (see Figure C.3), so too are equations (6.5b) and (6.8).

6.3 Thin-Walled, Closed Tubes

The shear flow in a closed tube under a single vertical force F_y applied at any position is written generally as $q = q_b + q_0$, where q_b is the flexural shear flow [1, 2],

$$q_b = \frac{F_y D_{xb}}{I_x} = \frac{F_y}{I_x}\int_A y\,\mathrm{d}A \qquad (6.9)$$

Unlike for an open section, we now have no free surface, where $q_0 = 0$, to take as the origin for s. So for a closed tube, a constant q_0 must be added to q_b. The sum $q_b + q_0$ ensures static equivalence between the net shear flow and the torque that arises when F_y does not pass through the shear centre. In the case of an asymmetric tube this equivalence is written as

$$F_y p = \oint q_b R\,\mathrm{d}s + q_0 \oint R\,\mathrm{d}s \qquad (6.10)$$

where p is the horizontal (perpendicular) distance between F_y and any convenient reference point P, and R is the perpendicular distance from P to the incremental force $q_b\delta s$ within the tube wall (see Figure 6.3).

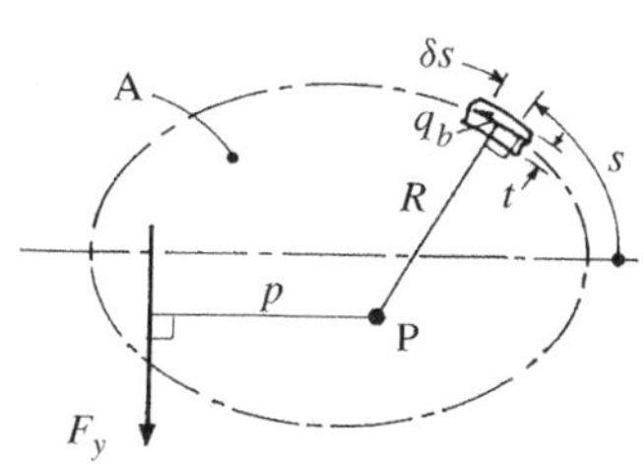

Figure 6.3 Static equivalence between torque ($F_y p$) and shear flow q_b

The second path integral in equation (6.10) is twice the area A enclosed by the mid-wall. Hence, the second term in equation (6.10) becomes the contribution from the St Venant torque $2Aq_0$. Here it is convenient to consider the shear flows from transverse shear and St Venant torsion together rather than to separate them, since we neither have pure flexure nor pure torsion under F_y. This allows q_0 to be identified with the shear flow at the origin chosen for the mid-wall path s. In § 5.3 we considered the special case where F_y passed through the shear centre E, in which the net shear flow was written as $q = q_b + q_E$, where equations (5.67a,b) and (5.68a,b) apply. As the q_b shear flows found from equation (5.67b) remain unaltered in regard to the position of F_y, we may use the q_b expressions derived previously in Chapter 5 for a given cross-section. However, q_0 now needs to be found to establish the modified net shear flow. It is convenient to treat the two effects together, but, if they need to be separated, the torsional effect amounts to subtracting the shear flows given in Chapter 5 from the net shear flows given here. That is,

$$(q_b + q_0) - (q_b + q_E) = q_0 - q_E$$

i.e. the torsional shear flow, arising from a transverse force displaced from the shear centre, is the difference between the 'initial' shear flows for each case.

6.3.1 *Thin-Walled Circular Tube*

Let a vertical force F_y be applied tangentially to the wall of a thin, circular tube of mean wall radius r and uniform wall thickness t, as shown in Figure 6.4. The q_b shear flow was found in § 5.4.1. That is, equation (5.70a) gives the dependence of the basic shear flow upon θ as

$$q_{\theta b} = \frac{F_y r^2 t(1 - \cos\theta)}{\pi r^3 t} = \frac{F_y}{\pi r}(1 - \cos\theta) \tag{6.11a}$$

which is distributed in the manner of Figure 6.5a.

Since F_y does not now lie at the shear centre (i.e. the circle's centre) q_0 must be added to q_b. To find q_0 we apply equations (6.10) and (6.11a) as follows:

$$F_y r = \oint_s q_b r \, \mathrm{d}s + 2Aq_0 = \frac{F_y r}{\pi} \oint_\theta (1 - \cos\theta)\, \mathrm{d}\theta + 2\pi r^2 q_0$$

$$= \frac{F_y r}{\pi} \left| \theta - \sin\theta \right|_0^{2\pi} + 2\pi r^2 q_0 = 2F_y r + 2\pi r^2 q_0, \quad \Rightarrow \quad q_0 = -\frac{F_y}{2\pi r} \tag{6.11b}$$

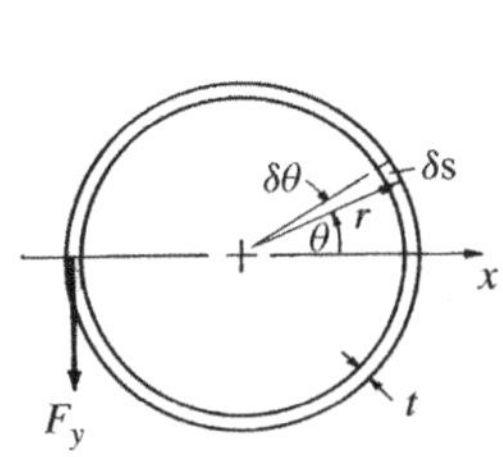

Figure 6.4 Thin-walled circular tube under transverse shear

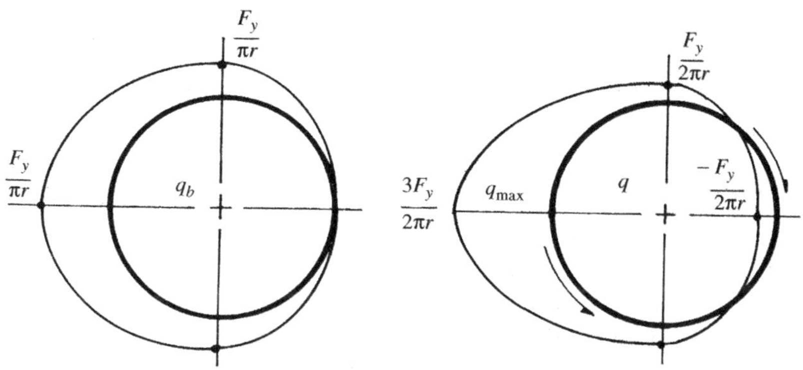

Figure 6.5 Shear flow in a thin circular tube

Hence, the net shear flow distribution, shown in Figure 6.5b, is the sum of equations (6.11a,b):

$$q_{\theta,\,\mathrm{net}} = \frac{F_y}{\pi r}(1-\cos\theta) - \frac{F_y}{2\pi r} = \frac{F_y}{2\pi r}(1-2\cos\theta) \tag{6.11c}$$

which has its maximum $q_{\max} = 3F_y/2\pi r$ at $\theta = 180°$, giving the maximum shear stress as

$$\tau_{\max} = \frac{3F_y}{2\pi r t} \tag{6.12}$$

When $\tau_{\max}$ is limited to the design stress τ_y we may identify from the product rt within equation (6.12) a shape factor f directly from the weight expression $W = 2\pi r t L\rho$ as follows:

$$\left(\frac{W}{L^3}\right) = \frac{2\pi L\rho}{L^3} rt = \frac{2\pi L\rho}{L^3}\frac{3F_y}{2\pi\tau_y} = 3\left(\frac{\rho}{\tau_y}\right)\left(\frac{F_y}{L^2}\right) \tag{6.13}$$

revealing a less efficient shape factor of 3, compared to 2, when F_y acts at the shear centre (see equation (5.73)).

For a long tube, $(L/r)^2(t/r) > 40$, the optimum dimensions follow from imposing the buckling condition, equation (4.53b) at the maximum shear stress:

$$\tau_y = \frac{3F_y}{2\pi r t} = K_p E_T \left(\frac{t}{r}\right)^{3/2} \tag{6.14a}$$

which gives

$$t_{\mathrm{opt}} = \frac{1}{K_p^{1/3}}\left(\frac{3}{2\pi}\right)^{1/2} \frac{F_y^{1/2}}{E_T^{1/3}\tau_y^{1/6}}, \quad r_{\mathrm{opt}} = K_p^{1/3}\left(\frac{3}{2\pi}\right)^{1/2} \frac{F_y^{1/2}E_T^{1/3}}{\tau_y^{5/6}} \tag{6.14b,c}$$

These optimum dimensions depend upon the force, material and geometrical instability. We have seen previously in § 5.4.1 how Gerard's plasticity reduction factor μ (see Figure 4.8) for long tubes [3] allows for plastic buckling through the use of E_T and K_p within equations (6.14b,c).

6.3.2 Uniform, Thin-Walled, Equilateral Triangular Tube

Let a vertical force F_y be applied along the vertical wall BC of an equilateral triangular tube section ABC in the manner of Figure 6.6a. With side length a and uniform wall-thickness t, the q_b shear flows within each wall were established previously in Figure 5.21a. These are given from equations (5.78a–c) as

$$q_{AB} = \frac{F_y s^2}{a^3} \tag{6.15a}$$

$$q_{BC} = \frac{2F_y}{a^3}(as - s^2) + \frac{F_y}{a} \tag{6.15b}$$

$$q_{CA} = -\frac{2F_y}{a^3}\left(as - \frac{s^2}{2}\right) + \frac{F_y}{a} \tag{6.15c}$$

where s has its origin at the respective corners A, B and C. The net shear flow is found from adding q_0, which we can now find from equation (6.10). When moments are taken about point A in Figure 6.6a there are no contributions to the shear flow from the sloping sides. This leaves only q_{BC} in equation (6.15b) to be substituted into equation (6.10) as follows:

$$F_y\left(\frac{\sqrt{3}a}{2}\right) = \frac{\sqrt{3}a}{2}\int_0^a q_{BC}\,\mathrm{d}s + 2Aq_0 \tag{6.16a}$$

$$= \frac{\sqrt{3}a}{2}\int_0^a \left[\frac{2F_y}{a^3}(as - s^2) + \frac{F_y}{a}\right]\mathrm{d}s + 2\left(\frac{a}{2}\times\frac{\sqrt{3}a}{2}\right)q_0$$

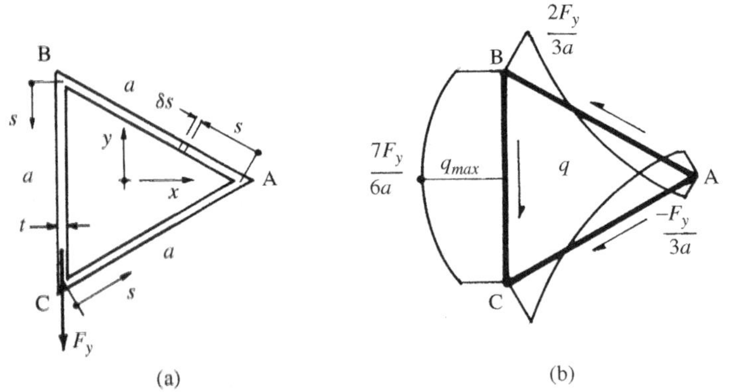

Figure 6.6 Shear flow in a thin-walled, triangular tube

in which q_0 follows from the integration as

$$q_0 = \frac{F_y}{a} - \frac{F_y}{a}\left|\frac{2}{a^3}\left(\frac{as^2}{2} - \frac{s^3}{3}\right) + \frac{s}{a}\right|_0^a = -\frac{F_y}{3a} \tag{6.16b}$$

Adding q_0 from (6.16b) in turn to q_{AB}, q_{BC} and q_{CA} in equations (6.15a–c) gives the net shear flow in each limb (see Figure 6.6b). It is apparent from this that the greatest net shear flow occurs at the mid-side BC (i.e. for $s = a/2$):

$$q_{\max} = \frac{3F_y}{2a} - \frac{F_y}{3a} = \frac{7F_y}{6a} \tag{6.17a}$$

Correspondingly, the maximum shear stress at this position is

$$\tau_{\max} = \frac{7F_y}{6at} \tag{6.17b}$$

upon which the optimum design is to be based for $\tau_{\max} \leq \tau_y$. Using $Q = 7/6$ to denote the numerical factor within the greatest shear stress, the objective function follows from the weight $W_{\text{opt}} = 3atL\rho$:

$$\left(\frac{W}{L^3}\right)_{\text{opt}} = 3Q\left(\frac{\rho}{\tau_y}\right)\left(\frac{F_y}{L^2}\right) \tag{6.18}$$

in which the shape factor $f = 3Q = 3.5$. With the force applied along side BC, the optimum dimensions t_{opt} and a_{opt} follow from allowing BC to buckle under the maximum shear stress. The condition for this is

$$\tau_y = \frac{QF_y}{at} = K_a E_T\left(\frac{t}{a}\right)^2 \tag{6.19a}$$

giving

$$t_{\text{opt}} = \left(\frac{Q^2}{K_a}\right)^{1/4}\left(\frac{F_y^2}{\tau_y E_T}\right)^{1/4}, \qquad a_{\text{opt}} = (Q^2 K_a)^{1/4}\left(\frac{F_y^2 E_T}{\tau_y^3}\right)^{1/4} \tag{6.19b,c}$$

where K_a refers to a plate in shear with both edges simply-supported (see Figure C.3). When, in Figure 5.20, F_y is displaced to the shear centre E, $Q = 5/6$ in equations (6.18) and (6.19a–c). This leads to the more favourable shape factor $f = 2.5$, found previously in equation (5.81), compared to $f = 3.5$ here.

6.3.3 *Non-Uniform, Equilateral, Triangular Tube*

The previous shear flows in equations (6.15a–c) again apply since they do not depend upon thickness. Given that less shear stress exists in the sloping sides AB and AC, a further reduction to the weight should be possible by reducing their thicknesses (see Figure 6.7).

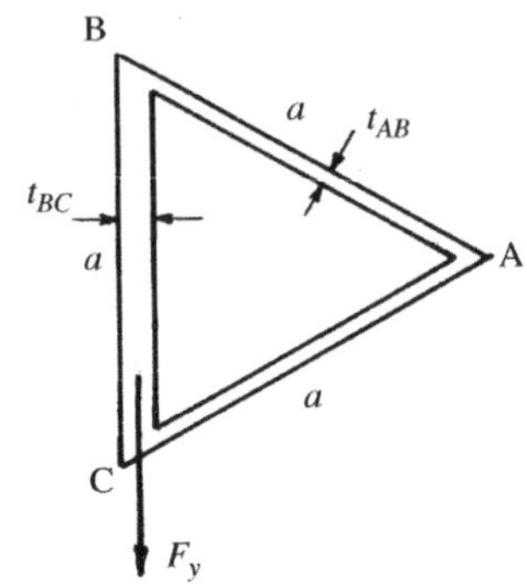

Figure 6.7 Variable thickness triangular tube

If we reduce thickness when F_y is aligned with BC, a simultaneous buckling of AB and BC occurs when

$$\left(\frac{\tau_{AB}}{\tau_{BC}}\right)_{\max} = \left(\frac{\tau_{AB}}{\tau_{BC}}\right)_{cr} \tag{6.20a}$$

where $(\tau_{BC})_{\max}$ is given by equation (6.17b). Now as $q_{AB} = \tau_{AB} t_{AB}$, it follows from equations (6.15a) and (6.16b) that $\tau_{AB} t_{AB} = F_y/a - F_y/3a$. Substituting the maximum net shear stress in each side along with their critical shear buckling stresses into equation (6.20a) leads to

$$\frac{2F_y}{3at_{AB}} \times \frac{6at_{BC}}{7F_y} = \frac{t_{AB}^2}{t_{BC}^2}, \quad \Rightarrow \quad \frac{t_{AB}}{t_{BC}} = \left(\frac{4}{7}\right)^{1/3} \tag{6.20b}$$

The objective function follows from the weight expression

$$W = a(t_{BC} + 2t_{AB})L\rho = at_{BC}\left[1 + 2\left(t_{AB}/t_{BC}\right)\right]L\rho \tag{6.21a}$$

Substituting from equation (6.20b), with the maximum stress limit $\tau_{\max} = 7F_y/6at_{BC} \le \tau_y$, leads to

$$\frac{W}{L^3} = \frac{7F_y}{6\tau_y}\left[1 + 2\left(\frac{4}{7}\right)^{1/3}\right]\frac{\rho}{L^2} = 3.1\left(\frac{\rho}{\tau_y}\right)\left(\frac{F_y}{L^2}\right) \tag{6.21b}$$

Hence, a lower shape factor of 3.1 applies, compared to a factor of 3.5 for the uniformly thin tube under similar loading. For this further weight reduction, equations (6.19b,c) will again optimise the dimensions provided t_{opt} is identified with t_{BC}. Note that, with $Q = 7/6$ in these equations, F_y is applied vertically along side BC.

6.3.4 *Non-Uniform, Thin-Walled, Rectangular Tube*

Let a vertical force F_y be applied to the left vertical side CD of a rectangular tube of side lengths b and d and wall thicknesses t_b and t_d, in the manner of Figure 6.8a. The q_b

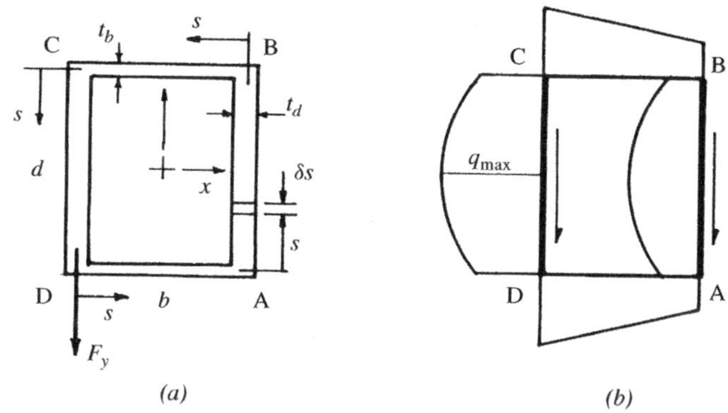

Figure 6.8 Non-uniform rectangular tube showing net shear flow

shear flows (see Figure 5.24a) were established previously, when F_y acted at the shear centre. These are, from equations (5.91a–d):

$$q_{AB} = \frac{F_y t_d}{2I_x}\left(s^2 - ds\right) \tag{6.22a}$$

$$q_{BC} = \frac{dF_y t_b s}{2I_x} \tag{6.22b}$$

$$q_{CD} = \frac{F_y t_d}{2I_x}\left(ds - s^2\right) + \frac{F_y t_b bd}{2I_x} \tag{6.22c}$$

$$q_{DA} = \frac{F_y t_b d}{2I_x}(-s + b) \tag{6.22d}$$

where the origin for s lies at the respective corners A, B, C and D.

It follows from equations (6.22b,c) that maxima in the linear and parabolic distributions lie along sides BC and CD (see Figure 5.24a). They are, for $s = b$ and $s = d/2$ respectively,

$$q_{BC} = \frac{F_y t_b bd}{2I_x}, \quad q_{CD} = \frac{F_y t_d d^2}{8I_x} + \frac{F_y t_b bd}{2I_x} \tag{6.23a,b}$$

where the second moment of area about the x-axis is

$$I_x = \frac{t_d d^3}{12}\left(1 + 3\frac{t_b}{t_d}\frac{b}{d}\right) \tag{6.23c}$$

The net shear flow in each side (see Figure 6.8b) is found by adding q_0 to q_b. Here q_0 is found by applying equation (6.10) in which moments are taken about corner A:

$$F_y b = \left[d\int_0^b q_{BC}\,\mathrm{d}s + b\int_0^d q_{CD}\,\mathrm{d}s\right] + 2Aq_0 \tag{6.24a}$$

Substituting into equation (6.24a) the two shear flows which contribute to this moment equation, i.e. q_{BC} and q_{CD} from equations (6.23a,b),

$$F_y b = \frac{F_y d^2 t_b}{2I_x} \int_0^b s\,\mathrm{d}s + \frac{F_y b t_d}{2I_x} \int_0^d (ds - s^2)\,\mathrm{d}s + \frac{F_y b^2 d t_b}{2I_x} \int_0^d \mathrm{d}s + 2Aq_0$$

$$= \frac{F_y d^2 t_b}{2I_x} \left| \frac{s^2}{2} \right|_0^b + \frac{F_y b t_d}{2I_x} \left| \frac{ds^2}{2} - \frac{s^3}{3} \right|_0^d + \frac{F_y b^2 d t_b}{2I_x} \left| s \right|_0^d + 2bdq_0 \tag{6.24b}$$

Equation (6.24b) leads to

$$q_o = \frac{F_y t_d d^2}{24 I_x} \left(1 - 3 \frac{b}{d} \frac{t_b}{t_d} \right) \tag{6.24c}$$

which may be checked by taking moments about another corner. Hence, Figure 6.8b shows that the net shear flow has its greatest value ($q_{\max}$) at the centre of side CD. This is found by adding q_0 from equation (6.24c) to equation (6.23b):

$$q_{\max} = (q_{CD})_{\max} = \frac{F_y t_d d^2}{6 I_x} \left(1 + \frac{9}{4} \frac{t_b}{t_d} \frac{b}{d} \right) \tag{6.25a}$$

Substituting I_x from equation (6.23c) and with $\tau_{\max} = q_{\max}/t_d$, the maximum shear stress in CD becomes

$$(\tau_{CD})_{\max} = \frac{2F_y[1 + (9/4)(t_b/t_d)(b/d)]}{t_d d[1 + 3(t_b/t_d)(b/d)]} = \frac{QF_y}{t_d d} \tag{6.25b}$$

Limiting $(\tau_{CD})_{\max}$ to τ_y, under which buckling of side CD is to occur, provides the design criterion

$$\tau_y = \frac{QF_y}{dt_d} = K_d E_T \left(\frac{t_d}{d} \right)^2 \tag{6.26a}$$

from which the optimised dimensions follow:

$$(t_d)_{\text{opt}} = \left(\frac{Q^2}{K_d} \right)^{1/4} \left(\frac{F_y^2}{\tau_y E_T} \right)^{1/4}, \qquad d_{\text{opt}} = (Q^2 K_d)^{1/4} \left(\frac{F_y^2 E_T}{\tau_y^3} \right)^{1/4} \tag{6.26b,c}$$

The optimum (minimum) weight is then

$$W_{\text{opt}} = 2(bt_b + dt_d)L\rho = 2dt_d[1 + (b/d)(t_b/t_d)]L\rho \tag{6.27a}$$

and substituting for the product dt_d from equations (6.26b,c) leads to the objective function

$$\frac{W}{L^3} = 2Q\left(1 + \frac{b}{d} \frac{t_b}{t_d} \right) \frac{\rho}{\tau_y} \frac{F}{L^2} = f\left(\frac{\rho}{\tau_y} \right) \left(\frac{F}{L^2} \right) \tag{6.27b}$$

In equation (6.27b) the shape factor f depends upon Q in equation (6.25b), to give

$$f = \frac{4[1 + (9/4)(t_b/t_d)(b/d)][1 + (t_b/t_d)(b/d)]}{1 + 3(t_b/t_d)(b/d)} \quad (6.27c)$$

A further relationship between t_b/t_d and b/d will follow when we arrange for a simultaneous buckling of the side BC (and AD). Note, from the asymmetric shear flow in Figure 6.8b that side AB cannot be made to buckle under its lower shear stress whilst its thickness remains at t_d. The maximum net shear stress in BC occurs at C and is found from $q = q_{BC} + q_0$, where q_{BC} and q_0 are given in equations (6.23a) and (6.24c). This leads to

$$(q_{BC})_{\max} = \frac{F_y t_d d^2}{24 I_x}\left(1 + 9\frac{t_b}{t_d}\frac{b}{d}\right) \quad (6.28a)$$

Substituting I_x from equation (6.23c) and with $\tau_{\max} = q_{\max}/t_b$, the maximum shear stress in BC is

$$(\tau_{BC})_{\max} = \frac{F_y\left[1 + 9(t_b/t_d)(b/d)\right]}{2t_d d(t_b/t_d)\left[1 + 3(t_b/t_d)(b/d)\right]} \quad (6.28b)$$

A simultaneous buckling of CD and BC (and DA) occurs when

$$(\tau_{CD}/\tau_{BC})_{\max} = (\tau_{CD}/\tau_{BC})_{cr} \quad (6.29a)$$

Substituting equations (6.25b) and (6.28b) into equation (6.29a), together with the critical buckling expressions, leads to the following relationship between the flat lengths and thickness ratios:

$$\frac{4(t_b/t_d)[1 + (9/4)(t_b/t_d)(b/d)]}{1 + 9(t_b/t_d)(b/d)} = \frac{K_d}{K_b}\left(\frac{t_d}{t_b}\right)^2\left(\frac{b}{d}\right)^2$$

$$4(K_b/K_d)(t_b/t_d)^3[1 + (9/4)(t_b/t_d)(b/d)] = (b/d)^2[1 + 9(t_b/t_d)(b/d)] \quad (6.29b)$$

The thickness ratios supplied by equation (6.29b) apply to the range of b/d ratios given in Table 6.1. Also calculated from equation (6.27c) are the shape factors at these ratios. The shape factor f, which has no minimum value, increases with b/d to attain highly inefficient values. Clearly, $b/d < 1$ is more desirable for lessening weight as we might expect since the side BC (and AD) suffers the least net stress.

Table 6.1 Length and thickness ratios for three-wall buckling in a rectangular tube

b/d	0.25	0.5	1	2	3	4	5
t_b/t_d	0.28	0.507	0.910	1.540	2.044	2.493	2.902
f	4.094	4.472	6.242	12.64	21.76	33.27	46.88

Note that these solutions apply to a ratio $K_b/K_d = 1$, which may be assumed for long plates (b/L and $d/L < 0.1$) with similar edge restraints (see Figure C.3). Comparing Tables 6.1 and 5.3, the latter shows some reduction in f-values when F_y acts at the shear centre. Moreover, under the latter loading condition the fourth side will also buckle under the balanced net shear flows. In contrast, one or, at most, two sides can be arranged to fail from loading either at the shear centre or along the vertical side when the thickness is uniform throughout.

6.3.5 *Uniform Hexagonal Tube*

We shall minimise the weight of a regular hexagonal tube with a force applied along one vertical side (see Figure 6.9) from the beginning. The second moment of this area about the x-axis is the sum of contributions from the two vertical and four sloping sides:

$$I_x = 2(ta^3/12) + 4\int_0^a (a/2 + s\sin 30^\circ)t\,\mathrm{d}s = 5ta^3/2 \tag{6.30a}$$

and the area enclosed by this tube is

$$A = 6 \times \frac{a}{2} \times \frac{\sqrt{3}a}{2} = \frac{3\sqrt{3}}{2}a^2 \tag{6.30b}$$

The q_b shear flow distributions within the sides follow from equation (6.9) as follows.

For side AB, with origin for s at A,

$$q_{AB} = \frac{F_y t}{I_x}\int_0^s (s - a/2)\,\mathrm{d}s = \frac{F_y t}{2I_x}s(s-a) \tag{6.31a}$$

giving $q_A = q_B = 0$ for $s = 0$ and $s = a$, respectively. The maximum in the parabolic distribution, $q = -F_y ta^2/(8I_x) = -F_y/(20a)$, occurs for $s = a/2$, in which the minus sign implies that q opposes the s-direction, as shown in Figure 6.10a.

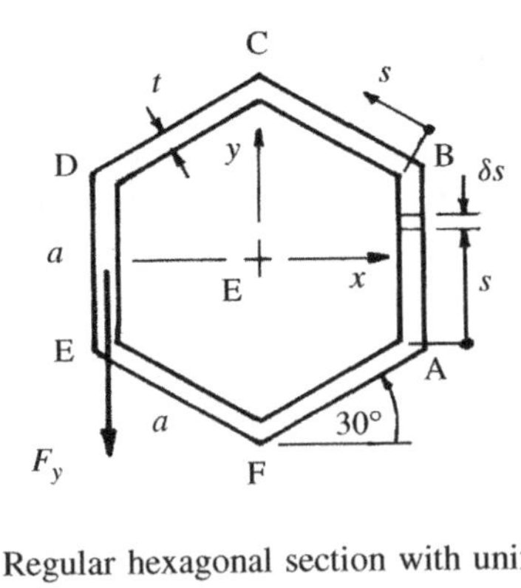

Figure 6.9 Regular hexagonal section with uniform thickness

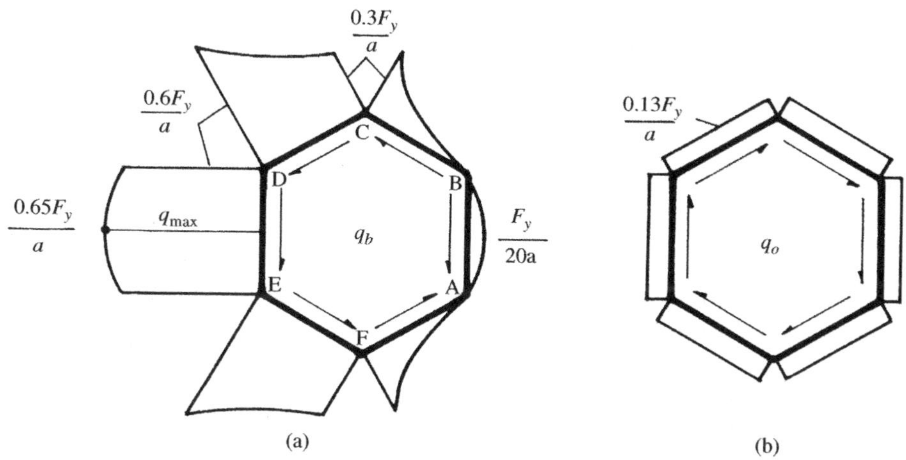

Figure 6.10 Shear flows q_b and q_0 in a uniform hexagonal tube

For side BC, with origin for s at B,

$$q_{BC} = \frac{F_y t}{2I_x}\int_0^s (a+s)\,ds + q_B = \frac{F_y t}{2I_x} s(a+s/2) \tag{6.31b}$$

Equation (6.31b) matches $q_B = 0$ for $s = 0$. The maximum in the parabolic shear flow distribution (see Figure 6.10a) is $q_C = 3F_y ta^2/(4I_x) = 0.3F_y/a$, which occurs at $s = a$.

For side CD, with origin for s at C,

$$q_{CD} = \frac{F_y t}{I_x}\int_0^s [a/2 + (a-s)/2]\,ds + q_C = \frac{F_y t}{I_x}[s(a-s/4) + 3a^2/4] \tag{6.31c}$$

Equation (6.31c) matches q_C for $s = 0$ and gives the maximum in the parabolic shear flow distribution as $q_D = 3F_y ta^2/(2I_x) = 0.6F_y/a$ for $s = a$ (see Figure 6.10a).

For side DE, with origin for s at D,

$$q_{DE} = \frac{F_y t}{I_x}\int_0^s (a/2 - s)\,ds + q_D = \frac{F_y t}{2I_x}[s(a-s) + 3a^2] \tag{6.31d}$$

which matches q_D for $s = 0$ and gives $q_E (= q_D) = 3F_y ta^2/(2I_x) = 0.6F_y/a$ for $s = a$. The maximum in the parabolic distribution, $q_{\max} = 13F_y ta^2/(8I_x) = 0.65F_y/a$, occurs for $s = a/2$, as shown in Figure 6.10a.

For side EF, with origin for s at E,

$$q_{EF} = \frac{-F_y t}{I_x}\int_0^s (a/2 + s/2)\,ds + q_E = \frac{F_y t}{2I_x}\left[3a^2 - s(a+s/2)\right] \tag{6.31e}$$

Equation (6.31e) matches q_E for $s = 0$, which is the maximum in this parabolic distribution. At F, $s = a$ when equation (6.31e) gives $q_F = 3F_y ta^2/(4I_x) = 0.3F_y/a$, as shown in Figure 6.10a.

For side FA, with origin for s at F,

$$q_{FA} = \frac{-F_y t}{I_x} \int_0^s (a - s/2)\,\mathrm{d}s + q_F = \frac{F_y t}{I_x}\left[3a^2/4 - s(a - s/4)\right] \tag{6.31f}$$

matching q_F for $s = 0$, which is the maximum in this parabolic distribution. Equation (6.31f) confirms $q_A = 0$ for $s = a$, as shown in Figure 6.10a.

The net shear flow in each side follows by adding the two distributions q_0 to q_b, given in Figure 6.10. Here, the constant q_0 is found by applying equation (6.10). Taking moments about corner A in Figure 6.10a,

$$F_y(\sqrt{3}a) = \left[\frac{\sqrt{3}a}{2}\int_0^a q_{BC}\,\mathrm{d}s + \sqrt{3}a\int_0^a q_{CD}\,\mathrm{d}s + \sqrt{3}a\int_0^a q_{DE}\,\mathrm{d}s + \frac{\sqrt{3}a}{2}\int_0^a q_{EF}\,\mathrm{d}s\right] + 2Aq_0 \tag{6.32a}$$

Substituting the four shear flows which contribute to this moment equation, i.e. q_{BC}, q_{CD}, q_{DE} and q_{EF} from equations (6.31b–e), each integral in equation (6.32a) becomes

$$\begin{aligned}
\frac{\sqrt{3}a}{2}\int_0^a q_{BC}\,\mathrm{d}s &= \frac{\sqrt{3}a}{2}\left(\frac{F_y t}{2I_x}\right)\int_0^a s(a + s/2)\,\mathrm{d}s = \frac{\sqrt{3}a^4 F_y t}{6I_x}\\
\sqrt{3}a\int_0^a q_{CD}\,\mathrm{d}s &= \sqrt{3}a\left(\frac{F_y t}{I_x}\right)\int_0^a \left[s(a - s/4) + 3a^2/4\right]\mathrm{d}s = \frac{7\sqrt{3}a^4 F_y t}{6I_x}\\
\sqrt{3}a\int_0^a q_{DE}\,\mathrm{d}s &= \sqrt{3}a\left(\frac{F_y t}{2I_x}\right)\int_0^a \left[s(a - s) + 3a^2\right]\mathrm{d}s = \frac{19\sqrt{3}a^4 F_y t}{12I_x}\\
\frac{\sqrt{3}a}{2}\int_0^a q_{EF}\,\mathrm{d}s &= \frac{\sqrt{3}a}{2}\left(\frac{F_y t}{2I_x}\right)\int_0^a \left[3a^2 - s(a + s/2)\right]\mathrm{d}s = \frac{7\sqrt{3}a^4 F_y t}{12I_x}
\end{aligned} \tag{6.32b-e}$$

Substituting equations (6.32b-e) into equation (6.32a) with $2A = 3\sqrt{3}a^2$ and $I_x = 5ta^3/2$ (see equations (6.30a,b)),

$$F_y(\sqrt{3}a) = 7\sqrt{3}F_y a/5 + 3\sqrt{3}a^2 q_0, \quad \Rightarrow \quad q_0 = -\frac{2F_y}{15a} \tag{6.32f}$$

The minus sign in equation (6.32f) indicates the contraflow shown in Figure 6.10b. It follows that the maximum in the net shear flow $q = q_b + q_0$, upon which the optimum design is based, occurs at the centre of DE with magnitude

$$q_{\max} = \frac{13F_y}{20a} - \frac{2F_y}{15a} = \frac{31F_y}{60a} \tag{6.33}$$

When buckling occurs in this side under the maximum shear stress $\tau_{\max} = q_{\max}/t$, we may limit this to our design stress τ_y and write

$$\tau_y = KE_T\left(\frac{t}{a}\right)^2 = \frac{QF_y}{at} \tag{6.34}$$

in which $Q = 31/60$. Solving equation (6.34) shows that the optimum dimensions a_{opt} and t_{opt} are again given by similar equations (6.16b,c) to those found for the triangular tube. The minimum weight follows from $W = 6atL\rho$ in which equation (6.34) gives the optimum product $at = QF_y/\tau_y$. Consequently, the objective function gives the minimum weight as

$$\left(\frac{W}{L^3}\right)_{\min} = 6Q\left(\frac{\rho}{\tau_y}\right)\left(\frac{F_y}{L^2}\right) \tag{6.35}$$

where the shape factor $f = 6Q = 3.1$. When the force F_y is displaced to the shear centre E, it can be shown that $Q = 7/20$ in equations (6.34) and (6.35). Hence, this position leads to a more favourable shape factor $f = 2.1$, compared to 3.1 found presently.

6.4 Concluding Remarks

The analysis of shear loading in beams with thin-walled cross-sections is complicated when the transverse force is offset from the shear centre of the section. Both torsion and flexural shear stress contribute to a raising of the beam's weight, thereby offsetting the greatest weight saving that can be achieved when the force is applied at the shear centre. Fortunately, it is not necessary to separate the two effects when determining the position and magnitude of the maximum shear stress upon which the optimum design is based. Whether the thin section be open or closed it is possible to optimise its dimensions by limiting the maximum stress to a design stress under which limb buckling also occurs. This ensures that the most highly stressed material is used to its full load bearing capacity. In the presence of stress gradients, the under-stressed material elsewhere is not used as efficiently unless limb dimensions can be adjusted to raise the stress to a similar design level and thereby save weight further. In this way the limbs concerned contribute more effectively in supporting the load and will buckle simultaneously at the load limit. In practice, of course, such failure is prevented by imposing a margin of safety such that the maximum stress, upon which the optimum design is based, is a fraction of the material's limiting stress. However, raising the safety factor will result in a weight increase and, therefore,

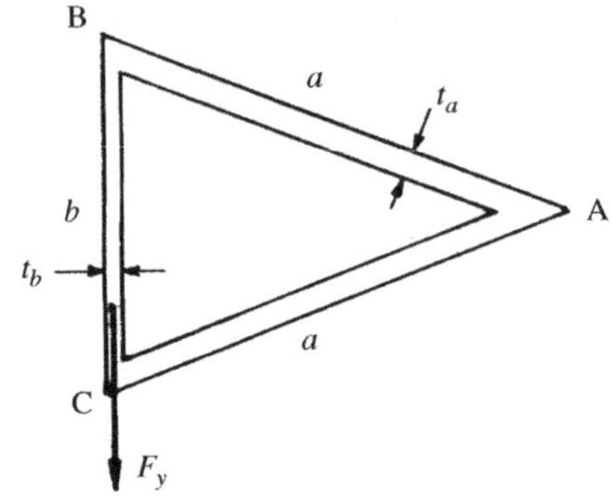

Figure 6.11 Non-uniform, isosceles triangular tube

deciding upon a suitable trade off between safety and weight remains a judgmental issue with each application.

References

[1] Heins, C. P. *Bending and Torsional Design in Structural Members*, Lexington, MA, 1975.
[2] Trahair, N. S. and Bradford, M. A. *The Behaviour and Design of Steel Structures*, Chapman & Hall, 1977.
[3] Gerard G. Compressive and torsional buckling of thin-walled cylinders in yield region, NACA Tech Note 3726, August 1956.

Exercises

6.1 A thin-walled isosceles triangle, with non-uniform thickness, is to be loaded with a transverse shear force F_y along its shortest, vertical side, as shown in Figure 6.11. Determine expressions for the optimum side lengths a and b and the respective thicknesses t_a and t_b which minimise the weight. Compare shape factors when F_y is displaced from the left vertical side to act at the shear centre of this section (*Hint*: see § 5.4.2).

6.2 Determine, from first principles, the optimum dimensions and the shape factor for a uniform rectangular tube of sides $b \times d$ and thickness t, when a vertical force F_y is applied along the left vertical side (as in Figure 6.8a). Check your answers by making the appropriate reductions to the corresponding non-uniform, rectangular tube equations given in § 6.3.4.

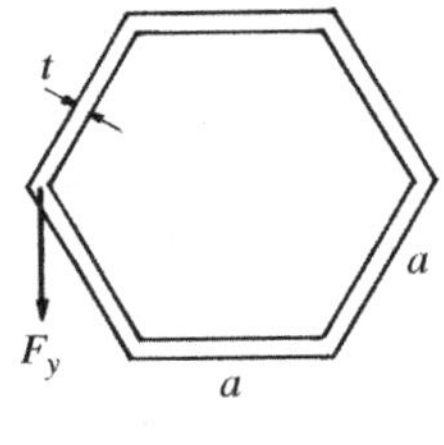

Figure 6.12 Uniform hexagonal tube under a corner force

6.3 What is the lowest shape factor that can be expected from the tube in Exercise 6.2 when the ratio b/d is allowed to vary in the range $0 < b/d \leq 1$?

6.4 Examine how best to reduce the wall thicknesses to lessen weight for the thin-walled, regular hexagonal tube in Figure 6.9.

6.5 Show that, when a vertical force acts on the shear centre E of the uniform hexagonal tube in Figure 6.9, the shape factor is at its minimum value of 2.1.

6.6 Determine the shape factor for the uniformly thin, regular, hexagonal tube shown in Figure 6.12. Compare with the shape factor when the vertical force is displaced from the corner to the shear centre.

7

Combined Shear and Bending in Idealised Sections

7.1 Introduction

When a beam carries a combination of distributed and concentrated loading, the material within any transverse cross-section is subjected to both bending and shear. The usual design of a solid beam section involves the calculation of the principal stresses at the positions in the length and depth where the bending and shear stresses attain their maximum values. However, the positions at which the respective stresses attain their maxima are not coincident. Within the cross-section, the maximum bending stress occurs at the outer surface, the maximum shear stress at the neutral axis. Thus, it may not be adequate to base the design upon one or other maximum by following the general rule: bending for long beams, shear for short 'beams', the latter including pins, rivets and gear teeth. This is especially true for materials where the limiting stresses for bending and shear differ widely. The design of thin-walled sections allows certain simplifying assumptions to be made. For example, in Chapter 5 it was observed that the flanges of an open cross-section carry most of the bending moment while the web carries most of the shear force. Therefore, in an ideal structure it becomes necessary to design the web to support the whole of the transverse shear force and the flanges to support all the bending moment. This is normally done within a web-boom idealisation in which the flanges are replaced with in-line concentrations of area A (the booms) whose second moment summation $I = \sum Ay^2$ about the neutral axis (NA) matches that of the original section (see Figure 7.1). The idealisation is used for both thin-walled open and closed (tubular) beam sections in order to simplify the precise shear flow calculations adopted in the previous two chapters. Recall that there the maximum shear stress in the wall governed the design. The idealisation provides a good estimate of this stress which can then be applied to the original section to optimise its dimensions and minimise weight as before.

7.2 Idealised Beam Sections

Figure 7.1 shows the idealisation for an I-beam that is subjected to both a bending moment M and a shear force F. Firstly, the tensile and compressive bending stresses σ in the

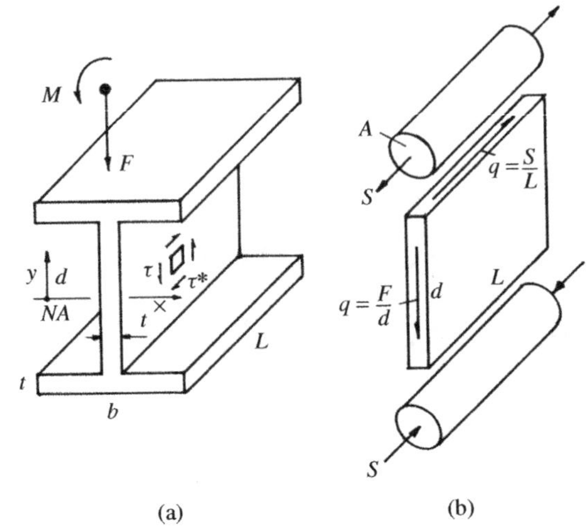

Figure 7.1 Forces in a web-boom idealisation for an I-section

booms are found. These stresses would normally be carried by the flanges in their support of the bending moment M (see Figure 7.1a). Bending theory gives M as [1]

$$M = \sigma I/y \tag{7.1a}$$

where $y = d/2$ is the distance between A and the neutral axis of bending. Within this idealisation, the tensile and compressive forces in the booms $S = \sigma A$ (see Figure 7.1b) also give this moment as a couple:

$$M = Sd = \sigma Ad \tag{7.1b}$$

Equations (7.1a,b) agree for $y = d/2$ and $I = 2A(d/2)^2$. In its support of the transverse shear force F, the shear stress in the web τ (see Figure 7.1a) is assumed to be constant over its depth d such that

$$F = \tau \times td = qd \tag{7.2a}$$

where $q = \tau t = F/d$ is the constant shear flow or force per unit web depth. Now q is found from the first moment summation of boom areas lying above the neutral axis. That is, from the first term in equation (5.35),

$$q = \frac{F}{I}\sum_{r=1}^{n}(Ay)_r \tag{7.2b}$$

The relationship between the boom and web forces follows from the nature of complementary shear in the web. That is, $\tau = \tau^*$ (see Figure 7.1a) where $\tau = F/(td)$ and with the web top and bottom areas supporting the boom forces S, it follows that $\tau^* = S/(Lt)$. Hence,

$$\frac{F}{td} = \frac{S}{tL}, \quad \Rightarrow \quad S = \frac{FL}{d} \tag{7.3}$$

Therefore, $q = S/L = F/d$ is also constant over the length of the web when S and F are constant (see Figure 7.1b), this being assumed for the idealised sections that follow. In idealising a cross-section in this way we shall see next how the boom area is calculated by adding to the flange area a contribution from the connecting webs. The simplified shear analysis also applies to thin plates placed in shear unaccompanied by bending, i.e. where S is a constant shear force applied directly along the top surface of a plate where an equal and opposite tangential force is applied to the bottom surface. We shall refer to this in Chapter 8 as the mode of loading in a shear web whose shape may be corrugated or stiffened to resist the shear force without buckling.

7.2.1 Calculation of Boom Areas

To calculate the two boom areas A_1 and A_2 for an I-section, appearing within the formulae (7.1) and (7.2), each flange area is added to a contribution from their adjacent web(s). When replacing the flange with a boom, the second moment of area must be preserved about the section's centroidal axis x. For a flange area A_f, whose individual centroid g lies at a distance y from x, the parallel axis theorem states (for the rth flange) that

$$I_{xr} = I_{gr} + A_{fr} y_r^2 \tag{7.4a}$$

Provided that the second term is dominant, with y_r being large and I_{gr} being small for a thin-walled flange, then $I_{xr} \approx A_{fr} y_r^2$. Thus, the flange contribution to the boom area A_{fr} is simply the product of the flange breadth and thickness, positioned at perpendicular distance y_r from x, i.e. along the web's centre-line. An example of the boom positioning for a double channel section is shown in Figure 7.2.

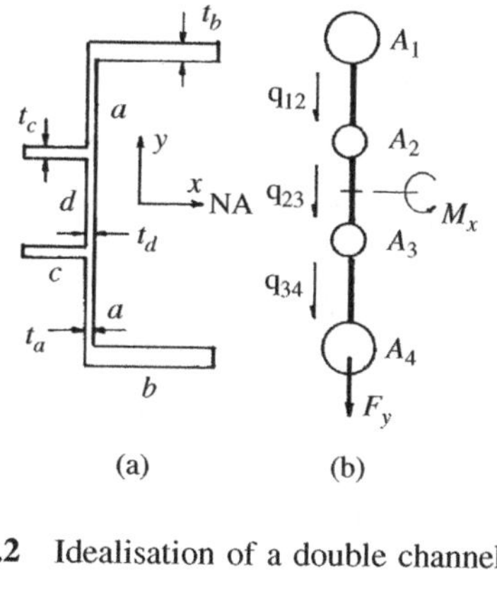

Figure 7.2 Idealisation of a double channel section

Allowing for the different thicknesses, the boom areas in the idealisation (Figure 7.2b) are

$$A_{f1} = A_{f4} = bt_b,\ A_{f2} = A_{f3} = ct_c \quad \Rightarrow \quad A_{fr} = (bt)_r \tag{7.5a}$$

The constant shear flows q_{12}, q_{23} and $q_{34} = q_{12}$ between the booms follow from equation (7.2b). Thicker webs contribute to the boom areas, adding to equation (7.5a) to ensure that the true bending stress and second moment of area are preserved within the idealisation. The web contribution to the rth boom is expressed as [2, 3]

$$A_{wr} = \frac{s_r t_r}{6}\left(2 + \frac{y_{r-1}}{y_r}\right) + \frac{s_{r+1} t_{r+1}}{6}\left(2 + \frac{y_{r+1}}{y_r}\right) \tag{7.5b}$$

Hence, the total boom areas are $A_r = A_{fr} + A_{wr}$, for $r = 1, \ldots, 4$. We shall see how to apply equations (7.5a,b) in this and other examples that follow. The idealisation given above simplifies the stress calculations for bending about the x-axis under a moment M_x in combination with a vertical shear force F_y. It is best employed for sections where stiffeners and corner reinforcement amount to large concentrations of area in which the contributions from webs are negligible, i.e. $A_{fr} > A_{wr}$

Where an additional bending moment M_y and shear force F_x exist, a second set of boom areas are required to preserve I_y when calculating their contributions to the net bending stress and shear flow. Consequently, there is no advantage in using an idealised section when the boom areas are derived from the web areas alone. This would require the use of equation (7.5b) and its counterpart in which x replaces y, as the following example shows.

Example 7.1 Find the areas of four- and six-boom idealisations of the lipped channel section shown in Figure 7.3, when bending occurs about both the x and y centroidal axes.

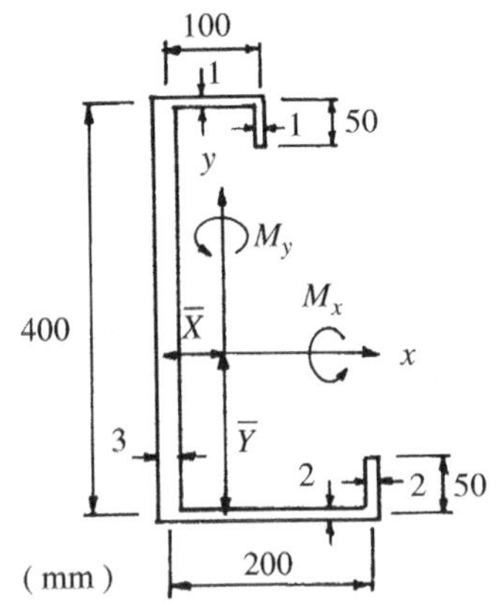

Figure 7.3 Lipped channel section

Firstly, the position of the centroid $(\overline{X}, \overline{Y})$ is found by taking first moments of area about each side:

$$(50 \times 1)(400 - 25) + (100 \times 1 \times 400) + (400 \times 3 \times 200) + (200 \times 2 \times 1)$$
$$+(50 \times 2 \times 25) = 1850\overline{Y}$$
$$\overline{Y} = 163.05 \text{ mm}$$
$$(50 \times 1 \times 100) + (100 \times 1 \times 50) + (400 \times 3 \times 1.5) + (200 \times 2 \times 100)$$
$$+(50 \times 2 \times 200) = 1850\overline{X}$$
$$\overline{X} = 38.81 \text{ mm}$$

The I_x and I_y values that are required to be matched within each of the following idealisations are found from the mean wall lengths as:

$$I_x = [1 \times 50^3/12 + 50(375 - 163.05)^2] + 100(400 - 163.05)^2$$
$$+ [3 \times 400^3/12 + (400 \times 3)(200 - 163.05)^2]$$
$$+ (200 \times 2)163.05^2 + [2 \times 50^3/12 + 100(163.05 - 25)^2]$$
$$= 38.1 \times 10^6 \text{ mm}^4$$
$$I_y = (50 \times 100^2) + [1 \times 100^3/12 + 100(50 - 38.81)^2] + (400 \times 3)(38.81 - 1.5)^2$$
$$+ [2 \times 200^3/12 + 400(100 - 38.81)^2] + 100(200 - 38.81)^2 = 7.69 \times 10^6 \text{ mm}^4$$

In Figure 7.4 the boom positions are numbered 0, 1, ..., 5, for each idealisation as shown.

Table 7.1 applies to the boom dimensions for both the four- and six-boom idealisations with respect to the x- and y-axes (see Figure 7.4).

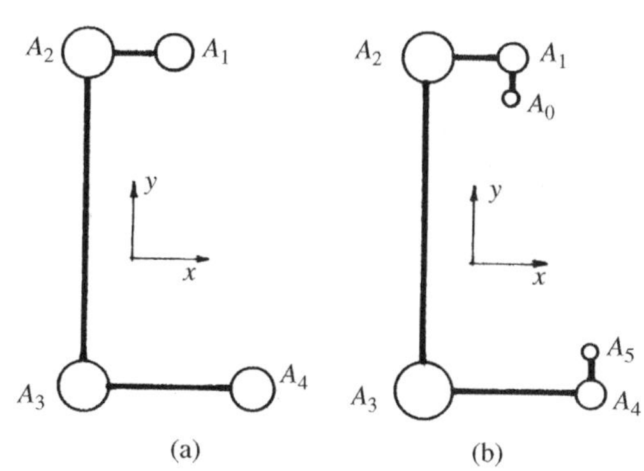

Figure 7.4 Four- and six-boom idealisations of a lipped channel

Table 7.1 Dimensions (mm) for boom area calculations

Boom	0	1	2	3	4	5
s	–	50	100	400	200	50
t	–	1	1	3	2	2
y	187	237	237	−163.1	−163.1	−113.1
x	62	62	−38.8	−38.8	161.2	161.2

(i) Boom areas referred to the x-axis
For the four-boom areas (see Figure 7.4a), the webs contribute as follows:

$$A_{w1} = (s_2t_2/6)(2 + y_2/y_1) = (100 \times 1/6)(2 + 1) = 50 \text{ mm}^2$$

$$A_{w2} = (s_2t_2/6)(2 + y_1/y_2) + (s_3t_3/6)(2 + y_3/y_2)$$
$$= (100 \times 1/6)(2 + 1) + (400 \times 3/6)(2 - 163.1/237) = 50 + 262.4 = 312.4 \text{ mm}^2$$

$$A_{w3} = (s_3t_3/6)(2 + y_2/y_3) + (s_4t_4/6)(2 + y_4/y_3)$$
$$= (400 \times 3/6)(2 - 237/163.1) + (200 \times 2/6)(2 + 1) = 109.4 + 200 = 309.4 \text{ mm}^2$$

$$A_{w4} = (s_4t_4/6)(2 + y_3/y_4) = (200 \times 2/6)(2 + 1) = 200 \text{ mm}^2$$

Adding the lip areas for booms 1 and 2, the areas of each boom are:

$$A_1 = A_{w1} + A_{f1} = 50 + 50 = 100 \text{ mm}^2$$
$$A_2 = 312.4 \text{ mm}^2$$
$$A_3 = 309.4 \text{ mm}^2$$
$$A_4 = A_{w4} + A_{f4} = 200 + 100 = 300 \text{ mm}^2$$

We may now check I_x based upon these boom areas:

$$I_x = (100 + 312.4)(400 - 163.05)^2 + (300 + 309.4)163.05^2 = 39.36 \times 10^6 \text{ mm}^4$$

For six booms (see Figure 7.4b), their areas are found from equation (7.1b) and Table 7.1 as follows:

$$A_{w0} = (s_1t_1/6)(2 + y_1/y_0) = (50 \times 1/6)(2 + 237/187) = 27.20 \text{ mm}^2$$

$$A_{w1} = (s_1t_1/6)(2 + y_0/y_1) + (s_2t_2/6)(2 + y_2/y_1)$$
$$= (50 \times 1/6)(2 + 187/237) + (100 \times 1/6)(2 + 1) = 23.2 + 50 = 73.2 \text{ mm}^2$$

$$A_{w2} = (s_2t_2/6)(2 + y_1/y_2) + (s_3t_3/6)(2 + y_3/y_2)$$
$$= (100 \times 1/6)(2 + 1) + (400 \times 3/6)(2 - 163.1/237) = 50 + 262.4 = 312.4 \text{ mm}^2$$

$$A_{w3} = (s_3t_3/6)(2 + y_2/y_3) + (s_4t_4/6)(2 + y_4/y_3)$$
$$= (400 \times 3/6)(2 - 237/163.1) + (200 \times 2/6)(2 + 1) = 109.4 + 200 = 309.4 \text{ mm}^2$$

$$A_{w4} = (s_4t_4/6)(2 + y_3/y_4) + (s_5t_5/6)(2 + y_5/y_4)$$
$$= (200 \times 2/6)(2 + 1) + (50 \times 2/6)(2 + 113.1/163.1) = 244.9 \text{ mm}^2$$
$$A_{w5} = (s_5t_5/6)(2 + y_4/y_5) = (50 \times 2/6)(2 + 163.1/113.1) = 57.37 \text{ mm}^2$$

Here the contributions from the lips appear within the additional areas A_0 and A_5. This gives a closer estimate of I_x:

$$I_x = 27.2(350 - 163.05)^2 + (73.2 + 312.4)(400 - 163.05)^2 + (309.4 + 244.9)163.05^2$$
$$+ 57.37(163.05 - 50)^2 = 38.07 \times 10^6 \text{ mm}^4$$

(ii) Boom areas referred to the y-axis
For the four-boom areas (see Figure 7.4a), the webs contribute as follows:

$$A_{w1} = (s_2t_2/6)(2 + x_2/x_1) = (100 \times 1/6)(2 - 38.8/62) = 22.9 \text{ mm}^2$$
$$A_{w2} = (s_2t_2/6)(2 + x_1/x_2) + (s_3t_3/6)(2 + x_3/x_2)$$
$$= (100 \times 1/6)(2 - 62/38.8) + (400 \times 3/6)(2 + 1) = 6.7 + 600 = 606.7 \text{ mm}^2$$
$$A_{w3} = (s_3t_3/6)(2 + x_2/x_3) + (s_4t_4/6)(2 + x_4/x_3)$$
$$= (400 \times 3/6)(2 + 1) + (200 \times 2/6)(2 - 161.2/38.8) = 600 - 143.64 = 456.4 \text{ mm}^2$$
$$A_{w4} = (s_4t_4/6)(2 + x_3/x_4) = (200 \times 2/6)(2 - 38.8/161.2) = 117.3 \text{ mm}^2$$

Adding the flange areas to booms 1 and 2 gives the areas of each boom as:

$$A_1 = A_{w1} + A_{f1} = 22.9 + 50 = 72.9 \text{ mm}^2$$
$$A_2 = 606.7 \text{ mm}^2$$
$$A_3 = 456.4 \text{ mm}^2$$
$$A_4 = A_{w4} + A_{f4} = 117.3 + 100 = 217.3 \text{ mm}^2$$

We may now check I_y based upon these boom areas:

$$I_y = 72.9(100 - 38.8)^2 + (606.7 + 456.4)38.8^2 + 217.3(200 - 38.8)^2$$
$$= 7.52 \times 10^6 \text{ mm}^4$$

Using six booms (Figure 7.4b), their areas are found from equation (7.1b) and Table 7.1 as follows:

$$A_{w0} = (s_1t_1/6)(2 + x_1/x_0) = (50 \times 1/6)(2 + 1) = 25 \text{ mm}^2$$
$$A_{w1} = (s_1t_1/6)(2 + x_0/x_1) + (s_2t_2/6)(2 + x_2/x_1)$$
$$= (50 \times 1/6)(2 + 1) + (100 \times 1/6)(2 - 38.8/62) = 25 + 22.9 = 47.9 \text{ mm}^2$$
$$A_{w2} = (s_2t_2/6)(2 + x_1/x_2) + (s_3t_3/6)(2 + x_3/x_2)$$
$$= (100 \times 1/6)(2 - 62/38.8) + (400 \times 3/6)(2 + 1) = 6.7 + 600 = 606.7 \text{ mm}^2$$

$$A_{w3} = (s_3t_3/6)(2 + x_2/x_3) + (s_4t_4/6)(2 + x_4/x_3)$$
$$= (400 \times 3/6)(2 + 1) + (200 \times 2/6)(2 - 161.2/38.8) = 600 - 143.64$$
$$= 456.43 \text{ mm}^2$$
$$A_{w4} = (s_4t_4/6)(2 + x_3/x_4) + (s_5t_5/6)(2 + x_5/x_4)$$
$$= (200 \times 2/6)(2 - 38.8/161.2) + (50 \times 2/6)(2 + 1) = 117.28 + 50 = 167.28 \text{ mm}^2$$
$$A_{w5} = (s_5t_5/6)(2 + x_4/x_5) = (50 \times 2/6)(2 + 1) = 50 \text{ mm}^2$$

Here the contributions from the lips appear within the additional areas A_0 and A_5. This agrees with the four-boom estimate of I_y:

$$I_y = (25 + 47.9)(100 - 38.8)^2 + (606.7 + 456.43)38.8^2 + (167.28 + 50)(200 - 38.8)^2$$
$$= 7.52 \times 10^6 \text{ mm}^4$$

7.2.2 *Calculation of Shear Flow and Bending Stress*

Unlike the previous example, a single set of boom areas will apply to both axes x and y if the booms are derived from concentration of area lying well away from the centroidal axes, as with stiffeners supporting a thin skin (see Figure 7.5). In this case the skin has a negligible effect upon the boom areas used in the stress calculations. The bending stress within the rth boom is written based on equation (7.1a) as

$$\sigma_r = \frac{M_x y_r}{I_x} + \frac{M_y x_r}{I_y} = \frac{M_x y_r}{\sum_{i=1}^{n} A_i y_i^2} + \frac{M_y x_r}{\sum_{i=1}^{n} A_i x_i^2} \tag{7.6}$$

in which the summation applies across all boom areas ($i = 1, 2, \ldots, r - 1, r, r + 1, \ldots, n$). The basic shear flow within their adjacent 'webs' is derived from equation (5.59):

$$q_{r-1,r,b} = \frac{F_y}{I_x} D_{r-1,x,b} + \frac{F_x}{I_y} D_{r-1,y,b} = \frac{F_y}{\sum_{i=1}^{n} A_i y_i^2} \sum_{i=1}^{r-1} A_i y_i + \frac{F_x}{\sum_{i=1}^{n} A_i x_i^2} \sum_{i=1}^{r-1} A_i x_i \tag{7.7}$$

to which q_0 must be added if the section is closed. In Figure 7.5, for example, equations (7.6) and (7.7) would be applied in an anticlockwise sense for $r = 4$, to give σ_4 and q_{34b} as:

$$\sigma_4 = \frac{M_x y_4}{I_x} + \frac{M_y x_4}{I_y}, \quad q_{34b} = \frac{F_y}{I_x}(A_1 y_1 + A_2 y_2 + A_3 y_3) + \frac{F_x}{I_y}(A_1 x_1 + A_2 x_2 + A_3 x_3)$$

in which the signs of moments and forces are shown positive and the signs of coordinates (x_i, y_i) follow the usual convention. Equivalent forces and moments are used within

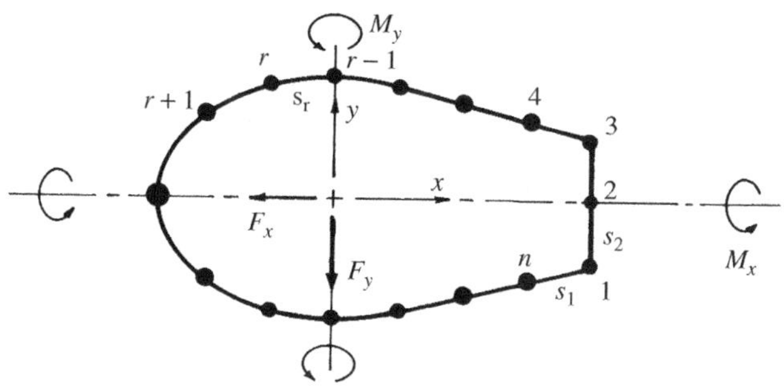

Figure 7.5 Idealised, stiffened closed tube

equations (7.6) and (7.7) when a section is asymmetric. Thus, M_x and M_y in equation (7.6) are replaced by the equivalent moments [4]:

$$\overline{M}_x = \frac{M_x - M_y(I_{xy}/I_y)}{1 - I_{xy}^2/(I_x I_y)}, \quad \overline{M}_y = \frac{M_y - M_x(I_{xy}/I_x)}{1 - I_{xy}^2/(I_x I_y)} \tag{7.8a,b}$$

and F_x, F_y in equation (7.7) are replaced by the equivalent forces:

$$\overline{F}_y = \frac{F_y - F_x(I_{xy}/I_y)}{1 - I_{xy}^2/(I_x I_y)}, \quad \overline{F}_x = \frac{F_x - F_y(I_{xy}/I_x)}{1 - I_{xy}^2/(I_x I_y)} \tag{7.9a,b}$$

As the shear force is the derivative of the bending moment with respect to length z, the correspondence between the moments and forces in equations (7.8a) and (7.9a) is

$$F_y = \mathrm{d}M_x/\mathrm{d}z, \quad F_x = \mathrm{d}M_y/\mathrm{d}z \tag{7.10a,b}$$

$$\overline{F}_y = \mathrm{d}\overline{M}_x/\mathrm{d}z, \quad \overline{F}_x = \mathrm{d}\overline{M}_y/\mathrm{d}z \tag{7.11a,b}$$

7.3 Idealised Open Sections

Bending and shear in any open section may be treated with the simplified web-boom idealisation given that the thicknesses of flats are small in comparison to their lengths. Otherwise, the web contribution to the boom area is required, as will now be shown.

7.3.1 I-Section

Figure 7.6a shows the idealisation for an I-beam of breadth b and depth d, with uniform thickness t, given earlier in Figure 7.1b. In this case equations (7.5a,b) give each boom area as $A = bt + dt/6$ at $y = \pm d/2$ from the neutral axis. Hence, $I_x = 2Ay^2$ restores our previous expression (3.45b) for this I-section:

$$I_x = 2A\left(\frac{d}{2}\right)^2 = \left(b + \frac{d}{6}\right)\frac{td^2}{2} \tag{7.12}$$

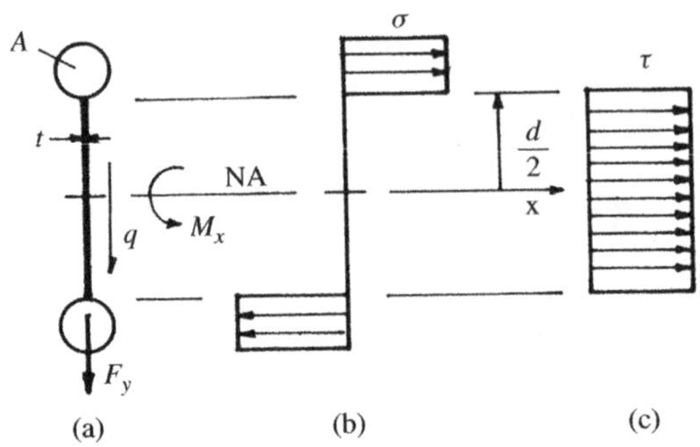

Figure 7.6 Web-boom representation for an I-section

The idealised section provides a constant bending stress $\sigma = M_x d/(2I)$ in the boom and a constant shear stress $\tau = F_y/(dt)$ within the web, as shown in Figure 7.6 b,c. These are used to optimise the dimensions as follows.

(a) Bending

When a bending moment M_x is applied about the x-axis, the constant bending stress within the boom follows from equation (7.12) as:

$$\sigma = \frac{M_x d}{2I_x} = \frac{M_x d}{2} \times \frac{4}{2Ad^2} = \frac{M_x}{dA} \tag{7.13a}$$

Thus, in its support of the moment, the distance between the booms, i.e. the depth of the web, in our idealised beam section must be set at

$$d_{\text{opt}} = \frac{M_x}{A\sigma_y} \tag{7.13b}$$

where σ_y is the maximum (design) stress allowed in the boom.

(b) Shear

Given that d cannot be altered because of the moment requirement in equation (7.13b), the optimum thickness for the web follows from the usual limit placed upon the shear stress $\tau \leq \tau_y$, which also becomes the critical stress for web buckling:

$$\tau_y = \frac{F}{dt} = K_d E_T \left(\frac{t}{d}\right)^2 \tag{7.14a}$$

Setting $F = qd$, where $q = \tau t$, the optimum thickness for the given depth in equation (7.13b) follows from equation (7.14a) as:

$$t_{\text{opt}} = \left(\frac{qd^2}{K_d E_T}\right)^{1/3} \tag{7.14b}$$

where K_d is a function of d/L and the degree of edge restraint. Figure C.3 shows, for example, that $K_d = 4.84$ as $d/L \to 0$ for a very long, simply supported web. Combining equations (7.14a,b) sets the design stress at

$$\tau_y = \left[K_d E_T \left(\frac{q}{d}\right)^2\right]^{1/3} \tag{7.15}$$

where q/d is the structural index. Thus, when τ_y is predetermined, the structural index follows from equation (7.15). The web's optimum (minimum) weight per unit length is found from $w = W/L = \rho d t_{\mathrm{opt}}$, when, from equation (7.14b),

$$w_{\mathrm{opt}} = \rho d \left(\frac{qd^2}{K_d E_T}\right)^{1/3} = \frac{\rho d^2}{(K_d E_T)^{1/3}} \left(\frac{q}{d}\right)^{1/3} \tag{7.16}$$

which applies where the lengthwise shear force S is constant over the given length, i.e. for the constant shear flow $q = F/d = S/L$ in Figure 7.1b.

In this design two allowable (design) stresses σ_y and τ_y appear, one direct stress and one shear stress. In fact, one can be derived from the other since it has been assumed there is no interaction between them. The von Mises and Tresca yield criteria provide $\sigma_y = \sqrt{3}\tau_y$ and $\sigma_y = 2\tau_y$, respectively. Tresca's relation is the more conservative when τ_y is to be found from σ_y. Note here that we have assumed that $\sigma_{\max} = \sigma_y$ and $\tau_{\max} = \tau_y$ in equations (7.13b) and (7.14a).

7.3.2 Lipped Channel Section

A four-boom idealisation of the symmetrical channel, shown in Figure 7.7a, becomes necessary with it having two additional, outer vertical lips. Here we shall calculate boom areas from equations (7.5a,b) knowing that areas $A_1 = A_4$ and $A_2 = A_3$ are separated by a and d, as shown in Figure 7.7b. This gives the boom areas A_1 and A_2 as a sum of flange and web contributions, where A_1 subsumes the whole length of the lip:

$$A_{f1} = ct_c, \quad A_{w1} = \frac{at_a}{6}\left(2 + \frac{d/2}{d/2}\right) = \frac{at_a}{2}, \Rightarrow A_1 = A_{f1} + A_{w1} = ct_c + \frac{at_a}{2} \tag{7.17a}$$

$$A_{f2} = 0, \quad A_{w2} = \frac{at_a}{6}\left(2 + \frac{d/2}{d/2}\right) + \frac{dt_d}{6}\left(2 + \frac{-d/2}{d/2}\right) = \frac{at_a}{2} + \frac{dt_d}{6} = A_2 \tag{7.17b}$$

(a) Bending under M_x

The second moment of the four areas, $2 \times A_1$ and $2 \times A_2$, about the x-axis is

$$I_x = 2\left[A_1\left(\frac{d}{2}\right)^2 + A_2\left(\frac{d}{2}\right)^2\right] = \frac{d^2}{2}(A_1 + A_2) \tag{7.18}$$

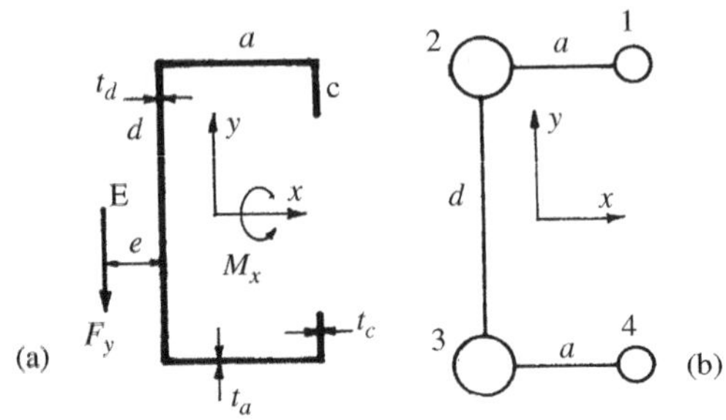

Figure 7.7 A 4-boom idealisation of a lipped channel

where A_1 and A_2 are identified in equations (7.17a,b). With a moment M_x applied about the x-axis (see Figure 7.7a), tension is produced in booms 1 and 2, compression in booms 3 and 4, as shown in Figure 7.8a. From these, the maximum allowable bending stress is

$$\sigma = \frac{M_x d}{2I_x} = \frac{M_x d}{2} \times \frac{2}{d^2(A_1 + A_2)} = \frac{M_x}{d(A_1 + A_2)} \tag{7.19a}$$

and from equation (7.19a), with $\sigma = \sigma_y$ as our design stress, the depth of the web is set at

$$d_{\text{opt}} = \frac{M_x}{\sigma_y(A_1 + A_2)} \tag{7.19b}$$

(b) Shear under F_y
The web shear flows follow from equation (7.7) as:

$$q_{12} = \frac{F_y}{I_x} A_1 \frac{d}{2}, \quad q_{23} = \frac{F_y}{I_x}\left(A_1 \frac{d}{2} + A_2 \frac{d}{2}\right),$$
$$q_{34} = \frac{F_y}{I_x}\left(A_1 \frac{d}{2} + A_2 \frac{d}{2} - A_2 \frac{d}{2}\right) = q_{12} \tag{7.20a,b,c}$$

Each q in equations (7.20a–c) is distributed uniformly between the booms, as shown in Figure 7.8b. Combining equations (7.18) and (7.20a–c), these shear flows become

$$q_{12} = q_{34} = \frac{F_y}{d(1 + A_2/A_1)}, \quad q_{23} = \frac{F_y}{d} \tag{7.21a,b}$$

and, as d has been fixed within equation (7.19b), an optimum thickness for the vertical web $(t_d)_{\text{opt}}$ follows from equation (7.14b) and (7.21b):

$$(t_d)_{\text{opt}} = \left(\frac{q_{23} d^2}{K_d E_T}\right)^{1/3} \tag{7.22}$$

It is possible to optimise dimension a for the horizontal webs by assuming a different thickness t_a initially. Let the ratio of the shear stresses in each web equal the ratio of

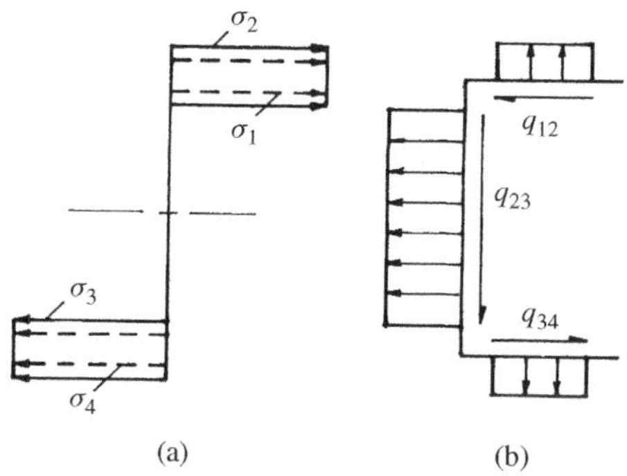

Figure 7.8 Bending stress and shear flow in an idealised lipped-channel section

their critical buckling stresses. A simultaneous buckling of all three webs will then occur when

$$\frac{q_{12}t_d}{q_{23}t_a} = \frac{K_a}{K_d}\left(\frac{t_a}{a}\right)^2\left(\frac{d}{t_d}\right)^2 \tag{7.23a}$$

and substituting q_{12} and q_{23} from equations (7.21a,b) leads to

$$\left(\frac{a}{d}\right)^2 = \frac{K_a}{K_d}\left(1 + \frac{A_2}{A_1}\right)\left(\frac{t_a}{t_d}\right)^3 \tag{7.23b}$$

Equation (7.23b) shows that the thickness ratio can remain at unity while a/d is adjusted. Figure C.3 indicates that the buckling coefficient for the web will differ from that for each flange by their edge restraints. The ratio a/d and $(t_d)_{\text{opt}}$ reappear in the optimum, uniformly distributed, weight expression $w = \rho d(t_d)_{\text{opt}}(1 + 2a/d)$. Substituting from equations (7.22) and (7.23b) leads to

$$\begin{aligned} w_{\text{opt}} &= \frac{\rho d^2}{(K_d E_T)^{1/3}}\left(1 + 2\frac{a}{d}\right)\left(\frac{q_{23}}{d}\right)^{1/3} \\ &= \frac{\rho d^2}{(K_d E_T)^{1/3}}\left[1 + 2\sqrt{\frac{K_a}{K_d}\left(1 + \frac{A_2}{A_1}\right)}\right]\left(\frac{q_{23}}{d}\right)^{1/3} \end{aligned} \tag{7.24}$$

in which $q_{23} = F/d$ expresses the known force to be supported over the fixed depth.

(c) Bending under M_x and M_y

With M_x and M_y acting about the centroidal axes x and y (see Figure 7.9a), the stresses within each boom may be found from equation (7.6). Example 7.1 showed that the boom areas for bending about x are different from those for bending about y when these areas are calculated from the adjacent webs. However, with L-shaped stiffeners placed at the corners 2 and 3 with lip 'attachments' at 1 and 4, their areas alone may be used to define similar boom areas for both axes when the connecting webs are thin. Hence, for similar

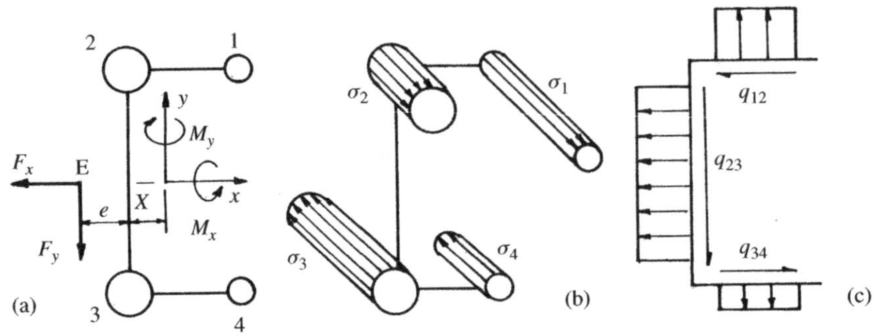

Figure 7.9 Boom stresses and web shear flow arising from bending and shear in both x- and y-axes

concentrations of area at positions 1 to 4, we may refer $A_1 = A_4$ and $A_2 = A_3$ to both axes x and y. The centroid position $\overline{X}$ from the vertical side is found to be

$$2A_1 a = 2(A_1 + A_2)\overline{X}, \quad \Rightarrow \quad \overline{X} = \frac{aA_1}{A_1 + A_2}, \quad a - \overline{X} = \frac{aA_2}{A_1 + A_2} \tag{7.25a,b}$$

Equations (7.25a,b) give the second moments of area:

$$I_x = 2(A_1 + A_2)\left(\frac{d}{2}\right)^2, \quad I_y = \frac{2A_1 A_2^2 a^2}{(A_1 + A_2)^2} + \frac{2A_2 A_1^2 a^2}{(A_1 + A_2)^2} = \frac{2A_1 A_2 a^2}{A_1 + A_2} \tag{7.26a,b}$$

With the appropriate sign of x, y, taken for each boom within equation (7.6), the bending stresses follow from equations (7.26a,b) as:

$$\sigma_1 = \frac{M_x(d/2)}{2(A_1 + A_2)(d/2)^2} + \frac{M_y a A_2(A_1 + A_2)}{(A_1 + A_2)2A_1 A_2 a^2} = \frac{M_x}{d(A_1 + A_2)} + \frac{M_y}{2aA_1} \tag{7.27a}$$

$$\sigma_2 = \frac{M_x(d/2)}{2(A_1 + A_2)(d/2)^2} - \frac{M_y a A_1(A_1 + A_2)}{(A_1 + A_2)2A_1 A_2 a^2} = \frac{M_x}{d(A_1 + A_2)} - \frac{M_y}{2aA_2} \tag{7.27b}$$

$$\sigma_3 = -\frac{M_x}{d(A_1 + A_2)} - \frac{M_y}{2aA_2}, \quad \sigma_4 = -\frac{M_x}{d(A_1 + A_2)} + \frac{M_y}{2aA_1} \tag{7.27c,d}$$

The stresses in equations (7.27a–d) act longitudinally, either in tension or compression, within the boom areas in the manner of Figure 7.9b. In placing a limit on the greatest of the boom stresses, i.e. $\sigma_1 \le \sigma_y$, equation (7.27a) gives

$$\sigma_1 = \sigma_y = \frac{M_x}{2dA_1}\left[\frac{2A_1}{A_1 + A_2} + \frac{dM_y}{aM_x}\right] \tag{7.28a}$$

Equation (7.28a) provides the web depth d in terms of the area, depth and moment ratios:

$$d = \frac{M_x}{2A_1\sigma_y}\left[\frac{2}{1 + A_2/A_1} + \frac{M_y/M_x}{a/d}\right] \tag{7.28b}$$

(d) Shear under F_x and F_y

Let the two shear forces act at the shear centre E whose x-coordinate, e in Figure 7.9a, may be found from the shear flow q_{12} when F_y acts alone at E. Hence, using the previous equation (7.20a) for q_{12}, we take moments about the web centre to give:

$$F_y e_x = q_{12}\,ad = \frac{F_y ad}{d(1 + A_2/A_1)}, \quad \Rightarrow \quad e_x = \frac{a}{1 + A_2/A_1} \tag{7.29}$$

When F_x is present, the shear flows follow from equation (7.7). Working anticlockwise from boom 1,

$$q_{12} = \frac{F_y A_1 y_1}{I_x} + \frac{F_x A_1 x_1}{I_y} = \frac{F_y A_1 d}{d^2(A_1 + A_2)} + \frac{F_x A_1 A_2 a}{A_1 + A_2} \times \frac{A_1 + A_2}{2A_1 A_2 a^2}$$

$$= \frac{F_y A_1}{d(A_1 + A_2)} + \frac{F_x}{2a} \tag{7.30a}$$

$$q_{23} = \frac{F_y}{I_x}\sum_{i=1}^{2} A_i y_i + \frac{F_x}{I_y}\sum_{i=1}^{2} A_i x_i$$

$$= \frac{F_y(d/2)(A_1 + A_2)}{(d_1^2/2)(A_1 + A_2)} + \frac{F_x[A_1(a - \overline{X}) - A_2\overline{X}](A_1 + A_2)}{2A_1 A_2 a^2} = \frac{F_y}{d} \tag{7.30b}$$

$$q_{34} = \frac{F_y}{I_x}\sum_{i=1}^{3} A_i y_i + \frac{F_x}{I_y}\sum_{i=1}^{3} A_i x_i$$

$$= \frac{F_y(d/2)(A_1 + A_2 - A_2)}{(d_1^2/2)(A_1 + A_2)} + \frac{F_x[A_1(a - \overline{X}) - A_2\overline{X} - A_2\overline{X}](A_1 + A_2)}{2A_1 A_2 a^2}$$

$$= \frac{F_y A_1}{d(A_1 + A_2)} - \frac{F_x}{2a} \tag{7.30c}$$

A check can be made on q_{34} in equation (7.30c) in working clockwise from boom 4:

$$q_{43} = \frac{F_y A_4 y_4}{I_x} + \frac{F_x A_4 x_4}{I_y} = -\frac{F_y A_1 d}{d^2(A_1 + A_2)} + \frac{F_x A_1(a - \overline{X})(A_1 + A_2)}{2A_1 A_2 a^2}$$

$$= -\frac{F_y A_1}{d(A_1 + A_2)} + \frac{F_x}{2a}$$

in which a sign change $q_{43} = -q_{34}$ accompanies the direction reversal. For a further check on the shear centre position e_x in the presence of both shear forces, take moments about the web centre, where equations (7.30a,c) apply:

$$F_y e_x = \frac{q_{12}ad}{2} + \frac{q_{34}ad}{2} = \frac{F_y A_1 ad}{2d(A_1 + A_2)}, \quad \Rightarrow \quad e_x = \frac{a}{1 + A_2/A_1}$$

which agrees with equation (7.29). Figure 7.9c shows how the shear flows are modified in the presence of the two forces. The q_{12} and q_{23} shear flows in equations (7.30a,b), being the greater contributors, may be used to provide the thicknesses t_a and t_d as a ratio. Employing a similar method to that which led to equation (7.23b), we now have

$$\left(\frac{t_a}{t_d}\right)^3 = \frac{K_d}{K_a}\left(\frac{a/d}{1 + A_2/A_1} + \frac{1}{2}\frac{F_x}{F_y}\right)\left(\frac{a}{d}\right)^2 \tag{7.31}$$

As equation (7.23b) is a reduction to equation (7.31) for $F_x = 0$, we may follow the earlier reasoning in which the thickness ratio is left at unity while a/d is adjusted to satisfy equation (7.31). The optimum web thickness is again given by equation (7.22) and the corresponding minimum weight distribution becomes

$$w_{\text{opt}} = \rho d t_{\text{opt}}\left(1 + 2\frac{a}{d}\right) = \frac{\rho d^2}{(K_d E_T)^{1/3}}\left(1 + 2\frac{a}{d}\right)\left(\frac{q_{23}}{d}\right)^{1/3} \tag{7.32}$$

In equation (7.32), d is defined by equation (7.28b) and q_{23} is given by equation (7.30b).

7.3.3 Re-entrant Channel Section

The x-symmetric, re-entrant channel in Figure 7.2a was given the in-line, four-boom idealisation shown in Figure 7.2b. Here we shall calculate both the flange and web contributions to each boom area $A_1, \ldots, A_4$, where these are separated by a and d, as shown in Figure 7.2a. The two outer flange areas are $A_{f1} = bt_b$ and the two inner flange areas are $A_{f2} = ct_c$. Table 7.2 assists with the calculation of the 'web areas' in which the length dimension s lies between booms and y is the vertical distance from the neutral axis (NA) to each boom position [5]. The web contributions are, from equation (7.5b),

$$A_{w1} = \frac{s_2 t_2}{6}\left(2 + \frac{y_2}{y_1}\right) = \frac{a t_a}{6}\left[2 + \frac{d/2}{(a + d/2)}\right] \tag{7.33a}$$

$$A_{w2} = \frac{s_2 t_2}{6}\left(2 + \frac{y_1}{y_2}\right) + \frac{s_3 t_3}{6}\left(2 + \frac{y_3}{y_2}\right) = \frac{a t_a}{6}\left(2 + \frac{a + d/2}{d/2}\right) + \frac{d t_d}{6}\left[2 + \frac{(-d/2)}{d/2}\right] \tag{7.33b}$$

$$A_{w3} = \frac{s_3 t_3}{6}\left(2 + \frac{y_2}{y_3}\right) + \frac{s_4 t_4}{6}\left(2 + \frac{y_4}{y_3}\right) = \frac{d t_d}{6}\left[2 + \frac{d/2}{(-d/2)}\right] + \frac{a t_a}{6}\left[2 + \frac{-(a + d/2)}{(-d/2)}\right] \tag{7.33c}$$

$$A_{w4} = \frac{s_4 t_4}{6}\left(2 + \frac{y_3}{y_4}\right) = \frac{a t_a}{6}\left[2 + \frac{-(d/2)}{-(a + d/2)}\right] \tag{7.33d}$$

Table 7.2 Dimensions for web contribution to boom areas

Boom	1	2	3	4
y	$(a + d/2)$	$d/2$	$-d/2$	$-(a + d/2)$
s	0	a	d	a
t	–	t_a	t_d	t_a

Adding to equations (7.33a–d) the flange areas, so that for the rth boom $A_r = A_{fr} + A_{wr} (r = 1, \ldots, 4)$, it follows that the total boom areas will conform to $A_1 = A_4$ and $A_2 = A_3$.

(a) Bending under M_x

For bending about the x-axis, the second moment of the pair of areas A_1 and A_2, lying above and below the x-axis, becomes

$$I_x = 2\left[A_1\left(a + \frac{d}{2}\right)^2 + A_2\left(\frac{d}{2}\right)^2\right] = \frac{A_1 d^2}{2}\left\{\left(1 + 2\frac{a}{d}\right)^2 + \frac{A_2}{A_1}\right\} \tag{7.34}$$

The longitudinal tensile stress in booms 1 and 2 follow from equation (7.34):

$$\sigma_1 = \frac{M_x y}{I_x} = \frac{M_x(1 + 2a/d)}{A_1 d\{(1 + 2a/d)^2 + A_2/A_1\}} \tag{7.35a}$$

$$\sigma_2 = \frac{M_x d/2}{I_x} = \frac{M_x}{A_1 d\{(1 + 2a/d)^2 + A_2/A_1\}} \tag{7.35b}$$

The longitudinal compressive stress for booms 3 and 4 have identical magnitudes to those in equations (7.35a,b). These are shown as σ_3 and σ_4 in Figure 7.10a. The greatest bending stress is $\sigma_1 (= -\sigma_4)$ which, when limited to σ_y (i.e. $\sigma_1 \leq \sigma_y$), gives the web depth d as:

$$d = \frac{M_x(1 + 2a/d)}{A_1 \sigma_y\{(1 + 2a/d)^2 + A_2/A_1\}} \tag{7.36}$$

in which a/d may be taken as an independent ratio, having an optimum value within the following consideration of the weight distribution under shear flow.

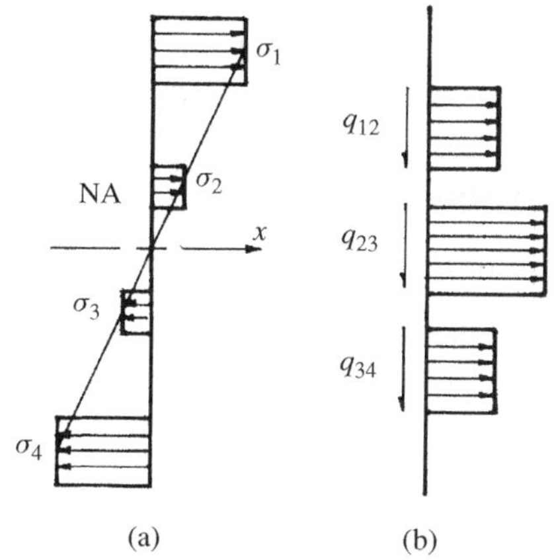

Figure 7.10 Bending stress and shear flow for an idealised re-entrant channel

(b) Shear Flow under F_y

Starting from the top free surface, the three web shear flows, shown in Figure 7.10b, follow from equation (7.2b):

$$q_{12} = \frac{F_y}{I_x} A_1 \left(a + \frac{d}{2}\right), \quad q_{23} = \frac{F_y}{I_x}\left[A_1\left(a + \frac{d}{2}\right) + A_2\frac{d}{2}\right], \tag{7.37a,b}$$

$$q_{34} = \frac{F_y}{I_x}\left[A_1\left(a + \frac{d}{2}\right) + A_2\frac{d}{2} - A_2\frac{d}{2}\right] = q_{12} \tag{7.37c}$$

Combining equations (7.37a–c) and (7.34), these shear flows become

$$q_{12} = q_{34} = \frac{F_y(1 + 2a/d)}{d\{(1 + 2a/d)^2 + A_2/A_1\}} \tag{7.38a}$$

$$q_{23} = \frac{F_y(1 + 2a/d + A_2/A_1)}{d\{(1 + 2a/d)^2 + A_2/A_1\}} \tag{7.38b}$$

Hence each shear flow shown in Figure 7.10b is constant. With q_{23} being the greatest shear flow, the buckling criterion applied to the central web shows that equation (7.22) will again provide its optimum thickness. Equation (7.23a) will also apply to the buckling of all three webs together. Substituting q_{12} and q_{23} from equations (7.37a,b) into equation (7.23a) leads to a relationship among four ratios:

$$\left(1 + 2\frac{a}{d}\right)\left(\frac{a}{d}\right)^2 = \left(1 + 2\frac{a}{d} + \frac{A_2}{A_1}\right)\frac{K_a}{K_d}\left(\frac{t_a}{t_d}\right)^3 \tag{7.39}$$

When $t_a/t_d = 1$, equation (7.39) permits an optimum a/d to be found within the distributed weight expression:

$$w_{\text{opt}} = \rho t_{\text{opt}}\, d\left(1 + 2\frac{a}{d}\right) = \frac{\rho d^2}{(K_d E_T)^{1/3}}\left(1 + 2\frac{a}{d}\right)\left(\frac{q_{23}}{d}\right)^{1/3} \tag{7.40}$$

where q_{23}/d is the structural index for the central web.

7.4 Idealised Closed Tubes

7.4.1 Rectangular Tube with F_y and M_x

Figure 7.11a shows a rectangular tube with horizontal flats $b \times t_b$ and vertical webs $d \times t_d$. The shear flow arising from a vertical force F_y applied to the left web (not the shear centre) may be found from the idealisation shown in Figure 7.11b.

With the origin of x and y at the centroid of the section, the booms are 'dimensioned' as shown in Table 7.3. The s-dimension refers to the web length connected behind each boom 1, 2, 3 and 4, when taken in the anticlockwise direction. That is, $s_1 = b$, $s_2 = d$

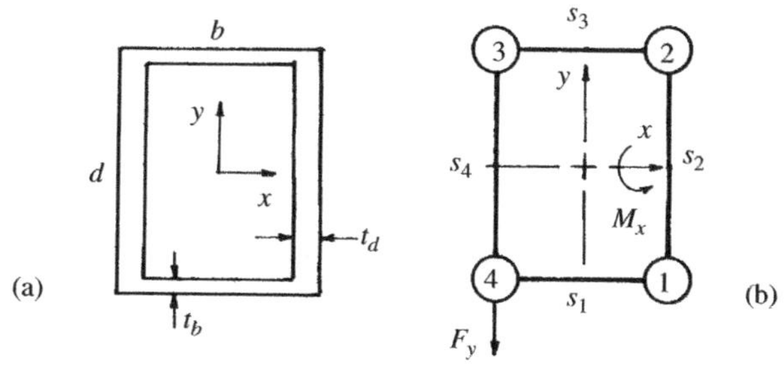

Figure 7.11 Rectangular tube with its web-boom idealisation

etc, as shown. The boom areas follow from applying equation (7.5b) to Table 7.3. Here there are no concentrations of area within the corners, so only the webs contribute to the boom areas:

$$A_1 = \frac{s_1t_1}{6}\left(2+\frac{y_4}{y_1}\right)+\frac{s_2t_2}{6}\left(2+\frac{y_2}{y_1}\right)=\frac{bt_b}{6}\left[2+\frac{(-d/2)}{(-d/2)}\right]+\frac{dt_d}{6}\left[2+\frac{d/2}{(-d/2)}\right]$$
$$=\frac{bt_b}{2}+\frac{dt_d}{6}=A_3 \tag{7.41a}$$
$$A_2 = \frac{s_2t_2}{6}\left(2+\frac{y_1}{y_2}\right)+\frac{s_3t_3}{6}\left(2+\frac{y_3}{y_2}\right)=\frac{dt_d}{6}\left[2+\frac{(-d/2)}{d/2}\right]+\frac{bt_b}{6}\left(2+\frac{d/2}{d/2}\right)$$
$$=\frac{dt_b}{6}+\frac{bt_b}{2}=A_4 \tag{7.41b}$$

Equations (7.41a,b) show that all four boom areas are equal, written as:

$$A = \frac{dt_d}{6}\left(1+3\frac{b}{d}\frac{t_b}{t_d}\right) \tag{7.41c}$$

Equation (7.41c) leads to $I_x = \sum Ay^2 = 4A(\mathrm{d}/2)^2$, agreeing with the finding from the original 'flat' areas in equation (3.34b):

$$I_x = \frac{t_d d^3}{6}\left(1+3\frac{b}{d}\frac{t_b}{t_d}\right) \tag{7.42a}$$

Table 7.3 Web dimensions for boom area contributions

Boom	1	2	3	4
y	$-d/2$	$d/2$	$d/2$	$-d/2$
s	b	d	b	d
t	t_b	t_d	t_b	t_d

(a) Bending under M_x
The second moment of area equation (7.42a) for the idealisation is more simply written as:

$$I_x = 4A(d/2)^2 = Ad^2 \tag{7.42b}$$

which gives the magnitude of the maximum bending stresses: tension in booms 2 and 3, compression in booms 1 and 4, as follows:

$$\sigma_2 = \sigma_3 = \frac{M_x d/2}{I_x} = M_x\left(\frac{d}{2} \times \frac{4}{4Ad^2}\right) = \frac{M_x}{2Ad} \tag{7.43a,b}$$

$$\sigma_1 = \sigma_4 = -\frac{M_x}{2Ad}$$

Limiting each stress to $\pm\sigma_y$ appropriately, the web depth is set at

$$d = \frac{M_x}{2A\sigma_y} \tag{7.44}$$

(b) Shear Flow under F_y
Writing each boom area as A and making use of Table 7.3, the q_b web shear flows become:

$$1\text{–}2: \quad q_{12b} = \frac{F_y}{I_x}(A_1 y_1) = -\frac{F_y A d}{2I_x} = -\frac{F_y}{2d} \tag{7.45a}$$

$$2\text{–}3: \quad q_{23b} = \frac{F_y}{I_x}(A_1 y_1 + A_2 y_2) = 0 \tag{7.45b}$$

$$3\text{–}4: \quad q_{34b} = \frac{F_y}{I_x}(A_1 y_1 + A_2 y_2 + A_3 y_3) = \frac{F_y A d}{2I_x} = \frac{F_y}{2d} \tag{7.45c}$$

$$4\text{–}1: \quad q_{41b} = \frac{F_y}{I_x}(A_1 y_1 + A_2 y_2 + A_3 y_3 + A_4 y_4) = 0 \tag{7.45d}$$

Figure 7.12a shows the two, constant, non-zero shear flows. The negative shear flow found in 1–2 infers an opposing direction as shown. The net shear flow $q_{\text{net}} = q_b + q_0$ is shown in Figure 7.12b where q_0 is found from equation (6.10). Taking moments about point 4:

$$F_y \times 0 = (q_{12b} \times d \times b) + (2bd)q_o, \quad \Rightarrow \quad q_0 = -\frac{q_{12b}}{2} = \frac{F_y A d}{4I_x} \tag{7.46a}$$

and substituting I_x from equations (7.42b):

$$q_0 = F_y/4d \tag{7.46b}$$

Comparing the net shear flows in Figure 7.12b to the exact distributions found previously for the non-uniform rectangular tube in Figure 6.8b, the nature of the idealisation becomes

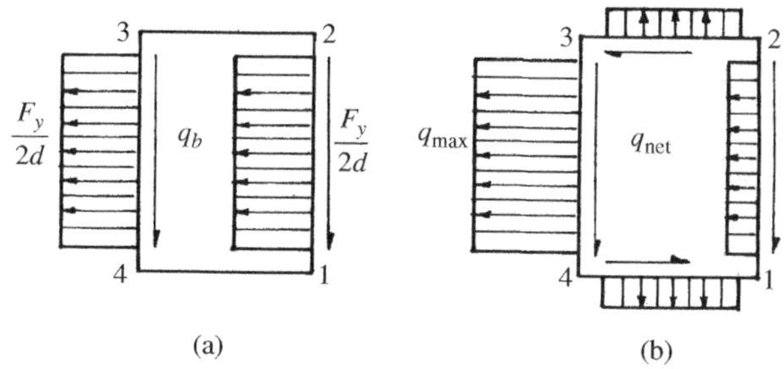

Figure 7.12 Shear flows in an idealised tube

apparent. Here, q_{max} in Figure 7.12b identifies with the constant net shear flow $q_{34} = q_{34b} + q_0$:

$$q_{34} = \frac{F_y}{2d} + \frac{F_y}{4d} = \frac{3F_y}{4d} \tag{7.47a}$$

In the remaining webs $q_{net} = F_y/4d$ with the directions shown. Now that the shear flows are established, we return to the original section and apply the optimum design criteria. Correspondingly, when the maximum shear stress q_{34}/t_d for this web is limited by buckling,

$$\frac{q_{34}}{t_d} = \frac{3F_y}{4dt_d} = K_d E_T \left(\frac{t_d}{d}\right)^2 \tag{7.47b}$$

Equation (7.47b) gives the optimum thickness for a depth d, set within equation (7.44), as

$$(t_d)_{opt} = \left(\frac{q_{34}d^2}{K_d E_T}\right)^{1/3} \tag{7.48}$$

The distributed weight is written as

$$w = \frac{W}{L} = 2(bt_b + dt_d)\rho = 2dt_d\left(1 + \frac{b}{d}\frac{t_b}{t_d}\right)\rho \tag{7.49a}$$

Substituting $t_d = (t_d)_{opt}$ from equation (7.48), the optimum (minimum) weight becomes

$$w_{opt} = \frac{2\rho d^2}{(K_d E_T)^{1/3}}\left(1 + \frac{b}{d}\frac{t_b}{t_d}\right)\left(\frac{q_{34}}{d}\right)^{1/3} \tag{7.49b}$$

in which the coefficient in parentheses on the structural index q_{34}/d contains both the web length and thickness ratios. We may connect these ratios when sides 1–2 and 2–3 buckle together under their respective maximum shear stresses:

$$\frac{q_{34}t_b}{q_{23}t_d} = \frac{K_d E_T}{K_b E_T}\left(\frac{t_d}{d}\right)^2\left(\frac{b}{t_b}\right)^2 \tag{7.50a}$$

Table 7.4 Web and thickness ratios in equation (7.49a) that optimise weight of a rectangular tube

b/d	0.25	0.33	0.5	1.0	2.0	3.0	4.0	5.0
t_b/t_d	0.275	0.331	0.436	0.693	1.101	1.442	1.747	2.027
$1 + (b/d)(t_b/t_d)$	1.069	1.109	1.218	1.693	3.201	5.326	7.988	11.135

where $q_{23} = q_0$ in equation (7.46b). Substituting equations (7.46b) and (7.47a) into equation (7.50a) leads to the required relationship:

$$\frac{t_b}{t_d} = \frac{1}{3^{1/3}} \left(\frac{K_d}{K_b} \right)^{1/3} \left(\frac{b}{d} \right)^{2/3} \tag{7.50b}$$

If the tube is long (i.e. $b/L < 0.1$) then we see from Figure C.3 that $K_a \approx K_d$ for plates (i.e. the tube's side walls) having identical edge restraints. Setting $K_a/K_d \approx 1$ in equation (7.50b) gives the thickness ratios in Table 7.4 that minimise weight over a range of b/d values.

While no minimum exists between the two relationships (7.49b) and (7.50b), Table 7.4 shows clearly that a narrow tube width reduces weight by more than an order of magnitude. We can improve on this further when the force F_y is displaced to act at the shear centre (through the tube centroid). Equation (5.68a) shows that $q_E = 0$, and hence the net shear flows are the q_b shear flows in equations (7.45a–d). These are $q_{34b} = F_y/2d = -q_{12b}$, as shown in Figure 7.12a. Equation (7.48) again optimises t_d but, as the sides 2–3 and 4–1 remain unstressed, the ratio t_b/t_d may then be made as small as possible to reduce weight further.

7.4.2 Rectangular Tube with F_x, F_y, M_x and M_y

(a) Bending under M_x and M_y

Let us assume that equal boom areas within a thin-walled rectangular tube arise solely from the concentration of area due to corner stiffeners. Shear forces F_x and F_y and bending moments M_x and M_y are applied in the manner of Figure 7.13a. With a negligible contribution to the boom areas from the thin walls, the second moments of area for such an idealisation are simply:

$$I_x = 4A(d/2)^2 = Ad^2 \quad \text{and} \quad I_y = 4A(b/2)^2 = Ab^2 \tag{7.51a,b}$$

Taking the appropriate signs for $x = \pm b/2$ and $y = \pm d/2$ in equation (7.6) gives the bending stresses within the four booms:

$$\sigma_1 = -\frac{M_x d}{2I_x} + \frac{M_y b}{2I_y} = \frac{1}{2A} \left(-\frac{M_x}{d} + \frac{M_y}{b} \right) \tag{7.52a}$$

$$\sigma_2 = \frac{M_x d}{2I_x} + \frac{M_y b}{2I_y} = \frac{1}{2A} \left(\frac{M_x}{d} + \frac{M_y}{b} \right) \tag{7.52b}$$

$$\sigma_3 = \frac{M_x d}{2I_x} - \frac{M_y b}{2I_y} = \frac{1}{2A}\left(\frac{M_x}{d} - \frac{M_y}{b}\right) \tag{7.52c}$$

$$\sigma_4 = -\frac{M_x d}{2I_x} - \frac{M_y b}{2I_y} = -\frac{1}{2A}\left(\frac{M_x}{d} + \frac{M_y}{b}\right) \tag{7.52d}$$

If we assume $M_y/b > M_x/d$ then stresses from equations (7.52a–d) are found to act in tension (outward) and compression (inward), as shown in Figure 7.13b. When the greatest tensile stress σ_2 is limited to σ_y, the depth is set at

$$d = \frac{M_x}{2A\sigma_y}\left(1 + \frac{d}{b}\frac{M_y}{M_x}\right) \tag{7.53}$$

in which the ratio d/b is provided by the following shear flow analysis.

(b) Shear Flow under F_x and F_y

The two shear forces act through the shear centre E which is coincident with the centroid for this tube (see Figure 7.13a). The basic shear flows, shown in Figure 7.14a, follow from equation (7.7):

$$q_{12b} = -\frac{F_y A_1 d}{2Ad^2} + \frac{F_x A_1 b}{2Ab^2} = -\frac{F_y}{2d} + \frac{F_x}{2b} \tag{7.54a}$$

$$q_{23b} = -\frac{F_y[A_1(-d/2) + A_2 d/2]}{Ad^2} + \frac{F_x[A_1 b/2 + A_2 b/2]}{Ab^2} = \frac{F_x}{b} \tag{7.54b}$$

$$q_{34b} = -\frac{F_y[A_1(-d/2) + A_2 d/2 + A_3 d/2]}{Ad^2} + \frac{F_x[A_1 b/2 + A_2 b/2 + A_3(-b/2)]}{Ab^2}$$

$$= \frac{F_y}{2d} + \frac{F_x}{2b} \tag{7.54c}$$

$$q_{41b} = -\frac{F_y[A_1(-d/2) + A_2 d/2 + A_3 d/2 + A_4(-d/2)]}{Ad^2}$$

$$+ \frac{F_x[A_1 b/2 + A_2 b/2 + A_3(-b/2) + A_4(-b/2)]}{Ab^2} = 0 \tag{7.54d}$$

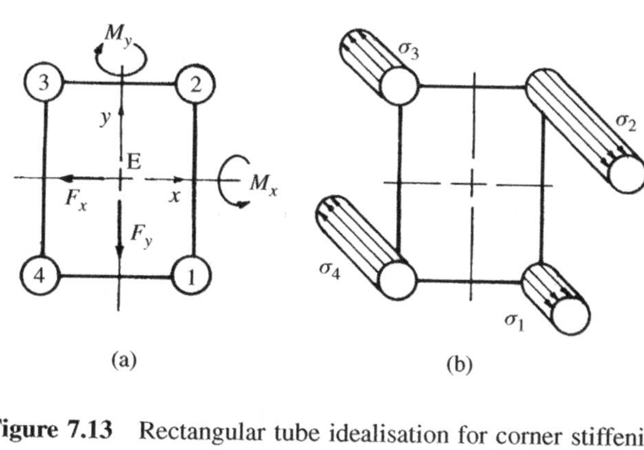

Figure 7.13 Rectangular tube idealisation for corner stiffening

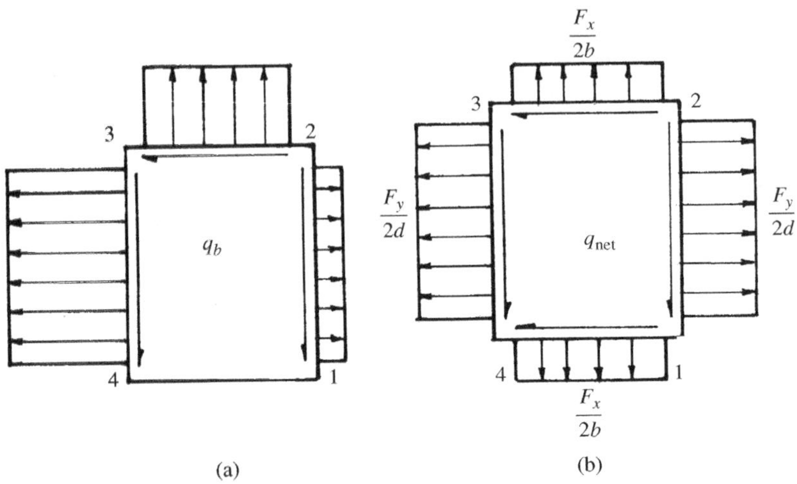

Figure 7.14 Shear flows in rectangular tube

With $A_1 = A_2 = A_3 = A_4 = A$, the initial shear flow q_E is found from equation (5.68a):

$$\left(q_{12b}\frac{d}{t_d} + q_{23b}\frac{b}{t_b} + q_{34b}\frac{d}{t_d} + q_{41b}\frac{b}{t_b}\right) + 2q_E\left(\frac{d}{t_d} + \frac{b}{t_b}\right) = 0 \tag{7.55a}$$

and substituting for the q_b shear flows from equations (7.54a–d),

$$q_E = -\frac{F_x}{2b} \tag{7.55b}$$

giving the net shear flows $q_{net} = q_b + q_E$, as

$$q_{12} = q_{12b} + q_E = -\frac{F_y}{2d}, \qquad q_{23} = q_{23b} + q_E = \frac{F_x}{2b} \tag{7.56a,b}$$

$$q_{34} = q_{34b} + q_E = \frac{F_y}{2d}, \qquad q_{41} = q_{41b} + q_E = -\frac{F_x}{2b} \tag{7.56c,d}$$

whose constant magnitudes and directions are shown within Figure 7.14b.

Working with the positive shear flows, the sides b and d will buckle when

$$\frac{q_{23}}{t_b} = K_b E_T\left(\frac{t_b}{b}\right)^2, \qquad \frac{q_{34}}{t_d} = K_d E_T\left(\frac{t_d}{d}\right)^2 \tag{7.57a,b}$$

Substituting for q_{23} and q_{34} from equations (7.56b,c) then dividing equations (7.57a,b), the following relationship is found when, for a long tube, $K_b \approx K_d$:

$$\frac{b}{d} = \frac{F_y}{F_x}\left(\frac{t_b}{t_d}\right)^3 \tag{7.58a}$$

Equation (7.58a) allows t_b/t_d to be selected arbitrarily, from which b/d follows. The vertical wall depth d_{opt} is given by equation (7.53), and if $F_y > F_x$ then the optimum thickness t_d will follow from allowing the greatest shear stress in the tube to attain its design stress τ_y as buckling occurs. Thus, with $q_{34} = \tau_y t_d$, equation (7.57b) gives this thickness as

$$(t_d)_{\text{opt}} = d_{\text{opt}} \left(\frac{\tau_y}{K_d E_T} \right)^{1/2} \tag{7.58b}$$

Note also from equation (7.56c) that the maximum vertical shear force allowed is $F_y = 2\tau_y (d t_d)_{\text{opt}}$.

The optimum values of b and t_b and the allowable F_x value may then be examined from equation (7.58a) in combination with the distributed weight expression:

$$w = 2(bt_b + dt_d)\rho = 2dt_d \left(1 + \frac{b}{d}\frac{t_b}{t_d}\right)\rho \tag{7.59}$$

By calculating b/d over a range of thickness ratios from equation (7.58a), the search for an acceptably low distributed weight within equation (7.59) may be carried out.

7.4.3 *Tube with Parallelogram Section under F_x, F_y, M_x and M_y*

Figure 7.15 show a large, thin-walled, parallelogram tube idealised to four boom areas arising from corner reinforcements, such that $A_1 = A_3 = A$ and $A_2 = A_4 = 2A$. The wall lengths a and d with different thicknesses t_a and t_d, respectively, are assumed to make a negligible contribution to the boom areas. Consequently, the second moments of the boom areas may be referred to both centroidal axes x, y with their origin at the shear centre E. Here, because the section is asymmetric, a product moment is also required. The second moments are:

$$I_x = A_2 d^2 + A_4 d^2 = (A_2 + A_4)d^2 = 4Ad^2 \tag{7.60a}$$

$$I_y = (A_1 + A_4)d^2 + (A_2 + A_3)d^2 = 6Ad^2 \tag{7.60b}$$

$$\begin{aligned} I_{xy} &= A_1 \times d \times 0 + A_2 \times (-d) \times d + A_3 \times (-d) \times 0 + A_4 \times d \times (-d) \\ &= -(A_2 + A_4)d^2 = -4Ad^2 \end{aligned} \tag{7.60c}$$

(a) Bending under M_x and M_y

Bending moments M_x and M_y act along both centroidal axes, as shown in Figure 7.15. Substituting equations (7.60a–c) into equations (7.8a,b) gives the equivalent moments:

$$\overline{M}_x = \frac{M_x + \frac{2}{3}M_y}{1 - \frac{16}{24}} = 3M_x + 2M_y, \quad \overline{M}_y = \frac{M_y + M_x}{1 - \frac{16}{24}} = 3(M_y + M_x) \tag{7.61a,b}$$

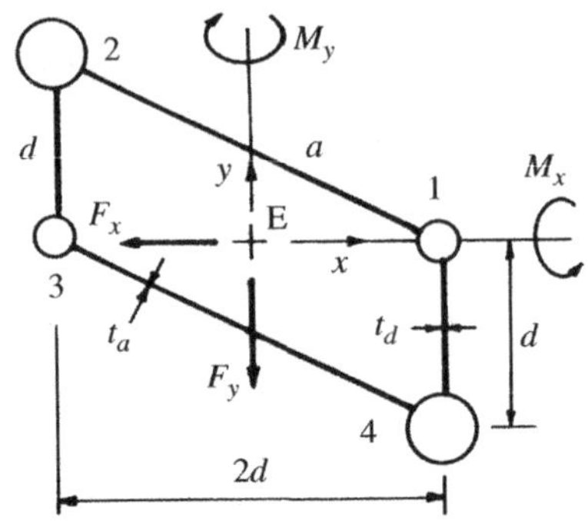

Figure 7.15 Idealised parallelogram tube

Substituting equations (7.61a,b) into equation (7.6) gives the bending stress expression

$$\sigma = \frac{\overline{M}_x y}{I_x} + \frac{\overline{M}_y x}{I_y} = \frac{(3M_x + 2M_y)y}{4Ad^2} + \frac{3(M_x + M_y)x}{6Ad^2} \tag{7.62}$$

In particular, using the (x, y) coordinates indicated for each boom, we find, from equation (7.62),

$$\text{at 1 } (d, 0), \quad \sigma_1 = \frac{(3M_x + 2M_y)0}{4Ad^2} + \frac{3(M_x + M_y)d}{6Ad^2} = \frac{M_x + M_y}{2Ad} \tag{7.63a}$$

$$\text{at 2 } (-d, d), \quad \sigma_2 = \frac{(3M_x + 2M_y)d}{4Ad^2} - \frac{3(M_x + M_y)d}{6Ad^2} = \frac{M_x}{4Ad} \tag{7.63b}$$

$$\text{at 3 } (-d, 0), \quad \sigma_3 = \frac{(3M_x + 2M_y)0}{4Ad^2} - \frac{3(M_x + M_y)d}{6Ad^2} = \frac{-(M_x + M_y)}{2Ad} \tag{7.63c}$$

$$\text{at 4 } (d, -d), \quad \sigma_4 = \frac{-(3M_x + 2M_y)d}{4Ad^2} + \frac{3(M_x + M_y)d}{6Ad^2} = \frac{-M_x}{4Ad} \tag{7.63d}$$

Limiting the greatest stress (i.e. equation (7.63a)) to $\sigma_1 \le \sigma_y$, the dimension d follows as

$$\frac{M_x + M_y}{2Ad} \le \sigma_y, \quad \Rightarrow \quad d \ge \frac{M_x}{2A\sigma_y}\left(1 + \frac{M_y}{M_x}\right) \tag{7.64}$$

Previously, the equality $\sigma_1 = \sigma_y$ has been taken when optimising depth, but the reader would be aware that σ_y may be lowered with a safety factor, a procedure that would normally be applied when seeking a safe weight reduction.

(b) Shear Flow under F_x and F_y

Here shear forces F_x and F_y pass through the origin for x and y (the centroid), assumed to be coincident with the shear centre E in Figure 7.15. The net shear flow is $q = q_b + q_E$ in which both q_b and q_E are to be found. The q_b shear flows follow from equation (7.7) in which the equivalent forces are the derivatives of the equivalent moments in equations (7.61a,b), equations (7.10 a,b) and (7.11a,b) show that:

$$\overline{F}_x = \frac{d\overline{M}_y}{dz} = 3(F_x + F_y), \quad \overline{F}_y = \frac{d\overline{M}_x}{dz} = 3F_y + 2F_x \tag{7.65a,b}$$

where $F_x = dM_y/dz$ and $F_y = dM_x/dz$. The q_b shear flows in each wall follow from equations (7.7) and (7.65a,b) as

$$\begin{aligned} q_b &= \frac{\overline{F}_y \sum_{r=1}^{n} A_r y_r}{I_x} + \frac{\overline{F}_x \sum_{r=1}^{n} A_r x_r}{I_y} \\ &= \frac{(3F_y + 2F_x) \sum_{r=1}^{n} A_r y_r}{4Ad^2} + \frac{3(F_x + F_y) \sum_{r=1}^{n} A_r x_r}{6Ad^2} \end{aligned} \tag{7.66}$$

Using the (x, y) coordinates, with I_x and I_y given above, equation (7.66) gives

$$q_{12b} = \frac{\overline{F}_y(A_1 y_1)}{I_x} + \frac{\overline{F}_x(A_1 x_1)}{I_y} = \frac{F_x + F_y}{2d} \tag{7.67a}$$

$$q_{23b} = \frac{\overline{F}_y(A_1 y_1 + A_2 y_2)}{I_x} + \frac{\overline{F}_x(A_1 x_1 + A_2 y_2)}{I_y} = \frac{F_x + 2F_y}{2d} \tag{7.67b}$$

$$q_{34b} = \frac{\overline{F}_y(A_1 y_1 + A_2 y_2 + A_3 y_3)}{I_x} + \frac{\overline{F}_x(A_1 x_1 + A_2 y_2 + A_3 x_3)}{I_y} = \frac{F_y}{2d} \tag{7.67c}$$

$$q_{41b} = \frac{\overline{F}_y(A_1 y_1 + A_2 y_2 + A_3 y_3 + A_4 y_4)}{I_x} + \frac{\overline{F}_x(A_1 x_1 + A_2 y_2 + A_3 x_3 + A_4 x_4)}{I_y} = 0 \tag{7.67d}$$

and q_E follows from equation (5.68a) as

$$\begin{aligned} &\frac{F_x + F_y}{2d}\left(\frac{a}{t_a}\right) + \frac{2F_y + F_x}{2d}\left(\frac{d}{t_d}\right) + \frac{F_y}{2d}\left(\frac{a}{t_a}\right) + 2q_E\left(\frac{a}{t_a} + \frac{d}{t_d}\right) = 0 \\ &q_E = -\left(\frac{F_x + 2F_y}{4d}\right) \end{aligned} \tag{7.68}$$

The net shear flow $q_{\text{net}} = q_b + q_E$ in each wall becomes

$$q_{12} = \frac{F_x + F_y}{2d} - \frac{F_x + 2F_y}{4d} = \frac{F_x}{4d} \tag{7.69a}$$

$$q_{23} = \frac{F_x + 2F_y}{2d} - \frac{F_x + 2F_y}{4d} = \frac{F_x + 2F_y}{4d} \tag{7.69b}$$

$$q_{34} = \frac{F_y}{2d} - \frac{F_x + 2F_y}{4d} = -\frac{F_x}{4d} \tag{7.69c}$$

$$q_{41} = 0 - \frac{F_x + 2F_y}{4d} = -\frac{F_x + 2F_y}{4d} \tag{7.69d}$$

The q_b and q_{net} shear flows are shown in Figure 7.16.

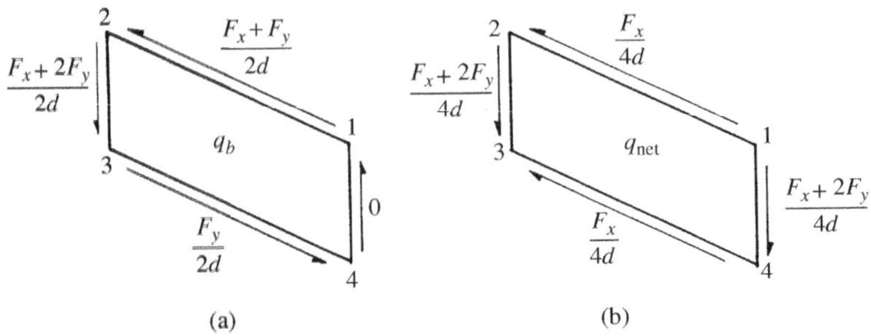

Figure 7.16 Constant q_b and net q_{net} shear flows in tube wall

We may now check that the origin for x, y coincides with the shear centre from the requirement that the twist rate at E for the tube is absent under q_{net}. That is, from equation (4.102a),

$$\frac{d\theta}{dz} = \frac{1}{2AG}\oint q_{\text{net}}\frac{ds}{t} = 0 \tag{7.70a}$$

in which it follows that

$$\oint q_{\text{net}}\frac{ds}{t} = q_{12}\frac{a}{t_a} + q_{23}\frac{d}{t_d} + q_{34}\frac{a}{t_a} + q_{41}\frac{d}{t_d} \tag{7.70b}$$

Substituting equations (7.69a–d) into the right-hand side of equation (7.70b),

$$\frac{F_x}{4d}\frac{a}{t_a} - \frac{F_x+2F_y}{4d}\frac{d}{t_d} - \frac{F_x}{4d}\frac{a}{t_a} - \frac{F_x+2F_y}{4d}\frac{d}{t_d} = 0 \tag{7.70c}$$

Equation (7.70c) confirms that the shear centre E lies at its assumed position. The sides a and d will buckle when their shear stresses attain the critical values:

$$\frac{q_{12}}{t_a} = K_a E_T\left(\frac{t_a}{a}\right)^2, \quad \frac{q_{23}}{t_d} = K_d E_T\left(\frac{t_d}{d}\right)^2 \tag{7.71a,b}$$

Substituting for q_{12} and q_{23} from equations (7.69a,b) and then dividing equations (7.71a,b), the following relationship is found:

$$\left(\frac{a}{d}\right)^2 = \left(1+\frac{2F_y}{F_x}\right)\frac{K_a}{K_d}\left(\frac{t_a}{t_d}\right)^3 \tag{7.72a}$$

Provided the tube is long, the left-hand side of Figure C.3 shows that $K_a \approx K_d$. Equation (7.72a) allows t_a/t_d to be selected arbitrarily, from which a/d follows. Here the vertical wall depth d is known from equation (7.64). Since this wall carries the greatest shear stress its optimum thickness is found from imposing the limiting shear stress. Thus, $(t_d)_{\text{opt}}$ will follow from equating (7.71b) to τ_y. This gives

$$(t_d)_{\text{opt}} = d_{\text{opt}}\left(\frac{\tau_y}{K_d E_T}\right)^{\frac{1}{2}} \tag{7.72b}$$

where E_T becomes the tangent modulus when τ_y is the proof stress at a given offset strain. Note also from equation (7.69b) that the shear force combination should conform to

$$F_x + 2F_y = 4\tau_y(dt_d)_{\text{opt}}.$$

Here a has been set at $\sqrt{5}d$ (see Figure 7.15) but, more generally, in the search for the optimum values of a and t_a/t_d, their role in the distributed weight expression should also be examined. That is,

$$w = 2(at_a + dt_d)\rho = 2dt_d\left(1 + \frac{a}{d}\frac{t_a}{t_d}\right)\rho \tag{7.73}$$

Hence, by applying the relationship between the geometric ratios, according to equation (7.72a), within equation (7.73), the section may be designed to have minimum weight.

References

[1] Rees, D. W. A. *Basic Solid Mechanics*, Macmillan, 1997.
[2] Megson, T. H. G. *Aircraft Structures for Engineering Students*, Arnold, 1972.
[3] Rees, D. W. A. *Mechanics of Solids and Structures*, Imperial College Press, 2000.
[4] Boresi, A.P., Schmidt, R. J. and Sidebottom, O. M. *Advanced Mechanics of Materials*, 5th edition, Wiley, 1993.
[5] Williams, D. *Theory of Aircraft Structures*, Arnold, 1960.

Exercises

7.1 Optimise the idealised parallelogram section in Figure 7.15 for weight when a single shear force is applied: (i) down the left-hand vertical wall (2–3), (ii) down the top sloping wall (2–1).

7.2 Examine whether a minimum exists in the distributed weight expression (7.73) by allowing the thickness ratio in equation (7.72a) to vary from 0.1 to 0.9. For this, take $K_a/K_d = 1$ and repeat this sequence of weight calculations over the range of thickness ratios.

7.3 Determine the shear flow distribution for the four-boom idealisation of the lipped channel section given in Example 7.1, when shear forces $F_x = 5$ kN and $F_y = 10$ kN are applied (i) at the shear centre and (ii) at the centroid, in the negative x- and y-directions (see Figure 7.3).

7.4 Compare the shear flow distributions found in Exercise 7.3 with that found for a six-boom idealisation of the same section under identical loading (see Example 7.1).

8

Shear in Stiffened Webs

8.1 Introduction

Chapter 7 showed that the design of any cross-section under flexural shear is based upon its requirement to support a vertical shear force F_y in a given depth d. The latter was fixed by the accompanying bending moment. A similar approach applies to the design of corrugated and reinforced shear webs. Here the objective is to minimise weight within the optimisation of all other dimensions that describe the long, stiffened web shape. For corrugated shear webs these dimensions include the thickness, pitch, flat length and inclination of the folded sloping sides. Similar considerations apply to the L-section and top-hat stiffening of shear webs and to how material should be removed effectively for buttress cut-outs from a web. The basic design criterion to be applied is that of matching a limiting, maximum shear stress for the web material with the critical stress for web buckling in shear. Additionally, we shall see that a condition of orthotropic buckling applies to the two axes of a corrugation and that further constraint is exerted by a stiffener upon a plate. The strategy for minimising weight allows for all such effects within the application of further failure criteria to optimise the following web designs.

8.2 Castellations in Shear

Buttress-like cut-outs along an aircraft wing and fuselage frame allow for the passage of transverse stringers [1]. In Figure 8.1 the skin, which is attached to the top rectangular, horizontal flat ($a \times b$), transmits a shear force $Q = \tau \times bt$ to the root of each vertical castellation. The latter is taken to have a rectangular shape, $b \times d \times t$, as shown. The sides may taper by up to 20°, becoming narrower at the top, where b refers to the root width. The horizontal, equal-angle fixing provides a rigid support along the root of the castellation such that with $b \approx d$, failures occur, mostly in the plastic range: (i) by root bending for low d/t, and (ii) by plastic buckling for moderately high d/t. Experiments [2] have shown that buckling occurs at a lower critical shear stress than can be predicted by applying a plasticity reduction factor to the elastic shear buckling formula (see Appendix C):

$$\tau_e = \frac{Q^*}{dt} = KE\left(\frac{t}{d}\right)^2 \tag{8.1a}$$

Mechanics of Optimal Structural Design: Minimum Weight Structures D. W. A. Rees

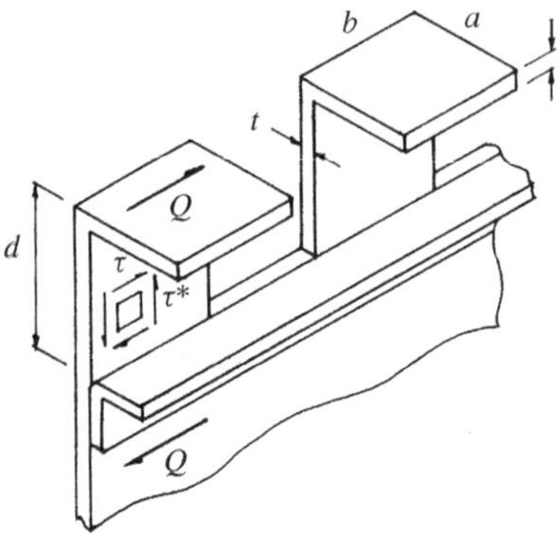

Figure 8.1 Shear force Q in a vertical castellation

where Q^* is the vertical induced shear force that follows from complementary shear $\tau = \tau^*$, in Figure 8.1:

$$\frac{Q}{bt} = \frac{Q^*}{dt}, \quad Q^* = \frac{d}{b}Q \tag{8.1b}$$

Combining equations (8.1a,b) gives the contribution to Q from elastic buckling,

$$Q_e = btKE\left(\frac{t}{d}\right)^2 \tag{8.2a}$$

A more dominant influence of plasticity can be found by combining equation (8.2a) with a limiting plastic stress τ_p. This allows the contribution to Q from plasticity to be found as

$$\tau_p = \frac{Q_p}{bt}, \quad \Rightarrow \quad Q_p = bt\tau_p \tag{8.2b}$$

Each contribution to Q is written in a combination of equations (8.2a,b):

$$\frac{1}{Q_{\mathrm{cr}}} = \frac{1}{Q_p} + \frac{1}{Q_e}, \quad \Rightarrow \quad Q_{\mathrm{cr}} = \frac{Q_p Q_e}{Q_e + Q_p} \tag{8.3a,b}$$

Equation (8.3a) gives $Q_{\mathrm{cr}} \approx Q_e$ for large d/t and $Q_{\mathrm{cr}} \approx Q_p$ for small b/t. Substituting equations (8.2a,b) into equation (8.3b) leads to a 'plastic' buckling stress $\tau_{\mathrm{cr}} = Q_{\mathrm{cr}}/bt$:

$$\tau_{\mathrm{cr}} = \frac{\tau_p}{1 + (\tau_p/KE)(d/t)^2} \tag{8.3c}$$

where K follows from Figure C.3 for a plate with one edge clamped. Here, the limiting shear stress τ_p is influenced by both bending and shear [2]. Consequently, we shall take both τ_p and $\tau_p/(KE)$ in equation (8.3c) to depend upon d/b to match the experimental

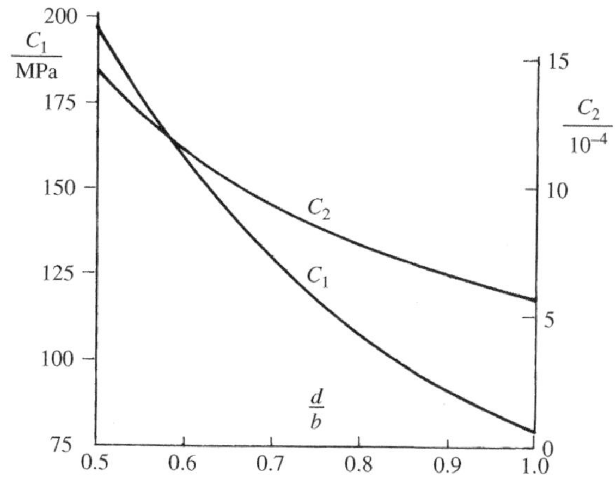

Figure 8.2 Dependence of C_1 and C_2 upon d/b in equation (8.4a)

data within the range $0.5 \leq d/b \leq 1$ [2]. Equation (8.3c) is therefore written in a more useful semi-empirical form:

$$\tau_{cr} = \frac{C_1}{1 + C_2(d/t)^2} \tag{8.4a}$$

where C_1(MPa = N/mm^2) and C_2 (no units) depend upon d/b, as shown in Figure 8.2. When we set $\tau_{cr} = \tau_{max}$, in which $\tau_{max} = q/t = Q/bt$ is the design stress, the optimum thickness follows from equation (8.4a) as the positive root of the cubic equation

$$t^3 - \left(\frac{q}{C_1}\right)t^2 - \left(\frac{C_1}{C_2}\right)qd^2 = 0 \tag{8.4b}$$

Example 8.1 An aluminium alloy castellation having a root breadth $b = 75$ mm and a depth $d = 55$ mm is required to support a shear force $Q = 10$ kN (see Figure 8.1). Calculate a safe thickness t required for the castellation based upon an optimum weight condition. Take $\rho = 2800$ kg/m^3 for aluminium alloy.

The shear flow is:

$$q = \frac{Q}{b} = \frac{10 \times 10^3}{75} = 133.33 \text{ N/mm}$$

and reading from Figure 8.2, for $d/b = 55/75 = 0.733$, we find:

$$C_1 = 120 \text{ N/mm}^2 \text{ and } C_2 = 8.7 \times 10^{-4}$$

Substituting into equation (8.4b) gives a cubic in t (mm):

$$t^3 - 1.111t^2 - 2.924 = 0$$

for which the solution is $t_{opt} = 1.91$ mm. At this thickness the theory predicts buckling to occur. So if t is raised to 2 mm the safe design stress is

$$\tau_{max} = \frac{Q}{bt} = \frac{10 \times 10^3}{75 \times 2} = 66.67 \text{ MPa}$$

this being just less than the critical buckling stress

$$\tau_{cr} = \frac{q}{t_{opt}} = \frac{133.33}{1.91} = 69.81 \text{ MPa}$$

Correspondingly, the optimum distributed weight for an aluminium alloy design becomes:

$$\frac{w}{b} = \rho\, dt = 2800 \times 9.81 \times (55 \times 2)10^{-6} = 3.02 \text{ N/m}$$

In this example the total weight is increased disproportionately within the horizontal flat supporting the skin. In principal, this flat's depth a and thickness t_a (see Figure 8.1) may also be optimised. For this, we shall assume that equation (8.1a) applies to the elastic buckling of this 'plate' where one edge is simply supported:

$$\tau_{cr} = K_a E \left(\frac{t_a}{a}\right)^2 \tag{8.5}$$

Imposing a similar buckling stress $\tau_{cr} = 68.91$ MPa for an aspect ratio $a/b = 0.5$ (say), equation (8.5) gives $t_a = 0.47$ mm. Note here that $K_a = 5.93$, for $a/b = 0.5$, is taken from Figure C.3.

8.3 Corrugated Web

Figure 8.3 shows the manner in which a shear web is corrugated over a given length. The depth is fixed at d from bending stress considerations in the manner outlined in Chapter 7. Other corrugation dimensions shown in Figure 8.3b are: land length e, sloping side length b, height h, inclination θ and thickness t. Each of these will need to be optimised in shear within one pitch length, $p = e + b\cos\theta$.

Referring to Figure 8.3b, the cross-sectional area of a corrugation within one pitch is

$$A_p = (e + b)t \tag{8.6}$$

The second moment of this area about the centroidal axis of the corrugation is the sum of contributions from the horizontal and sloping flats. Using the parallel axis theorem, with the sloping length dimension s as shown, gives

$$I_x = \left[\frac{et^3}{12} + et\left(\frac{h}{2}\right)^2\right] + \left[\int_0^b (s\sin\theta)^2 t\, \mathrm{d}s - bt\left(\frac{b}{2}\sin\theta\right)^2\right] \tag{8.7a}$$

$$\approx \frac{h^2 te}{4} + \frac{b^3 t\sin^2\theta}{12}$$

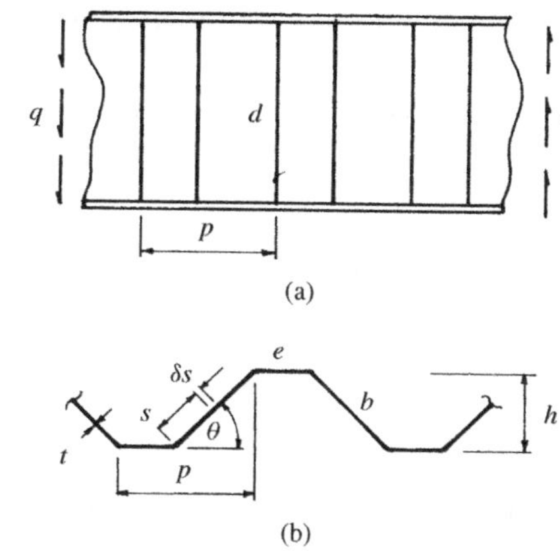

Figure 8.3 Geometry of a corrugated shear web

Substituting $\sin\theta = h/b$ into (8.7a) gives

$$I_x = \frac{h^2te}{4} + \frac{h^2tb}{12} = \frac{h^2t}{12}(3e + b) \tag{8.7b}$$

(i) Objective Function

The weight per unit length (the pitch) defines our objective function:

$$w = \frac{W}{p} = \frac{\rho A_p d}{p} = \frac{\rho(e + b)td}{(e + b\cos\theta)} \tag{8.8a}$$

and since $q = \tau t =$ constant within the depth of a corrugation (see Figure 8.3a), we may write equation (8.8a) as

$$w = \frac{\rho dq(e/b + 1)}{\tau(e/b + \cos\theta)} \tag{8.8b}$$

where $\tau \le \tau_y$ is limited to a predetermined design stress at which both local and global buckling of the corrugation are arranged to occur as in the following manner.

(ii) Local Buckling

Local plate buckling refers to the critical, elastic buckling stress within the flat lengths b and e with edge restraints described by Figure C.3. When the thickness t is uniform and $b > e$, then the local buckling criterion, based upon the longer sloping side, becomes

$$\tau_L = K_b E\left(\frac{t}{b}\right)^2 \tag{8.9a}$$

where, for example, Figure C.3 shows that $K_b = 4.84$ for a long plate ($b/d < 0.1$) with simply supported edges. Substituting $h = b\sin\theta$ into equation (8.9a) gives

$$\tau_L = (K_b \sin^2\theta)E\left(\frac{t}{h}\right)^2 \tag{8.9b}$$

If $b < e$, then e replaces b in equation (8.9a) such that equation (8.9b) becomes

$$\tau_L = \frac{K_e}{(e/b)^2} \sin^2\theta E \left(\frac{t}{h}\right)^2 \tag{8.9c}$$

(iii) Orthotropic Buckling

Buckling of a corrugation as a whole must be considered about the two perpendicular axes aligned with d and p in Figure 8.3a [3–5]. The critical shear stress τ_0 is given by

$$\tau_0 = \frac{4K_0(D_1 D_2^3)^{1/4}}{d^2 t} \tag{8.10}$$

where K_0 is an orthotropic buckling coefficient. The flexural rigidities, D_1 and D_2, parallel and perpendicular to the corrugations respectively, are

$$D_1 = \left(\frac{t}{t'}\right)\frac{Et^3}{12}, \qquad D_2 = Et'k^2 \tag{8.11a,b}$$

where t' is an equivalent thickness,

$$t' = \frac{(e+b)t}{e + b\cos\theta} \tag{8.12a}$$

and k is the radius of gyration, which we find from equations (8.6) and (8.7b) as

$$k^2 = \frac{I_x}{A_p} = \frac{h^2(3e+b)}{12(e+b)} \tag{8.12b}$$

Substituting equations (8.11a,b) and (8.12a,b) into equation (8.10) gives

$$\tau_0 = \frac{K_0}{3}\left\{\frac{[3(e/b)+1]^{3/4}}{(e/b+1)^{1/4}(e/b+\cos\theta)^{1/2}}\right\}\frac{Eh^{3/2}t^{1/2}}{d^2} \tag{8.13}$$

(iv) Optimisation

With $b > e$ we may equate the shear stresses within equations (8.9b) and (8.13) to ensure a simultaneous local and global buckling at the design stress $\tau_y = q/t$. The design criterion is

$$\tau_y = \frac{q}{t} = K_b E \sin^2\theta \left(\frac{t}{h}\right)^2 = \frac{K_0}{3}\left\{\frac{[3(e/b)+1]^{3/4}}{(e/b+1)^{1/4}(e/b+\cos\theta)^{1/2}}\right\}\frac{Eh^{3/2}t^{1/2}}{d^2} \tag{8.14}$$

Here, τ_y is taken as the shear stress at the limit of proportionality so that E is the elastic modulus. If τ_y is taken to exceed the elastic limit then the allowance for plasticity is made through the use of a tangent modulus and the plasticity reduction factor (see Figure 4.8) in the usual way [6]. Abbreviating the quantity in braces in equation (8.14) above as { ■ }, a manipulation of the three parts of this equation gives the optimum thickness in two forms:

$$\frac{t_{\text{opt}}}{d} = \left(\frac{\tau_y}{K_b E \sin^2\theta}\right)^{7/8} \left\{\frac{[\bullet]}{\{\blacksquare\}}\right\}^{1/2}, \quad t_{\text{opt}} = \left(\frac{q}{K_b E \sin^2\theta}\right)^{7/15} \left\{\frac{[\bullet]}{\{\blacksquare\}}\right\}^{4/15} d^{8/15} \tag{8.15a,b}$$

where $[\bullet] = [3(K_b/K_0)\sin^2\theta]$. The optimum height is expressed in a similar manner:

$$\frac{h_{\text{opt}}}{d} = \left(\frac{\tau_y}{K_b E \sin^2\theta}\right)^{3/8} \left\{\frac{[\bullet]}{\{\blacksquare\}}\right\}^{1/2}, \quad h_{\text{opt}} = \left(\frac{q}{K_b E \sin^2\theta}\right)^{1/5} \left\{\frac{[\bullet]}{\{\blacksquare\}}\right\}^{2/5} d^{4/5} \tag{8.16a,b}$$

where $q = \tau_y t$ connects each pair of expressions (8.15a,b) and (8.16a,b). Using equation (8.15b) to eliminate t from the distributed weight expression, equation (8.8b) leads to

$$\left(\frac{w}{d^2}\right)_{\text{opt}} = \left(\frac{3}{K_b^{3/4} K_0}\right)^{4/15} \left[\frac{(e/b+1)^{16}}{(e/b+\cos\theta)^{13}(3e/b+1)^3 \sin^6\theta}\right]^{1/15} \frac{\rho}{E^{7/15}} \left(\frac{q}{d}\right)^{7/15} \tag{8.17}$$

in which the usual three controlling quantities appear within the objective function: (a) the shape factor $f = (\,)^{4/15} \times [\,]^{1/15}$; (b) the material factor $m = \rho/E^{7/15}$; and (c) the structural index $S = q/d$. To achieve minimum weight the shape factor must be minimised. Alternatively, minimum weight corresponds to a maximum in the reciprocal of the shape factor. Recall from Chapters 1 and 2 that $1/f$ is called the shape efficiency factor F. This gives

$$F = \left[\frac{(e/b+\cos\theta)^{13}(3e/b+1)^3 \sin^6\theta}{(e/b+1)^{16}}\right]^{1/15}, \quad \text{for } e \le b \tag{8.18a}$$

With $b < e$ we see, by making the comparison between equations (8.9b,c), that K_b in the four optimal expressions above should be replaced by $K_e/(e/b)^2$. Consequently, this modifies F in equation (8.18a):

$$F = \left[\frac{(e/b+\cos\theta)^{13}(3e/b+1)^3 \sin^6\theta}{(e/b)^6(e/b+1)^{16}}\right]^{1/15}, \quad \text{for } b \le e \tag{8.18b}$$

The two independent variables e/b and θ will control the weight distribution. This is demonstrated with the following two conditions.

(a) Equal Flat Lengths $e/b = 1$

Here equations (8.18a,b) coincide and F depends upon θ alone:

$$2^{10} F^{15} = (1+\cos\theta)^{13} \sin^6\theta \tag{8.19a}$$

Hence, we may differentiate F in equation (8.19a) with respect to θ for the required maximum in F:

$$\frac{\mathrm{d}F}{\mathrm{d}\theta} = 6(1+\cos\theta)^{13}\sin^5\theta\cos\theta - 13\sin^6\theta(1+\cos\theta)^{12}\sin\theta = 0$$

$$19\cos^2\theta + 6\cos\theta - 13 = 0 \tag{8.19b}$$

The solution to equation (8.19b) is $\theta = 46.83°$, whence equation (8.19a) yields $F_{max} = 0.872$. This solution appears as the upper $e/b = 1.0$ plot in Figure 8.4. Correspondingly, the optimum dimensions follow from equations (8.15a) and (8.16a):

$$\frac{t}{d} = \frac{1.621}{K_b^{3/8}K_0^{1/2}}\left(\frac{\tau_y}{E}\right)^{7/8}, \quad \frac{h}{d} = \frac{1.182K_b^{1/8}}{K_0^{1/2}}\left(\frac{\tau_y}{E}\right)^{3/8} \tag{8.20a,b}$$

(b) Unequal Flat Lengths $e \neq b$

Where equations (8.18a,b) apply, we may plot F versus θ for constant ratios e/b in the manner of Figure 8.4. These ratios, which include the case above ($e/b = 1$), end with a V-corrugation where $e/b = 0$. Each e/b plot in Figure 8.4 reveals the θ-value that gives F_{max}. Of these maxima, none is greater than that found for case (a).

The equivalent analytical expression giving the maximum in F follows from a partial differentiation of equations (8.18a,b) in which θ is found for a given e/b. This leads to

$$19\cos^2\theta + 6(e/b)\cos\theta - 13 = 0 \tag{8.21}$$

Hence, F_{max} corresponds to the positive root ($\cos\theta$) of equation (8.21). With $e/b = 2$, for example, equation (8.21) gives the required root as

$$\cos\theta = \frac{-12 + \sqrt{12^2 + (4\times 19\times 13)}}{2\times 19} = 0.570$$

Hence $\theta = 55.3°$ and $F_{max} = 0.726$, in agreement with Figure 8.4.

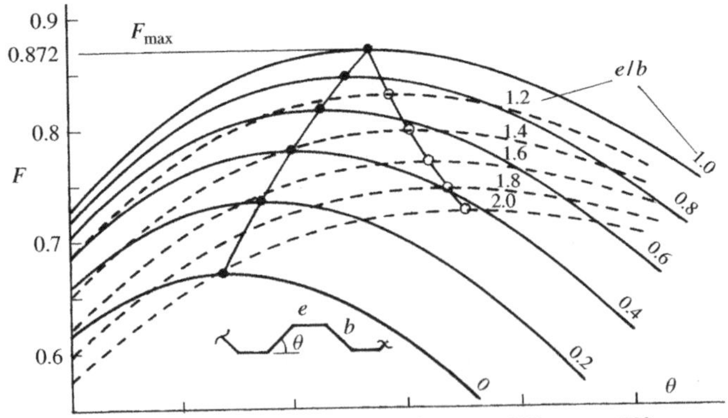

Figure 8.4 Shape efficiency factors F whose maxima give minimum weight of a corrugated shear web

8.4 Flat Web with Stiffeners

Figure 8.5 shows the elevation and plan for regular spaced, top-hat stiffening of a flat plate with thickness t_p, to give a shear web of depth d and pitch b with $b < d$. The thickness and height of the stiffener (see Figure 8.5b) need not be specified. Instead, its cross-sectional area A_s within one pitch is employed. This area enables other regular spaced reinforcements, both single- and double-sided, to be dealt with in a similar manner. The following analysis is similar to those where L- and Z-section stiffener attachments are used.

(i) Objective Function

The objective function is the distributed weight per unit length. With web length L the number of stiffeners is L/b, each of section area A_s. The total weight of the assembly is the sum of the plate and stiffener weights,

$$W = W_p + W_s = \rho[dt_pL + A_s(L/b)d] \tag{8.22a}$$

which gives the objective function

$$w = \frac{W}{L} = w_p + w_s = \rho\left(dt_p + \frac{A_sd}{b}\right) \tag{8.22b}$$

For a given number of stiffeners, as the pitch b and length L increase the two contributions w_p and w_s to the total weight in equation (8.22b) will oppose (see Example 8.2). It follows that the minimum weight of the assembly will correspond to an optimum pitch spacing b in Figure 8.5b, which we wish to find.

(ii) Shear Buckling

Shear buckling of the flat panel, arising from the shear flow q, is expressed in terms of a critical shear stress

$$\tau_p = \frac{q}{t} = KE_T\left(\frac{t_p}{b}\right)^2 \tag{8.23}$$

where E_T is the tangent modulus for the plate material to be taken with a correction within K to allow for plate plasticity (see Appendix C). Otherwise, when buckling is elastic, E and K_e replace E_T and K. When the web has to bear a shear force Q, we substitute $q = Q/d$ into equation (8.23) to give the plate thickness required:

$$t_p = \left(\frac{Qb^2}{K_eEd}\right)^{1/3} \tag{8.24}$$

Given $b < d$, Figure C.3 shows that $K_p = 4.84$ for a 'long' plate (i.e. $b/d < 0.1$) with simply supported edges. It will be seen from Figure 8.6 how this buckling coefficient is modified in the presence of stiffeners, depending upon the d/b ratio.

The criterion for choosing the minimum area A_s is based upon the stiffener's role in preventing the spread of buckling from panel to panel within the flat plate. A non-dimensional

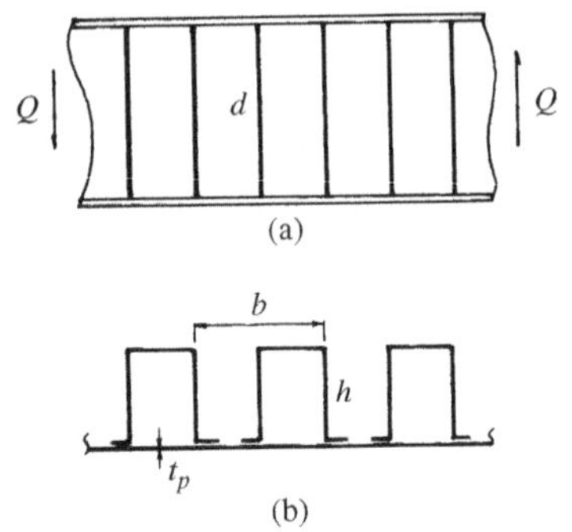

Figure 8.5 Stiffened shear web

coefficient μ provides the constraint exerted by the stiffener upon the plate [7]:

$$\mu = \left(\frac{E_s I_s b}{E_p d^2 t_p^3} \right)^{1/2} \tag{8.25}$$

where subscripts s and p refer to the stiffener and plate respectively. When the plate and stiffener are of the same material $E_p = E_s$ then μ appears as the abscissa within Figures 8.6 and 8.7. In Figure 8.6 the top and bottom horizontal edges of the panel are simply supported over length L and in Figure 8.7 these edges are clamped. Both figures assume that the stiffeners have no torsional stiffness. Hence, the constraint measure μ in equation (8.25) determines the buckling stress coefficient K, for a stiffened plate within the pitch area's aspect ratio d/b.

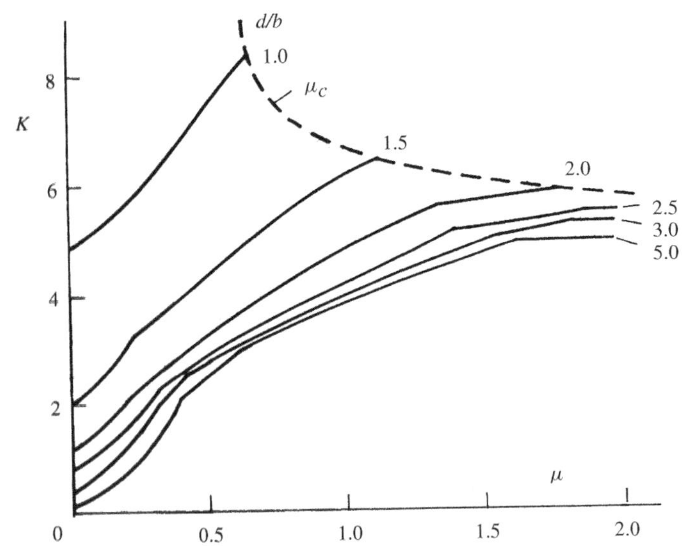

Figure 8.6 Influence of constraint factor upon buckling coefficient for a panel with simply supported edges

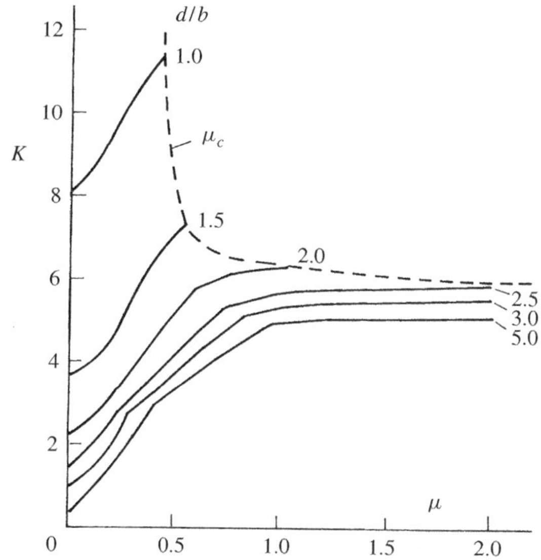

Figure 8.7 Influence of constraint factor upon buckling coefficient for a panel with edges clamped

Figures 8.6 and 8.7 show that as μ increases beyond its critical value μ_c the coefficient K approaches the K_e-value for an unsupported plate. Hence, the stiffeners then become ineffective in providing support to the plate depth. The intersection curve (dashed line in each figure) provides the critical μ_c-value for a stiffener to impart simple support to the plate. This support prevents shear buckling in the unsupported panel between stiffeners. In what follows the coefficient K, as read from each figure, is used within equations (8.23) and (8.24). Strictly, for a closed-section reinforcement, the buckling equation (8.23) should refer to the plate's unsupported plate region, i.e. the stiffener base length should be subtracted from the pitch [5]. Setting $I_s = A_s k_s^2$ in equation (8.25), this is rearranged to give

$$\frac{A_s d}{b} = \frac{\mu^2 d^3 t_p^3 E_p}{k_s^2 b^2 E_s} \tag{8.26a}$$

Substituting t_p from equation (8.24) into equation (8.26a)

$$\frac{A_s d}{b} = \frac{\mu^2 d^3 E_p}{k_s^2 b^2 E_s} \times \frac{Qb^2}{dKE_p} = \left(\frac{\mu d}{k_s}\right)^2 \frac{Q}{KE_s} \tag{8.26b}$$

Then, substituting equations (8.24) and (8.26b) into the distributed weight expression (8.22b),

$$w = \rho\left[\left(\frac{Qb^2d^2}{KE_p}\right)^{1/3} + \left(\frac{\mu d}{k_s}\right)^2 \frac{Q}{KE_s}\right] = \rho\left[\left(\frac{Qd^2}{E_p}\right)^{1/3}\left(\frac{b^2}{K}\right)^{1/3} + \frac{Qd^2}{k_s^2 E_s}\left(\frac{\mu^2}{K}\right)\right] \tag{8.27a,b}$$

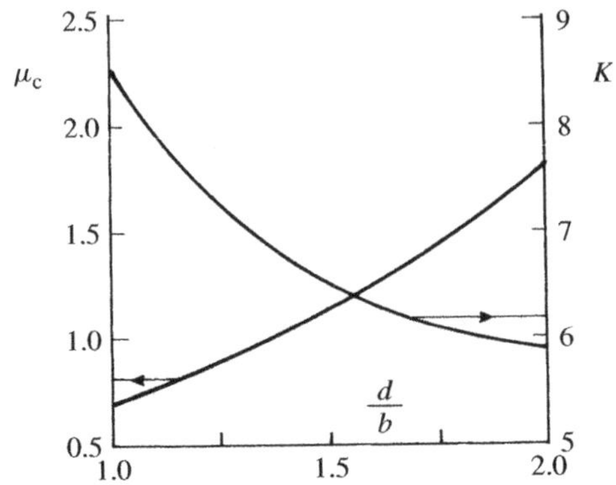

Figure 8.8 Dependence of μ and K upon d/b with plate simply supported by stiffener

In equation (8.27b) the variable terms b^2/K and μ^2/K, which lead to w_{min}, have been separated. The breadth b, corresponding to w_{min}, is found from a graphical application of equation (8.27b) in which $E_p = E_s$ is implied by Figure 8.6. For example, with the edges of each stiffener providing simple supports over the web depth d, we may read both $\mu(=\mu_c)$ and K from Figure 8.6. The plot derived in Figure 8.8 is more convenient to use within the range $1.0 \leq d/b \leq 2.0$. Here the top and bottom edges of the reinforced panel are also simply supported. Thus, for each value of b selected, d/b is calculated (or vice versa). Then μ_c and K follow directly from Figure 8.8, which allow w to be found from equation (8.27b). When w is plotted against b, the minimum condition is established. The calculation of k is open to a number of interpretations [5]. Normally, I and hence $k = \sqrt{(I/A_s)}$ may be based upon the stiffener alone about the mid-plane of the plate. Alternatively, with a contribution from the plate, k should be referred to their combined centroid. The following example shows how these techniques are applied and the effect that each k-value has upon w.

Example 8.2 Regular spacing of an L-shaped stiffener section is to be used to reinforce a flat vertical web in its support of a shear force of $Q = 60$ kN over a depth $d = 0.5$ m (see inset figure within Figure 8.9). Both web and stiffener are of similar steel grade with an elastic modulus $E = 200$ GPa and density $\rho = 7835$ kg/m^3. At what pitch b should the stiffeners be spaced to minimise the combined weight distribution of the plate and stiffener?

Firstly, identify A_s with a section area that includes stiffener and plate, i.e. the section dimensioned in Figure 8.9, assuming $t_p = 1.5$ mm:

$$A_s = 3(30 \times 1.5) = 135 \text{ mm}^2$$

The second moment of area is referred to the centroidal axis, with distance $\bar{y}$ from the base,

$$A\bar{y} = (30 \times 1.5)0.75 + (30 \times 1.5)16.5, \quad \Rightarrow \quad \bar{y} = 776.25/135 = 5.75 \text{ mm}$$

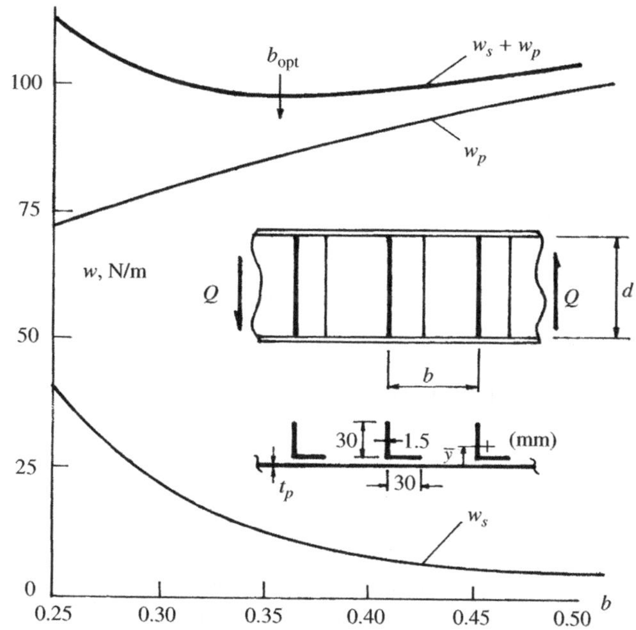

Figure 8.9 Single-sided, L-shaped stiffener attachment showing the contribution to distributed weight from each term in equation (8.22b)

and I is taken to be the second moment of area for the stiffener, plus a contribution from the plate, about the centroidal axis. Using the parallel axis theorem,

$$I = \frac{30(3)^3}{12} + (30 \times 3)(5.75 - 1.5)^2 + \frac{1.5(30)^3}{12} + (30 \times 1.5)(15 - 4.25)^2 = 10268 \text{ mm}^4$$

which provides the radius of gyration for this section as

$$k_s^2 = \frac{I}{A_s} = \frac{10268}{135} = 76.1 \text{ mm}^4$$

Substituting all appropriate values into equation (8.27b) in N and m units gives w in units of N/m:

$$w = 76.86\left[4.217\left(\frac{b^2}{K}\right)^{1/3} + 0.9856\left(\frac{\mu^2}{K}\right)\right] \tag{8.28}$$

With the edges of each stiffener providing simple supports over the plate depth, we may read both $\mu = \mu_c$ and K from Figure 8.8 within its range $1.0 \le d/b \le 2.0$ (see Table 8.1).

Each term in equation (8.28) contributes to the total weight in the manner of Figure 8.9 (the separated w_s and w_p plots). In the fourth row of Table 8.1 the distributed weights w have been calculated from the sum $(w_s + w_p)$. Figure 8.9 shows that the optimum stiffener pitch for minimum weight is $b_{opt} = 0.357$ m. The plate thickness follows from

Table 8.1 Distributed weight calculations from equation (8.28)

d/b	1.0	1.1	1.2	1.3	1.4	1.5	1.75	2.0
b, m	0.5	0.455	0.417	0.385	0.357	0.333	0.286	0.25
$\mu = \mu_c$	0.68	0.75	0.83	0.92	1.02	1.13	1.45	1.80
K	8.5	7.95	7.47	7.07	6.75	6.50	6.07	5.90
w, N/m	104.2	101.4	99.5	98.3 ($k_s^2 = 76.1$ mm^2)	98.0	98.3	103.3	112.8
w^*, N/m	104.7	102.1	100.4	99.4 ($k_s^2 = 89.29$ mm^2)	99.4	100.1	106.5	117.7

equation (8.24), using N and mm units, as

$$t_p = \left[\frac{(60 \times 10^3) \times 357^2}{6.75 \times 500 \times (200 \times 10^3)}\right]^{1/3} = 2.25 \text{ mm}$$

Since this differs from the 1.5 mm plate thickness assumed in the I-calculation for the stiffener section, the I-calculation could be revised and the solution refined at this stage. However, this is unlikely to matter since the stiffener dominates the I-value. The refinement is unnecessary when, in accordance with the derivation given above, A_s remains the area of the stiffener. More importantly here, there are two alternative calculations for I. In the first, I refers to the centroid axis of the stiffener, its position from the base being

$$(28.5 \times 1.5 \times 0.75) + (30 \times 1.5 \times 15) = 1.5(28.5 + 30)\bar{y}, \quad \Rightarrow \quad \bar{y} = 8.058 \text{ mm}$$

and I about this axis is

$$I = \frac{28.5(1.5)^3}{12} + (28.5 \times 1.5)(8.058 - 0.75)^2 + \frac{1.5(30)^3}{12} + (30 \times 1.5)(15 - 8.058)^2$$
$$= 7834.77 \text{ mm}^4$$

giving $k^2 = 89.29$ mm^2. In the second, I is calculated for the stiffener about the mid plane of the plate [5]. This may be found from applying parallel axes to the first result directly:

$$I = 7834.77 + 1.5(28.5 + 30)\left(8.058 + \frac{1.5}{2}\right)^2 = 14642.5 \text{ mm}^2$$

or, in a separate calculation, where $t_p = 1.5$ mm is again assumed:

$$I = \frac{28.5(1.5)^3}{12} + (28.5 \times 1.5)1.5^2 + \frac{1.5(30)^3}{12} + (30 \times 1.5)\left(15 + \frac{1.5}{2}\right)^2$$
$$= 14642 \text{ mm}^4$$

giving, apparently, $k^2 = 166.86$ mm^2. The first calculation, giving $k_s^2 = 89.29$ mm^2, when based upon the section's centroidal axis, appears to be more sound. In taking this value for k_s^2 the revised weights w^*, given in Table 8.1, are found to be only slightly different from the previous values.

References

[1] ESDU 02.03.24, Strength of castellations in shear, September 1989.
[2] Leach, J. F. Strength tests on fuselage frame castellations, British Aeroplane Co, Rpt No. C.R. 530 September 1948.
[3] Peterson, J. P. and Card, M. F. Buckling strength of corrugated webs in shear, NACA Tech Note D-424, June 1960.
[4] Rothwell, A. The buckling of shallow corrugated webs in shear, October 1968, *Journal of the Royal Aeronautical Society*, 72, 883–886.
[5] Darley, K. Static shear tests on corrugated web panels, British Aerospace Co, Rpt No TN 2030, October 1964.
[6] ESDU 71005, Buckling of flat plates in shear, September 1976
[7] ESDU 02.03.02, Flat panels in shear. Buckling of long panels with transverse stiffeners, April 1983.

Exercises

8.1 Figure 8.10 shows a V-corrugation with equal side lengths b and included angles θ that is to be used as a shear web. Derive: (i) an expression for its efficiency in resisting simultaneous local and orthotropic buckling and (ii) the optimum θ, corresponding to minimum weight of the corrugation.

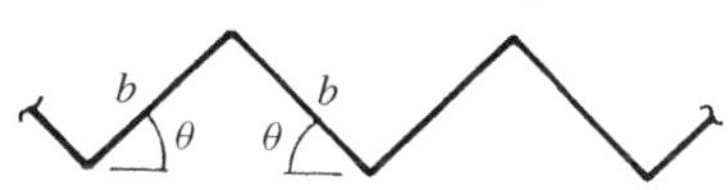

Figure 8.10 V-corrugated shear web

8.2 A shear web of depth 0.6 m and thickness 1 mm is subjected to a constant shear force of 45 kN. The web is to be reinforced on one side with Z-stiffeners having the dimensions shown in Figure 8.11. Both the web and the stiffener are made from a similar grade of aluminium alloy. The long, horizontal edges of the reinforced panel are clamped against the shear force. Find the optimum pitch of the stiffeners that will minimise the net weight of the assembly. Take $E = 70$ GPa and $\rho = 2800$ kg/m^3.

8.3 What would the optimum spacing be for a double-sided attachment of L-stiffeners to the flat plate given in Example 8.2? The stiffener and plate geometry are to remain as shown in Figure 8.9. Take the panel long edges to be simply supported.

8.4 Referring to Example 8.2, what should the optimum spacing be for a single-sided attachment of L-stiffeners to the flat web when the long edges of the panel are clamped? The stiffener and plate geometry are to remain as those shown in Figure 8.9.

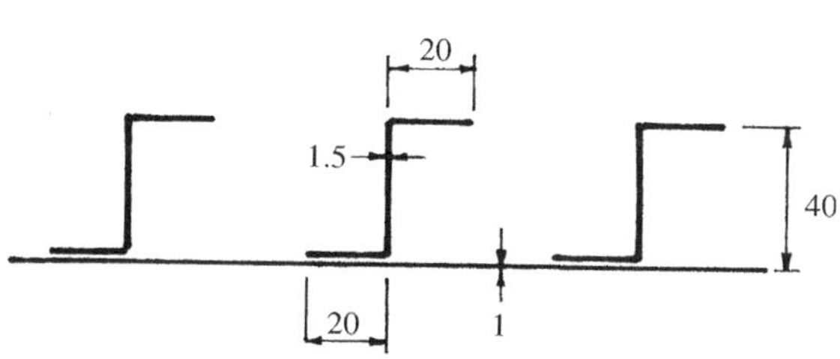

Figure 8.11 Z-stiffener reinforcement

Figure 8.11

9

Frame Assemblies

9.1 Introduction

The weight of struts and ties within pin-jointed frames may be optimised from the loading that they carry [1–12]. Either the corresponding compressive and tensile stresses in each member may be allowed to reach a similar limit, or a lesser elastic stress may be imposed upon strut buckling. The minimum weight solutions provided by these criteria extend across a wide range of structural indices from which an optimum weight is chosen to dictate the frame geometry. The method is first illustrated for relatively simple, two-bar plane frames and then extended to three-bar space frames. In certain cases the members may be entirely under compression (struts), but more usually both tension and compression arise within the bars of a frame under load. Frames with many struts and tie bars may be treated in a similar manner by generalising their respective optimising procedures.

9.2 Double-Strut Assembly

Here the problem is to span a fixed distance $2B$ with two struts, each of length L, that are required to support a central vertical load F with an optimum inclination θ (see Figure 9.1). Since $B = L\cos\theta$, it is only necessary to optimise the inclination θ when minimising weight.

Each vertical reaction is $F/2$, this being in equilibrium with the compressive bar force P:

$$0.5F - P\sin\theta = 0 \tag{9.1a}$$

Equation (9.1a) gives the compressive strut force P and its stress σ as

$$P = \frac{F}{2\sin\theta}, \quad \sigma = \frac{P}{A} = \frac{F}{2A\sin\theta} \tag{9.1b,c}$$

The weight of this double-strut assembly follows as

$$W = 2\rho AL = \frac{2\rho FB}{2\sigma\sin\theta\cos\theta} = \frac{\rho}{\sigma}\frac{FB}{\sin\theta\cos\theta} \tag{9.2}$$

Mechanics of Optimal Structural Design: Minimum Weight Structures D. W. A. Rees

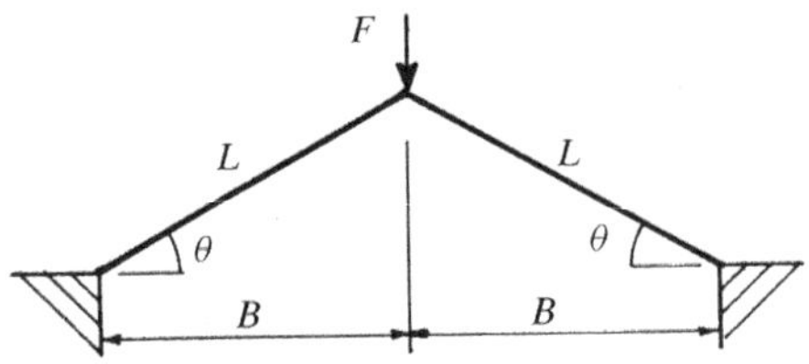

Figure 9.1 Double-strut assembly supporting a vertical load

Clearly, the denominator, $\sigma \sin\theta \cos\theta$, in equation (9.2) should be maximised to give minimum weight ($W_{\min}$) within the criteria that follow.

9.2.1 *Failure Criteria*

We may derive two objective functions from equation (9.2): one based upon flexural buckling and the other upon a limiting compressive stress.

(i) Flexural Buckling

Taking each strut in isolation, our previous analysis of a slender strut with circular cross-section (see § 1.3.1) gave the objective function in equation (1.8a). Thus, the stress that is required to be maximised is

$$\sigma_{\text{opt}} = \frac{0.886}{L}\sqrt{E_T P} = \frac{0.886 \cos\theta}{B}\sqrt{\frac{E_T F}{2 \sin\theta}} \tag{9.3}$$

Substituting equation (9.3) into equation (9.2) leads to

$$W = 1.596\rho \left(\frac{B}{\cos\theta}\right)^2 \sqrt{\frac{F}{E_T \sin\theta}} \tag{9.4a}$$

and dividing equation (9.4a) by B^3 gives the objective function

$$\frac{W}{B^3} = \frac{1.596}{\sqrt{\sin\theta}\cos^2\theta}\left(\frac{\rho}{E_T^{1/2}}\right)\left(\frac{F}{B^2}\right)^{1/2} \tag{9.4b}$$

The shape factor within equation (9.4b) will minimise the weight when its denominator

$$f_{(i)} = \sqrt{\sin\theta} \times \cos^2\theta \tag{9.4c}$$

is a maximum. Thus, applying the condition $\mathrm{d}f_{(i)}/\mathrm{d}\theta = 0$ leads to $\theta_{\text{opt}} = 26.57°$. Substituting $\theta = \theta_{\text{opt}}$ into equation (9.4c) gives $f_{(i)} = 0.535$. The minimum weight required follows from equation (9.4b):

$$\left(\frac{W}{B^3}\right)_{\min} = 2.983\left(\frac{\rho}{E_T^{1/2}}\right)\left(\frac{F}{B^2}\right)^{1/2} \tag{9.5}$$

Correspondingly, the optimum diameter for the circular-section strut is found by substituting $L = B/\cos\theta_{opt}$ into equation (1.7a):

$$d_{opt} = \left(\frac{4}{\pi}\right)^{3/4}\left(\frac{PL^2}{E_T}\right)^{1/4} = 1.304\left(\frac{FB^2}{E_T}\right)^{1/4} \tag{9.6}$$

(ii) Limiting Stress

When the compressive stress in equation (9.2) is limited to $\sigma \le \sigma_y$, then

$$\frac{\rho}{W}\frac{FB}{\sin\theta\cos\theta} \le \sigma_y \tag{9.7a}$$

Taking the equality in equation (9.7a) gives the lowest limit for W/B^3 within an alternative objective function:

$$\frac{W}{B^3} = \frac{1}{\sin\theta\cos\theta}\left(\frac{\rho}{\sigma_y}\right)\left(\frac{F}{B^2}\right) \tag{9.7b}$$

Comparing equations (9.4b) and (9.7b) shows that both the shape function and the material factor differ between the two criteria. Equation (9.7b) shows that the weight is minimised within its shape factor when

$$f_{(ii)} = \sin\theta\cos\theta \tag{9.7c}$$

is a maximum. Thus, applying the condition $\mathrm{d}f_{(ii)}/\mathrm{d}\theta = 0$ to equation (9.7c) leads to $\theta_{opt} = 45^\circ$. Substituting $\theta = \theta_{opt}$ into equations (9.7b,c) gives $f_{(ii)} = 1/2$ and a minimum weight

$$\left(\frac{W}{B^3}\right)_{min} = 2\left(\frac{\rho}{\sigma_y}\right)\left(\frac{F}{B^2}\right) \tag{9.8}$$

The optimum diameter now follows from σ_y and equation (9.1c) as

$$\sigma_y = \frac{P}{A} = \frac{F}{2\sin\theta} \times \frac{4}{\pi d^2} \tag{9.9a}$$

giving d_{opt} for $\theta = \theta_{opt}$ as

$$d_{opt} = 0.949\left(\frac{F}{\sigma_y}\right)^{1/2} \tag{9.9b}$$

9.2.2 Structural Index

The critical value of the structural index F/B^2 applies when W/B^3 is the same within criteria (i) and (ii) above. Hence, equating (9.5) and (9.8),

$$\left(\frac{W}{B^3}\right)_{min} = 2.983\left(\frac{\rho}{E_T^{1/2}}\right)\left(\frac{F}{B^2}\right)^{1/2} = 2\left(\frac{\rho}{\sigma_y}\right)\left(\frac{F}{B^2}\right) \tag{9.10a}$$

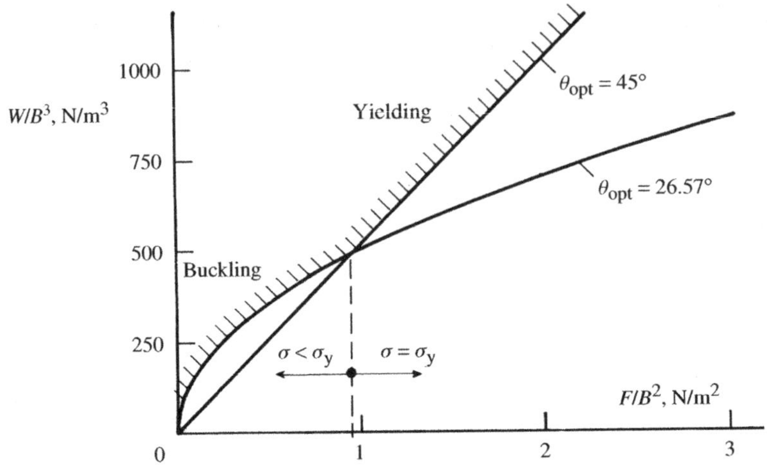

Figure 9.2 Dependence of weight upon structural index from yielding and buckling criteria

It follows from equation (9.10a) that

$$\left(\frac{F}{B^2}\right)_{cr} = 2.225\frac{\sigma_y^2}{E_T} \tag{9.10b}$$

Figure 9.2 shows this solution graphically as the intersection point between the plots of equations (9.5) and (9.8). Also revealed within this plot are the *greater* minimum weight regions (shaded) where the buckling and yielding criteria are to apply to optimum designs.

Alternative, non-dimensional forms of equation (9.5) and (9.8) are written as

$$n = 2.983\left(\frac{\sigma_y}{E}\right)^{1/2} s^{1/2} \quad \text{and} \quad n = 2s \tag{9.11a,b}$$

respectively, where

$$n = \frac{1}{\rho}\frac{W}{B^3} \quad \text{and} \quad s = \frac{1}{\sigma_y}\frac{F}{B^2} \tag{9.12a,b}$$

Equations (9.11a,b) and (9.12a,b) admit a graphical interpretation of equation (9.10b) for a range of materials. Here we take σ_y at the elastic limit for which E_T becomes the elastic modulus. Equations (9.12a,b) allow an examination of the weight parameter n across six logarithmic decades (say) of the structural index parameter s. Figure 9.3 shows that within the first two decades of s, where B is large, the optimum weight is controlled by buckling at a lower stress, i.e. equation (9.11a) applies with $\theta_{opt} = 26.57°$. The relationship (9.11a) between n and s, which depends upon both σ_y and E, is shown for brass, aluminium, steel and glass-fibre reinforced epoxy struts. In the decades beyond the intersection point, the minimum weight of each material is dictated by the stress imited (yielding) criterion (9.11b) with $\theta_{opt} = 45°$. Here a single line $n = 2s$ applies to all bar materials operating at their stress limit. Table 9.1 gives s_{cr} at the intersection between equations (9.11a,b)

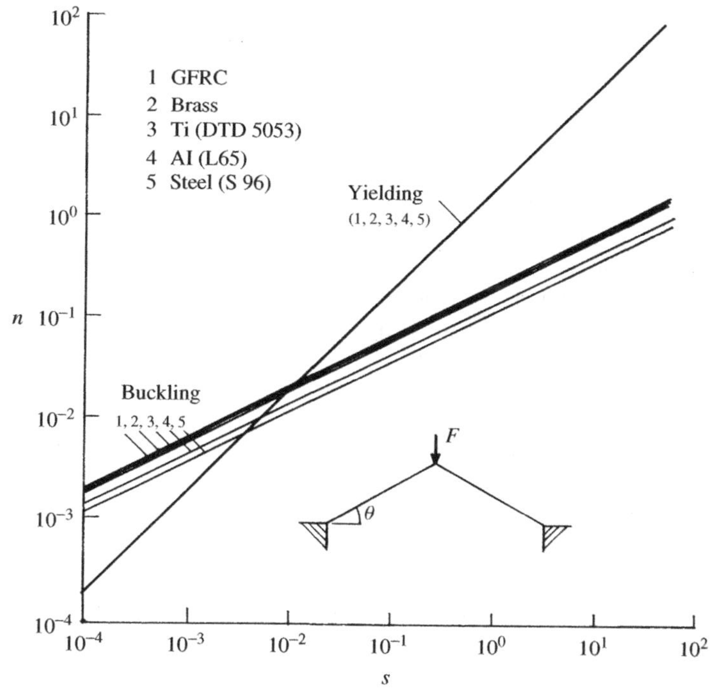

Figure 9.3 Weight dependence upon structural index. (Full key in Table 9.1)

for each material, and Appendix A provides the material properties necessary for the conversion to weight W.

Given that the bar inclination differs between the two criteria, it follows that, when choosing between them, both the length and diameter must be altered accordingly to ensure the minimum weight condition. At the intersections between the buckling and yielding limit plots, shown in Figure 9.2, the critical structural indices may be written in terms of two material properties, since it follows from equation (9.10b) that

$$s_{cr} = \frac{1}{\sigma_y}\left(\frac{F}{B^2}\right)_{cr} = 2.225\frac{\sigma_y}{E_T} \tag{9.12c}$$

Table 9.1 Material properties and critical structural indices for a two-bar frame

Material	σ_y (MPa)	E (GPa)	Equation (9.11a)	$s_{cr}(\times 10^{-3})$
(1) GFRC	100	20	$n = 0.211\sqrt{s}$	11.13
(2) Brass	450	100	$n = 0.2\sqrt{s}$	10.01
(3) Titanium	480	118	$n = 0.19\sqrt{s}$	9.051
(4) Aluminium	150	75	$n = 0.133\sqrt{s}$	4.45
(5) Steel	300	200	$n = 0.113\sqrt{s}$	3.338

Values found from equation (9.12c) are listed in Table 9.1. In reading n from Figure 9.3, the weight $W = n\rho B^3$ in kg can be found from equation (9.12a) given the base dimension B and the material densities ρ in Appendix A.

9.3 Multiple-Strut Assembly

The structure solved in § 9.2 may be extended to one spanning a plane horizontal diameter $2B$ with N identical, equi-spaced struts. Each strut is inclined at θ in its support of a vertical load F at height H, as shown in Figure 9.4.

Since $B = L\cos\theta$, it is again only necessary to optimise the inclination θ when minimising weight. The vertical equilibrium equation for the top joint is

$$-F + NP\sin\theta = 0 \tag{9.13a}$$

Equation (9.13a) gives the compressive force P and the stress σ in each strut as

$$P = \frac{F}{N\sin\theta}, \quad \sigma = \frac{P}{A} = \frac{F}{NA\sin\theta} \tag{9.13b}$$

The weight of the assembly depends upon the strut length L and area A in equation (9.13b) as

$$W = N\rho AL = \frac{N\rho FB}{N\sigma\sin\theta\cos\theta} = \frac{\rho}{\sigma}\frac{FB}{\sin\theta\cos\theta} \tag{9.14}$$

in which the denominator $\sigma\sin\theta\cos\theta$ is to be maximised for $W_{\min}$ by the following criteria.

9.3.1 *Failure Criteria*

(i) Flexural Buckling

The optimum stress level for a slender strut with circular cross-section has already been derived. We write this from equation (9.3) and (9.13b) as

$$\sigma_{\text{opt}} = \frac{\sqrt{\pi EP}}{2L} = \frac{\cos\theta}{2B}\sqrt{\frac{\pi EF}{N\sin\theta}} \tag{9.15}$$

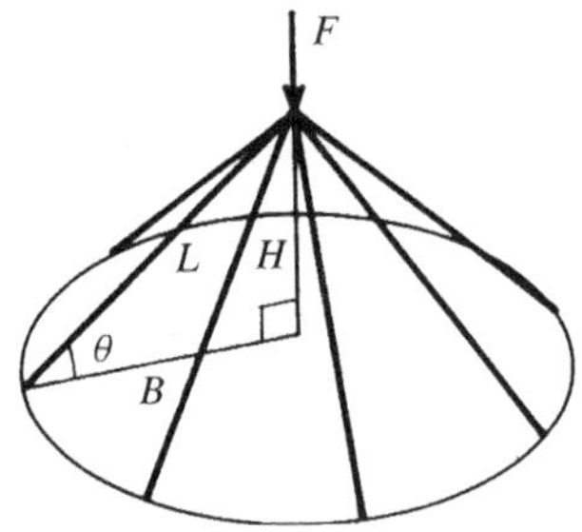

Figure 9.4 Multiple-struts in support of vertical load

Substituting equation (9.15) into equation (9.14) leads to

$$W = 2\rho\left(\frac{B}{\cos\theta}\right)^2\sqrt{\frac{NF}{\pi E\sin\theta}} \tag{9.16a}$$

and dividing equation (9.16a) by B^3 gives the objective function

$$\frac{W}{B^3} = \frac{2}{\sqrt{\sin\theta\cos^2\theta}}\left(\frac{N}{\pi}\right)^{1/2}\left(\frac{\rho}{\sqrt{E}}\right)\left(\frac{F}{B^2}\right)^{1/2} \tag{9.16b}$$

which is a minimum for a given number of struts N, when the denominator is a maximum:

$$f = \sqrt{\sin\theta} \times \cos^2\theta \tag{9.16c}$$

Thus, applying the condition $\mathrm{d}f/\mathrm{d}\theta = 0$ to equation (9.16c) leads to $\sin\theta_{\text{opt}} = 1/\sqrt{5}$, $\cos\theta_{\text{opt}} = 2/\sqrt{5}$, from which $\theta_{\text{opt}} = 26.57°$. Substituting $\theta = \theta_{\text{opt}}$ into equation (9.16b) gives the minimum weight:

$$\left(\frac{W}{B^3}\right)_{\min} = 3.738\left(\frac{N}{\pi}\right)^{1/2}\left(\frac{\rho}{\sqrt{E}}\right)\left(\frac{F}{B^2}\right)^{1/2} \tag{9.17}$$

Substituting $L = B/(\cos\theta_{\text{opt}}) = \sqrt{5}B/2$ and $P = F/(N\sin\theta_{\text{opt}}) = \sqrt{5}F/N$ into equation (1.7a) provides the corresponding optimum diameter for each of the N circular-section struts

$$d_{\text{opt}} = \left(\frac{4}{\pi}\right)^{3/4}\left(\frac{PL^2}{E}\right)^{1/4} = \frac{1.55}{N^{1/4}}\left(\frac{FB^2}{E}\right)^{1/4} \tag{9.18}$$

(ii) Limiting Stress

When the stress in equation (9.14) is limited to $\sigma \le \sigma_y$, then

$$\frac{F}{N\sin\theta} \times \frac{N\rho B}{W\cos\theta} \le \sigma_y \tag{9.19a}$$

Taking the equality in equation (9.19a) gives the lowest limit for W/B^3:

$$\frac{W}{B^3} = \frac{1}{\sin\theta\cos\theta}\left(\frac{\rho}{\sigma_y}\right)\left(\frac{F}{B^2}\right) \tag{9.19b}$$

which is independent of the number of struts. The minimum weight follows for a maximum in the denominator

$$f = \sin\theta\cos\theta \tag{9.19c}$$

Applying the condition $\mathrm{d}f/\mathrm{d}\theta = 0$ to equation (9.19c) leads to $\sin\theta_{opt} = \cos\theta_{opt} = 1/\sqrt{2}$, from which $\theta_{opt} = 45°$. Substituting $\theta = \theta_{opt}$ into equation (9.19b) gives

$$\left(\frac{W}{B^3}\right)_{min} = 2\left(\frac{\rho}{\sigma_y}\right)\left(\frac{F}{B^2}\right) \tag{9.20}$$

The optimum diameter now follows from the limiting stress σ_y within each strut:

$$\sigma_y = \frac{P}{A} = \frac{F}{N\sin\theta} \times \frac{4}{\pi d^2} \tag{9.21a}$$

Equation (9.21a) gives d_{opt} for $\theta = \theta_{opt}$ as

$$d_{opt} = \left(\frac{1.8}{N}\right)^{1/2}\left(\frac{F}{\sigma_y}\right)^{1/2} \tag{9.21b}$$

The critical value of F/B^2 applies when W/B^3 is the same within each criterion (i) and (ii) above. Hence, equating (9.17) and (9.20),

$$\left(\frac{W}{B^3}\right)_{min} = 3.738\left(\frac{N}{\pi}\right)^{1/2}\frac{\rho}{\sqrt{E}}\left(\frac{F}{B^2}\right)^{1/2} = 2\left(\frac{\rho}{\sigma_y}\right)\left(\frac{F}{B^2}\right) \tag{9.22a}$$

and, from equation (9.22a), the critical structural index is

$$\left(\frac{F}{B^2}\right)_{cr} = 3.493\frac{N}{\pi}\frac{\sigma_y^2}{E} \tag{9.22b}$$

Using equations (9.12a,b) to introduce non-dimensional parameters n and s into equations (9.22a,b),

$$n = 2s = 3.738\left(\frac{N\sigma_y}{\pi E}\right)^{1/2} s^{1/2}, \quad s_{cr} = 3.493\frac{N\sigma_y}{\pi E} \tag{9.22c,d}$$

The graphical interpretation of the solution (9.22c,d) appears at the intersection between the buckling limit plots and the single stress limit plot $n = 2s$, as shown in Figure 9.5. The plots refer to steel where $\sigma_y = 300$ MPa, $E = 210$ GPa and $\rho = 7850$ kg/m^3. The intersection coordinates appear in Table 9.2 for $N = 2, 3, \ldots, 6$ bars, with the former value confirming our earlier equation (9.10b) for the two bar frame.

Given that the bar inclination differs between the two criteria we may choose between them at the critical structural index. These will give the same weight when the length and diameter match the chosen inclination. Otherwise, in taking non-critical structural indices in support of a given load, the weight of an optimally designed, N-bar assembly is to be taken from Figure 9.5 at the greater n-value.

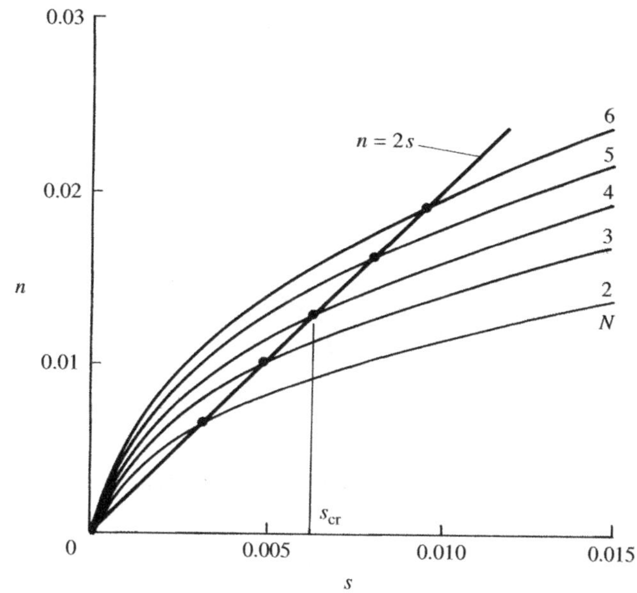

Figure 9.5 Critical structural indices for an N-bar frame

9.4 Cantilevered Framework

In Figure 9.6, two bars of different lengths L_1 and L_2, are connected from a wall to the load point with inclinations α and β as shown. The frame supports a force F at a perpendicular distance B from the connecting wall. Assuming joint tension at the load point, the equilibrium condition that exists between F and the bar forces P_1 and P_2 is expressed as

$$\uparrow \quad P_1 \sin\alpha - P_2 \sin\beta - F = 0 \tag{9.23a}$$

$$\rightarrow \quad -P_1 \cos\alpha - P_2 \cos\beta = 0 \tag{9.23b}$$

Solving equations (9.23a,b) leads to the tensile force in bar 1 (a tie)

$$P_1 = \frac{F \cos\beta}{\sin(\alpha + \beta)} \tag{9.24a}$$

Table 9.2 Critical structural indices for a multiple, steel strut assembly

N, bars	2	3	4	5	6
$3.493(N/\pi)$	2.224	3.336	4.448	5.56	6.671
s_{cr}	3.18×10^{-3}	4.77×10^{-3}	6.35×10^{-3}	7.94×10^{-3}	9.53×10^{-3}
$(F/B^2)_{cr}$ (MPa)	0.953	1.430	1.906	2.383	2.859
W/B^3(kg/m^3)	49.93	74.89	99.70	124.66	149.62

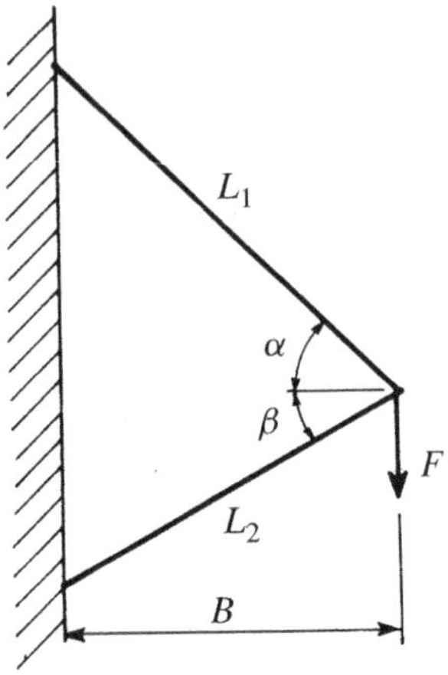

Figure 9.6 Two-bar cantilevered frame

and the compressive force within bar 2 (a strut)

$$P_2 = -\frac{F\cos\alpha}{\sin(\alpha+\beta)} \tag{9.24b}$$

Writing the bar section areas as

$$A_1 = \frac{P_1}{\sigma_1}, \quad A_2 = \frac{|P_2|}{\sigma_2} \tag{9.25a,b}$$

and their lengths as

$$L_1 = \frac{B}{\cos\alpha}, \quad L_2 = \frac{B}{\cos\beta} \tag{9.26a,b}$$

the total weight of differing bar materials follows as

$$W = \rho_1 A_1 L_1 + \rho_2 A_2 L_2 \tag{9.27a}$$

Substituting equations (9.25a,b) and (9.26a,b) into equation (9.27a),

$$W = FB\left[\frac{\rho_1\cos\beta}{\sigma_1\sin(\alpha+\beta)\cos\alpha} + \frac{\rho_2\cos\alpha}{\sigma_2\sin(\alpha+\beta)\cos\beta}\right] \tag{9.27b}$$

Dividing equation (9.27b) by B^3 introduces the structural index to each bar's weight contribution:

$$\frac{W}{B^3} = \frac{F}{B^2}\left[\frac{\rho_1}{\sigma_1}\frac{\cos\beta}{\sin(\alpha+\beta)\cos\alpha} + \frac{\rho_2}{\sigma_2}\frac{\cos\alpha}{\sin(\alpha+\beta)\cos\beta}\right] \tag{9.27c}$$

9.4.1 Failure Criteria

(i) Limiting Stress

For the tie, the limiting stress criterion $\sigma_1 \le \sigma_y$ applies. If, for a single material frame, we impose a limiting stress condition upon the strut ($\sigma_2 \le \sigma_y$), equation (9.27c) may be written as

$$\frac{W}{B^3} = \left[\frac{\cos\beta}{\sin(\alpha+\beta)\cos\alpha} + \frac{\cos\alpha}{\sin(\alpha+\beta)\cos\beta}\right]\left(\frac{\rho}{\sigma_y}\right)\left(\frac{F}{B^2}\right) \tag{9.28}$$

Equation (9.28) has a minimum value when $\alpha = \beta = 45°$. Here strut buckling has been offset with the selection of a bar diameter to give a slenderness ratio L_2/k that ensures the buckling stress lies well into the plastic regime. That is $\sigma_{cr} > \sigma_y$ which, for pinned ends, gives

$$\frac{\pi^2 E}{(L_2/k)^2} > \sigma_y, \quad \Rightarrow \quad \frac{L_2}{k} < \sqrt{\frac{\pi^2 E}{\sigma_y}} \tag{9.29a,b}$$

Since $k^2 = d^2/16$, the strut diameter required becomes

$$d > \frac{4L_2}{\pi}\sqrt{\frac{\sigma_y}{E}} \tag{9.30a}$$

An alternative diameter for the strut, when fully stressed, follows from $\sigma_y = P_2/A_2$, where P_2 is given by equation (9.24b). This leads to

$$d = \sqrt{\frac{4F\cos\alpha}{\pi\sigma_y \sin(\alpha+\beta)}} \tag{9.30b}$$

If we now equate (9.30a,b), then buckling occurs under σ_y for the condition

$$\frac{\cos\alpha}{\sin(\alpha+\beta)} = \frac{4L_2^2\sigma_y^2}{\pi E F} \tag{9.31a}$$

Substituting equation (9.31a) into equation (9.27c) gives an alternative objective function

$$\frac{W}{B^3} = \frac{4}{\cos\beta}\left(\frac{1}{\cos^2\alpha} + \frac{1}{\cos^2\beta}\right)\frac{\rho\sigma_y}{\pi E} \tag{9.31b}$$

for which the orientations $\alpha = \beta = 45°$ again provide a minimum weight for the frame.

(ii) Strut Buckling

In this criterion the strut is at the point of buckling when the stress is at its maximum elastic level, i.e. $\sigma_2 = \sigma_{2y}$. Taking σ_2 from equation (1.8a) and combining with equations (9.24b) and (9.26b)

$$(\sigma_2)_{opt} = \frac{0.886}{L_2}\sqrt{E_2 P_2} = \frac{0.886}{B}\sqrt{E_2 F}\cos\beta\sqrt{\frac{\cos\alpha}{\sin(\alpha+\beta)}} \tag{9.32a}$$

Substituting equation (9.32a) into equation (9.27c), gives an objective function for different bar materials

$$\frac{W}{B^3} = \left[\frac{\cos\beta}{\sin(\alpha+\beta)\cos\alpha}\right]\frac{F}{B^2}\frac{\rho_1}{\sigma_{1y}} + \frac{1.129}{\cos^2\beta}\sqrt{\frac{\cos\alpha}{\sin(\alpha+\beta)}}\left(\frac{F}{B^2}\right)^{1/2}\frac{\rho_2}{E_2^{1/2}} \tag{9.32b}$$

Equations (9.32a,b) allow the limiting stress to be reached in the tie when buckling of the strut occurs simultaneously for bars of different materials. For similar bar materials $\sigma_{1y} = \sigma_{2y}$ and $\rho_1 = \rho_2$ when equation (9.32b) simplifies to

$$\frac{W}{B^3} = \left[\frac{\cos\beta}{\sin(\alpha+\beta)\cos\alpha}\right]\frac{F}{B^2}\frac{\rho}{\sigma_y} + \frac{1.129}{\cos^2\beta}\sqrt{\frac{\cos\alpha}{\sin(\alpha+\beta)}}\left(\frac{F}{B^2}\right)^{1/2}\frac{\rho}{E^{1/2}} \tag{9.32c}$$

To minimise the weight under this condition, the material properties ρ, E and σ_y are inserted into equation (9.32c) for a given F/B^2. Values of α and β are then sought to minimise W/B^3. This process may be repeated over a range of F/B^2, equivalent to equation (9.32c) at α_{opt} and β_{opt}.

9.4.2 *Special Case for $\beta = 0$*

(i) Limiting stress condition

With both bars of similar material, fully stressed to the limit σ_y, equation (9.28) becomes

$$\frac{W}{B^3} = \left(\frac{1}{\sin\alpha\cos\alpha} + \frac{\cos\alpha}{\sin\alpha}\right)\frac{\rho}{\sigma_y}\frac{F}{B^2} = \left(\frac{1+\cos^2\alpha}{\sin\alpha\cos\alpha}\right)\frac{\rho}{\sigma_y}\frac{F}{B^2} \tag{9.33a}$$

Equation (9.33a) is written in the non-dimensional form

$$n = \left(\frac{1+\cos^2\alpha}{\sin\alpha\cos\alpha}\right)s \tag{9.33b}$$

where s and n are as defined in equation (9.12a,b) . The minimum weight follows from applying the condition $\partial n/\partial\alpha = 0$ to equation (9.33b), which leads to

$$\cos\alpha = 1/\sqrt{3}, \quad \Rightarrow \quad \alpha = 54.74^\circ, \quad n = 2.8283s$$

The non-dimensional relationship $n = 2.8283s$ is material-independent, though the specific properties ρ and σ_y do appear within s and n, as defined in equations (9.12a,b)

(ii) Strut Buckling

When bar 1 is yield stress limited and bar 2 at its optimum buckling condition, equation (9.32c) is written as

$$\frac{W}{B^3} = \frac{1}{\sin\alpha\cos\alpha}\frac{F}{B^2}\frac{\rho}{\sigma_y} + 1.129\left(\frac{\sigma_y}{E}\right)^{1/2}\left(\frac{\cos\alpha}{\sin\alpha}\right)^{1/2}\left(\frac{F}{B^2}\right)^{\frac{1}{2}}\frac{\rho}{\sigma_y^{1/2}} \tag{9.34a}$$

Table 9.3 Dependence of optimum weight of a cantilever frame upon α when $\beta = 0°$

s	0.001	0.01	0.10	1.0	10	100
α	53.17°	47.9°	45.95°	45.3°	45.02	45
n	0.003253	0.0242	0.2134	2.0426	20.13	200.4

Equation (9.34a) is written in terms of our non-dimensional parameters as

$$n = \frac{s}{\sin\alpha\cos\alpha} + 1.129\left(\frac{\cos\alpha}{\sin\alpha}\right)^{1/2}\left(\frac{\sigma_y}{E}\right)^{1/2} s^{1/2} \tag{9.34b}$$

Here the relation $n = n(s)$ is material-specific. In steel, for example, with its properties taken from Table 9.1, equation (9.34b) becomes

$$n = \frac{s}{\sin\alpha\cos\alpha} + 0.0427\left(\frac{\cos\alpha}{\sin\alpha}\right)^{1/2} s^{1/2} \tag{9.35a}$$

Minimum weight follows from applying the condition $\partial n/\partial\alpha = 0$ to equation (9.35a). This leads to

$$s(\sin^2\alpha - \cos^2\alpha) - 0.02135 s^{1/2}\cos^{3/2}\alpha\sin^{1/2}\alpha = 0 \tag{9.35b}$$

Table 9.3 gives the trial solutions for α and n from equations (9.35a,b) , over five decades of the structural index parameter s.

The minimum weight solutions from the two criteria (i) and (ii) in § 9.4.2 above appear in Figure 9.7. This shows that condition (ii) applies to the first decade in s and condition (i) applies to the remaining decades. Thus, to achieve minimum weight for a given structural index, the geometry of the frame and the section diameter must be adjusted accordingly, as shown later in Figure 9.8.

9.4.3 Special Case for $\alpha = 0$

(i) Limiting stress condition

With both bars fully stressed to the yield limit σ_y, equation (9.27c) becomes

$$\frac{W}{B^3} = \left(\frac{\cos\beta}{\sin\beta} + \frac{1}{\sin\beta\cos\beta}\right)\frac{\rho}{\sigma_y}\frac{F}{B^2} = \frac{1+\cos^2\beta}{\sin\beta\cos\beta}\frac{\rho}{\sigma_y}\frac{F}{B^2} \tag{9.36a}$$

which we write as

$$n = \left(\frac{1+\cos^2\beta}{\sin\beta\cos\beta}\right)s \tag{9.36b}$$

where s and n are as defined in equation (9.12a,b). Minimum weight follows from applying the condition $\partial n/\partial\beta = 0$ to equation (9.36b). This leads to

$$\cos\beta = 1/\sqrt{3}, \quad \Rightarrow \beta = 54.74°, \quad n = 2.8283s$$

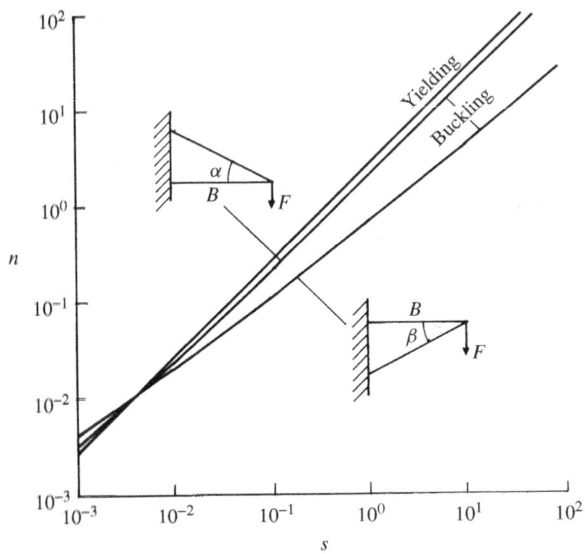

Figure 9.7 Weight parameter versus structural index when $\alpha = 0$ and when $\beta = 0$

The relationship $n = 2.8283s$ is again material-independent, though the specific properties ρ and σ_y do appear within the s and n definitions in equations (9.12a,b).

(ii) Strut Buckling

With bar 1 at its stress limit and bar 2 at its optimum buckling condition, equation (9.32c) becomes

$$\frac{W}{B^3} = \frac{\cos\beta}{\sin\beta}\frac{F}{B^2}\frac{\rho}{\sigma_y} + \frac{1.129}{\cos^2\beta\sin^{1/2}\beta}\left(\frac{\sigma_y}{E}\right)^{1/2}\left(\frac{F}{B^2}\right)^{1/2}\frac{\rho}{\sigma_y^{1/2}} \tag{9.37a}$$

which is written in the non-dimensional form as

$$n = \frac{\cos\beta}{\sin\beta}s + \frac{1.129}{\cos^2\beta\sin^{1/2}\beta}\left(\frac{\sigma_y}{E}\right)^{1/2}s^{1/2} \tag{9.37b}$$

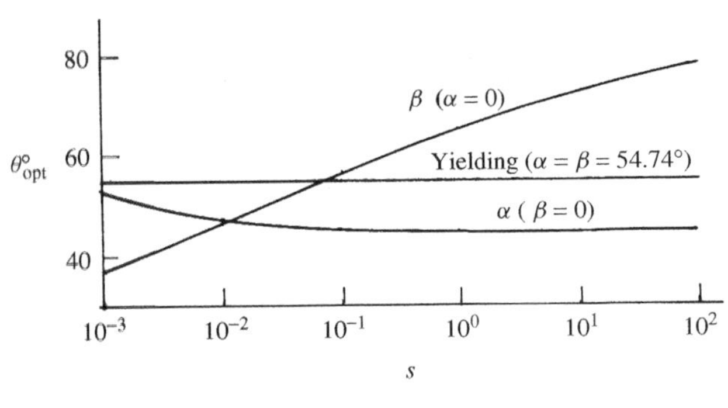

Figure 9.8 Bar orientations versus structural index when $\alpha = 0$ and when $\beta = 0$

Table 9.4 Dependence of optimum weight of upon frame geometry for $\alpha = 0°$

s	0.001	0.01	0.10	1.0	10	100
β	36.89°	46.11°	56.27°	65.44°	72.71	78.022
n	0.00406	0.0201	0.115	0.716	4.677	31.24

For steel bars, with $\sigma_y = 300\,\text{MPa}$ and $E = 200$ GPa, equation (9.37b) becomes

$$n = \frac{\cos\beta}{\sin\beta}s + \frac{0.0427}{\cos^2\beta\sin^{1/2}\beta}s^{1/2} \tag{9.38a}$$

Minimum weight follows from applying the condition $\partial n/\partial\beta = 0$ to equation (9.38a). This gives

$$-s\cos^3\beta - 0.02135s^{1/2}\cos^2\beta\sin^{1/2}\beta + 0.0954s^{1/2}\sin^{5/2}\beta = 0 \tag{9.38b}$$

The solutions to β, found by trial from equation (9.38b), appear in Table 9.4 over five decades of s.

The minimum weight solutions from the two criteria (i) and (ii), in § 9.4.3 above, appear in Figure 9.7. As with $\beta = 0$, considered earlier in § 9.4.2, when $\alpha = 0°$ this again shows that condition (ii) applies to the first decade in s while condition (i) applies to the remaining decades. Thus, to achieve minimum weight for a given structural index, the geometry of each frame must be adjusted according to Figure 9.8.

The buckling plots show the inclination of the sloping bar over a similar range of indices s in each case, i.e. when $\alpha = 0°$ and when $\beta = 0°$. Note that only the buckling criterion is sensitive to each special case. In the stress limited (yielding) criterion it matters not whether the inclined bar lies above or below the horizontal bar, the inclination remains fixed at 54.74° and the weight of the structure remains the same at a given s-value.

9.5 Tetrahedron Framework

We saw earlier in § 9.3 how a number of struts were arranged within a three-dimensional space frame to support a vertical load. That analysis is now extended to minimise the weight of a space frame consisting of three inclined struts connected to three base ties. Let this frame be formed from three struts of equal lengths L_1 and equal inclinations θ, based upon three further tie bars of equal lengths L_2 set in an equilateral triangle (see Figure 9.9).

With a base radius B, the sloping length L_1 and the base lengths are L_2 are given by

$$L_1 = \frac{B}{\cos\theta}, \quad L_2 = \sqrt{3}B \tag{9.39a,b}$$

When in support of a load F, applied at the apex, the bar forces are resolved to

$$P_1 = -\frac{F}{3\sin\theta}, \quad P_2 = \frac{F}{3\sqrt{3}\tan\theta} \tag{9.40a,b}$$

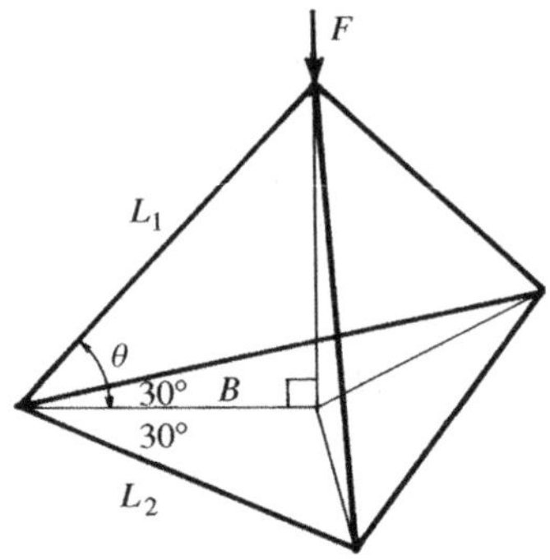

Figure 9.9 Tetrahedron frame

Equations (9.40a,b) confirm that the three 1-bars are struts and that the three 2-bars are ties. It follows that the areas of these bars are

$$A_1 = \frac{|P_1|}{\sigma_1} = \frac{F}{3\sigma_1 \sin\theta}, \qquad A_2 = \frac{P_2}{\sigma_2} = \frac{F}{3\surd\sigma_2 \tan\theta} \tag{9.41a,b}$$

Consequently, the weight of the 6 bars in the same material may be written as

$$W = 3(\rho A_1 L_1 + \rho A_2 L_2) \tag{9.42a}$$

Substituting equations (9.39a,b) and (9.41a,b) into equation (9.42a) and then dividing by B^3 provides this frame's objective function:

$$\frac{W}{B^3} = \left(\frac{1}{\sin\theta\cos\theta}\frac{\rho}{\sigma_1} + \frac{1}{\tan\theta}\frac{\rho}{\sigma_2}\right)\frac{F}{B^2} \tag{9.42b}$$

The following two criteria may be applied to equation (9.42b) to minimise its weight.

(i) All bars stress limited
When at the stress limit, $\sigma_1 = \sigma_2 = \sigma_y$, equation (9.42b) becomes

$$\frac{W}{B^3} = \frac{1+\cos^2\theta}{\sin\theta\cos\theta}\frac{\rho}{\sigma_y}\frac{F}{B^2}, \quad \text{or} \quad n = \frac{1+\cos^2\theta}{\sin\theta\cos\theta}s \tag{9.43a,b}$$

where s and n are as defined in equation (9.12a,b) . The minimum weight follows from applying the condition $\partial n/\partial\theta = 0$ to equation (9.43b). This gives

$$\cos\theta = 1/\sqrt{3} \quad \Rightarrow \quad \theta = 54.74^\circ, \quad n = 2\sqrt{2}s$$

The optimum diameters of the three struts follow from equation (9.40a) as

$$(d_1)_{\text{opt}} = \sqrt{\frac{4P_1}{\pi\sigma_y}} = 0.721\left(\frac{F}{\sigma_y}\right)^{1/2} \tag{9.44a}$$

and those of the three ties, from equation (9.40b),

$$(d_2)_{\text{opt}} = \sqrt{\frac{4P_2}{\pi\sigma_y}} = 0.4163\left(\frac{F}{\sigma_y}\right)^{1/2} \tag{9.44b}$$

(ii) Base bars stress limited with inclined bars at optimum buckling

Equation (1.8a) provides the optimum buckling stress for the inclined bars. Substituting P_1 from equation (9.40a) and L_1 from equation (9.39a) into equation (1.8a) gives this stress as

$$(\sigma_1)_{\text{opt}} = \frac{0.866E^{1/2}P_1^{1/2}}{L_1} = 0.866E^{1/2}\left(\frac{F}{3\sin\theta}\right)^{1/2}\frac{\cos\theta}{B} = \frac{0.5}{B}\sqrt{EF}\frac{\cos\theta}{\sqrt{\sin\theta}} \tag{9.45a}$$

where, from (9.41a), the optimum diameter of each strut follows as

$$d_{\text{opt}} = \left(\frac{4}{\pi}\right)^{3/4}\left(\frac{P_1L_1^2}{E}\right)^{1/4} = \left(\frac{4}{\pi}\right)^{3/4}\left[\frac{F}{3E\sin\theta}\left(\frac{B}{\cos\theta}\right)^2\right]^{1/4} \tag{9.45b}$$

Substituting $\sigma_1 = (\sigma_1)_{\text{opt}}$ from equation (9.45a) with $\sigma_2 = \sigma_y$ in equation (9.42b), the objective function is

$$\frac{W}{B^3} = \frac{\cos\theta}{\sin\theta}\frac{F}{B^2}\frac{\rho}{\sigma_y} + \frac{2}{\cos^2\theta\sin^{1/2}\theta}\left(\frac{\sigma_y}{E}\right)^{1/2}\left(\frac{F}{B^2}\right)^{1/2}\frac{\rho}{\sigma_y^{1/2}} \tag{9.46a}$$

which appears in the non-dimensional form as

$$n = \frac{\cos\theta}{\sin\theta}s + \frac{2}{\cos^2\theta\sin^{1/2}\theta}\left(\frac{\sigma_y}{E}\right)^{1/2}s^{1/2} \tag{9.46b}$$

Again, for steel bars, with $\sigma_y = 300$ MPa and $E = 210$ GPa, equation (9.46b) becomes

$$n = \frac{\cos\theta}{\sin\theta}s + \frac{0.0756}{\cos^2\theta\sin^{1/2}\theta}s^{1/2} \tag{9.47a}$$

Minimum weight follows from applying the condition $\partial n/\partial\theta = 0$ to equation (9.47a). This leads to

$$-s\cos^3\theta - 0.0378s^{1/2}\cos^2\theta\sin^{1/2}\theta + 0.1512s^{1/2}\sin^{5/2}\theta = 0 \tag{9.47b}$$

The solutions to θ, found from equation (9.47b) by trial, appear in Table 9.5 over five decades of s.

Table 9.5 Dependence of optimum weight n of upon frame geometry θ and structural index s

s	0.001	0.01	0.10	1.0	10	100
θ	34.6°	42.5°	52.4°	62.0°	70.08	76.1
n	0.00613	0.028	0.15	0.897	5.75	38.04

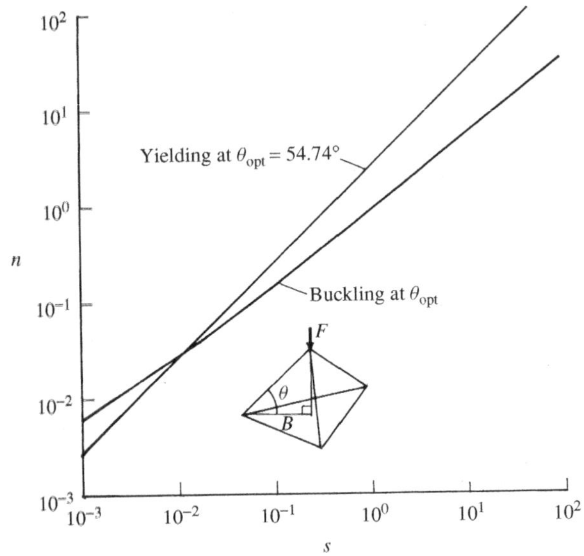

Figure 9.10 Weight parameter and bar inclinations versus structural index for steel frame

The minimum weight solutions from the two criteria (i) and (ii) above appear in Figure 9.10 which shows that the buckling condition (ii) applies to the first decade in s where the frame geometry is variable. The yielding condition (i) provides the greater weight for the remaining decades where the frame geometry is constant at $\theta = 54.74°$. Figure 9.11 indicates the strut inclination provided by each criterion. This shows that a buckling-based, optimum design can only be achieved when θ is adjusted to match the given structural index s.

9.6 Cantilever Frame with Two Struts

The wall-mounted space frame in Figure 9.12 has two struts of equal length L_2 in a horizontal plane making an equilateral triangle with the wall surface. These two struts are

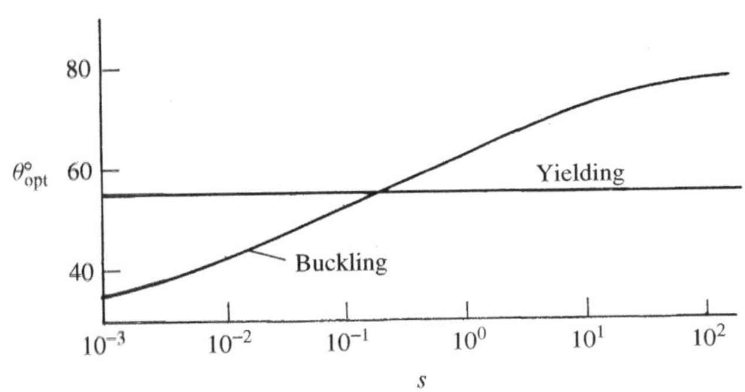

Figure 9.11 Bar orientations versus structural index for steel

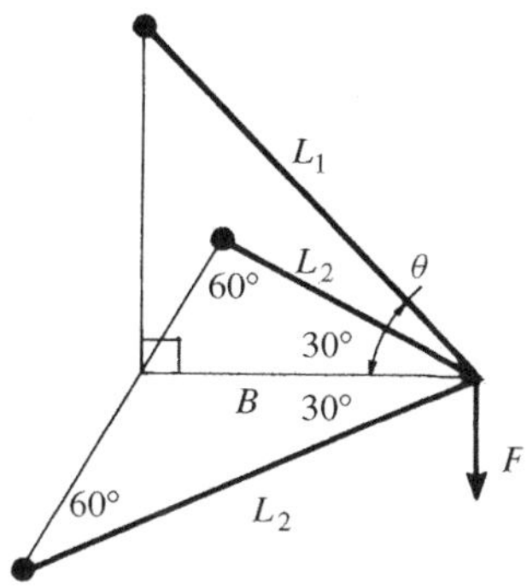

Figure 9.12 Cantilever frame with two struts and one tie

supported by a single tie bar of length L_1 and inclination θ in support of a force F applied as shown. The bars remain in vertical and horizontal force equilibrium with F when

$$(+\uparrow) \qquad P_1 \sin\theta - F = 0 \tag{9.48a}$$

$$(+\rightarrow) \qquad -P_1 \cos\theta - 2P_2 \cos\theta = 0 \tag{9.48b}$$

from which the bar forces are

$$P_1 = \frac{F}{\sin\theta}, \qquad P_2 = -\frac{F\cos\theta}{\sqrt{3}\sin\theta} \tag{9.49a,b}$$

where the minus sign indicates that each 'bar 2' acts as a strut. It follows from equations (9.49a,b) that the areas of the bars are

$$A_1 = \frac{P_1}{\sigma_1} = \frac{F}{3\sigma_1 \sin\theta}, \qquad A_2 = \frac{|P_2|}{\sigma_2} = \frac{F}{3\sqrt{3}\sigma_2 \tan\theta} \tag{9.50a,b}$$

The base lengths L_2 and the sloping length L_1 are expressed in terms of the structural index dimension B as

$$L_1 = \frac{B}{\cos\theta}, \qquad L_2 = \frac{2B}{\sqrt{3}} \tag{9.51a,b}$$

The weight of the three bars may be written as

$$W = \rho(A_1 L_1 + 2A_2 L_2) \tag{9.52a}$$

Substituting equations (9.50a,b) and (9.51a,b) into equation (9.52a) and dividing by B^3 gives the objective function

$$\frac{W}{B^3} = \left(\frac{1}{\sin\theta\cos\theta}\frac{\rho}{\sigma_1} + \frac{4\cos\theta}{3\sin\theta}\frac{\rho}{\sigma_2}\right)\frac{F}{B^2} \tag{9.52b}$$

The following two criteria may be applied to equation (9.52b) to minimise the weight of this structure.

(i) Bars 1 and 2 stress limited
When $\sigma_1 = \sigma_2 = \sigma_y$, equation (9.52b) becomes

$$\frac{W}{B^3} = \frac{3 + 4\cos^2\theta}{3\sin\theta\cos\theta}\frac{\rho}{\sigma_y}\frac{F}{B^2}, \quad \text{or} \quad n = \frac{3 + 4\cos^2\theta}{3\sin\theta\cos\theta}s \tag{9.53a,b}$$

where s and n are as defined in equation (9.12a,b) . The minimum weight follows from equation (9.53b) by applying the condition $\partial n/\partial\theta = 0$. This gives

$$\cos\theta_{opt} = 3/10 \quad \Rightarrow \quad \theta_{opt} = 56.78^\circ, \quad n = 3.055s$$

The optimum diameter of the tie then follows from setting $\theta = \theta_{opt}$ in equation (9.50a):

$$(d_1)_{opt} = \sqrt{\frac{4P_1}{\pi\sigma_y}} = 1.234\left(\frac{F}{\sigma_y}\right)^{1/2} \tag{9.54a}$$

The optimum diameters of the two struts follow from equation (9.50b)

$$(d_2)_{opt} = \sqrt{\frac{4P_2}{\pi\sigma_y}} = 0.694\left(\frac{F}{\sigma_y}\right)^{1/2} \tag{9.54b}$$

(ii) Stress limited tie bar with each strut under optimum buckling
Equation (1.8a) applies to the optimum buckling stress condition for each bar 2. Substituting for $|P_2|$ from equation (9.49b) and L_2 from equation (9.51b) gives this stress as

$$(\sigma_2)_{opt} = \frac{0.866E^{1/2}P_2^{1/2}}{L_2} = 0.866E^{1/2}\left(\frac{F\cos\theta}{\sqrt{3}\sin\theta}\right)^{1/2}\frac{\sqrt{3}}{2B}$$
$$= \frac{0.57}{B}\sqrt{EF}\left(\frac{\cos\theta}{\sin\theta}\right)^{1/2} \tag{9.55a}$$

where, from equation (1.7a), the optimum diameter becomes

$$(d_2)_{opt} = \left(\frac{4}{\pi}\right)^{3/4}\left(\frac{P_2L_2^2}{E}\right)^{1/4} = \left(\frac{4}{\pi}\right)^{3/4}\left[\frac{F\cos\theta}{\sqrt{3}E\sin\theta}\left(\frac{2B}{\sqrt{3}}\right)^2\right]^{1/4} \tag{9.55b}$$

Substituting $\sigma_2 = (\sigma_2)_{opt}$ from equation (9.55a) and $\sigma_1 = \sigma_y$ into equation (9.52b) leads to the objective function for these two conditions:

$$\frac{W}{B^3} = \frac{1}{\sin\theta\cos\theta}\frac{F}{B^2}\frac{\rho}{\sigma_y} + 2.34\left(\frac{\cos\theta}{\sin\theta}\right)^{1/2}\left(\frac{\sigma_y}{E}\right)^{1/2}\left(\frac{F}{B^2}\right)^{1/2}\frac{\rho}{\sigma_y^{1/2}} \tag{9.56a}$$

which appears in our non-dimensional form as

$$n = \frac{s}{\sin\theta\cos\theta} + 2.34\left(\frac{\cos\theta}{\sin\theta}\right)^{1/2}\left(\frac{\sigma_y}{E}\right)^{1/2} s^{1/2} \tag{9.56b}$$

where s and n are as defined in equations (9.12a,b) . For steel, with $\sigma_y = 300$ MPa and $E = 210$ GPa, equation (9.56b) becomes

$$n = \frac{s}{\sin\theta\cos\theta} + 0.0884\left(\frac{\cos\theta}{\sin\theta}\right)^{1/2} s^{1/2} \tag{9.57a}$$

Minimum weight follows from applying the condition $\partial n/\partial\theta = 0$ to equation (9.57a). This leads to

$$s(\sin^2\theta - \cos^2\theta) - 0.0442 s^{1/2}\sin^{1/2}\theta\cos^{3/2}\theta = 0 \tag{9.57b}$$

Table 9.6 gives the trial solutions for θ from equation (9.57b) over five decades of s.

The minimum weight solutions from the two criteria (i) and (ii) above appear in Figure 9.13. These show that condition (ii) applies to the first decade in s where the tie's inclination θ is variable. The yielding condition (i), when applied to all three bars, provides the greater weight for the remaining decades at a constant geometry ($\theta = 56.78°$).

In the narrow range of structural indices, $s < 10^{-2}$, where the buckling design applies, θ must be selected from Figure 9.13 and Table 9.6 at the corresponding n-value. Note that in this design, as in other frame designs, the geometries are *not* identical at the intersection ($s_{cr} \approx 10^{-2}$) between the two plots in Figure 9.13, despite their having the same weight. Both the geometry parameter θ and the bar diameters differ in each design to provide a similar weight. Figure 9.14 shows the optimum orientations for the two designs. This shows how θ_{opt} differs between the 'yielding' and 'buckling' designs for $s_{cr} \approx 10^{-2}$.

Table 9.6 Dependence of the optimum weight parameter n upon index s and frame geometry θ

s	0.001	0.01	0.10	1.0	10	100
θ	59.18°	50.66°	46.93°	45.68°	45.20	45.06
n	0.00443	0.0284	0.228	2.088	20.28	200.88

9.7 Cantilever Frame with One Strut

The three-bar, wall mounted support shown in Figure 9.15 has a single strut of length L_2 and inclination θ with two connecting ties, each of length L_1. The two ties and the wall surface form an equilateral triangle lying in the horizontal plane. The three bars remain in vertical and horizontal equilibrium with the applied force F when

$$\uparrow \quad -P_2\sin\theta - F = 0 \tag{9.58a}$$

$$\rightarrow \quad -P_2\cos\theta - 2P_1\cos 30° = 0 \tag{9.58b}$$

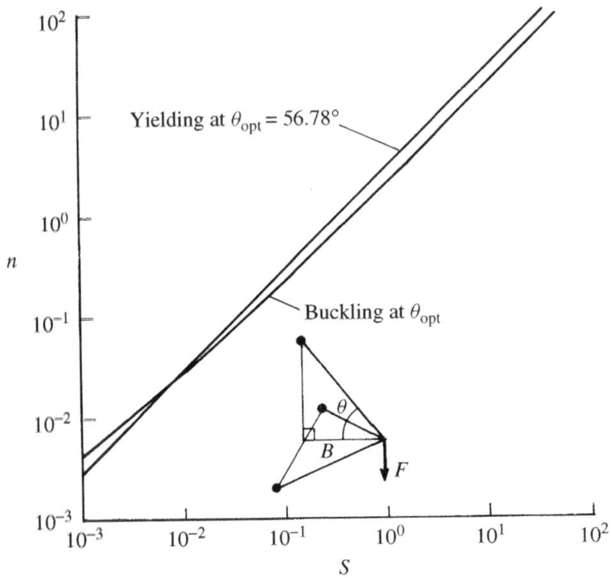

Figure 9.13 Weight parameter versus structural index at optimum frame geometry

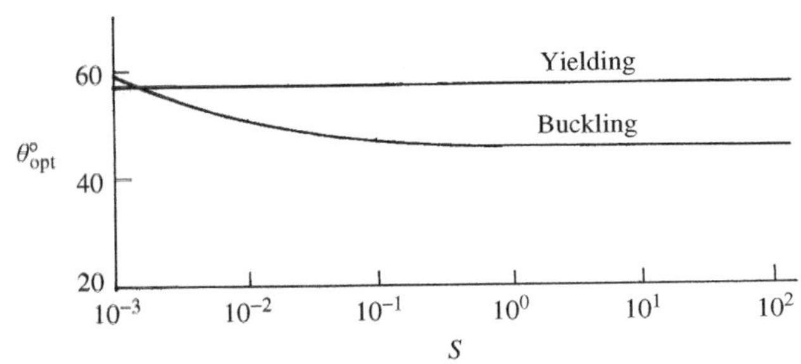

Figure 9.14 Bar inclination versus structural index for steel

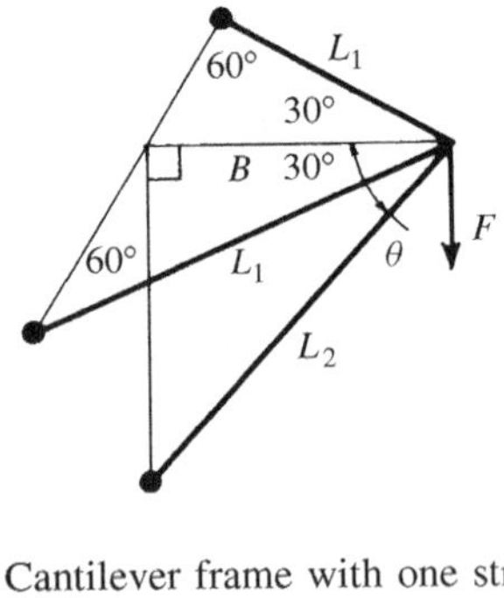

Figure 9.15 Cantilever frame with one strut and two ties

from which the bar forces are

$$P_1 = \frac{F\cos\theta}{\sqrt{3}\sin\theta}, \quad P_2 = \frac{-F}{\sin\theta} \tag{9.59a,b}$$

The minus sign confirms that bar 2 acts as a strut. It follows that the areas of these bars are

$$A_1 = \frac{P_1}{\sigma_1} = \frac{F\cos\theta}{\sqrt{3}\sigma_1\sin\theta}, \quad A_2 = \frac{|P_2|}{\sigma_2} = \frac{F}{\sigma_2\sin\theta} \tag{9.60a,b}$$

The bar lengths L_1 and L_2 are written in terms of the structural index dimension B as

$$L_1 = \frac{2B}{\sqrt{3}}, \quad L_2 = \frac{B}{\cos\theta} \tag{9.61a,b}$$

Consequently, the weight of this three-bar frame may be written as

$$W = \rho(2A_1L_1 + A_2L_2) \tag{9.62a}$$

Substituting equations (9.60a,b) and (9.61a,b) into equation (9.62a) and then dividing by B^3 gives the objective function required to optimise this frame:

$$\frac{W}{B^3} = \left(\frac{4\cos\theta}{3\sin\theta}\frac{\rho}{\sigma_1} + \frac{1}{\sin\theta\cos\theta}\frac{\rho}{\sigma_2}\right)\frac{F}{B^2} \tag{9.62b}$$

The following two criteria may be applied to equation (9.62b) to minimise its weight.

(i) Bars 1 and 2 stress limited

When $\sigma_1 = \sigma_2 = \sigma_y$, equation (9.62b) becomes

$$\frac{W}{B^3} = \frac{3+4\cos^2\theta}{3\sin\theta\cos\theta}\frac{\rho}{\sigma_y}\frac{F}{B^2}, \quad \Rightarrow \quad n = \frac{3+4\cos^2\theta}{3\sin\theta\cos\theta}s \tag{9.63a,b}$$

which agrees with equation (9.53b) for the inverted frame arrangement given in § 9.6. Again, the minimum weight follows from applying the condition $\partial n/\partial\theta = 0$ to equation (9.63b). This gives

$$\cos\theta = 3/10 \quad \Rightarrow \quad \theta = 56.78^\circ, \quad n = 3.055s$$

The optimum diameters of the two ties follow from equation (9.60a,b) as

$$(d_1)_{\text{opt}} = \sqrt{\frac{4P_1}{\pi\sigma_y}} = 0.694\left(\frac{F}{\sigma_y}\right)^{1/2} \tag{9.64a}$$

and that of the strut as

$$(d_2)_{\text{opt}} = \sqrt{\frac{4P_2}{\pi\sigma_y}} = 1.234\left(\frac{F}{\sigma_y}\right)^{1/2} \tag{9.64b}$$

(ii) Stress limited ties with strut at optimum buckling

Equation (1.8a) applies to the optimum buckling stress for the strut (bar 2). Substituting equations (9.59b) and (9.61b) into equation (1.8a) gives

$$(\sigma_2)_{opt} = \frac{0.866E^{1/2}P_2^{1/2}}{L_2} = 0.866E^{1/2}\left(\frac{F}{\sin\theta}\right)^{1/2}\frac{\cos\theta}{B} = \frac{0.886}{B}\sqrt{EF}\frac{\cos\theta}{\sqrt{\sin\theta}} \tag{9.65a}$$

where, from equation (1.7a), the optimum diameter for this strut becomes

$$(d_2)_{opt} = \left(\frac{4}{\pi}\right)^{3/4}\left(\frac{P_2L_2^2}{E}\right)^{1/4} = \left(\frac{4}{\pi}\right)^{3/4}\left[\frac{F\cos\theta}{\sqrt{3}E\sin\theta}\left(\frac{B}{\cos\theta}\right)^2\right]^{1/4} \tag{9.65b}$$

Substituting $\sigma_1 = \sigma_y$ and $\sigma_2 = (\sigma_2)_{opt}$, from equation (9.65a), into the objective function (9.62b),

$$\frac{W}{B^3} = \frac{4\cos\theta}{3\sin\theta}\frac{F}{B^2}\frac{\rho}{\sigma_y} + \frac{1.129}{\sqrt{\sin\theta\cos^2\theta}}\left(\frac{\sigma_y}{E}\right)^{1/2}\left(\frac{F}{B^2}\right)^{1/2}\frac{\rho}{\sigma_y^{1/2}} \tag{9.66a}$$

which appears in the non-dimensional form as

$$n = \frac{4\cos\theta}{3\sin\theta}s + \frac{1.129}{\sqrt{\sin\theta\cos^2\theta}}\left(\frac{\sigma_y}{E}\right)^{1/2}s^{1/2} \tag{9.66b}$$

For steel, with $\sigma_y = 300$ MPa and $E = 210$ GPa, equation (9.66b) becomes

$$n = \frac{4\cos\theta}{3\sin\theta}s + \frac{0.0427}{\sqrt{\sin\theta\cos^2\theta}}s^{1/2} \tag{9.67a}$$

Minimum weight follows from applying the condition $\partial n/\partial\theta = 0$ to equation (9.67a). This gives

$$-4s\cos^3\theta + 0.256s^{1/2}\sin^{5/2}\theta - 0.064s^{1/2}\sin^{1/2}\theta\cos^2\theta = 0 \tag{9.67b}$$

Table 9.7 gives the trial solutions for θ from equation (9.67b) for five decades of s.

Table 9.7 Dependence of optimum weight parameter upon structural index and frame geometry

s	0.001	0.01	0.10	1.0	10	100
θ	40.29°	49.83°	59.72°	68.24°	74.77	79.51
n	0.00446	0.0230	0.135	0.854	5.621	37.58

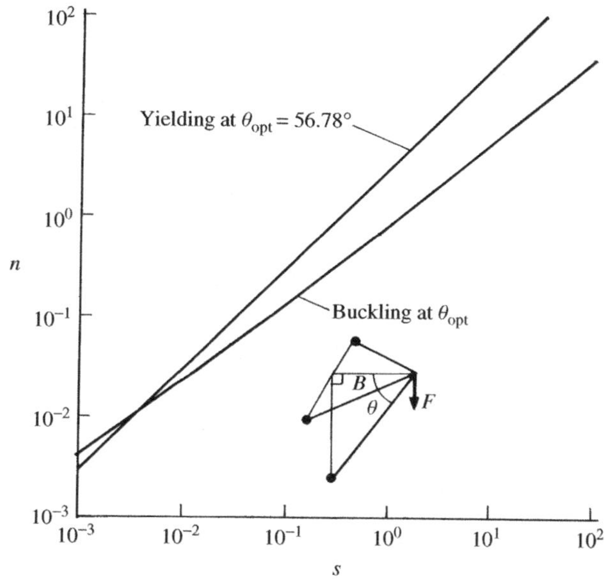

Figure 9.16 Weight parameter versus structural index for steel

The minimum weights from the two criteria (i) and (ii), given above, appear in Figure 9.16. This figure shows that the buckling condition (ii) applies to the first decade in s, within which the frame geometry is variable. The yielding condition (i) provides the greater weight for the remaining decades in a frame where $\theta = 56.78°$ is constant. Figure 9.17 shows the particular orientations required within these two optimum designs.

The orientations are to be chosen from Figure 9.17 (or Table 9.7) such that the greater weight n dictates the design at a given structural index s. At the critical s-value, for which the two criteria give equal weight, two frame geometries are permissible. This is made possible by the differing inclination and optimum bar diameters that the buckling and yielding criteria provide. Thus, we may read from Figure 9.16 at the intersection between these two criteria $s_{cr} \approx 6 \times 10^{-3}$. Correspondingly, for this critical structural

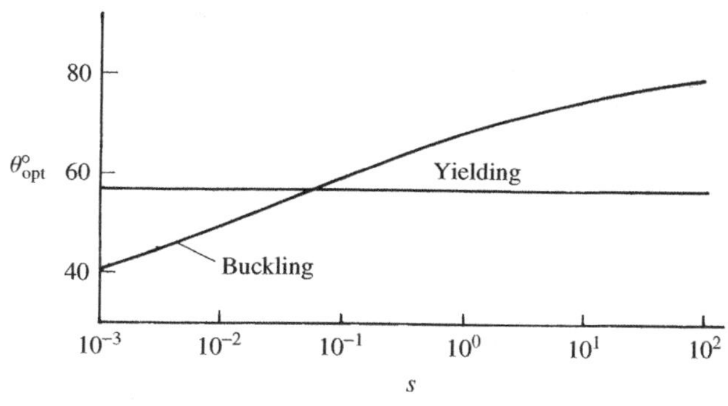

Figure 9.17 Strut orientation versus structural index for steel

index Figure 9.17 gives $\theta \approx 44^\circ$ for the buckling criterion, compared to a constant orientation $\theta = 56.78^\circ$ for the stress limited yielding criterion. The structure may conform to either geometry when bearing load at this critical index. However, when $s < s_{cr}$ the buckling design applies and for $s > s_{cr}$ the yielding design applies. Each design applies to the greater weight domain.

References

[1] Arya, C. *Design of Structural Elements*, 2nd edition, Spon Press, 1994
[2] Cox, H. L. *The Design of Structures of Least Weight*, Pergamon, 1965.
[3] Farkas, J. *Optimum Design of Metal Structures*, Ellis-Horwood 1984.
[4] Gallagher, R. H. and Zienkiewicz, O. C. *Optimum Structural Design*, Wiley, 1973.
[5] Hemp, W. S. *Optimum Structures*, Clarendon Press, 1973.
[6] Kirsch, U. *Optimum Structural Design*, McGraw-Hill, 1981.
[7] Lïsborg, N. *Principles of Structural Design*, Batsford, 1967.
[8] Majid, K. I. (ed.) *Optimum Design of Structures*, Newnes-Butterworth, 1974.
[9] Save, M. and Prager, W. *Structural Optimisation, Optimality Criteria*, Volume 1, Plenum Press, 1985.
[10] Shanley, *Weight-Strength Analysis of Aircraft Structures*, McGraw-Hill, 1952.
[11] Spunt, L. *Optimum Structural Design*, Prentice Hall, 1971.
[12] Timoshenko, S. J. and Gere, J. M. *Theory of Elastic Stability*, McGraw-Hill, 1961.

Exercises

9.1 Examine the optimum solutions to the plane frames in Figures 9.1 and 9.6 when a square section bar is used for each member (see § 1.3.2).

9.2 How do the solutions to the space frames in Figures 9.12 and 9.15 differ when, instead of the vertical force F, each is in support of a horizontal pull at the junction of the three bars?

9.3 Calculate the optimum dimensions of the bar section and the geometry of the tetrahedron steel frame in Figure 9.9 when in support of a vertical force of 100 kN. Take $E = 200$ GPa and $\sigma_y = 300$ MPa.

9.4 A four-bar multiple-strut assembly (as in Figure 9.4) is to support a load of 50 kN at its critical structural index. Calculate the optimum alternative dimensions of circular, aluminium alloy bar sections and the geometry of the frame according to the designs provided by the yielding and buckling criteria. Take $E = 70$ GPa and $\sigma_y = 150$ MPa.

9.5 Examine the possibility of optimum solutions to the cantilevered frame in Figure 9.6 when (i) $\alpha = \beta$, (ii) $\alpha = 45^\circ$ and (iii) $\beta = 45^\circ$.

10

Simply Supported Beams and Cantilevers

10.1 Introduction

In our earlier consideration of slender beams (Chapter 3) the bending moment was taken to be unchanging along the length. Whilst uniform moments do arise in practice, it is more likely that the moment M varies with the length z of a beam where it has to bear a combination of concentrated and distributed loads. Here, we will consider alongside M a variable, transverse shear force given by $F = \mathrm{d}M/\mathrm{d}z$. When we design for minimum weight both the moment and the shear force will influence the beam's cross-sectional dimensions. In particular, the section area may be reduced in low stressed regions within the beam's length. Firstly, it is shown that it would not be necessary to maintain a uniform section for longer beams where the effect of shear force may be ignored. However, for short beams where the length is of similar dimension to the section depth, shear will govern the design. The procedure for attaining minimum weight beams is demonstrated for four standard cases of loading with the more common shapes of solid cross-section. The particular sections and the loading have been chosen to provide closed solutions. A similar procedure may be implemented numerically for any manner of loading a beam for these and other cross-sections with combinations of concentrated and distributed loads.

10.2 Variable Bending Moments

When shear force effects can be ignored in longer beams the optimum solution to the section geometry may be based upon the varying moments that arise in each of the following four standard beams.

10.2.1 Simply Supported Beam with Central Load (Figure 10.1a)

The magnitude of the ordinate in the bending moment diagram in Figure 10.1b may be expressed as $|M(z)| = Pz/2$ within the half region $0 \le z \le L/2$. When the cross-section

Mechanics of Optimal Structural Design: Minimum Weight Structures D. W. A. Rees

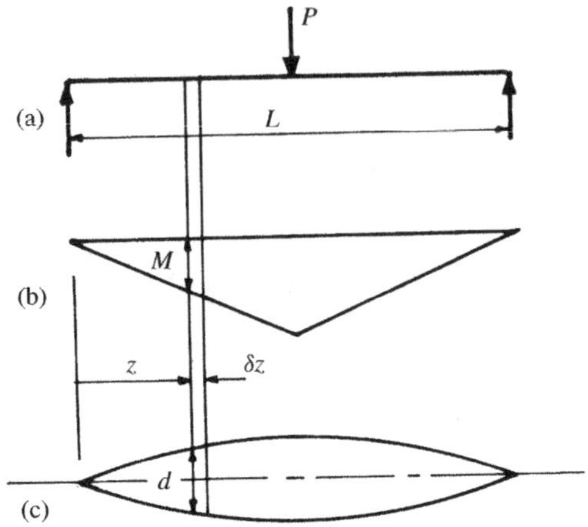

Figure 10.1 Moment diagram for a centrally loaded, simply supported beam

is circular, the optimum diameter is found by combining $|M(z)|$ with equation (3.3b):

$$d = \left[\frac{32|M(z)|}{\pi\sigma_y}\right]^{1/3} = \left(\frac{16P}{\pi\sigma_y}\right)^{1/3} z^{1/3} \tag{10.1}$$

Figure 10.1c shows how d varies with z according to equation (10.1). For an elemental length of beam δz, with this variable section, the weight is $\delta W = \rho(\pi/4)d^2\delta z$. Substituting from equation (10.1) and integrating over each half gives W:

$$W = 2\frac{\rho\pi}{4}\left(\frac{16P}{\pi\sigma_y}\right)^{2/3} \int_0^{L/2} z^{2/3}\mathrm{d}z = \frac{3\rho\pi^{1/3}P^{2/3}L^{5/3}}{5\sigma_y^{2/3}} \tag{10.2a}$$

Setting $M_{\max} = PL/4$ in equation (10.2a) and dividing by L^3 optimises the weight as

$$\left(\frac{W}{L^3}\right)_{\text{opt}} = 2.2143\,\frac{\rho}{\sigma_y^{2/3}}\left(\frac{M_{\max}}{L^3}\right)^{2/3} \tag{10.2b}$$

in which the shape factor $f = 2.2143$ has been reduced from 3.691 for a uniform circular section under a constant moment (see equation (3.4b) and § 3.2.1).

10.2.2 Cantilever with End-Load (Figure 10.2a)

The moment diagram in Figure 10.2b may be expressed as $M(z) = Pz$ within the full length ($0 \le z \le L$). When the cross-section is circular, the optimum diameter is found by combining $M(z)$ with equation (3.3b):

$$d = \left[\frac{32M(z)}{\pi\sigma_y}\right]^{1/3} = \left(\frac{32P}{\pi\sigma_y}\right)^{1/3} z^{1/3} \tag{10.3}$$

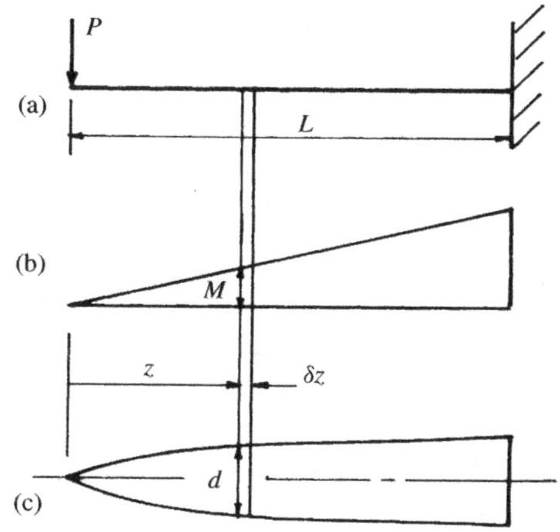

Figure 10.2 Moment diagram for an end-loaded cantilever beam

Figure 10.2c shows how d varies with z according to equation (10.3). For the elemental length δz, the weight is $\delta W = \rho(\pi/4)d^2\delta z$. Substituting from equation (10.3), the integration gives

$$W = \frac{\rho\pi}{4}\left(\frac{32P}{\pi\sigma_y}\right)^{2/3}\int_0^{L/2} z^{2/3}\mathrm{d}z = \frac{3\times 2^{4/3}\rho\pi^{1/3}P^{2/3}L^{5/3}}{5\sigma_y^{2/3}} \qquad (10.4a)$$

Setting $M_{\max} = PL$ in equation (10.4a) and dividing by L^3 optimises the weight to

$$\left(\frac{W}{L^3}\right)_{\text{opt}} = 2.2143\frac{\rho}{\sigma_y^{2/3}}\left(\frac{M_{\max}}{L^3}\right)^{2/3} \qquad (10.4b)$$

in which the shape factor is identical to the previous beam loading in § 10.2.1.

10.2.3 Simply Supported Beam with Distributed Load (Figure 10.3a)

Within the half length $0 \le z \le L/2$, the moment ordinate in Figure 10.3b may be expressed as

$$|M(z)| = \left|-\frac{wLz}{2} + \frac{wz^2}{2}\right| = \frac{wz}{2}(L - z) \qquad (10.5)$$

For a circular section, the optimum diameter is found by combining $|M(z)|$ in equation (10.5) with equation (3.2b):

$$d = \left[\frac{32|M(z)|}{\pi\sigma_y}\right]^{1/3} = \left(\frac{16w}{\pi\sigma_y}\right)^{1/3}[z(L - z)]^{1/3} \qquad (10.6)$$

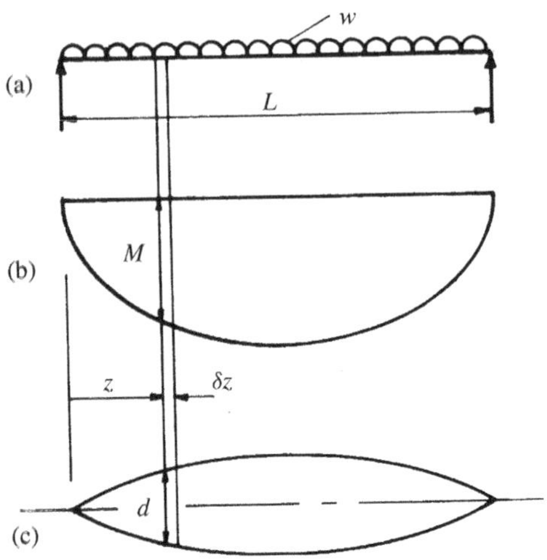

Figure 10.3 Moment diagram for a simply supported beam under distributed loading

Figure 10.3c shows the variation in d with z according to equation (10.6). For the elemental length δz, the beam weight is $\delta W = \rho(\pi/4)d^2\delta z$. Substituting from equation (10.6) and integrating gives

$$W = 2\frac{\rho\pi}{4}\left(\frac{16wL}{\pi\sigma_y}\right)^{2/3}\int_0^{L/2} z^{2/3}(1 - z/L)^{2/3}\,\mathrm{d}z \tag{10.7a}$$

We may integrate equation (10.7a) numerically or, in this case, expand the integrand within equation (10.7a) as a series before integration. This leads to

$$W = \frac{\rho\pi}{2}\left(\frac{16wL}{\pi\sigma_y}\right)^{2/3}\left|\frac{3z^{5/3}}{5} - \frac{2z^{8/3}}{8L} - \frac{z^{11/3}}{33L^2} - \frac{2z^{14/3}}{189L^3} - \ldots\right|_0^{L/2} \tag{10.7b}$$

Equation (10.7b) optimises the weight to

$$W = 0.466\,\frac{\rho\pi}{2}\left(\frac{16wL}{\pi\sigma_y}\right)^{2/3}\left(\frac{L}{2}\right)^{5/3} \tag{10.8a}$$

Setting $M_{\max} = wL^2/8$ and dividing equation (10.8a) by L^3 gives the objective function

$$\left(\frac{W}{L^3}\right)_{\text{opt}} = 2.73\left(\frac{\rho}{\sigma_y^{2/3}}\right)\left(\frac{M_{\max}}{L^3}\right)^{2/3} \tag{10.8b}$$

Again, the shape factor value of 2.73 is correspondingly reduced from a value of 3.691 for a uniform circular section under a constant moment (see § 3.2.1).

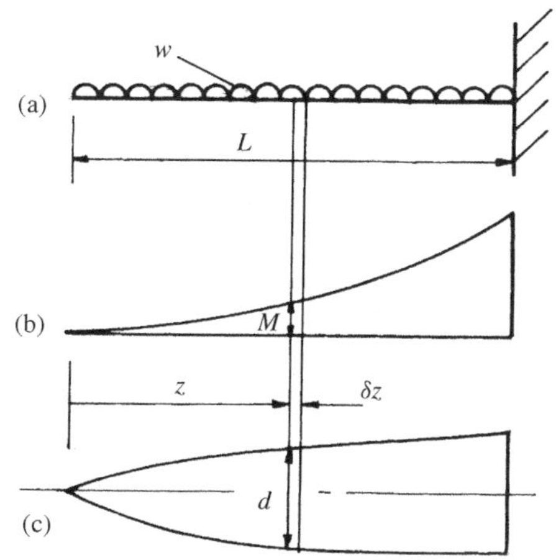

Figure 10.4 Moment diagram for a cantilever with distributed loading

10.2.4 Cantilever with Distributed Load (Figure 10.4a)

Within the full length, $0 \leq z \leq L$, the moment M at position z in Figure 10.4b is expressed as

$$M(z) = wz^2/2 \tag{10.9}$$

For a circular section, the optimum diameter is found by combining the $M(z)$ expression in equation (10.9) with equation (3.3b):

$$d = \left[\frac{32M(z)}{\pi\sigma_y}\right]^{1/3} = \left(\frac{16w}{\pi\sigma_y}\right)^{1/3} z^{2/3} \tag{10.10}$$

Figure 10.4c shows how d varies with z according to equation (10.10). For an elemental length δz the weight is $\delta W = \rho(\pi/4)d^2\delta z$. Substituting from equation (10.10) and integrating gives

$$W = \frac{\rho\pi}{4}\left(\frac{32F}{\pi\sigma_y}\right)^{2/3}\int_0^L z^{4/3}\mathrm{d}z = \frac{3\times 2^{2/3}\rho\pi^{1/3}w^{2/3}L^{7/3}}{7\sigma_y^{2/3}} \tag{10.11a}$$

Setting $M_{\text{max}} = wL^2/2$ and dividing by L^3 optimises weight within the objective function:

$$\left(\frac{W}{L^3}\right)_{\text{opt}} = 1.582\left(\frac{\rho}{\sigma_y^{2/3}}\right)\left(\frac{M_{\text{max}}}{L^3}\right)^{2/3} \tag{10.11b}$$

Again, this shape factor value $f = 1.582$ shows a considerable reduction from the $f = 3.691$ value that applies to a uniform circular section under a constant moment (see § 3.2.1).

10.2.5 *Cantilever with End- and Distributed Loads (Figure 10.5a)*

Within the full length $0 \le z \le L$, the moment diagrams in Figures 10.2b and 10.4b may be added to give, for the combined loading,

$$M(z) = Pz + wz^2/2, \quad M_{max} = PL + wL^2/2 \tag{10.12a,b}$$

where M_{max} applies to $z = L$. Dividing equations (10.12a,b) gives

$$\frac{M(z)}{M_{max}} = \frac{z}{L}\left[1 + \frac{wL}{2P}\left(\frac{z}{L}\right)\right] \Big/ \left(1 + \frac{wL}{2P}\right) \tag{10.13}$$

For a circular cross-section, the optimum diameter is found by combining $M(z)$ in equation (10.12a), or equation (10.13), with equation (3.3b).

$$d = \left[\frac{32M(z)}{\pi\sigma_y}\right]^{1/3} = \left(\frac{32}{\pi\sigma_y}\right)^{1/3}\left(Pz + \frac{wz^2}{2}\right)^{1/3} \tag{10.14}$$

Figure 10.5b shows how d varies with z according to equation (10.14). For the elemental length δz the weight is $\delta W = \rho(\pi/4)d^2\delta z$. Substituting from equation (10.14) leads to the integral for this beam's total weight:

$$W = \frac{\rho\pi L}{4}\left(\frac{32PL}{\pi\sigma_y}\right)^{2/3}\int_0^1 \left(\frac{z}{L}\right)^{2/3}\left[1 + \frac{wL}{2P}\left(\frac{z}{L}\right)\right]^{2/3} \mathrm{d}\left(\frac{z}{L}\right) \tag{10.15a}$$

The integral may be obtained numerically for a given load ratio wL/P. Writing I for the integral value, setting $PL = M_{max}/(1 + wL/2P)$ in equation (10.15a) and then dividing by L^3 optimises the weight:

$$\left(\frac{W}{L^3}\right)_{opt} = \frac{3.691I}{(1 + wL/2P)^{2/3}}\left(\frac{\rho}{\sigma_y^{2/3}}\right)\left(\frac{M_{max}}{L^3}\right)^{2/3} \tag{10.15b}$$

If, as an approximation, we assume the diameter to be the average of the optimum diameters found when each load is applied independently, then the optimum weights in

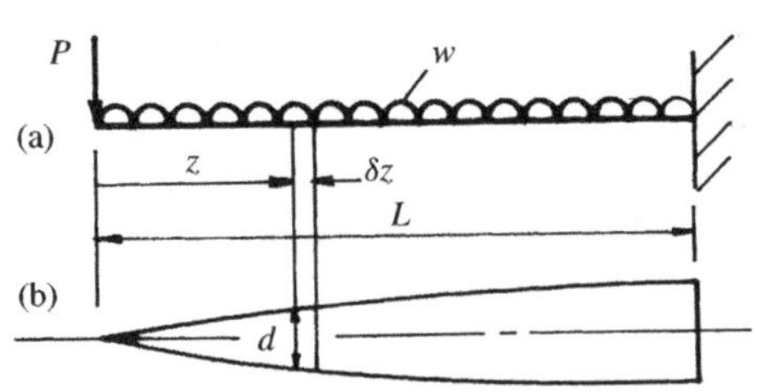

Figure 10.5 Cantilever profile for combined loading

equations (10.4b) and (10.11b) might be averaged between each load:

$$\left(\frac{W}{L^3}\right)_{\text{opt}} = 1.107\left(\frac{\rho}{\sigma_y^{2/3}}\right)\left(\frac{M_{F\max}}{L^3}\right)^{2/3} + 0.791\left(\frac{\rho}{\sigma_y^{2/3}}\right)\left(\frac{M_{w\max}}{L^3}\right)^{2/3} \quad (10.16)$$

where $M_{F\max} = PL$ and $M_{w\max} = wL^2/2$. Strictly, however, it is apparent from the weight optimisation integral (10.15a) that superposition would not be applicable to this mixed loading. That is, the diameter does not derive from a linear sum of their respective moments. An alternative, exact solution to this problem appears in Chapter 12 (see § 12.7). It is left as an exercise for the reader to examine the influence upon bending from a change in section, say, from circular to square. Certain solid sections in which shear is also accounted for are given in § 10.3. It becomes a feature in all these beam designs that the cross-section should be variable. This would not pose any difficulty to modern machining methods which should be adopted where the design requires that weight be minimised. Thus, all superfluous material may be removed to ensure that what material remains becomes fully stressed in its support of the loading. This is the essential principle of weight reduction for beams bearing loads, but, as we shall see, it is made more complicated when the design needs to alternate between the more severe fully stressed conditions from bending and shear.

10.3 Cantilever with End-Load

When we return to the standard cantilever (see Figure 10.6a) and take the shear force into consideration a structural index P/L^2 is required. Regions will now be established within this index to show where the shear force and bending moment each contribute to an optimum design, or where one effect is dominant over the other. Three solid cross-sections will be considered – circular, square and rectangular – whose dimensions must again be allowed to vary with the length dimension z if minimum weight is to be achieved. The respective force and moment diagrams for this loading are given in Figure 10.6b,c. The two diagrams may be expressed as $F = P =$ constant and $M(z) = Pz$ within the full length ($0 \le z \le L$). Note that each diagram is independent of the shape of beam cross-section.

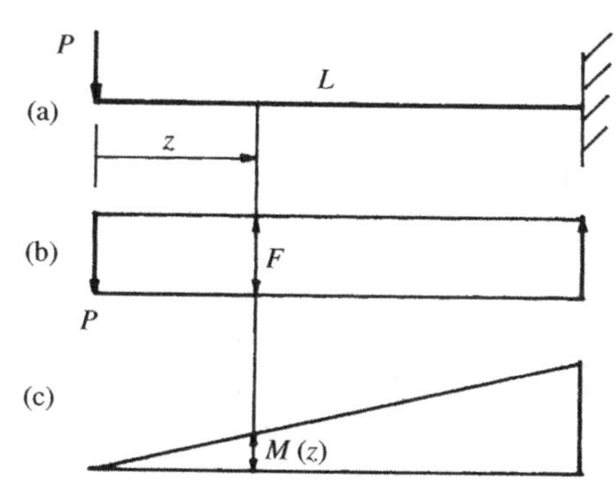

Figure 10.6 Force and moment diagrams for an end-loaded cantilever

10.3.1 Circular Section, Diameter d

When the section is circular, the optimum diameter in bending is again found by combining $M(z) = Pz$ with equation (3.3b). Dividing by $2L$ reveals the variation in normalised radius with normalised length:

$$\frac{d}{2L} = \left(\frac{4}{\pi}\right)^{1/3}\left(\frac{1}{\sigma_y}\frac{P}{L^2}\right)^{1/3}\left(\frac{z}{L}\right)^{1/3} = 1.0838\, s^{1/3}\left(\frac{z}{L}\right)^{1/3} \tag{10.17a}$$

in which $s = P/(\sigma_y L^2)$ is the *normalised structural index*. Note that this index is to be identified with short beams when s is large and long beams when s is small.

For transverse shear loading of a circular section beam it was shown previously that equation (5.15b) gave the optimum diameter (see § 5.2.5). Dividing by $2L$ and writing the relationship between the tensile and shear yield stresses as $\sigma_y = Q\tau_y$ leads to

$$\frac{d}{2L} = 2\left(\frac{Q}{3\pi}\right)^{1/2}\left(\frac{1}{\sigma_y}\frac{P}{L^2}\right)^{1/2} = 0.8574\, s^{1/2} \tag{10.17b}$$

where the numerical coefficient $Q = \sqrt{3}$ applies to the von Mises yield criterion. For a given structural index s, equation (10.17b) shows that the diameter is constant in shear while for bending equation (10.17a) shows that diameter varies with the length. Equating (10.17a,b) gives the critical position in the length where the diameter from a shear design is displaced by the diameter from a bending design:

$$\left(\frac{z}{L}\right)_{cr} = \left(\frac{Q}{3}\right)^{3/2}\left(\frac{4}{\pi}\right)^{1/2} s^{1/2} \tag{10.18a}$$

Setting $(z/L)_{cr} = 1$ in equation (10.18a) provides the lowest limiting s for an all-shear design:

$$s_{FM} \geq \frac{\pi}{4}\left(\frac{3}{Q}\right)^3 \tag{10.18b}$$

This limit occurs at the division in structural indices between shear and bending designs. With $Q = \sqrt{3}$ (von Mises) we find this limit to be $s_{FM} = 4.081$. Thus, if we select $s = 1$, for example, equation (10.18a) gives $(z/L)_{cr} = 0.495$. Then, in the region of shear, $0 \leq z/L \leq 0.495$, the diameter is held constant according to equation (10.17b) and in the remaining region of bending, $0.495 \leq z/L \leq 1$, the diameter varies by equation (10.17a). Figure 10.7 shows the beam radius profile by this combined design. Given that $d/2L$ is the normalised radius, Figure 10.7 identifies the half-section of the beam's length. In the shear region this radius is $d/2L = 0.857$, and in the bending region it increases to 1.084 at the fixed end. This particular design applies to a short beam or pin whose radius is similar to its length.

With $s \leq s_{FM}$, the beam's weight follows from the two contributions. For an elemental length δz, within the shear and bending regions, the weight is $\delta W = \rho(\pi/4)d^2\delta z$. Thus, for $0 \leq z/L \leq z_{cr}/L$ (shear) we substitute $d/2L$ from equation (10.17b), when the integration

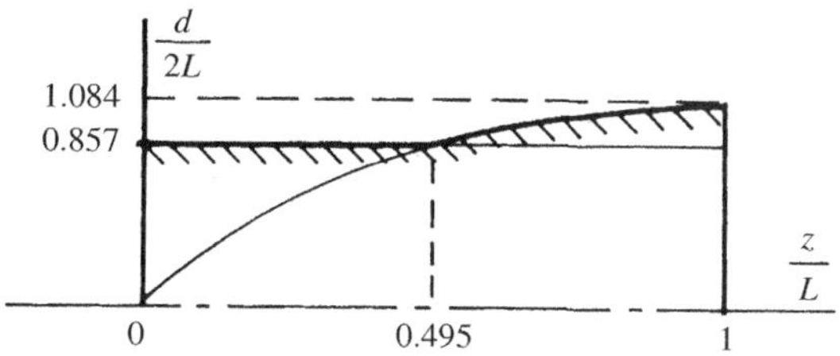

Figure 10.7 Length profile for a circular-section cantilever

gives

$$\frac{W_F}{L^3} = \rho\pi \int_0^{(z/L)_{cr}} \left(\frac{d}{2L}\right)^2 d\left(\frac{z}{L}\right) = \rho\pi \frac{4Q}{3\pi} s \int_0^{(z/L)_{cr}} d\left(\frac{z}{L}\right) = \frac{4Q\rho s}{3}\left(\frac{z}{L}\right)_{cr} \qquad (10.19a)$$

For $z_{cr}/L \leq z/L \leq 1$ (bending) we substitute $d/2L$ from equation (10.17a), when the integration gives

$$\frac{W_M}{L^3} = \rho\pi \int_{(z/L)_{cr}}^{1} \left(\frac{d}{2L}\right)^2 d\left(\frac{z}{L}\right) = \rho\pi\left(\frac{4s}{\pi}\right)^{2/3} \int_{(z/L)_{cr}}^{1} \left(\frac{z}{L}\right)^{2/3} d\left(\frac{z}{L}\right)$$

$$= \frac{3\rho\pi}{5}\left(\frac{4s}{\pi}\right)^{2/3}\left[1-\left(\frac{z}{L}\right)_{cr}^{5/3}\right] \qquad (10.19b)$$

Adding equations (10.19a,b) and substituting $(z/L)_{cr}$ from equation (10.18a) leads to the total weight

$$\frac{W}{L^3} = 4\rho\left(\frac{Q}{3}\right)^{5/2}\left(\frac{4}{\pi}\right)^{1/2} s^{3/2} + \frac{3\rho\pi}{5}\left(\frac{4s}{\pi}\right)^{2/3}\left[1-\left(\frac{Q}{3}\right)^{5/2}\left(\frac{4}{\pi}\right)^{5/6} s^{5/6}\right] \qquad (10.20a)$$

Putting $Q = \sqrt{3}$ (from von Mises) and introducing the *non-dimensional weight parameter* $n = (1/\rho)(W/L^3)$, equation (10.20a) becomes

$$n = 1.1432\, s^{3/2} + 2.2143\, s^{2/3}\left(1 - 0.3098\, s^{5/6}\right) = 2.2143\, s^{2/3} + 0.4572\, s^{3/2} \qquad (10.20b)$$

For $s \geq s_{FM}$, the diameter is constant, with the beam's weight determined by shear only. Here, the upper limit to the integral in equation (10.19a) becomes unity:

$$\frac{W}{L^3} = \rho\pi \int_0^1 \left(\frac{d}{2L}\right)^2 d\left(\frac{z}{L}\right) = \rho\pi\left(\frac{4Q}{3\pi}\right) s \int_0^1 d\left(\frac{z}{L}\right) = \frac{4Q\rho s}{3} \qquad (10.21a)$$

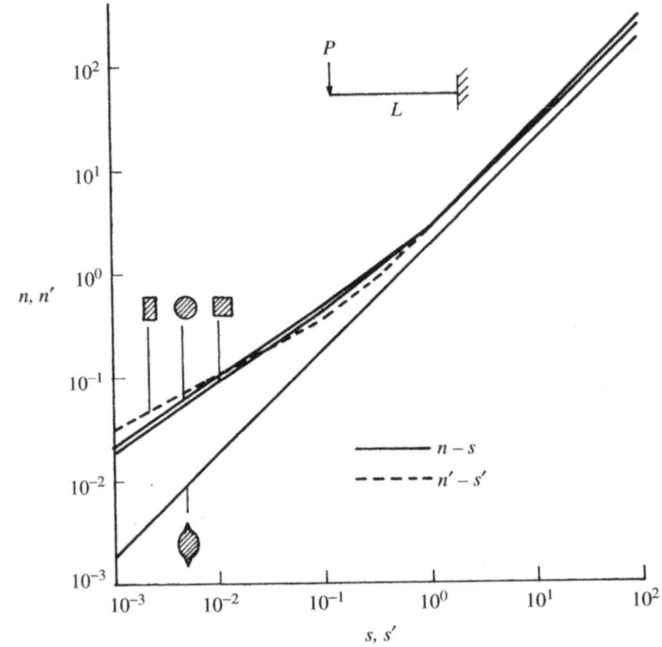

Figure 10.8 Weight parameter versus structural index for end-loaded cantilevers of different cross-sections (note : $n = (1/\rho)(W/L^3)$; $s = P/(\sigma_y L^2)$; $n' = n/(\sigma_y/E)$; and $s' = s/(\sigma_y/E)$)

which, on making similar substitutions for n and Q, has the linear form

$$n = 2.3094\, s \tag{10.21b}$$

Figure 10.8 plots n versus s from equations (10.20b) and (10.21b) over five decades, to reveal the regions within which bending, combined shear and bending, and shear designs apply with increasing s. As the beam's length increases, i.e. as s decreases, the contribution from shear diminishes. Shear may be neglected here for $s < 0.01$ where $z_{cr}/L < 0.05$. However, strictly speaking, in these optimum bending designs the free end should not start with a point as suggested by equation (10.17a). In this regard shear will always contribute, e.g. with $s = 0.01$, equation (10.17b) gives $d/2L = 0.0857$ at the free end. Figure 10.8 shows that both shear and bending control the optimum design for an index s in the region $0.001 < s < 4.081$ where equation (10.20b) supplies the normalised weight n. For $s \geq 4.081$, shear controls the optimum design and equation (10.21b) applies to the linear n versus s relationship.

When the effect of shear can be ignored, the first term in equation (10.20b) applies to bending, i.e. $n = 2.2143 s^{2/3}$ and, with the substitutions for s and n, this returns to equation (10.4b).

10.3.2 Square Section, Side Length a

The force and moment diagrams in Figure 10.6b,c again apply as these are independent of the cross-section. When the section is square, the optimum side length a in bending is given by combining $M(z)$ with equation (3.6b). Dividing by $2L$ reveals the variation in semi-depth ($a/2$), normalised with the beam length L:

$$\frac{a}{2L}=\left(\frac{3}{4}\right)^{1/3}\left(\frac{1}{\sigma_y}\frac{P}{L^2}\right)^{1/3}\left(\frac{z}{L}\right)^{1/3}=0.9086\,s^{1/3}\left(\frac{z}{L}\right)^{1/3} \tag{10.22a}$$

Equation (5.4c) gives a_{opt} for a transverse shear force applied to a square section. Dividing by $2L$ and writing $\sigma_y = Q\tau_y$ leads to

$$\frac{a}{2L}=\left(\frac{3Q}{8}\right)^{1/2}\left(\frac{1}{\sigma_y}\frac{P}{L^2}\right)^{1/2}=0.806\,s^{1/2} \tag{10.22b}$$

where the numerical coefficient applies to $Q=\sqrt{3}$ (von Mises). Equating (10.22a,b) gives the critical position in the length dividing the shear and bending designs:

$$\left(\frac{z}{L}\right)_{\text{cr}}=\left(\frac{3Q^3}{2^5}\right)^{1/2}s^{1/2} \tag{10.23a}$$

and, since at its limit $(z/L)_{\text{cr}}=1$, the lowest limiting index s for an all-shear design occurs at the shear–bending boundary:

$$s_{FM}\geq\frac{2^5}{3Q^3} \tag{10.23b}$$

For example, with $Q=\sqrt{3}$ we find this limit to be $s_{FM}=2.053$. For $s\leq s_{FM}$, say $s=1$, equation (10.23a) gives $(z/L)_{\text{cr}}=0.462$. Thus, for $z/L\leq 0.462$, a shear design from equation (10.22b) maintains the side length constant at $a/2L=0.806$. For $0.462\leq z/L\leq 1$, the bending design (equation (10.22a)) shows that the side length must increase to 0.909 at the fixed end. Figure 10.9 shows the half-depth beam profile by this combined design. The shaded profile indicates where material may be removed from a uniform bar, thereby reducing weight within a combined shear–bending design.

For $s\leq s_{FM}$ the total weight is found from the two contributions by taking an elemental length δz within each region where the weight is $\delta W=\rho a^2\delta z$. Thus, for $0\leq z/L\leq z_{\text{cr}}/L$ (shear), we substitute $a/2L$ from equation (10.22b), when the integration gives

$$\frac{W_F}{L^3}=4\rho\int_0^{(z/L)_{\text{cr}}}\left(\frac{a}{2L}\right)^2\mathrm{d}\left(\frac{z}{L}\right)=4\rho s\frac{3Q}{8}\left(\frac{z}{L}\right)_{\text{cr}} \tag{10.24a}$$

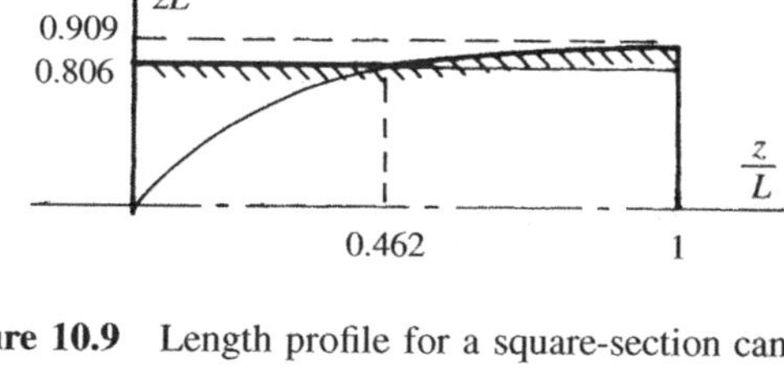

Figure 10.9 Length profile for a square-section cantilever

For $z_{cr}/L \leq z/L \leq 1$ (bending), we substitute $a/2L$ from equation (10.22a), when the integration gives

$$\frac{W_M}{L^3} = 4\rho \int_{(z/L)_{cr}}^{1} \left(\frac{a}{2L}\right)^2 \mathrm{d}\left(\frac{z}{L}\right) = 4\rho \left(\frac{3s}{4}\right)^{2/3} \int_{(z/L)_{cr}}^{1} \left(\frac{z}{L}\right)^{2/3} \mathrm{d}\left(\frac{z}{L}\right)$$

$$= \frac{(4 \times 3^5)^{1/3} \rho s^{2/3}}{5} \left[1 - \left(\frac{z}{L}\right)_{cr}^{5/3}\right] \qquad (10.24b)$$

Adding equations (10.24a,b) and substituting $(z/L)_{cr}$ from equation (10.23a) leads to

$$\frac{W}{L^3} = 4\rho \frac{3Q}{8} s \left(\frac{3Q^3}{2^5}\right)^{1/2} s^{1/2} + \frac{(4 \times 3^5)^{1/3} \rho}{5} s^{2/3} \left[1 - \left(\frac{3Q^3}{2^5}\right)^{5/6} s^{5/6}\right] \qquad (10.25a)$$

Setting $n = (1/\rho)(W/L^3)$ and $Q = \surd 3$, equation (10.25a) becomes

$$n = 1.8133\, s^{3/2} + 1.9811\, s^{2/3}(1 - 0.5492\, s^{5/6}) \qquad (10.25b)$$

For $s \geq s_{FM}$ the diameter is constant and the beam's weight follows from shear only. Thus, the upper limit to equation (10.24a) becomes unity:

$$\frac{W_F}{L^3} = 4\rho \int_0^1 \left(\frac{a}{2L}\right)^2 \mathrm{d}\left(\frac{z}{L}\right) = \rho s \frac{3Q}{2} \qquad (10.26a)$$

which, upon making similar substitutions for n and Q, has the linear form

$$n = 2.598\, s \qquad (10.26b)$$

Figure 10.8 plots equations (10.25b) and (10.26b) over five decades in s to reveal, with increasing s, regions in which the bending, combined shear and bending, and shear designs apply. For $s_{FM} = 2.053$, both of these equations give $n = 5.334$, at the division between the shear and combined shear and bending designs. In a pure shear design ($s \geq s_{FM}$), where the depth is held constant by equation (10.22b), the linear relationship (10.26b) applies. Equation (10.25b) applies to $s < s_{FM}$, where observations upon the varying depth by equation (10.22a) are similar to those made for the circular section above. Figure 10.8 shows the weight of a square-section beam to be less than that of the circular section for a shear–bending design but greater for a shear design.

10.3.3 *Rectangular Section, Breadth b, Depth d*

With the possibility of buckling arising in a thin rectangular section, where $b \ll d$, we could employ equations (3.13a,b) to optimise these dimensions separately in bending.

Alternatively, since the product $(bd)_{\text{opt}}$ will appear in the weight expression, we shall work with this product as our variable. Hence, for bending, equation (3.11) becomes

$$bd = \left(\frac{6|M(z)|L}{2.61E}\right)^{1/2} \tag{10.27a}$$

Substituting $|M(z)| = Pz$ from Figure 10.6c and dividing b by L and d by $2L$ leads to

$$\frac{b}{L} \times \frac{d}{2L} = \left(\frac{0.5747Pz}{L^3E}\right)^{1/2} = 0.7851\left(\frac{\sigma_y}{E}\right)^{1/2}\left(\frac{1}{\sigma_y}\frac{P}{L^2}\right)^{1/2}\left(\frac{z}{L}\right)^{1/2} \tag{10.27b}$$

Equation (5.3c) gives $(bd)_{\text{opt}}$ for a transverse shear force $F = P$ (Figure 10.6b) applied to a rectangular section. Dividing b by L, d by $2L$ and writing $\sigma_y = \sqrt{3}\tau_y$ leads to

$$\frac{b}{L} \times \frac{d}{2L} = 1.299\frac{1}{\sigma_y}\frac{P}{L^2} \tag{10.28}$$

Setting $s = (1/\sigma_y)(P/L^2)$ and equating (10.27b) and (10.28), gives the position in the length dividing the shear and bending designs:

$$\left(\frac{z}{L}\right)_{\text{cr}} = \frac{2.9361s}{\sigma_y/E} \tag{10.29a}$$

With $(z/L)_{\text{cr}} \leq 1$, for a combined shear and bending design, s_{FM} follows from equation (10.29a) as

$$s_{FM} \leq 0.3406\frac{\sigma_y}{E} \tag{10.29b}$$

In the region $s < s_{FM}$ the weight W is found from the two contributions by taking an elemental length δz within each region, where $\delta W = \rho bd \times \delta z$. Thus, for $0 \leq z/L \leq z_{\text{cr}}/L$ (shear), we substitute from equation (10.28) for the integration:

$$\frac{W_F}{L^3} = 2\rho \times 1.299\,s\int_0^{(z/L)_{\text{cr}}} \mathrm{d}\left(\frac{z}{L}\right) = 2.598\rho s\left(\frac{z}{L}\right)_{\text{cr}} \tag{10.30a}$$

For $z_{\text{cr}}/L \leq z/L \leq 1$ (bending), the weight integral is formed from equation (10.27b):

$$\frac{W_M}{L^3} = 2\rho \times 0.7581\left(\frac{\sigma_y}{E}\right)^{1/2} s^{1/2}\int_{(z/L)_{\text{cr}}}^{1}\left(\frac{z}{L}\right)^{1/2}\mathrm{d}\left(\frac{z}{L}\right)$$

$$= 1.0109\rho\left(\frac{\sigma_y}{E}\right)^{1/2} s^{1/2}\left[1 - \left(\frac{z}{L}\right)_{\text{cr}}^{3/2}\right] \tag{10.30b}$$

Adding equations (10.30a,b) and then substituting $(z/L)_{cr}$ from equation (10.29a) leads to the total weight

$$\frac{W}{L^3} = \frac{7.628\,\rho s^2}{\sigma_y/E} + 1.0109\rho\left(\frac{\sigma_y}{E}\right)^{1/2} s^{1/2}\left[1 - \frac{5.031}{(\sigma_y/E)^{3/2}}s^{3/2}\right] \tag{10.31a}$$

Setting $n = (1/\rho)(W/L^3)$ simplifies equation (10.31a) to

$$n = 1.0109\left(\frac{\sigma_y}{E}\right)^{1/2} s^{1/2} + 2.5422\left(\frac{\sigma_y}{E}\right)^{-1} s^2 \tag{10.31b}$$

For $s \geq s_{FM}$ the product bd is constant, with the beam's weight following from shear only. Thus, with an upper limit of unity, equation (10.30a) becomes

$$\frac{W_F}{L^3} = 2\rho \times 1.299\,s\int_0^1 \mathrm{d}\left(\frac{z}{L}\right) = 2.598\rho s, \quad \Rightarrow \quad n = 2.598s \tag{10.32}$$

Here, equations (10.31b) and (10.32) are normalised further by dividing each side by σ_y/E to give

$$n' = 1.0109\,s'^{1/2} + 2.5422\,s'^2 \tag{10.33a}$$

$$n' = 2.598s' \tag{10.33b}$$

where $n' = n/(\sigma_y/E)$ and $s' = s/(\sigma_y/E)$. This change to the non-dimensional parameters has influenced the bending design only. Note that the shear coefficient in equation (10.32) is identical to that which provided an optimum weight for the square section in equation (10.26b).

Figure 10.8 plots equations (10.33a,b) over five decades in s' to reveal, with increasing s', regions in which the bending, combined shear and bending, and shear designs apply. We see from equation (10.29b) that for $s'_{FM} = 0.3406$, both equations (10.33a,b) give $n' = 0.885$, at the division between the shear and combined shear and bending designs. In overlaying the weight plots for the three cross-sections in Figure 10.8 there appears to be little between them across the full range of structural indices shown. However, given the new definition s' and n' for a rectangular section and the logarithmic scale, the weight of an optimum rectangular section is much less than beams with optimum circular and square sections (see Example 10.1). The lesser weight applies both to shear ($s \geq s'_{FM}$) and bending ($s < s'_{FM}$) designs in short and long rectangular-section beams, respectively. Note that the lowest weight plot in Figure 10.8 refers to a fully optimised cross-section in which weight reductions apply to combined bending and shear designs across the full range of indices. The derivation of this cross-section geometry is given in Chapter 11 (see § 11.6).

The optimum dimensions b and d for a rectangular section, with its relationship between minimum weight and structural index (n' versus s') in Figure 10.8, are found from the buckling criterion subsumed within equation (10.27a). For the constant bending moment, the breadth and depth were given earlier in equations (3.13a,b) . Here, the section showed a constant breadth:

$$\left(\frac{b}{L}\right)_{\text{opt}} = 0.383\,\frac{\sigma_y}{E} \tag{10.34a}$$

accompanied by a variable depth (at a given s)

$$\left(\frac{d}{2L}\right)_{\text{opt}} = \frac{3.957}{2}\frac{E^{1/2}}{\sigma_y}\left[\frac{M(z)}{L^3}\right]^{1/2} = 1.979\left(\frac{E}{\sigma_y}\right)^{1/2}\left(\frac{1}{\sigma_y}\frac{P}{L^2}\right)^{1/2}\left(\frac{z}{L}\right)^{1/2}$$
$$= 1.979\left(\frac{\sigma_y}{E}\right)^{-1/2} s^{1/2}\left(\frac{z}{L}\right)^{1/2} \tag{10.34b}$$

where $M(z) = Pz$. Note that the product of equations (10.34a,b) restores equation (10.27b). The depth variation $d \propto z^{1/2}$ in equation (10.34b) differs from the cube root dependence of the square side length ($d \propto z^{1/3}$) in equation (10.22a). In a further contrast between these two sections, the optimum shear design for a beam with rectangular section requires both constant breadth and depth. Here the separation of b and d from within equation (10.28) involves the buckling coefficient $K = K(d/L)$ for $0 \le d/L \le 1$, i.e. lying in the range $4.84 \le K \le 8.4$, from Figure C.3. Equation (5.5c) may be shaped to give the depth in terms of the present parameters as

$$\left(\frac{d}{2L}\right)_{\text{opt}} = \frac{1}{2}\left(\frac{9K}{4}\right)^{1/4}\left(\frac{F^2E}{\tau_y^3L^4}\right)^{1/4} = \frac{1}{2}\left(\frac{9K}{4}\right)^{1/4}\left(\frac{F^2EQ^3}{\sigma_y^3L^4}\right)^{1/4} \tag{10.35a}$$

Substituting $Q = \sqrt{3}$ with $F = P =$ constant into equation (10.35a) gives the depth as

$$\left(\frac{d}{2L}\right)_{\text{opt}} = \frac{Q^{3/4}}{2}\left(\frac{9K}{4}\right)^{1/4}\left(\frac{E}{\sigma_y}\right)^{1/4}\left(\frac{1}{\sigma_y}\frac{P}{L^2}\right)^{1/2} = 0.755\left(\frac{9K}{4}\right)^{1/4}\left(\frac{\sigma_y}{E}\right)^{-1/4}s^{1/2} \tag{10.35b}$$

which is not dependent upon z. Equation (5.5b) gives the breadth as

$$\left(\frac{b}{L}\right)_{\text{opt}} = \left(\frac{9}{4K}\right)^{1/4}\left(\frac{P^2Q}{E\sigma_yL^4}\right)^{1/4} = Q^{1/4}\left(\frac{9}{4K}\right)^{1/4}\left(\frac{\sigma_y}{E}\right)^{1/4}\left(\frac{P^2}{\sigma_y^2L^4}\right)^{1/4} \tag{10.36a}$$

Substituting $Q = \sqrt{3}$ with $F = P =$ constant, gives a constant breadth

$$\left(\frac{b}{L}\right)_{\text{opt}} = 1.147\left(\frac{9}{4K}\right)^{1/4}\left(\frac{\sigma_y}{E}\right)^{1/4}s^{1/2} \tag{10.36b}$$

and, to check, the product of equations (10.35b) and (10.36b) will restore equation (10.28). It is seen that equations (10.35b) and (10.36b) are both independent of z. The resulting optimum beam shape within a combined design ($s < s_{FM}$) is shown in the following example.

Example 10.1 Find the optimum dimensions of a rectangular section, steel cantilever, its weight and the end-load it carries when $s' = 0.1$ and $L = 1$m. Find the weights of optimum circular- and square-section beams at the same load. The properties for steel are $\rho = 7860$ kg/m^3, $E = 210$ GPa and $\sigma_y = 300$ MPa.

With $s' = 0.1$, Figure 10.8 (or equation (10.33a)) gives $n' = 0.3451$. These (s', n') parameters require a combined shear and bending design since $s' < s'_{FM}$ where, from equation (10.29b), $s'_{FM} = 0.3406$. The division between the two designs occurs at the critical length position given by equation (10.29a) as

$$(z/L)_{\text{cr}} = 2.9361\, s' = 0.2936$$

Hence, in the region where $0 \le z/L \le 0.2936$, the shear design applies. The optimum breadth and depth then follow from equations (10.35b) and (10.36b):

$$\left(\frac{b}{L}\right)_{\text{opt}} = 0.3627\left(\frac{\sigma_y}{E}\right)^{3/4}\left(\frac{9}{4K}\right)^{1/4} \tag{i}$$

$$\left(\frac{d}{2L}\right)_{\text{opt}} = 0.2388\left(\frac{\sigma_y}{E}\right)^{1/4}\left(\frac{9K}{4}\right)^{1/4} \tag{ii}$$

Since $K = K(d/L)$, equation (ii) will requires an iterative solution to $d/2L$. For steel, we find that equation (ii) is satisfied by $d/L = 0.17$ when $K = 5.03$ from Figure C.3. Equation (i) then gives the normalised breadth as $(b/L)_{\text{opt}} = 0.0022$. Neither optimum dimension varies with the length and so equations (i) and (ii) describe a uniform rectangular section for the relevant portion of the length (rounded to 0.294 m), as shown in Figure 10.10.

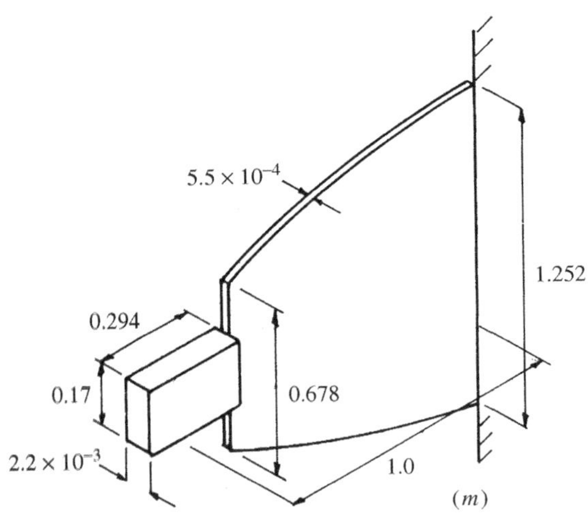

Figure 10.10 Optimum design of an end-loaded cantilever beam under combined shear and bending

In the region where $0.294 \le z/L \le 1.0$, bending controls the design. The constant, optimum breadth follows from equation (10.34a) as

$$\left(\frac{b}{L}\right)_{\text{opt}} = 0.383\,\frac{\sigma_y}{E} = 0.000547 \qquad \text{(iii)}$$

and, from equation (10.34b), the optimum depth will vary with z/L:

$$\left(\frac{d}{2L}\right)_{\text{opt}} = 1.979\,s'^{1/2}\left(\frac{z}{L}\right)^{1/2} = 0.6258\left(\frac{z}{L}\right)^{1/2} \qquad \text{(iv)}$$

Equation (iv) gives particular depths: $d/L = 0.678$ for $z/L = 0.294$, $d/L = 0.885$ for $z/L = 0.50$ and $d/L = 1.252$ for $z/L = 1$. Elsewhere within this length region the square root dependence $d \propto z^{1/2}$ gives a convex curvature to the beam edges (see Figure 10.10). Given a 1 m length for this beam, Figure 10.10 shows the beam's length and depth dimensions in metres. The weight of the beam is found directly from $n' = 0.3451$, as follows:

$$n' = \frac{1}{\rho}\left(\frac{W}{L^3}\right) \Big/ \left(\frac{\sigma_y}{E}\right) = 0.3451, \quad \Rightarrow \quad W = 0.3451\,\frac{\sigma_y}{E}\rho L^3 = 3.876 \text{ kg}$$

By comparison, the weights of square- and circular-section beams correspond to $s = s'\sigma_y/E = 1.43 \times 10^{-4}$. From Figure 10.8 we read $n \approx 10^{-2}$, giving their approximately equal weights as $W = n\rho L^3 = 78.6$ kg. The common load supported by each beam follows from $s' = 0.1$:

$$s' = \frac{1}{\sigma_y}\left(\frac{P}{L^2}\right) \Big/ \left(\frac{\sigma_y}{E}\right) = 0.1, \quad P = \frac{0.1(\sigma_y L)^2}{E} = 42.85 \text{ kN}$$

It may surprise the reader that such a low weight beam is predicted to support such a large load, but this is a consequence of the failure criteria adopted. In fact, it is most unlikely that the 0.55 mm breadth is capable of supporting this end-load without twist, flexural or lateral buckling occurring, none of which has been considered in optimising this design. To offset the likelihood of these failures we could accept the depth as given and then forfeit some weight by adding vertical stiffeners whose optimum spacing and geometry are found by the methods outlined in Chapter 8. A general discussion upon additional failure modes arising within thin rectangular beam sections is given in § 10.7.

10.4 Cantilever with Distributed Loading

Regions of structural indices will again be established to show where both shear force and bending moment contribute to the optimum design of the cantilever beam in Figure 10.11a. The force and moment diagrams in Figure 10.11b,c are expressed positively as $|F(z)| = wz$ and $M(z) = wz^2/2$ within the full length $0 \le z \le L$. These diagrams apply to any cross-section. In particular, a minimum weight will be achieved in each of three solid

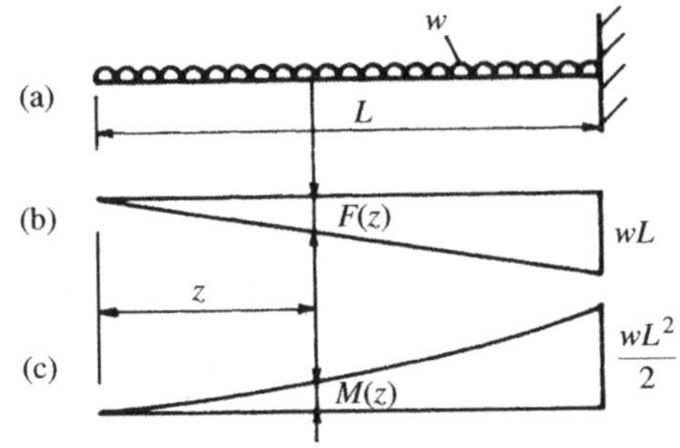

Figure 10.11 Force and moment diagrams for a cantilever with uniformly distributed loading

cross-sections – circular, square and rectangular – by allowing the section dimensions to vary with the length dimension z in a manner now to be determined.

10.4.1 Circular Section, Diameter d

When the cantilever cross-section is circular, the optimum diameter in bending is again given by combining $M(z) = wz^2/2$ with equation (3.3b). Dividing by $2L$ reveals the variation in radius with normalised length as

$$\frac{d}{2L} = \left(\frac{2}{\pi}\right)^{1/3}\left(\frac{1}{\sigma_y}\frac{F_{\max}}{L^2}\right)^{1/3}\left(\frac{z}{L}\right)^{2/3} = 0.8603\, s^{1/3}\left(\frac{z}{L}\right)^{2/3} \tag{10.37a}$$

in which $s = (1/\sigma_y)(F_{\max}/L^2)$ is the normalised structural index, with $F_{\max} = wL$. Note that a large index s identifies a short beam and a small index identifies a long beam.

For a transverse shear force applied to a circular-section beam it was shown previously that equation (5.15b) gave the optimum diameter. Writing $F = |F(z)| = wz$, dividing by $2L$ and writing the relationship between the tensile and shear yield stresses as $\sigma_y = Q\tau_y$ leads to

$$\frac{d}{2L} = 2\left(\frac{Q}{3\pi}\right)^{1/2}\left(\frac{1}{\sigma_y}\frac{F_{\max}}{L^2}\right)^{1/2}\left(\frac{z}{L}\right)^{1/2} = 0.8574\, s^{1/2}\left(\frac{z}{L}\right)^{1/2} \tag{10.37b}$$

in which we may choose between $Q = \sqrt{3}$ (taken here) and $Q = 2$ by the von Mises and Tresca criteria of yielding, respectively. For a given s, equations (10.37a,b) show that the optimum diameters vary with length both in shear and in bending. Equating (10.37a,b) gives the critical length position where the diameter from shear switches to the diameter from bending:

$$\left(\frac{z}{L}\right)_{\text{cr}} = \left(\frac{Q}{3}\right)^3\frac{2^4}{\pi}s = 0.9801\, s \tag{10.38a}$$

At its limit equation (10.38a) gives $(z/L)_{\text{cr}} = 1$ which provides the minimum structural index for an all-shear design. This index (s_{FM}) lies at the division between shear and

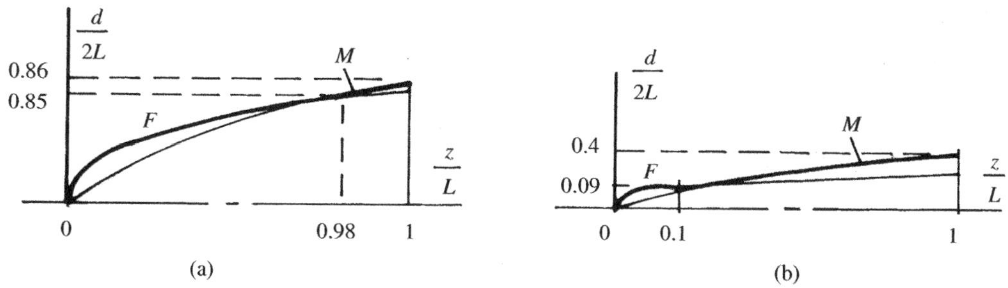

Figure 10.12 Near, all-shear (F) and all-bending (M) designs in short-length beams

bending designs:

$$s_{FM} \geq \frac{\pi}{2^4}\left(\frac{3}{Q}\right)^3 \tag{10.38b}$$

With $Q = \sqrt{3}$ we obtain $s_{FM} = 1.0203$. Thus, if we select $s = 1$, for example, equation (10.38a) gives $(z/L)_{cr} = 0.98$. For $0.98 \leq z/L \leq 1$, the diameter varies according to equation (10.37a) and for $0 \leq z/L \leq 0.98$, the diameter varies according to equation (10.37b), i.e. a shear design for most of the length. At the other extreme, if we select $s = 0.1$, equation (10.38a) gives $(z/L)_{cr} = 0.098$ (say, 0.1). Hence, for $0 \leq z/L \leq 0.1$, the diameter varies according to equation (10.37b) and for $0.1 \leq (z/L) \leq 1$, the diameter varies according to equation (10.37a), i.e. a bending design for most of the length. Figure 10.12 shows the beam profiles according to these two combined designs. The ordinates show that both plots apply to short beams or pins whose maximum radii are 0.86 and 0.4 of their lengths.

For $s \leq s_{FM}$, the beam's weight follows from bending and shear within their respective regions. Thus, for an elemental length δz within each region, the weight of the thin disc is $\delta W = \rho(\pi/4)d^2\,\delta z$. Thus, for $0 \leq z/L \leq z_{cr}/L$ (shear) we substitute $d/2L$ from equation (10.37b), when the integration gives

$$\frac{W_F}{L^3} = \rho\pi \int_0^{(z/L)_{cr}} \left(\frac{d}{2L}\right)^2 \mathrm{d}\left(\frac{z}{L}\right) = \rho\pi\left(\frac{4Q}{3\pi}\right)s \int_0^{(z/L)_{cr}} \left(\frac{z}{L}\right)\mathrm{d}\left(\frac{z}{L}\right) = \frac{2Q\rho s}{3}\left(\frac{z}{L}\right)^2_{cr} \tag{10.39a}$$

For $z_{cr}/L \leq z/L \leq 1$ (bending) we substitute $d/2L$ from equation (10.37a), when the integration gives

$$\begin{aligned}\frac{W_M}{L^3} &= \rho\pi \int_{(z/L)_{cr}}^{1} \left(\frac{d}{2L}\right)^2 \mathrm{d}\left(\frac{z}{L}\right) = \rho\pi\left(\frac{2s}{\pi}\right)^{2/3} \int_{(z/L)_{cr}}^{1} \left(\frac{z}{L}\right)^{4/3} \mathrm{d}\left(\frac{z}{L}\right) \\ &= \frac{3\rho\pi}{7}\left(\frac{2s}{\pi}\right)^{2/3}\left[1 - \left(\frac{z}{L}\right)^{7/3}_{cr}\right]\end{aligned} \tag{10.39b}$$

Adding equations (10.39a,b) and substituting $(z/L)_{cr}$ from equation (10.38a) leads to

$$\frac{W}{L^3} = \frac{2\rho Q s}{3}\left(\frac{2^4}{\pi}\right)^2\left(\frac{Q}{3}\right)^6 s^2 + \frac{3\rho\pi}{7}\left(\frac{2s}{\pi}\right)^{2/3}\left[1 - \left(\frac{Q}{3}\right)^7\left(\frac{2^4}{\pi}\right)^{7/3} s^{7/3}\right] \quad (10.40a)$$

Putting $n = (1/\rho)(W/L^3)$ and $Q = \sqrt{3}$, equation (10.40a) becomes

$$n = 1.1093\,s^3 + 0.9964\,s^{2/3}(1 - 0.9542\,s^{7/3}) = 0.9964\,s^{2/3} - 0.1585\,s^3 \quad (10.40b)$$

For $s \geq s_{FM}$ the diameter is constant and the beam's weight follows from shear only:

$$\frac{W}{L^3} = \rho\pi\int_0^1\left(\frac{d}{2L}\right)^2 d\left(\frac{z}{L}\right) = \rho\pi\left(\frac{4Q}{3\pi}\right)s\int_0^1\left(\frac{z}{L}\right)d\left(\frac{z}{L}\right) = \frac{2Q\rho s}{3} \quad (10.41a)$$

which, with similar substitutions for n and Q, has the linear form

$$n = 1.1547\,s \quad (10.41b)$$

Figure 10.13 plots equations (10.40b) and (10.41b) over five decades in ascending s to reveal those regions in which the bending, combined shear and bending, and all-shear

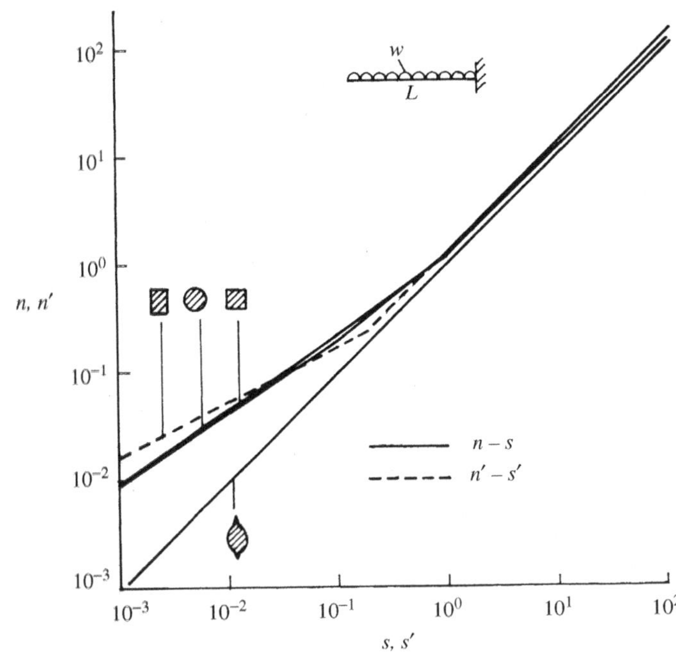

Figure 10.13 Weight versus structural index for uniformly distributed loaded cantilevers with various sections. (note: $n = (1/\rho)(W/L^3)$; $s = (1/\sigma_y)(F_{max}/L^2)$, where $F_{max} = wL$; $s' = s/(\sigma_y/E)$; $n' = n/(\sigma_y/E)$)

designs apply. With s-values in the middle region, $0.1 < s < 1.0203$, Figure 10.13 shows that both shear and bending control an optimum design in which equation (10.40b) supplies the normalised weight n. Shear controls the design for $s > 1.0203$ where equation (10.41b) applies to n. Bending dominates the section profile design for long beams ($s < 0.01$).

10.4.2 *Square Section, Side Length a*

As in § 10.4.1 the shear force and bending moment diagrams in Figure 10.11b,c may be expressed as $|F(z)| = wz$ and $M(z) = wz^2/2$ over the full length ($0 \le z \le L$). When the section is square, its optimum depth a under bending is found by combining $M(z)$ with equation (3.3b). Dividing by $2L$ reveals the variation between the normalised half-depth ($a/2L$) and the normalised length (z/L) as

$$\frac{a}{2L} = \left(\frac{3}{8}\right)^{1/3} \left(\frac{1}{\sigma_y}\frac{F_{max}}{L^2}\right)^{1/3} \left(\frac{z}{L}\right)^{2/3} = 0.721 s^{1/3} \left(\frac{z}{L}\right)^{2/3} \tag{10.42a}$$

in which $s = (1/\sigma_y)(F_{max}/L^2)$ is the normalised structural index and $F_{max} = wL$. Note that the index s is to be identified with short beams when s is large and long beams when s is small.

For a transverse shear force applied to a square section beam it was shown previously that equation (5.4c) gave the optimum depth of section. Dividing equation (5.4c) by $2L$ and writing the relationship between the tensile and shear yield stresses as $\sigma_y = Q\tau_y$ gives

$$\frac{a}{2L} = \left(\frac{3Q}{8}\right)^{1/2} \left(\frac{1}{\sigma_y}\frac{F_{max}}{L^2}\right)^{1/2} \left(\frac{z}{L}\right)^{1/2} = 0.8059\, s^{1/2} \left(\frac{z}{L}\right)^{1/2} \tag{10.42b}$$

where the numerical coefficient $Q = \sqrt{3}$ by the von Mises yield criterion. For a given s, equations (10.42a,b) show how a varies with the length dimension z in both regions. Equating (10.42a,b) gives the critical position in the length where the transition between shear and bending designs occurs:

$$\left(\frac{z}{L}\right)_{cr} = \frac{3}{8} Q^3 s = 1.949\, s \tag{10.43a}$$

At its limit $(z/L)_{cr} = 1$ in equation (10.43a). Hence, the lowest limiting s for an all-shear design corresponds to the index dividing the shear and bending designs:

$$s_{FM} \ge \frac{8}{3Q^3} \tag{10.43b}$$

With $Q = \sqrt{3}$, we obtain $s_{FM} = 0.5132$. Thus, if we select $s = 1$, for example, equation (10.43a) gives $(z/L)_{cr} > 1.949$. As this lies outside the length the design is to be based entirely upon shear. That there will be no intersection between equations (10.42a,b) for $s = 1$ is shown in Figure 10.14a. Shear (F) provides the safer, larger depth over the full length. If we select $s = 0.1$, equation (10.43a) gives $(z/L)_{cr} = 0.195$, when shear plays

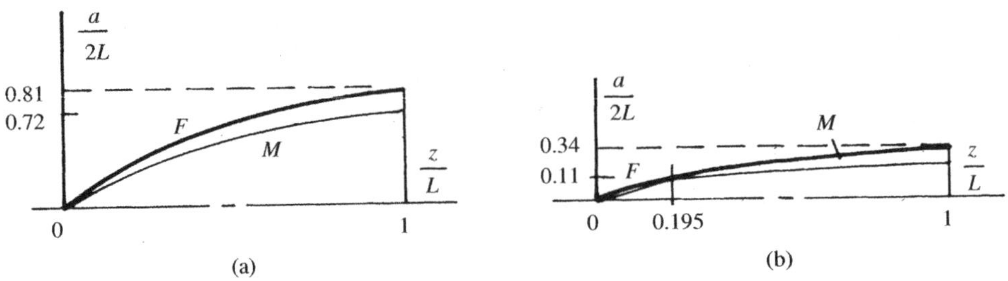

Figure 10.14 All-shear and near, all-bending designs in short length beams

a lesser role in the design. That is, for $0 \le z/L \le 0.195$, Figure 10.14b shows that the depth in shear (F) varies as equation (10.42b). For the remaining length, $0.195 < z/L \le 1$, bending (M) controls the depth by equation (10.42a). Figures 10.14a,b are in fact semi-depth profiles over the length. The normalised ordinates in these figures show that both plots apply to short beams whose depths are similar to their length. As s decreases further and $(z/L)_{\text{cr}} \approx 0$, bending dominates the design. The fact that $(z/L)_{\text{cr}}$ is never zero shows an ever present influence of shear, but this effect is less important to distributed loads where the free-end may start as a point.

Consider $s \le s_{FM}$, where the beam's weight follows from the two contributions. For an elemental length δz within each region, the weight is $\delta W = \rho a^2 \delta z$. That is, for the shear design $0 \le z/L \le z_{\text{cr}}/L$, we substitute $a/2L$ from equation (10.42b), when the integration gives

$$\frac{W_F}{L^3} = 4\rho \int_0^{(z/L)_{\text{cr}}} \left(\frac{a}{2L}\right)^2 \mathrm{d}\left(\frac{z}{L}\right) = 4\rho \left(\frac{3Q}{8}\right) s \int_0^{(z/L)_{\text{cr}}} \left(\frac{z}{L}\right) \mathrm{d}\left(\frac{z}{L}\right) = \frac{3Q\rho s}{4} \left(\frac{z}{L}\right)_{\text{cr}}^2 \tag{10.44a}$$

For $z_{\text{cr}}/L \le z/L \le 1$ (bending) we substitute $a/2L$ from equation (10.42a), when the integration gives

$$\frac{W_M}{L^3} = 4\rho \int_{(z/L)_{\text{cr}}}^{1} \left(\frac{a}{2L}\right)^2 \mathrm{d}\left(\frac{z}{L}\right) = 4\rho \left(\frac{3s}{8}\right)^{2/3} \int_{(z/L)_{\text{cr}}}^{1} \left(\frac{z}{L}\right)^{4/3} \mathrm{d}\left(\frac{z}{L}\right)$$

$$= \frac{12\rho}{7} \left(\frac{3s}{8}\right)^{2/3} \left[1 - \left(\frac{z}{L}\right)_{\text{cr}}^{7/3}\right] \tag{10.44b}$$

Adding equations (10.44a,b) and substituting $(z/L)_{\text{cr}}$ from equation (10.43a) leads to

$$\frac{W}{L^3} = \frac{3\rho Q s}{4} \left(\frac{3}{8}\right)^2 Q^6 s^2 + \frac{12\rho}{7} \left(\frac{3s}{8}\right)^{2/3} \left[1 - \left(\frac{3}{8}\right)^{7/3} Q^7 s^{7/3}\right] \tag{10.45a}$$

Putting $n = (1/\rho)(W/L^3)$ and $Q = \sqrt{3}$, equation (10.45a) becomes

$$n = 4.9323\, s^3 + 0.8914\, s^{2/3}(1 - 4.7424\, s^{7/3}) \tag{10.45b}$$

Now consider $s \geq s_{FM}$, where the depth is given by equation (10.42b) and the beam's weight follows from shear only:

$$\frac{W}{L^3} = 4\rho \int_0^1 \left(\frac{a}{2L}\right)^2 \mathrm{d}\left(\frac{z}{L}\right) = 4\rho \left(\frac{3Q}{8}\right) s \int_0^1 \left(\frac{z}{L}\right) \mathrm{d}\left(\frac{z}{L}\right) = \frac{3Q\rho s}{4} \tag{10.46a}$$

which, for similar substitutions for n and Q, takes the linear form

$$n = 1.299\, s \tag{10.46b}$$

Figure 10.13 plots equations (10.45b) and (10.46b) over five decades in ascending s to reveal the regions in which bending, combined shear and bending, and all-shear designs apply. This shows that both shear and bending control the optimum design for mid-range indices $0.1 < s < 0.5132$, where equation (10.45b) supplies the normalised weight n. Shear controls the optimum design for $s \geq 0.5132$, where the linear equation (10.46b) applies to n. Bending dominates for $s < 0.01$.

10.4.3 Rectangular Section, Breadth b, Depth d

The possibility of lateral buckling arises in thin rectangular sections where $b \ll d$. We could employ equations (3.13a,b) to optimise b and d separately in bending. However, since the product $(bd)_{\text{opt}}$ will appear in the weight expression, we shall work with this product as our variable. That is, for bending, equation (3.11) becomes

$$bd = \left[\frac{6M(z)L}{2.61E}\right]^{1/2} \tag{10.47a}$$

Substituting $M(z) = wz^2/2$ from Figure 10.11 and dividing b by L and d by $2L$ leads to

$$\frac{b}{L}\frac{d}{2L} = \left(\frac{6wz^2L}{5.22E \times 4L^4}\right)^{1/2} = 0.536\left(\frac{\sigma_y}{E}\right)^{1/2}\left(\frac{1}{\sigma_y}\frac{|F_{\max}|}{L^2}\right)^{1/2}\frac{z}{L}$$

$$= 0.536\, s^{1/2}\left(\frac{\sigma_y}{E}\right)^{1/2}\frac{z}{L} \tag{10.47b}$$

Equation (5.3c) gives $(bd)_{\text{opt}}$ for a transverse shear force applied to a rectangular section. Substituting $|F(z)| = wz$, dividing b by L, d by $2L$ and setting $\sigma_y = \sqrt{3}\tau_y$ and $F_{\max} = wL$ within equation (5.3c) leads to

$$\frac{b}{L}\frac{d}{2L} = \frac{3QF_{\max}z}{4\sigma_y L^3} = 1.299\left(\frac{1}{\sigma_y}\frac{F_{\max}}{L^2}\right)\frac{z}{L} = 1.299\, s\frac{z}{L} \tag{10.48a}$$

where $s = (1/\sigma_y)(F_{\max}/L^2)$. Clearly, if we equate (10.47b) and (10.48a) the position in the length which divides a bending design from a shear design is indeterminate. Instead, this equality provides a critical index:

$$s_{FM} \leq 0.1703\frac{\sigma_y}{E} \tag{10.48b}$$

Thus, for $s \leq s_{FM}$, the safe design become entirely bending-based, when from equation (10.47b) the weight becomes

$$\frac{W_M}{L^3} = 2\rho \times 0.536\left(\frac{\sigma_y}{E}\right)^{1/2} s^{1/2} \int_0^1 \left(\frac{z}{L}\right) \mathrm{d}\left(\frac{z}{L}\right) = 0.536\rho\left(\frac{\sigma_y}{E}\right)^{1/2} s^{1/2} \tag{10.49a}$$

For $s \geq s_{FM}$, the safe design is entirely shear-based, where the beam weight follows from equation (10.48a):

$$\frac{W_F}{L^3} = 2\rho \times 1.299\, s \int_0^1 \left(\frac{z}{L}\right) \mathrm{d}\left(\frac{z}{L}\right) = 1.299\rho s \tag{10.49b}$$

Dividing each side of equations (10.49a,b) by $\rho(\sigma_y/E)$ gives the respective weights in the normalised coordinates $n' = n/(\sigma_y/E)$ and $s' = s/(\sigma_y/E)$ as

$$n' = 0.536\, s'^{1/2}, \quad n' = 1.299\, s' \tag{10.50a,b}$$

at which the transition, $S'_{FM} = 0.1703$, follows from equation (10.48b). Figure 10.13 plots equations (10.50a,b) over five decades in s' to reveal, with increasing s', the two regions in which the all-bending and all-shear designs apply (the dashed line). We see that for $s'_{FM} = 0.1703$, both equations (10.50a,b) give $n' = 0.2212$, at the division between the two designs. Again, as explained previously in Figure 10.8, by overlaying the weights for the three cross-sections in Figure 10.13, there appears little to distinguish them across five decades in s. However, the revised parameters s' and n', used for the optimum design of a thin, rectangular section, will result in a much lighter beam (see Example 10.2).

The optimum dimensions of the rectangular section, with weight and structural indices n' and s' in Figure 10.13, admit additional lateral buckling criteria within equations (10.47b) and (10.48a). Separating b from within bd for the bending design (see equation (3.13b)) gives a constant breadth

$$\frac{b_{\text{opt}}}{L} = 0.383\,\frac{\sigma_y}{E} \tag{10.51a}$$

and, dividing equation (10.47b) by (10.51a), a linear varying depth

$$\left(\frac{d}{2L}\right)_{\text{opt}} = \frac{0.536(\sigma_y/E)^{1/2} s^{1/2}(z/L)}{0.383(\sigma_y/E)} = 1.4\left(\frac{\sigma_y}{E}\right)^{-1/2} s^{1/2}\frac{z}{L} \tag{10.51b}$$

In contrast, the optimum shear design requires both variable breadth and depth. Equations (5.5b) and (5.5c) show that the separation of b from d within equation (10.48a) involves the buckling coefficient $K = K(d/L)$, where, from Figure C.3, K lies in the range $4.84 \leq K \leq 8.4$ for $0 \leq d/L \leq 1$. Equation (5.5c) may be arranged to express the variable depth in the present parameters as follows:

$$\frac{d}{2L} = \frac{1}{2L}\left(\frac{9K}{4}\right)^{1/4}\left(\frac{|F(z)|^2 E}{\tau_y^3}\right)^{1/4} = \frac{1}{2}\left(\frac{9K}{4}\right)^{1/4}\left(\frac{|F(z)|^2 E Q^3}{\sigma_y^3 L^4}\right)^{1/4} \tag{10.52a}$$

Substituting $Q = \sqrt{3}$ and $|F(z)| = wz = F_{\max}z/L$ (where $F_{\max} = wL$) into equation (10.52a)

$$\left(\frac{d}{2L}\right)_{\text{opt}} = \frac{Q^{3/4}}{2}\left(\frac{9K}{4}\right)^{1/4}\left(\frac{1}{\sigma_y}\frac{F_{\max}}{L^2}\right)^{1/2}\left(\frac{z^2 E}{\sigma_y L^2}\right)^{1/4}$$

$$= 0.925K^{1/4}s^{1/2}\left(\frac{\sigma_y}{E}\right)^{-1/4}\left(\frac{z}{L}\right)^{1/2} \tag{10.52b}$$

Equation (5.5b) gives the variable breadth in terms of these parameters

$$\frac{b}{L} = \frac{1}{L}\left(\frac{9}{4K}\right)^{1/4}\left(\frac{|F(z)|^2}{E\tau_y}\right)^{1/4} = \left(\frac{9}{4K}\right)^{1/4}\left(\frac{Q|F(z)|^2}{E\sigma_y L^4}\right)^{1/4} \tag{10.53a}$$

With similar substitutions, $Q = \sqrt{3}$ and $|F(z)| = wz = F_{\max}z/L$, equation (10.53a) becomes

$$\left(\frac{b}{L}\right)_{\text{opt}} = \left(\frac{9}{4K}\right)^{1/4}\left(\frac{QF_{\max}^2 z^2}{E\sigma_y L^6}\right)^{1/4} = \frac{1.405}{K^{1/4}}s^{1/2}\left(\frac{\sigma_y}{E}\right)^{1/4}\left(\frac{z}{L}\right)^{1/2} \tag{10.53b}$$

and, to check, the product of equations (10.52b) and (10.53b) restores equation (10.48a). At the intersection between equations (10.50a,b), where $s' = 0.1703$ and $n' = 0.221$, the solutions to the optimum b and d agree, giving

$$\frac{b_{\text{opt}}}{L} = 0.383\,\frac{\sigma_y}{E},\quad \frac{d_{\text{opt}}}{L} = 2 \times 0.253\left[\frac{K(d_{\text{opt}}/L)\sigma_y}{E}\right]^{1/2},\quad \left(\frac{z}{L}\right)_{\text{cr}} = 0.437\left(\frac{K\sigma_y}{E}\right)^{1/2} \tag{10.54a,b,c}$$

Equations (10.54a,b) correspond to s_{FM} where the optimum b and d from the bending and shear designs in equations (10.51)–(10.53) become the same. Here, equation (10.54b) requires an iterative solution to $(d/L)_{\text{opt}}$ using Figure C.3.

Example 10.2 Design optimum cantilevers to carry 10 kN distributed uniformly over a 1 m length in aluminium alloy when the cross-sections are: (a) circular, (b) square and (c) rectangular. Take $E = 75$ GPa, $\sigma_y = 150$ MPa and $\rho = 2710$ kg/m^3.

Here we identify $F_{\max} = 10$ kN and $L = 1$ m, so that the normalised structural index for all three sections becomes

$$s = \frac{1}{\sigma_y} \times \frac{F_{\max}}{L^2} = \frac{1}{150 \times 10^6} \times \frac{10 \times 10^3}{1^2} = 6.67 \times 10^{-5}$$

(a) In the case of a circular section, equations (10.38a,b) show that, with $s_{FM} = 1.0203$,

$$(z/L)_{cr} = 0.9801s = 6.534 \times 10^{-5}$$

and since a bending design applies to $(z/L)_{cr} \leq z/L \leq 1$, it follows that here the beam design may be based upon bending alone. Thus, equation (10.37a) gives the normalised radius as

$$\left(\frac{d}{2L}\right)_{opt} = 0.8603\left(6.67 \times 10^{-5}\right)^{1/3}\left(\frac{z}{L}\right)^{2/3} = 0.03489\left(\frac{z}{L}\right)^{2/3}$$

showing that

(i) when $z = L, d/2L = 0.03489$, giving $d = 0.03489 \times 2 \times 1$
$= 0.06978$ m $= 69.78$ mm,
(ii) when $z = L/2, d/2L = 0.02198$, giving $d = 0.02198 \times 2 \times 1$
$= 0.04396$ m $= 43.96$ mm.

Figure 10.15a shows the rounded diameters of an optimum beam design corresponding to this s-value. Here the section may start as a point at the free end given that the shear force is zero at this position (see Figure 10.11b). Neglecting the shear contribution, it follows that (10.39b) provides the normalised weight of this design:

$$\left(\frac{W}{L^3}\right)_M = \rho\pi\left(\frac{2}{\pi}\right)^{2/3} s^{2/3}\int_0^1\left(\frac{z}{L}\right)^{4/3} d\left(\frac{z}{L}\right)$$

Substituting $s = 6.67 \times 10^{-5}$, the integration reveals the minimum weight

$$\frac{1}{\rho}\left(\frac{W_{opt}}{L^3}\right) = 1.6377 \times 10^{-3}, \quad \Rightarrow \quad W_{opt} = 1.6377 \times 10^{-3} \times 2710 \times 1^3 = 4.438 \text{ kg}$$

(b) In the case of a square section, equation (10.43a,b) shows that, with $s_{FM} = 0.5131$,

$$(z/L)_{cr} = 1.949s = 1.3 \times 10^{-4}$$

and since the bending design applies to $(z/L)_{cr} \leq z/L \leq 1$, it follows that here the beam design may be based upon bending alone. Thus, equation (10.42a) gives the normalised half-depth

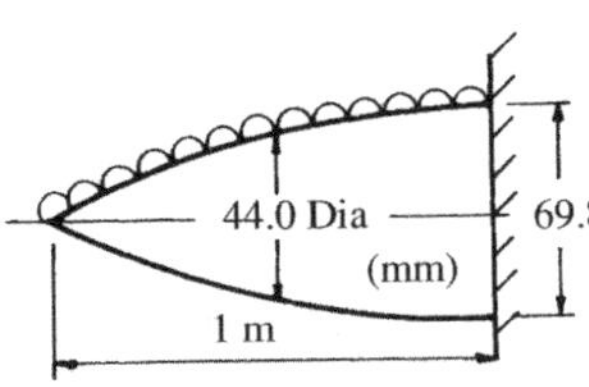

(a)

Figure 10.15a Circular-section, length profile for $s = 6.67 \times 10^{-5}$

$$\left(\frac{a}{2L}\right)_{\rm opt} = \left(\frac{3}{8}\right)^{1/3} (6.67 \times 10^{-5})^{1/3} \left(\frac{z}{L}\right)^{2/3} = 0.02924 \left(\frac{z}{L}\right)^{2/3}$$

showing that

(i) when $z = L, a/2L = 0.02924$, giving $a = 0.02924 \times 2 \times 1 = 0.05848$ m $= 58.48$ mm,
(ii) when $z = L/2, a/2L = 0.01842$, giving $a = 0.01842 \times 2 \times 1 = 0.03684$ m $= 36.84$ mm.

Figure 10.15b shows the optimum beam design for this s-value. The smaller depth dimensions shown suggest a weight saving over the circular section (see Figure 10.15a). To show this, when we neglect shear, it follows from equation (10.44b) that the normalised weight of this design becomes

$$\left(\frac{W}{L^3}\right)_M = 4\rho \left(\frac{3}{8}\right)^{2/3} s^{2/3} \int_0^1 \left(\frac{z}{L}\right)^{4/3} \mathrm{d}\left(\frac{z}{L}\right)$$

Substituting $s = 6.67 \times 10^{-5}$, the integration reveals that

$$\frac{1}{\rho}\left(\frac{W_{\rm opt}}{L^3}\right) = 1.4656 \times 10^{-3}, \quad \Rightarrow \quad W_{\rm opt} = 1.4656 \times 10^{-3} \times 2710 \times 1^3 = 3.972 \text{ kg}$$

(c) In the case of a rectangular section, we have seen that the design is either all-shear or all-bending. In this case, the latter applies since $s < s_{FM}$, where from equation (10.48b),

$$s_{FM} = 0.1703\frac{\sigma_y}{E} = 0.1703\frac{150}{75 \times 10^3} = 0.3406 \times 10^{-3}$$

Therefore, the cross-section dimensions follow from equations (10.51a,b):

$$\left(\frac{b}{L}\right)_{\rm opt} = 0.383\frac{\sigma_y}{E} = 0.383\frac{150}{75 \times 10^3} = 0.766 \times 10^{-3}$$

$$\left(\frac{d}{2L}\right)_{\rm opt} = 1.4(6.67 \times 10^{-5})^{1/2} \left(\frac{150}{75 \times 10^3}\right)^{-1/2} \left(\frac{z}{L}\right) = 0.2566\frac{z}{L}$$

from which the constant breadth is found as $b = 0.77$ mm and the depth varies linearly with the length as shown in Figure 10.15c. In particular, at the fixed end and at mid-span, the depths are far greater than in the previous two designs:

(i) when $z = L, d/2L = 0.2556$, giving $d = 0.2556 \times 2 \times 1 = 0.5112$ m $=$ 511.2 mm,
(ii) when $z = L/2, d/2L = 0.1278$, giving $d = 0.1278 \times 2 \times 1 = 0.2556$ m $= 255.6$ mm

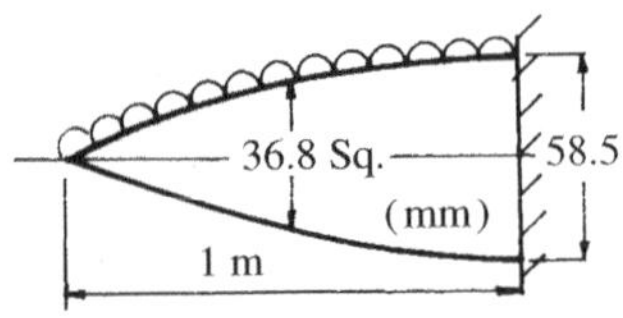

(b)

Figure 10.15b Square-section, length profile for $s = 6.67 \times 10^{-5}$

This design, which could easily be fabricated from sheet metal, presents a considerable weight saving over the other designs. Equation (10.49a) provides the normalised weight for this beam design:

$$\left(\frac{W}{L^3}\right)_M = 2\rho \times 0.536 \times \left(\frac{\sigma_y}{E}\right)^{1/2} s^{1/2} \int_0^1 \left(\frac{z}{L}\right) \mathrm{d}\left(\frac{z}{L}\right)$$

Substituting $s = 6.67 \times 10^{-5}$, the integration reveals that

$$\frac{1}{\rho}\frac{W_{\text{opt}}}{L^3} = 1.9572 \times 10^{-4}, \quad \Rightarrow \quad W_{\text{opt}} = 1.9572 \times 10^{-4} \times 2710 \times 1^3 = 0.5304 \text{ kg}$$

This design raises the question as to whether such a thin section is stable. Unless the load is perfectly evenly distributed upon the top surface there is the possibility of further torsion and lateral buckling arising, for which additional failure criteria would be required (see § § 10.7.4–10.7.6). However, the tapered depth, distributed loading and encastré fixing is a far more stable combination under these modes than is the simply supported, thin beam of constant depth in support of concentrated loading. One such example follows.

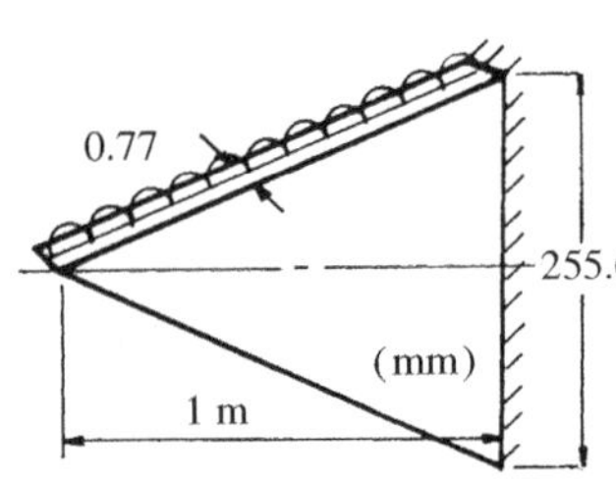

(c)

Figure 10.15c Rectangular-section, length profile for $s = 6.67 \times 10^{-5}$

10.5 Simply Supported Beam with Central Load

As before (see § 10.2), this standard case, shown in Figure 10.16a, refers to a beam with a central concentrated load P. With a discontinuity arising in the F- and M-diagrams

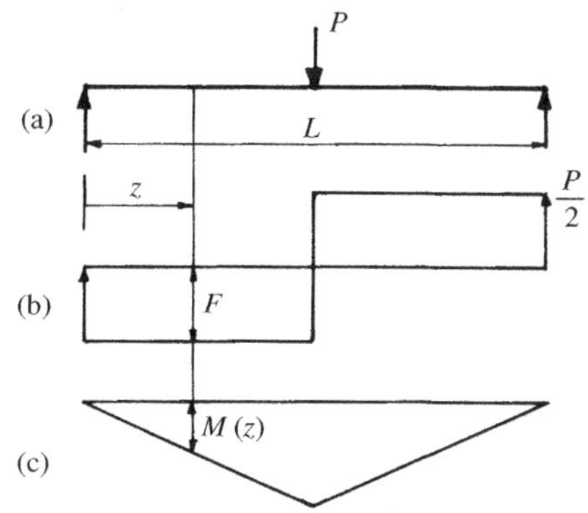

Figure 10.16 Force and moment diagrams in a centrally loaded, simply supported beam

beneath the load (Figure 10.16b,c) it becomes necessary to work on each side of the beam with z taken from each end towards the centre. The force and moment diagram in Figure 10.16a,b show magnitudes $|F| = P/2 =$ constant and $|M(z)| = Pz/2$ within the half length $0 \le z \le L/2$. It follows that with a central symmetry in each diagram, the total, optimum weight will be twice that of one half.

10.5.1 Circular Section, Diameter d

When the section is circular, the optimum diameter in bending is found by combining $|M(z)|$ with equation (3.3b). Dividing by $2L$ reveals the variation in radius with normalised length,

$$\left(\frac{d}{2L}\right)_{opt} = \left(\frac{2}{\pi}\right)^{1/3}\left(\frac{1}{\sigma_y}\frac{P}{L^2}\right)^{1/3}\left(\frac{z}{L}\right)^{1/3} = 0.8603\, s^{1/3}\left(\frac{z}{L}\right)^{1/3} \qquad (10.55a)$$

in which $s = (1/\sigma_y)(P/L^2)$ is the normalised structural index that varies inversely with the square of the beam length.

For a transverse shear force F applied to a circular section beam it was shown previously that equation (5.15b) gave the optimum diameter. Dividing by $2L$ and using the relationship between the tensile and shear yield stresses, $\sigma_y = Q\tau_y$,

$$\left(\frac{d}{2L}\right)_{opt} = \left(\frac{2Q}{3\pi}\right)^{1/2}\left(\frac{1}{\sigma_y}\frac{P}{L^2}\right)^{1/2} = 0.6063 s^{1/2} \qquad (10.55b)$$

where $Q = \sqrt{3}$ from the von Mises yield criterion. For a given s, equation (10.55b) shows that the diameter is constant in shear while, for bending, equation (10.55a) shows how the design diameter should vary with the length. Equating (10.55a,b) gives the critical position in the half-length where the shear-based design diameter is displaced by a bending-based diameter:

$$\left(\frac{z}{L}\right)_{cr} = \left(\frac{Q}{3}\right)^{3/2}\left(\frac{2}{\pi}\right)^{1/2} s^{1/2} \qquad (10.56a)$$

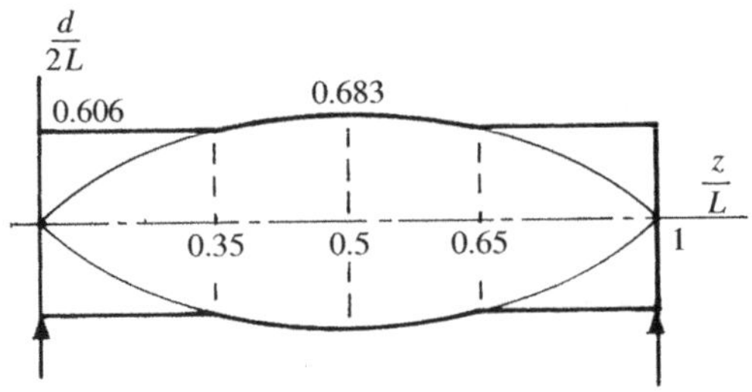

Figure 10.17 Length profile for a circular-section beam

At its limit $(z/L)_{cr} = 0.5$, when equation (10.56a) gives the lowest index s for an all-shear design,

$$s_{FM} \geq \frac{\pi}{8}\left(\frac{3}{Q}\right)^3 \tag{10.56b}$$

With $Q = \sqrt{3}$ in equation (10.56b), we find this index to be $s_{FM} = 2.041$. Smaller indices require a combined shear and bending design. Thus, if we select $s = 1$, for example, equation (10.56a) gives $(z/L)_{cr} = 0.35$. Then, for $0 \leq z/L \leq 0.35$, the diameter is held constant ($d/2L = 0.606$) in shear according to equation (10.55b). For $0.35 \leq z/L \leq 0.5$, the diameter increases to a maximum at the centre ($d/2L = 0.683$) in bending according to equation (10.55a). Figure 10.17 shows the beam's profile for $s = 1$ according to the combined design, in which the two halves mirror one another. Given that $d/2L$ identifies the normalised radius, Figure 10.17 replicates the section profile over the beam's length. This particular design applies to a short beam whose diameter is similar to its length.

For $s \leq s_{FM}$, the beam's weight follows from two contributions: shear and bending. For an elemental length δz within each region the weight is $\delta W = \rho(\pi/4)d^2\delta z$. Thus, for $0 \leq z/L \leq z_{cr}/L$ (shear), we substitute $d/2L$ from equation (10.55b) for the integration:

$$\frac{W_F}{L^3} = \rho\pi \int_0^{(z/L)_{cr}} \left(\frac{d}{2L}\right)^2 \mathrm{d}\left(\frac{z}{L}\right) = \rho\pi\frac{2Q}{3\pi}s \int_0^{(z/L)_{cr}} \mathrm{d}\left(\frac{z}{L}\right) = \frac{2Q\rho s}{3}\left(\frac{z}{L}\right)_{cr} \tag{10.57a}$$

For $z_{cr}/L \leq z/L \leq 0.5$ (bending) we substitute $d/2L$ from equation (10.55a) into the weight integral:

$$\frac{W_M}{L^3} = \rho\pi \int_{(z/L)_{cr}}^{0.5} \left(\frac{d}{2L}\right)^2 \mathrm{d}\left(\frac{z}{L}\right) = \rho\pi\left(\frac{2s}{\pi}\right)^{2/3} \int_{(z/L)_{cr}}^{0.5} \left(\frac{z}{L}\right)^{2/3} \mathrm{d}\left(\frac{z}{L}\right)$$

$$= \frac{3\rho\pi}{5}\left(\frac{2s}{\pi}\right)^{2/3}\left[0.5^{5/3} - \left(\frac{z}{L}\right)_{cr}^{5/3}\right] \tag{10.57b}$$

Doubling and adding equations (10.57a,b) and then substituting $(z/L)_{cr}$ from equation (10.56a) leads to

$$\frac{W}{L^3} = 2\pi\rho\left(\frac{Q}{3}\right)^{5/2}\left(\frac{2}{\pi}\right)^{3/2}s^{3/2} + \frac{6\rho\pi}{5}\left(\frac{2s}{\pi}\right)^{2/3}\left[0.5^{5/3} - \left(\frac{Q}{3}\right)^{5/2}\left(\frac{2}{\pi}\right)^{5/6}s^{5/6}\right] \quad (10.58a)$$

Setting $n = (1/\rho)(W/L^3)$ and $Q = \sqrt{3}$, equation (10.58a) becomes

$$n = 0.8084\,s^{3/2} + 2.79\,s^{2/3}(0.315 - 0.1739\,s^{5/6}) = 0.8789\,s^{2/3} - 0.3232\,s^{3/2} \quad (10.58b)$$

For $s \geq s_{FM}$, the diameter is constant and the beam's weight follows from shear only. This modifies the upper limit to the integral in equation (10.57a). Doubling to cover the full length,

$$\frac{W}{L^3} = 2\rho\pi\int_0^{0.5}\left(\frac{d}{2L}\right)^2 \mathrm{d}\left(\frac{z}{L}\right) = 2\rho\pi\frac{2Q}{3\pi}s\int_0^{0.5}\mathrm{d}\left(\frac{z}{L}\right) = \frac{2Q\rho s}{3} \quad (10.59a)$$

which, for similar substitutions n and Q, has a linear, non-dimensional form

$$n = 1.1547\,s \quad (10.59b)$$

Figure 10.18 plots equations (10.58b) and (10.59b) over five decades in ascending s to contain the three regions in which bending, combined shear and bending, and shear designs apply to this beam. As the beam's length increases, i.e. s decreases, the contribution from shear diminishes. Shear may neglected here for $s < 0.01$ where $z_{cr}/L < 0.035$. However, strictly speaking, in these optimum bending designs the ends of the beam should never start at the point given by equation (10.56a). In this regard shear always plays a role, e.g. with $s = 0.01$, equation (10.55b) gives $d/2L = 0.0606$ for the end sections. Given $s_{FM} = 2.041$, Figure 10.18 shows: (i) that both shear and bending control the optimum design with s-values in the region $0.01 < s < 2.041$, where equation (10.58b) supplies the normalised weight n; and (ii) shear controls the optimum design for $s > 2.041$, where equation (10.59b) will provide n.

10.5.2 Square Section, Side Length a

We have seen in § 10.5.1 that the shear force F and the bending moment M in Figure 10.16 may again be expressed as $|F| = P/2 =$ constant and $|M(z)| = Pz/2$, within the half length $0 \leq z \leq L/2$. When the section is square, its optimum depth a in bending is again found by combining $|M(z)|$ with equation (3.6b). Dividing this combination by $2L$ reveals the variation in half-depth with normalised length as

$$\left(\frac{a}{2L}\right)_{opt} = \left(\frac{3}{8}\right)^{1/3}\left(\frac{1}{\sigma_y}\frac{P}{L^2}\right)^{1/3}\left(\frac{z}{L}\right)^{1/3} = 0.7211s^{1/3}\left(\frac{z}{L}\right)^{1/3} \quad (10.60a)$$

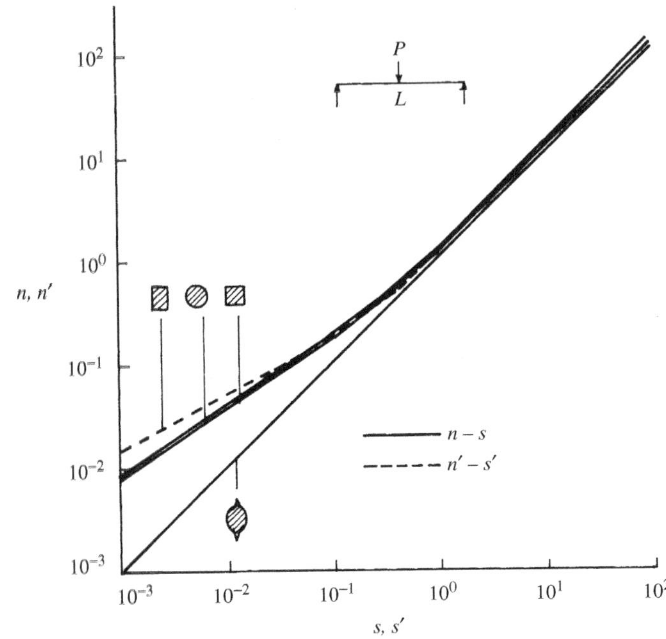

Figure 10.18 Optimum weights for simply supported beams with different cross-sections under central concentrated loading (note: $n = (1/\rho)(W/L^3)$; $s = (1/\sigma_y)(P/L^2)$; $s' = s/(\sigma_y/E)$; $n' = n/(\sigma_y/E)$)

in which $s = (1/\sigma_y)(P/L^2)$ is the normalised structural index, identified with short beams when s is large and vice versa.

For a transverse shear force $|F|$ applied to a square-section beam it was shown previously that equation (5.4c) gave the optimum depth. Dividing by $2L$ and writing the relationship between the tensile and shear yield stresses as $\sigma_y = Q\tau_y$ leads to

$$\left(\frac{a}{2L}\right)_{\text{opt}} = \left(\frac{3Q}{16}\right)^{1/2}\left(\frac{1}{\sigma_y}\frac{P}{L^2}\right)^{1/2} = 0.570\,s^{1/2} \tag{10.60b}$$

when $Q = \sqrt{3}$ from the von Mises yield criterion. For a given s, equation (10.60b) shows that the depth a is constant in shear while for bending equation (10.60a) shows that a varies with the length. Equating (10.60a,b) gives the critical position in the half-length where the depth from a shear design is displaced by the depth from a bending design:

$$\left(\frac{z}{L}\right)_{\text{cr}} = \left(\frac{3}{8}\right)^{1/2}\left(\frac{Q}{2}\right)^{3/2} s^{1/2} = 0.494\,s^{1/2} \tag{10.61a}$$

and since, at its limit $(z/L)_{\text{crit}} = 0.5$, equation (10.61a) gives the s-value dividing the shear and bending designs. Thus, for a pure shear design,

$$s_{FM} \geq \frac{2}{3}\left(\frac{2}{Q}\right)^3 \tag{10.61b}$$

Smaller s-values than s_{FM} imply a combined shear and bending design. For example, taking $Q = \sqrt{3}$ in equation (10.61b), we find this limit to be $s_{FM} = 1.0265$. If we select $s = 1$, for example, equation (10.61a) gives $(z/L)_{\mathrm{cr}} = 0.494$. For $0 \le z/L \le 0.494$, the semi-depth is held constant at $a/2L = 0.57$ according to equation (10.60b). This design is almost entirely shear-based except for the region $0.494 \le z/L \le 0.5$, where the depth increases slightly to a maximum $a/2L = 0.572$ in a bending design, given by equation (10.60a).

Figure 10.19 shows the beam's length profile according to the combined design, in which the left half's geometry derived above mirrors its adjacent half in both length and depth. Here the section dimension $a/2L$ identifies the normalised half-depth and the normalised length dimension z/L applies to the central axis for the beam's profile shown. This design applies to a short beam whose depth is similar to its length.

For $s \le s_{FM}$, the beam's weight follows from both shear and bending contributions. For an elemental length δz within their respective length regions, the weight is $\delta W = \rho a^2 \delta z$. Thus, for $0 \le z/L \le z_{\mathrm{cr}}/L$ (shear) we substitute $a/2L$ from equation (10.60b). The integration gives

$$\frac{W_F}{L^3} = 4\rho \int_0^{(z/L)_{\mathrm{cr}}} \left(\frac{a}{2L}\right)^2_{\mathrm{opt}} \mathrm{d}\left(\frac{z}{L}\right) = 4\rho \frac{3Qs}{16}\left(\frac{z}{L}\right)_{\mathrm{cr}} \tag{10.62a}$$

For $z_{\mathrm{cr}}/L \le z/L \le 0.5$ (bending) we substitute $a/2L$ from equation (10.60a) for the integration:

$$\frac{W_M}{L^3} = 4\rho \int_{(z/L)_{\mathrm{cr}}}^{0.5} \left(\frac{a}{2L}\right)^2_{\mathrm{opt}} \mathrm{d}\left(\frac{z}{L}\right) = 4\rho\left(\frac{3s}{8}\right)^{2/3} \int_{(z/L)_{\mathrm{cr}}}^{0.5} \left(\frac{z}{L}\right)^{2/3} \mathrm{d}\left(\frac{z}{L}\right)$$

$$= \frac{12\rho}{5}\left(\frac{3s}{8}\right)^{2/3}\left[0.5^{5/3} - \left(\frac{z}{L}\right)^{5/3}_{\mathrm{cr}}\right] \tag{10.62b}$$

Doubling and adding equations (10.62a,b) and then substituting $(z/L)_{\mathrm{cr}}$ from equation (10.61a) leads to

$$\frac{W}{L^3} = 3\rho\left(\frac{Q}{2}\right)^{5/2}\left(\frac{3}{8}\right)^{1/2} s^{3/2} + \frac{24}{5}\left(\frac{3s}{8}\right)^{2/3}\left[0.5^{5/3} - \left(\frac{3}{8}\right)^{5/6}\left(\frac{Q}{2}\right)^{5/2} s^{5/6}\right] \tag{10.63a}$$

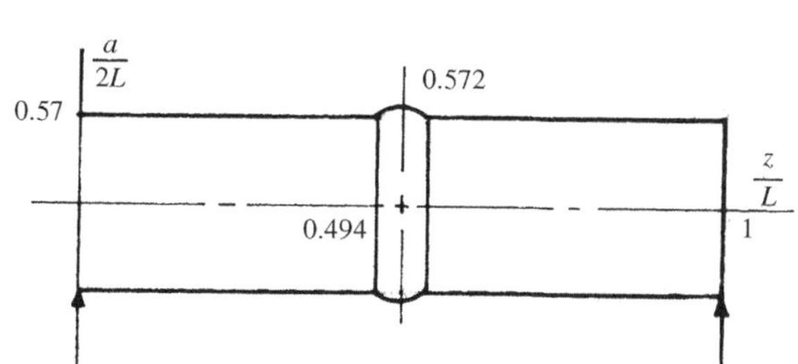

Figure 10.19 Length profile for a square-section, simply supported beam

Setting $n = (1/\rho)(W/L^3)$ and $Q = \sqrt{3}$, equation (10.63a) becomes

$$n = 1.2822\, s^{3/2} + 2.4961\, s^{2/3}(0.315 - 0.3082\, s^{5/6}) \tag{10.63b}$$

For $s \geq s_{FM}$ the depth is constant and the beam's weight follows from shear only:

$$\frac{W}{L^3} = 8\rho \int_0^{0.5} \left(\frac{a}{2L}\right)^2 \mathrm{d}\left(\frac{z}{L}\right) = 8\rho\left(\frac{3Q}{16}\right) s \int_0^{0.5} \mathrm{d}\left(\frac{z}{L}\right) = \frac{3Q\rho s}{4} \tag{10.64a}$$

which, for similar substitutions, has linear form

$$n = 1.299\, s \tag{10.64b}$$

Figure 10.18 plots equations (10.63b) and (10.64b) over five decades in ascending s to reveal the regions in which the bending, combined shear and bending, and all-shear designs apply. As the beam length increases, i.e. as s decreases, the contribution from shear diminishes. Shear may be neglected here for $s < 0.01$ where, from equation (10.61a) $z_{cr}/L < 0.0494$. However, strictly speaking, in these optimum bending designs the ends should never start at the point given by equation (10.60a). In this regard shear will always contribute, e.g. with $s = 0.01$, equation (10.60b) gives $a/2L = 0.057$ for the end-sections. Figure 10.18 applies to indices the regions: (i) $0.01 < s < 1.0265$, where both shear and bending control the optimum design and for which equation (10.63b) supplies the normalised weight n; and (ii) $s > 1.0265$, in which shear controls the optimum design for which equation (10.64b) provides n. Apart from the difference in s_{FM} between the circular and square sections, there is little difference between their optimum weights.

10.5.3 Rectangular Section, Breadth b, Depth d

When the section is rectangular, the optimum product $(bd)_{\text{opt}}$ in bending is found by combining $M(z)$ with equation (3.11). Dividing b by L and d by $2L$ reveals the variation in the product of these normalised dimensions with s and z/L:

$$\begin{aligned}\frac{b}{L}\left(\frac{d}{2L}\right)_{\text{opt}} &= 0.536\left(\frac{1}{\sigma_y}\frac{P}{L^2}\right)^{1/2}\left(\frac{\sigma_y}{E}\right)^{1/2}\left(\frac{z}{L}\right)^{1/2} \\ &= 0.536 s^{1/2}\left(\frac{\sigma_y}{E}\right)^{1/2}\left(\frac{z}{L}\right)^{1/2}\end{aligned} \tag{10.65a}$$

Equation (3.13b) shows that the breadth is constant:

$$\left(\frac{b}{L}\right)_{\text{opt}} = 0.383\,\frac{\sigma_y}{E} \tag{10.65b}$$

Hence, it follows, from dividing equations (10.65a,b) , that only the depth will vary with z/L:

$$\left(\frac{d}{2L}\right)_{\text{opt}} = 1.399 s^{1/2}\left(\frac{\sigma_y}{E}\right)^{-1/2}\left(\frac{z}{L}\right)^{1/2} \tag{10.65c}$$

For a transverse shear force F applied to a rectangular section beam it was shown previously that equation (5.3c) gave the optimum sectional dimensions within the product bd. Writing this product in its normalised form, i.e. $b/L \times d/2L$, and setting $\tau_y = \sigma_y/\sqrt{3}$ leads to

$$\frac{b}{L}\left(\frac{d}{2L}\right)_{\text{opt}} = 0.65\left(\frac{1}{\sigma_y}\frac{P}{L^2}\right) = 0.65\,s \tag{10.66a}$$

In both equations (10.65a) and (10.66a), $s = (1/\sigma_y)(P/L^2)$ is the normalised structural index. In equations (5.5b,c) the optimum sectional dimensions b and d were separated using the buckling coefficient K in equation (4.10). In the shear design it is found that both the optimum dimensions b and d remain constant in the length as follows:

$$\left(\frac{b}{L}\right)_{\text{opt}} = 0.811\left(\frac{\sigma_y}{E}\right)^{1/4}\left(\frac{9}{4K}\right)^{1/4} s^{1/2} \tag{10.66b}$$

$$\left(\frac{d}{2L}\right)_{\text{opt}} = 0.534\left(\frac{\sigma_y}{E}\right)^{-1/4}\left(\frac{9K}{4}\right)^{1/4} s^{1/2} \tag{10.66c}$$

in which the product of equations (10.66b,c) restores equation (10.66a). Equating (10.65a) and (10.66a) gives the critical position in the half-length where the product bd from a shear design is displaced by one from bending,

$$\left(\frac{z}{L}\right)_{\text{cr}} = 1.468\,s\left(\frac{\sigma_y}{E}\right)^{-1} \tag{10.67a}$$

and since, at its limit, $(z/L)_{\text{cr}} = 0.5$, equation (10.67a) gives the index s at the division between shear and bending designs,

$$s_{FM} = 0.3406\left(\frac{\sigma_y}{E}\right) \tag{10.67b}$$

When $s < s_{FM}$, the weight $\delta W = \rho(bd)\delta z$ is found by taking an elemental length δz to lie within the shear and bending regions. Thus, for $0 \leq z/L \leq z_{\text{cr}}/L$ (shear) we substitute from equation (10.66a) to form the weight integral

$$\frac{W_F}{L^3} = 2\rho\int_0^{(z/L)_{\text{cr}}} \frac{b}{L}\frac{d}{2L}\,\mathrm{d}\left(\frac{z}{L}\right) = 1.30\rho s\left(\frac{z}{L}\right)_{\text{cr}} \tag{10.68a}$$

For $z_{cr}/L \leq z/L \leq 1/2$ (bending), equation (10.65a) applies and the integration gives

$$\frac{W_M}{L^3} = 2\rho \times 0.536 \left(\frac{\sigma_y}{E}\right)^{1/2} s^{1/2} \int_{(z/L)_{cr}}^{1/2} \left(\frac{z}{L}\right)^{1/2} \mathrm{d}\left(\frac{z}{L}\right)$$

$$= 0.715\rho \left(\frac{\sigma_y}{E}\right)^{1/2} s^{1/2} \left[0.5^{3/2} - \left(\frac{z}{L}\right)_{cr}^{3/2}\right] \tag{10.68b}$$

Adding equations (10.68a,b) then doubling and substituting $(z/L)_{cr}$ from equation (10.67a) leads to

$$\frac{W}{L^3} = \frac{3.8168\rho s^2}{\sigma_y/E} + 1.4294\,\rho \left(\frac{\sigma_y}{E}\right)^{1/2} s^{1/2} \left[0.3536 - \frac{1.7786}{(\sigma_y/E)^{3/2}} s^{3/2}\right] \tag{10.69a}$$

Setting $n = (1/\rho)(W/L^3)$, equation (10.69a) simplifies to

$$n = 3.8168 \left(\frac{\sigma_y}{E}\right)^{-1} s^2 + 1.4294\, s^{1/2} \left(\frac{\sigma_y}{E}\right)^{1/2} \left[0.3536 - 1.7786\, s^{3/2} \left(\frac{\sigma_y}{E}\right)^{-3/2}\right] \tag{10.69b}$$

For $s \geq s_{FM}$, the beam's total weight follows from shear only. With $(z/L)_{cr} = 1/2$ as the upper limit in equation (10.68a), this gives

$$\frac{W_F}{L^3} = 2 \times 0.65\rho s = 1.3\rho s, \quad \Rightarrow \quad n = 1.30\, s \tag{10.69c}$$

for which the product bd is constant. Equations (10.69b,c) may be normalised further by writing $n' = n/(\sigma_y/E)$ and $s' = s/(\sigma_y/E)$. This gives

$$n' = 3.8168\, s'^2 + 1.4294\, s'^{1/2}(0.3536 - 1.7786\, s'^{3/2}) \tag{10.70a}$$

$$= 0.5054\, s'^{1/2} + 1.2745\, s'^{1/2} \quad \text{for} \quad s' \leq 0.3406$$

$$\text{and } n' = 1.30\, s' \quad \text{for} \quad s' > 0.3406 \tag{10.70b}$$

Figure 10.18 plots n' from equations (10.70a,b) over 5 decades in s' as a dashed line. This plot reveals, with increasing s', those regions in which the bending, combined shear and bending, and all-shear designs apply. We see that for $s'_{FM} = 0.3406$, both equations (10.70a,b) give $n' = 0.443$, at the division between the shear and combined shear and bending designs. As the beam length increases, i.e. as s decreases, the contribution from shear diminishes. The shear weight may be neglected here for $s < 0.01$ where $z_{cr}/L < 0.05$. However, strictly speaking, in these optimum bending designs the free end should never start at the point suggested by equation (10.65c). Here shear continues to play its part since the product $bd \neq 0$ for all z in equation (10.66c), as the following example will show.

Equations (10.65a,b) and (10.66b,c) show that for a given index s, only the depth varies within a region designed by bending. A similar observation was made for the solid circular and square sections above. The reader should not be deceived by Figure 10.18 into believing that the optimum rectangular section carries the greatest weight. In fact, with the revised coordinates s' and n' used in this figure, as with the cantilever plots in Figures 10.8 and 10.13, the rectangular section is the lightest. The following example will serve to demonstrate this (see also Examples 10.1 and 10.2).

Example 10.3 Find the optimum rectangular section dimensions, the self-weight and the load that can be carried by simply supported steel beams loaded at their centres (a) when $s' = 1$, $L = 100$ mm and (b) when $s' = 0.1$, $L = 1$ m. Take $\rho = 7860\ \text{kg/m}^3$, $E = 210$ GPa and $\sigma_y = 300$ MPa.

(a) With $s' = 1$ (i.e. 10^0 in Figure 10.18) this gives $n' = 1.3$. Alternatively, n' may be calculated directly from equation (10.70b) for an all-shear design ($s' \le s'_{FM}$) in which it follows from equation (10.67b) that $s'_{FM} = 0.3406$. The optimum breadth and depth follow from equations (10.66b,c):

$$\left(\frac{b}{L}\right)_{\text{opt}} = 0.811\left(\frac{\sigma_y}{E}\right)^{3/4}\left(\frac{9}{4K}\right)^{1/4} s'^{1/2} = 5.96 \times 10^{-3}\left(\frac{9}{4K}\right)^{1/4} s'^{1/2} \qquad \text{(i)}$$

$$\left(\frac{d}{2L}\right)_{\text{opt}} = 0.534\left(\frac{\sigma_y}{E}\right)^{1/4}\left(\frac{9K}{4}\right)^{1/4} s'^{1/2} = 0.1038\left(\frac{9K}{4}\right)^{1/4} s'^{1/2} \qquad \text{(ii)}$$

in which $K = K(d/L)$ and, thereby, equation (ii) requires an iterative solution to $d/2L$. For steel, we find that equation (ii) is satisfied by $d/2L \approx 0.2$, where $K = 5.5$ from Figure C.3. Equations (i) and (ii) then give the normalised breadth and depth as $(b/L)_{\text{opt}} = 0.0048$ and $(d/L)_{\text{opt}} \approx 0.39$, showing that the section's dimensions remain constant for the full length in Figure 10.20a. Given $L = 100$ mm, then $b = 0.48$ mm and $d = 39$ mm. These dimensions give the very low weight of this uniformly thin beam as

$$W = (0.48 \times 39 \times 100)10^{-9} \times (7860 \times 10^3) = 14.7\ \text{g}$$

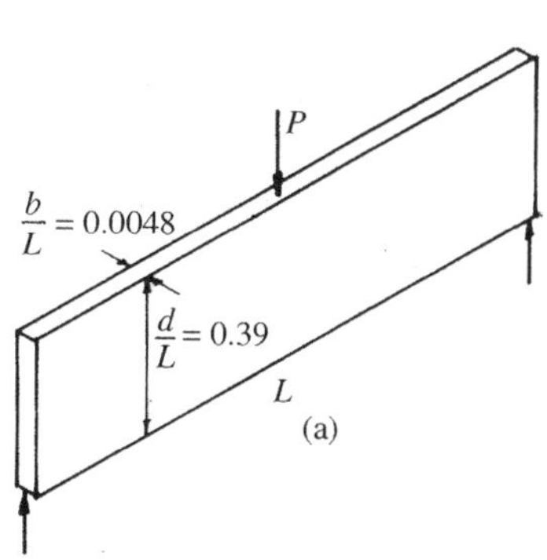

Figure 10.20a All shear design of a centrally loaded, simply supported beam

Alternatively, we may find the exact weight directly from $n' = 1.3$, as follows

$$n' = \frac{1}{\rho}\left(\frac{W}{L^3}\right)\Big/\left(\frac{\sigma_y}{E}\right) = 1.3, \quad \Rightarrow \quad W = 1.3\left(\frac{\sigma_y}{E}\right)\rho L^3 = 14.6 \text{ g}$$

In fact, $d = 38.7$ mm corresponds to the exact weight. The load supported for $s' = 1$ is

$$s' = \frac{1}{\sigma_y}\left(\frac{P}{L^2}\right)\Big/\left(\frac{\sigma_y}{E}\right) = 1, \quad \Rightarrow \quad P = \frac{(\sigma_y L)^2}{E} = 4.29 \text{ kN}$$

(b) With $s' = 0.1$, Figure 10.18 (or equation (10.70a)) gives $n' = 0.1726$. These parameters require a combined shear and bending design since $s' < s'_{FM}$, where, from equation (10.67b), $s'_{FM} = 0.3406$. The division between the two designs occurs at the critical length position given by equation (10.67a):

$$(z/L)_{cr} = 1.468s' = 0.147 \text{ (rounded)}$$

Hence, in the region $0 \le z/L \le 0.147$, the shear design applies. The optimum breadth and depth follow from setting $s' = 0.1$ in equations (i) and (ii) above:

$$\left(\frac{b}{L}\right)_{opt} = 0.2565\left(\frac{\sigma_y}{E}\right)^{3/4}\left(\frac{9}{4K}\right)^{1/4} \tag{iii}$$

$$\left(\frac{d}{2L}\right)_{opt} = 0.1688\left(\frac{\sigma_y}{E}\right)^{1/4}\left(\frac{9K}{4}\right)^{1/4} \tag{iv}$$

As $K = K(d/L)$, equation (iv) thereby requires an iterative solution to $d/2L$. For steel, we find that equation (iv) is satisfied for $d/L = 0.12$, where $K = 4.91$ from Figure C.3. Equation (iii) then gives the normalised breadth as $(b/L)_{opt} = 0.001552$ (say, 0.0016). Neither of these optimum section-dimensions vary with the length and so they describe a uniform rectangular section for a part of the length (0.147 m), as shown in Figure 10.20b. In the region $0.147 \le z/L \le 0.50$, where bending controls the design, the constant, optimum breadth follows from equation (10.65b),

$$\left(\frac{b}{L}\right)_{opt} = 0.383\frac{\sigma_y}{E} = 0.00055 \tag{v}$$

and, from equation (10.65c), the optimum depth will vary with z/L,

$$\left(\frac{d}{2L}\right)_{opt} = 1.399s'^{1/2}\left(\frac{z}{L}\right)^{1/2} = 0.4424\left(\frac{z}{L}\right)^{1/2} \tag{vi}$$

Equation (vi) gives particular depths: $d/L = 0.34$ for $z/L = 0.147$ and $d/L = 0.63$ for $z/L = 0.50$. Between these extremes the square root dependence $d \propto z^{1/2}$ gives

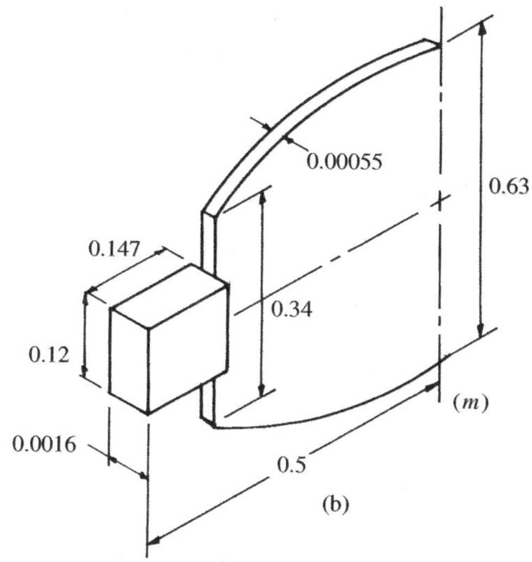

Figure 10.20b Optimum design of a centrally loaded, simply supported beam (one half shown)

a convex curvature to the beam edges in this region (see Figure 10.20b). For a beam length $L = 1$ m, these normalised section dimensions transpose directly to m. The weight of this length is found from $n' = 0.1726$, as follows:

$$n' = \frac{1}{\rho}\left(\frac{W}{L^3}\right) \Big/ \left(\frac{\sigma_y}{E}\right) = 0.1726, \quad \Rightarrow \quad W = 0.1726\frac{\sigma_y}{E}\rho L^3 = 1.938 \text{ kg}$$

The load supported follows from $s' = 0.1$, where

$$s' = \frac{1}{\sigma_y}\left(\frac{P}{L^2}\right) \Big/ \left(\frac{\sigma_y}{E}\right) = 0.1, \quad P = \frac{0.1(\sigma_y L)^2}{E} = 42.85 \text{ kN}$$

Similar comments apply here to those made earlier for the rectangular section designs in Examples 10.1 and 10.2. Again, low weighted beams found in (a) and (b) above are seen to be capable of supporting large central loads. However, the reader is warned of other influences that could arise from loading thin rectangular beam sections. For example, additional buckling modes, considered in § 10.7, would require the application of further failure criteria to ensure a safe design.

10.6 Simply Supported Beam with Uniformly Distributed Load

This standard case (Figure 10.21a) refers to beam loading that is uniformly distributed along the length. With a central symmetry in the F- and M-diagrams (see Figure 10.21b,c), it is only necessary to work on one half of the beam with z taken from the left end

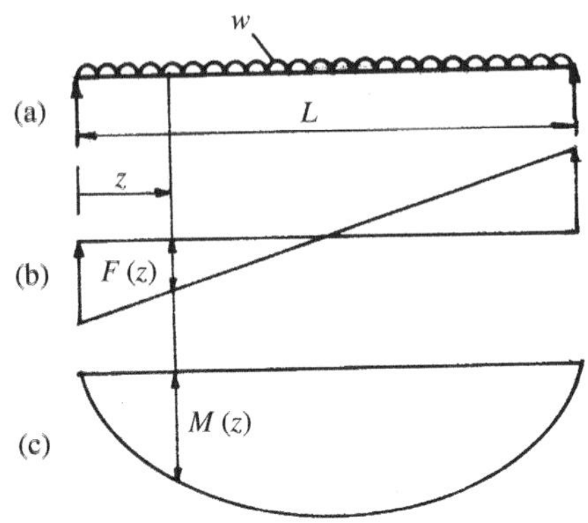

Figure 10.21 Force and moment diagrams for distributed loading

towards the centre. The total weight becomes twice that of one half. Within the half length $0 \leq z \leq L/2$, the ordinates in the force and moment diagrams in Figure 10.21b,c provide their magnitudes:

$$|F(z)| = w\left(\frac{L}{2} - z\right) = \frac{2F_{\max}}{L}\left(\frac{L}{2} - z\right) \tag{10.71a}$$

$$|M(z)| = \frac{wz}{2}(L - z) = \frac{F_{\max}z}{L}(L - z) \tag{10.71b}$$

where $F_{\max} = wL/2$ is the magnitude of each supporting reaction.

10.6.1 Circular Section, Diameter d

For a circular cross-section, the optimum diameter in bending is found by combining $|M(z)|$ in equation (10.71b) with equation (3.3b). Dividing by $2L$ reveals the variation in radius with length in the normalised form

$$\left(\frac{d}{2L}\right)_{\text{opt}} = \left(\frac{4}{\pi}\right)^{1/3}\left(\frac{1}{\sigma_y}\frac{F_{\max}}{L^2}\right)^{1/3}\left[\frac{z}{L}\left(1 - \frac{z}{L}\right)\right]^{1/3} = 1.084\,s^{1/3}\left[\frac{z}{L}\left(1 - \frac{z}{L}\right)\right]^{1/3} \tag{10.72a}$$

in which $s = (1/\sigma_y)(F_{\max}/L^2)$ is the normalised structural index. Consistent with all other standard beams, index s is to be identified with short lengths when large and vice versa.

For a transverse shear force applied to a circular section beam it was shown previously that equation (5.15b) gave the optimum diameter. Substituting from equation (10.71a), dividing by $2L$ and writing the relationship between the tensile and shear yield stresses as $\sigma_y = Q\tau_y$ leads to

$$\left(\frac{d}{2L}\right)_{\text{opt}} = \left(\frac{8Q}{3\pi}\right)^{1/2}\left(\frac{1}{\sigma_y}\frac{F_{\max}}{L^2}\right)^{1/2}\left(\frac{1}{2} - \frac{z}{L}\right)^{1/2} = 1.2125s^{1/2}\left(\frac{1}{2} - \frac{z}{L}\right)^{1/2} \tag{10.72b}$$

Table 10.1 Division between optimum shear and optimum bending designs

s	0.001	0.01	0.10	1.0	10	100
q_c	0.0151	0.0452	0.1197	0.2426	0.360	0.4306
n_F	3.4345×10^{-5}	9.967×10^{-4}	2.433×10^{-2}	0.4243	5.3208	56.623
n_M	5.3446×10^{-3}	2.4325×10^{-2}	0.1027	0.3529	0.9329	2.1756
n	0.0108	0.0506	0.2539	1.5545	12.508	117.6

where $Q = \sqrt{3}$ according to the von Mises yield criterion. For a given s, both equations (10.72a,b) show that the diameter varies with the length (see Example 10.3). Equating (10.72a,b) gives the critical position $q_c = (z/L)_{cr}$ in the half-length where the diameter from a shear design is displaced by the diameter from a bending design:

$$\frac{[q_c(1-q_c)]^{1/3}}{(1-2q_c)^{1/2}} = 0.791s^{1/6} \tag{10.72c}$$

Now equation (10.72c) gives $s = \infty$ for $q_c = 0.5$. Hence, the design of this beam can never be purely shear-based, unlike the previous beam loadings where a finite structural index s_{FM} divided the two design criteria. Here, equation (10.72c) has been employed to find q_c numerically in the half-region $0 \le q_c \le 1/2$ at each s-value. It follows that for the first sub-region of the half-length, $0 \le q \le q_c$, the design is to be shear-based and for the remaining, adjacent sub-region, $q_c < q \le 1/2$, the design is to be moment-based. The results are given in Table 10.1.

Correspondingly, the weight of the beam from each region becomes twice the sum of these two contributions. For an elemental length δz within each sub-region ($q < q_c$ and $q > q_c$) the weight is $\delta W = \rho(\pi/4)d^2\delta z$. Thus, for $0 \le q \le q_c$ (shear), we substitute $d/2L$ from equation (10.72b). The integration gives

$$\frac{W_F}{L^3} = \rho\pi \int_0^{(z/L)_{cr}} \left(\frac{d}{2L}\right)^2 \mathrm{d}\left(\frac{z}{L}\right) = \frac{8Qs}{3}\int_0^{q_c}\left(\frac{1}{2} - q\right)\mathrm{d}q = \frac{4Q}{3}\rho s q_c(1-q_c) \tag{10.73a}$$

For $q_c \le q \le 0.5$ (bending), we substitute $d/2L$ from equation (10.72a). The integration gives

$$\frac{W_M}{L^3} = \rho\pi \int_{q_c}^{1/2} \left(\frac{d}{2L}\right)^2 \mathrm{d}\left(\frac{z}{L}\right) = \rho(16\pi)^{1/3}s^{2/3}\int_{q_c}^{1/2}\left[\frac{z}{L}\left(1-\frac{z}{L}\right)\right]^{2/3}\mathrm{d}\left(\frac{z}{L}\right)$$

$$= 3.6905\rho s^{2/3}\int_{q_c}^{1/2}[q(1-q)]^{2/3}\,\mathrm{d}q \tag{10.73b}$$

Putting $n = (1/\rho)(W/L^3)$ and $Q = \sqrt{3}$, equations (10.73a,b) become

$$n_F = 2.3094 s q_c(1-q_c) \tag{10.74a}$$

$$n_M = 3.6905s^{2/3} \int_{q_c}^{0.5} [q(1-q)]^{2/3}\, dq \tag{10.74b}$$

Substituting q_c and s from Table 10.1, equations (10.74a,b) may be evaluated numerically to give the n_F and n_M values entered in that table. Finally, doubling and adding these n-values gives the total normalised weight as $n = 2(n_F + n_M)$. Figure 10.22 presents equations (10.74a,b) over five decades in s to reveal the regions in which bending and combined shear and bending designs apply (no all-shear design). As the beam length increases, i.e. as s decreases, the contribution from shear diminishes. Table 10.1 shows that the shear weight contribution (n_F) may be ignored here for $s < 0.01$ where $q_c(= z/L)_{cr} < 0.0452$. Strictly speaking, in applying these conditions to an optimum bending design, the free end should never start as a point given by equation (10.72a). In this regard shear will always contribute, e.g. with $s = 0.01$, equation (10.72b) gives $d/2L = 0.086$ for each end diameter. The influence of shear becomes greater for $s > 0.01$ and here bending will also continue to influence the design over the full range of s.

Example 10.4 Reading from Figure 10.22 (or Table 10.1), determine the radius profile and the weight of a 10 mm long, circular-section steel pin when $s = 10$. What is the distributed load being supported? Take $\sigma_y = 300$ MPa and $\rho = 7860$ kg/m^3.

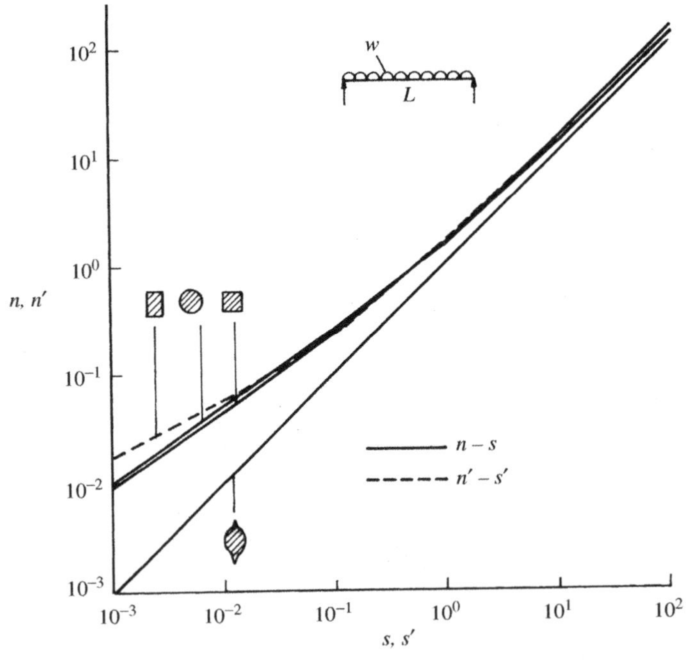

Figure 10.22 Optimum weights of simply supported beams with different cross-sections under distributed loading (note: $n = (1/\rho)(W/L^3)$; $s = (1/\sigma_y)(F_{max}/L^2)$ where $F_{max} = wL/2$; $n' = n/(\sigma_y/E)$ and $s' = s/(\sigma_y/E)$)

With $q_c = 0.36$, taken from Table 10.1, equation (10.72b) gives the variation in optimum diameter with length from a shear design as

$$\left(\frac{d}{2L}\right)_{\text{opt}} = 0.8574 \times 10^{1/2}(1-2q)^{1/2} = 2.7113(1-2q)^{1/2}, \quad \text{for } 0 \le q \le 0.36 \quad \text{(i)}$$

Particular values at the ends and mid-span for this region are: for $q = 0$, $d/2L = 2.711$; for $q = 0.18$, $d/2L = 2.170$; and for $q = 0.36$, $d/2L = 1.435$.

Equation (10.72a) gives the variation in optimum diameter with length from bending as

$$\left(\frac{d}{2L}\right)_{\text{opt}} = 1.084 \times 10^{1/3}[q(1-q)]^{1/3} = 2.335q(1-q)^{1/3}, \quad \text{for } 0.36 \le q \le 0.50 \quad \text{(ii)}$$

Particular values are: for $q = 0.36, d/2L = 1.432$; for $q = 0.43, d/2L = 1.461$; and for $q = 0.50, d/2L = 1.471$.

The half-length radius profile (with rounded values), given by equations (i) and (ii) above, is shown in Figure 10.23a. Figure 10.23b shows how the diameter varies over the full length. The greatest diameter is 54.23 mm, which lies at the ends. The smallest diameter is 28.6 mm and occurs at positions 3.6 mm in from each end. Figure 10.22 (or Table 10.1) gives the normalised weight of the pin as $n = 12.508$, from which

$$\frac{1}{\rho}\frac{W}{L^3} = 12.508, \quad \Rightarrow \quad W = 12.508 \times 7860 \times 0.01^3 = 0.0983 \text{ kg}$$

Given $F_{\max} = wL/2$, the distributed loading follows from the normalised structural index s as

$$\frac{1}{\sigma_y}\left(\frac{F_{\max}}{L^2}\right) = 10, \quad \Rightarrow \quad w = 20\sigma_y L = 20 \times 300 \times 0.01 = 60 \text{ MN/m (kN/mm)}$$

10.6.2 Square Section, Side Length a

For a square cross-section, the optimum depth a in bending is found by combining $M(z)$ from equation (10.71b) with equation (3.6b). Dividing a by $2L$ reveals the variation in

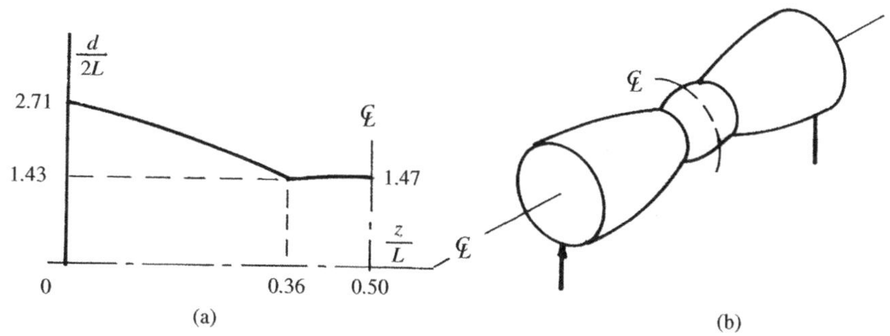

Figure 10.23 Radius and diameter profiles for a circular section

semi-depth with length in a normalised form:

$$\left(\frac{a}{2L}\right)_{\text{opt}} = \left(\frac{3}{4}\right)^{1/3} \left(\frac{1}{\sigma_y}\frac{F_{\max}}{L^2}\right)^{1/3} \left[\frac{z}{L}\left(1 - \frac{z}{L}\right)\right]^{1/3} = 0.9086 s^{1/3}[q(1-q)]^{1/3} \tag{10.75a}$$

in which $s = (1/\sigma_y)(F_{\max}/L^2)$ and $q = z/L$. Again, the index is to be identified with short beams when s is large and long beams when s is small.

With a transverse shear force applied to a square section beam it was shown previously that equation (5.4c) gave the optimum size. Substituting $|F(z)|$ from equation (10.71a), dividing by $2L$ and using the relationship between the tensile and shear yield stresses, $\sigma_y = Q\tau_y$, leads to

$$\left(\frac{a}{2L}\right)_{\text{opt}} = \left(\frac{3Q}{4}\right)^{1/2} \left(\frac{1}{\sigma_y}\frac{F_{\max}}{L^2}\right)^{1/2} \left(\frac{1}{2} - \frac{z}{L}\right)^{1/2} = 1.1398\, s^{1/2} \left(\frac{1}{2} - q\right)^{1/2} \tag{10.75b}$$

where $Q = \sqrt{3}$. For a given s, equations (10.75a,b) both show that the diameter varies with the length (see Example 10.4). Equating (10.75a,b) gives the critical position $q_c = (z/L)_{\text{cr}}$ in the half-length where a shear-based depth changes to a moment-based depth:

$$\frac{[q_c(1-q_c)]^{1/3}}{(1-2q_c)^{1/2}} = 0.8871\, s^{1/6} \tag{10.75c}$$

Equation (10.75c) shows that $s = \infty$ for $q_c = 0.5$, meaning that the section depth cannot be shear-based for beam's full length. Equation (10.75c) allows q_c to be found numerically in the half-length ($0 \le q_c \le 0.5$) for the s-values given in Table 10.2. For $0 \le q \le q_c$, the diameter is shear-based and for $q_c \le q \le 0.5$, the diameter is moment-based.

Correspondingly, the weight of the beam is the sum of the two regions within which the weight of an elemental length δz is $\delta W = \rho a^2 \delta z$. For the shear contribution to δW we substitute $a/2L$ from equation (10.75b). Integration gives

$$\frac{W_F}{L^3} = 4\rho \int_0^{(z/L)_{\text{cr}}} \left(\frac{a}{2L}\right)^2 \mathrm{d}\left(\frac{z}{L}\right) = 5.1962\rho s \int_0^{q_c} \left(\frac{1}{2} - q\right) \mathrm{d}q = 2.5981\rho s q_c(1-q_c) \tag{10.76a}$$

Table 10.2 Square section beam weights showing division between optimum shear and bending designs

s	0.001	0.01	0.10	1.0	10	100
q_c	0.0212	0.060	0.151	0.2811	0.3862	0.4454
n_F	5.3912×10^{-5}	1.4653×10^{-3}	3.313×10^{-2}	0.5250	6.1588	64.198
n_M	4.7629×10^{-3}	0.0215	0.0866	0.2734	0.6832	1.5158
n	9.6336×10^{-3}	0.0459	0.2394	1.5969	13.684	131.43

For the bending contribution to δW, substitute $a/2L$ from equation (10.75a). Integration gives

$$\frac{W_M}{L^3} = 4\rho \int_{z/L_{cr}}^{1/2} \left(\frac{a}{2L}\right)^2 \mathrm{d}\left(\frac{z}{L}\right) = 3.302\rho s^{2/3} \int_{q_c}^{1/2} \left[\frac{z}{L}\left(1 - \frac{z}{L}\right)\right]^{2/3} \mathrm{d}\left(\frac{z}{L}\right)$$

$$= 3.302\rho s^{2/3} \int_{q_c}^{1/2} [q(1-q)]^{2/3} \mathrm{d}q \tag{10.76b}$$

Setting $n = (1/\rho)(W/L^3)$, equations (10.76a,b) become

$$n_F = 2.5981 s q_c(1 - q_c) \tag{10.77a}$$

$$n_M = 3.302\, s^{2/3} \int_{q_c}^{0.5} [q(1-q)]^{2/3} \mathrm{d}q \tag{10.77b}$$

Substituting q_c and s from Table 10.2, equations (10.77a,b) were evaluated numerically to give the n_F and n_M values there entered. Finally, doubling and adding these gives the total normalised weights $n = 2(n_F + n_M)$ in the final row of Table 10.2. Figure 10.22 plots n versus s over five decades to reveal the regions in which these combined shear and bending designs apply. As the beam length increases, i.e. as s decreases, the contribution from shear diminishes. The shear weight may be neglected for $s < 0.01$, where $q_c (= z/L)_{cr} < 0.06$. However, in applying only the optimum bending design, the free end should never start as a point suggested by equation (10.75a). In this regard shear will always contribute, e.g. with $s = 0.01$, equation (10.75b) gives $a/2L = 0.0805$ for the each end-section. The influence of shear becomes greater for $s > 0.01$, but again bending continues to influence the design over the full range of s in Table 10.2.

Example 10.5 Reading from Figure 10.22 (or Table 10.2), determine the length profile and the weight of a 100 mm long, square-section steel beam when $s = 0.10$. What is the distributed load being supported? Take $\sigma_y = 300$ MPa and $\rho = 7860$ kg/m^3.

Firstly, we read $q_c = 0.151$ from Table 10.2. For the shear design, equation (10.75b) gives the variation in the optimum size of the square in the length region, $0 \le q \le 0.151$, as

$$\left(\frac{a}{2L}\right)_{\text{opt}} = \left(\frac{3\sqrt{3}}{4}\right)^{1/2} \times \left(\frac{0.10}{2}\right)^{1/2} (1 - 2q)^{1/2} = 0.255(1 - 2q)^{1/2} \tag{i}$$

Particular values at the ends and at mid-span for this region are: for $q = 0$, $a/2L = 0.255$; for $q = 0.0755$, $a/2L = 0.235$; and for $q = 0.151$, $a/2L = 0.213$.

From the bending design, equation (10.75a) gives the variation in the optimum size of the square in the length region: $0.151 \le q \le 0.50$, as

$$\left(\frac{a}{2L}\right)_{\text{opt}} = \left(\frac{3}{4}\right)^{1/3} \times 0.10^{1/3} [q(1-q)]^{1/3} = 0.422[q(1-q)]^{1/3} \tag{ii}$$

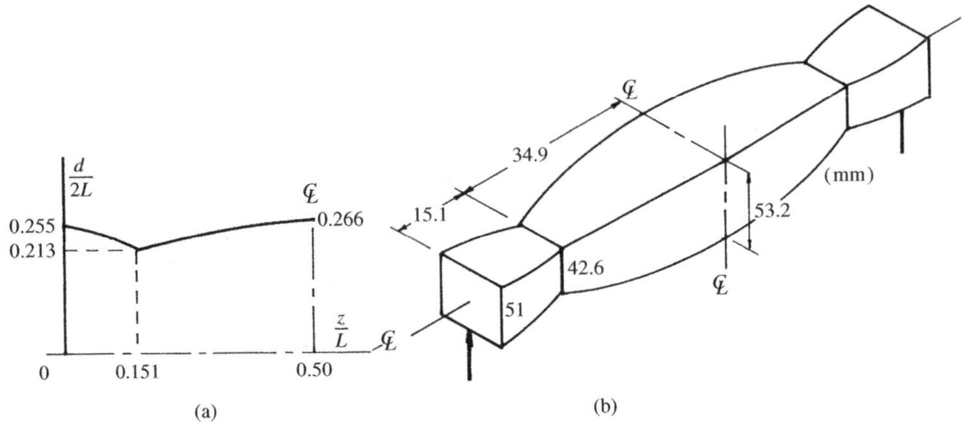

Figure 10.24 Depth and length profiles for a square section beam carrying distributed loading

Particular values are: for $q = 0.151$, $a/2L = 0.213$; for $q = 0.25$, $a/2L = 0.242$; and for $q = 0.50$, $a/2L = 0.266$.

The section profile for the half-length, given by equations (i) and (ii), is shown with these particular values indicated in Figure 10.24a. Figure 10.24b shows how the section profile would appear over the full length. The greatest, 53.2 mm, square section lies at the centre and the smallest, 42.6 mm, occurs at positions 15.1 mm in from each end.

Figure 10.22 (or Table 10.2) gives the beam's normalised weight as $n = 0.2394$, from which

$$\frac{1}{\rho}\frac{W}{L^3} = 0.2394, \quad \Rightarrow \quad W = 0.2394 \times 7860 \times 0.1^3 = 1.882 \text{ kg}$$

With $F_{\max} = wL/2$, the distributed load follows from the normalised structural index s as

$$\frac{1}{\sigma_y}\frac{F_{\max}}{L^2} = 0.10, \quad \Rightarrow \quad w = 0.20\sigma_y L = 0.20 \times 300 \times 0.1 = 6 \text{ MN/m (kN/mm)}$$

10.6.3 Rectangular Section, Breadth b, Depth d

For a rectangular cross-section, the optimum product $(bd)_{\text{opt}}$ in bending is found by combining equation (3.11) with $|M(z)|$ in equation (10.71b). Dividing b by L and d by $2L$ reveals the variation in this product with length in normalised form:

$$\frac{b}{L}\left(\frac{d}{2L}\right)_{\text{opt}} = 0.758\left(\frac{1}{\sigma_y}\frac{F_{\max}}{L^2}\right)^{1/2}\left(\frac{\sigma_y}{E}\right)^{1/2}\left[\frac{z}{L}\left(1-\frac{z}{L}\right)\right]^{1/2}$$

$$= 0.758\, s^{1/2}\left(\frac{\sigma_y}{E}\right)^{1/2}[q(1-q)]^{1/2} \qquad (10.78\text{a})$$

in which $q = z/L$ and the index $s = (1/\sigma_y)(F_{max}/L^2)$ identifies short beams when s is large and vice versa. Equation (3.13b) shows that the breadth in bending is constant at

$$\left(\frac{b}{L}\right)_{opt} = 0.383\frac{\sigma_y}{E} \tag{10.78b}$$

Dividing equations (10.78a,b), the depth is seen to vary with $q(= z/L)$:

$$\left(\frac{d}{2L}\right)_{opt} = 1.979\, s^{1/2}\left(\frac{\sigma_y}{E}\right)^{-1/2}[q(1-q)]^{1/2} \tag{10.78c}$$

Equation (5.3c) gives the required product $(bd)_{opt}$ for a rectangular-section beam under a transverse shear force. Taking $Q = \sqrt{3}$ and combining equations (5.3c) and (10.71a) leads to

$$\frac{b}{L}\left(\frac{d}{2L}\right)_{opt} = \frac{3\sqrt{3}}{2}\frac{1}{\sigma_y}\frac{F_{max}}{L^2}\left(\frac{1}{2}-\frac{z}{L}\right) = 2.5981s\left(\frac{1}{2}-q\right) \tag{10.79a}$$

We may separate b from d using the relationship involving the buckling coefficient K in equation (5.5a). In this case, the optimum dimensions b and d are found from equations (5.5b,c)

$$\left(\frac{b}{L}\right)_{opt} = 1.147\left(\frac{\sigma_y}{E}\right)^{1/4}\left(\frac{9}{4K}\right)^{1/4} s^{1/2}(1-2q)^{1/2} \tag{10.79b}$$

$$\left(\frac{d}{2L}\right)_{opt} = 0.755\left(\frac{\sigma_y}{E}\right)^{-1/4}\left(\frac{9K}{4}\right)^{1/4} s^{1/2}(1-2q)^{1/2} \tag{10.79c}$$

in which the product of equations (10.79b,c) agrees with equation (10.79a). For a given s, equations (10.79b,c) show that dimensions b and d vary with the length parameter $q = z/L$ (similar to the square section in Example 10.5). Equating (10.78a) and (10.79a) gives the critical position $q_c = (z/L)_{cr}$ in the half-length ($0 \le q \le 1/2$), where the optimum, shear-based product $(bd)_{opt}$ changes to a moment-based product. That is, for the region $0 \le q \le q_c$, the section size is shear-based, and for $q_c \le q \le 1/2$, the section size is moment-based:

$$\frac{[q_c(1-q_c)]^{1/2}}{1-2q_c} = 1.7138\, s'^{1/2} \tag{10.80}$$

where $s' = s/(\sigma_y/E)$. The dimensions b and d cannot be entirely shear-based over the full length, because $s' = \infty$ for $q_c = 0.5$ in equation (10.80). Recall, that this was also shown for short beams, with circular and square sections, under a similar distributed loading. Thus, both shear and bending, in which q_c divides their designs, contribute to the rectangular section's optimum shape. Equation (10.80) enables q_c in the half-length

Table 10.3 Division between optimum shear and bending designs for a rectangular section beam

s'	0.001	0.01	0.10	1.0	10	100
q_c	0.0029	0.027	0.161	0.360	0.454	0.4854
$n_F{}'$	7.5126×10^{-6}	6.8255×10^{-4}	3.510×10^{-2}	0.5986	6.4403	64.897
$n_M{}'$	9.1757×10^{-3}	0.02862	0.0735	0.1045	0.1101	0.1107
n'	0.01837	0.0586	0.2172	1.4062	13.101	130.02

($0 < q_c < ½$), to be found numerically for the s'-values given in Table 10.3. For $0 \le q \le q_c$ the depth is shear-based, and for $q_c \le q \le ½$ the depth is moment-based.

Taking an elemental length δz within each of these regions, the weight is $\delta W = \rho(bd)\delta z$. For the shear contribution to this weight, we substitute the product term bd from equation (10.79a) into the δW expression, whereupon the integration gives

$$\frac{W_F}{L^3} = 2\rho \int_0^{(z/L)_{cr}} \frac{b}{L}\frac{d}{2L} \mathrm{d}\left(\frac{z}{L}\right) = 5.1962\rho s \int_0^{q_c} \left(\frac{1}{2} - q\right) \mathrm{d}q$$

or, in the equivalent non-dimensional form,

$$n_F = 2.5981 s q_c(1 - q_c) \tag{10.81a}$$

where $n_F = (1/\rho)(W_F/L^3)$. For $q_c \le q \le 0.5$ (bending), we substitute bd from equation (10.78a) to provide the elemental weight. The integration gives

$$\frac{W_M}{L^3} = 2\rho \int_{z/L_{cr}}^{1/2} \frac{b}{L}\frac{d}{2L} \mathrm{d}\left(\frac{z}{L}\right) = 1.516\,\rho s^{1/2}\left(\frac{\sigma_y}{E}\right)^{1/2} \int_{z/L_{cr}}^{1/2} \left[\frac{z}{L}\left(1 - \frac{z}{L}\right)\right]^{1/2} \mathrm{d}\left(\frac{z}{L}\right)$$

$$n_M = 1.516 s^{1/2}\left(\frac{\sigma_y}{E}\right)^{1/2} \int_{q_c}^{1/2} [q(1-q)]^{1/2} \mathrm{d}q \tag{10.81}$$

where $n_M = (1/\rho)(W_M/L^3)$. Setting $s' = s/(\sigma_y/E)$ and $n' = n/(\sigma_y/E)$, equations (10.81a,b) appear in terms of these parameters as

$$n'_F = 2.5981 s' q_c(1 - q_c) \tag{10.82a}$$

$$n'_M = 1.516 s'^{1/2} \int_{q_c}^{0.5} [q(1-q)]^{1/2} \mathrm{d}q \tag{10.82b}$$

The total weight of the beam is summed from within each region. Substituting q_c and s' from Table 10.3, equations (10.82a,b) have been evaluated numerically to give the n'_F

and n'_M values entered therein. Finally, adding n'_F and n'_M and then doubling the result gives the total normalised weight n'. Figure 10.22 plots n' versus s' over five decades to reveal the relative amounts by which shear and bending contribute to the weight of this optimum design. This shows that as the beam's length increases, i.e. as s' decreases, the contribution from shear diminishes. Shear may be neglected for $s' < 0.01$ where $q_c(= z/L)_{cr} < 0.027$. Again, in applying these conditions to an optimum bending design, the free-end depth should never start as a point given by equation (10.78c). In this regard shear will always contribute in that equations (10.79b,c) give $b/L > 0$ and $d/2L > 0$ at the end where $q = 0$. The end-breadth should be selected as the greater of equations (10.78b) and (10.79b). Table 10.3 shows that the influence of shear becomes greater for $s' > 0.01$ and is dominant for $s' > 100$. The bending contribution to the weight (n'_M) may be ignored for $s' \geq 10$.

Example 10.6 Reading from Figure 10.22 (or Table 10.3), determine the length profile and the weight of a 1 m long, simply supported steel beam with rectangular section at a structural index $s' = 1.0$. What is the distributed load being supported? For steel take $\sigma_y = 300$ MPa, $E = 210$ GPa and $\rho = 7860$ kg/m^3.

With $q_c = 0.36$, taken from Table 10.3, the shear design applies to the region $0 \leq q \leq 0.36$ where the optimum breadth and depth follow from equations (10.79b,c). The latter appear in terms of the modified structural index $s' = s/(\sigma_y/E)$ to give the optimum breadth b as

$$\left(\frac{b}{L}\right)_{\text{opt}} = 1.147\left(\frac{\sigma_y}{E}\right)^{3/4}\left(\frac{9}{4K}\right)^{1/4} s'^{1/2}(1-2q)^{1/2} \tag{i}$$

and the optimum depth d as

$$\left(\frac{d}{2L}\right)_{\text{opt}} = 0.755\left(\frac{\sigma_y}{E}\right)^{1/4}\left(\frac{9K}{4}\right)^{1/4} s'^{1/2}(1-2q)^{1/2} \tag{ii}$$

Equations (i) and (ii) describe the variation in optimum breadth and depth with length. Particular values for b/L and $d/2L$ at the end $q = 0$ are based upon the required match with the buckling coefficient from Figure C.3. Thus, when $s' = 1$, equations (i) and (ii) become

$$\left(\frac{d}{2L}\right)_{\text{opt}} = 0.755\left(\frac{9K}{4}\right)^{1/4}\left(\frac{\sigma_y}{E}\right)^{1/4}, \quad \left(\frac{b}{L}\right)_{\text{opt}} = 1.147\left(\frac{9}{4K}\right)^{1/4}\left(\frac{\sigma_y}{E}\right)^{3/4} \tag{iii, iv}$$

Since $K = K(d/L)$ we must solve for d/L within equation (iii) iteratively. For example, if we assume $d/L = 0.5$, then the buckling coefficient $K = 6.05$ is read from Figure C.3 for a plate with unsupported edges under shear. Substituting $K = 6.05$, $\sigma_y = 300$ MPa and $E = 210$ GPa into the right-hand side of equation (iii) gives $d/2L = 0.281$, which disagrees with the assumed value of d/L. Fortunately, equation (iii) soon converges to the correct (rounded) solution $d/L = 0.56$. Equation (iv) may then be employed to give $b/L = 6.6 \times 10^{-3}$.

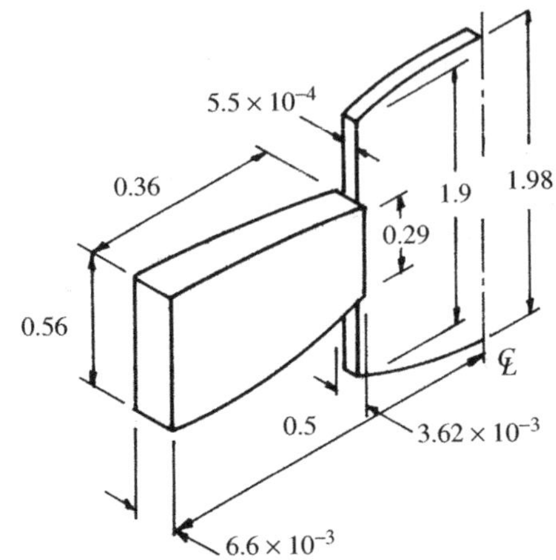

Figure 10.25 Breadth and depth profiles within the half-length for a rectangular-section, simply supported beam carrying uniformly distributed loading

At the other end of the shear region, where $q = 0.36$, equations (i) and (ii) become

$$\left(\frac{b}{L}\right)_{\text{opt}} = 1.147 \times 0.529 \left(\frac{9}{4K}\right)^{1/4} \left(\frac{\sigma_y}{E}\right)^{3/4}, \tag{v}$$

$$\left(\frac{d}{2L}\right)_{\text{opt}} = 0.755 \times 0.529 \left(\frac{9K}{4}\right)^{1/4} \left(\frac{\sigma_y}{E}\right)^{1/4} \tag{vi}$$

for which the solutions are $b/L = 3.62 \times 10^{-3}$, $d/L = 0.29$ with $K = 5.21$ from Figure C.3. Equations (iii)–(vi) describe the shape that the shear design imposes upon the beam for the region $0 \le q \le 0.36$, within the half-length, as shown in Figure 10.25. Both the breadth and the depth's dependencies upon $\sqrt{(1-2q)}$ appear within their convex edge curvatures shown.

Within the remaining region ($0.36 \le q \le 0.5$) of the half-length, bending controls the design. Equation (10.78b) shows that the breadth remains constant at

$$\left(\frac{b}{L}\right)_{\text{opt}} = 0.383\frac{\sigma_y}{E} = 5.471 \times 10^{-4} \tag{vii}$$

Equation (10.78c) gives the variation in depth with length for a bending design

$$\left(\frac{d}{2L}\right)_{\text{opt}} = 1.979 s'^{1/2} [q(1-q)]^{1/2} \tag{viii}$$

Particular values, found from equation (viii), are: for $q = 0.36$, $d/2L = 0.95$; for $q = 0.43$, $d/2L = 0.98$; and for $q = 0.50$, $d/2L = 0.99$. The section profile, described by equations (vii) and (viii), for this region, is shown in Figure 10.25. With $L = 1$ m length, the particular $d/2L$ values given above equate to the section's half-depths in metres directly. Of course, it may not be feasible to employ this stepped shape where the thinner, deeper central region is at risk of buckling as a wide strut. Here, additional flexural and local buckling criteria would be required to limit the section dimensions further (see § 10.7.6). Alternatively, as discussed earlier in Example 10.2, we could apply the wide strut design to optimise the reinforcement applied to the 'shear depth' that the present analysis has provided (see Chapter 8). Reinforcement would be a more appropriate weight increase to consider where the shear loading is more severe under concentrated loading. Adding weight to the beam through reinforcement is less essential when, as here, the applied loading is uniformly distributed. The basic shape of beam, shown in Figure 10.25, carries the minimum weight possible in its support of the distributed load. It is a consequence of the failure criteria that have been applied to its design. Accounting for other instabilities (see § 10.7) would alter this shape and weight.. The optimum design given carries a minimum weight which follows from $n' = 1.4062$ (see Table 10.3):

$$n' = \frac{1}{\rho}\left(\frac{W}{L^3}\right) \Big/ \left(\frac{\sigma_y}{E}\right) = 1.4062 \qquad \text{(ix)}$$

Equation (ix) gives the weight W of an $L = 1$ m long beam as

$$W = 1.4062\rho L^3 \frac{\sigma_y}{E} = 1.4062 \times 7860 \times 1^3 \frac{300 \times 10^6}{210 \times 10^9} = 15.79 \text{ kg}$$

The load supported follows from $s' = 1$, where

$$s' = \frac{1}{\sigma_y}\left(\frac{F_{\max}}{L^2}\right) \Big/ \left(\frac{\sigma_y}{E}\right) = 1 \qquad \text{(x)}$$

Substituting $F_{\max} = wL/2$ into equation (x) gives the distributed loading w as

$$w = 2L\frac{\sigma_y^2}{E} = 2 \times 1 \times \frac{(300 \times 10^6)^2}{210 \times 10^9} = 857.14 \text{ kN/m} \qquad \text{(xi)}$$

The exercises given at the end of this chapter have been selected to demonstrate how to use the weight versus structural index plots derived here for standard beam loadings. A number of exercises, which require solutions in a similar form to the worked Examples 10.1–10.6 given here, are intended to raise familiarity with the optimisation process. In practice, the question of how to achieve an optimum design would not be posed in quite the same way. Normally, one would be required to optimise the section dimensions to minimise weight given that a certain beam section has to support a specified load in a given length. This is a common design requirement, for which additional exercises are given.

10.7 Additional Failure Criteria

Our examples on the optimum design of rectangular beam sections have promoted, invariably, the thin, deep section that is known from elementary bending theory to provide the required strength at low weight. When coupled with a shear design, the interface with a bending design is stepped in both the breadth and the depth, though their products are equal at this position (see Examples 10.1 and 10.3). Each solution given is one of a number of possibilities that depend upon the failure criteria adopted. Those that were employed limited the bending and shear stress to the design value in the region of the highest bending moment and shear force. Under the latter a shear buckling mode was also considered at the position of maximum shear stress along the beam's neutral axis. An equivalent compression at this position would ripple the sheet in a 45° direction as shown in Figure 10.26a. Alternative design criteria should admit the possibility of torsion, flexural, lateral and local buckling. Figures 10.26b–e respectively show how each effect would appear in a thin, rectangular section. It is convenient to examine which of these are the most likely modes for the concentrated and distributed loadings into which our standard beam was divided.

10.7.1 Inclined End Force

In the case of a concentrated load P this need only be offset or inclined slightly to the mid-vertical plane to activate one or more of these additional buckling modes. Figure 10.27a shows the first of these in which an inclined end force P is applied to a cantilever.

The force P is resolved into its vertical component $P_V = P\cos\alpha$ (Figure 10.27b) and a horizontal component $P_H = P\sin\alpha$ that may be replaced by the statically equivalent system consisting of the force P_H and an initial torque T_0, both acting at the centroid

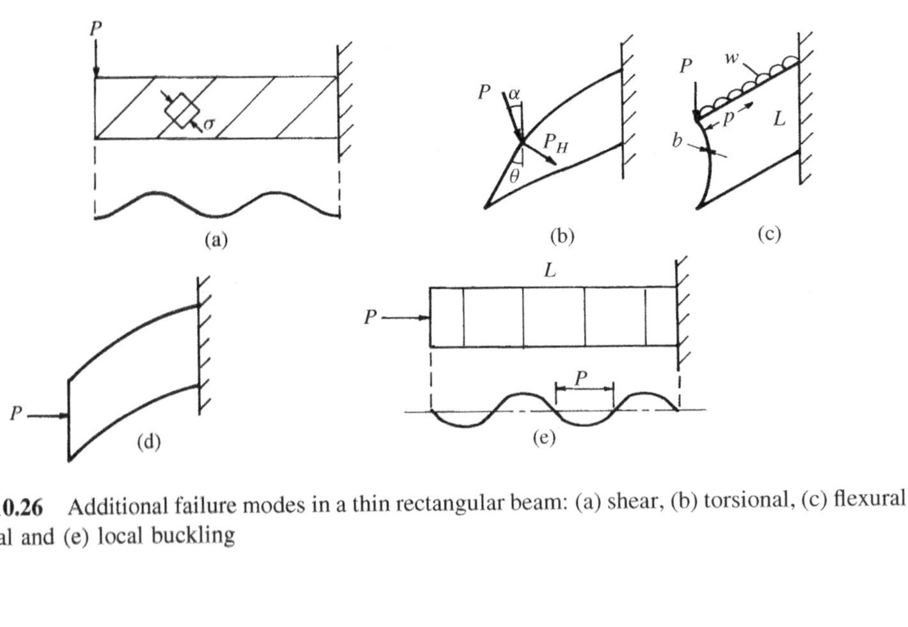

Figure 10.26 Additional failure modes in a thin rectangular beam: (a) shear, (b) torsional, (c) flexural, (d) lateral and (e) local buckling

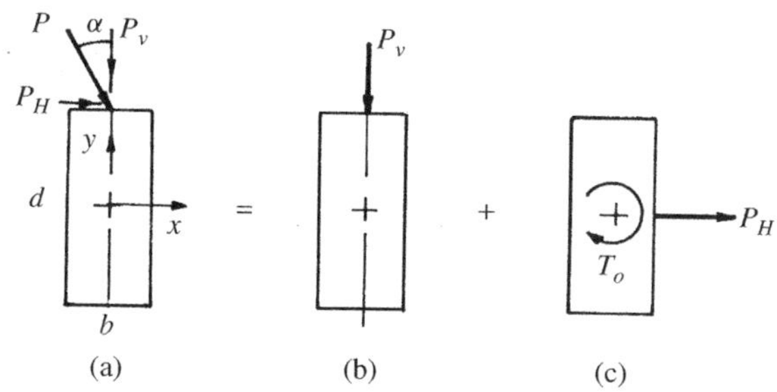

Figure 10.27 Inclined end-force for a rectangular section cantilever

(see Figure 10.27c),

$$T_0 = P_H \times \frac{d}{2} = \frac{Pd}{2}\sin\alpha$$

As the free-end rotates with an angular twist θ (as shown in Figure 10.26b) the torque is modified by the current tangential force $P\sin(\alpha+\theta)$ to

$$T = \frac{Pd}{2}\sin(\alpha+\theta) = \frac{T_0\sin(\alpha+\theta)}{\sin\alpha} \tag{10.83}$$

The maximum shear stress at the section's edges and the angular twist within the thin strip are given by the standard formulae [1]:

$$\tau_{\max} = \frac{3T}{db^2}, \quad \theta = \frac{3TL}{Gdb^3} \tag{10.84a,b}$$

Combining equations (10.83) and (10.84b) allows θ to be found as the solution to

$$\theta = \frac{3T_0L}{Gdb^3} \times \frac{\sin(\alpha+\theta)}{\sin\alpha} \tag{10.84c}$$

Equation (10.84a) is limited to a design stress τ_y and equated to the critical shear stress for the plate buckling mode shown in Figure 10.26a. Combining with equation (4.10), we have a new torsional buckling criterion from which b and d may be optimised:

$$\tau_y = \frac{3T}{db^2} = KE\left(\frac{b}{d}\right)^2 \tag{10.85}$$

where $K = K(d/L)$ is read from Figure C.3 for the aspect ratio d/L. Equation (10.85) leads to the optimum dimensions

$$b_{\text{opt}} = \left(\frac{9T^2}{KE\tau_y}\right)^{1/6}, \quad d_{\text{opt}} = (3K)^{1/3}\left(\frac{ET}{\tau_y^2}\right)^{1/3} \tag{10.86a,b}$$

where T and θ are given by equations (10.83) and (10.84c) and $E = 2G(1+\nu)$ applies.

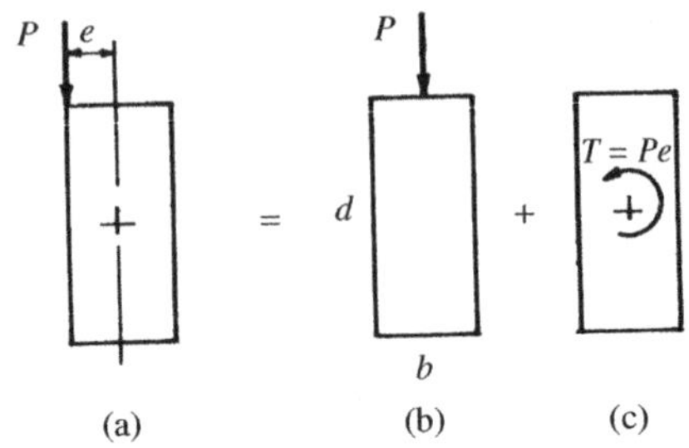

Figure 10.28 Off-set force applied to the end of a rectangular section cantilever

10.7.2 Eccentric End-Force

Figure 10.28a shows a vertical end-force P offset by an amount e from the beam's vertical, mid-plane. In the equivalent system this force is displaced to the mid-plane (Figure 10.28b) where an axial torque $T = Pe$ applies to the centroidal axis (Figure 10.28c). Equations (10.84a,b) again apply. Taking the worst case to be when the end force lies at the plate edge, we may substitute $T = Pb/2$ into equation (10.84a) to give the maximum shear stress arising from torsion:

$$\tau_{\max} = \frac{3P}{2db} \tag{10.87a}$$

Applying similar design criteria, which led to equation (10.85), we now have, for shear buckling,

$$\tau_y = \frac{3P}{2db} = KE\left(\frac{b}{d}\right)^2 \tag{10.87b}$$

The section's optimum dimensions are found from equation (10.87b) to be

$$b_{\text{opt}} = \left(\frac{3}{2}\right)^{1/2}\left(\frac{P^2}{KE\tau_y}\right)^{1/4}, \quad d_{\text{opt}} = \left(\frac{3}{2}\right)^{1/2}\left(\frac{KEP^2}{\tau_y^3}\right)^{1/4} \tag{10.88a,b}$$

in which the limiting design stress τ_y may be set to correspond with a given offset strain. This design stress allows a tangent modulus E_T to replace the elastic modulus E where plate buckling involves plasticity.

10.7.3 Transverse Shear Effects

The previous two solutions have ignored the transverse shear stress distribution arising from the vertical force acting within the mid-vertical plane in each statically equivalent system (see Figures 10.27b and 10.28b).

(a) Inclined End-Force

Transverse flexural shear stress arises from the vertical component P_V of the inclined force, shown in Figure 10.27b. Assuming a small angular twist, the maximum transverse shear stress across the neutral plane can never exceed

$$\tau_{\max} = \frac{3P_V}{2db} = \frac{3P\cos\alpha}{2db} \tag{10.89a}$$

Equation (10.89a) assumes that a small angular twist arises from the horizontal force component P_H in Figure 10.27a, from which the axial torque T_0 (Figure10.27c) is written as

$$T_0 = \frac{P_H d}{2} = \frac{Pd}{2}\sin\alpha \tag{10.89b}$$

Hence, the maximum shear stress from torsion follows from equations (10.84a) and (10.89b):

$$\tau_{\max} = \frac{3P\sin\alpha}{2db} \tag{10.89c}$$

Adding equation (10.89c) to (10.89a) gives the total shear stress at the edges of the neutral plane:

$$\sum \tau_{\max} = \frac{3P}{2db}(\sin\alpha + \cos\alpha) \tag{10.90a}$$

The design criterion corresponding to equation (10.85) is now modified,

$$\tau_y = \frac{3P}{2db}(\sin\alpha + \cos\alpha) = KE\left(\frac{b}{d}\right)^2 \tag{10.99b}$$

which provides the optimum dimensions for the section,

$$b_{opt} = \left[\frac{3}{2}(\sin\alpha + \cos\alpha)\right]^{1/2}\left(\frac{P^2}{KE\tau_y}\right)^{1/4},$$

$$d_{opt} = \left[\frac{3}{2}(\sin\alpha + \cos\alpha)\right]^{1/2}\left(\frac{KEP^2}{\tau_y^3}\right)^{1/4} \tag{10.91a,b}$$

In general, equations (10.91a,b) will increase the section dimensions compared to a design based only upon torsion (i.e. from equations (10.86a,b)).

(b) Eccentric End-Load

A similar approach to that in (a) above can be applied to the vertical force P acting at the centroid within the equivalent system, shown in Figure 10.28b,c. The maximum in the parabolic shear stress distribution lies across the mid-horizontal (neutral) plane, the magnitude being given by equation (5.3b):

$$\tau_{\max} = \frac{3P}{2db} \tag{10.92a}$$

If we now add the two shear stress maxima from torsion and shear (equations (10.87a) and (10.92a), respectively), this gives the greatest shear stress at one edge of the neutral plane:

$$\Sigma\tau_{\text{max}} = \frac{3P}{db} \tag{10.92b}$$

The buckling criterion within equation (10.87b) is thus modified to

$$\tau_y = \frac{3P}{db} = KE\left(\frac{b}{d}\right)^2 \tag{10.93a}$$

which increases the section's optimum dimensions,

$$b_{\text{opt}} = \sqrt{3}\left(\frac{P^2}{KE\tau_y}\right)^{1/4}, \qquad d_{\text{opt}} = \sqrt{3}\left(\frac{KEP^2}{\tau_y^3}\right)^{1/4} \tag{10.93b,c}$$

whose product $(bd)_{\text{opt}} = 3P/\tau_y$ is consistent with equation (10.93a). Equations (10.93b,c) have increased the section dimensions by a factor of $\sqrt{2}$ compared to the design based only upon torsion (see equations (10.88a,b)).

10.7.4 Lateral Buckling

In long, thin rectangular sections, the possibility of lateral buckling (see Figure 10.26d) arising from the horizontal force component P_H, shown in Figure 10.27a, should also be considered. Moreover, two-axis bending occurs from the vertical and horizontal components, P_V and P_H, of a concentrated force P applied with an inclination α to the mid-vertical plane. Here we shall equate the critical, lateral buckling stress to the net stress found from two-axis bending and then set these both to our limiting design stress σ_y. The buckling stress, equation (3.10), is [2]

$$\sigma = 2.61E\left(\frac{b}{L}\right) \tag{10.94}$$

The force components P_H and P_V in Figure 10.27a bend the beam about the section's two centroidal axes x and y, in which the maximum moments are

$$M_x = P_V L = PL\cos\alpha, \quad M_y = P_H L = PL\sin\alpha \tag{10.95a,b}$$

Each moment produces a bending stress that alternates in sign on either side of their reference axes. It follows that the greatest net stress arises at points where the stress from each bending moment are additive in tension and in compression. Such points lie on opposite corners of the section's diagonal. Their magnitudes are

$$\sigma_{\text{net}} = \frac{M_x d/2}{I_x} + \frac{M_y b/2}{I_y} \tag{10.96a}$$

and, substituting $I_x = bd^3/12$ and $I_y = db^3/12$, equation (10.96a) gives

$$\sigma_{\text{net}} = \frac{6}{bd}\left(\frac{M_y}{d} + \frac{M_x}{b}\right) \tag{10.96b}$$

Equating (10.94) and (10.96b) and setting the limiting net stress as $\sigma_{\text{net}} \leq \sigma_y$ leads to the required combination of these design criteria:

$$\sigma_y = 2.61E\left(\frac{b}{L}\right) = \frac{6}{bd}\left(\frac{M_y}{d} + \frac{M_x}{b}\right) \tag{10.97a}$$

The first two parts of equation (10.97a) confirm the optimum, constant breadth of the section given in equation (3.13b) as

$$b_{\text{opt}} = 0.383\left(\frac{\sigma_y}{E}\right)L \tag{10.97b}$$

Combining equations (10.97a,b) provides a quadratic for the optimum depth in normalised form

$$\left(\frac{d}{L}\right)^2 - \frac{40.873E^2M_x}{\sigma_y^3L^3}\frac{d}{L} - \frac{15.66EM_y}{\sigma_y^2L^3} = 0 \tag{10.97c}$$

Note that equation (10.97b) is the same expression as used in our original analyses of rectangular sections that led to the very thin section that prompted this search for further failure criteria.

10.7.5 Combined Effects

In § § 10.7.1–10.7.3 the effects of shear, torsional and flexural buckling were combined to provide the optimum section dimensions. Direct stress effects –bending and lateral buckling –were employed separately in § 10.7.4. An inclination and an eccentricity of a concentrated force have been taken separately in each of Figures 10.27 and 10.28. There remains the possibility of a loading in which their effects are combined and therefore a corresponding failure criterion is required. Firstly, we should note that while the shear criteria apply to the maximum shear stress at the neutral axis, the direct stress criteria refer to the outer edges. This complicates the matter of basing the design upon the major principal stress since its position would lie somewhere between these two positions. A non-uniform distribution in the major principal stress applies to the depth within all common, solid shapes, including the rectangular section. Therefore, the strategy recommended is to determine the optimum section dimensions separately from both shear and direct stress effects and then select the greatest values of b_{opt} and d_{opt} that the various criteria provide. For this, some estimation of the offset and inclination of the end-load applied to the cantilever, however small, would need to be made. For a simply supported beam, carrying a central point load, the detailed analysis of shear and direct stress effects is similar to that given above with the appropriate change made to the maximum shear force and bending moment (refer here to Figures 10.6 and 10.16). In the following chapter an optimum cross-sectional shape is determined on the basis of equalising an effective stress within the depth from shear and direct stress effects.

10.7.6 Distributed Loading

The question as to which additional failure criteria might apply to a rectangular section beam when loading is uniformly distributed along the length was raised earlier. A number of possibilities arise from Figure 10.26 in the case of a thin, deep rectangular section.

(i) Flexural Buckling (see Figure 10.26c)
For a wide strut of pitch length p (see Chapter 2) the Euler flexural buckling stress is

$$\sigma_F = \frac{\pi^2 EI}{AL^2} = \frac{\pi^2 E(pb^3/12)}{pbL^2} = \frac{\pi^2 Ep^2}{12L^2} \tag{10.98a}$$

The direct compressive stress within the pitch length p is

$$\sigma = \frac{wp}{bp} = \frac{w}{p} \tag{10.98b}$$

(ii) Local Buckling (see Figure 10.26e)
The possibility of local compressive buckling in thin sheet should not be discounted, e.g. in the walls of a tube or in the web and flanges of I-, T- and U-sections. This requires the use of local buckling formulae for the plate-like flats of thin-walled cross sections [3]. In the case of a thin rectangular section this buckling mode is interpreted in Figure 10.26e. There is not a direct axial force, but compression arises from bending at a critical stress:

$$\sigma_L = KE\left(\frac{b}{p}\right)^2 \tag{10.99}$$

where $K = 3.62$ when b/p lies within the standard range of deep plates with simply supported edges interconnecting cells of regular pitch p. Combining equations (10.98a,b) with equation (10.99) gives the optimum breadth and pitch:

$$b_{\text{opt}} = 0.761\left(\frac{wL}{E}\right)^{1/2}, \quad p_{\text{opt}} = 3.62\frac{Eb^2}{w} \tag{10.100a,b}$$

For this analysis to apply, the pitch, given by equation (10.100b), should not exceed the length (i.e. $p \le L$). Preferably, the beam length should be an integral number of optimum pitch lengths.
This mode is also likely to occur where an inclined, concentrated force has a component that exerts axial compression. When this mode is combined with other direct stress modes of failure, e.g. lateral buckling, the optimum breadth should be taken as the greater of equations (10.97b) and (10.100a).

References

[1] Rees, D. W. A. *Mechanics of Solids and Structures*, Imperial College Press, 2000.
[2] Timoshenko, S. and Gere, J. M. *Theory of Elastic Stability*, McGraw-Hill, 1961.
[3] ESDU 72019, Buckling of flat isotropic plates under uniaxial and biaxial loading, September 1972.

Exercises

10.1 Why do equations (10.26b) and (10.32), pertaining to the optimum shear designs of end-loaded cantilever beams, with square and rectangular section, have identical coefficients?

10.2 Design optimum cantilevers required to carry 25 kN distributed uniformly in a 2 m length of steel when the cross-sections are: (i) circular, (ii) square and (iii) rectangular. Take, for the beam material, $E = 210$ GPa, $\sigma_y = 300$ MPa and $\rho = 7860$ kg/m^3.

10.3 Find the rectangular section dimensions, the weight and the central load carried by simply supported, aluminium alloy beams when: (a) $s' = 1$, $L = 100$ mm; (b) $s' = 0.1$, $L = 1$ m. Take, for the beam material, $E = 75$ GPa, $\sigma_y = 150$ MPa and $\rho = 2710$ kg/m^3.

10.4 Reading from Figure 10.22 (or Table 10.1), determine the optimum diameter profile and the minimum weight of a circular-section brass pin 10 mm long, when $s = 10$. What is the distributed load being supported? Take $\sigma_y = 130$ MPa and $\rho = 8900$ kg/m^3.

10.5 Reading from Figure 10.22 (or Table 10.2), determine the optimum length profile and the minimum weight of a 100 mm long, square-section, stainless steel pin, when $s = 0.10$. What is the distributed load being supported? Take $\sigma_y = 300$ MPa and $\rho = 7860$ kg/m^3.

10.6 Reading from Figure 10.22 (or Table 10.3), determine the optimum length profile and the minimum weight of a 1 m long, rectangular-section, simply supported, cast iron beam when $s' = 1.0$. What is the distributed load being supported? Take $\sigma_y = 150$ MPa, $E = 96.5$ GPa and $\rho = 7860$ kg/m^3.

10.7 Design optimum cantilevers required to carry 15 kN distributed uniformly in a 0.5 m length of aluminium alloy when the cross-sections are: (i) circular, (ii) square and (iii) rectangular. Take, for the beam material, $E = 75$ GPa, $\sigma_y = 150$ MPa and $\rho = 2710$ kg/m^3.

10.8 Find the rectangular section dimensions, the weight and load carried by simply supported steel beams loaded at their centres when: (a) $s' = 0.5$, $L = 50$ mm; (b) $s' = 0.01$, $L = 1$ m. Take, for the beam material, $\rho = 7860$ kg/m^3, $E = 210$ GPa and $\sigma_y = 300$ MPa.

10.9 Reading from Figure 10.22 (or Table 10.1), determine the diameter profile and minimum weight of a 25 mm long, circular-section, silver steel pin when $s = 20$. What is the distributed load being supported? Take $\sigma_y = 350$ MPa and $\rho = 7860$ kg/m^3.

10.10 Reading from Figure 10.22 (or Table 10.2), determine the length profile and minimum weight of a 100 mm long, square-section, chromium steel axle when $s = 0.2$. What is the distributed load being supported? Take $\sigma_y = 500$ MPa and $\rho = 7860$ kg/m^3.

10.11 Reading from Figure 10.22 (or Table 10.3), determine the length profile and the minimum weight of a 1 m long, rectangular-section, simply supported titanium beam when $s' = 5$. What distributed load is this beam capable of supporting? Take $\sigma_y = 850$ MPa, $E = 118$ GPa and $\rho = 4540$kg/m^3

10.12 Determine optimum dimensions of a rectangular-section, glass-fibre reinforced composite cantilever beam required to support an end-load of 15 kN over a length of 2 m. The breadth is built up in layers of cross-ply reinforcement whose fibres are aligned with the length and depth of the beam. Take, for both length and depth, $E = 20$ GPa, $\sigma_y = 100$ MPa and $\rho = 1800$ kg/m^3.

10.13 Examine the effect upon equations (10.14) and (10.15) when the cross-section of a cantilever under combined loading (see Figure 10.5) is changed from circular to square. Compare the variable geometry and minimum weight with those given for the circular section cantilever at $s = 1$ and $s = 0.01$.

11

Optimum Cross-Sections for Beams

11.1 Introduction

So far we have considered how the dimension of a solid section, whether it be circular, square or rectangular, should vary with the length of a beam if we are to minimise its weight. In fact, a closer inspection of stress arising from bending and shear reveals that the optimum cross section will depart from these standard shapes. It may not always be necessary to machine a beam's cross-section to vary with its length, but it would become essential when weight removal is crucial. Here, it is possible to produce the desired shape of the cross-section and the length profile required to minimise weight (i.e. an optimised beam) from numerically controlled machining. The program to do this would incorporate dimensional details based upon the combined shear and bending design that can be found here. The principle behind this is that all the material within the section's depth is stressed to its limit and not just the material lying at its outer edges and central (neutral) axis. Previously, in Chapter 10, the maximum stress at each of these positions was limited to a design stress in separate criteria for bending and shear. Recall here that the maximum bending stress occurs at the top and bottom surfaces while the maximum shear stress occurs at the neutral axis. The respective stress distributions are linear and parabolic and so, for any intermediate position in the section's depth, an effective (von Mises) stress is required to combine them. Hence, when the von Mises stress has attained the design stress value at all depths this defines the optimised section. The problem lies in knowing how to establish the shape of cross-section that meets this requirement when none of the common beam cross-sections do. The solution given here is developed in stages, from initial approximations where the depth is preset, to evaluations of the more common, axisymmetric beam sections, before a numerical scheme identifying the required shape is given. Worked examples illustrate the optimum beam design for our four standard cases of transverse loading, namely, simply supported beams and cantilevers carrying concentrated and distributed loadings. Though the complex section shapes and the length profiles that the analysis provides may appear to be impractical, they are perfectly capable of manufacture with modern machining methods. The range of structural indices covered

Mechanics of Optimal Structural Design: Minimum Weight Structures D. W. A. Rees

allows for the optimum design of short pins under shear and long beams under bending in each of the transverse loadings considered.

11.2 Approaching Optimum Sections

Let us establish, firstly, how the optimum section shape of a cantilever appears when carrying a uniformly distributed load (see Figure 10.11a). The shear force and bending moment diagrams shown in Figure 10.11b,c were used previously to establish how the breadth, depth or diameter of the chosen section should vary with beam length. Here we shall fix the depth of the section but allow the force and moment diagram to control the section's breadth.

11.2.1 Optimum Section for Shear

Equation (5.3a) shows that the shear stress, due to a transverse shear force $F(z)$, varies with depth y from the centre of a rectangular section according to

$$\tau = \frac{6|F(z)|}{bd^3}\left(\frac{d^2}{4} - y^2\right) \tag{11.1}$$

where $I = bd^3/12$, $\tau \leq \tau_y$ is the design stress and y is the distance above or below the horizontal centre line (the neutral axis NA) in Figure 5.2. It follows from equation (11.1) that if depth d is constant, then the breadth b must vary in its non-dimensional form as

$$\frac{b}{L} = \frac{6\sqrt{3}}{(d/L)^3}\left(\frac{1}{\sigma_y} \times \frac{|F(z)|}{L^2}\right)\left[\left(\frac{d}{2L}\right)^2 - \left(\frac{y}{L}\right)^2\right] \tag{11.2a}$$

where $\sigma_y = \sqrt{3}\tau_y$ is the tensile yield stress of the beam material. For the cantilever loading in Figure 11.1a, $|F(z)| = wz = F_{\max}(z/L)$, where $F_{\max} = wL$, and so equation (11.2a) becomes

$$\frac{b}{L} = \frac{6\sqrt{3}}{(d/L)^3}\left(\frac{1}{\sigma_y} \times \frac{F_{\max}}{L^2}\right)\left(\frac{z}{L}\right)\left[\left(\frac{d}{2L}\right)^2 - \left(\frac{y}{L}\right)^2\right] = C_1\frac{z}{L}\left[\left(\frac{d}{2L}\right)^2 - \left(\frac{y}{L}\right)^2\right] \tag{11.2b}$$

where C_1 is a constant. Thus, the section breadth b/L varies with $(y/L)^2$ at any given position z/L in the length in the manner shown in Figure 11.1b. This design results in the paraboloid cross-section shown that 'grows' linearly with z/L, from a point at the free end to its maximum size at the fixed end.

11.2.2 Optimum Section for Bending

The stress due to bending may be written, from equation (3.9), as

$$\sigma = \frac{12|M(z)|y}{bd^3} \tag{11.3}$$

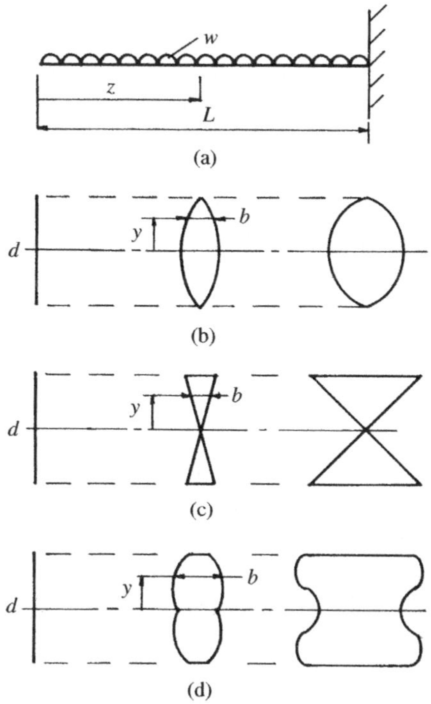

Figure 11.1 Approaches to optimum sections for a cantilever in shear and bending

where $\sigma \leq \sigma_y$ is the design stress and y is the distance above or below the horizontal centre line (the neutral axis). It follows that when the depth d is constant, the breadth b must vary in its non-dimensional form as

$$\frac{b}{L} = \frac{12}{(d/L)^3}\left(\frac{1}{\sigma_y} \times \frac{|M(z)|}{L^3}\right)\frac{y}{L} \tag{11.4a}$$

For the cantilever in Figure 11.1a, $|M(z)| = wz^2/2 = M_{\max}(z/L)^2$, where $M_{\max} = wL^2/2$. Hence equation (11.4a) becomes

$$\frac{b}{L} = \frac{12}{(d/L)^3}\left(\frac{1}{\sigma_y} \times \frac{M_{\max}}{L^3}\right)\left(\frac{z}{L}\right)^2\frac{y}{L} = C_2\left(\frac{z}{L}\right)^2\frac{y}{L} \tag{11.4b}$$

where C_2 is a constant. Equation (11.4b) shows that the section breadth b/L varies linearly with y/L at any given position z/L in the length in the manner shown in Figure 11.1c. This design results in a cross-section of mirrored, isosceles triangles as shown. These 'grow' with z^2, from an edge at the free end to attain their greatest dimensions at the fixed end.

11.2.3 Optimum Section for Combined Bending and Shear

When the two basic shapes, determined by shear and bending above, are added there results the waisted shape (see Figure 11.1d) that will support the design stresses through the depth. The section under the maximum stress approximates to the traditional I-section in which the flange supports bending and the web supports shear. There is no universal solution to finding the shape variation by this method since both $F(z)$ and $M(z)$ in equations (11.1) and (11.3) will depend upon the loading. We should also note the approximate nature of the above analysis that begins with a rectangle and ends with another shape whose I-value will have been altered. Consequently, an improved approach to the optimum section design should begin with the two basic section shapes similar to those established above (triangle and parabola). This analysis is given in the following section.

11.3 Generalised Optimum Sections

11.3.1 Bending

In bending we begin with a concave, asymmetric section in which the yield stress σ_y has been reached in all fibres of width b and thickness δy (see Figure 11.2). The precise dependence of the breadth upon depth, i.e. $b = b(y)$, is to be established for this condition when it is assumed that $b(y) = Ky^n$ describes the concave curvature.

The force upon this fibre is $\sigma_y b(y)\delta y$ and the moment exerted by this force is $\delta M = \sigma_y b(y)y\delta y$. Integrating to the half-depth $d_o/2$ on both sides of the neutral axis (NA),

$$M(z) = 2\sigma_y \int_0^{d_o/2} b(y)y\mathrm{d}y \tag{11.5a}$$

If we let $b(y) = Ky^n$, equation (11.5a) integrates as

$$M(z) = 2K\sigma_y \int_0^{d_o/2} y^{n+1}\mathrm{d}y = \frac{2K\sigma_y}{n+2}\left(\frac{d_o}{2}\right)^{n+2} \tag{11.5b}$$

from which K and then $b(y)$ follow:

$$K = \frac{(n+2)M(z)}{2\sigma_y(d_o/2)^{n+2}}, \qquad b(y) = \frac{(n+2)M(z)}{2\sigma_y(d_o/2)^{n+2}}y^n \tag{11.6a,b}$$

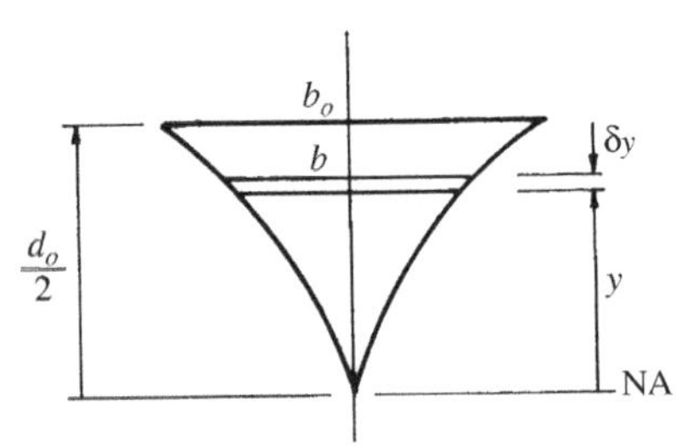

Figure 11.2 Beam section determined solely by bending

If, for the edges ($y = d_o/2$), we set $b(d_o/2) = b_o$, then n follows from equation (11.6b):

$$n = 2\left[\frac{b_o d_o^2 \sigma_y}{4M(z)} - 1\right] \tag{11.6c}$$

within the simpler form for $b(y)$,

$$b(y) = \frac{b_o}{(d/2)^n} y^n \tag{11.6c}$$

where, from equation (11.6c), n appears to be controlled by the outer dimensions b_o and d_o of the section and the moment it carries at the given position z in the beam length.

11.3.2 Shear

In shear we begin with a convex asymmetric section in which the shear yield stress τ_y has been reached in all fibres of width b and thickness δy (see Figure 11.3). The dependence of the breadth upon depth, i.e. $b = b(y)$, is to be established.

The vertical force upon this fibre is $\delta F = \tau_y b(y)\delta y$. The total force carried by the section is found from integrating to the half-depth $d_o/2$ on both sides of the neutral axis:

$$F(z) = 2\tau_y \int_0^{d_o/2} b(y)dy \tag{11.7a}$$

Guided by the approximate analysis above, we let $b(y) = C[(d_o/2)^m - y^m]$, when equation (11.7a) integrates to

$$F(z) = 2C\tau_y \int_0^{d_o/2} [(d_o/2)^m - y^m]dy = \frac{2Cm\tau_y}{m+1}\left(\frac{d_o}{2}\right)^{m+1} \tag{11.7b}$$

from which C and, hence, $b(y)$ follow as

$$b(y) = \left[\frac{(m+1)F(z)}{2m\tau_y (d_o/2)^{m+1}}\right][(d_o/2)^m - y^m] \tag{11.8a}$$

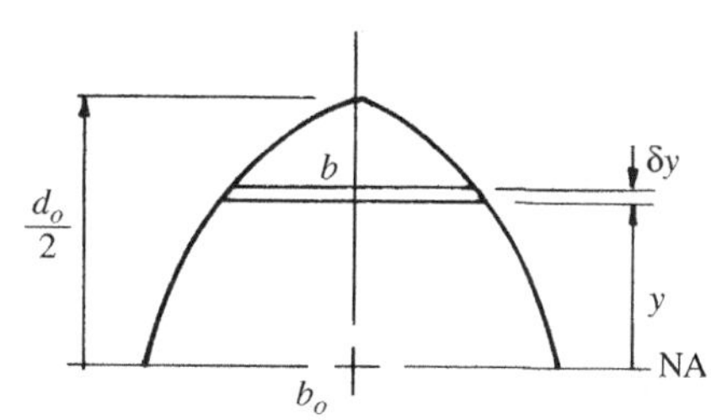

Figure 11.3 Beam section determined by shear

If, for $y = 0$, we set $b(0) = b_o$, then m follows as:

$$m = \left[\frac{b_o d_o \tau_y}{F(z)} - 1\right]^{-1} \tag{11.8b}$$

within the simpler form for $b(y)$,

$$b(y) = \frac{b_o d_o}{2} \frac{(d_o/2)^m - y^m}{(d_o/2)^{m+1}} \tag{11.8c}$$

where m in equation (11.8b) appears to be controlled by the outer dimensions b_o and d_o of the section and the force it carries at a given position z in the length. A differential relationship $F(z) = \mathrm{d}M(z)/\mathrm{d}z$ allows the connection between n and m here and in § 11.3.1 to be made.

Once again, these independent analyses of fully stressed beam sections in bending and shear, when combined, lead to similar, optimised shapes as found previously in Figure 11.1b–d by the approximate method given in § 11.2.3.

11.4 Optimum Section, Combined Bending and Shear

The above analyses are simplified in that they ignore the fact that both σ and τ contribute to yielding at intermediate depths $0 < y < d_o/2$ for the arbitrary cross-section in Figure 11.4a. Hence, a yield criterion is required when allowing for elastic variations in the two component stresses $\sigma(y)$ and $\tau(y)$. Let this stress combination satisfy the von Mises yield criterion [1, 2]:

$$[\sigma(y)]^2 + 3[\tau(y)]^2 = \sigma_y^2 \tag{11.9}$$

Referring to Figure 11.4b,c, the following two stresses apply to position y within any cross-section at a position in the length, where the shear force and bending moment are F and M:

$$\sigma(y) = \frac{My}{I}, \quad \tau(y) = \frac{FA\bar{y}}{Ib} \tag{11.10a,b}$$

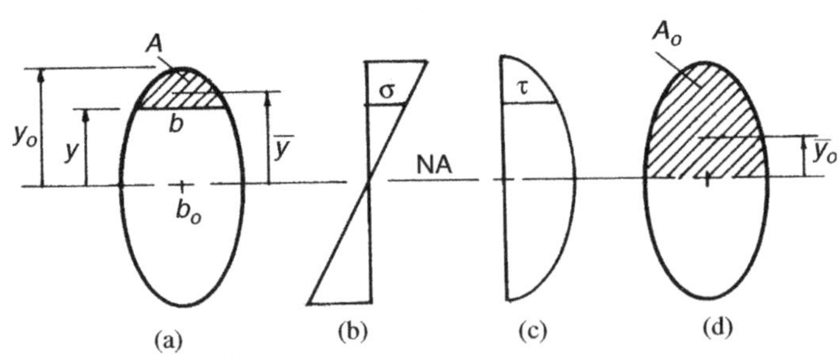

Figure 11.4 Bending and shear stress in the depth of an arbitrary section

These stresses are distributed in the manner of Figure 11.4b,c. Now, when we allow their maxima to attain the design stress, equations (11.10a,b) become

$$\sigma(y_o) = \frac{My_o}{I} = \sigma_y, \quad \tau(0) = \frac{FA_o\overline{y}_o}{Ib_o} = \tau_y = \frac{\sigma_y}{\sqrt{3}} \tag{11.11a,b}$$

where $A_o\overline{y}_o$ is the first moment of the half-section area in Figure 11.4d. Eliminating M/I between equations (11.10a) and (11.11a) and F/I between (11.10b) and (11.11b):

$$\sigma(y) = \frac{\sigma_y y}{y_o}, \quad \tau(y) = \frac{\sigma_y b_o A\overline{y}}{\sqrt{3}bA_o\overline{y}_o} \tag{11.12a,b}$$

Substituting equations (11.12a,b) into equation (11.9) leads to the condition for which σ_y is reached at all y according to the von Mises combination of $\sigma(y)$ and $\tau(y)$:

$$\left(\frac{y}{y_o}\right)^2 + \left[\frac{(A/A_o)(\overline{y}/\overline{y}_o)}{b/b_o}\right]^2 = 1, \quad \sum = \left(\frac{y}{y_o}\right)^2 + \left[\frac{(A/A_o)(\overline{y}/\overline{y}_o)}{b/b_o}\right]^2 \tag{11.13a,b}$$

In principle, equation (11.13a) could be solved to define the optimum section, but this is no easy matter. Later, a numerical solution to equation (11.13a) will be given but first we shall apply the summation within its left-hand side (written as $\sum$ in equation (11.13b)) to see what the most common beam cross-sections can provide compared to the ideal sum $\sum = 1$.

11.5 Solid, Axisymmetric Sections

Equation (11.13a) is the von Mises condition for which the yield stress σ_y is reached within any fibre at depth y from the neutral axis of a solid, axisymmetric cross-section. The square, rectangle, circle and ellipse often appear in beam cross-sections. In what follows, expressions for their three geometric ratios, $A/A_o, b/b_o$ and $\overline{y}/\overline{y}_o$, appearing within the second term on the right-hand side of equation (11.13b), are determined at a normalised depth y/y_o.

11.5.1 Rectangle

Equation (11.13b) is simple to apply to the rectangular section $b \times d$, shown in Figure 11.5. This section shows $b/b_o = 1, A/A_o = 1 - y/y_o$ and $\overline{y}/\overline{y}_o = 1 + y/y_o$ where $A_o = by_o$. Table 11.1 provides each ratio for normalised depths $0 \le y/y_o \le 1$. The final row gives the required summation $\sum$, which may also be found from y/y_o directly as

$$\sum = \left(\frac{y}{y_o}\right)^2 + \left[1 - \left(\frac{y}{y_o}\right)^2\right]^2 \tag{11.14}$$

We see from Table 11.1 that each summation for $0 < y/y_o < 1$ falls short of the unity value required of an optimum section. The results are plotted in Figure 11.6, where, from

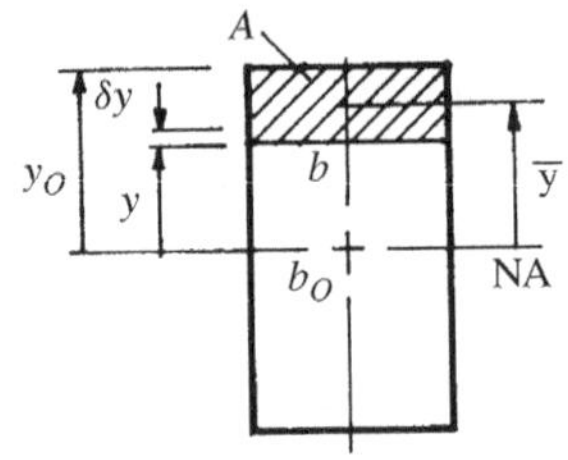

Figure 11.5 Dimensions of a rectangular section

Table 11.1 Stress summation $\sum$ in equation (11.13b), applied to a solid, rectangular section

y/y_o	0	0.2	0.4	0.6	0.8	0.9	1.0
A/A_o	1	0.8	0.6	0.4	0.2	0.1	0
$\overline{y}/\overline{y}_o$	1	1.2	1.4	1.6	1.8	1.9	2.0
$\sum$	1	0.9616	0.8656	0.7696	0.7696	0.8461	1

the analyses that follow, the rectangular distribution in $\sum$ appears identical to those of both circular and elliptical sections. It will be shown that even though the individual ratios b/b_o, A/A_o and $\overline{y}/\overline{y}_o$, at a given y/y_o, differ between these sections, they do so in a way as to preserve the summation given in equation (11.14). However, it will be seen that other sections given in Figure 11.6 have different summations.

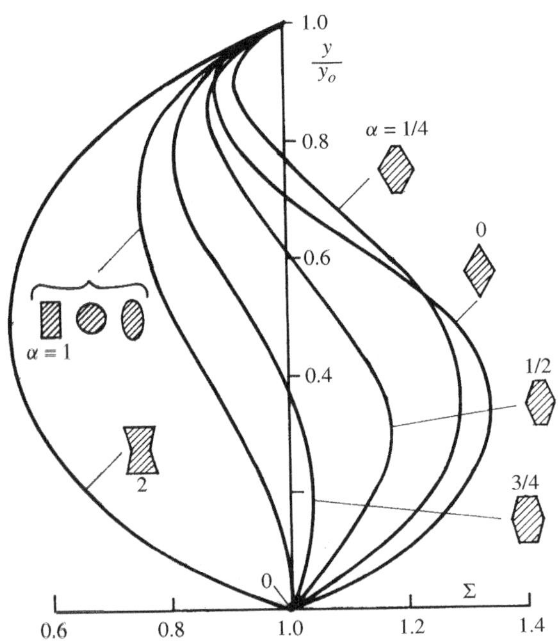

Figure 11.6 Stress summation $\sum$ distribution with depth y/y_o for various axisymmetric sections

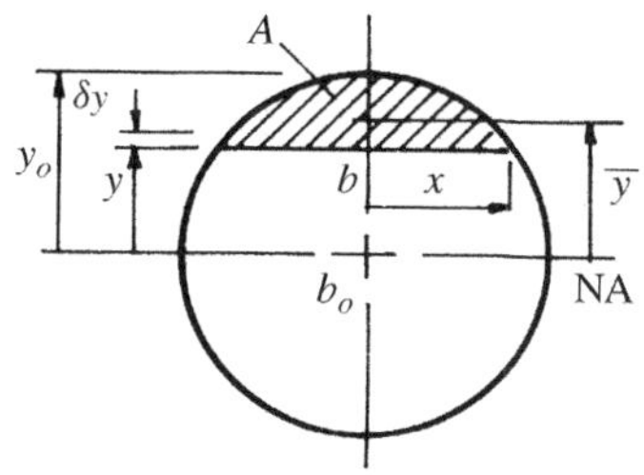

Figure 11.7 Dimensions of a circular section

11.5.2 Circular Section

The dimensions are referred to the upper-half of the circular section, as shown in Figure 11.7. The first moment of the shaded area A about the x-axis (NA) may be expressed as

$$A\bar{y} = 2\int_{y}^{y_o} xy\,dy = 2\int_{y}^{y_o} (y_o^2 - y^2)^{1/2} y\,dy = \frac{2}{3}(y_o^2 - y^2)^{3/2} \tag{11.15a}$$

and, with the lower limit in equation (11.15a) set at $y = 0$, the first moment of the semi-circular area becomes:

$$A_o\bar{y}_o = \frac{2}{3}y_o^3 \tag{11.15b}$$

Hence, the product ratio within equation (11.13b) follows from equations (11.15a,b) as

$$\frac{A\bar{y}}{A_o\bar{y}_o} = \left[1 - \left(\frac{y}{y_o}\right)^2\right]^{3/2} \tag{11.16a}$$

The breadth ratio becomes

$$\frac{b}{b_o} = \frac{x}{y_o} = \left[1 - \left(\frac{y}{y_o}\right)^2\right]^{1/2} \tag{11.16b}$$

Substituting equations (11.16a,b) into the right-hand side of equation (11.13b), gives the summation

$$\sum = \left(\frac{y}{y_o}\right)^2 + \left[1 - \left(\frac{y}{y_o}\right)^2\right]^2 \tag{11.17}$$

which is identical to equation (11.14), indicating a similar summation to the rectangular section. Again, equation (11.17) satisfies the unity values required for $y/y_o = 0$ and 1. For intermediate depths ($0 < y/y_o < 1$) the stress summations fall short of unity, as shown in Table 11.2.

Table 11.2 Stress summation equation (11.17) applied to solid, circular and elliptical sections

y/y_o	0	0.2	0.4	0.6	0.8	0.9	1.0
Σ	1.0	0.9616	0.8656	0.7696	0.7696	0.8461	1.0

11.5.3 Elliptical Section

The dimensions given in Figure 11.8 refer to the upper-half of a solid, elliptical section. The equation for the right-side 'quadrant', with coordinates (x_o, y_o), is

$$(x/x_o)^2 + (y/y_o)^2 = 1 \tag{11.18a}$$

The breadth ratio follows from equation (11.18a):

$$\frac{b}{b_o} = \frac{x}{x_o} = \left[1 - \left(\frac{y}{y_o}\right)^2\right]^{1/2} \tag{11.18b}$$

The first moment of the shaded area A about the x-axis (NA) is expressed as

$$A\overline{y} = 2\int_{y}^{y_o} xydy = 2x_o\int_{y}^{y_o}[1-(y/y_o)^2]^{1/2}ydy = \frac{2}{3}x_o y_o^2[1-(y/y_o)^2]^{3/2} \tag{11.19a}$$

and, with the lower limit in equation (11.19a) set at $y = 0$, the first moment of the half area becomes

$$A_o\overline{y}_o = \frac{2}{3}x_o y_o^2 \tag{11.19b}$$

Hence, the product ratio required for equation (11.13b) follows from equations (11.18b) and (11.19a,b):

$$\frac{A\overline{y}}{A_o\overline{y}_o} = \left[1 - \left(\frac{y}{y_o}\right)^2\right]^{3/2} \tag{11.19c}$$

The ratios (11.18b) and (11.19c) are identical to those found for the circular section, and so too will be their stress summations. That is, Table 11.2 also applies to an ellipse,

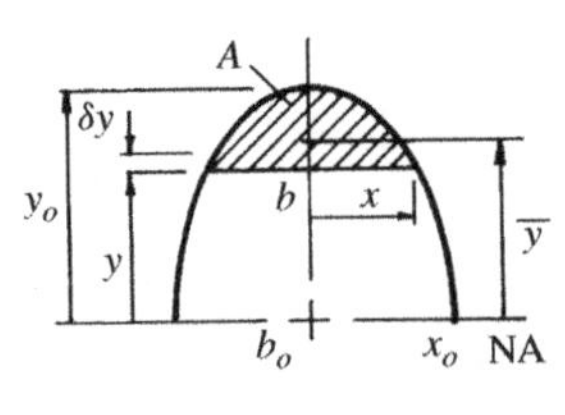

Figure 11.8 Dimensions of an elliptical section (upper half shown)

irrespective of the ratio x_o/y_o between the lengths of the ellipse's major and minor axes.

11.5.4 Double Triangular Section

The dimensions given in Figure 11.9 refer to the upper half of an axisymmetric, double triangular, solid section that is reflected at the neutral axis (NA). The equation of the right-hand sloping side is given as

$$x/x_o + y/y_o = 1 \tag{11.20a}$$

The breadth ratio follows from equation (11.20a):

$$b/b_o = x/x_o = 1 - y/y_o \tag{11.20b}$$

The first moment of the shaded area A about the x-axis (NA) may be expressed as

$$A\overline{y} = 2\int_y^{y_o} xy\,dy = 2x_o\int_y^{y_o}\left(1 - \frac{y}{y_o}\right)y\,dy = \frac{x_o y_o^2}{3}\left[1 - 3\left(\frac{y}{y_o}\right)^2 + 2\left(\frac{y}{y_o}\right)^3\right] \tag{11.21a}$$

and, with the lower limit as $y = 0$, the first moment of the triangular area above the NA is

$$A_o\overline{y}_o = x_o y_o^2/3 \tag{11.21b}$$

Substituting equations (11.20b) and (11.21a,b) into the following left-hand-side term in equation (11.13b),

$$\frac{(A\overline{y})/(A_o\overline{y}_o)}{(b/b_o)} = 1 + \frac{y}{y_o} - 2\left(\frac{y}{y_o}\right)^2 \tag{11.22}$$

An alternative approach to forming equation (11.22) is to write for the shaded area A, in Figure 11.9, the following normalised relationships:

$$\frac{A}{A_o} = \left(1 - \frac{y}{y_o}\right)^2, \quad \frac{\overline{y}}{\overline{y}_o} = 1 + 2\frac{y}{y_o}, \quad \frac{b}{b_o} = 1 - \frac{y}{y_o}, \tag{11.23a,b,c}$$

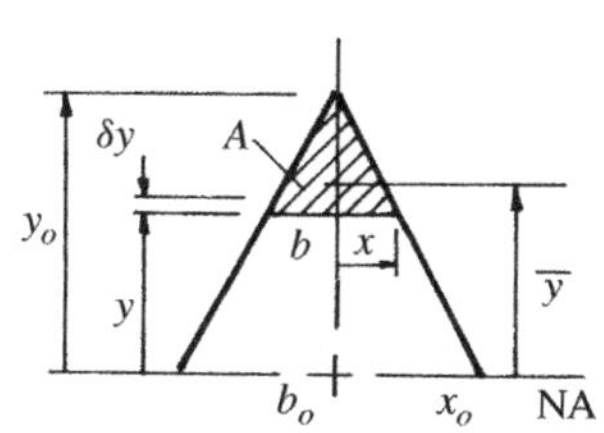

Figure 11.9 Dimensions of the upper-half of a double-triangular section

Table 11.3 Stress summation equation (11.13) applied to a double-triangular section

y/y_o	0	0.2	0.4	0.6	0.8	0.9	1.0
Σ	1.0	1.2944	1.3264	1.1344	0.9104	0.8884	1.0

When the product of equations (11.23a,b) is divided by equation (11.23c), this returns to equation (11.22). Substituting equation (11.22) into equation (11.13b) the summation is seen to satisfy the unity values required for $y/y_o = 0$ and 1. Elsewhere within the depth (i.e. $0 < y/y_o < 1$), the individual summations given in Table 11.3 are found to be either greater or less than unity.

This summation contrasts with that for the rectangle, circle and ellipse (see Figure 11.6) whose interiors were understressed. For a double-triangular section, Figure 11.6 shows that material on either side of its NA is overstressed, indicating that greater width is required.

11.5.5 Double Trapezoidal Section

Referring again to Table 11.1, the rectangular section suggests that the breadth of the cross-section should be reduced between its top and bottom edges. In contrast, the double triangular section (Table 11.3) suggests an optimum section with a greater breadth over most of its depth. The double trapezoidal section (i.e. an irregular hexagon) modifies breadth as suggested within its interior region. Dimensions are referred to the half sections shown in Figure 11.10.

(a) Method 1

For the coordinates indicated in Figure11.10, the equation for the right-hand sloping side is

$$y - y_o = \frac{y_o(x - \alpha x_o)}{x_o(1 - \alpha)} \tag{11.24a}$$

from which the breadth ratio $b/b_o(= x/x_o)$ in equation (11.13b) becomes

$$\frac{x}{x_o} = 1 - \frac{y}{y_o}(1 - \alpha) \tag{11.24b}$$

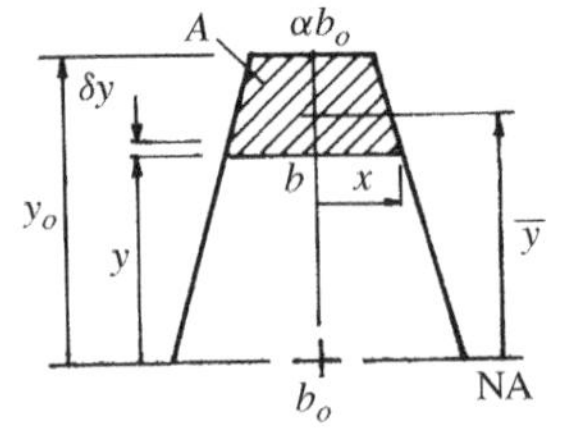

Figure 11.10 Dimensions of the upper half of a double trapezoidal section

The first moment of the shaded area A about the x-axis (NA) may be expressed as

$$A\overline{y} = 2\int_{y}^{y_o} xydy = 2x_o \int_{y}^{y_o} \left[1 - \frac{y}{y_o}(1-\alpha)\right] ydy$$

$$= \frac{x_o y_o^2}{3}\left\{(1+2\alpha) - \left[3\left(\frac{y}{y_o}\right)^2 - 2(1-\alpha)\left(\frac{y}{y_o}\right)^3\right]\right\} \quad (11.25a)$$

and, with $y = 0$, the first moment of the full, half-section area becomes

$$A_o\overline{y}_o = \frac{x_o y_o^2}{3}(1+2\alpha) \quad (11.25b)$$

Substituting equations (11.24b) and (11.25a,b) into the right-hand-side term in equation (11.13b):

$$\frac{(A\overline{y})/(A_o\overline{y}_o)}{(b/b_o)} = \frac{(1+2\alpha) - 3(y/y_o)^2 + 2(1-\alpha)(y/y_o)^3}{(1+2\alpha)[1-(1-\alpha)(y/y_o)]} \quad (11.26)$$

which reduces to equation (11.22) for the triangular section ($\alpha = 0$). For a rectangular section $\alpha = 1$ in equation (11.26), which confirms the summation given in equation (11.14). We may choose other values of α to examine its effect upon the summation. For example, when $\alpha = 1/2$ this provides the closest approach to the required $\sum = 1$ for each y/y_o value (see Table 11.4).

(b) Method 2

We may check the results for $\sum$ in Table 11.4 from the shaded area A in Figure 11.10, whose geometrical properties are normalised as

$$\frac{b}{b_o} = \frac{2x}{b_o} = 1 - (1-\alpha)\frac{y}{y_o} \quad (11.27a)$$

$$\frac{A}{A_o} = \left(1 - \frac{1-\alpha}{1+\alpha}\frac{y}{y_o}\right)\left(1 - \frac{y}{y_o}\right) \quad (11.27b)$$

$$\frac{\overline{y}}{\overline{y}_o} = \frac{3(1+y/y_o) - 2(1-\alpha)[1 + y/y_o + (y/y_o)^2]}{(1+2\alpha)\{1 - [(1-\alpha)/(1+\alpha)](y/y_o)\}} \quad (11.27c)$$

Table 11.4 Stress summation equation (11.13b) applied to a double trapezoidal section for $\alpha = 1/2$

y/y_o	0	0.2	0.4	0.6	0.8	0.9	1.0
$\sum$	1.0	1.1402	1.1401	1.0184	0.8834	0.8839	1.0

Table 11.5 Geometrical ratios within stress summations for a trapezoidal section with $\alpha = 1/2$

y/y_o	0	0.2	0.4	0.6	0.8	0.9	1.0
b/b_o	1	0.9	0.8	0.7	0.6	0.55	0.5
A/A_o	1	0.746	0.52	0.32	0.146	0.07	0
$\bar{y}/\bar{y}_o$	1	1.2643	1.5231	1.775	2.0182	2.1357	2.25
$\sum$	1	1.1402	1.1401	1.0184	0.8834	0.8839	1

Again, the constant α allows the gradient of the sloping side to be altered. When $\alpha = 1/2$, the ratios entered into Table 11.5 follow from equations (11.27a–c). The final row gives the summation required in equation (11.13b), which agrees with that entered in Table 11.4.

Clearly, as this section is more nearly stressed to the optimum amount we might seek to improve upon it by examining the influence of α upon $\sum$. Values of $\alpha = 0$, $1/4$, $1/2$, $3/4$, 1 and 2 are given in Table 11.6 The previous results for the rectangular ($\alpha = 1$) and double triangular ($\alpha = 0$) sections are added for completeness. Table 11.6 shows that the interior of a trapezoidal section, whose breadth is greatest at the top ($\alpha = 2$), remains unstressed, as with the rectangle ($\alpha = 1$). However, the interior of a trapezoidal section with $\alpha < 1$ becomes both under- and overstressed by varying amounts. At the extreme, the most severe overstressing occurs in the NA region of the double triangular section ($\alpha = 0$). These results are plotted together in Figure 11.6. Note, that the summation plot for the rectangle is coincident with that for both the circular and elliptical sections (see §§ 11.5.1–11.5.3).

Figure 11.6 suggests that if a summation of unity is to apply at all depths then α should fall (from $\alpha \approx 3/4$ to $\alpha \approx 1/4$) with increasing y before the topmost layers assume a near triangular shape ($\alpha = 1/4$). The curved shape of a section's sides that matches the requirement $\sum = 1$ will be derived in § 11.6, but first we shall examine $\sum$ values for uniform I-sections in a similar manner, given that these are believed to be 'efficient' as beam cross-sections.

11.5.6 I-Section

Consider an I-section with uniform thickness t in which α expresses the ratio between half the web height ($= \alpha b$) and the flange length b, within the half-section shown in Figure 11.11a.

Table 11.6 Influence of α upon $\sum$ for trapezoidal and triangular half-sections

y/y_o	0	0.2	0.4	0.6	0.8	0.9	1.0
$\sum(\alpha = 0)$	1	1.2944	1.3264	1.1344	0.9104	0.8884	1
$\sum(\alpha = 1/4)$	1	1.232	1.2897	1.1733	0.9764	0.9225	1
$\sum(\alpha = 1/2)$	1	1.1402	1.1401	1.0184	0.8834	0.8839	1
$\sum(\alpha = 3/4)$	1	1.0323	0.9916	0.877	0.8147	0.8603	1
$\sum(\alpha = 1)$	1	0.9616	0.8656	0.7696	0.7696	0.8461	1
$\sum(\alpha = 2)$	1	0.6672	0.5537	0.5501	0.6920	0.8236	1

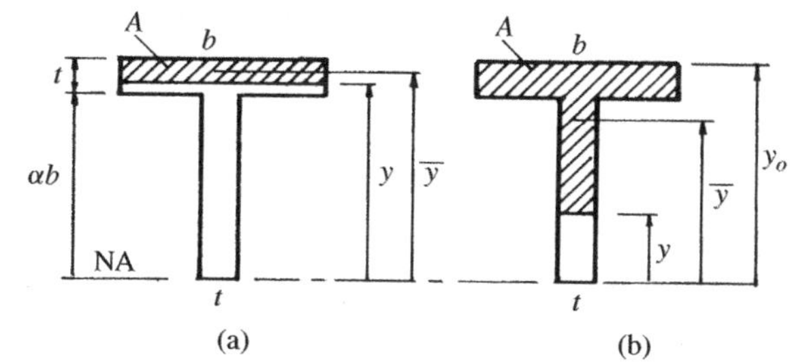

Figure 11.11 Uniform I-section (top half shown)

The half-section area A_o and its first moment about the neutral axis (NA) are

$$A_o = bt(1+\alpha) \tag{11.28a}$$

$$A_o\bar{y}_o = bt\left(\alpha b + \frac{t}{2}\right) + \alpha bt\left(\frac{\alpha b}{2}\right) \tag{11.28b}$$

The centroid position $\bar{y}_o$ for the half-section follows from the division of equations (11.28a,b):

$$\bar{y}_o = \frac{b}{1+\alpha}\left[\alpha\left(1+\frac{\alpha}{2}\right) + \frac{1}{2}\frac{t}{b}\right] \tag{11.28c}$$

Now, the ratios required within equation (11.13b) must be referred separately to the position y lying within the flange and the web areas as follows. Within the flange (see Figure 11.11a):

$$\frac{y}{y_o} = \frac{y/b}{\alpha + t/b} \quad \text{for } \alpha \le y/b \le \alpha + t/b \tag{11.29a}$$

$$\frac{A}{A_o} = \frac{b}{t(1+\alpha)}\left(\alpha + \frac{t}{b} - \frac{y}{b}\right) \tag{11.29b}$$

$$\frac{\bar{y}}{\bar{y}_o} = \frac{(1+\alpha)(\alpha + t/b + y/b)}{2[\alpha(1+\alpha/2) + t/2b]} \tag{11.29c}$$

Within the web (see Figure 11.11b):

$$\frac{y}{y_o} = \frac{y/b}{\alpha + t/b} \quad \text{for } 0 \le y/b \le \alpha \tag{11.30a}$$

$$\frac{A}{A_o} = \frac{1+\alpha - y/b}{1+\alpha} \tag{11.30b}$$

$$\frac{\bar{y}}{\bar{y}_o} = \frac{(1+\alpha)\{(\alpha + t/2b) + [\alpha^2 - (y/b)^2]/2\}}{(1+\alpha - y/b)[\alpha(1+\alpha/2) + t/2b]} \tag{11.30c}$$

Equations (11.29a–c) and (11.30a–c) allow I-sections, with various α and t/b values, to be examined for the optimum condition $\sum = 1$, from equation (11.13a). Nine combinations

Table 11.7 Optimum design analysis for an I-section flange with $\alpha = 1, b/t = 20$

y/b	1.0	1.02	1.04	1.05
y/y_o	0.9542	0.9715	0.9905	1.0
A/A_o	0.5	0.30	0.10	0
$\overline{y}/\overline{y}_o$	1.3442	1.3573	1.3704	1.377
$\sum$	0.9082	0.9442	0.9811	1.0

$(\alpha, b/t)$ were chosen, for $\alpha = 0.5$, 1, 1.5 and $b/t = 10$, 20, 30. As an example, consider the mid-range pair $\alpha = 1$ and $b/t = 20$. The flange equations (11.29a–c) become

$$y/y_o = 0.9524y/b, \quad A/A_o = 10(1.05 - y/b), \quad \overline{y}/y_o = 0.6557(1.05 + y/b) \tag{11.31a,b,c}$$

Equations (11.31a–c) provide the ratios and stress summation given in Table 11.7. The web equations (11.30a–c) become

$$\frac{y}{y_o} = 0.9524\frac{y}{b}, \quad \frac{A}{A_o} = 1 - \frac{y}{2b}, \quad \frac{\overline{y}}{\overline{y}_o} = \frac{6.1 - 2(y/b)^2}{3.05(2 - y/b)} \tag{11.32a,b,c}$$

Equations (11.32a–c) give the ratios and the required summation entered in Table 11.8.

The summations $\sum$ in Tables 11.7 and 11.8 are the result of applying the right-hand side of equation (11.13b) to their respective rows. Similar calculations are made for two further I-section designs with $\alpha = 0.5$ and $\alpha = 1.5$ in Figure 11.11a. These lead to the $\sum$ versus y/y_o plots shown in Figure 11.12.

Figure 11.12 shows that the divergence from a stress summation of unity is greatest at the web–flange interface, where a consistent discontinuity occurs. The discontinuity is due to the effect that the sudden change in breadth at the web–flange interface has upon the transverse shear stress. The thickest section ($b/t = 10$), having web to flange length ratios α of (a) 0.5, (b) 1.0 and (c) 1.5, shows the lowest divergence from unity on the web side of the interface, but this divergence is at its highest on the flange side. Figure 11.12 shows that converse observations apply to the thinnest section ($b/t = 30$) for each geometry $\alpha = 0.5$, 1.0 and 1.5. Given that the web carries more than 90% of the shear force, it follows that an optimum design is more closely approximated by decreasing the ratio b/t, i.e. by thickening the web. Coupled to this is the influence of α, where the greatest value ($\alpha = 1.5$) is seen to be more desirable, i.e. by lengthening

Table 11.8 Optimum design analysis for an I-section web with $\alpha = 1, b/t = 20$

y/b	0	0.2	0.4	0.6	0.8	0.9	1.0
y/y_o	0	0.1905	0.381	0.5714	0.7619	0.8572	0.9524
A/A_o	1	0.90	0.80	0.70	0.60	0.55	0.5
$\overline{y}/\overline{y}_o$	1	1.0965	1.1844	1.2599	1.3169	1.3353	1.3443
$\sum$	1	1.0102	1.043	1.1043	1.2048	1.2742	1.3589

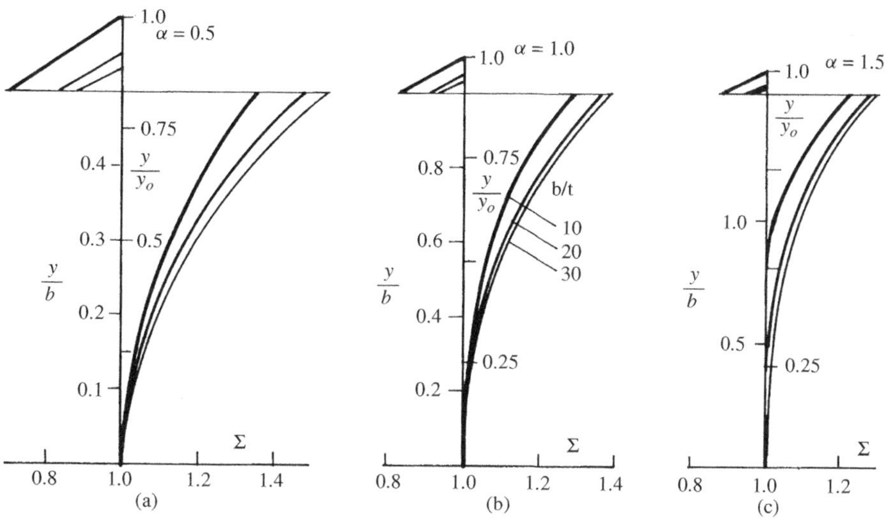

Figure 11.12 Search for an optimum I-section by varying its geometric ratios (a) $\alpha = 0.5$, (b) 1.0 and (c) 1.5 in Figure 11.11a. Note: stress summations $\sum$ apply to two ordinates y/y_o and y/b within (a), (b) and (c)

the web. Thus, the greater the web area is, compared to the flange area, the closer to an optimum design the I-section becomes. While some further minor improvement may be found for a non-uniform I-section, by altering its web–flange thickness ratio, an I-section can never be fully optimised with its inherent discontinuity. To find the true optimum section requires that equation (11.13a) is upheld throughout the depth. The following numerical analysis provides a good approximation to the optimum section shape.

11.6 Fully Optimised Section

Up to now we have seen that it has not been possible to match equation (11.13a) precisely within our examination of various beam cross-sections. If we are to do this then the sides of the section must be allowed to vary in a non-linear manner. The observations made upon Figure 11.6 suggest that the gradient of the sides should initially be steep and then fall before becoming steeper again at the top (as in Figure 11.13a). We shall

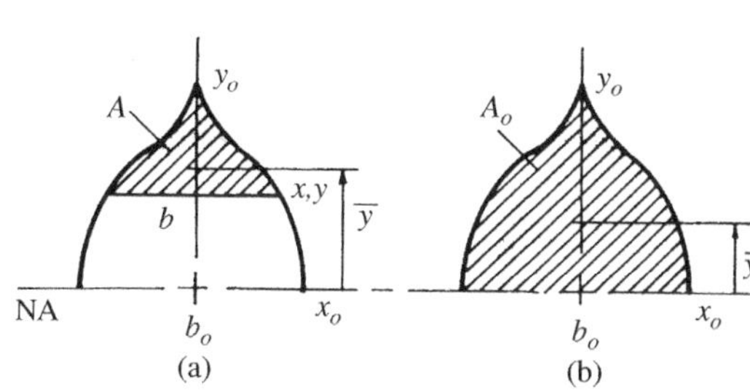

Figure 11.13 Fully optimised half-section

express the optimum section shape in non-dimensional coordinates $X = x/x_o$ $(= b/b_o)$ and $Y = y/y_o$. Correspondingly, the condition required in equation (11.13a) is rearranged as

$$\frac{A\overline{y}}{A_o\overline{y}_o} = X(1 - Y^2)^{1/2} \tag{11.33a}$$

The left-hand side of equation (11.33a) is identified with the first moments of the shaded areas, A and A_o, in Figure 11.13a and Figure 11.13b, respectively. Thus, equation (11.33a) becomes

$$\frac{\int_Y^1 XYdY}{\int_0^1 XYdY} = X(1 - Y^2)^{1/2} \tag{11.33b}$$

We shall express the side of the shape required as a polynomial:

$$X = 1 + A_1Y + A_2Y^2 + A_3Y^3 + A_4Y^4 + A_5Y^5 \tag{11.34a}$$

in which the condition $X = x/x_o = 1$ applies to $Y = 0$ (i.e. at the neutral axis NA). If, at the top of the section where $Y = 1$, we are to assume that $X = 0$ then it follows from equation (11.34a) that $A_5 = -(1 + A_1 + A_2 + A_3 + A_4)$. Hence, X in equation (11.34a) becomes

$$\begin{aligned} X &= 1 + A_1Y + A_2Y^2 + A_3Y^3 + A_4Y^4 - (1 + A_1 + A_2 + A_3 + A_4)Y^5 \\ &= (1 - Y^5) + A_1(Y - Y^5) + A_2(Y^2 - Y^5) + A_3(Y^3 - Y^5) + A_4(Y^4 - Y^5) \end{aligned} \tag{11.34b}$$

We shall fit equation (11.34a) to depths $Y = 0.2$, 0.4, 0.6 and 0.8. It follows from equation (11.34b) that the top integral in equation (11.33b) becomes:

$$\begin{aligned} \int_Y^1 XYdY &= \int_Y^1 (Y + A_1Y^2 + A_2Y^3 + A_3Y^4 + A_4Y^5 + A_5Y^6)dY \\ &= \left(\frac{1}{2} + \frac{A_1}{3} + \frac{A_2}{4} + \frac{A_3}{5} + \frac{A_4}{6} + \frac{A_5}{7}\right) \\ &\quad - \left(\frac{Y^2}{2} + \frac{A_1Y^3}{3} + \frac{A_2Y^4}{4} + \frac{A_3Y^5}{5} + \frac{A_4Y^6}{6} + \frac{A_5Y^7}{7}\right) \end{aligned} \tag{11.35a}$$

and setting $Y = 0$ in equation (11.35a) identifies the lower integral in equation (11.33b) as

$$\int_0^1 XYdY = \frac{1}{2} + \frac{A_1}{3} + \frac{A_2}{4} + \frac{A_3}{5} + \frac{A_4}{6} + \frac{A_5}{7} \tag{11.35b}$$

Substituting equations 11.34b, and equations (11.35a,b) into equation (11.33b) eventually leads to an equation in the four independent constants $A_1, \ldots, A_4$:

$$\left[\left(\frac{1}{2}-\frac{1}{7}\right)+\left(\frac{1}{3}-\frac{1}{7}\right)A_1+\left(\frac{1}{4}-\frac{1}{7}\right)A_2+\left(\frac{1}{5}-\frac{1}{7}\right)A_3+\left(\frac{1}{6}-\frac{1}{7}\right)A_4\right]$$
$$\times\{1-[(1-Y^5)+A_1(Y-Y^5)+A_2(Y^2-Y^5)+A_3(Y^3-Y^5)+A_4(Y^4-Y^5)](1-Y^2)^{1/2}\}$$
$$=\left(\frac{Y^2}{2}-\frac{Y^7}{7}\right)+\left(\frac{Y^3}{3}-\frac{Y^7}{7}\right)A_1+\left(\frac{Y^4}{4}-\frac{Y^7}{7}\right)A_2+\left(\frac{Y^5}{5}-\frac{Y^7}{7}\right)A_3+\left(\frac{Y^6}{6}-\frac{Y^7}{7}\right)A_4 \quad (11.36a)$$

The constants $A_1 = 0.069$, $A_2 = -4.222$, $A_3 = 2.561$, $A_4 = 2.580$ in equation (11.36a) were found by trial from the solution to the four simultaneous equations that follow from setting $Y = 0.2$, 0.4, 0.6 and 0.8. Equation (11.36a) becomes

$$X = 1 + 0.069Y - 4.222Y^2 + 2.561Y^3 + 2.580Y^4 - 1.988Y^5 \quad (11.36b)$$

Figure 11.14 shows one quadrant of the fully optimised profile in normalised coordinates as described by equation (11.36b). It can be seen that this optimum section modifies the double triangular section considered above (see Figure 11.9) in its thickening of the breadth where lower region material was overstressed and in its removal of upper region material from the understressed region. Consequently, the straight sides of the triangular section require modification, for example by numerical machining or by extrusion. As the shear force $F(z)$ and bending moment $M(z)$ will vary with the length z of a beam, this optimum section shape will need to inflate and contract along the length to achieve the maximum weight reduction possible. Given the section's equation (11.36b)

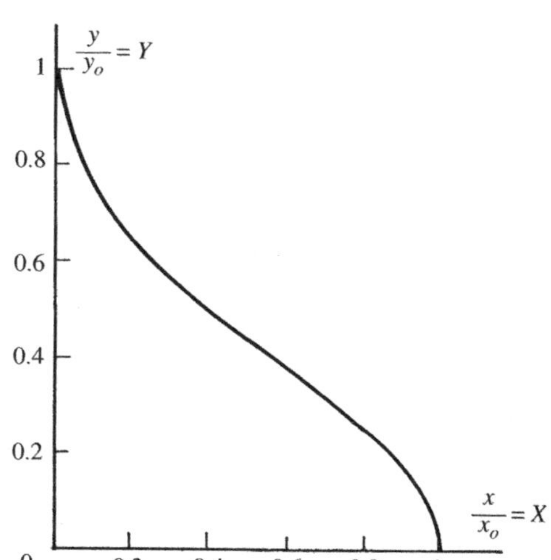

Figure 11.14 Profile of one quadrant of a fully optimised beam section

in non-dimensional coordinates $X = x/x_o$ and $Y = y/y_o$, we need only specify x_o and y_o when applying the bending and shear stress formulae. These will ensure that the yield stress (or design stress) will be reached at the section's top and at its neutral axis together. These two yield stresses are written respectively as

$$\sigma_y = \frac{M(z)y_o}{I}, \quad \tau_y = \frac{\sigma_y}{\sqrt{3}} = \frac{F(z)(A_o\overline{y}_o)}{I(2x_o)} \tag{11.37a,b}$$

where $\sigma_y = \sqrt{3}\tau_y$ is consistent with the von Mises yield criterion [1], $A_o\overline{y}_o$ is the first moment of the half-section area and I is the second moment of area for the whole section. These are:

$$A_o\overline{y}_o = 2x_o y_o^2 \int_0^1 XY\,dY$$
$$= 2x_o y_o^2\left[\left(\frac{1}{2}-\frac{1}{7}\right) + A_1\left(\frac{1}{3}-\frac{1}{7}\right) + A_2\left(\frac{1}{4}-\frac{1}{7}\right) + A_3\left(\frac{1}{5}-\frac{1}{7}\right) + A_4\left(\frac{1}{6}-\frac{1}{7}\right)\right] \tag{11.38a}$$

$$I = 4x_o y_o^3 \int_0^1 XY^2\,dY$$
$$= 4x_o y_o^3\left[\left(\frac{1}{3}-\frac{1}{8}\right) + A_1\left(\frac{1}{4}-\frac{1}{8}\right) + A_2\left(\frac{1}{5}-\frac{1}{8}\right) + A_3\left(\frac{1}{6}-\frac{1}{8}\right) + A_4\left(\frac{1}{7}-\frac{1}{8}\right)\right] \tag{11.38b}$$

when X is given by equation (11.34b). Using the constants $A_1, \ldots, A_4$ stated above, equations (11.38a,b) become

$$A_o\overline{y}_o = 0.2512x_o y_o^2, \quad I = 0.2123x_o y_o^3 \tag{11.39a,b}$$

Substituting equations (11.39a,b) into equations (11.37a,b) leads to the section's base dimensions at each position in the length where the shear force and bending moment are known:

$$x_o = 0.2229\frac{[F(z)]^2}{\sigma_y M(z)}, \quad y_o = 4.5967\frac{M(z)}{F(z)} \tag{11.40a,b}$$

Equations (11.40a,b) would appear to present a problem where $M(z) = 0$, giving $x_o = \infty$ and $y_o = 0$ and also where $F(z) = 0$, giving $x_o = 0$ and $y_o = \infty$. However, if we work with the product $x_o y_o$, equations (11.40a,b) give this as

$$x_o y_o = 1.0247\frac{F(z)}{\sigma_y} \tag{11.40c}$$

Equation (11.40c) implies that we may base the design upon $F(z)$ at position z. However, when we apply equations (11.40a,b) to separate the optimum section's base dimensions for the four standard cases of beam loading (see §§ 10.3 –10.6), both bending and shear will contribute to the optimum section dimensions, as will now be shown.

11.7 Fully Optimised Weight

Previously, minimum weight was established for standard sections in which length regimes were identified with structural indices where shear and bending dominated. Also identified was that narrow transition range of indices where the interaction between bending and shear provided matching contributions to their optimum designs. Shear and bending contribute to the weight of a fully optimised section in a similar manner. Here, however, we find that the weight function n is simplified to one having a linear dependence upon the index s.

11.7.1 Simply Supported Beam with Central Concentrated Load

In the case of a simply supported beam of length L, carrying a central concentrated load P (see §10.5 and Figure 10.16), equations (11.40a–c) are applied to the half-length $0 \le z/L \le 0.5$, where $|F(z)| = P/2$ and $|M(z)| = Pz/2$. These give

$$x_o = \frac{0.1115P}{\sigma_y z}, \quad y_o = 4.5967z, \quad x_o y_o = \frac{0.5125P}{\sigma_y} \tag{11.41a,b,c}$$

The total weight of the full length of this beam becomes

$$\frac{W}{L^3} = \rho \frac{x_o y_o}{L^2}\left(4\int_0^1 X\mathrm{d}Y\right)\left[2\int_0^{1/2} \mathrm{d}\left(\frac{z}{L}\right)\right] \tag{11.42}$$

in which the term within first bracket is the section area ($\times x_o y_o$), one quarter of which is provided by the four constants $A_1, \ldots, A_4$ within equation (11.36b), as

$$\int_0^1 X\mathrm{d}Y = \int_0^1 (1 + 0.069Y - 4.222Y^2 + 2.561Y^3 + 2.580Y^4 - 1.988Y^5)\mathrm{d}Y = 0.4521 \tag{11.43}$$

As the product $x_o y_o$ in equation (11.41c) does not depend upon z, the value of integral within the second bracket [ ] in equation (11.42) is simply unity in this case. Substituting equations (11.41c) and (11.43) into equation (11.42) leads to the objective function

$$\frac{W}{L^3} = 0.9266\frac{\rho}{\sigma_y}\frac{P}{L^2}, \quad \Rightarrow \quad \frac{1}{\rho}\frac{W}{L^3} = 0.9266\frac{1}{\sigma_y}\frac{P}{L^2} \tag{11.44a,b}$$

in which the shape factor is now at its lowest possible value for this loading. The advantage of the ideal section is seen by applying the alternative expression (11.44b) to Figure 10.18.

The normalised weight of the beam having an optimum section takes a simple linear form $n = 0.9266s$. The weight is seen to be lower than the weight of standard section beams of all lengths. However, it may only be practical to machine this optimum section shape into longer beams ($s < 0.1$). For shorter beams ($s > 0.1$), where the weight continues to be slightly less than that of the standard shapes, the latter may be preferred for convenience. Here, the interior of standard sections (circle, square and rectangle) would be overstressed (see Figure 11.6), a fact not considered earlier when applying bending and shear criteria separately to outer dimension designs. As a rule of thumb, to avoid overstressing by the percentage amounts shown in Figure 11.6, the design stress should be lowered by a similar percentage.

Example 11.1 Find the optimum section dimensions, the weight and load carried by a simply supported steel beam, loaded at its centre, when the normalised index $s = 1$ in a length $L = 100$ mm. Compare with the weights of circular- and square-section beams when operating under the same conditions. Take $\rho = 7860$ kg/m^3, $E = 210$ GPa and $\sigma_y = 300$ MPa.

Firstly, we find the central load carried at a structural index value $s = 1$,

$$\frac{1}{\sigma_y}\left(\frac{P}{L^2}\right) = 1, \quad \Rightarrow \quad P = \sigma_y L^2 = 300 \times 0.1^2 = 3 \text{ MN}$$

and from equation (11.44b) the weight follows as

$$\frac{1}{\rho}\left(\frac{W}{L^3}\right) = 0.9266 \times 1, \quad \Rightarrow \quad W = 0.9266\,\rho L^3 = 0.9266 \times 7860 \times 0.1^3 = 7.283 \text{ kg}$$

which compares with the greater weights of 9.45 kg and 10.21 kg (taken from Figure 10.18) for the overstressed circular and square sections respectively under the same conditions (see Figure 11.6). Equations (11.41a,b) give the two base dimensions for the optimum section as

$$x_o = \frac{0.1115P}{\sigma_y z} = \frac{0.1115 \times 3 \times 10^6}{300\,z} = \frac{1115}{z}, \qquad y_o = 4.5967z \qquad \text{(i, ii)}$$

with z in mm. The constant product of these dimensions is $x_o y_o = 5125$ mm^2. Figure 11.15 shows the manner in which x_o and y_o vary with z/L within the beam's normalised half length ($0 \leq z/L \leq 0.5$).

To avoid an infinite x_o, when $z = 0$, we shall design the initial region of the beam such that $x_o = y_o$, so maintaining the product $x_o y_o$ constant, as shown in Figure 11.15. Equating (11.41a,b) gives the critical length position at the intersection point:

$$\left(\frac{z}{L}\right)_{\text{cr}} = 0.1558\left(\frac{1}{\sigma_y}\frac{P}{L^2}\right)^{1/2} = 0.1558 \qquad \text{(iii)}$$

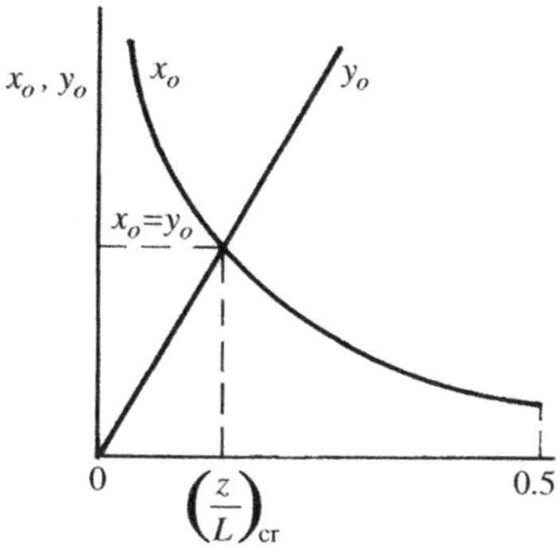

Figure 11.15 Base dimension variation within the half-length

Hence, it follows from equations (ii) and (iii) that, within the region $0 \le z/L \le (z/L)_{cr}$, the dimensions $x_o = y_o$ will remain constant at

$$\frac{y_o}{L} = \frac{x_o}{L} = 4.5967\left(\frac{z}{L}\right)_{cr} = 4.5967 \times 0.1558 = 0.7162 \qquad \text{(iv)}$$

and with $L = 100$ mm, equation (iv) gives $x_o = y_o = 71.62$ mm. For the remaining region within the half-length, i.e. $(z/L)_{cr} \le z/L \le 0.5$, the base dimensions may vary with z according to equations (i) and (ii), reaching their mid-span values $x_o = 22.3$ mm, $y_o =$ 229.84 mm. These are doubled to give outer dimensions in orthographic projection of the

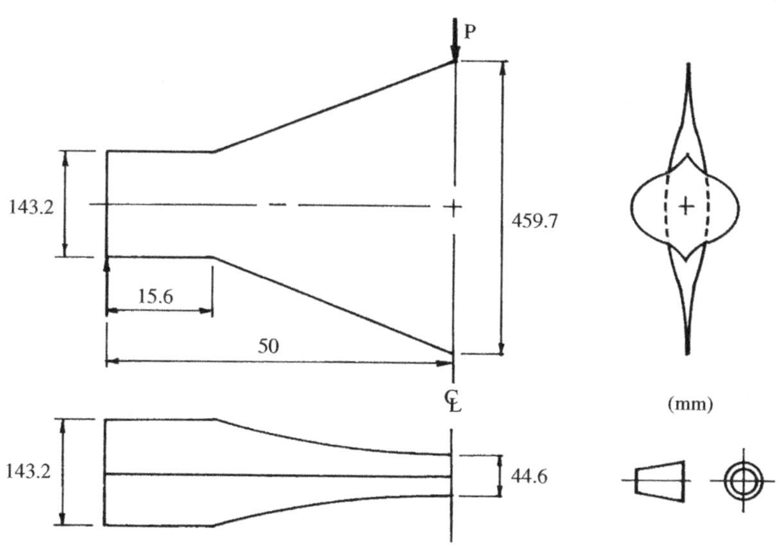

Figure 11.16 Orthographic projection of an optimum, half-length beam design

beam's half-length in Figure 11.16. Within both length regions the section's fractional breadths $X = x/x_o$ follow from equation (11.34b) at fractional depths $Y = y/y_o = 0.2$, 0.4, 0.6 and 0.8 using the constant values $A_1, \ldots, A_4$ given earlier.

When the intersection between shear and bending designs is to coincide with the beam's mid-span, the limiting structural index is found from equation (iii):

$$0.1558\left(\frac{1}{\sigma_y}\frac{P}{L^2}\right)^{1/2}_{FM} = 0.5, \quad \Rightarrow \quad s_{FM} = \left(\frac{1}{\sigma_y}\frac{P}{L^2}\right)_{FM} = 10.3 \qquad \text{(v)}$$

Equation (v) shows that when $s_{FM} < 10.3$ the section's base dimensions may be taken to be equal within the region $0 \leq z/L \leq (z/L)_{\mathrm{cr}}$ (as is the case in this example). When $s_{FM} \geq 10.3$, the intersection lies outside the half-length and we may design according to equation (11.40c). Here there is flexibility in the choice of the two base dimensions (x_o, y_o) but with the restriction that their product remains constant.

11.7.2 Simply Supported Beam with Distributed Loading

In the case of a simply supported beam of length L, carrying a uniformly distributed load w per unit length (see §10.6 and Figure 10.21), equations (11.40a,b,c) are applied to the beam's half length $0 \leq z/L \leq 0.5$, where $|F(z)| = w(L/2 - z)$ and $|M(z)| = (wz/2)(L - z)$. These give

$$x_o = \frac{0.1115w(L-2z)^2}{\sigma_y z(L-z)}, \quad y_o = \frac{4.5968z(L-z)}{(L-2z)}, \quad x_o y_o = \frac{0.5124w(L-2z)}{\sigma_y} \qquad \text{(11.45a,b,c)}$$

As the product $x_o y_o$ in equation (11.45c) now depends upon z, the weight of the full length of this beam is derived from

$$\frac{W}{L^3} = \rho\left(4\int_0^1 X\mathrm{d}Y\right)\left[2\int_0^{1/2}\frac{x_o y_o}{L^2}\mathrm{d}\left(\frac{z}{L}\right)\right] \qquad \text{(11.46a)}$$

in which the first bracketed integral on the right-hand side is the normalised section area, evaluated in §11.7.1 as 4 × 0.4521. The second bracketed integral is found by substituting equation (11.45c) into equation (11.46a). Making these substitutions,

$$\frac{W}{L^3} = 2 \times 4 \times 0.4521 \times 0.5124\frac{\rho w}{L\sigma_y}\int_0^{1/2}\left(1 - 2\frac{z}{L}\right)\mathrm{d}\left(\frac{z}{L}\right) \qquad \text{(11.46b)}$$

Equation (11.46b) leads to the objective function

$$\frac{W}{L^3} = 0.4633\frac{\rho}{\sigma_y}\frac{w}{L}, \Rightarrow \frac{1}{\rho}\frac{W}{L^3} = 0.9266\frac{1}{\sigma_y}\frac{F_{\max}}{L^2} \qquad \text{(11.47a,b)}$$

where $F_{\max} = wL/2$ is the maximum shear force in this beam. The comparison between equations (11.47a,b) and (11.44a,b) shows that this beam has half the weight of a simply supported beam carrying the same total load ($P = wL$) applied at its centre. The shape factor value is now the lowest possible for each type of loading. The alternative, non-dimensional expression (11.47b) is applied to Figure 10.22 for comparison with other non-optimal shapes (circle, square and rectangle). In the axes of this figure the weight of the beam having the optimum section takes a simple linear form, $n = 0.9266s$. The comparison shows that the weight has been lowered across the full range of s. However, it may only be practical to machine the optimum shape into longer beams, say for $s < 0.3$ where the weight saving is more significant. Here we should note that if a greater weight is to be tolerated in any of the other standard sections shown in Figure 10.22, their interiors would become overstressed in the manner of Figure 11.6, a fact not considered previously in their 'outer dimension' designs.

Example 11.2 Reading from Figure 10.22, determine the length profile and the weight of a simply supported, 1 m long, steel beam with an optimised cross-section, when $s = 0.01$. What is the maximum distributed load that can be supported? Compare with the weights of circular, square and rectangular sections when operating under similar conditions. Take, for steel, $\sigma_y = 300$ MPa, $E = 210$ GPa and $\rho = 7860$ kg/m^3.

Firstly, we find the load carried at the structural index value $s = 0.01$ from equation (11.47b):

$$s = \frac{1}{\sigma_y}\left(\frac{F_{\max}}{L^2}\right) = 0.01, \quad \Rightarrow \quad F_{\max} = 0.01\,\sigma_y L^2 = 0.01 \times 300 \times 1^2 = 3 \text{ MN}$$

and hence the distributed loading is $w = 2F_{\max}/L = 6$ MN/m. The beam's weight follows from equation (11.47b):

$$n = \frac{1}{\rho}\frac{W}{L^3} = 0.9266 \times 0.01, \quad \Rightarrow \quad W = 0.9266 \times 0.01 \times 7860 \times 1^3 = 72.83 \text{ kg}$$

Reading from Figure 10.22, this contrasts with far greater weights of 398.05 kg and 360.8 kg for the optimum circular and square sections, respectively. For the rectangular section, both the abscissa and the ordinate in Figure 10.22 have been divided by σ_y/E. To find its weight at a similar structural index, we approximate that part of the plot in the region where bending dominates with the relationship $n' = 0.75s'^{0.54}$. Hence, when $s = 0.01$ then $s' = s/(\sigma_y/E) = 7$, from which $n' = 2.145$, and this gives $n = n'\sigma_y/E = 0.3064 \times 10^{-2}$. From this, the weight $W = n\rho L^3 = 24.08$ kg, which is much less than the weight of a beam with the optimum section. However, additional buckling modes are likely for thin rectangular sections in bending which can render an unsafe design (see §10.7).

Equations (11.40a,b) give the two base dimensions for the optimum cross-section.

$$\frac{x_o}{L} = \frac{0.1115(1-2q)^2}{q(1-q)}\frac{w}{L\sigma_y} \tag{i}$$

$$\frac{y_o}{L} = \frac{4.5968q(1-q)}{1-2q} \tag{ii}$$

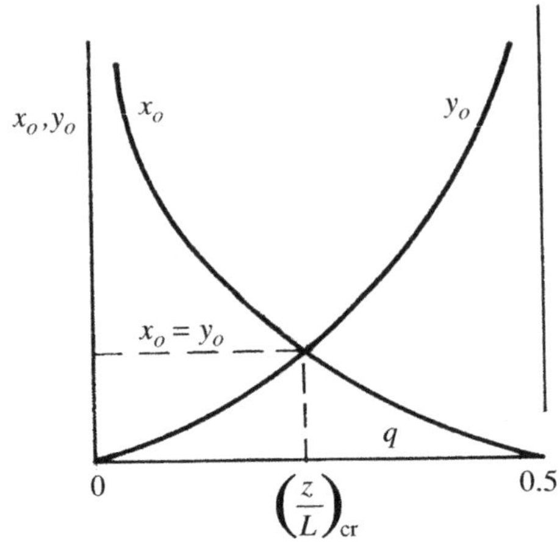

Figure 11.17 Variation in base dimensions within the half length $0 \leq q \leq 0.5$

where $q = z/L$. Equations (i) and (ii) show an infinite breadth and depth for when $q = 0$ and $q = 0.5$, respectively (see Figure 11.17). To avoid these infinities the design uses the product $x_o y_o$, written from equations (i) and (ii) as

$$\frac{x_o}{L}\frac{y_o}{L} = 0.5124(1 - 2q)\frac{w}{L\sigma_y} \tag{iii}$$

When $x_o = y_o$ we may equate (i) and (ii) to find q_c within the half length $0 \leq q \leq 0.5$. This corresponds to the intersection between the x_o and y_o plots, shown in Figure 11.17, at

$$q_c^2(1 - q_c)^2 = 0.02426\frac{w}{L\sigma_y}(1 - 2q_c)^3 \tag{iv}$$

where $q_c = (z/L)_{cr}$. Substituting $w = 6$ MN/m, $L = 1$ m and $\sigma_y = 300$ MN/m^2, the solution to equation (iv) is found by trial as $q_c = 0.021$. Correspondingly, equations (i) and (ii) give $x_o/L = y_o/L = 0.099$, so that $x_o = y_o = 99$ mm. Elsewhere, the product $x_o y_o$, must vary as in equation (iii) this allowing one dimension to be held constant for the full length. The weight is not altered by this procedure as it depends solely upon $x_o y_o$. Nor is the effective stress altered since any combination of the base dimensions satisfying equation (iii) ensures an optimum design.

11.7.3 Cantilever with Concentrated End-Load

In the case of a cantilever of length L, carrying a concentrated load P at its end (see §10.3 and Figure 10.6, equations (11.40a,b,c) are applied to the full length $0 \leq z/L \leq 1$, where $|F(z)| = P$ and $|M(z)| = Pz$. These give

$$x_o = \frac{0.2229P}{\sigma_y z}, \quad y_o = 4.5968z, \quad x_o y_o = \frac{1.0246P}{\sigma_y} \tag{11.48a,b,c}$$

Since equation (11.48c) shows that the product $x_o y_o$ is independent of z, the total weight of the full length of this beam becomes

$$\frac{W}{L^3} = \rho \frac{x_o y_o}{L^2}\left(4\int_0^1 X\mathrm{d}Y\right)\left[\int_0^1 \mathrm{d}\left(\frac{z}{L}\right)\right] = 4 \times 0.4521 \times 1.0246 \frac{\rho}{\sigma_y}\frac{F}{L^2} \qquad (11.49a)$$

Equation (11.49a) may be written in the following two ways:

$$\frac{W}{L^3} = 1.8532 \frac{\rho}{\sigma_y}\frac{F}{L^2}, \Rightarrow \frac{1}{\rho}\frac{W}{L^3} = 1.8532 \frac{1}{\sigma_y}\frac{F}{L^2} \qquad (11.49b,c)$$

The second expression (11.49c) may be applied to Figure 10.8 for comparison with other non-optimal section shapes (circular, square and rectangular). The weight of the beam having an optimum section has a simple linear form, $n = 1.8532s$. This shows that a greater weight saving applies to longer beams where it would be more practical to machine the optimum section shape. For shorter beams, the lesser weight saving may not justify the machining expense, though we should note that if a standard section is employed its interior would become overstressed when the outer surface material attains the allowable design stress.

Example 11.3 Find the optimum section dimensions, the weight and the end-load carried by a steel cantilever beam loaded at its end when $s = 0.1$ and $L = 1$ m. Compare with the weights of solid circular-, square- and rectangular-section cantilever beams when operating under similar conditions. Take $\rho = 7860$ kg/m^3, $E = 210$ GPa and $\sigma_y = 300$ MPa.

Firstly, we find the load carried from the structural index value $s = 0.1$,

$$s = \frac{1}{\sigma_y}\frac{P}{L^2} = 0.1, \quad \Rightarrow \quad P = s\sigma_y L^2 = 0.1 \times 300 \times 1^2 = 30 \text{ MN}$$

and from equation (11.49c) the weight follows from $n = 1.8532s$ as

$$\frac{1}{\rho}\frac{W}{L^3} = 1.8532 \times 0.1, \quad \Rightarrow \quad W = 0.18532\,\rho L^3 = 0.18532 \times 7860 \times 1^3 = 1456.6 \text{ kg}$$

which contrasts with more than a twofold increase in the weights (3863.2 kg and 3535.4 kg) of the overstressed circular and square sections respectively under the same conditions (see Figure 10.8). For the rectangular section both the abscissa and the ordinate in Figure 10.8 have been divided by σ_y/E. To find its weight at a similar structural index we approximate that part of the plot in the region where bending dominates with the relationship $n' = 1.13s'^{0.52}$. Hence, when $s = 0.1$ then $s' = s/(\sigma_y/E) = 70$, from which $n' = 10.293$, and this gives $n = n'\sigma_y/E = 1.47 \times 10^{-2}$. Thus the beam weight is $W = n\rho L^3 = 115.54$ kg, but note here the comments made on thin rectangular sections in Example 11.2.

Equations (11.48a,b) give the two base dimensions for the optimum section as

$$x_o = \frac{0.2229P}{\sigma_y z} = \frac{0.2229 \times 30 \times 10^6}{300\,z} = \frac{22.29 \times 10^3}{z}, \qquad y_o = 4.5968\,z \qquad \text{(i,ii)}$$

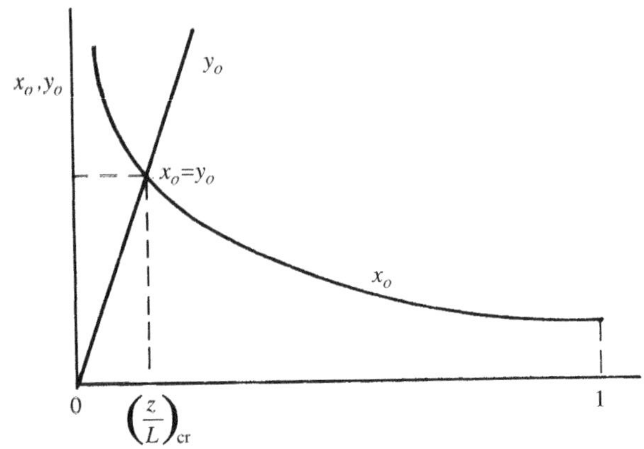

Figure 11.18 Variation in base dimensions within the full length

for z in mm. The constant product of these dimensions is $x_o y_o = 102.46 \times 10^3\ \mathrm{mm}^2$. The manner in which x_o and y_o vary with z within the beam's full length ($0 \le z/L \le 1.0$) is shown in Figure 11.18.

To avoid an infinite x_o when $z = 0$ we shall design the initial region of the beam so that base dimensions $x_o = y_o$, whilst retaining this product $x_o y_o =$ constant, as shown. Equating (i) and (ii) gives the critical length position at the intersection point in Figure 11.18:

$$\left(\frac{z}{L}\right)_{cr} = 0.2202 \left(\frac{1}{\sigma_y}\frac{P}{L^2}\right)^{1/2} = 0.06964 \tag{iii}$$

Hence, it follows from equation (ii) and (iii) that within the region $0 \le z/L \le (z/L)_{cr}$ the normalised dimensions $x_o = y_o$ remain constant at

$$\frac{y_o}{L} = \frac{x_o}{L} = 4.5968 \left(\frac{z}{L}\right)_{cr} = 4.5968 \times 0.06964 = 0.3201 \tag{iv}$$

and with $L = 1000$ mm, equation (iii) and (iv) give $z_{cr} = 69.64$ mm and $x_o = y_o = 320.1$ mm. For the remaining length $(z/L)_{cr} \le z/L \le 1.0$, the base dimensions may vary with z according to equations (i) and (ii), reaching end values of $x_o = 22.29$ mm and $y_o = 4596.8$ mm. Within both these regions of the length the section's fractional breadths x/x_o follow from equation (11.34b) at fractional depths $y/y_o = 0.2, 0.4, 0.6$ and 0.8 using the constants $A_1, \ldots, A_4$ as given earlier.

When the intersection $(z/L)_{cr}$ is made to coincide with the beam end, the limiting structural index is found from

$$0.2202 \left(\frac{1}{\sigma_y}\frac{P}{L^2}\right)^{1/2}_{FM} = 1.0, \quad \Rightarrow \quad s_{FM} = \left(\frac{1}{\sigma_y}\frac{P}{L^2}\right)_{FM} = 20.62 \tag{v}$$

Equation (v) shows that when $s_{FM} < 20.62$ the section's base dimensions may be taken to be equal (as in this example). When $s_{FM} \ge 20.62$ the intersection lies outside the half-length and we may design according to equation (11.48c) in which there is flexibility in the choice of one base dimension within the requirement that their product $(x_o y_o)$ is constant.

The reader should note here that the large sectional dimensions and weight found in this example correspond to the enormous load limit at which the cantilever can perform optimally at the given structural index. In normal design practice, where loads and lengths are specified, the optimum design will dictate the cantilever's structural index and, consequently, its section shape and self-weight (see Example 11.4).

11.7.4 Cantilever with Distributed Loading

In the case of a cantilever of length L, carrying a distributed load w (see §10.4 and Figure 10.11), equations (11.40a,b,c) are applied to the full length $0 \leq z/L \leq 1$ where $|F(z)| = wz$ and $|M(z)| = wz^2/2$. These give

$$x_o = \frac{0.4458w}{\sigma_y}, \quad y_o = 2.2984z, \quad x_o y_o = 1.0247\frac{wz}{\sigma_y} \tag{11.50a,b,c}$$

As the product $x_o y_o$ now depends upon z, the objective function for the weight of the full length of this beam is derived from

$$\frac{W}{L^3} = \rho\left(4\int_0^1 X\mathrm{d}Y\right)\left[\int_0^1 \frac{x_o y_o}{L^2}\mathrm{d}\left(\frac{z}{L}\right)\right]$$

$$= 4 \times 0.4521 \times 1.0247\frac{\rho}{\sigma_y}\frac{w}{L}\int_0^1 \frac{z}{L}\mathrm{d}\left(\frac{z}{L}\right) \tag{11.51a}$$

Equation (11.51a) leads to the following two forms for the objective function:

$$\frac{W}{L^3} = 0.9266\frac{\rho}{\sigma_y}\frac{w}{L}, \quad \Rightarrow \quad \frac{1}{\rho}\frac{W}{L^3} = 0.9266\frac{1}{\sigma_y}\frac{F_{\max}}{L^2} \tag{11.51b,c}$$

where $F_{\max} = wL$ is the maximum shear force in this beam. Comparison between equations (11.49c) and (11.51c) shows that this cantilever has half the weight of a cantilever carrying the same total load (i.e. wL) applied to its end. The shape factor value of 0.9266 is at its lowest possible for this distributed loading. Equation (11.51c) is applied to Figure 10.13 for comparison with other non-optimal shapes (circular, square and rectangular). The weight of the beam having this optimum section has the simple linear form $n = 0.9266s$. When applied to Figure 10.13, this equation shows that the weight is lowered more effectively for longer beams where it would be beneficial to machine its optimum shape. For shorter beams the lesser weight saving may not warrant the machining cost. If a standard, solid section is preferred in this instance Figure 11.6 has revealed that some of its interior material would be stressed beyond the design stress value assigned to the outer and centroidal positions, this being the procedure adopted in Chapter 10 for separated bending and shear designs.

Example 11.4 Design optimally an aluminium alloy cantilever required to carry 50 kN distributed uniformly in a 1 m length. Compare beam weights when the cross-sections are circular, square and rectangular. Take $E = 75$ GPa, $\sigma_y = 150$ MPa and $\rho = 2710$ kg/m^3.

Firstly, we find the structural index s from the load to be carried. Equation (10.37a) gives

$$s = \frac{1}{\sigma_y}\frac{F_{\max}}{L^2} = \frac{1}{150 \times 10^6}\frac{50 \times 10^3}{1^2} = 3.3 \times 10^{-4}$$

Hence, the non-dimensional weight follows from equation (11.51c) as

$$n = \frac{1}{\rho}\frac{W}{L^3} = 0.9266s = 0.9266 \times 3.3 \times 10^{-4} = 3.089 \times 10^{-4}$$

from which the beam weight is

$$W = n\rho L^3 = (3.089 \times 10^{-4}) \times 2710 \times 1^3 = 0.8370 \text{ kg}$$

The weights of cantilevers with optimum circular and square sections operating at similar structural indices fall outside the range of Figure 10.13. However, in the lower s range, where the bending design applies, their linear plots may be extrapolated to $s = 3.3 \times 10^{-4}$, where $n = 46 \times 10^{-4}$ and 43×10^{-4} for the optimum circular and square sections respectively. Thus, the weights associated with each section are increased considerably by factors 7.45 and 6.96 to 12.47 kg and 11.65 kg.

For the rectangular section beam, both the abscissa and the ordinate in Figure 10.13 have been divided by σ_y/E. To find its weight at a similar structural index we approximate that part of the plot $s' < 0.1$, where bending dominates, with the relationship $n' = 0.58s'^{0.52}$. Hence, when $s = 3.3 \times 10^{-4}$ then $s' = s/(\sigma_y/E) = 0.165$, from which $n' = 0.227$, and this gives $n = n'\sigma_y/E = 4.545 \times 10^{-4}$. It follows that this beam's weight, $W = n\rho L^3 = 1.232$ kg, is comparable to that of the optimum section, but additional buckling modes that arise with thin rectangular sections in bending could render this design unsafe (see §10.7).

For the optimum section shape, equations (11.50a,b) give the two base dimensions required. Substituting $w = 50$ kN/m(= N/mm) and $\sigma_y = 150$ MN/m^2(= N/mm^2) leads to

$$x_o = \frac{0.4458w}{\sigma_y} = \frac{0.4458 \times 50}{150} = 0.1486 \text{ mm}, \quad y_o = 2.2984z$$

which show that the optimum section has a constant width of $2x_o \approx 0.3$mm at the neutral axis with a depth $(2y_o)$ that increases linearly from zero at the free end ($z = 0$) to 4.6 m at the fixed end ($z = 1$m). Figure 11.19 shows the optimum beam shape provided by the theory, which appears to be unrealistic in this case. As we have come to expect, the optimum beam design reveals a thin, deep cross-section that tapers in its length. Here, the contrasting dimensions found raise the question of the appropriate choice of design criteria. Similar comments were applied to the thin, deep rectangular section beam designs found previously in Examples 11-1–11.3. Thus, additional criteria, to offset both flexural (lateral) and torsional buckling in thin section, would be required here (see again §10.7).

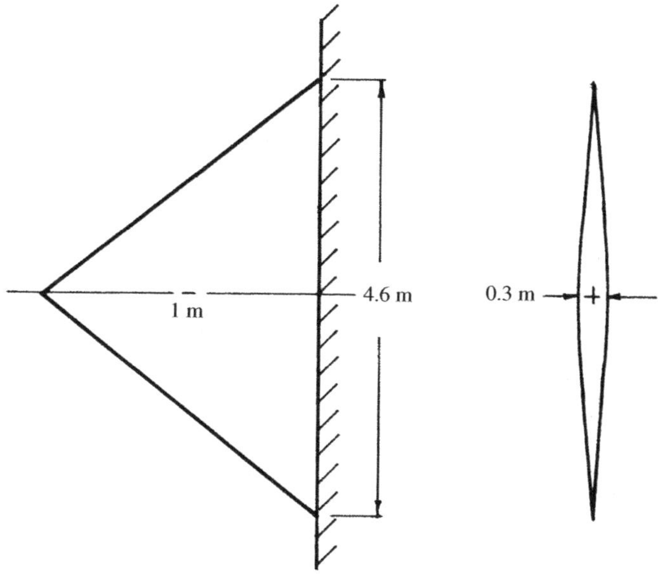

Figure 11.19 Optimum cantilever for carrying uniformly distributed loading

11.8 Summary

For the optimised cross-section, the weight function n shows a simple linear relation to the structural index s. That is $n = fs$, where the shape factor f becomes the constant of proportionality. For each of the four standard cases of loading beams having optimal cross-sections, we can interpret $n = fs$ as follows:

$$\frac{W}{L^3} = f\frac{\rho}{\sigma_y}\frac{P}{L^2}, \quad \Rightarrow \quad \frac{1}{\rho}\frac{W}{L^3} = f\frac{1}{\sigma_y}\frac{P}{L^2}, \quad \Rightarrow \quad n = fs \qquad (11.52a,b,c)$$

The structural index s is defined in terms of the *total* applied load, either P or $P = wL$ in equation (11.52b), for when the load is either concentrated or distributed. The shape factors f for these beams were found to be:

(1) simply supported beam with central concentrated load, $f = 0.9266$;
(2) simply supported beam with uniformly distributed load, $f = 0.4633$;
(3) cantilever concentrated end-load, $f = 1.8532$;
(4) cantilever distributed load, $f = 0.9266$.

Alternatively, the following comparisons apply between shape factors for simply supported and cantilever beams supporting similar total loading:

Concentrated loads P:
simply supported with load at centre, $f = 0.9266$; cantilever with load at end, $f = 1.8532$.

Distributed loads w per unit length $(P = wL)$:
simply supported, $f = 0.4633$; cantilever, $f = 0.9266$.

This comparison shows that the self-weight of both simply supported and cantilever beams is halved when loading is distributed. A total load P may be carried by a beam of least possible weight when it is simply supported provided P is uniformly distributed.

References

[1] Chakrabarty, J. *Theory of Plasticity*, Butterworth-Heinemann, 2006.
[2] Rees, D. W. A. *Basic Engineering Plasticity*, Butterworth-Heinemann, 2006.

Exercises

11.1 Find the optimum section dimensions, the weight and the maximum load that can be carried by a simply supported steel beam, loaded at its centre, when the structural index is $s = (1/\sigma_y)(F/L^2) = 1$ in a length $L = 500$ mm. Take $\rho = 7850$ kg/m^3, $E = 207$ GPa and $\sigma_y = 280$ MPa.

11.2 Reading from Figure 10.22, determine the length profile and the weight of a 5 m long, simply supported steel beam having an optimised cross-section, when its structural index is $s = 0.02$. What is the maximum distributed load that can be supported? Compare the self-weight of this beam with the self-weights of beams with circular, square and rectangular cross-sections when operating under the same conditions. Take $\sigma_y = 300$ MPa, $E = 210$ GPa and $\rho = 7860$ kg/m^3.

11.3 Find the optimum section dimensions, the self-weight and the end-load carried by a steel cantilever beam, loaded at its end, when the structural index and length are $s = 0.2$ and $L = 2$ m, respectively. Take $\rho = 7860$ kg/m^3, $E = 210$ GPa and $\sigma_y = 310$ MPa.

11.4 Design optimally an aluminium alloy cantilever required to carry 20 kN distributed uniformly in a 5 m length. Compare the weight of this cantilever with the minimum weights of cantilevers whose cross-sections are circular, square and rectangular, operating under otherwise similar conditions. Take $E = 78$ GPa, $\sigma_y = 130$ MPa and $\rho = 2700$ kg/m^3.

11.5 A simply supported beam carrying a uniformly distributed load w per unit length is to have an optimum cross-section design. Show from equations (11.45a,b) that the critical position $q = z/L$ within the half-length $(0 \le z \le L/2)$ at which this section's base dimensions (x_o and y_o) are similar is found from the solution to the following equation:

$$41.22q_c^2(1 - q_c)^2 = \frac{w}{L\sigma_y}(1 - 2q_c)^3$$

Hence, show that at this position the product of these two dimensions is given by

$$x_o y_o = 0.5124(1 - 2q_c)\left(\frac{wL}{\sigma_y}\right)$$

11.6 Design optimally an aluminium alloy cantilever required to carry 30 kN distributed uniformly in a 2 m length. Compare beam weights when the cross-sections are circular, square and rectangular. Take, for aluminum, $E = 75$ GPa, $\sigma_y = 150$ MPa and $\rho = 2710$ kg/m^3.

12

Structures under Combined Loading

12.1 Introduction

Up to now we have looked at isolated 'actions' upon structures as found in beams, ties, struts, torsion bars and shear webs. We were able to separate bending from shear in beams because their greatest stresses acted at different positions in the cross-section –the edges and the centre, respectively. However, many structures are subjected to combinations of these loadings in which the stress state which governs the section design is complex [1–8]. Particular applications to the optimum design of such structures under combinations of bending, torsion and shear will be given here. A yield criterion [1–4] is required to allow for the combined effects of the loading upon the limiting design stress. The approach to optimising a structure subjected to combined loading is to employ free-body diagrams within its separated limbs, minimise their weight and then reassemble to provide an overall minimum weight. Within this procedure it is usually possible to derive an objective function for which the shape function is at set at its minimum.

12.2 Combined Bending and Torsion

This common combination of loading arises in drive shafts and other rotating parts within transmission systems. Let the torque T and the bending moment M be uniform along the length L of a shaft of diameter d (see Figure 12.1). As with the beams and torsion bars considered in previous chapters, the maximum bending stress and the maximum shear stress occur at the surface of the shaft. They are given as

$$\sigma = \frac{32M}{\pi d^3}, \quad \tau = \frac{16T}{\pi d^3} \tag{12.1a,b}$$

Arising from the combination of these two stresses, the major and minor principal surface stresses (σ_1, σ_2), are given by [4]

$$\sigma_{1,2} = \frac{\sigma}{2} \pm \frac{1}{2}\sqrt{\sigma^2 + 4\tau^2} \tag{12.2a,b}$$

Mechanics of Optimal Structural Design: Minimum Weight Structures D. W. A. Rees

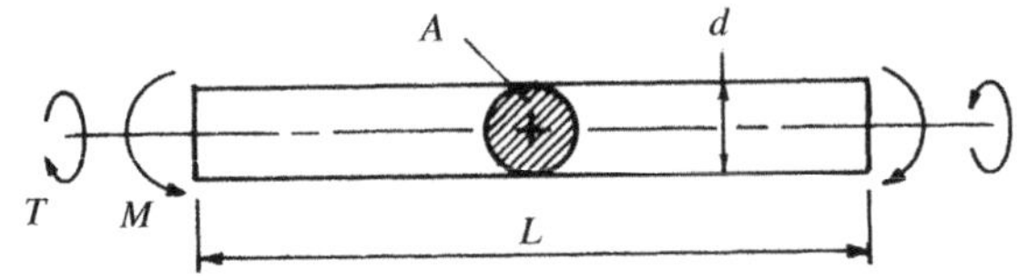

Figure 12.1 Uniform shaft subjected to combined bending and torsion

If the limiting design stress $\sigma_1 \leq \sigma_y$ is placed upon the greater (major) principal stress, an 'equivalent moment' M_E may be used to optimise the shaft diameter. Substituting equations (12.1a,b) into equation (12.2a) (positive discriminant), this moment appears in

$$\sigma_1 = \frac{32}{\pi d^3} \times \frac{1}{2}\{M + [M^2 + T^2]^{1/2}\} = \frac{32 M_E}{\pi d^3} \tag{12.3}$$

from which

$$d_{\text{opt}} = \left(\frac{32 M_E}{\pi \sigma_y}\right)^{1/3}, \quad M_E = \frac{1}{2}\{M + [M^2 + T^2]^{1/2}\} \tag{12.4a,b}$$

This approach corresponds to Rankine's yield criterion [2], which is known to be deficient in that it ignores the minor principal stress σ_2 in equation (12.2b). A better choice is to set the stress limit by the Tresca or von Mises yield criterion. We can write these as [1–4]

$$\sigma^2 + 4\tau^2 = \sigma_y^2 \quad \text{and} \quad \sigma^2 + 3\tau^2 = \sigma_y^2 \tag{12.5a,b}$$

respectively. Substituting from equations (12.1a,b) in turn, both criteria (12.5a,b) appear in a common form,

$$\frac{32}{\pi d^3}(M^2 + QT^2)^{1/2} = \sigma_y \tag{12.6}$$

where $Q = 1$ for Tresca and $Q = 3/4$ for von Mises. Equation (12.6) identifies a modified equivalent moment M_E' with a consequent modification to the optimum diameter as follows:

$$d_{\text{opt}} = \left(\frac{32 M_E'}{\pi \sigma_y}\right)^{1/3}, \quad M_E' = (M^2 + QT^2)^{1/2} \tag{12.7a,b}$$

The weight $W = \rho A L = \rho(\pi d_{\text{opt}}^2/4)L$ follows from equations (12.7a) as

$$W = \frac{\pi \rho L}{4}\left(\frac{32}{\pi \sigma_y}\right)^{2/3}(M^2 + QT^2)^{1/3} \tag{12.8}$$

and dividing equation (12.8) by L^3 provides the objective function

$$\frac{W}{L^3} = (16\pi)^{1/3}\frac{\rho}{\sigma_y^{2/3}}\frac{(M^2 + QT^2)^{1/3}}{L^2} = 3.691\frac{\rho}{\sigma_y^{2/3}}\left(\frac{M_E'}{L^3}\right)^{2/3} \tag{12.9}$$

12.3 Cranked Cantilever

A further combination of bending (with shear) and torsion appears when loading the cranked cantilevers as shown in Figure 12.2. This problem is more complicated than the previous one as the moment varies with the length z within both limbs. Moreover, the varying moment introduces a shear force $F = \mathrm{d}M/\mathrm{d}z$, but if we assume that the limb lengths L_1 and L_2 are large compared to their diameters then transverse shear effects may be ignored.

12.3.1 End-Load

As a free-body diagram, limb 1 in Figure 12.2a appears as a cantilever with end-load P. Using equation (3.3b), the optimum diameter for a pure bending design of limb 1 becomes

$$(d_1)_{\mathrm{opt}} = \left(\frac{32M(z)}{\pi\sigma_y}\right)^{1/3} = \left(\frac{32Pz}{\pi\sigma_y}\right)^{1/3} \tag{12.10}$$

which shows that d_1 varies with length z. The minimum weight is given by equation (10.4b)

$$\left(\frac{W_1}{L_1^3}\right)_{\mathrm{opt}} = 2.2143\frac{\rho}{\sigma_y^{2/3}}\left(\frac{M_{\mathrm{max}}}{L_1^3}\right)^{2/3} \tag{12.11}$$

where $M_{\mathrm{max}} = PL_1$. Limb 2 is subjected to a constant torque $T = PL_1$ and a bending moment $M(z) = Pz$ that again varies with z, having its origin in the position shown. Using the optimum diameter expression (12.7a), the diameter d_2 follows as

$$(d_2)_{\mathrm{opt}} = \left(\frac{32}{\pi\sigma_y}\right)^{1/3}[(Pz)^2 + Q(PL_1)^2]^{1/6} = \left(\frac{32PL_1}{\pi\sigma_y}\right)^{1/3}\left[Q + \left(\frac{z}{L}\right)^2\right]^{1/6} \tag{12.12}$$

The weight, integrated over the length L_2 with this varying diameter, becomes

$$W_2 = \rho\frac{\pi}{4}\left(\frac{32PL_1}{\pi\sigma_y}\right)^{2/3}\int_0^{L_2}\left[Q + \left(\frac{z}{L_1}\right)^2\right]^{1/3}\mathrm{d}z \tag{12.13a}$$

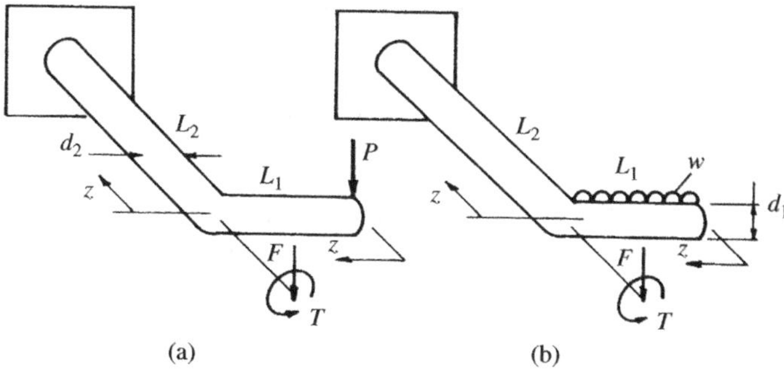

Figure 12.2 Loading of a cranked cantilever

The integral I in equation (12.13a) may be evaluated numerically from a five-term series expansion (say), where $L_2/L_1 < 1$, as

$$I = Q^{1/3}L_2\left\{1 + \frac{1}{9Q}\left(\frac{L_2}{L_1}\right)^2 - \frac{1}{45Q^2}\left(\frac{L_2}{L_1}\right)^4 + \frac{5}{567Q^3}\left(\frac{L_2}{L_1}\right)^6 - \frac{10}{3187Q^4}\left(\frac{L_2}{L_1}\right)^8 + \ldots\right\} \tag{12.13b}$$

or, where $L_1/L_2 < 1$, as

$$I = Q^{1/3}L_2\left\{\left[\left(\frac{L_1}{L_2}\right)^8 + \frac{1}{9Q}\left(\frac{L_1}{L_2}\right)^6 - \frac{1}{45Q^2}\left(\frac{L_1}{L_2}\right)^4 + \frac{5}{567Q^3}\left(\frac{L_1}{L_2}\right)^2 - \frac{10}{3187Q^4}\right]\Big/\left(\frac{L_1}{L_2}\right)^8\right\} \tag{12.13c}$$

We may take either expansion in braces in equations (12.13b,c) for the value of the integral in equation (12.13a). This leads to the common objective function

$$\left(\frac{W_2}{L_2^3}\right)_{\text{opt}} = (2^4\pi Q)^{1/3}\{\ \}\frac{\rho}{\sigma_y^{2/3}}\left(\frac{T}{L_2^3}\right)^{2/3} \tag{12.14a}$$

where {} refers to either of the five-term series expansions (12.13b,c). When $L_1/L_2 = 1$, $Q = 3/4$ and $\sigma_y = \sqrt{3}\tau_y$ (von Mises), the series term $\{\} = 1.1151$ and equation (12.14a) leads to

$$\left(\frac{W_2}{L_2^3}\right)_{\text{opt}} = 2.592\frac{\rho}{\tau_y^{2/3}}\left(\frac{T}{L_2^3}\right)^{2/3} \tag{12.14b}$$

Note that the shape factor $f = 2.592$ in equation (12.14b) differs from the shape factor $f = 2.325$ found from the application of a pure torque to a circular bar (see equation (4.4b), § 4.2). Taking $Q = 3/4$ in equation (12.14a) and adding to equation (12.11), gives the total weight of this structure:

$$\left(\frac{\sum W}{L_1^3}\right) = 2.214\frac{\rho}{\sigma_y^{2/3}}\left(\frac{M_{\max}}{L_1^3}\right)^{2/3} + 3.353\{\ \}\frac{\rho}{\sigma_y^{2/3}}\left(\frac{T}{L_2^3}\right)^{2/3}\left(\frac{L_2}{L_1}\right)^3 \tag{12.15a}$$

Substituting $M_{\max} = T = PL_1$ and introducing a non-dimensional index $s_1 = (1/\sigma_y)(P/L_1^2)$ together with the normalised weight parameter $n_1 = (1/\rho)(\sum W/L_1^3)$ simplifies equation (12.15a), for when $L_2/L_1 < 1$:

$$n_1 = [2.214 + 3.353f(L_2/L_1)]s_1^{2/3} \tag{12.15b}$$

where $f(L_2/L_1)$ is the product of L_2/L_1 and the series expansion {} in equation (12.13b). This is evaluated as

$$f\left(\frac{L_2}{L_1}\right) = \frac{L_2}{L_1}\left[1 + 0.148\left(\frac{L_2}{L_1}\right)^2 - 0.0395\left(\frac{L_2}{L_1}\right)^4 + 0.0209\left(\frac{L_2}{L_1}\right)^6 - 0.01445\left(\frac{L_2}{L_1}\right)^8 + \ldots\right] \quad (12.15c)$$

When $L_2/L_1 > 1$, the length ratio in equations (12.15b,c) may be inverted to give the net weight of the structure as:

$$n_2 = \left[2.214\left(\frac{L_1}{L_2}\right)^{5/3} + 3.353\left(\frac{L_1}{L_2}\right)^{2/3} g\left(\frac{L_1}{L_2}\right)\right] s_2^{2/3} \quad (12.16a)$$

where $s_2 = (1/\sigma_y)(P/L_2^2)$, $n_2 = (1/\rho)(\sum W/L_2^3)$ and $g(L_1/L_2)$ is the series expansion in equation (12.13c),

$$g\left(\frac{L_1}{L_2}\right) = \left[\left(\frac{L_1}{L_2}\right)^8 + 0.148\left(\frac{L_2}{L_1}\right)^6 - 0.0395\left(\frac{L_2}{L_1}\right)^4 + 0.0209\left(\frac{L_2}{L_1}\right)^2 - 0.01445 + \ldots\right] \Big/ \left(\frac{L_1}{L_2}\right)^8 \quad (12.16b)$$

12.3.2 Distributed Loading

If the load is uniformly distributed over limb 1, as shown in Figure 12.2b, then equation (10.10) is used to replace equation (12.10):

$$(d_1)_{\text{opt}} = \left(\frac{32M(z)}{\pi\sigma_y}\right)^{1/3} = \left(\frac{16wz^2}{\pi\sigma_y}\right)^{1/3} \quad (12.17a)$$

Comparing equations (12.17a) and (12.10) shows that d_1 varies with length z differently from a beam with a concentrated end-load. The minimum weight is given by equation (10.11b),

$$\left(\frac{W}{L_1^3}\right)_{\text{opt}} = 1.582\frac{\rho}{\sigma_y^{2/3}}\left(\frac{M_{\max}}{L_1^3}\right)^{2/3} \quad (12.17b)$$

where $M_{\max} = wL_1^2/2$. Hence, when equation (12.17b) replaces the first term within equation (12.15a), for $P = wL_1$ and $T = M_{\max}$, we may proceed as before to derive an optimum weight expression n_1 in terms of s_1. This is left for the reader to do (see Exercise 12.5).

12.4 Cranked Strut with End-Load

Figure 12.3 shows free-body diagrams for the two limbs of a cranked strut of circular cross-section. From these we recognise the horizontal limb as a cantilever of length L_1 with an end-load P for which the previous analysis, given in § 12.3.1, applies.

When dealing with a circular section, we saw in Chapter 10 that the optimum diameter for the cantilever beam is variable as both the shear force and bending moment influence its profile. Depending upon its length L_2, the vertical limb may be identified either as a column or a strut carrying an eccentric end-load. As a column its optimum diameter will be uniform according to the following analysis.

12.4.1 Stress Limited Column

When L_2 is short enough to offset buckling, the diameter must be chosen so that the net stress in the section does not exceed the yield stress σ_y. Figure 12.3b shows that the net stress is the sum of direct and bending stress contributions, which we write as

$$\sigma = \sigma_1 + \sigma_2 = -\frac{P}{A_2} \pm \frac{My}{I} \tag{12.18a}$$

Clearly, equation (12.18a) shows, from the additive compression, that the greatest net stress lies at the outer diameter $y = d_2/2$. We shall take this sum to be positive and limit it to σ_y:

$$\sigma_y = \frac{P}{A_2} + \frac{Md_2}{2I} \tag{12.18b}$$

Substituting $A_2 = \pi d_2^2/4$, $I = \pi d_2^4/64$ and $M = PL_1$ into equation (12.18b) leads to a cubic in d_2:

$$d_2^3 - \left(\frac{4P}{\pi\sigma_y}\right)d_2 - \left(\frac{32PL_1}{\pi\sigma_y}\right) = 0 \tag{12.19a}$$

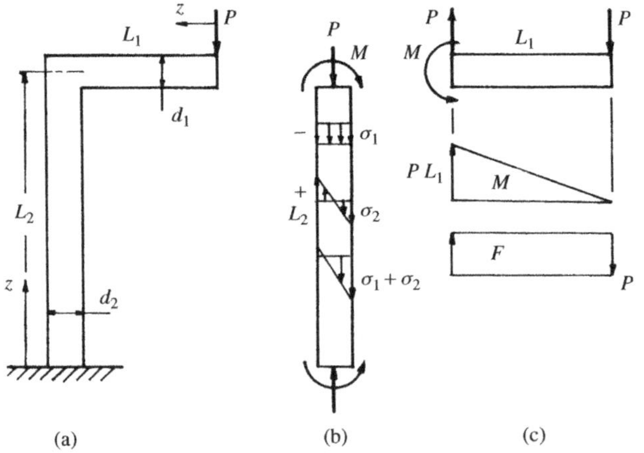

Figure 12.3 Cranked strut showing each limb as free-body diagrams

When this equation is normalised to include the structural index $s_1 = P/(\sigma_y L_1^2)$, used earlier for the beam design, it becomes

$$\left(\frac{d_2}{L_2}\right)^3 - 1.273\left(\frac{L_1}{L_2}\right)^2\left(\frac{d_2}{L_2}\right)s_1 - 10.186\left(\frac{L_1}{L_2}\right)^3 s_1 = 0 \qquad (12.19\text{b})$$

which allows the two limbs to be optimised at similar values of s_1. Equation (12.19a) shows that d_2 is independent of L_2, this restricting the design to short supporting columns which are not likely to buckle, i.e. $L_2 < L_{cr}$, where L_{cr} is the minimum length of an Euler strut with similar end conditions, as demonstrated in the following example.

Example 12.1 A circular steel column is to support a concentrated load of $P = 15$kN with an overhang $L_1 = 1$ m from its longitudinal axis (see Figure 12.3a). Find the diameter d_2 of the vertical column when $L_2 = 500$ mm and show that this length is valid for the short column theory employed. Take $E = 210$ GPa and $\sigma_y = 300$ MPa.
Firstly, we evaluate the structural index value s_1 for the cantilever overhang as

$$s_1 = \frac{1}{\sigma_y}\frac{P}{L_1^2} = \frac{15 \times 10^3}{(300 \times 10^6)1^2} = 50 \times 10^{-6}$$

As this value lies outside the range given in Figure 10.8, the implication is that the cantilever profile design is dominated by bending. Hence, equation (12.10) will provide the variation in optimum diameter d_1 with the length z taken from the free end (see Figure 12.3a). For the 500 mm long column, we substitute $L_1/L_2 = 2$ with $s_1 = 50 \times 10^{-6}$ into equation (12.19b):

$$\left(\frac{d_2}{L_2}\right)^3 - 2.546 \times 10^{-4}\frac{d_2}{L_2} - 4.0744 \times 10^{-3} = 0$$

for which the solution is $d_2/L_2 \approx 0.16$, giving $d_2 = 0.16 \times 500 = 80$ mm. The slenderness ratio, $L_2/k = 500/(80/4) = 25$, provides a check that the critical length of the column has not been exceeded, i.e. where strut theory would apply. This equates the Euler buckling stress for a strut with 'fixed-free' ends to the material's yield stress, thereby providing a critical slenderness ratio below which column theory applies:

$$\sigma_y = \frac{\pi^2 E}{4(L_2/k)^2}, \quad \Rightarrow \quad \left(\frac{L_2}{k}\right)_{cr} = \frac{\pi}{2}\sqrt{\frac{E}{\sigma_y}} = \frac{\pi}{2}\sqrt{\frac{210 \times 10^3}{300}} = 41.56$$

Hence our ratio of 25 would be valid if the load were to be applied along the column axis, but as it is not, the effect of the eccentricity is likely to be more restrictive (see § 12.4.2). For slenderness ratios greater than this critical value the following strut theory applies.

12.4.2 Optimum Strut Design

The vertical member in Figure 12.3a may be treated as a strut having one end fixed with an eccentric load applied with its free-end connection to the cantilever beam. This is a standard case in strut theory, where the bending moment at position z from the fixed end is [1–4]

$$M(z) = PL_1 \sec(\alpha L_2)\cos(\alpha z) \tag{12.20a}$$

where the 'eccentricity' is L_1 and $\alpha = \sqrt{(P/EI_2)}$. Equation (12.20a) allows for the deflection in the strut, the greatest and least moments being those at the fixed and free ends respectively, where $z = 0$ and $z = L_2$. These are

$$M_{\max} = M(0) = PL_1 \sec(\alpha L_2) \tag{12.20b}$$

$$M_{\min} = M(L_2) = PL_1 \tag{12.20c}$$

Equation (12.18b) again gives the net compressive stress in the strut. At position z we write this as a positive summation of the direct and bending compressive stresses:

$$\sigma(z) = \frac{P}{A_2} + \frac{M(z)}{M_{\max}}\frac{M_{\max}y}{I_2} \tag{12.21a}$$

Substituting for $M(z)$ and $M_{\max}$ from equations (12.20a,b) into equation (12.21a),

$$\sigma(z) = \frac{P}{A_2} + \frac{PL_1 y}{I_2}\cos(\alpha z)\sec(\alpha L_2) \tag{12.21b}$$

where $A_2 = \pi d_2^2/4$, $I_2 = \pi d_2^4/64$ and $y = d_2/2$. Equation(12.21b) shows that d_2 will vary with z for an optimum design where, say, $\sigma(z) = \sigma_y =$ constant. In particular, at the free end, where $z = L_2$, equation (12.21b) gives

$$\sigma_y = \frac{4P}{\pi d_2^2}\left(1 + \frac{8L}{d_2}\right) \tag{12.22a}$$

which allows the end diameter d_2 to be found. Elsewhere in the length the trial solution to $d_2 = d_2(z)$ from equation (12.21b) is assisted by Webb's approximation to the sec (αL_2) term [2],

$$\sec(\alpha L_2) = \frac{(\pi/2)^2 + (\pi^2/8 - 1)(\alpha L_2)^2}{(\pi/2)^2 - (\alpha L_2)^2} = \frac{P_E + 4P(\pi^2/8 - 1)}{P_E - 4P} \tag{12.22b}$$

where $P_E = \pi^2 EI_2/L_2^2$ is Euler's buckling load for a pinned-end strut of similar dimensions. For example, at the fixed end, where $z = 0$, the combination of equations (12.21b) and (12.22b) leads to an equation including the maximum, fixed-end diameter for this strut:

$$\sigma_y = \frac{4P}{\pi d_2^2}\left\{1 + \frac{8L_1}{d_2}\left[\frac{\pi^3 Ed_2^4 + 256P(\pi^2/8 - 1)L_2^2}{\pi^3 Ed_2^4 - 256PL_2^2}\right]\right\} \tag{12.22c}$$

In the optimum design, equations (12.22a,c) supply the end-diameters for a tapered length profile. In a safe but non-optimum design, the maximum diameter, found from equation (12.22c), may be held constant for the full length of the strut. Equation (12.22c) can also be rearranged to give the critical length of a uniform strut, at any given diameter d_2, bearing a load P offset by an amount L_1. Thus, for lengths L_2 less than the following critical value, the column theory given above applies:

$$(L_2)^2_{\mathrm{cr}} = \frac{2\left(\frac{\pi}{8}\right)^3 E d_2^4 \left\lfloor \left(\sigma_y - \frac{4P}{\pi d_2^2}\right)\frac{\pi d_2^3}{32PL_1} - 1 \right\rfloor}{P\left\lceil \left(\frac{\pi^2}{8} - 1\right) + \left(\sigma_y - \frac{4P}{\pi d_2^2}\right)\frac{\pi d_2^3}{32PL_1} \right\rceil} \tag{12.22d}$$

12.5 Cranked Bracket with End-Load

Figure 12.4 shows a vertical bracket arrangement carrying an end-force P. Each circular-section limb experiences different actions. The horizontal limb 1 is subjected to a bending moment $M(z) = Pz$ (ignoring shear). The vertical limb 2 is placed under combined bending $M = PL_1$ and axial tension P, as shown (not additional forces).

12.5.1 Stress Limited Tie

The free-body diagrams for this problem are similar to those for Figure 12.3, as described in § 12.4, except that buckling does not arise in the vertical tension member (tie bar 2). In this case equation (12.18b) refers to the net stress in the tie bar, found by adding the stresses from axial tension and bending effects. With a stress limited tension member it will be shown that equation (12.19a) may again be used to find its optimum diameter. In certain materials, known to be weaker in tension than in compression ($\sigma_{yt} < \sigma_{yc}$), the design stress appearing in these equations will need to be reduced. Thus, a stress limited condition will optimise the tensile bar such that the net stress in its cross-section does not exceed the tensile yield stress σ_{yt}. The greatest net stress, which is the sum of direct and

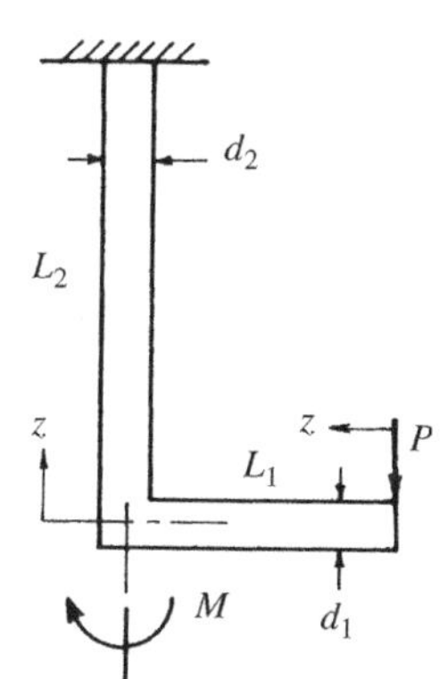

Figure 12.4 Cranked bracket with end-load

bending contributions, lies at the outer surface ($y = d_2/2$) on the tensile side, which we limit to σ_{yt}:

$$\sigma_{yt} = \frac{P}{A_2} + \frac{Md_2}{2I} \tag{12.23a}$$

Substituting $A_2 = \pi d_2^2/4$, $I = \pi d_2^4/64$ and $M = PL_1$ into equation (12.23a) leads to a cubic in d_2:

$$d_2^3 - \frac{4P}{\pi\sigma_{yt}}d_2 - \frac{32PL_1}{\pi\sigma_{yt}} = 0 \tag{12.23b}$$

which is identical to equation (12.19a) but without the previous problem of having a length restriction imposed upon L_2 from compressive buckling. Hence, to support a load of $P = 15$ kN at $L_1 = 1$m from the axis of a steel tie bar, whose limiting tensile strength is 300 MPa, then equation (12.23b) will again give a minimum uniform diameter of 80 mm irrespective of length L_2 (see Example 12.1).

12.5.2 Effect of Shear Force

So far the shear force effect upon the design of limb 1 (Figures 12.3 and 12.4) has been neglected. This would be a reasonable assumption to make when bending dominates in long limbs where the length L_1 is comparable with L_2. However, with an eccentric loading of the strut in Figure 12.3a or the tie bar in Figure 12.4, where the offset L_1 from the axis of each is reckoned to be small, then the effect of shear in the short beam must be considered.

The free-body diagram for this limb (see Figure 12.3c) identifies with the loading imposed upon an end-loaded cantilever, for which a solution to the optimum design under combined bending and shear is given in § 10.3. It is sufficient to call upon the results given there to provide this account of shear in which the F-diagram shows that the shear force is constant at $|F| = P$. Thus, for a short, cantilever of circular cross-section with index $s_1 \geq 4.081$, shear governs the design. The diameter d_1 is held constant throughout the length at $d_1/2L_1 = 0.8574\sqrt{s_1}$ and, from equation (10.21b), the minimum weight is linear in s_1,

$$n_1 = 2.3094s_1 \tag{12.24a}$$

where $s_1 = (1/\sigma_y)(P/L_1^2)$. On the other hand, when $s_1 < 4.08$, both shear and bending govern the design of limb 1. A critical position in its length $(z/L)_{\text{cr}} = 0.495\sqrt{s}$ divides the two designs such that shear applies to the sub-region $0 \leq z/L \leq (z/L)_{\text{cr}}$ across which the radius is held constant at $d/2L = 0.8574\sqrt{s}$, as for the all-shear design. A bending design applies to the adjacent region $(z/L)_{\text{cr}} \leq z/L \leq 1$, where the radius varies as $d/2L = 1.0838(sz/L)^{1/3}$. The contributions to the minimum weight from within the two regions are additive (see equation (10.20b)):

$$n_1 = 2.2143s_1^{2/3} + 0.4572\,s_1^{3/2} \tag{12.24b}$$

Equations (12.24a,b) agree, giving $n_1 = 9.42$, at the lowest structural index $s_1 = 4.081$, for which an all-shear design applies. Note that bending dominates the design for $s_1 \ll 4.081$, but shear is never entirely eliminated, if we are to uphold the requirement that the beam should never taper to a point at its ends. The graphical presentation of equations (12.24a,b) allows n to be found across five decades of s (see Figure 10.8), this providing for designs based entirely upon shear ($s > 4.081$), those based upon dominant bending ($s \ll 4.08$) and other interactive combined designs ($s < 4.081$). Note, that here $<$ refers to less than by approximately one order of magnitude and $\ll$ is less than by two or more orders of magnitude. When the shear design ($s > 4.081$) applies to the cantilever limb carrying the end-load in Figure 12.4, then the total weight of the whole structure $n = n_1 + n_2$ is written as

$$\frac{1}{\rho}\frac{W}{L_1^3} = 2.3094 s_1 + \pi\left(\frac{d_2}{2L_2}\right)^2\left(\frac{L_2}{L_1}\right)^3 \tag{12.25a}$$

where n_1 is given by equation (12.24a). For the interactive design ($s < 4.08$) the total weight becomes

$$\frac{1}{\rho}\frac{W}{L_1^3} = 2.214 s_1^{2/3} + 0.457 s_1^{3/2} + \pi\left(\frac{d_2}{2L_2}\right)^2\left(\frac{L_2}{L_1}\right)^3 \tag{12.25b}$$

where n_1 is given by equation (12.24b). In each of equations (12.25a,b) $d_2/2L_2$ for the tie-bar will depend upon the structural index s_1 for the cantilever. This dependence follows from equation (12.23b) as the solution to

$$\left(\frac{d_2}{2L_2}\right)^3 - \frac{s_1}{\pi}\frac{d_2}{2L_2}\left(\frac{L_1}{L_2}\right)^2 - \frac{4s_1}{\pi}\left(\frac{L_1}{L_2}\right)^3 = 0 \tag{12.25c}$$

Example 12.2 Optimise the cranked bracket shown in Figure 12.4 when the 1 m long tie is cranked to support a tensile load of 50 kN offset by 25 mm from the tie's longitudinal axis. The bar is to have a circular section in steel with $\sigma_{yt} = 280$ MPa and $\rho = 7860\text{kg/m}^3$.

The cantilever is to operate under the following structural index:

$$s_1 = \frac{1}{\sigma_{yt}}\frac{P}{L_1^2} = \frac{1}{280}\frac{50 \times 10^3}{25^2} = 0.286$$

and, since $s_1 < 4.08$, by approximately one order of magnitude, this limb will require an interactive design. Firstly, the tie's diameter is found from equation (12.25c),

$$\left(\frac{d_2}{2L_2}\right)^3 - 56.9 \times 10^{-6}\frac{d_2}{2L_2} - 5.69 \times 10^{-6} = 0$$

from which $d_2/2L_2 = 0.01892$($d_2/2 = 18.92$ mm). Substituting into equation (12.25b) gives the normalised weight parameter for this structure,

$$\frac{1}{\rho}\frac{W}{L_1^3} = 2.214 \times 0.286^{2/3} + 0.457 \times 0.286^{3/2} + \pi \times 0.01892^2(1000/25)^3 = 73.0$$

from which the total weight follows as

$$W = 73\rho L_1^3 = 73 \times 7860 \times 0.025^3 = 8.97 \text{ kg}$$

Referring to § 12.5.2, the diameter of the cantilever $d_1/2L_1 = 0.8574\sqrt{s_1} = 0.4585$ is constant at $d_1 = 22.93$ mm within the region $0 \leq z/L_1 \leq (z/L_1)_{cr}$, where $(z/L_1)_{cr} = 0.495\sqrt{s_1} = 0.265$. In the remaining region of its length $(z/L_1)_{cr} \leq z/L_1 \leq 1$, the normalised radius varies as $d_1/2L_1 = 1.0838(s_1 z/L_1)^{1/3} = 0.714(z/L_1)^{1/3}$, i.e. from 22.93 mm to 35.7 mm at its junction with the tie bar.

12.6 Portal Frame with Central Load

The portal frame in Figure 12.5 may be treated as a horizontal beam carrying a central load P, resting upon two simple end-supports provided by long vertical struts. Figure 10.18 dealt with this optimum beam design in three common cross-sections and so we may read its weight parameter n for any index value s within five decades. Each support may be designed as a minimum length strut, with one end hinged and the other end fixed, in any of these sections. The optimum design of the individual elements of this structure is best demonstrated within the following example.

Example 12.3 Find the dimensions of the portal frame in Figure 12.5 when it is required to carry a central load of $P = 50$ kN operating at structural indices: (i) $s = 1$ and (ii) $s = 0.01$. Assume that each stanchion is an optimally designed, minimum length, circular-section strut, fixed at its base and hinged to each end of the beam. Take both beam and strut to be made of steel with $\sigma_y = 300$ MPa, $\rho = 7860$ kg/m^3 and $E = 210$ GPa.

(a) Beam Design

We showed in § 10.5.1 that a short-length beam design may be based entirely upon shear when $s \geq 2.041$. Here, however, $s < 2.041$ for both the indices given and so each of the required designs involves both bending and shear. The bending contribution leads to beam designs with central, variable cross-sections, as illustrated in Figures 10.17 and 10.19. In

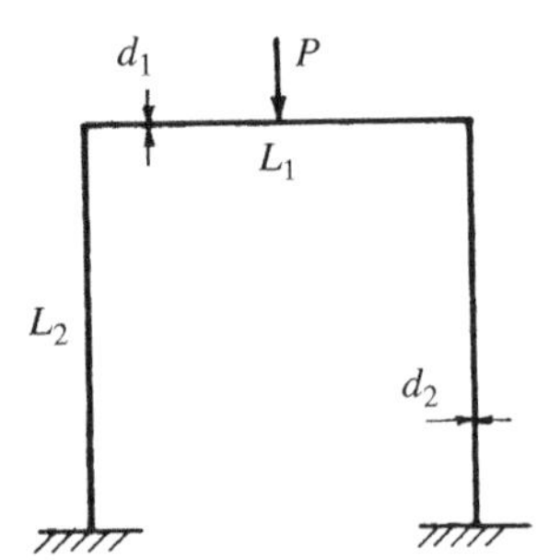

Figure 12.5 Portal frame carrying a central-concentrated load

the case of a circular cross-section, for example, Figure 10.17 shows the optimum beam design for $s = 1$. The length L_1, used to normalise its diameter as $d/2L_1$, follows from

$$s = \frac{1}{\sigma_y} \times \frac{P}{L_1^2}, \quad \Rightarrow \quad L_1 = \left(\frac{P}{s\sigma_y}\right)^{1/2} \tag{i}$$

and with $P = 50$ kN, we find an optimum length $L_1 = 12.91$ mm for the beam when $s_1 = 1$. Referring to Figure 10.17, the maximum and minimum diameters within the beam profile are $0.606(2 \times 12.91) = 15.66$ mm and $0.683(2 \times 12.91) = 17.63$ mm. The lesser diameter terminates at the critical length positions $(z/L_1)_{cr} = 0.35$ (i.e. 4.52 mm) from each end. The greater diameter applies to the beam centre (see Figure 12.6a).
For $s = 0.01$, equation (i) gives a tenfold increase in the length, $L_1 = 129.1$mm. A similar optimum beam profile applies (see Figure 12.6b), where the minimum end-diameter is found from a shear design (see equation 10.55b):

$$\left(\frac{d}{2L_1}\right)_{opt} = \left(\frac{2\sqrt{3}}{3\pi}\right)^{1/2} (0.01)^{1/2} = 0.0603, \Rightarrow d_{opt} = 0.0603 \times 2 \times 129.1 = 15.56 \text{ mm}$$

The end-diameter extends over a parallel length $0 \le z \le z_{cr}$, where z_{cr} is given by equation (10.56a):

$$\left(\frac{z}{L_1}\right)_{cr} = \left(\frac{\sqrt{3}}{3}\right)^{3/2} \left(\frac{2}{\pi}\right)^{1/2} 0.01^{1/2} = 0.035, \Rightarrow z_{cr} = 0.035 \times 129.1 = 4.52 \text{ mm}$$

For $z_{cr} \le z \le L_1$, the bending design controls the diameter, which increases according to equation (10.55a), reaching its maximum at the beam centre:

$$\left(\frac{d}{2L_1}\right)_{opt} = \left(\frac{2}{\pi}\right)^{1/3} (0.01)^{1/3} (0.5)^{1/3} = 0.147, \Rightarrow d_{opt} = 0.147 \times 2 \times 129.1 = 37.96 \text{ mm}$$

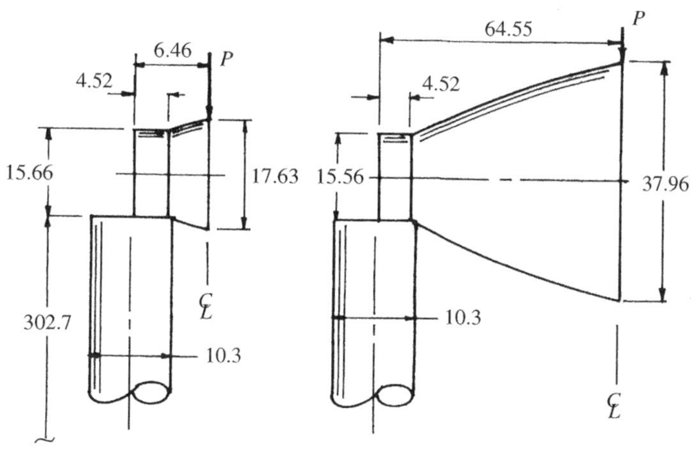

Figure 12.6 Optimum portal-frame designs

The short lengths, found for each s_1-value, reflect the influence of the shear contribution to each design at their optimum weights. The latter are calculated from n as follows:

$$n = \frac{1}{\rho}\frac{W}{L_1^3}, \quad \Rightarrow \quad W = \rho L_1^3 n \tag{ii}$$

where the n-value is read from Figure 10.18, or calculated from equation (10.58b), corresponding to $s = 0.01$. This gives $n = 0.041$, when from equation (ii),

$$W = 7860 \times 129.1^3 \times 10^{-9} \times 0.041 = 0.693 \text{ kg}$$

For $s = 1, n = 1.2021$, when equation (ii) gives an almost negligible weight for the short beam:

$$W = 7860 \times 12.91^3 \times 10^{-9} \times 1.2021 = 0.0203 \text{ kg}$$

(b) Minimum Length of Fixed-Pinned Euler Struts

When the yield stress σ_y is reached in each strut's circular-section (see Figure 12.5) they are required to buckle under their compressive reactions $F = P/2$. The strut's slenderness ratio follows from the Euler theory as

$$F = \frac{2\pi^2 EI}{L_2^2} = \frac{2\pi^2 EAk^2}{L_2^2}, \quad \Rightarrow \quad \left(\frac{L_2}{k}\right)^2 = \frac{2\pi^2 E}{\sigma_y} \tag{iii}$$

Now as $\sigma_y = F/A = 2P/\pi d_2^2$ and $k^2 = d_2^2/16$, we find from equation (iii) the minimum diameter and length expressions for an Euler supporting strut with these end conditions:

$$d_2 = \left(\frac{2P}{\pi\sigma_y}\right)^{1/2} = \left(\frac{2 \times 50 \times 10^3}{\pi \times 300}\right)^{1/2} = 10.3 \text{ mm}$$

$$L_2 = \left(\frac{\pi EP}{4\sigma_y^2}\right)^{1/2} = \left(\frac{\pi \times 210 \times 10^3 \times 50 \times 10^3}{4 \times 300^2}\right)^{1/2} = 302.7 \text{ mm}$$

giving the total weight of the two struts as

$$W = 2 \times \frac{\rho\pi d_2^2 L_2}{4} = 2 \times 7860 \times 10^{-9} \times \frac{\pi}{4} \times 10.3^2 \times 302.7 = 0.397 \text{ kg}$$

Figure 12.6 shows the optimum dimensions calculated for the frame at each structural index. The beam and strut designs when assembled show that the connection may be made conveniently between the uniform diameter struts and the constant diameter beam ends. We see from this that the beam with the greater s-value requires support over most of its length, in which the combined weight of the assembly becomes $0.397 + 0.0203 = 0.417$ kg. The longer beam requires identical supporting struts, giving a combined weight to the frame of $0.397 + 0.697 = 1.094$ kg. In practice, we would not allow the strut to operate at the yield stress. Therefore, the need arises to increase the strut diameter beyond 10.3 mm to lessen its internal stress and thus provide the required safety factor.

Alternatively, if we require supporting struts of greater length we should adopt the optimum design of a circular-section strut, discussed in § 1.3.1, in which the buckling stress and limiting compressive design stress σ_d are made equal. That is,

$$\sigma_d = \frac{2\pi^2 E}{(L_2/k)^2} = \frac{4F}{\pi d_2^2} \tag{iv}$$

and with $k^2 = d_2^2/16$, $F = P/2$, the strut diameter is expressed from equation (iv) as

$$d_2 = \frac{2}{\pi^{3/4}}\left(\frac{PL_2^2}{E}\right)^{1/4}, \quad \Rightarrow \quad \frac{d_2}{2L_2} = \left(\frac{\sigma_d}{\pi^3 E}\right)^{1/4} s^{1/4} \tag{v, vi}$$

Here, the slenderness ratio for the Euler strut must lie in its valid range, i.e. greater than the critical value $L_2/k = 117.55$, found from equation (iii). Suppose, for example, that the length is 500 mm. Then from equation (v), $d_2 = 13.24$ mm, giving $L_2/k = 4L_2/d_2 = 151.07$ and $\sigma_d = 181.6$ MPa. On the other hand, for an invalid length of 200 mm, equation (v) gives $d_2 = 8.373$ mm, $L_2/k = 95.55$ and $\sigma_d = 454.03$ MPa, which exceeds the yield stress.

The length calculated from equation (vi) would allow each strut to operate under a safe design stress at a similar structural index to the beam within the range $s \leq s_c$. Equating (iii) and (v) gives the critical value $s_c = 4\sigma_d/\pi E$, below which this can be arranged. When $\sigma_d = \sigma_y$, the limiting index becomes $s_c = 1.82 \times 10^{-3}$. Hence, this design cannot be achieved for greater s-values, including the two values given. For example, with $s = 0.01$, equation (vi) becomes $d_2/(2L_2) = 0.0693 s^{1/4} = 0.022$, which corresponds to a slenderness ratio of 91.25, compared to a ratio of 117.55 required for a valid Euler strut.

12.7 Cantilever with End and Distributed Loading

Consider a cantilever beam of circular section carrying a uniformly distributed load w per unit length together with an end-load P (see Figure 12.7a). The shear force and bending

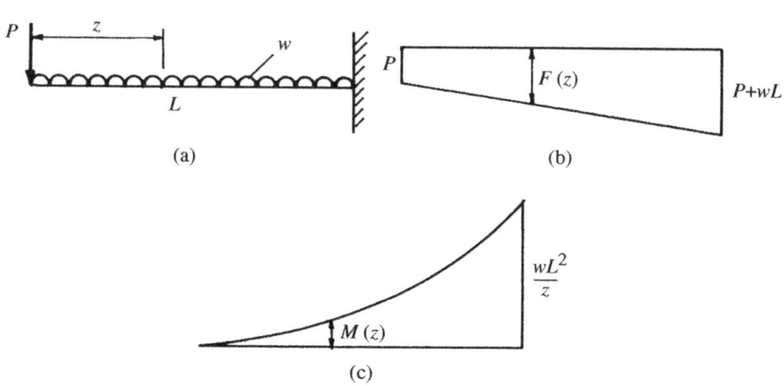

Figure 12.7 F- and M-diagrams for a cantilever under combined loads

moment diagrams under the combined loads are as shown in Figure 12.7b,c. From these diagrams the beam diameter is to be the greater of the two values found by applying moment and force criteria (see equations (3.3b) and (5.15b)) to the ordinates in each figure. Setting $F_1 = P$ and $F_2 = wL$, we have $F_2/F_1 = wL/P$ with the magnitudes of the ordinates in the F- and M-diagrams:

$$|F(z)| = F_1 + F_2 z/L \tag{12.26a}$$

$$|M(z)| = F_1 z + F_2 z^2/2L \tag{12.26b}$$

The various sign conventions [1–8], used for F- and M-diagrams, should not alter equations (12.26a,b). Correspondingly, the normalised beam diameters required to withstand the shear force and moment, when taken separately, follow from equations (12.26a,b) as

$$\left(\frac{d}{2L}\right)_F = \left(\frac{4}{\sqrt{3}\pi}\right)^{1/2} s_1^{1/2}\left[1 + \left(\frac{s_2}{s_1}\right) q\right]^{1/2} \tag{12.27a}$$

$$\left(\frac{d}{2L}\right)_M = \left(\frac{4}{\pi}\right)^{1/3} s_1^{1/3}\left[q + \frac{1}{2}\left(\frac{s_2}{s_1}\right) q^2\right]^{1/3} \tag{12.27b}$$

where $q = z/L, s_1 = (1/\sigma_y)(F_1/L^2)$ and $s_2 = (1/\sigma_y)(F_2/L^2)$ are positive variables which we shall vary in the ratio $s_2/s_1 (= wL/P)$ across several decades of s_1 from 0.001 to 100. Taking this range for s_1 will embrace both very long and short beams in which the design varies according to the relative contributions from bending and shear. Note that s_2/s_1 is also the force ratio $F_2/F_1 = wL/P$. Thus, if we equate (12.27a,b), the critical position q_c in the length is found at which the moment-based design becomes shear-based, with s increasing:

$$\left[q_c\left(1 + \frac{q_c}{2}\frac{s_2}{s_1}\right)\right]^{1/3} = 0.7911 s_1^{1/6}\left(1 + \frac{s_2}{s_1} q_c\right)^{1/2} \tag{12.28}$$

The total weight W is found by summing $\int \rho(\pi d^2/4)\mathrm{d}z$ expressions within each region. The regions refer to a shear design for $0 \le q \le q_c$ and a bending design for $q_c \le q \le 1$. This allows us to write the normalised weight parameter across the two regions as a sum:

$$n = \frac{1}{\rho}\frac{W}{L^3} = \pi \int_0^{q_c} \left(\frac{d}{2L}\right)^2 \mathrm{d}q + \pi \int_{q_c}^{1} \left(\frac{d}{2L}\right)^2 \mathrm{d}q \tag{12.29a}$$

Substituting $d/2L$ from equations (12.27a,b) into equation (12.29a), n then appears in terms of q, s_1 and s_2/s_1, with contributions to the design diameter from both shear and bending:

$$n = \frac{4}{\sqrt{3}} s_1 \int_0^{q_c} \left(1 + \frac{s_2}{s_1} q\right) \mathrm{d}q + \pi \left(\frac{4}{\pi}\right)^{2/3} s_1^{2/3} \int_{q_c}^{1} \left[q\left(1 + \frac{1}{2}\frac{s_2}{s_1} q\right)\right]^{2/3} \mathrm{d}q \tag{12.29b}$$

Table 12.1 Critical length positions, with $s_2/s_1 = 0.1$, for shear- and moment-based designs

s_1	1	0.1	0.01	0.001
q_c	0.5216	0.1597	0.050	0.0157

The design becomes wholly shear-based at a critical value of $s_1 = s_{1FM}$, which is found by setting $q_c = 1$ in equation (12.28). This gives

$$s_{1FM} = 4.081\left(1 + \frac{1}{2}\frac{s_2}{s_1}\right)^2 \Big/ \left(1 + \frac{s_2}{s_1}\right)^3 \tag{12.30a}$$

For example, taking three ratios $s_2/s_1 = 0.1$, 1 and 10, equation (12.30a) gives the critical indices as $s_{1FM} = 3.3804$, 1.1478 and 0.1104, respectively. Here the wholly shear-based weight is found by altering the limits within the first integral in equation (12.29b). That is,

$$n = \frac{4}{\sqrt{3}}s_1 \int_0^1 \left(1 + \frac{s_2}{s_1}q\right)\mathrm{d}q = 1.1547s_1\left(2 + \frac{s_2}{s_1}\right) \tag{12.30b}$$

With the index (or 'force') ratio $s_2/s_1 = 0.1$, equation (12.30a) gives $s_{1FM} = 3.3804$. Thus, for $s_2/s_1 = 0.1$ we must have $s_1 < 3.3804$ if $q_c < 1$ for the design to be both shear- and moment-based. Table 12.1 gives q_c values across three decades of s_1 that satisfy this requirement. The q_c values in Table 12.1 were found by trial as the solution to equation (12.28).

Once q_c is known, the total weight of the beam, at each s_1-value, is found from equation (12.29b):

$$n = 2.3094s_1 \int_0^{q_c} (1 + 0.1q)\mathrm{d}q + 3.6905s_1^{2/3} \int_{q_c}^1 [q(1 + 0.05q)]^{2/3}\mathrm{d}q \tag{12.31}$$

The second integral in equation (12.31) is evaluated numerically for each combination (s_1, q_c) in Table 12.1. When $s_1 \geq 3.3804$, then $q_c = 1$, and the section diameter becomes a wholly shear-based design. Correspondingly, equation (12.30b) applies to give the normalised weight $n = 2.4249s_1$. Table 12.2 relates the index s_1 to the weight parameter n over five decades.

Table 12.2 Optimised weight parameters over the full range of s_1 for $s_2/s_1 = 0.1$

s_1	0.001	0.01	0.1	1	3.3804	10	100
n	0.0223	0.1042	0.49645	2.73	8.197	24.249	242.49

Table 12.3 Critical length positions, with $s_2/s_1 = 1.0$, for shear- and moment-based designs

s_1	1	0.1	0.01	0.001
q_c	0.890	0.185	0.0521	0.0159

With a force (index) ratio $s_2/s_1 = 1.0$, equation (12.30a) gives $s_{1FM} = 1.1478$. Hence, for $s < s_{1FM}$, we find, by trial from equation (12.28), the q_c-values entered in Table 12.3. The optimised weights in Table 12.4 apply to the full range of s_1.

Table 12.4 Optimised weight parameters for the full range of s_1 with $s_2/s_1 = 1.0$

s_1	0.001	0.01	0.1	1	1.1478	10	100
n	0.02634	0.1224	0.5836	3.476	3.9761	34.641	346.41

Finally, when $s_2/s_1 = 10$, equation (12.30a) gives $s_{1FM} = 0.1104$. With $s < s_{1FM}$ we find from equation (12.28) the q_c-values in Table 12.5. The optimised weights in Table 12.6 apply to the full range of s_1.

Table 12.5 Critical length positions, with $s_2/s_1 = 10$, for shear- and moment-based designs

s_1	0.1	0.01	0.001
q_c	0.90	0.089	0.0186

The results given in Tables 12.2, 12.4 and 12.6, are presented graphically in Figure 12.8. Also included in Figure 12.8 are the results for ratios (i) $s_2/s_1 = 0$, when P acts alone upon a cantilever (taken from Figure 10.8) and (ii) $s_2/s_1 = -0.1$, where the distributed loading is propped at its end. Any intermediate ratio s_2/s_1 may be interpolated within the figure to cover optimum beam designs in circular-section. Negative ratios s_2/s_1, which arise in the case of end-propping this cantilever, are given separate consideration in § 12.9.

Table 12.6 Optimised weight parameters over the full range of s_1 with $s_2/s_1 = 10$

s_1	0.001	0.01	0.1	0.1104	1	10	100
n	0.0592	0.2756	1.39	1.53	13.856	138.56	1385.6

Example 12.4 Find the geometry and the minimum weight of a 0.25 m long, circular-section, steel cantilever when designed optimally to support a uniformly distributed load of $w = 80$ kN/m together with a concentrated end-load of $P = 20$ kN. Take $\sigma_y = 320$ MPa and $\rho = 7850$ kg/m^3.

The load (or index) ratio for this load combination is

$$\frac{s_2}{s_1} = \frac{wL}{P} = \frac{80 \times 0.25}{20} = 1$$

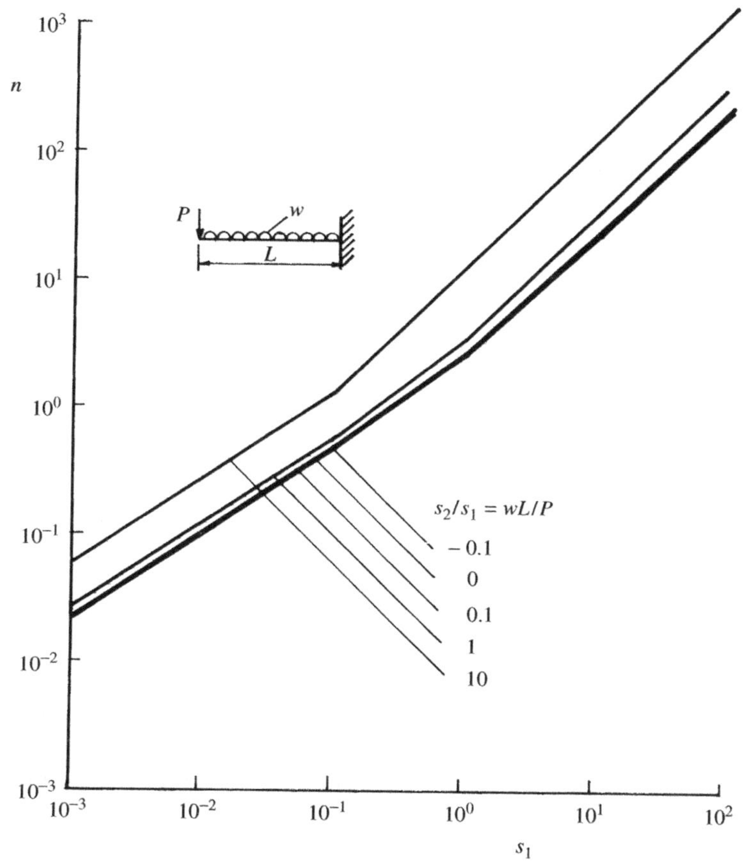

Figure 12.8 Minimum weight of circular section cantilever under combined loading (note: $s_1 = (1/\sigma_y)(P/L^2)$)

Correspondingly, equation (12.30a) gives $s_{1FM} = 1.1478$ for an all-shear design. Here the structural index s_1 follows from the concentrated load P as

$$s_1 = \frac{1}{\sigma_y}\frac{P}{L^2} = \frac{1}{320 \times 10^6}\frac{20 \times 10^3}{0.25^2} = 0.001$$

Clearly, $s_1 < s_{1FM}$ and so both bending and shear contribute to the optimum design. With s_1 at the lower end of the range, Table 12.3 gives the critical length position at $q_c = 0.0159$ (i.e. $z_{cr} = q_c L = 3.98$ mm). Hence, shear applies to the length region $0 \le q \le q_c$ and bending to the region $q_c \le q \le 1$. Thus, bending will dominate this design but, in what follows, we shall see that shear is important to sizing the beam at its free end.

For $0 \le q \le 0.0159$, equation (12.27a) gives the shear-based diameters as

$$\left(\frac{d}{2L}\right)_F = \left(\frac{4}{\sqrt{3}\pi}\right)^{1/2} s_1^{1/2}\left(1 + \frac{s_2}{s_1}q\right)^{1/2} = 0.02711(1+q)^{1/2} \qquad \text{(i)}$$

Equation (i) gives the two limiting diameters

$$q = 0, \quad \Rightarrow \quad \left(\frac{d}{2L}\right)_F = 0.02711, \quad \Rightarrow \quad d = 0.02711 \times 2 \times 250 = 13.56 \text{ mm}$$

$$q = 0.0159, \quad \Rightarrow \quad \left(\frac{d}{2L}\right)_F = 0.02733, \quad \Rightarrow \quad d = 0.02733 \times 2 \times 250 = 13.67 \text{ mm}$$

For $0.0159 \leq q \leq 1.0$, equation (12.27b) gives the bending-based diameters

$$\left(\frac{d}{2L}\right)_M = \left(\frac{4}{\pi}\right)^{1/3} s_1^{1/3}\left[q\left(1 + \frac{1}{2}\frac{s_2}{s_1}q\right)\right]^{1/3} = 0.1084[q(1 + 0.5q)]^{1/3} \tag{ii}$$

Equation (ii) gives two limiting diameters and one intermediate diameter:

$$q = 0.0159, \quad \Rightarrow \quad \left(\frac{d}{2L}\right)_M = 0.02733, \quad \Rightarrow \quad d = 0.0233 \times 2 \times 250 = 13.67 \text{ mm}$$

$$q = 0.50, \quad \Rightarrow \quad \left(\frac{d}{2L}\right)_M = 0.0927, \quad \Rightarrow \quad d = 0.0927 \times 2 \times 250 = 46.34 \text{ mm}$$

$$q = 1.0, \quad \Rightarrow \quad \left(\frac{d}{2L}\right)_M = 0.1241, \quad \Rightarrow \quad d = 0.1241 \times 2 \times 250 = 62.04 \text{ mm}$$

Figure 12.9 shows the cantilever profile based upon these optimised dimensions. The role of shear ensures that the beam does not taper to a point where the free-end bears load. The corresponding minimum weight of this design follows from equation (12.29b) as

$$\sum n = \pi\left(\frac{4}{\sqrt{3\pi}}\right) s_1 \int_0^{0.0159} (1 + q)\mathrm{d}q + \pi\left(\frac{4}{\pi}\right)^{2/3} s_1^{2/3} \int_{0.0159}^{1} [q(1 + 0.5q)]^{2/3}\mathrm{d}q$$

which was evaluated as $\sum n = 0.02634$ in Table 12.4. Hence, the weight of this beam is found from

$$\sum n = \frac{1}{\rho}\frac{W}{L^3} = 0.02634$$

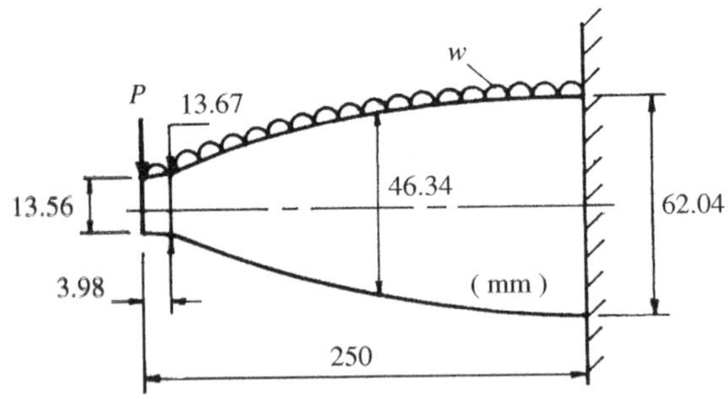

Figure 12.9 Diameter profile for cantilever under combined loading

which gives the beam's weight as

$$W = 0.02634 \times 7850 \times 0.25^3 = 3.23 \text{ kg}$$

This calculation indicates that many practical examples involving long beams may fall within the range of s_1 given in Tables 12.1–12.6. Where $s_1 < 0.001$, the equations given for combined shear and bending designs remain valid. Though bending will dominate these designs, shear will continue to dictate the end-diameter.

12.8 Centrally Propped Cantilever with End-Load

Figure 12.10a shows a cantilever carrying an end-load P with a prop placed at mid-span to eliminate the deflection there. In the presence of this prop the shear force and bending moment diagrams are as shown in Figure 12.10b–d.

12.8.1 Prop Reaction

The moment area theorem [1, 2], applied to separated bending moment diagrams due to P and to the prop reaction Q (see Figure 12.10c), enables Q to be found. Thus, for the two bending moment diagrams between the encastré fixing and the prop position, this theorem states:

$$\text{Moment of } M\text{-diagram due to } P - \text{Moment of } M\text{-diagram due to } Q = 0 \quad (12.32a)$$

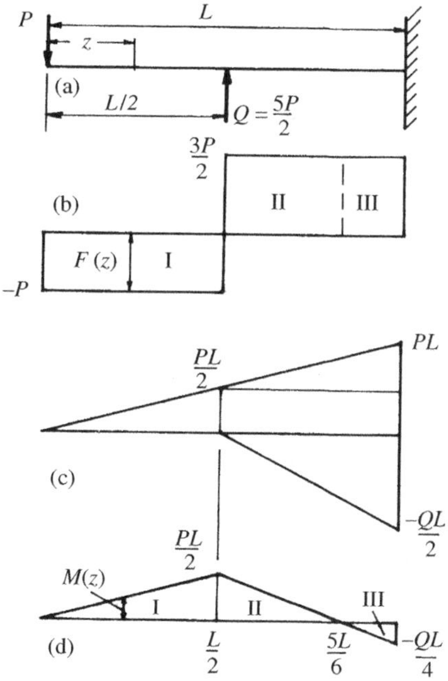

Figure 12.10 *F*- and *M*-diagrams for a centrally propped cantilever carrying an end-load

With the dimensions indicated in Figure 12.10c, equation (12.32a) becomes

$$\left(\frac{1}{2}\times\frac{L}{2}\times\frac{PL}{2}\times\frac{2}{3}\times\frac{L}{2}\right)+\left(\frac{PL}{2}\times\frac{L}{2}\times\frac{1}{2}\times\frac{L}{2}\right)-\left(\frac{1}{2}\times\frac{L}{2}\times\frac{QL}{2}\times\frac{2}{3}\times\frac{L}{2}\right)=0 \tag{12.32b}$$

Cancelling L^3 from within each term in equation (12.32b) leads to $Q = 5P/2$. Consequently, the ordinates $F(z)$ and $M(z)$ in the net F- and M-diagrams (Figure 12.10b,d) may be found. Within these diagrams their magnitudes $|F(z)|$ and $|M(z)|$ for the regions I, II and III shown are used to determine optimum diameters for the beam's solid circular section.

12.8.2 *Beam Profile*

At a given position in the length the beam diameter is to be the greater of the two values found by applying equations (3.3b) and (5.15b) to the ordinates $M(z)$ and $F(z)$ in Figure 12.10b,d. Firstly, on setting $F_1 = P$ and $F_2 = -Q = -5P/2$, we have $F_1/F_2 = -0.40$. Three regions I, II and III, are identified within the full length, $0 \le q \le 1.0$, where $q = z/L$. Correspondingly, the magnitudes of the ordinates in the F- and M-diagrams become the following.

In region I ($0 \le q \le \frac{1}{2}$),

$$|F(z)| = F_1 \tag{12.33a}$$

$$|M(z)| = F_1 z \tag{12.33b}$$

Hence, the normalised beam diameters required to withstand the shear forces and bending moments in this region follow from equations (12.33a,b):

$$\left(\frac{d}{2L}\right)_F = \left(\frac{4}{\sqrt{3}\pi}\right)^{1/2}\left(\frac{1}{\sigma_y}\times\frac{F_1}{L^2}\right)^{1/2} = \left(\frac{4}{\sqrt{3}\pi}\right)^{1/2} s_1^{1/2} \tag{12.34a}$$

$$\left(\frac{d}{2L}\right)_M = \left(\frac{4}{\pi}\right)^{1/3}\left(\frac{1}{\sigma_y}\times\frac{F_1 z}{L^3}\right)^{1/3} = \left(\frac{4}{\pi}\right)^{1/3}(s_1 q)^{1/3} \tag{12.34b}$$

where $s_1 = (1/\sigma_y)(F_1/L^2)$ is a defined as a *positive*, non-dimensional structural index which we shall vary across five decades from 0.001 to 100. This range will embrace both very long and very short beams, where the design diameter varies with the relative contributions from bending and shear.

In region II ($\frac{1}{2} \le q \le \frac{5}{6}$), the upper limit follows from $M(z) = 0 = Pz - (5P/2)(z - L/2)$, giving $z = 5L/6$. In this region, the magnitudes of the ordinates to the F- and M-diagrams are written as

$$|F(z)| = -F_1 - F_2 \tag{12.35a}$$

$$|M(z)| = F_1 z + F_2(z - L/2) \tag{12.35b}$$

Equation (12.35b) confirms the upper limit $q = z/L = 5/6$ for this region by substituting $|M(z)| = 0$ and $F_2/F_1 = -5/2$:

$$z = \frac{F_2 L}{2(F_1 + F_2)} = \frac{5}{6}L \tag{12.35c}$$

Correspondingly, the beam diameters required to withstand the shear forces and moments in this region become

$$\left(\frac{d}{2L}\right)_F = \left(\frac{4}{\sqrt{3}\pi}\right)^{1/2} s_1^{1/2}\left(-1 - \frac{s_2}{s_1}\right)^{1/2} = \left(\frac{4}{\sqrt{3}\pi}\right)^{1/2} s_1^{1/2}\left(\frac{3}{2}\right)^{1/2} \tag{12.36a}$$

$$\left(\frac{d}{2L}\right)_M = \left(\frac{4}{\pi}\right)^{1/3} s_1^{1/3}\left[q + \frac{s_2}{s_1}\left(q - \frac{1}{2}\right)\right]^{1/3} = \left(\frac{4}{\pi}\right)^{1/3} s_1^{1/3}\left(\frac{5}{4} - \frac{3q}{2}\right)^{1/3} \tag{12.36b}$$

where $s_2/s_1 = F_2/F_1 = -5/2$.

In region III ($\frac{5}{6} \le q \le 1$),

$$|F(z)| = -F_1 - F_2 \tag{12.37a}$$

$$|M(z)| = -F_1 z - F_2(z - L/2) \tag{12.37b}$$

The beam diameters required to withstand the shear forces and moments in this region follow from equations (12.37a,b) as

$$\left(\frac{d}{2L}\right)_F = \left(\frac{4}{\sqrt{3}\pi}\right)^{1/2} s_1^{1/2}\left(\frac{3}{2}\right)^{1/2} \tag{12.38a}$$

$$\left(\frac{d}{2L}\right)_M = \left(\frac{4}{\pi}\right)^{1/3} s_1^{1/3}\left(-\frac{5}{4} + \frac{3q}{2}\right)^{1/3} \tag{12.38b}$$

Table 12.7 gives the limiting diameters within each region, calculated from equations (12.34a,b), (12.36a,b) and (12.38a,b), for s_1 positive, taken over five decades. It is clear

Table 12.7 Normalised beam diameters ($d/2L$) and weight parameter n within the three zones of $q(= z/L)$

$d/2L$	s_1	$0 \le q \le 0.50$	$0.50 \le q \le 0.8\dot{3}$	$0.8\dot{3} \le q \le 1.0$	$\sum n$
M	100	0–3.9928	3.9928–0	0–3.1692	
F	100	8.574–8.574	10.501–10.501	10.501–10.501	288.68
M	10	0–1.8534	1.8534–0	0–1.4710	
F	10	2.7113–2.7113	3.3207–3.3207	3.3207–3.3207	28.868
M	1	0–0.8603	0.8603–0	0–0.6828	
F	1	0.8574–0.8574	1.0501–1.0501	1.0501–1.0501	2.8868
M	0.1	0–0.3993	0.3993–0	0–0.3169	
F	0.1	0.2711–0.2711	0.33207–0.33207	0.3321–0.3321	0.3606
M	0.01	0–0.1853	0.1853–0	0–0.1471	
F	0.01	0.0857–0.0857	0.10501–0.10501	0.1051–0.1051	0.0629
M	0.001	0–0.08603	0.08603–0	0–0.06828	
F	0.001	0.0271–0.0271	0.03321–0.03321	0.03321–0.03321	0.0132

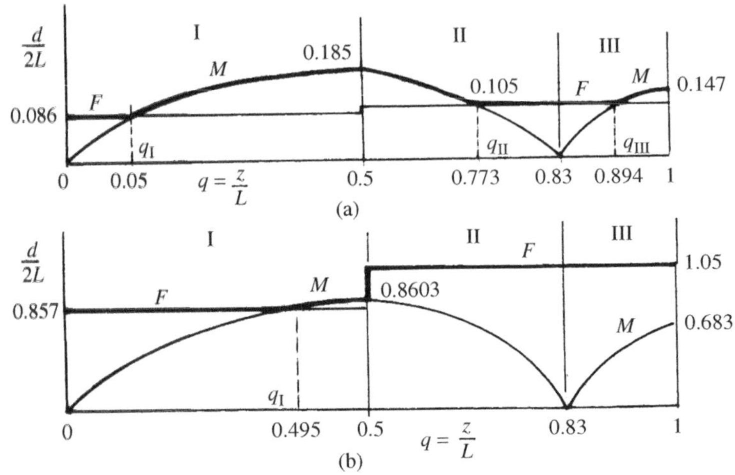

Figure 12.11 Cantilever radius profiles, for structural indices (a) $s_1 = 0.01$ and (b) $s_1 = 1$

from Table 12.7 that the diameter required must be based upon both F and M across these regions. Taking the greater diameter within each length region determines the beam's profile. Two examples of long and short cantilever beam profiles, extracted from Table 12.7 for $s_1 = 0.01$ and $s_1 = 1$, are given in Figure 12.11. It is apparent that, when minimising weight across each region I, II and II, applying their $F(z)$ and $M(z)$ expressions produce a variable beam diameter. The design rests with the selection of the greater diameter between the force F and moment M plots within regions I, II and III, as shown. Consequently, in Figure 12.11b it is seen that the cantilever's diameter is stepped at the central prop position. Specific diameters within each design follow from multiplying the ordinates shown by $2L$ (see Example 12.5).

12.8.3 *F and M Transitions*

Where intersections occur between the F and M plots within a particular region of Figure 12.11a,b, the corresponding length positions, q_{I}, q_{II} and q_{III}, are found by equating the paired expressions $(d/2L)_F$ and $(d/2L)_M$ in equations (12.34a,b), (12.36a,b) and (12.38a,b). These calculations now follow for each index s_1.

In region I,

$$\left(\frac{4}{\pi}\right)^{1/3}(s_1 q_{\mathrm{I}})^{1/3} = \left(\frac{4}{\sqrt{3}\pi}\right)^{1/2} s_1^{1/2}$$

from which $q_{\mathrm{I}} = 0.495\sqrt{s_1}$, giving the values at each s_1 in Table 12.8. The value $q_{\mathrm{I}} = 0.5$ for $s_1 \geq 10$ in Table 12.8 indicates no intersection and hence the design is purely shear.

In region II,

$$\left(\frac{4}{\pi}\right)^{1/3} s_1^{1/3}\left(\frac{5}{4} - \frac{3q_{\mathrm{II}}}{2}\right)^{1/3} = \left(\frac{2\sqrt{3}}{\pi}\right)^{1/2} s_1^{1/2}$$

Table 12.8 Positions of F and M transitions in region I

s_1	0.001	0.01	0.10	1.0	10	100
q_I	0.01565	0.0495	0.1565	0.4950	0.50	0.50

from which $q_{II} = 2/3(1.25 - 0.9096\sqrt{s_1})$, giving the values in Table 12.9. The value $q_{II} = 0.5$ for $s_1 \geq 1$ within Table 12.9 indicates no intersection and hence the design is purely shear.

In region III,

$$\left(\frac{4}{\pi}\right)^{1/3} s_1^{1/3} \left(-\frac{5}{4} + \frac{3q_{III}}{2}\right)^{1/3} = \left(\frac{2\sqrt{3}}{\pi}\right)^{1/2} s_1^{1/2}$$

from which $q_{III} = 2/3(1.25 + 0.9096\sqrt{s_1})$, giving the values in Table 12.10. The value $q_{III} = 1.0$ for $s_1 \geq 0.1$ within Table 12.10 indicates that no intersection occurs and hence the design is purely shear-based. We see that three intersections q_I, q_{II} and q_{III} appear in Figure 12.11a for $s_1 = 0.01$, but only one ($q_I = 0.495$) in Figure 12.11b when $s_1 = 1$.

12.8.4 Optimal Weights

(a) Shear and Bending Design

Once the intersection positions $q_I, q_{II}, \ldots$ are known, the weight of each region, and hence the total weight, is found from summing integrated weight expressions $W = \int \rho(\pi d^2/4)\mathrm{d}z$. For example, within region I we can write a normalised weight expression $n = (1/\rho)(W/L^3)$ as the sum of the shear and bending contributions. From equations (12.34a,b),

$$n_I = \frac{4}{\sqrt{3}} s_1 \int_0^{q_1} \mathrm{d}q + \pi \left(\frac{4}{\pi}\right)^{2/3} s_1^{2/3} \int_{q_1}^{0.5} q^{2/3}\mathrm{d}q$$

$$= 2.3094 s_1 q_1 + 2.2143 s_1^{2/3}(0.315 - q_1^{5/3}) \quad (12.39a)$$

Table 12.9 Positions of F and M transitions in region II

s_1	0.001	0.01	0.10	1.0	10	100
q_{II}	0.8142	0.7727	0.6416	0.50	0.50	0.50

Table 12.10 Positions of F and M transitions in region III

s_1	0.001	0.01	0.10	1.0	10	100
q_{III}	0.853	0.894	1.0	1.0	1.0	1.0

Similarly, within region II, we have from equations (12.36a,b),

$$n_{\text{II}} = \pi \left(\frac{4}{\pi}\right)^{2/3} s_1^{2/3} \int_{0.50}^{q_{\text{II}}} \left(\frac{5}{4} - \frac{3q}{2}\right)^{2/3} \mathrm{d}q + \frac{4}{\sqrt{3}} s_1 \times \frac{3}{2} \int_{q_{\text{II}}}^{5/6} \mathrm{d}q$$

$$= 1.4762 s_1^{2/3} \left[0.315 - \left(\frac{5}{4} - \frac{3q_{\text{II}}}{2}\right)^{5/3}\right] + 3.4641 s_1 \left(\frac{5}{6} - q_{\text{II}}\right) \tag{12.39b}$$

Within region III (where an intersection in the F and M plots occurs for $s_1 \le 0.01$) equations (12.38a,b) apply:

$$n_{\text{III}} = \frac{4}{\sqrt{3}} s_1 \times \frac{3}{2} \int_{5/6}^{q_{\text{III}}} \mathrm{d}q + \pi \left(\frac{4}{\pi}\right)^{2/3} s_1^{2/3} \int_{q_{\text{III}}}^{1.0} \left(\frac{3q}{2} - \frac{5}{4}\right)^{2/3} \mathrm{d}q$$

$$= 3.4641 s_1 \left(q_{\text{III}} - \frac{5}{6}\right) + 1.4762 s_1^{2/3} \left[0.09921 - \left(\frac{3q_{\text{III}}}{2} - \frac{5}{4}\right)^{5/3}\right] \tag{12.39c}$$

Summing equations (12.39a–c) gives the normalised weights $\sum n = n_{\text{I}} + n_{\text{II}} + n_{\text{III}}$ for each s_1-value within the combined designs. Table 12.11 gives $\sum n$, so calculated, over five decades of s_1.

(b) Shear Design

Where, as indicated in Tables 12.8–12.10, there is no intersection between the F and M plots within the three regions I–III, then the design is to be based entirely upon shear. Hence, we take the first integral within each of the three expressions (12.39a–c) with an appropriate upper limit. Thus, for region I,

$$n_{\text{I}} = \frac{4}{\sqrt{3}} s_1 \int_0^{0.5} \mathrm{d}q = 1.1547 s_1 \tag{12.40a}$$

within region II,

$$n_{\text{II}} = \frac{4}{\sqrt{3}} s_1 \times \frac{3}{2} \int_{0.5}^{5/6} \mathrm{d}q = 1.1547 s_1 \tag{12.40b}$$

Table 12.11 Normalised weight parameter sum $\sum n$ for a centrally propped cantilever carrying an end-load

s_1	0.001	0.01	0.1	1.0	10	100
$\sum n$	0.0132	0.0629	0.3606	2.8868	28.868	288.68

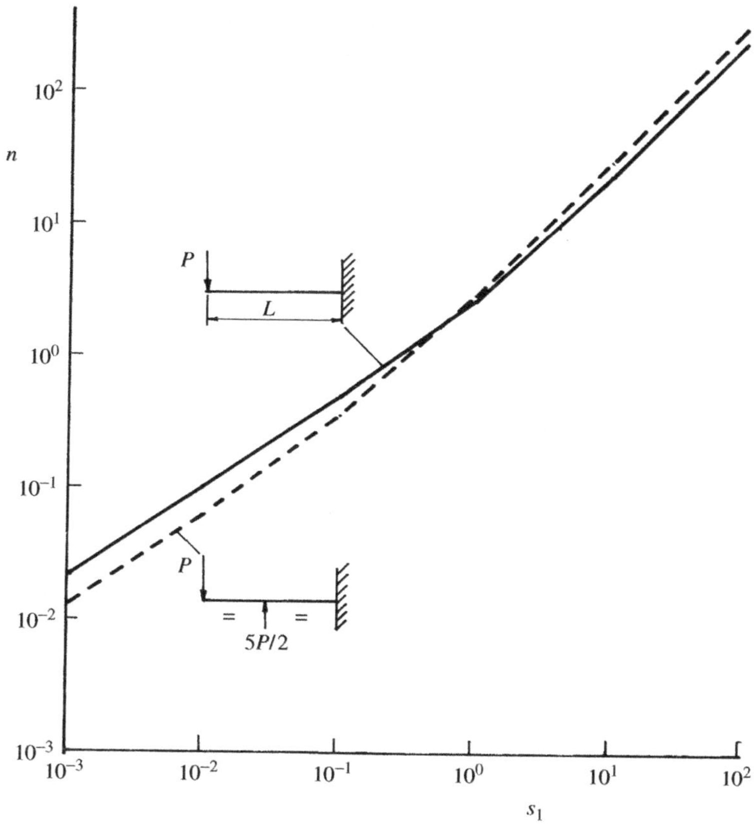

Figure 12.12 Minimum weights of propped and unpropped cantilevers. (note: $s_1 = (1/\sigma_y)(P/L^2)$)

and within region III,

$$n_{III} = \frac{4}{\sqrt{3}} s_1 \times \frac{3}{2} \int_{5/6}^{1.0} dq = 0.5774 s_1 \tag{12.40c}$$

Summing equations (12.40a–c) gives the normalised weight $\sum n = n_I + n_{II} + n_{III}$ at each s_1-value. Table 12.11 gives $\sum n$, calculated for shear designs where $s_1 \geq 1$ lies within the five decades of s_1 chosen. In Figure 12.12 the weight parameter n is plotted (dashed line) against s_1 to allow for an optimum design of any length of centrally propped, circular-section cantilever carrying a concentrated end-load. Figure 12.12 shows that for longer beams, i.e in the region $s_1 < 0.1$, where bending dominates the design, a propped cantilever has less weight than one without a prop. However, it appears from this figure that shear-based designs for $s_1 \geq 1$ do not save the beam weight, rather, they increase weight marginally. If we consider the weight to include the beam plus the prop, then the weight will be increased over the full range of s_1, as the following example shows.

Example 12.5 Find the dimensions of the centrally propped, circular-section cantilever, shown in Figure 12.10a, when it is to carry an end load of $P = 50$ kN for structural index values $s_1 = 0.01$ and $s_1 = 1$. Assume that the prop is to be an optimally designed, minimum length, Euler strut of circular cross-section, fixed at its base and hinged to the underside of the cantilever. Which index s_1 provides a more realistic design? Both beam and strut are made of steel, with $\sigma_y = 300$ MPa, $\rho = 7860$ kg/m^3 and $E = 210$ GPa.

(a) Beam Design

Figure 12.11 shows the beam profiles $d/2L$ at each s_1-value, for which the respective lengths follow:

$$s_1 = \frac{1}{\sigma_y} \times \frac{F_1}{L_1^2}, \quad \Rightarrow \quad L_1 = \left(\frac{F_1}{s_1 \sigma_y}\right)^{1/2} \tag{i}$$

Setting $F_1 = P = 50$ kN, we find from equation (i) that the optimum lengths for the cantilever are $L_1 = 129.1$ mm for $s_1 = 0.01$ and $L_1 = 12.91$ mm for $s_1 = 1$. Correspondingly, the beam diameter profiles are as shown in Figure 12.11 in which the greatest diameters are: (a) $2 \times 129.1 \times 0.185 = 47.8$ mm and (b) and $2 \times 12.91 \times 1.05 = 27.1$ mm. Both short length beams reveal the influence of the shear contribution to each beam design at their optimum weights. The latter follow from $\sum n$, given in Tables 12.7 and 12.11, as

$$\sum n = \frac{1}{\rho} \frac{W}{L_1^3}, \quad \Rightarrow \quad W = \rho L_1^3 \sum n \tag{ii}$$

Equation (ii) gives, for $s_1 = 0.01$,

$$W = 7860 \times 129.1^3 \times 10^{-9} \times 0.0629 = 1.064 \text{ kg}$$

and for $s_1 = 1$,

$$W = 7860 \times 12.91^3 \times 10^{-9} \times 2.8868 = 0.0488 \text{ kg}$$

which is the low weight we should expect of a shear pin or a rivet.

(b) Minimum Length of a Fixed-Pinned Euler Strut

Let d_2 and L_2 be the minimum diameter and length for the supporting prop (see Figure 12.13). These dimensions are found when the prop behaves as an Euler strut where the yield stress σ_y is reached in its circular-section as it buckles under the compressive reaction $Q = 5P/2$. This reaction sets a limiting condition for Euler strut buckling in which the slenderness ratio follows from

$$Q = \frac{2\pi^2 EI}{L_2^2} = \frac{2\pi^2 EAk^2}{L_2^2}, \quad \Rightarrow \quad \left(\frac{L_2}{k}\right)^2 = \frac{2\pi^2 E}{\sigma_y} \tag{iii}$$

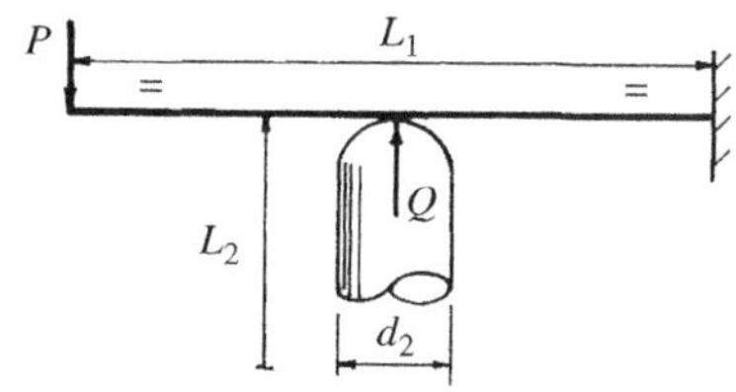

Figure 12.13 Centrally propped cantilever

Now as $\sigma_y = Q/A = 4Q/\pi d_2^2 = 10P/\pi d_2^2$ and $k^2 = d_2^2/16$, we find from equation (iii) the minimum diameter and length expressions for an Euler strut with these end-fixings:

$$d_2 = \left(\frac{10P}{\pi \sigma_y}\right)^{1/2} = \left(\frac{10 \times 50 \times 10^3}{\pi \times 300}\right)^{1/2} = 23.03 \text{ mm}$$

$$L_2 = \left(\frac{5\pi EP}{4\sigma_y^2}\right)^{1/2} = \left(\frac{5\pi \times 210 \times 10^3 \times 50 \times 10^3}{4 \times 300^2}\right)^{1/2} = 676.87 \text{ mm}$$

The strut weight remains independent of each s_1-value when applied to the beam design:

$$W = \frac{\rho \pi d^2 L}{4} = 7860 \times 10^{-9} \times \frac{\pi}{4} \times 23.033^2 \times 676.87 = 2.217 \text{ kg}$$

Obviously, only the beam with the lesser s_1-value (= 0.01) provides a cantilever with sufficient length $L_1 = 129.1$ mm to admit a prop at half-length with a diameter $d_2 = 23.03$ mm (see Figure 12.13). Hence, the total weight of this more realistic assembly becomes $1.064 + 2.217 = 3.281$ kg.

12.9 End-Propped Cantilever with Distributed Load

Consider a cantilever beam propped at its end when carrying a uniformly distributed load w per unit length (see Figure 12.14a). The shear force and bending moment diagrams appear in Figure 12.14b–d. When the prop is used to eliminate the end-displacement, the prop reaction P can be found by one of the following two methods.

12.9.1 Prop Reaction

The moment area theorem [1–3], applied to the separated bending moment diagrams due to w and to the prop reaction P (see Figure 12.14c), enables P to be found. The theorem states:

$$\text{Moment of } M\text{-diagram due to } w - \text{Moment of } M\text{-diagram due to } P = 0 \quad (12.41a)$$

where each M-diagram refers to the region between the encastré fixing and the prop. For the dimensions given in each M-diagram in Figure 12.14c, equation (12.41a) becomes

$$\left(\frac{wL^2}{2} \times \frac{L}{3}\right)\frac{3L}{4} - \left(PL \times \frac{L}{2}\right)\frac{2L}{3} = 0 \qquad (12.41b)$$

Equation (12.41b) gives $P = 3wL/8$. Alternatively, we may equate the end-displacements from each load acting separately upon the cantilever using the standard expressions [1, 2]

$$\frac{PL^3}{3EI} = \frac{wL^4}{8EI} \qquad (12.41c)$$

Equation (12.41c) also gives $P = 3wL/8$. In the presence of a prop with the reaction P, the net shear force and bending moment diagrams are as shown in Figure 12.14b,d. Three regions (I, II and III) are identified within each diagram as shown.. The F and M ordinates for each region will now be used to determine the section's diameter profile.

12.9.2 Beam Profile

The optimum beam diameter is to be the greater of the two values found by applying the usual force and moment equations (3.3b) and (5.15b) to the ordinates in Figure 12.14b,d. On setting $F_1 = -P$ and $F_2 = wL$, we have $F_1/F_2 = -0.375$. Three regions are identified within the full length, $0 \le q \le 1.0$, where $q = z/L$, and the magnitudes of the ordinates in the F- and M-diagrams 12.14b,d are as follows.

In region I ($0 \le q \le 0.375$),

$$|F(z)| = -F_1 - F_2 z/L, \qquad \Rightarrow |F(q)| = F_2(0.375 - q) \qquad (12.42a)$$

$$|M(z)| = -F_1 z - F_2 z^2/2L \qquad \Rightarrow |F(q)| = F_2 L q(0.375 - q/2) \qquad (12.42b)$$

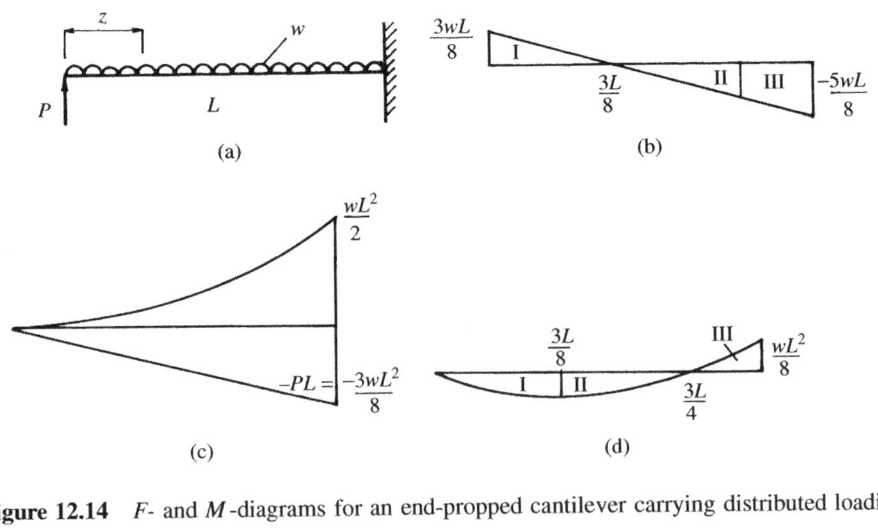

Figure 12.14 F- and M-diagrams for an end-propped cantilever carrying distributed loading

Correspondingly, the normalised beam diameters required to withstand the shear forces and moments in this region follow from equations (12.42a,b):

$$\left(\frac{d}{2L}\right)_F = \left(\frac{4}{\sqrt{3}\pi}\right)^{1/2} s_2^{1/2}(0.375 - q)^{1/2} \tag{12.43a}$$

$$\left(\frac{d}{2L}\right)_M = \left(\frac{4}{\pi}\right)^{1/3} s_2^{1/3}[q(0.375 - q/2)]^{1/3} \tag{12.43b}$$

where $s_2 = (1/\sigma_y)(F_2/L^3)$ is defined as a *positive* structural index for the loading ($F_2 = wL$). The index s_2 will be varied over five decades, from 0.001 to 100, to include both very long and short beams whose profiles depend upon the relative contributions from bending and shear.

In region II ($0.375 \leq q \leq 0.75$),

$$|F(z)| = F_1 + F_2 z/L, \qquad \Rightarrow |F(q)| = F_2(q - 0.375) \tag{12.44a}$$

$$|M(z)| = -F_1 z - F_2 z^2/2L, \qquad \Rightarrow |F(q)| = F_2 Lq(0.375 - q/2) \tag{12.44b}$$

Correspondingly, the beam diameters required to withstand the shear force and bending moment in equations (12.44a,b) become

$$\left(\frac{d}{2L}\right)_F = \left(\frac{4}{\sqrt{3}\pi}\right)^{1/2} s_2^{1/2}(q - 0.375)^{1/2} \tag{12.45a}$$

$$\left(\frac{d}{2L}\right)_M = \left(\frac{4}{\pi}\right)^{1/3} s_2^{1/3}[q(0.375 - q/2)]^{1/3} \tag{12.45b}$$

In region III ($0.75 \leq q \leq 1$),

$$|F(z)| = F_1 + F_2 z/L, \qquad \Rightarrow |F(q)| = F_2(q - 0.375) \tag{12.46a}$$

$$|M(z)| = F_1 z + F_2 z^2/2L, \qquad \Rightarrow |F(q)| = F_2 Lq(q/2 - 0.375) \tag{12.46b}$$

The beam diameters required to withstand the shear forces and moments in this region follow from equations (12.46a,b) as

$$\left(\frac{d}{2L}\right)_F = \left(\frac{4}{\sqrt{3}\pi}\right)^{1/2} s_2^{1/2}(q - 0.375)^{1/2} \tag{12.47a}$$

$$\left(\frac{d}{2L}\right)_M = \left(\frac{4}{\pi}\right)^{1/3} s_2^{1/3}[q(q/2 - 0.375)]^{1/3} \tag{12.47b}$$

Table 12.12 gives the limiting diameters within each region of q with s_2 spanning five decades from 0.001 to 100. Selecting the greater diameter from Table 12.12, it is clear how the section size $d/2L$ must be based upon both the F and M designs across the three regions. Two examples of beam profiles, taken from Table 12.12, for $s_2 = 0.01$ and $s_2 = 1$, are given in Figure 12.15. Figure 12.15a shows that both shear and bending

Table 12.12 Normalised beam diameters ($d/2L$) and weight parameter n for three regions of length $q(=z/L)$

$d/2L$	s_2	$0 \le q \le 0.375$	$0.375 \le q \le 0.75$	$0.75 \le q \le 1.0$	$\sum n$
M	100	0–2.0761	2.0761–0	0–2.515	
F	100	5.2504–0	0–5.2504	5.2504–6.7782	62.126
M	10	0–0.9638	0.9638–0	0–1.1675	
F	10	1.6603–0	0–1.6603	1.6603–2.1435	6.5142
M	1	0–0.4474	0.4474–0	0–0.5419	
F	1	0.5250–0	0–0.5250	0.5250–0.6778	0.7551
M	0.1	0–0.2879	0.2879–0	0–0.2515	
F	0.1	0.16603–0	0–0.16603	0.16603–0.2143	0.1121
M	0.01	0–0.09638	0.09638–0	0–0.1168	
F	0.01	0.0525–0	0–0.0525	0.0525–0.06778	0.0223
M	0.001	0–0.04474	0.04474–0	0–0.05419	
F	0.001	0.0166–0	0–0.0166	0.0166–0.02143	0.00475

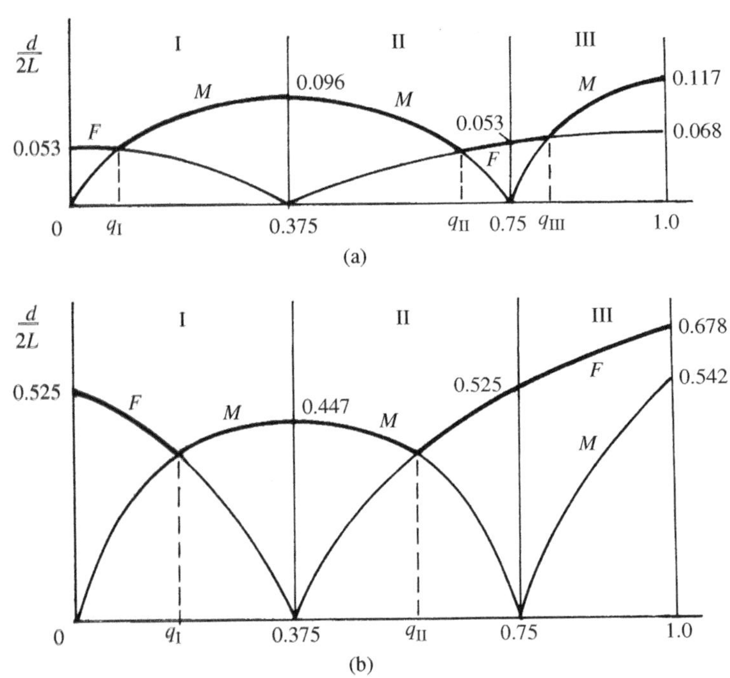

Figure 12.15 Beam profiles, for structural indices (a) $s_2 = 0.01$ and (b) $s_2 = 1$

play an interactive role within each region at the lesser index. Interactions also occur in Figure 12.15b within regions I and II for the higher index but not in region III where the design is wholly shear-based.

12.9.3 Minimum Weights

If we are to minimise weight across the three regions I, II and III, the $F(z)$ and $M(z)$ expressions (12.42a,b), (12.44a,b) and (12.46a,b) will lead to a variable diameter. Where intersections occur within a particular region, their length positions $q_{\mathrm{I}}, q_{\mathrm{II}}$ and q_{III} are found (by trial) from equating the respective paired $(d/2L)_F$ and $(d/2L)_M$ expressions (12.43a,b), (12.45a,b) and (12.47a,b). Once these transition positions are known, the weight of each region, and hence the total weight, is found by summing expressions $W = \int \rho(\pi d^2/4)\mathrm{d}z$. For example, within region I we can write from equations (12.43a,b) normalised weight expression $n = (1/\rho)(W/L^3)$ from the shear and bending contributions as

$$n_{\mathrm{I}} = \frac{4}{\sqrt{3}}s_2 \int_0^{q_{\mathrm{I}}} (0.375 - q)\mathrm{d}q + \pi\left(\frac{4}{\pi}\right)^{2/3} s_2^{2/3} \int_{q_{\mathrm{I}}}^{0.375} [q(0.375 - q/2)]^{2/3}\mathrm{d}q \qquad (12.48\mathrm{a})$$

Similarly, within region II, we have from equations (12.45a,b),

$$n_{\mathrm{II}} = \pi\left(\frac{4}{\pi}\right)^{2/3} s_2^{2/3} \int_{0.375}^{q_{\mathrm{II}}} [q(0.375 - q/2)]^{2/3}\mathrm{d}q + \frac{4}{\sqrt{3}}s_2 \int_{q_{\mathrm{II}}}^{0.75} (q - 0.375)\mathrm{d}q \qquad (12.48\mathrm{b})$$

Within region III (where an intersection in the F and M plots occurs for $s_2 < 1$), equations (12.47a,b) give

$$n_{\mathrm{III}} = \frac{4}{\sqrt{3}}s_2 \int_{0.75}^{q_{\mathrm{III}}} (q - 0.375)\mathrm{d}q + \pi\left(\frac{4}{\pi}\right)^{2/3} s_2^{2/3} \int_{q_{\mathrm{III}}}^{1.0} [q(q/2 - 0.375)]^{2/3}\mathrm{d}q \qquad (12.48\mathrm{c})$$

For $s_2 \geq 1$, Table 12.12 shows the design and hence the weight in region III is to be based wholly upon shear:

$$n_{\mathrm{III}} = \frac{4}{\sqrt{3}}s_2 \int_{0.75}^{1} (q - 0.375)\mathrm{d}q \qquad (12.48\mathrm{d})$$

A numerical integration of the bending contribution within equations (12.48a–c) facilitates their addition to the shear contribution within each region. Summing the weights across the regions gives the total, normalised weight parameter $\sum n = n_{\mathrm{I}} + n_{\mathrm{II}} + n_{\mathrm{III}}$ in Table 12.12. The total 'weight' $\sum n$ may then be plotted against s_2 in the manner of Figure 12.16, to enable an optimum design of any length of propped cantilever carrying uniformly

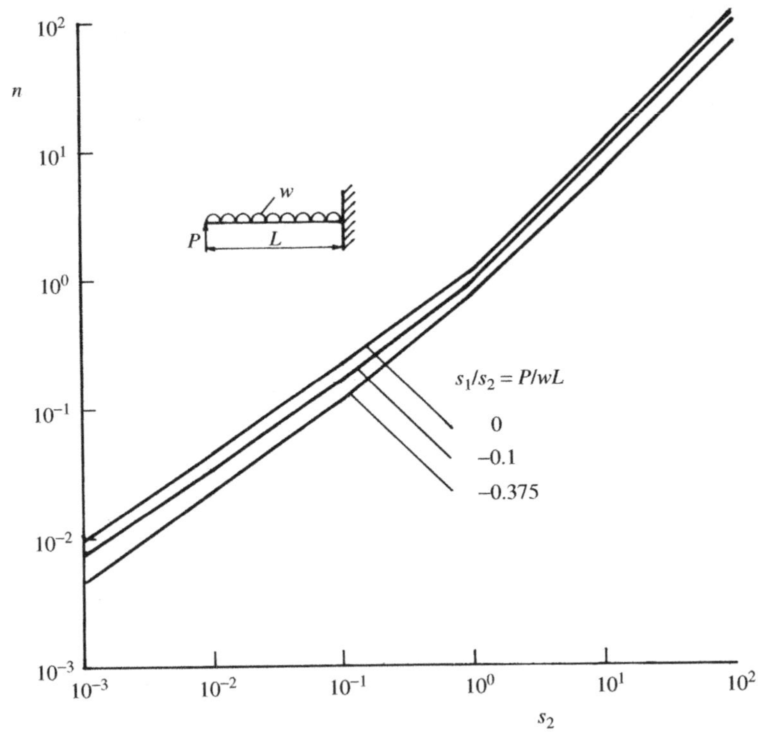

Figure 12.16 Normalised weights of a propped cantilever carrying distributed loading. (Note $s_1 = (1/\sigma_y)(P/L^2)$; $s_2 = (1/\sigma_y)(F_2/L^2)$, where $F_2 = wL$, refers to the total applied loading.)

distributed loading. This weight (i.e. the plot for $s_1/s_2 = P/wL = -0.375$), which does not include the prop's weight, is seen to be less than the weights of:

(i) a propped cantilever under a higher load ratio $s_1/s_2 = -0.1$;
(ii) a free-ended cantilever ($s_1/s_2 = 0$) with circular cross-section (taken from Figure 10.13).
The analysis of the higher load ratio ($P/wL = -0.1$) is left as an exercise for the reader.

12.10 Simply Supported Beam with Central-Concentrated and Distributed Loadings

Consider a simply supported beam of circular cross-section carrying a uniformly distributed load w per unit length in addition to a central concentrated load P (see Figure 12.17a). The shear force and bending moment diagrams are shown in Figure 12.17b,c. The beam diameter is to be the greater of the two values found by applying the force and moment criteria (see equations (3.3b) and (5.15b)) to the ordinates in Figure 12.17b,c. Setting $F_1 = P/2$ and $F_2 = wL/2$ gives a load ratio $F_2/F_1 = wL/P$.

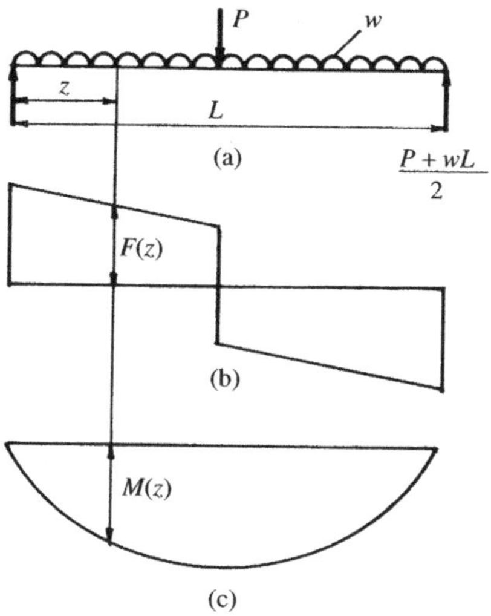

Figure 12.17 *F*- and *M*-diagrams for a simply supported beam under combined loading

Within the half-length ($0 \leq z/L \leq 0.50$), the ordinates to the *F*- and *M*-diagrams become

$$|F(z)| = F_1 + F_2(1 - 2z/L) \tag{12.49a}$$

$$|M(z)| = F_1 z + F_2 z(1 - z/L) \tag{12.49b}$$

Correspondingly, the normalised beam diameters required to withstand the shear force and bending moment in equations (12.49a,b) follow from equations (3.3b) and (5.15b) as

$$\left(\frac{d}{2L}\right)_F = \left(\frac{4}{\sqrt{3}\pi}\right)^{1/2} s_1^{1/2}\left[1 + \frac{s_2}{s_1}(1 - 2q)\right]^{1/2} \tag{12.50a}$$

$$\left(\frac{d}{2L}\right)_M = \left(\frac{4}{\pi}\right)^{1/3} s_1^{1/3} q^{1/3}\left[1 + \frac{s_2}{s_1}(1 - q)\right]^{1/3} \tag{12.50b}$$

where $q = z/L$, $s_1 = (1/\sigma_y)(F_1/L^2)$ and $s_2 = (1/\sigma_y)(F_2/L^2)$ are positive variables which we shall vary in the ratio s_2/s_1 across five decades of the index s_1 from 0.001 to 100. Note that s_2/s_1 is also the force ratio $F_2/F_1 = wL/P$. The range of s_1 includes both very long and very short beams, whose optimum designs vary with the relative contributions from bending and shear. Thus, when we equate (12.50a,b), the critical position q_c in the half-length is found at which a shear-based design becomes moment-based. This gives

$$q_c^{1/3}\left[1 + \frac{s_2}{s_1}(1 - q_c)\right]^{1/3} = 0.7911 s_1^{1/6}\left[1 + \frac{s_2}{s_1}(1 - 2q_c)\right]^{1/2} \tag{12.50c}$$

That is, a shear-based design applies to the region $0 \leq q \leq q_c$ and a moment-based design to $q_c \leq q \geq 0.5$. The total weight is found by summing the integral $W = \int \rho(\pi d^2/4)\mathrm{d}z$.

This integral is applied to each region of the half-length to give its normalised weight as

$$n = \frac{1}{\rho}\left(\frac{W}{L^3}\right) = \pi \int_0^{q_c} \left(\frac{d}{2L}\right)_F^2 dq + \pi \int_{q_c}^{0.5} \left(\frac{d}{2L}\right)_M^2 dq \tag{12.51a}$$

Substituting from equations (12.50a,b) and doubling for the two halves, n appears in terms of q, s_1 and s_2/s_1 with contributions to the minimum weight from both shear and bending:

$$n = \frac{8}{\sqrt{3}} s_1 \int_0^{q_c} \left[1 + \frac{s_2}{s_1}(1 - 2q)\right] dq + 2\pi \left(\frac{4}{\pi}\right)^{2/3} s_1^{2/3} \int_{q_c}^{0.5} \left\{q\left[1 + \frac{s_2}{s_1}(1 - q)\right]\right\}^{2/3} dq \tag{12.51b}$$

The design becomes wholly shear-based at a critical value of $s_1 = s_{1FM}$, which is found by setting $q_c = 0.5$ in equation (12.50c). This gives

$$s_{1FM} = 0.255(2 + s_2/s_1)^2 \tag{12.52a}$$

Thus, when $s_1 \geq s_{1FM}$, the total weight is given by a change to the limit for the first integral in equation (12.51b):

$$n = \frac{8}{\sqrt{3}} s_1 \int_0^{0.5} \left[1 + \frac{s_2}{s_1}(1 - 2q)\right] dq = 1.1547 s_1 \left[2 + \left(\frac{s_2}{s_1}\right)\right] \tag{12.52b}$$

(a) $s_2/s_1 = 0.1$
Taking the structural index (force) ratio as $s_2/s_1 = 0.1$, equation (12.52a) gives $s_{1FM} = 1.1246$. Thus, for $s_2/s_1 = 0.1$ we must have $s_1 < 1.1246$ if $q_c < 0.5$ for the design to be both shear- and moment-based. Table 12.13 gives q_c values across three decades of s_1 that satisfy this requirement. The q_c values in Table 12.13 were found from a trial solution to equation (12.50c). Once q_c is known, the total weight of a beam at each index s_1 is found from equation (12.51b). This becomes

$$n = 4.619 s_1 \int_0^{q_c} (1.1 - 0.2q) dq + 7.381 s_1^{2/3} \int_{q_c}^{0.5} [q(1.1 - 0.1q)]^{2/3} dq \tag{12.53a}$$

Table 12.13 Critical length positions with $s_2/s_1 = 0.1$ for shear- and moment-based designs

s_1	1	0.1	0.01	0.001
q_c	0.475	0.160	0.0514	0.0164

Equation (12.53a) was integrated for each pair of values (q_c, s_1) given in Table 12.13. Note that the second integral was evaluated numerically using Simpson's rule. When $s_1 \geq 1.1246$, then $q_c = 0.5$ and the section diameter becomes wholly shear-based, i.e. equation (12.52b) gives the normalised weight as

$$n = 1.1547 s_1 (2 + 0.1) = 2.4249 s_1 \tag{12.53b}$$

The optimised weights n, found from equations (12.53a,b), over five decades in s_1 are given in Table 12.14.

Table 12.14 Optimised weights over the full range of s_1 with $s_2/s_1 = 0.1$

s_1	0.001	0.01	0.1	1	1.1246	10	100
n	0.0143	0.0674	0.3413	2.4273	2.727	24.249	242.49

(b) $s_2/s_1 = 1$
With an index (force) ratio $s_2/s_1 = 1.0$, equation (12.52a) gives $s_{1FM} = 2.294$. For $s_1 < s_{1FM}$ we find from equation (12.50c) the q_c-values given in Table 12.15. The optimised weights in Table 12.16 apply to the full range of s_1, found from the integration in equations (12.51b) and (12.52b). The latter applies when $s_1 \geq 2.294$.

Table 12.15 Critical length positions with $s_2/s_1 = 1.0$ for shear- and moment-based designs

s_1	1	0.1	0.01	0.001
q_c	0.402	0.180	0.065	0.022

Table 12.16 Optimised weights over the full range of s_1 with $s_2/s_1 = 1.0$

s_1	0.001	0.01	0.1	1	2.294	10	100
n	0.0195	0.0926	0.489	3.5348	7.9466	34.641	346.41

(c) $s_2/s_1 = 10$
Finally, when $s_2/s_1 = 10$, equation (12.52a) gives $s_{1FM} = 36.72$. With $s_1 < s_{1FM}$, we find from equation (12.50c) the q_c-values given in Table 12.17. With $s_1 \geq (s_1)_{FM}$ and $q_c = 0.5$, the design is entirely shear-based. The optimised (minimum) weights in Table 12.18, apply to the full range of s_1 as found from the integration in equations (12.51b) and (12.52b).

The n versus s_1 results in Tables 12.14, 12.16 and 12.18 are plotted in Figure 12.18, to which we may add the previous results for the ratio $s_2/s_1 = 0$ in respect of a central force acting alone upon a beam with circular cross-section (taken from Figure 10.18). This

Table 12.17 Critical length positions with $s_2/s_1 = 10$ for shear- and moment-based designs

s_1	10	1	0.1	0.01	0.001
q_c	0.473	0.3919	0.2603	0.1259	0.048

special case ($s_2/s_1 = 0$) was considered in §10.5. Note that we cannot add the results for a distributed load acting alone (see Figure 10.22) without changing the abscissa from s_1 to s_2. Figure 12.18 allows an interpolation for any positive ratio s_2/s_1 to provide an optimum circular-section design. Negative ratios s_2/s_1, which apply to this beam in the presence of a central prop, require separate consideration, given in the following section.

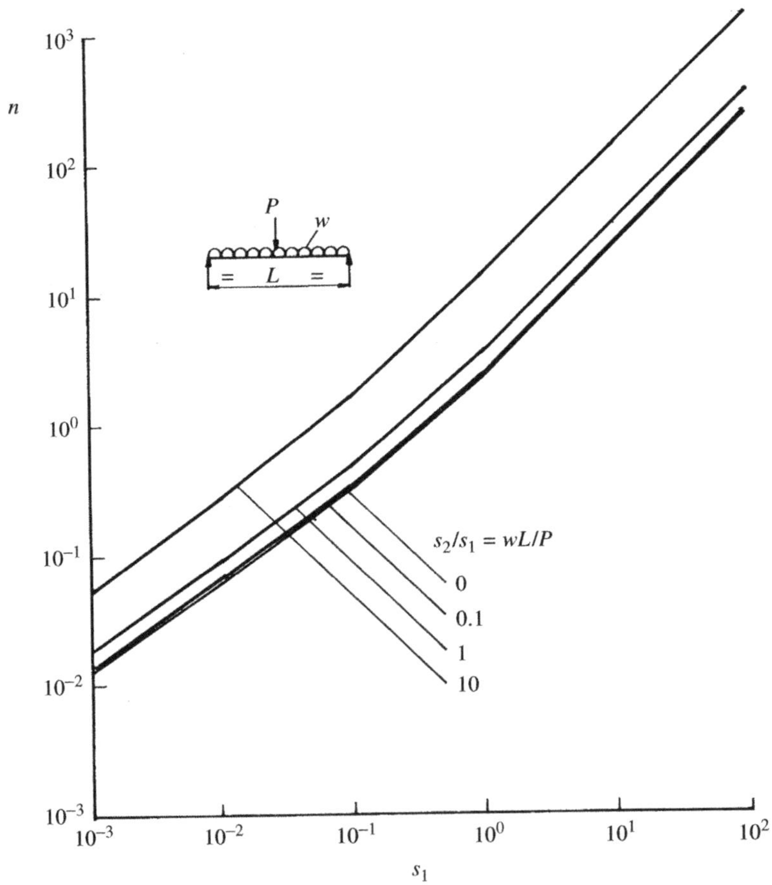

Figure 12.18 Minimum weights of simply supported beams under combined loading. (Note: $s_1 = (1/\sigma_y)(F_1/L^2)$; $s_2 = (1/\sigma_y)(F_2/L^2)$ where $F_1 = P/2$ and $F_2 = wL/2$.)

Table 12.18 Optimised weights over the full range of s_1 with $s_2/s_1 = 10$

s_1	0.001	0.01	0.1	1	10	36.72	100
n	0.0556	0.280	1.742	14.44	138.9	508.81	1385.64

12.11 Centrally Propped, Simply Supported Beam with Distributed Load

Consider a beam resting upon simple supports at its ends when carrying a uniformly distributed load w per unit length with a prop placed at its centre to eliminate the displacement there (see Figure 12.19a). The prop reaction P can be found by equating the cental vertical displacement from each load according to the standard expressions [1, 2]

$$\frac{PL^3}{48EI} = \frac{5wL^4}{384EI} \tag{12.54}$$

from which $P = wL/1.6$. The reader may wish to confirm this prop reaction by applying the moment area theorem. The net shear force and bending moment diagrams in the presence of this prop are as shown in Figure 12.19b,c.

There is a central symmetry, so we need only work upon one half of the beam to optimise its geometry and thereby minimise weight. Hence, on setting $F_1 = -P/2$ and $F_2 = wL/2$, we have $F_1/F_2 = -P/wL = -0.625$. Note also that $F_1/F_2 = s_1/s_2$, where $s_1 = (1/\sigma_y)(F_1/L^2)$ and $s_2 = (1/\sigma_y)(F_2/L^2)$. For the three regions I, II and III, within the left half-length $(0 \le q \le 0.5)$ in Figure 12.19b,c, the magnitudes of the ordinates in the F- and M-diagrams are as follows.

For region I $(0 \le q \le 0.1875$, where $q = z/L)$,

$$|F(z)| = F_1 + F_2(1 - 2z/L), \qquad \Rightarrow |F(q)| = F_2(0.375 - 2q) \tag{12.55a}$$

$$|M(z)| = F_1 z + F_2 z(1 - z/L), \qquad \Rightarrow |M(q)| = F_2 Lq(0.375 - q) \tag{12.55b}$$

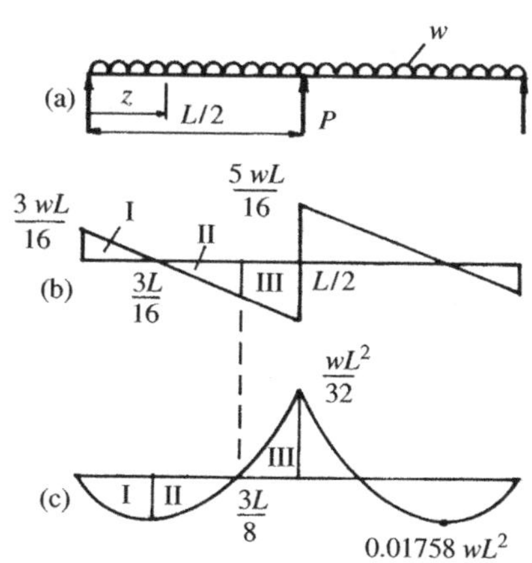

Figure 12.19 F- and M-diagrams for a propped beam carrying uniformly distributed loading

giving $|F(0.1875)| = 0$ and $|M(0.1875)| = 0.01758wL^2$. Correspondingly, within this region, the normalised beam diameters required to withstand the shear forces and moments in equations (12.55a,b) follow from equations (3.3b) and (5.15b) as

$$\left(\frac{d}{2L}\right)_F = \left(\frac{4}{\sqrt{3}\pi}\right)^{1/2} s_2^{1/2}(0.375 - 2q)^{1/2} \tag{12.56a}$$

$$\left(\frac{d}{2L}\right)_M = \left(\frac{4}{\pi}\right)^{1/3} s_2^{1/3}[q(0.375 - q)]^{1/3} \tag{12.56b}$$

where $s_2 = (1/\sigma_y)(F_2/L^2)$ is defined as a *positive* structural index which we shall vary across five decades from 0.001 to 100. This range includes both long and short beams with optimum profiles that vary depending upon the interaction between bending and shear.

For region II ($0.1875 \le q \le 0.375$),

$$|F(z)| = -F_1 - F_2(1 - 2z/L), \qquad \Rightarrow |F(q)| = F_2(2q - 0.375) \tag{12.57a}$$

$$|M(z)| = F_1 z + F_2 z(1 - z/L) \qquad \Rightarrow |F(q)| = F_2 Lq(0.375 - q) \tag{12.57b}$$

The beam diameters required to withstand the shear forces and moments in this region are

$$\left(\frac{d}{2L}\right)_F = \left(\frac{4}{\sqrt{3}\pi}\right)^{1/2} s_2^{1/2}(2q - 0.375)^{1/2} \tag{12.58a}$$

$$\left(\frac{d}{2L}\right)_M = \left(\frac{4}{\pi}\right)^{1/3} s_2^{1/3}[q(0.375 - q)]^{1/3} \tag{12.58b}$$

For region III ($0.375 \le q \le 0.5$),

$$|F(z)| = -F_1 - F_2(1 - 2z/L), \qquad \Rightarrow |F(q)| = F_2(2q - 0.375) \tag{12.59a}$$

$$|M(z)| = -F_1 z - F_2 z(1 - z/L), \qquad \Rightarrow |F(q)| = F_2 Lq(q - 0.375) \tag{12.59b}$$

The corresponding beam diameters for this region become

$$\left(\frac{d}{2L}\right)_F = \left(\frac{4}{\sqrt{3}\pi}\right)^{1/2} s_2^{1/2}(2q - 0.375)^{1/2} \tag{12.60a}$$

$$\left(\frac{d}{2L}\right)_M = \left(\frac{4}{\pi}\right)^{1/3} s_2^{1/3}[q(q - 0.375)]^{1/3} \tag{12.60b}$$

Table 12.19 gives the limiting diameters within each region for s_2 taken over five decades. It is clear from Table 12.19 that the optimum diameter required must be based upon both $F(z)$ and $M(z)$ across these regions. We take, for each region I, II and III within the half-length, the greater normalised radius ($d/2L$) from the paired equations (12.56a,b), (12.58a,b) and (12.60a,b) to yield a variable radius profile that minimises beam weight.

Table 12.19 Normalised radii ($d/2L$) and weight parameters $\sum n$ for three regions of $q(= z/L)$ within a half-length

$d/2L$	s_2	$0 \leq q \leq 0.1875$	$0.1875 \leq q \leq 0.375$	$0.375 \leq q \leq 0.50$	$\sum n$
M	100	0–1.648	1.648–0	0–1.9965	
F	100	5.25–0	0–5.25	5.25–6.778	61.65
M	10	0–0.765	0.765–0	0–0.9266	
F	10	1.6603–0	0–1.6603	1.6603–2.1435	6.90
M	1	0–0.3551	0.3551–0	0–0.4302	
F	1	0.525–0	0–0.525	0.525–0.6778	0.809
M	0.1	0–0.1648	0.1648–0	0–0.19966	
F	0.1	0.166–0	0–0.166	0.166–0.2144	0.114
M	0.01	0–0.0765	0.0765–0	0–0.09267	
F	0.01	0.0525–0	0–0.0525	0.0525–0.06778	0.021
M	0.001	0–0.03552	0.03552–0	0–0.04303	
F	0.001	0.0166–0	0–0.0166	0.0166–0.02143	4.34×10^{-3}

Two examples of long and short beam profiles at $s_2 = 0.01$ and $s_2 = 1$, from Table 12.19, appear in Figure 12.20. In Figure 12.20a the design radius divides between the F and M criteria within each region as shown. Figure 12.20b differs in that the radius is entirely shear-based within region III. Where intersections occur within regions I, II and III, their positions $q_{\mathrm{I}}, q_{\mathrm{II}}$ and q_{III} are found (by trial) by equating the respective $d/2L$ equations (12.56a,b)–(12.60a,b).

Once the diameters are known, the total weight W across each region within the half-length is found by summing weight integrals $W = \int \rho(\pi d^2/4)\mathrm{d}z$. For example,

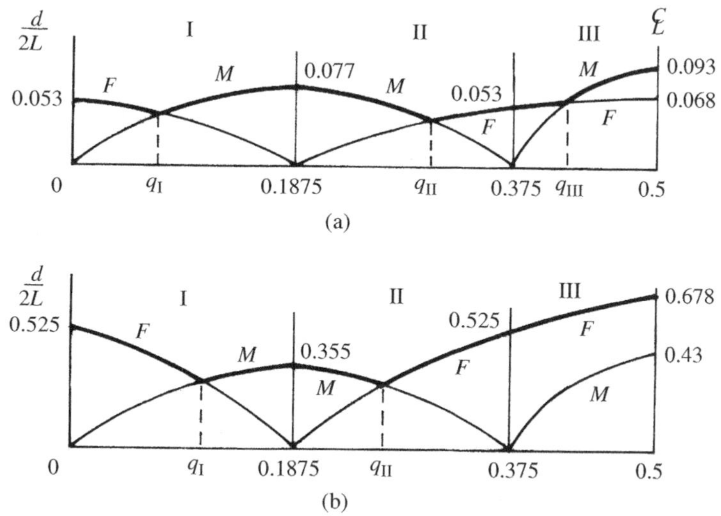

Figure 12.20 Beam radius profiles with a half-length for (a) $s_2 = 0.01$ and (b) $s_2 = 1$

within region I we can write a normalised weight expression from the shear and bending radii in equations (12.56a,b):

$$n_{\mathrm{I}} = \frac{4}{\sqrt{3}} s_2 \int_0^{q_{\mathrm{I}}} (0.375 - 2q)\mathrm{d}q + \pi \left(\frac{4}{\pi}\right)^{2/3} s_2^{2/3} \int_{q_{\mathrm{I}}}^{0.1875} [q(0.375 - q)]^{2/3} \mathrm{d}q \qquad (12.61)$$

where $n = (1/\rho)(W/L^3)$. Similarly, within region II we have from equations (12.58,b),

$$n_{\mathrm{II}} = \frac{4}{\sqrt{3}} s_2 \int_{q_{\mathrm{II}}}^{0.375} (2q - 0.375)\mathrm{d}q + \pi \left(\frac{4}{\pi}\right)^{2/3} s_2^{2/3} \int_{0.1875}^{q_{\mathrm{II}}} [q(0.375 - q)]^{2/3} \mathrm{d}q \qquad (12.62)$$

Within region III (where a q_{III} intersection occurs only for $s_2 \le 0.1$), equations (12.60a,b) apply:

$$n_{\mathrm{III}} = \frac{4}{\sqrt{3}} s_2 \int_{0.375}^{q_{\mathrm{III}}} (2q - 0.375)\mathrm{d}q + \pi \left(\frac{4}{\pi}\right)^{2/3} s_2^{2/3} \int_{q_{\mathrm{III}}}^{0.50} [q(q - 0.375)]^{2/3} \mathrm{d}q \qquad (12.63a)$$

For $s_2 > 0.1$, Table 12.19 shows that the design, and hence the weight, in region III is based wholly upon shear. Here the upper limit to the first integral in equation (12.63a) is altered to give

$$n_{\mathrm{III}} = \frac{4}{\sqrt{3}} s_2 \int_{0.375}^{0.5} (2q - 0.375)\mathrm{d}q \qquad (12.63b)$$

Where appropriate, a numerical integration of the bending contributions within equations (12.61)–(12.63a,b) enables the summation $n_{\mathrm{I}} + n_{\mathrm{II}} + n_{\mathrm{III}}$, to provide the normalised half-weight. Hence, the total optimised beam weight becomes $\sum n = 2(n_{\mathrm{I}} + n_{\mathrm{II}} + n_{\mathrm{III}})$. The results, given in the final column of Table 12.19, are used to plot $\sum n$ against s_2 in Figure 12.21. With s_2 as its abscissa, Figure 12.21 allows for an optimum design of any length of beam carrying uniformly distributed loading when resting upon three equi-spaced, level supports. It is seen that the weight is less than that of a beam without a central prop. Within the shaded region of this figure lie the weights of propped beams with negative load ratios in the range $-0.625 < P/wL < 0$.

In order to make a comparison with the greater weights of unpropped beams carrying both distributed and central loading (see Figure 12.18), it is necessary to do so at similar, positive indices s_2. This requires that the abscissa s_1 in Figure 12.18 be converted to s_2 using the indicated load ratios $s_2/s_1 = wL/P$. This is done in Table 12.20 to facilitate the comparison between this propped beam ($s_2/s_1 = -1.6$) and beams with combined load ratios $s_2/s_1 = 0.1$, 1 and 10 (see Tables 12.14, 12.16 and 12.18) across a similar range of s_2. Figure 12.22 has been constructed from the weights given Table 12.20. This shows that beams under increasing positive load ratios $s_2/s_1 = wL/P$ offer reduced weight on the basis of similar s_2. This applies to the three combined loadings $wL/P = 0.1$, 1 and 10 in

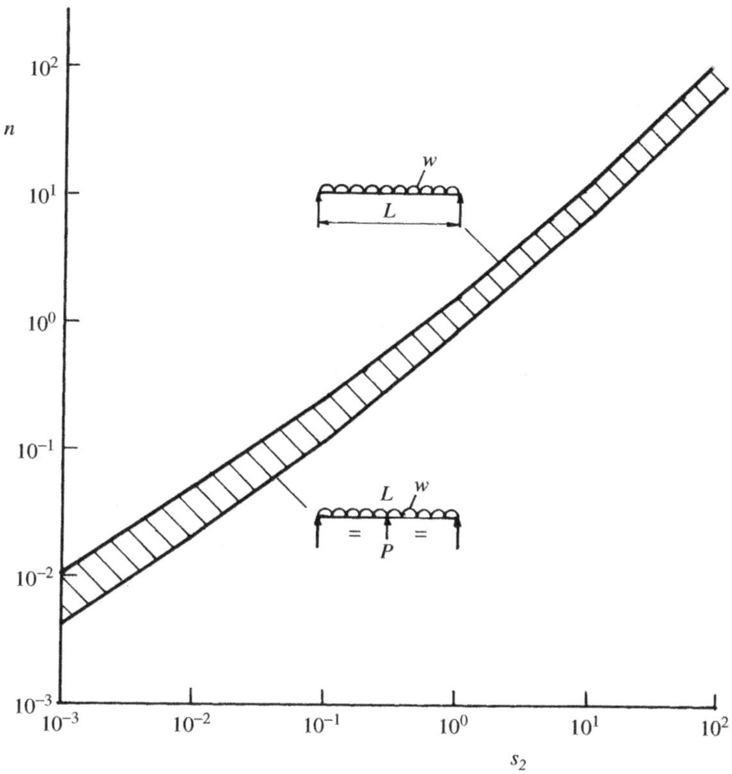

Figure 12.21 Minimum weights of propped and simply supported beams under distributed loading. (Note: $s_2 = (1/\sigma_y)(F_2/L^2)$ where $F_2 = wL/2$; $P = wL/1.6$.)

Table 12.20 Comparison between optimised weights for similar s_2 with $s_2/s_1 = -1.6$, 0.1, 1 and 10

$s_2 = -1.6s_1$	0.001	0.01	0.1	1	10	100
$\sum n$	0.0043	0.021	0.114	0.81	6.90	61.65
$s_2 = 0.1s_1$	0.001	0.01	0.1	1	10	100
$\sum n$	0.0674	0.3413	2.4273	24.249	242.49	2424.9
$s_2 = s_1$	0.001	0.01	0.1	1	10	100
$\sum n$	0.0195	0.0926	0.489	3.5348	34.64	346.41
$s_2 = 10s_1$	0.001	0.01	0.1	1	10	100
$\sum n$	0.012	0.0556	0.280	1.742	14.44	138.9

Figure 12.18. Also shown in Figure 12.22 are the weights of a simply supported beam carrying only a distributed load ($s_2/s_1 = \infty$) and a centrally propped beam ($s_2/s_1 = -1.6$), each taken from Figure 12.21. The former beam ($s_2/s_1 = \infty$) defines the lower band for positive load ratios. The greatest weight reduction applies to the negative load ratio ($s_2/s_1 = -1.6$) that arises from centrally propping the distributed loading to eliminate all central deflection.

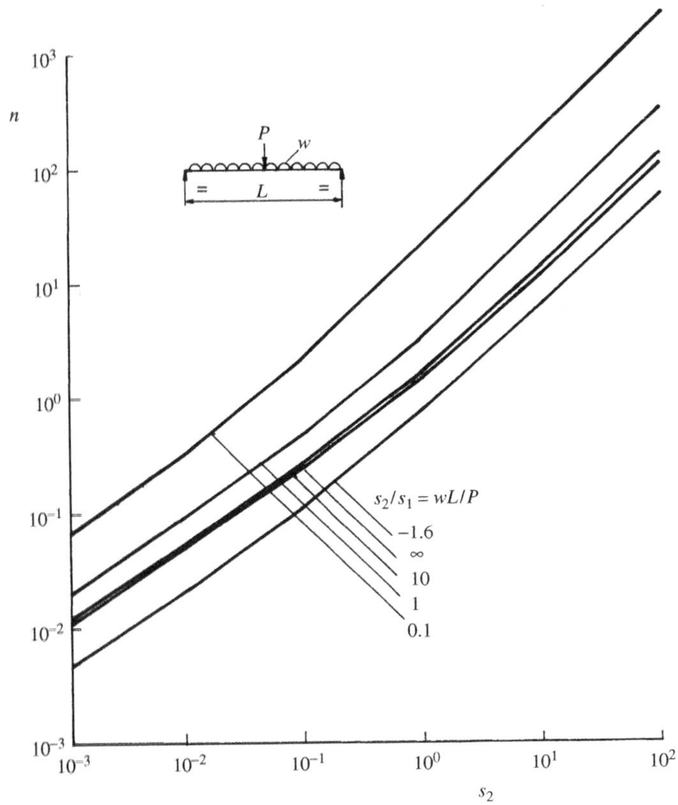

Figure 12.22 Weights for optimum, simply supported beams under combined loadings. (Note $s_1 = (1/\sigma_y)(F_1/L^2)$; $s_2 = (1/\sigma_y)(F_2/L^2)$ where $F_1 = P/2$ and $F_2 = wL/2$.)

References

[1] Case, J. and Chilver, A. H. *Strength of Materials and Structures*, 4th edition, Arnold, 1999
[2] Rees, D. W. A. *Basic Solid Mechanics*, Macmillan, 1997.
[3] Lancaster, P. R. and Michell, D. *The Mechanics of Materials*, McGraw-Hill, 1967.
[4] Timoshenko, S. and Young, D. *Elements of Strength of Materials*, van Nostrand, 1962.
[5] Megson, T. H. G. *Strength of Materials for Civil Engineers*, Nelson, 1980.
[6] Neěmec, J. *Rigidity and Strength of Steel Parts*, Academia Prague, 1966
[7] Pilkey, W. D. and Pilkey, O. H. *Mechanics of Solids*, Quantum, 1974.
[8] Shigley, J. E. *Mechanical Engineering Design*, McGraw-Hill, 1986.

Exercises

12.1 Find the optimum diameters for each limb of the cranked strut shown in Figure 12.3a when a load of 50 kN is applied at a distance of 0.25 m from the central axis of the strut. Assume that the strut is optimally designed, having a minimum length fixed at its base and hinged to the end of the beam. The material is aluminium alloy with yield stress $\sigma_y = 150$ MPa, elastic modulus $E = 70$ GPa and density $\rho = 2700$ kg/m^3.

12.2 A square-section steel column is to support a concentrated load of $P = 15$ kN with an overhang of $L_1 = 1$ m from its longitudinal axis (see Figure 12.3a). Find the square dimension for the vertical column when its length is $L_2 = 500$ mm and show that this length is valid for the column theory employed. Take $E = 210$ GPa and $\sigma_y = 300$ MPa.

12.3 Derive an expression for the diameter d_2 of limb 2 for the cranked cantilevers in Figure 12.2. Hence, derive the integrals that express their optimum weights and consider various methods by which these integrals could be evaluated.

12.4 Find the optimum diameters for each limb of the cranked bracket shown in Figure 12.4 when the load $P = 50$ kN is applied at a distance $L_1 = 0.25$ m from the axis of the tie, whose length is $L_2 = 1$ m. The material is aluminium alloy with yield stress $\sigma_y = 150$ MPa, elastic modulus $E =$ 70 GPa and density $\rho = 2700$ kg/m^3.

12.5 Optimise the cranked bracket shown in Figure 12.4 when the 1 m long tie is cranked to support a tensile load of 50 kN offset by 25 mm from its longitudinal axis. The bar is to be square section in steel with $\sigma_y = 280$ MPa, $E = 210$ GPa and $\rho = 7860$ kg/m^3.

12.6 When, for the cranked cantilever in Figure 12.2b, the loading is distributed uniformly along the crank so that equation (12.17b) replaces the first term in equation (12.15a), derive an expression for its optimum (minimum) weight for $L_2/L_1 < 1$, given $P = wL_1$ and $T = M_{\max} = wL_1^2/2$.

12.7 If both ends of each strut, supporting the beam within the portal frame in Figure 12.5, is built in, find their dimensions when $P = 50$ kN. What is the weight of the assembly when the beam indices are: (i) $s = 0.01$; (ii) $s = 1$? Take $\sigma_y = 300$ MPa, $E = 210$ GPa and $\rho = 7860$ kg/m^3. [Answers: $d_2 = 10.3$ mm and $L_2 = 428.1$ mm, (i) 0.9771 kg, (ii) 0.3007 kg]

12.8 Find the dimensions of the portal frame in Figure 12.5 when it is required to carry a central load of $P = 50$ kN operating at structural indices (i) $s = 0.1$ and (ii) $s = 0.001$. Assume that each stanchion is an optimally designed, minimum length, circular-section strut, fixed at its base and hinged to each end of the beam. Both beam and strut are made of steel with $\sigma_y = 300$ MPa, $\rho = 7860$ kg/m^3 and $E = 210$ GPa.

12.9 A cantilever of length L carries a uniformly distributed load w whilst a prop eliminates some free-end deflection. If the prop reaction is P, derive the minimum weight of the cantilever when it is to operate at a load ratio $P/wL(= s_1/s_2) = -0.1$. Show that the beam weight is greater than one where the prop eliminates the end-deflection entirely. [Answers appear in Figure 12.16.]

12.10 Find the critical structural index s_2 within region III of Table 12.19 at and beyond which the design diameter must be wholly shear-based.

12.11 A cantilever carries a uniformly distributed load w whilst a prop supports a load P at its centre. Show that if the prop is to eliminate all the central, free deflection then the load carried by the prop is $P = -15wL/16$. Hence, derive the weight of the propped cantilever when operating at a load ratio $P/wL(= s_1/s_2) = -15/16$. Compare with the weight of a cantilever carrying distributed loading in the absence of a prop.

12.12 Design the optimum geometry and find the minimum weight of a 0.5 m long, circular-section, aluminium alloy cantilever required to support a uniformly distributed load of 100 kN/m, together with a concentrated end-load of 5 kN. Take $\sigma_y = 200$ MPa and $\rho = 2700$ kg/m^3.

12.13 Design the optimum geometry and find the minimum weight of a 10 mm long, circular-section, cantilevered steel pin required to support a uniformly distributed load of 30 kN/m, together with a concentrated end-load of 3 kN. Take $\sigma_y = 300$ MPa and $\rho = 7850$ kg/m^3.

12.14 Determine the normalised weights of an end-propped cantilever having a circular cross-section, when carrying distributed loading at a load ratio $P/wL(= s_1/s_2) = -0.1$ over a range of structural indices $0.001 \leq s_2 \leq 100$ (see Figure 12.16). Compare graphically with the weight of a cantilever in the absence of a prop ($P/wL = 0$) over a similar range of structural indices, taken from Figure 10.13.

13

Encastré Beams

13.1 Introduction

In the previous two chapters simply supported beams and cantilevers with various cross-sections were optimised for minimum weight independently and as elements within more complex structures. The fully encastré beam, being a statically indeterminate structure, is optimised for minimum weight separately here. Standard encastré beams have their ends built in and thereby provide a stiffer, lighter structure than simply supported beams when the loading they are required to bear is similar. It is shown that encastré beams can operate at minimum weight with a variable profile in a circular cross-section for central point load and uniformly distributed loading acting separately and in combination. It is also shown how to extend the analysis to encastré beams with optimised square sections when loaded in a similar manner. Though the analysis of other non-circular solid sections will, in general, follow a similar procedure to that adopted here, additional local buckling criteria would be required in the case of open and closed sections with thin walls. We have seen how to marry local buckling with the limiting stress from bending and shear taken separately within Chapters 3 and 5. The encastré beam section, when loaded as described, is subjected to shear force and bending moment that vary with the length, in a manner that is well established [1–16]. Therefore, both bending and shear effects should be combined within a true optimum design, as we saw for simply supported beams in Chapter 10. When this is done, regions of the length appear over which the beam section is optimised in shear and in bending depending upon which design criterion dominates. It will be seen that this procedure for minimising beam weight can lead to some rather unusual length profiles as one design basis is displaced by another across each length region. The profiles may appear impracticable but they do show where and how much material in the length can be removed safely from a uniform-section beam when it becomes essential to reduce weight.

13.2 Central-Concentrated Load

The first standard case refers to a central point load (see Figure 13.1a). The shear force and bending moments vary with length in the manner of Figure 13.1b,c, irrespective of the shape of the beam's cross-section. The net moment diagram in Figure 13.1d is found

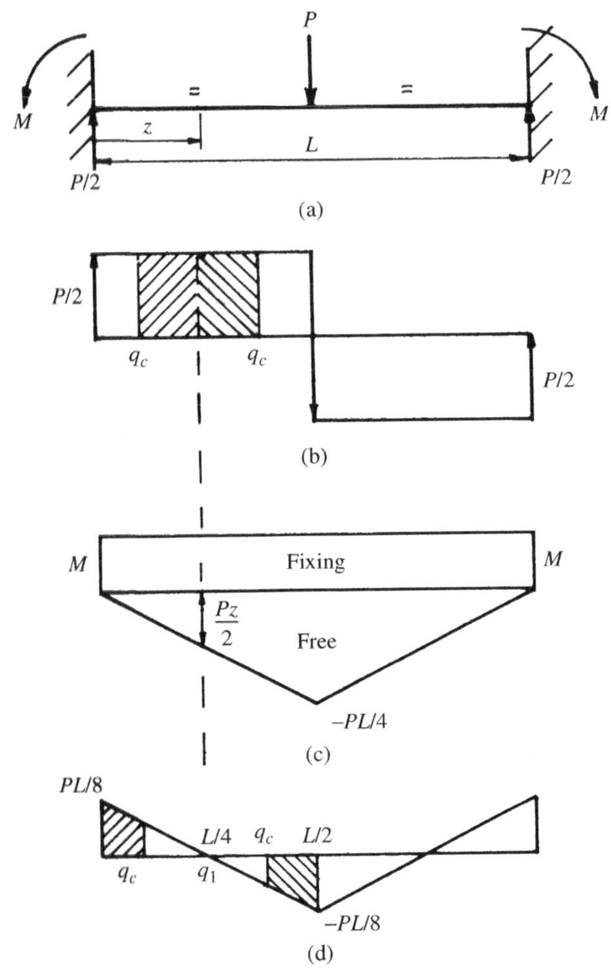

Figure 13.1 Force and moment diagrams for an encastré beam with central load

by taking the difference between ordinates in the 'free' and 'fixing' moment diagrams, whose separated areas, shown in Figure 13.1c, are equal. With a discontinuity arising in the net F- and M-diagrams at mid-span beneath the load, it becomes necessary to work within each half of the beam with z taken from each end towards the centre. With the symmetry in these diagrams (see Figure 13.1b,d) it is sufficient to confine the analysis to one half. Thus, for the left half-length ($0 \leq z \leq L/2$) in Figure 13.1a, the ordinates in the shear force diagram (Figure 13.1b) may be expressed simply as

$$|F(z)| = P/2 = \text{constant} \tag{13.1}$$

That part of the net M-diagram in Figure 13.1d which is positive within the first half-length of this beam gives the bending moment ordinates as

$$|M(z)| = \frac{P}{2}\left(\frac{L}{4} - z\right), \quad \text{for } 0 \le z \le \frac{L}{4} \tag{13.2a}$$

where clearly the net moment becomes zero when $z = L/4$. For that part of the net M-diagram within the first half-length, which is negative in Figure 13.1d, we write the magnitudes of the moment ordinates as

$$|M(z)| = \frac{P}{2}\left(z - \frac{L}{4}\right), \quad \text{for } \frac{L}{4} \le z \le \frac{L}{2} \tag{13.2b}$$

Equations (13.1) and (13.2a,b) will reveal how the optimum section should vary within each region of the half-length. With a central symmetry in each diagram, the total weight is twice that of one half, which will be established here for circular and square cross-sections.

13.2.1 Optimum Circular Section

For a transverse shear force applied to a circular section beam, it was shown previously that equation (5.15b) gave the optimum diameter. Substituting from equation (13.1), dividing by $2L$ and writing the relationship between the tensile and shear yield stresses as $\sigma_y = Q\tau_y$ leads to

$$\left(\frac{d}{2L}\right)_F = \left(\frac{2Q}{3\pi}\right)^{1/2}\left(\frac{1}{\sigma_y}\frac{P}{L^2}\right)^{1/2} = 0.6063\, s^{1/2} \quad \text{for } 0 \le \frac{z}{L} \le \frac{1}{2} \tag{13.3}$$

where P is the applied load. The numerical coefficient $Q = \sqrt{3}$ (used here) or $Q = ½$ by the von Mises or Tresca yield criterion, respectively. For a given structural index $s = (1/\sigma_y)(P/L^2)$, equation (13.3) shows that the normalised radius $d/2L$ remains constant in a shear design. Note that the index s is to be identified with short beams when it is large and long beams when it is small, i.e. an increasing beam length is associated with a decreasing index.

With the bending of a circular section, the optimum diameter is found by combining each $|M(z)|$ expression in (13.2a,b) in turn with equation (3.3b). Dividing by $2L$ reveals the variation in the radius (normalised with length), for two regions that define the half-length: for $0 \le z/L \le 1/4$,

$$\left(\frac{d}{2L}\right)_M = \left(\frac{2}{\pi}\right)^{1/3}\left(\frac{1}{\sigma_y}\frac{P}{L^2}\right)^{1/3}\left(\frac{1}{4} - \frac{z}{L}\right)^{1/3} = 0.8603\, s^{1/3}\left(\frac{1}{4} - \frac{z}{L}\right)^{1/3} \tag{13.4a}$$

and for $1/4 \le z/L \le 1/2$,

$$\left(\frac{d}{2L}\right)_M = \left(\frac{2}{\pi}\right)^{1/3}\left(\frac{1}{\sigma_y}\frac{P}{L^2}\right)^{1/3}\left(\frac{z}{L} - \frac{1}{4}\right)^{1/3} = 0.8603\, s^{1/3}\left(\frac{z}{L} - \frac{1}{4}\right)^{1/3} \tag{13.4b}$$

in which $s = (1/\sigma_y)(P/L^2)$ is a normalised structural index. Firstly, equating (13.4a) and (13.3) gives the critical position in the first quarter-length where the diameter from a bending design is displaced by one from shear:

$$\left(\frac{z}{L}\right)_{cr} = \frac{1}{4} - \left(\frac{Q}{3}\right)^{3/2}\left(\frac{2}{\pi}\right)^{1/2} s^{1/2} \quad \Rightarrow \quad q_c = 0.25 - 0.35s^{1/2} \qquad (13.5a,b)$$

Setting $q_c = 0$ in equation (13.5b) gives $s_{MF} = 0.51$. Hence, an all-shear design for the first quarter length $0 \leq q \leq 1/4$ applies when $s \geq s_{MF}$. When $s < s_{MF}$, *bending* dictates the diameter within the region $0 < q < q_c$ and *shear* for $q_c < q < 1/4$. Here $q_c(0 < q_c < 1/4)$ is found from equation (13.5b) at the particular s-value. Similarly, equating (13.4b) and (13.3) gives the critical position in the second quarter-length where the design diameter criterion changes from one of shear to one of bending:

$$\left(\frac{z}{L}\right)_{cr} = \frac{1}{4} + \left(\frac{Q}{3}\right)^{3/2}\left(\frac{2}{\pi}\right)^{1/2} s^{1/2} \quad \Rightarrow \quad q_c = 0.25 + 0.35s^{1/2} \qquad (13.6a,b)$$

Setting $q_c = 0.5$ in equation (13.6b) again gives $s_{FM} = 0.51$. An all-shear design applies to the second quarter length $1/4 \leq q \leq 1/2$ when $s \geq s_{FM}$. When $s < s_{FM}$, *shear* dictates the diameter within the region $1/4 < q < q_c$ and *bending* for $q_c < q < 1/2$. Here $q_c(1/4 < q_c < 1/2)$ is found from equation (13.6b) for a given index s. The particular regions of the shear force and bending moment diagram that dictate the optimum design are shaded in Figure 13.1b,d. Values of s are taken to span five decades from 0.001 to 100 for the specific cross-sections that follow.

13.2.2 Minimum Weight of Circular-Section Beam

With $s \geq s_{MF}$ (where $s_{MF} = 0.51$) the beam's diameter and weight within the quarter region $0 \leq q \leq 1/4$ follow solely from an all-shear design. Thus, for a normalised elemental length $\delta q = \delta z/L$ in this region, the normalised weight $n_F = (1/\rho)(W_F/L^3)$ from shear becomes

$$\frac{W_F}{L^3} = \rho\pi \int_0^{1/4} \left(\frac{d}{2L}\right)^2 \mathrm{d}\left(\frac{z}{L}\right) \quad \Rightarrow \quad n_F = \pi \int_0^{1/4} \left(\frac{d}{2L}\right)^2 \mathrm{d}q \qquad (13.7a,b)$$

where $q = z/L$. Substituting equation (13.3) into equation (13.7b) leads to the weight of the quarter-length:

$$n_F = \left(\frac{2Q}{3}\right) s \int_0^{1/4} \mathrm{d}q = \frac{s}{2\sqrt{3}} \qquad (13.8)$$

When $s < s_{MF}$, bending and then shear contribute to the optimal weight. From bending:

$$\frac{W_M}{L^3} = \rho\pi \int_0^{q_c} \left(\frac{d}{2L}\right)^2 \mathrm{d}\left(\frac{z}{L}\right) \quad \Rightarrow \quad n_M = \pi \int_0^{q_c} \left(\frac{d}{2L}\right)^2 \mathrm{d}q \qquad (13.9a,b)$$

Substituting equation (13.4a) into equation (13.9b),

$$n_M = \pi \left(\frac{2}{\pi}\right)^{2/3} s^{2/3} \int_0^{q_c} \left(\frac{1}{4} - q\right)^{2/3} \mathrm{d}q$$

$$= 1.395 s^{2/3} \left[\left(\frac{1}{4}\right)^{5/3} - \left(\frac{1}{4} - q_c\right)^{5/3}\right] = 1.395 s^{2/3}(0.0992 - 0.1738 s^{5/6}) \quad (13.9\text{c})$$

From shear,

$$\frac{W_F}{L^3} = \rho\pi \int_{q_c}^{1/4} \left(\frac{d}{2L}\right)^2 \mathrm{d}\left(\frac{z}{L}\right) \quad \Rightarrow \quad n_F = \pi \int_{q_c}^{1/4} \left(\frac{d}{2L}\right)^2 \mathrm{d}q \qquad (13.10\text{a,b})$$

Substituting equation (13.3) into equation (13.10b),

$$n_F = \frac{2}{\sqrt{3}} s \int_{q_c}^{1/4} \mathrm{d}q$$

$$= \frac{2}{\sqrt{3}} s \left(\frac{1}{4} - q_c\right) = \frac{2}{\sqrt{3}} s \left[\frac{1}{4} - \left(\frac{1}{4} - 0.35\, s^{1/2}\right)\right] = 0.4042\, s^{3/2} \qquad (13.10\text{c})$$

in which equation (13.5b) has defined q_c. Adding equations (13.9c) and (13.10c) gives the weight of the first quarter length:

$$n = n_M + n_F = 1.395\, s^{2/3}(0.0992 - 0.1738\, s^{5/6}) + 0.4042\, s^{3/2}$$

$$= 0.1617\, s^{3/2} + 0.1384\, s^{2/3} \qquad (13.11)$$

Within the symmetry of the net M-diagram in Figure 13.1d we should expect that the weight of the second quarter length ($1/4 \le q \le 1/2$) to be identical to those of equations (13.8) and (13.11). That is, for an all-shear design, $s \ge s_{FM}$ (where $s_{FM} = 0.51$), the normalised weight in our abbreviated notation becomes

$$n_F = \left(\frac{2Q}{3}\right) s \int_{1/4}^{1/2} \mathrm{d}q = \frac{s}{2\sqrt{3}} \qquad (13.12)$$

which agrees with equation (13.8). For $s < s_{FM}$, where shear and then bending contribute to the weight,

$$n_F = \frac{2}{\sqrt{3}} s \int_{1/4}^{q_c} \mathrm{d}q = \frac{2}{\sqrt{3}} s \left(q_c - \frac{1}{4}\right)$$

$$= \frac{2}{\sqrt{3}} s \left[\left(\frac{1}{4} + 0.35 s^{1/2}\right) - \frac{1}{4}\right] = 0.4042 s^{3/2} \qquad (13.13\text{a})$$

$$n_M = \pi \left(\frac{2}{\pi}\right)^{2/3} s^{2/3} \int_{q_c}^{1/2} \left(q - \frac{1}{4}\right)^{2/3} dq$$

$$= 1.395 s^{2/3} \left[\left(\frac{1}{2} - \frac{1}{4}\right)^{5/3} - \left(q_c - \frac{1}{4}\right)^{5/3} \right] = 1.395 s^{2/3} (0.0992 - 0.1738 s^{5/6}) \tag{13.13b}$$

in which equation (13.6b) has defined q_c. The total weight in this region, as found from the sum $n_F + n_M$ in equations (13.13a,b), agrees with equation (13.11).

The weight of the beam's half-length is found by adding the weights of its first two quarter lengths at a given s-value, i.e. $\sum n_F$ for $s \geq 0.51$ and $\sum(n_F + n_M)$ for $s < 0.51$ (as shown above). Finally, by doubling this, Table 13.1 gives the full-length beam weights across five decades of s according to these summations. Figure 13.2 plots $\sum n$ from Table 13.1 against s to provide the total weight from regions in which bending, combined shear and bending, and pure shear designs apply as the index s increases to cross its transitional value ($s_{MF} = s_{FM} = 0.51$). Note that this plot is identified later with a combined load ratio $s_2/s_1 = wL/P = 0$, thereby enabling comparisons with the minimum weights of encastré beams under concentrated P and distributed w loads. The comparison made in the last line of Table 13.1 with the weights of simply supported beams (from Figure 10.18) shows the latter to be far heavier for similar s in the region of bending. However, the weights of encastré and simply supported beams become identical in the region of shear.

13.2.3 Circular Section, Length Profile

As this encastré beam length increases, i.e. as s decreases, the contribution to the design diameter from shear diminishes. The following shows this within the beam's half-length profiles for three s-values, 1, 0.1 and 0.01. The ordinates $d/2L$ are the normalised radii, these being given by equations (13.3) and (13.4a,b), appropriate to each s-value.

(a) $s = 1$ ($s > s_{MF}$, where $s_{MF} = 0.51$)
The diameter for an all-shear design follows from equation (13.3) as:

$$\left(\frac{d}{2L}\right)_F = \left(\frac{2}{\sqrt{3\pi}}\right)^{1/2} \times 1^{1/2} = 0.6063 = \text{constant}, \quad \text{for } 0 \leq q \leq 0.5$$

Table 13.1 Normalised, circular-section beam weights across five decades of structural index s

s	0.001	0.01	0.10	0.51	1.0	10	100
$n(0 < q < {}^1\!/\!_4)$	1.3891×10^{-3}	6.585×10^{-3}	0.03494	0.1472	0.2887	2.887	28.87
$n({}^1\!/\!_4 < q < {}^1\!/\!_2)$	1.3891×10^{-3}	6.585×10^{-3}	0.03494	0.1472	0.2887	2.887	28.87
$\sum n$	5.5563×10^{-3}	0.02634	0.13974	0.5888	1.1548	11.548	115.48
Simply supported	8.8×10^{-3}	0.0412	0.1996	–	1.2021	11.547	115.47

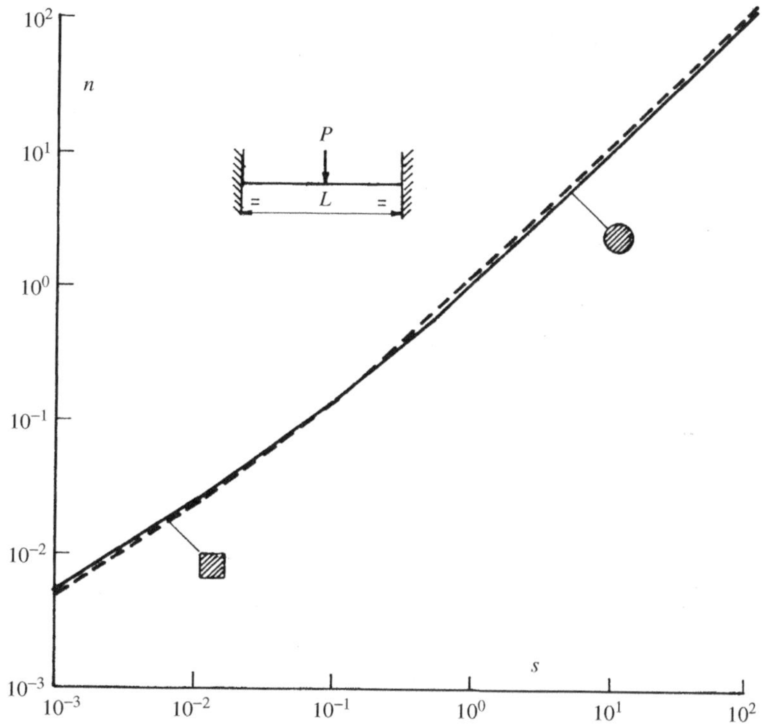

Figure 13.2 Effect of structural index upon weight of circular- and square-section encastré beams under central concentrated loading (note: $s = (1/\sigma_y)(P/L^2)$; $n = (1/\rho)(W/L^3)$)

(b) $s = 0.1 (s < s_{MF})$
Bending (subscript M) and then shear (subscript F) designs apply within the region $0 \leq q \leq 0.25$. Equation (13.5b) gives $q_c = 0.139$, and when equations (13.4a) and (13.3) are applied to this region they give the respective normalised radii

$$\left(\frac{d}{2L}\right)_M = \left(\frac{2}{\pi}\right)^{1/3} 0.1^{1/3} \left(\frac{1}{4} - q\right)^{1/3} = 0.3993 \left(\frac{1}{4} - q\right)^{1/3}, \quad \text{for } 0 \leq q \leq 0.139$$

$$\left(\frac{d}{2L}\right)_F = \left(\frac{2}{\sqrt{3}\pi}\right)^{1/2} \times 0.1^{1/2} = 0.1917 = \text{constant}, \quad \text{for } 0.139 \leq q \leq 0.25$$

Shear and then bending designs apply within the region $0.25 \leq q \leq 0.50$. Equation (13.6b) gives $q_c = 0.361$, and equations (13.3) and (13.4b) apply to this region as follows:

$$\left(\frac{d}{2L}\right)_F = \left(\frac{2}{\sqrt{3}\pi}\right)^{1/2} \times 0.1^{1/2} = 0.1917 = \text{constant}, \quad \text{for } 0.25 \leq q \leq 0.361$$

$$\left(\frac{d}{2L}\right)_M = \left(\frac{2}{\pi}\right)^{1/3} 0.1^{1/3} \left(q - \frac{1}{4}\right)^{1/3} = 0.3993 \left(q - \frac{1}{4}\right)^{1/3}, \quad \text{for } 0.361 \leq q \leq 0.5$$

(c) $s = 0.01(s < s_{MF})$
Bending and then shear designs apply within the quarter length $0 \leq q \leq 0.25$. Equation (13.5b) gives $q_c = 0.215$, for when equations (13.4a) and (13.3) are applied:

$$\left(\frac{d}{2L}\right)_M = \left(\frac{2}{\pi}\right)^{1/3} 0.01^{1/3}\left(\frac{1}{4} - q\right)^{1/3} = 0.1854\left(\frac{1}{4} - q\right)^{1/3}, \quad \text{for } 0 \leq q \leq 0.215$$

$$\left(\frac{d}{2L}\right)_F = \left(\frac{2}{\sqrt{3}\pi}\right)^{1/2} \times 0.01^{1/2} = 0.0607 = \text{constant}, \quad \text{for } 0.215 \leq q \leq 0.25$$

Shear and then bending designs apply within $0.25 \leq q \leq 0.50$. Equation (13.6b) gives $q_c = 0.285$, for when equations (13.3) and (13.4b) are applied:

$$\left(\frac{d}{2L}\right)_F = \left(\frac{2}{\sqrt{3}\pi}\right) \times 0.01^{1/2} = 0.0607 = \text{constant}, \quad \text{for } 0.25 \leq q \leq 0.285$$

$$\left(\frac{d}{2L}\right)_M = \left(\frac{2}{\pi}\right)^{1/3} 0.01^{1/3}\left(q - \frac{1}{4}\right)^{1/3} = 0.1854\left(q - \frac{1}{4}\right)^{1/3}, \quad \text{for } 0.285 \leq q \leq 0.5$$

Figure 13.3 shows diameters corresponding to the application of each pair of equations to their respective regions. These reveal: (a) an all-shear pin design diameter for $s = 1$, (b) a combined bending and shear design for $s = 0.1$ and (c) an almost all bending design for $s = 0.01$. The diameter is maintained constant in the case of each shear contribution to the optimum design, but for a bending design the diameter must always be varied in

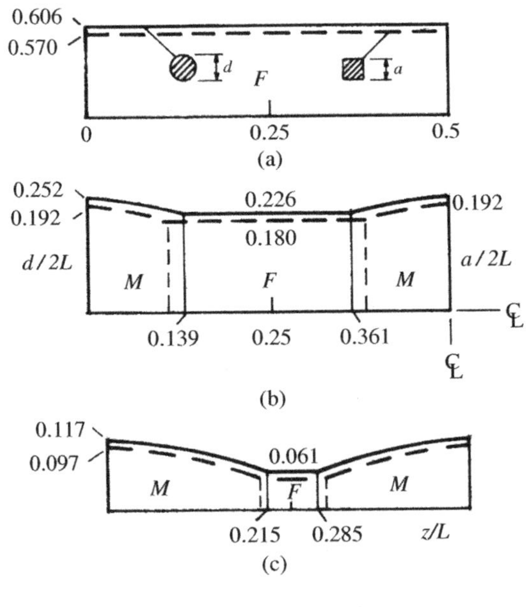

Figure 13.3 Dependence of optimum design diameter profiles upon structural index, s. Comparisons are made with an optimum square section for: (a) $s = 1$, (b) $s = 0.1$ and (c) $s = 0.01$

the manner shown to achieve the minimum weight condition. In contrast, the diameter of an encastré beam with uniformly distributed loading, which follows in § 13.3, will vary within both the shear and bending regions in an optimum design.

Example 13.1 If the weight W and length L of an encastré steel beam are to be restricted to 1 kg and 250 mm respectively, what is the greatest central-concentrated force P that a circular section beam is capable of supporting when: (a) the section is uniform; (b) the section is allowed to vary optimally? Determine the diameter for (a) and the shape of the beam contour for (b). Take $\rho = 7850$ kg/m^3 and $\sigma_y = 300$ MPa.

(a) Uniform Section
When the diameter is constant for the entire length and the weight is set at 1 kg, the diameter may be found independently of the loading as

$$W = \frac{\pi}{4} d^2 L\rho$$

from which

$$d = \sqrt{\frac{4W}{\pi L\rho}} = \sqrt{\frac{4 \times 1 \times 10^6}{\pi \times 0.25 \times 7850}} = 25.47 \text{ mm}$$

The greatest central force that this beam can carry is the lesser value found by applying separately the following shear- and moment-based design criteria. These criteria ensure that the respective stresses arising from the maximum shear force and bending moment in Figure 13.1b,d do not exceed the yield stress for the beam material.

Shear-Based Design. Figure 13.1b shows that the maximum shear stress will remain constant in each half length since the shear force $F(z) = P/2 =$ constant, for $0 \le z < L/2$. The maximum shear stress τ_{max}, which occurs at the horizontal diameter (neutral axis), is given by equation (5.1b),

$$\tau_{max} = \frac{[F(z)]A\bar{y}}{Ix} = \frac{(P/2)(\pi d^2/8)(2d/3\pi)}{(\pi d^4/64)d} = \frac{8P}{3\pi d^2}$$

and setting $\tau_{max} = \sigma_y/\sqrt{3}$ (von Mises), the allowable central force is found to be

$$P = \frac{\sqrt{3}}{8}\pi d^2 \sigma_y = \frac{\sqrt{3}}{8}\pi \times 25.47^2 \times 300 = 132.37 \text{ kN}$$

This shows that a bar of 25.47 mm diameter is capable of supporting a very high shear force when the bar material operates at its limiting yield stress. As a check, assuming a uniform shear stress distribution through the depth, we may approximate the maximum transverse shear force = stress × area, for the central section in single shear. This gives

$$\frac{P}{2} = \frac{\sigma_y}{\sqrt{3}}A = \frac{300}{\sqrt{3}}\pi\frac{25.47^2}{4} = 88.25 \text{ kN}, \quad \Rightarrow \quad P = 176.5 \text{ kN}$$

Moment-Based Design. Figure 13.1d shows that the maximum bending stress is reached simultaneously at the ends and the centre, where the ordinates of the net moment diagram are of similar magnitudes: $|M(0)| = |M(L/2)| = PL/8$. The opposing sign between these bending moments at each position indicates that the sense of the maximum surface stress alternates, e.g. when the top surface at the left end is under maximum tension the centre top surface is under maximum compression. The maximum bending stress may be expressed using equation (3.3a),

$$\sigma_{\max} = \frac{|M(0)|y}{I} = \frac{(PL/8)(d/2)}{(\pi d^4/64)} = \frac{4PL}{\pi d^3}$$

and setting $\sigma_{\max} = \sigma_y$, the allowable central force is found:

$$P = \frac{\pi d^3 \sigma_y}{4L} = \frac{\pi \times 25.47^3 \times 300}{4 \times 250} = 15.573 \text{ kN}$$

which limits the central load that can be applied to this much reduced value, i.e. shear plays no part in the design.

(b) Optimum Design

In general, the optimum analysis in § 13.2.1 will allow a contribution to the design from both the bending and shear criteria by allowing the diameter to vary. Hence, for the given length and weight, the non-dimensional weight parameter n required is

$$n = \frac{W_{\text{opt}}}{\rho L^3} = \frac{1}{7850 \times 0.25^3} = 8.1529 \times 10^{-3}$$

We may read the corresponding index s from Figure 13.2 directly, or interpolate from within Table 13.1. For the interpolation we must take logarithms for the s–n extract lying on either side of the present s-value, as shown in Table 13.2. A linear interpolation between the logarithms of s and n leads to an equation in the unknown $\log_{10} s$:

$$\frac{\log_{10} s - (-2)}{-2.08869 - (-1.5794)} = \frac{-3 - (-2)}{-2.2552 - (-1.5794)}$$

from which $\log_{10} s = -2.7536$, giving $s = 10^{-2.7536} = 1.7636 \times 10^{-3}$. Hence the central load is found to be

$$P = s\sigma_y L^2 = 1.7636 \times 10^{-3} \times 300 \times 250^2 = 33.07 \text{ kN}$$

Table 13.2 Extract for interpolation taken from Figure 13.2

s	0.001	?	0.010	...
n	5.5563×10^{-3}	8.1529×10^{-3}	0.02634	...
$\log_{10} s$	−3	?	−2	...
$\log_{10} n$	−2.2552	−2.08869	−1.5794	...

which doubles the allowable load (15.573 kN) found in (a) for a uniform section beam of similar 1 kg weight.

Contour Shape. Equations (13.5b) and (13.6b) provide the positions q_c within the first two quarter lengths, $0 \le q \le 0.25$ and $0.25 \le q \le 0.50$, at which the interchange between moment-based and shear-based design diameters occurs. These positions are

$$q_c = 0.25 - 0.35\,s^{1/2} = 0.25 - 0.35\,(1.7636 \times 10^{-3}) = 0.2353$$

$$q_c = 0.25 + 0.35\,s^{1/2} = 0.25 + 0.35\,(1.7636 \times 10^{-3}) = 0.2647$$

Hence, in the region $0 \le q \le 0.2353$, the moment-based diameters are provided by equation (13.4a):

$$\left(\frac{d}{2L}\right)_M = \left(\frac{2}{\pi}\right)^{1/3} s^{1/3}(0.25 - q)^{1/3}$$

$$= 0.8603(1.7636 \times 10^{-3})^{1/3}(0.25 - q)^{1/3} = 0.10394(0.25 - q)^{1/3} \qquad \text{(i)}$$

Equation (i) gives the normalised diameters at three selected positions:

$$q = 0, \quad (d/2L)_M = 0.06548, \quad d = 32.74 \text{ mm}$$

$$q = 0.125, (d/2L)_M = 0.05197, \quad d = 25.99 \text{ mm}$$

$$q = 0.2353, (d/2L)_M = 0.02546, \quad d = 12.73 \text{ mm}$$

Note that when $q = 0.25$, $(d/2L)_M = 0$, which indicates that shear has taken over. Shear applies to the region $0.2353 \le q \le 0.2647$, within which equation (13.3) applies:

$$\left(\frac{d}{2L}\right)_F = \left(\frac{2}{\sqrt{3}\pi}\right)^{1/2} s^{1/2} = 0.6063 \times (1.7636 \times 10^{-3})^{1/2} = 0.02546 \text{ (constant)} \qquad \text{(ii)}$$

A single diameter $d = 12.73$ mm is found from equation (ii). Finally, for the region $0.2647 \le q \le 0.50$, the moment-based diameters are provided by equation (13.4b):

$$\left(\frac{d}{2L}\right)_M = \left(\frac{2}{\pi}\right)^{1/3} s^{1/3}(q - 0.25)^{1/3}$$

$$= 0.8603(1.7636 \times 10^{-3})^{1/3}(q - 0.25)^{1/3} = 0.10394(q - 0.25)^{1/3} \qquad \text{(iii)}$$

Equation (iii) gives the normalised diameters at three selected positions:

$$q = 0.2647, (d/2L)_M = 0.2546, \quad d = 12.73 \text{ mm}$$

$$q = 0.375, \quad (d/2L)_M = 0.05197, \quad d = 25.99 \text{ mm}$$

$$q = 0.50, \quad (d/2L)_M = 0.06548, \quad d = 32.74 \text{ mm}$$

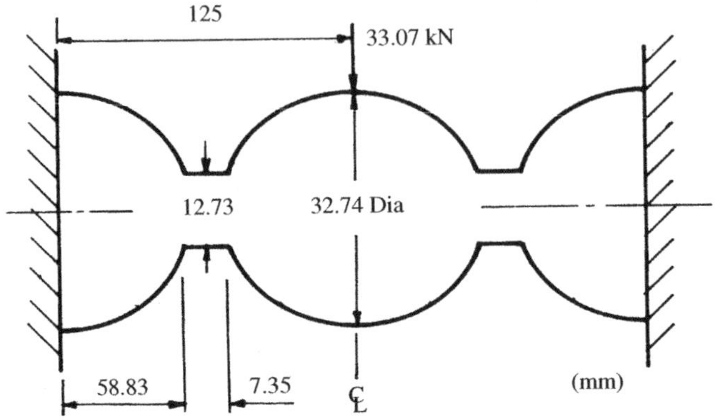

Figure 13.4 Diameter profile for a 1 kg encastré beam (structural index $s = 1.7636 \times 10^{-3}$)

The diameter calculations from equations (i)–(iii) reveal the length profile for each region shown in Figure 13.4. Here the half-length profile is mirrored into the second half and shows where material can be removed from the beam without impairing its load carrying capacity. The largest diameter d, which determines the sock size, is that at its ends and centre. This is found from $d/2L = 0.06548$ to be $d = 32.74$ mm, given that $L = 250$ mm. The smallest d is seen from the above shear calculations to be a constant 12.73 mm, this spanning 7.35 mm, beginning at $(0.2353 \times 250) = 58.83$ mm from each end-fixing.

13.2.4 Optimum Square Section

It was shown previously that, for a transverse shear force applied to a square-section beam, equation (5.4c) gave the optimum section size $a \times a$. Substituting $|F(z)|$ from equation (13.1), dividing a by $2L$ and writing the relationship between the tensile and shear yield stresses as $\sigma_y = \sqrt{3}\tau_y$ leads to

$$\left(\frac{a}{2L}\right)_{\text{opt}} = \left(\frac{3\sqrt{3}}{16}\right)^{1/2}\left(\frac{1}{\sigma_y}\frac{P}{L^2}\right)^{1/2} = 0.5699\, s^{1/2} \quad \text{for } 0 \le \frac{z}{L} \le \frac{1}{2} \tag{13.14}$$

The optimum side length a of a square section under bending is found by combining $|M(z)|$ from equations (13.2a,b) with equation (3.6b). Dividing a by $2L$ reveals the variation in the normalised half-depth within two regions of the half-length: for $0 \le z/L \le 1/4$,

$$\left(\frac{a}{2L}\right)_{\text{opt}} = \left(\frac{3}{8}\right)^{1/3}\left(\frac{1}{\sigma_y}\frac{P}{L^2}\right)^{1/3}\left(\frac{1}{4} - \frac{z}{L}\right)^{1/3} = 0.7211\, s^{1/3}\left(\frac{1}{4} - \frac{z}{L}\right)^{1/3} \tag{13.15a}$$

and for $1/4 \le z/L \le 1/2$,

$$\left(\frac{a}{2L}\right)_{\text{opt}} = \left(\frac{3}{8}\right)^{1/3}\left(\frac{1}{\sigma_y}\frac{P}{L^2}\right)^{1/3}\left(\frac{z}{L} - \frac{1}{4}\right)^{1/3} = 0.7211\, s^{1/3}\left(\frac{z}{L} - \frac{1}{4}\right)^{1/3} \tag{13.15b}$$

Again, the normalised structural index $s = (1/\sigma_y)(P/L^2)$ is to be identified with short beams when s is large and long beams when s is small.

For a given s, equation (13.14) shows that the side length is constant in shear, while for bending equations (13.15a,b) show how a varies with z within each region. Equating (13.14) and (13.15a) gives the critical position $(z/L)_{cr}$ in the first quarter-length where the depth from a shear design is displaced by the depth from bending. Writing $q = z/L$, this becomes

$$0.7211s^{1/3}(1/4 - q)^{1/3} = 0.5699\,s^{1/2} \quad \Rightarrow \quad q_c = 0.25 - 0.4936\,s^{1/2} \tag{13.16a}$$

Setting $q_c = 0$ in equation (13.16a) gives $s_{MF} = 0.2566$. When $s \geq 0.2566$, an all-shear design applies to the quarter length $0 \leq q \leq 1/4$. When $s < 0.2566$, bending will dictate a within the region $0 < q < q_c$ and shear for $q_c < q < 1/4$. Here $q_c (0 < q_c < 1/4)$ is found from equation (13.16a) at the particular s-value. Similarly, equating (13.14) and (13.15b) gives the critical position in the second quarter-length where the design criteria change:

$$0.7211\,s^{1/3}(q - 1/4)^{1/3} = 0.5699\,s^{1/2} \quad \Rightarrow \quad q_c = 0.4936\,s^{1/2} + 0.25 \tag{13.16b}$$

Setting $q_c = 0.5$ in equation (13.16b) again gives $s_{FM} = 0.2566$. Hence, an all-shear design applies to the quarter length $1/4 \leq q \leq 1/2$, when $s \geq 0.2566$. When $s < 0.2566$, shear dictates the diameter within the region $1/4 < q < q_c$ and bending for $q_c < q < 1/2$. Here $q_c (1/4 < q_c < 1/2)$ is found from equation (13.16b) at a particular s-value, chosen to span five decades from 0.001 to 100 for the minimum weight calculations that follow.

13.2.5 Minimum Weight of Square-Section Beam

With $s \geq s_{MF}$, the depth and weight within the region $0 < q < 1/4$ follow an all-shear design. Thus, for an elemental, normalised length δq within this region, the normalised weight for a shear design, $n_F = (1/\rho)(W_F/L^3)$, becomes

$$\frac{1}{\rho}\frac{W_F}{L^3} = 4\int_0^{1/4} \left(\frac{a}{2L}\right)^2 \mathrm{d}\left(\frac{z}{L}\right) = 4\left(\frac{a}{2L}\right)^2 \int_0^{1/4} \mathrm{d}q \tag{13.17a}$$

Substituting (13.14) into equation (13.17a) leads to the weight of the beam's quarter-length:

$$n_F = 4(0.5699\,s^{1/2})^2 \int_0^{1/4} \mathrm{d}q = 0.3248\,s \tag{13.17b}$$

With $s < s_{MF}$, bending and then shear will contribute to the optimal weight within the region $0 < q < 1/4$. From bending:

$$\frac{1}{\rho}\frac{W_M}{L^3} = 4\int_0^{(z/L)_{cr}} \left(\frac{a}{2L}\right)^2 \mathrm{d}\left(\frac{z}{L}\right) \quad \Rightarrow \quad n_M = 4\int_0^{q_c} \left(\frac{a}{2L}\right)^2 \mathrm{d}q \tag{13.18a}$$

Substituting equations (13.15a) into equation (13.18a),

$$n_M = 4(0.7211\,s^{1/3})^2\,s^{2/3}\int_0^{q_c}(1/4-q)^{2/3}\mathrm{d}q$$

$$= 1.248s^{2/3}\left[\left(\frac{1}{4}\right)^{5/3}-\left(\frac{1}{4}-q_c\right)^{5/3}\right] = 1.248\,s^{2/3}(0.0992-0.3083\,s^{5/6}) \qquad (13.18\text{b})$$

From shear:

$$\frac{W_F}{L^3} = 4\rho\int_{(z/L)_c}^{1/4}\left(\frac{a}{2L}\right)^2\mathrm{d}\left(\frac{z}{L}\right) \Rightarrow n_F = 4\int_{q_c}^{1/4}\left(\frac{a}{2L}\right)^2\mathrm{d}q \qquad (13.19\text{a})$$

Substituting equation (13.14) into equation (13.19a),

$$n_F = 4(0.5699\,s^{1/2})^2\int_{q_c}^{1/4}\mathrm{d}q = 1.2991\,s(1/4-q_c)$$

$$= 1.2991\,s\left[\frac{1}{4}-\left(\frac{1}{4}-0.4936\,s^{1/2}\right)\right] = 0.6412\,s^{3/2} \qquad (13.19\text{b})$$

Adding equations (13.18b) and (13.19b) gives the total weight of the first quarter-length:

$$n = n_M + n_F = 1.248\,s^{2/3}(0.0992-0.3083\,s^{5/6}) + 0.6412\,s^{3/2} \qquad (13.20)$$

$$= 0.2564\,s^{3/2} + 0.1238s^{2/3}$$

We should expect identical weights for the second quarter length $1/4 \le q \le 1/2$, given that the shear force remains constant and that the moment diagram mirrors that for the first quarter length. To show this, the limits in equations (13.17a), (13.18a) and (13.19a) must be changed with the substitutions from equations (13.15a,b) and (13.14). When $s \ge s_{FM}$, the weight of an all-shear design follows as

$$n_F = 4(0.5699\,s^{1/2})^2\int_{1/4}^{1/2}\mathrm{d}q = 0.3248\,s$$

in agreement with equation (13.17b). With $s < s_{FM}$, the weight contributions from shear and then bending within this region are

$$n_F = 4(0.5699\,s^{1/2})^2\int_{1/4}^{q_c}\mathrm{d}q = 1.2991\,s(q_c-1/4)$$

$$= 1.2991\,s\left[\left(\frac{1}{4}+0.4936\,s^{1/2}\right)-\frac{1}{4}\right] = 0.6412\,s^{3/2}$$

$$n_M = 4(0.7211\, s^{1/3})^2\, s^{2/3} \int_{q_c}^{1/2} (q - 1/4)^{2/3} \mathrm{d}q$$

$$= 1.248\, s^{2/3} \left[\left(\frac{1}{4}\right)^{5/3} - \left(q_c - \frac{1}{4}\right)^{5/3} \right] = 1.248\, s^{2/3} (0.0992 - 0.3083\, s^{5/6})$$

such that $n_F + n_M$ agrees with equation (13.20). Thus, we may construct Table 13.3 showing minimum weights for the full beam length simply by doubling n-values found from equation (13.17b) and (13.20) at each s-value and then doubling to give $\sum n$.

Figure 13.2 plots $\sum n$ from Table 13.3 against s to reveal regions in which bending and combined shear and bending designs apply as the index s increases to cross the two transitional values ($s_{MF} = s_{FM} = 0.2566$). It is seen that a beam with square section (dashed line) is lighter than one with a circular section for a bending design ($s < 0.2566$) but becomes heavier for a shear design when $s > 0.2566$. Example 13.2 illustrates the weight saving that can be achieved with a change from a circular to a square section. The comparison made in the bottom line of Table 13.3 shows that the weights of simply supported beams (taken from Figure 10.18) are heavier for similar indices s in in the region of bending. The weights of encastré and simply supported beams bearing central loads become identical in the region of shear.

13.2.6 Square-Section Profiles

As the present beam's length increases, i.e. as s decreases, the contribution to the design diameter from shear diminishes. The following shows this within the beam's half-length profiles for three s-values, 1, 0.1 and 0.01. The normalised section dimension $a/2L$, corresponding to each s-value, is found from equations (13.14) and (13.15a,b), as follows:

(a) $s = 1$
Here $s > s_{MF} (s_{MF} = 0.2566)$ when equation (13.14) gives the all-shear design dimension:

$$\left(\frac{a}{2L}\right)_F = 0.5699 \times 1^{1/2} = 0.5699 = \text{constant}, \quad \text{for } 0 \leq q \leq 0.5$$

(b) $s = 0.1$
Here $s < s_{MF}$ and, therefore, a bending design followed by a shear design provides the square dimensions within the first quarter region, $0 \leq q \leq 0.25$. Equation (13.16a) shows

Table 13.3 Normalised square-section beam weights across five decades of index s

s	0.001	0.01	0.10	0.2566	1.0	10	100
$n(0 < q < 1/4)$	1.2461×10^{-3}	6.003×10^{-3}	0.03478	0.0833	0.3248	3.248	32.48
$n(1/4 < q < 1/2)$	1.2461×10^{-3}	6.003×10^{-3}	0.03478	0.0833	0.3248	3.248	32.48
$\sum n$	4.984×10^{-3}	0.024	0.1391	0.3333	1.2992	12.992	129.92
Simply supported	7.88×10^{-3}	0.037	0.1856	0.3333	1.2992	12.992	129.92

that the division between the two designs occurs at the position $q_c = 0.25 - 0.4936 \times 0.1^{1/2} = 0.0939$. The respective square side-lengths follow from equations (13.16a) and (13.15):

$$\left(\frac{a}{2L}\right)_M = 0.3347\left(\frac{1}{4} - q\right)^{1/3}, \quad \text{for } 0 \leq q \leq 0.0939$$

$$\left(\frac{a}{2L}\right)_F = 0.5699 \times 0.1^{1/2} = 0.1802 = \text{constant}, \quad \text{for } 0.0939 \leq q \leq 0.25$$

In the second quarter region, $0.25 \leq q \leq 0.50$, shear and then bending designs apply. Equation (13.16b) shows that the division between them occurs at $q_c = 0.25 + 0.4936 \times 0.1^{1/2} = 0.4061$. The respective square side-lengths follow from equations (13.15) and (13.16b):

$$\left(\frac{a}{2L}\right)_F = 0.5699 \times 0.1^{1/2} = 0.1802 = \text{constant}, \quad \text{for } 0.25 \leq q \leq 0.4061$$

$$\left(\frac{a}{2L}\right)_M = 0.3347\left(q - \frac{1}{4}\right)^{1/3}, \quad \text{for } 0.4061 \leq q \leq 0.5$$

(c) $s = 0.01$
This is similar to (b) where bending and then shear designs are divided at position $q_c = 0.201$ within the first quarter length, $0 \leq q \leq 0.25$. Equations (13.16a) and (13.15) give the squares' normalised side-lengths:

$$\left(\frac{a}{2L}\right)_M = 0.1554\left(\frac{1}{4} - q\right)^{1/3}, \quad \text{for } 0 \leq q \leq 0.201$$

$$\left(\frac{a}{2L}\right)_F = 0.5699 \times 0.01^{1/2} = 0.05699 = \text{constant}, \quad \text{for } 0.201 \leq q \leq 0.25$$

For the second quarter length, $0.25 \leq q \leq 0.50$, shear and bending designs divide at position $q_c = 0.2994$. From equations (13.15) and (13.16b), the side-lengths become

$$\left(\frac{a}{2L}\right)_F = 0.5699 \times 0.01^{1/2} = 0.05699 = \text{constant}, \quad \text{for } 0.25 \leq q \leq 0.2994$$

$$\left(\frac{a}{2L}\right)_M = 0.1554 \times 0.01^{1/3}\left(q - \frac{1}{4}\right)^{1/3}, \quad \text{for } 0.2994 \leq q \leq 0.5$$

Figure 13.3 shows how the equations above describe the depth versus length profile of a square section beam at each index, i.e. s-values of 1, 0.1 and 0.01. These profiles (dashed lines) are similar to those found for the circular section. Though the depths of the square section are less than the diameters in each case, we should refer to Figure 13.2 to see if there is a weight saving (or gain) at each s-value.

13.3 Uniformly Distributed Load

The second, standard encastré beam refers to one that carries a uniformly distributed load w per unit length, as shown in Figure 13.5a. With the symmetry present at the centre of

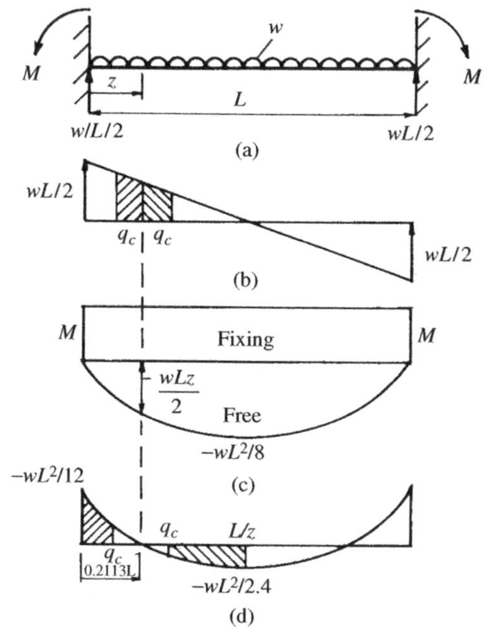

Figure 13.5 F- and M-diagrams for an encastré beam carrying uniformly distributed loading

the F- and M-diagrams, z is taken from one end towards the centre, given that the total weight will be twice that of one half. Thus, within the half-length $0 \le z/L \le 0.5$, the shear force diagram in Figure 13.5b may be expressed as

$$\left|F\left(\frac{z}{L}\right)\right| = \frac{wL}{2}\left(1 - 2\frac{z}{L}\right) = F_{\max}\left(1 - 2\frac{z}{L}\right) \tag{13.21}$$

where $F_{\max} = wL/2$ refers to the maximum ordinate within the F-diagram. The net moment diagram (see Figure 13.5d) is found by taking the difference between ordinates in the 'free' and 'fixing' moment diagrams in Figure 13.5c (whose areas are equal). Where this difference is positive, the net moments in Figure 13.5d are expressed as

$$\left|M\left(\frac{z}{L}\right)\right| = \frac{wL^2}{2}\left[\frac{1}{6} - \frac{z}{L}\left(1 - \frac{z}{L}\right)\right] = F_{\max}L\left[\frac{1}{6} - \frac{z}{L}\left(1 - \frac{z}{L}\right)\right] \tag{13.22a}$$

Setting $|M(z/L)| = 0$ in equation (13.22a) gives a quadratic equation whose roots provide the two length positions $z/L = 0.2113$ and 0.7887 in each half where the M-diagram crosses its axis as shown. Thus, equation (13.22a) applies to the sub-region $0 \le z/L \le 0.2113$ within the first half of the beam. Within the remaining sub-region, $0.2113 \le z/L \le 0.5$, where the net M-diagram is negative, we write

$$\left|M\left(\frac{z}{L}\right)\right| = \frac{wL^2}{2}\left[\frac{z}{L}\left(1 - \frac{z}{L}\right) - \frac{1}{6}\right] = F_{\max}L\left[\frac{z}{L}\left(1 - \frac{z}{L}\right) - \frac{1}{6}\right] \tag{13.22b}$$

13.3.1 Optimum Circular Section

For a transverse shear force applied to a circular section beam, it was shown previously that equation (5.15b) gave the optimum diameter. Dividing by $2L$, substituting from equation (13.21) and taking the relationship between the tensile and shear yield stresses as $\sigma_y = Q\tau_y$ leads to

$$\left(\frac{d}{2L}\right)_{\text{opt}} = \left(\frac{4Q}{3\pi}\right)^{1/2}\left(\frac{1}{\sigma_y}\frac{F_{\max}}{L^2}\right)^{1/2}\left(1 - 2\frac{z}{L}\right)^{1/2} = 0.8574\, s^{1/2}(1 - 2q)^{1/2} \quad (13.23)$$

Equation (13.23) applies to the half length $0 \leq q \leq \frac{1}{2}$ when $Q = \sqrt{3}$ by the von Mises yield criterion. Alternatively, we may take $Q = \frac{1}{2}$ by the Tresca yield criterion.

The optimum diameter of a circular-section in bending is again found by combining each $|M(z/L)|$ expression (13.22a,b) with equation (3.3b). Hence, the dependence of radius with length for each sub-region of the half length becomes, in normalised form, for $0 \leq q \leq 0.2113$,

$$\left(\frac{d}{2L}\right)_{\text{opt}} = \left(\frac{4}{\pi}\right)^{1/3}\left(\frac{1}{\sigma_y}\frac{F_{\max}}{L^2}\right)^{1/3}\left[\frac{1}{6} - \frac{z}{L}\left(1 - \frac{z}{L}\right)\right]^{1/3}$$

$$= 1.0838\, s^{1/3}\left[\frac{1}{6} - q(1 - q)\right]^{1/3} \quad (13.24a)$$

and for $0.2113 \leq q \leq 0.5$,

$$\left(\frac{d}{2L}\right)_{\text{opt}} = \left(\frac{4}{\pi}\right)^{1/3}\left(\frac{1}{\sigma_y}\frac{F_{\max}}{L^2}\right)^{1/3}\left[\frac{z}{L}\left(1 - \frac{z}{L}\right) - \frac{1}{6}\right]^{1/3}$$

$$= 1.0838 s^{1/3}\left[q(1 - q) - \frac{1}{6}\right]^{1/3} \quad (13.24b)$$

in which $q = z/L$ and $s = (1/\sigma_y)(F_{\max}/L^2)$ is the normalised structural index, identifying short beams when s is large and long beams when s is small.

For a given s, equations (13.23) and (13.24a,b) show that the optimum diameter is variable within their respective lengths by both shear and bending designs. Equating (13.23) and (13.24a) gives the critical position $q = q_c$ in the first region where the design diameter from bending is displaced by one from shear:

$$\left[\frac{1}{6} - q_c(1 - q_c)\right]^{1/3} = 0.7911\, s^{1/6}(1 - 2q_c)^{1/2} \quad (13.25a)$$

Setting $q_c = 0$ in equation (13.25a) gives $s = s_{MF} = 0.1133$. Thus, an all-shear design of diameters in the region $0 \leq q \leq 0.2113$ applies when $s \geq s_{MF}$. When $s < s_{MF}$, bending determines the diameter within the sub-region $0 < q < q_c$ and shear for the sub-region $q_c < q < 0.2113$. Here the q_c values ($0 < q_c < 0.2113$), given in Table 13.4 have been found by trial from equation (13.25a) for diminishing indices s, where bending becomes

Table 13.4 Division between bending and combined shear and bending designs ($0 < q_c < 0.2113$)

s	0.1133	0.1	0.01	0.001
q_c	0	0.0192	0.168	0.1987

increasingly dominant. Where the design diameter criteria change within the second region of the half-length ($0.2113 < q < 0.5$), the critical position q_c follows from equating (13.23) and (13.24b):

$$[q_c(1 - q_c) - 1/6]^{1/3} = 0.7911s^{1/6}(1 - 2q_c)^{1/2} \qquad (13.25b)$$

Setting $q_c = 0.5$ in equation (13.25b) gives $s = \infty$, showing that an all-shear design does not apply across this region. Hence, the diameter is provided by shear within a limited sub-region $0.2113 < q < q_c$ and by bending for $q_c < q < ½$. Table 13.5 gives the trial solutions to $q_c(0.2113 < q_c < ½)$, found from equation (13.25b), at each of six indices s spanning five decades. Figures 13.5b,d shade the corresponding regions of the shear force and bending moment diagrams that dictate the optimum bending–shear designs within the four sub-regions of the half-length.

Table 13.5 Normalised lengths q_c where design changes from shear to bending in five decades of index s

s	100	10	1	0.1	0.01	0.001
q_c	0.4658	0.4303	0.3696	0.2955	0.2448	0.2232

13.3.2 Minimum Weight of Circular-Section Beam

If the design in not wholly shear-based then intermediate regions, divided by q_c, will define the limits to the respective weight integrals for combined shear and bending designs.

Taking the region $0 < q < 0.2113$ first, we have seen, for $s \geq s_{MF}$, the beam's diameter and weight follow from an all-shear design. Thus, for an elemental, normalised length δq within this region, the normalised weight from shear, $n_F = (1/\rho)(W_F/L^3)$, becomes

$$\frac{W_F}{L^3} = \rho\pi \int_0^{0.2113} \left(\frac{d}{2L}\right)^2 \mathrm{d}\left(\frac{z}{L}\right) \quad \Rightarrow \quad n_F = \pi \int_0^{0.2113} \left(\frac{d}{2L}\right)^2 \mathrm{d}q \qquad (13.26a)$$

Substituting equation (13.23) into equation (13.26a) leads to the beam weight within this region

$$n_F = \left(\frac{4Q}{3}\right) s \int_0^{0.2113} (1 - 2q)\mathrm{d}q = 0.385s \qquad (13.26b)$$

With $s < s_{MF}$, both bending and shear contribute to the optimum weight. For bending:

$$\frac{W_M}{L^3} = \rho\pi \int_0^{q_c} \left(\frac{d}{2L}\right)^2 \mathrm{d}\left(\frac{z}{L}\right) \quad \Rightarrow \quad n_M = \pi \int_0^{q_c} \left(\frac{d}{2L}\right)^2 \mathrm{d}q \tag{13.27a}$$

Substituting equations (13.24a) into equation (13.27a), the weight parameter becomes

$$n_M = \pi \left(\frac{4}{\pi}\right)^{2/3} s^{2/3} \int_0^{q_c} \left[\frac{1}{6} - q(1-q)\right]^{2/3} \mathrm{d}q = 3.6905 s^{2/3} \int_0^{q_c} \left[\frac{1}{6} - q(1-q)\right]^{2/3} \mathrm{d}q \tag{13.27b}$$

Equation (13.27b) must be evaluated numerically for the values of s and q_c in Table 13.4. For shear, the s- and q_c-values will appear in a closed solution to the integration

$$\frac{W_F}{L^3} = \rho\pi \int_{q_c}^{1/4} \left(\frac{d}{2L}\right)^2 \mathrm{d}\left(\frac{z}{L}\right) \quad \Rightarrow \quad n_F = \pi \int_{q_c}^{1/4} \left(\frac{d}{2L}\right)^2 \mathrm{d}q \tag{13.28a}$$

Substituting from equation (13.23) into equation (13.28a)

$$n_F = \frac{4}{\sqrt{3}} s \int_{q_c}^{0.2113} (1-2q)\mathrm{d}q = 2.30994 s \left[\frac{1}{6} - q_c(1-q_c)\right] \tag{13.28b}$$

Then, taking the sum $n_M + n_F$ from equations (13.27b) and (13.28b) gives the total weight of this first region.

Turning now to the region $0.2113 < q < 0.50$, we need only alter the limits and interchange terms within equations (13.28b) and (13.27b) to give the weight contributions from shear and bending as follows:

$$n_F = \frac{4}{\sqrt{3}} s \int_{0.2113}^{q_c} (1-2q)\mathrm{d}q = 2.30994 s \left[q_c(1-q_c) - \frac{1}{6}\right] \tag{13.29}$$

$$\begin{aligned} n_M &= \pi \left(\frac{4}{\pi}\right)^{2/3} s^{2/3} \int_{q_c}^{0.5} \left[q(1-q) - \frac{1}{6}\right]^{2/3} \mathrm{d}q \\ &= 3.6905 s^{2/3} \int_{q_c}^{0.5} \left[q(1-q) - \frac{1}{6}\right]^{2/3} \mathrm{d}q \end{aligned} \tag{13.30}$$

Again, to achieve the sum $n_F + n_M$, equation (13.30) must first be integrated numerically for each pair of s- and q_c-values in Table 13.5. Finally, the total weight $\sum n$ for the beam requires that these sums be added at similar s-values over five decades within the two regions and then doubled, as shown in Table 13.6.

Table 13.6 Normalised beam weights across five decades of index s

s	0.001	0.01	0.10	1.0	10	100
$(0 < q < 0.2113)$						
$n_F + n_M$	1.3112×10^{-3}	6.291×10^{-3}	0.03844	0.38496	3.8496	38.496
$(0.2113 < q < 1/2)$						
$n_F + n_M$	1.4452×10^{-3}	7.095×10^{-3}	0.03675	0.2405	2.0334	19.4973
$\sum n$	5.513×10^{-3}	0.02677	0.1504	1.251	11.766	115.39
Simply supported	10.8×10^{-3}	0.0506	0.2539	1.5545	12.508	117.6

Figure 13.6 plots the total weight $\sum n$ from Table 13.6 against s for where shear and combined shear and bending designs apply, i.e. in crossing the transitional value $s_{MF} = 0.1133$. While Figures 13.2 and 13.6 appear similar, these do not imply that the minimum weight of a circular-section, encastré beam, carrying a central-concentrated load, will be similar to the minimum weight of a circular-section beam where the same load is distributed uniformly. In fact, for similar values of s and L, the total load that can be carried by the latter is twice that of the former for the definitions of s used (given in each figure caption). Hence, when the total load carried by each beam is to be the same, s must be doubled in Figure 13.2. The corresponding n-values show a greater

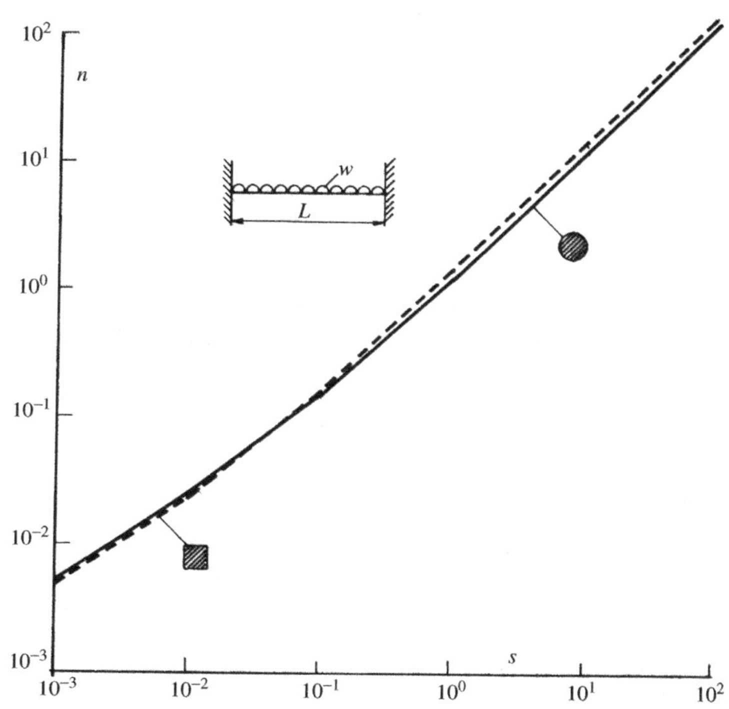

Figure 13.6 Minimum weights of circular and square encastré beams under distributed loading. (note: $s = (1/\sigma_y)(F_{max}/L^2)$ where $F_{max} = wL/2$; $n = (1/\rho)(W/L^3)$)

weight of beam required to support the concentrated load, as the following Example 13.2 will show.

The comparison made in the last two lines of Table 13.3 shows that the weights of simply supported beams (from Table 10.1) are heavier for similar s in the region of bending. It also appears here that the weights of encastré and simply supported beams bearing similar distributed loads are only similar for designs dominated by shear.

Example 13.2 Compare the weights of 500 mm lengths of encastré, circular-section, aluminium alloy beams required to support 50 kN when this load is: (i) applied at the centre; (ii) uniformly distributed over the length. Repeat the comparison for square-section aluminium alloy beams bearing the same loading. Take $\rho = 2700$ kg/m^3 and $\sigma_y = 100$ MPa.

The structural index for each beam follows from the definitions:

$$s_{(i)} = \frac{1}{\sigma_y}\frac{P}{L^2} = \frac{1}{100 \times 10^6}\frac{50 \times 10^3}{0.5^2} = 2 \times 10^{-3}$$

$$s_{(ii)} = \frac{1}{\sigma_y}\frac{wL/2}{L^2} = \frac{1}{100 \times 10^6}\frac{(50 \times 10^3)/2}{0.5^2} = 1 \times 10^{-3}$$

(a) Circular-Section Beams

Corresponding to $s_{(i)}$ and $s_{(ii)}$, the n-values are read from Figures 13.2 and 13.6 (or Tables 13.1 and 13.6) as $n_{(i)} = 9 \times 10^{-3}$ and $n_{(ii)} = 5.513 \times 10^{-3}$. Hence the respective weights of beams with concentrated and distributed loadings follow from the definition of n as:

$$W_{(i)} = n_{(i)}\rho L^3 = 9 \times 10^{-3} \times 2700 \times 0.5^3 = 3.038 \text{ kg}$$

$$W_{(ii)} = n_{(ii)}\rho L^3 = 5.513 \times 10^{-3} \times 2700 \times 0.5^3 = 1.861 \text{ kg}$$

(b) Square-Section Beams

Corresponding to the same indices $s_{(i)}$ and $s_{(ii)}$, the n-values are read from Figures 13.2 and 13.6 (or Tables 13.3 and 13.9) as $n_{(i)} = 8 \times 10^{-3}$ and $n_{(ii)} = 5.06 \times 10^{-3}$. Hence, the respective minimum weights of beams with concentrated and distributed loading (to follow) are

$$W_{(i)} = n_{(i)}\rho L^3 = 8 \times 10^{-3} \times 2700 \times 0.5^3 = 2.7 \text{ kg}$$

$$W_{(ii)} = n_{(ii)}\rho L^3 = 5.06 \times 10^{-3} \times 2700 \times 0.5^3 = 1.708 \text{ kg}$$

The calculations show that square section beams are lighter and that the weight of beams in each section, when required to support the concentrated loads, is approximately 60% greater than when the same load is distributed. Moreover, each of these beams will be seen to be lighter than optimised encastré beams bearing combinations of these two loadings in the ratios $wL/P = 0.1$, 1 and 10, cases to be considered in § 13.4. The particular profile for each beam is left to the reader to determine (see Exercise 13.2). In §§ 13.3.3 and 13.3.6 the method for calculating their profiles is shown.

13.3.3 Circular-Section Profiles

As the encastré beam length increases, i.e. as s decreases, the contribution from shear diminishes within both regions of the half-length. We shall demonstrate this for the beam's half-length radius profiles at three s-values 1, 0.1 and 0.01. The ordinates $d/2L$ are the normalised radii, these being given from equations (13.23) and (13.24,b) as follows.

(a) $s = 1$
An all-shear design applies to $0 \le q \le 0.2113$,

$$\left(\frac{d}{2L}\right)_F = \left(\frac{4}{\sqrt{3}\pi}\right)^{1/2} \times 1^{1/2} \times (1-2q)^{1/2} = 0.8574(1-2q)^{1/2}$$

Shear then bending designs apply to $0.2113 \le q \le 0.5$ ($q_c = 0.3696$).
For $0.2113 \le q \le 0.3696$,

$$\left(\frac{d}{2L}\right)_F = \left(\frac{4}{\sqrt{3}\pi}\right)^{1/2} \times 1^{1/2} \times (1-2q)^{1/2} = 0.8574(1-2q)^{1/2}$$

and for $0.3696 \le q \le 0.5$,

$$\left(\frac{d}{2L}\right)_M = \left(\frac{4}{\pi}\right)^{1/3} 1^{1/3}\left[q(1-q)-\frac{1}{6}\right]^{1/3} = 1.0839\left[q(1-q)-\frac{1}{6}\right]^{1/3}$$

(b) $s = 0.1$
Bending and then shear designs apply to $0 \le q \le 0.2113$ ($q_c = 0.0192$).
For $0 \le q \le 0.0192$,

$$\left(\frac{d}{2L}\right)_M = \left(\frac{4}{\pi}\right)^{1/3} 0.1^{1/3}\left[\frac{1}{6}-q(1-q)\right]^{1/3} = 0.5031\left[\frac{1}{6}-q(1-q)\right]^{1/3}$$

and for $0.0192 \le q \le 0.2113$,

$$\left(\frac{d}{2L}\right)_F = \left(\frac{4}{\sqrt{3}\pi}\right)^{1/2} \times 0.1^{1/2}(1-2q)^{1/2} = 0.2711(1-2q)^{1/2}$$

Shear and then bending designs apply to $0.2113 \le q \le 0.5$ ($q_c = 0.2955$).
For $0.2113 \le q \le 0.2955$,

$$\left(\frac{d}{2L}\right)_F = \left(\frac{4}{\sqrt{3}\pi}\right)^{1/2} \times 0.1^{1/2}(1-2q)^{1/2} = 0.2711(1-2q)^{1/2}$$

and for $0.2955 \le q \le 0.50$,

$$\left(\frac{d}{2L}\right)_M = \left(\frac{4}{\pi}\right)^{1/3} 0.1^{1/3}\left[q(1-q)-\frac{1}{6}\right]^{1/3} = 0.5031\left[q(1-q)-\frac{1}{6}\right]^{1/3}$$

(c) $s = 0.01$
Bending and then shear designs apply to $0 \le q \le 0.2113$ ($q_c = 0.168$).
For $0 \le q \le 0.168$,

$$\left(\frac{d}{2L}\right)_M = \left(\frac{4}{\pi}\right)^{1/3} 0.01^{1/3}\left[\frac{1}{6} - q(1-q)\right]^{1/3} = 0.2355\left[\frac{1}{6} - q(1-q)\right]^{1/3}$$

and for $0.168 \le q \le 0.2113$,

$$\left(\frac{d}{2L}\right)_F = \left(\frac{4}{\sqrt{3}\pi}\right)^{1/2} \times 0.01^{1/2}(1-2q)^{1/2} = 0.08574\,(1-2q)^{1/2}$$

Shear and then bending designs apply to $0.2113 \le q \le 0.5$ ($q_c = 0.2448$).
For $0.2113 \le q \le 0.2448$,

$$\left(\frac{d}{2L}\right)_F = \left(\frac{4}{\sqrt{3}\pi}\right)^{1/2} \times 0.01^{1/2}(1-2q)^{1/2} = 0.08574(1-2q)^{1/2}$$

and for $0.2448 \le q \le 0.50$,

$$\left(\frac{d}{2L}\right)_M = \left(\frac{4}{\pi}\right)^{1/3} 0.01^{1/3}\left[q(1-q) - \frac{1}{6}\right]^{1/3} = 0.2335\left[q(1-q) - \frac{1}{6}\right]^{1/3}$$

Figure 13.7 shows the beam's half-length radius profiles as continuous lines, these being described by the equations above, for the three s-values. The figure shows the interplay between shear and bending design criteria within one quarter of these beams' volumes (half-length and half-depth). No single design criterion applies to the three structural indices chosen, though it is seen that shear (F) plays the major role for $s = 1$ (Figure 13.7a) and bending (M) for $s = 0.01$ (Figure 13.7c). The criteria are equally influential for $s = 0.1$ (Figure 13.7b). Both the shear and bending minimum weight designs dictate a changing diameter in the regions to which they apply. The equations given above, which control the pin and beam profiles, could be programmed for numerical machining. Alternatively, for short length beams, a centre lathe could be used with a plate copy attachment machined with the chosen profile along one edge for the stylus to follow.

13.3.4 Optimum Square Section

Equation (5.4c) provides the optimum side-length a of a square-section beam under a transverse shear force $|F(z/L)|$ in equation (13.21). Combining these, dividing by $2L$ and taking the von Mises relationship between the tensile and shear yield stresses, $\sigma_y = \sqrt{3}\tau_y$, leads to

$$\left(\frac{a}{2L}\right)_{\text{opt}} = \left(\frac{3\sqrt{3}}{8}\right)^{1/2}\left(\frac{1}{\sigma_y}\frac{F_{\max}}{L^2}\right)^{1/2}\left(1 - 2\frac{z}{L}\right)^{1/2} = 0.8059\,s^{1/2}(1-2q)^{1/2} \quad (13.31)$$

Equation (13.31) apples to the half length $0 \le q \le 1/2$, where $F_{\max} = wL/2$. Equation (3.6b) provides the optimum side-length a of a square-section beam under the bending

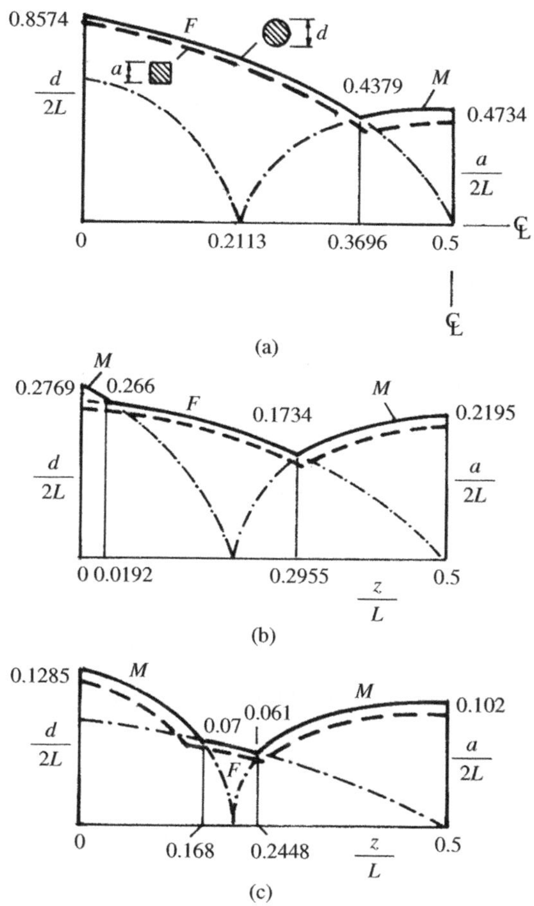

Figure 13.7 Beam radius profiles within half-length for three structural indices, (a) $s = 1$, (b) 0.1 and (c) 0.01, in circular and square sections

moments $|M(z/L)|$ given in equations (13.22a,b). Combining these and dividing a by $2L$ reveals the variation in the normalised side-length within the two regions of the half length: for $0 \leq q \leq 0.2113$,

$$\left(\frac{a}{2L}\right)_{\text{opt}} = \left(\frac{3}{4}\right)^{1/3} \left(\frac{1}{\sigma_y}\frac{F_{\max}}{L^2}\right)^{1/3} \left[\frac{1}{6} - \frac{z}{L}\left(1 - \frac{z}{L}\right)\right]^{1/3} = 0.9086 s^{1/3} \left[\frac{1}{6} - q(1-q)\right]^{1/3} \tag{13.32a}$$

and for $0.2113 \leq q \leq 0.5$,

$$\left(\frac{a}{2L}\right)_{\text{opt}} = \left(\frac{3}{4}\right)^{1/3} \left(\frac{1}{\sigma_y}\frac{F_{\max}}{L^2}\right)^{1/3} \left[\frac{z}{L}\left(1 - \frac{z}{L}\right) - \frac{1}{6}\right]^{1/3} = 0.9086 s^{1/3} \left[q(1-q) - \frac{1}{6}\right]^{1/3} \tag{13.32b}$$

Table 13.7 Division between bending and shear designs within region $0 < q_c < 0.2113$

s	0.05704	0.01	0.001
q_c	0	0.1465	0.1919

in which $q = z/L$ and $s = (1/\sigma_y)(F_{\max}/L^2)$ is a normalised structural index which identifies short beams when s is large and long beams when s is small. For a given s, equations (13.31) and (13.32a,b) depend upon q, showing that the optimum size of section is variable within their respective lengths provided by both bending and shear designs. Equating (13.31) and (13.32a) gives the critical position $q = q_c$ in the first region, where the depth from a bending criterion switches to one from a shear criterion:

$$\left[\frac{1}{6} - q_c(1 - q_c)\right]^{1/3} = 0.8870\, s^{1/6}(1 - 2q_c)^{1/2} \tag{13.33a}$$

Setting $q_c = 0$ in equation (13.33a) gives $s_{MF} = 0.05704$. Hence, an all-shear design follows when $s \geq s_{MF}$ in the region $0 \leq q \leq 0.2113$. When $s < s_{MF}$, the side-length is determined by bending for the region $0 < q < q_c$ and by shear for the region $q_c < q < 0.2113$. Here q_c-values $(0 < q_c < 0.2113)$, given in Table 13.7, have been found by trial from equation (13.33a) at the three indices within the range required. The critical position q_c at which design criteria change within the second region of the half length $(0.2113 < q < 0.5)$ follows from equating (13.31) and (13.32b):

$$\left[q_c(1 - q_c) - \frac{1}{6}\right]^{1/3} = 0.8870\, s^{1/6}(1 - 2q_c)^{1/2} \tag{13.33b}$$

Firstly, note that on setting $q_c = 0.5$ in equation (13.33b) this gives $s = \infty$, implying that an all-shear design does not apply to this region. We saw previously for the circular section how shear and then bending contributed to the design for all s-values in the second region. Then, shear provided the diameter within the sub-region $0.2113 < q < q_c$ and bending for $q_c < q < 1/2$. Similarly, for the square section the q_c-values $(0.2113 < q_c < 0.5)$ given in Table 13.8 are found by trial from equation (13.33b) to span five decades in the index s.

13.3.5 Minimum Weight of Square-Section Beam

With $s \geq s_{MF}$, the beam's section size and weight within the region $0 < q < 0.2113$ follow from the all-shear design. Thus, for an elemental, normalised length δq within this

Table 13.8 Division between shear and bending designs within region $0.2113 < q < 0.5$

s	100	10	1	0.1	0.01	0.001
q_c	0.4739	0.445	0.3906	0.3158	0.2562	0.2276

region, the normalised weight $n_F = (1/\rho)(W_F/L^3)$ from shear becomes

$$\frac{W_F}{L^3} = 4\rho \int_0^{0.2113} \left(\frac{a}{2L}\right)^2 \mathrm{d}\left(\frac{z}{L}\right) \quad \Rightarrow \quad n_F = 4 \int_0^{0.2113} \left(\frac{a}{2L}\right)^2 \mathrm{d}q \tag{13.34a}$$

Substituting equation (13.31) into equation (13.34a) leads to the beam weight within this region,

$$n_F = 4(0.8059\, s^{1/2})^2 \int_0^{0.2113} (1 - 2q)\mathrm{d}q = 0.43295\, s \tag{13.34b}$$

With $s < s_{MF}$, both bending and shear contribute to the optimal weight. From bending:

$$\frac{W_M}{L^3} = 4\rho \int_0^{q_c} \left(\frac{a}{2L}\right)^2 \mathrm{d}\left(\frac{z}{L}\right) \quad \Rightarrow \quad n_M = 4 \int_0^{q_c} \left(\frac{a}{2L}\right)^2 \mathrm{d}q \tag{13.35a}$$

Substituting equation (13.32a) into equation (13.35a),

$$n_M = 4\left(\frac{3s}{4}\right)^{2/3} \int_0^{q_c} \left[\frac{1}{6} - q(1-q)\right]^{2/3} \mathrm{d}q = 3.3022\, s^{2/3} \int_0^{q_c} \left[\frac{1}{6} - q(1-q)\right]^{2/3} \mathrm{d}q \tag{13.35b}$$

Equation (13.35b) is to be evaluated numerically for the s- and q_c-values in Table 13.8. For the shear contribution within this region,

$$\frac{W_F}{L^3} = 4\rho \int_{q_c}^{0.2113} \left(\frac{a}{2L}\right)^2 \mathrm{d}\left(\frac{z}{L}\right) \quad \Rightarrow \quad n_F = 4 \int_{q_c}^{0.2113} \left(\frac{a}{2L}\right)^2 \mathrm{d}q \tag{13.36a}$$

Substituting equation (13.31) into (13.36a) , the s- and q_c-values appear within a closed solution to the integration

$$n_F = 4(0.8059 s^{1/2})^2 \int_{q_c}^{0.2113} (1 - 2q)\mathrm{d}q = 2.5979 s \left[\frac{1}{6} - q_c(1 - q_c)\right] \tag{13.36b}$$

Taking the sum $n_M + n_F$ from equations (13.35b) and (13.36b), gives the total weight of this region. We now turn to the region $0.2113 < q < 0.50$. The intermediate regions, divided by q_c, define the limits within the respective weight integrals for the shear and bending designs. We need only alter the integration limits and thereby interchange terms within equations (13.36b) and (13.35b) respectively as follows:

$$n_F = 4(0.8059\, s^{1/2})^2 \int_{0.2113}^{q_c} (1 - 2q)\mathrm{d}q = 2.5979\, s \left[q_c(1 - q_c) - \frac{1}{6}\right] \tag{13.37}$$

$$n_M = 4\left(\frac{3s}{4}\right)^{2/3}\int_{q_c}^{0.5}\left[q(1-q)-\frac{1}{6}\right]^{2/3}dq = 3.3022s^{2/3}\int_{q_c}^{0.5}\left[q(1-q)-\frac{1}{6}\right]^{2/3}dq \tag{13.38}$$

Again, to achieve the sum $n_F + n_M$, equation (13.38) must first be integrated numerically for each pair of s- and q_c-values in Table 13.8. Finally, the total weight of the beam requires that each sum $n_M + n_F$ be added at similar indices s over five decades within the two regions and then doubled, as shown in Table 13.9.

Table 13.9 Normalised beam weights with distributed loading across five decades of index s

s	0.001	0.01	0.10	1.0	10	100
$(0 < q < 0.2113)$						
$n_M + n_F$	1.1881×10^{-3}	5.8617×10^{-3}	0.043295	0.43295	4.3295	43.295
$(0.2113 < q < ½)$						
$n_F + n_M$	1.34191×10^{-3}	6.4622×10^{-3}	0.03533	0.2519	2.2455	21.825
$\sum n$	5.060×10^{-3}	0.02465	0.15725	1.3697	13.15	130.24
Simply supported	9.634×10^{-3}	0.0459	0.2394	1.5969	13.684	131.43

Figure 13.6 plots $\sum n$ versus s from Table 13.9 to reveal where the shear and combined shear and bending designs apply, crossing the transitional value $s_{MF} = 0.05704$ (for the region $0 \le q \le 0.2113$). The comparison made with the circular-section plot shows sightly less weight in the region of a bending design (i.e. for $s > 0.05704$) but greater weight in the region of a shear design ($s > 0.05704$). While Figures 13.2 and 13.6 appear similar, these do not imply that the minimum weight of a square-section, encastré beam, carrying a central-concentrated load, will be similar to the minimum weight of a square-section beam where the same load is distributed uniformly. With the appropriate definition for s in these figures, Example 13.2 demonstrated that the weight of the latter is typically 60% less than the former. It will be seen in § 13.4 that both beams are lighter than optimised encastré beams under combined loading.

A comparison between the last two lines of Table 13.3 shows that the weights of simply supported beams (from Table 10.2) are greater than encastré beams for equal distributed loads in the region of bending. However, the weights of the two beams are similar in the region of a shear-dominated design.

13.3.6 Square-Section Profiles

The contributions to the square-section design from bending and shear depend upon the index s. We shall demonstrate this by designing depth profiles over the beam's half-length at three s-values, 1, 0.1 and 0.01. The normalised semi-depth $a/2L$ is found from equations (13.31) and (13.32a,b) as follows:

(a) $s = 1$
An all shear design apples to $0 \leq q \leq 0.2113$,

$$\left(\frac{a}{2L}\right)_F = 0.8059 \times 1^{1/2} \times (1 - 2q)^{1/2} = 0.8059(1 - 2q)^{1/2}$$

Shear and then bending designs apply to $0.2113 \leq q \leq 0.5$ within two sub-regions ($q_c = 0.3906$): for $0.2113 \leq q \leq 0.3906$,

$$\left(\frac{a}{2L}\right)_F = 0.8059 \times 1^{1/2} \times (1 - 2q)^{1/2} = 0.8059(1 - 2q)^{1/2}$$

and for $0.3906 \leq q \leq 0.5$,

$$\left(\frac{a}{2L}\right)_M = 0.9086 \times 1^{1/3}\left[q(1-q) - \frac{1}{6}\right]^{1/3} = 0.9086\left[q(1-q) - \frac{1}{6}\right]^{1/3}$$

(b) $s = 0.1$
An all shear design apples to $0 \leq q \leq 0.2113$,

$$\left(\frac{a}{2L}\right)_F = 0.8059 \times 0.1^{1/2} \times (1 - 2q)^{1/2} = 0.2549\,(1 - 2q)^{1/2}$$

Shear then bending designs apply to $0.2113 \leq q \leq 0.5$ within two sub-regions ($q_c = 0.3158$): for $0.2113 \leq q \leq 0.3158$,

$$\left(\frac{a}{2L}\right)_F = 0.8059 \times 0.1^{1/2} \times (1 - 2q)^{1/2} = 0.2549(1 - 2q)^{1/2}$$

and for $0.3158 \leq q \leq 0.5$,

$$\left(\frac{a}{2L}\right)_M = 0.9086 \times 0.1^{1/3}\left[q(1-q) - \frac{1}{6}\right]^{1/3} = 0.4217\left[q(1-q) - \frac{1}{6}\right]^{1/3}$$

(c) $s = 0.01$
Bending then shear designs apply to $0 \leq q \leq 0.2113$ within two sub-regions ($q_c = 0.1465$): for $0 \leq q \leq 0.1465$,

$$\left(\frac{a}{2L}\right)_M = 0.9086 \times 0.01^{1/3}\left[\frac{1}{6} - q(1-q)\right]^{1/3} = 0.1958\left[\frac{1}{6} - q(1-q)\right]^{1/3}$$

and for $0.1465 \leq q \leq 0.2113$,

$$\left(\frac{a}{2L}\right)_F = 0.8059 \times 0.01^{1/2}(1 - 2q)^{1/2} = 0.08059(1 - 2q)^{1/2}$$

Shear and then bending designs apply for $0.2113 \le q \le 0.5$ in two sub-regions ($q_c = 0.2562$): for $0.2113 \le q \le 0.2562$,

$$\left(\frac{a}{2L}\right)_F = 0.8059 \times 0.01^{1/2}(1 - 2q)^{1/2} = 0.08059(1 - 2q)^{1/2}$$

and for $0.2562 \le q \le 0.50$,

$$\left(\frac{a}{2L}\right)_M = 0.9086 \times 0.01^{1/3}\left[q(1 - q) - \frac{1}{6}\right]^{1/3} = 0.1958\left[q(1 - q) - \frac{1}{6}\right]^{1/3}$$

The square-section, beam profile plots, as derived from the equations above, appear as the inner, dashed lines in Figure 13.7. Each quarter-profile lies beneath that for a circular section beam at each index. Here we should read from Figure 13.6 whether the weight of an optimum square-section beam will be greater or less than a circular-section beam at the given s-value. The equations providing side-lengths a above could be implemented in a numerically controlled milling of each profile. We shall compare the optimum circular- and square-section beam designs further under distributed loading within the following example.

Example 13.3 Determine the optimum circular- and square-section profiles for minimum weight encastré, aluminium alloy beams required to support a load of 10 kN uniformly distributed over a length of 1 m. Determine the weight saving over a non-optimum design in each case. For aluminium alloy, take $\sigma_y = 150$ MPa and $\rho = 2710$ kg/m^3.
The maximum shear force at the end-fixing is $F_{max} = wL/2 = 5$ kN. Hence, irrespective of the section shape, the index s for each beam is

$$s = \frac{1}{\sigma_y}\frac{F_{max}}{L^2} = \frac{1}{150 \times 10^6}\frac{5 \times 10^3}{1^2} = 3.\dot{3} \times 10^{-5}$$

(a) Circular Section
The transition q_c in the first region of the beam $0 \le q_c \le 0.2113$, where a bending design switches to one of shear, follows from the solution to equation (13.25a):

$$\left[\frac{1}{6} - q_c(1 - q_c)\right]^{1/3} = 0.14192(1 - 2q_c)^{1/2}$$

giving, by trial, $q_c = 0.2091$.
For the region $0 \le q \le 0.2091$, the normalised radius is supplied by the bending criterion (13.24a),

$$\left(\frac{d}{2L}\right)_M = 1.08385\,(3.\dot{3} \times 10^{-5})^{1/3}\left[\frac{1}{6} - q(1 - q)\right]^{1/3} = 0.03488\left[\frac{1}{6} - q(1 - q)\right]^{1/3}$$

giving the extreme dimensions $(d/2L)_M = 0.0192$ and 3.796×10^{-3} for $q = 0$ and 0.2091, respectively.

For the region $0.2091 \leq q \leq 0.2113$, the normalised radius is supplied by the shear criterion (13.23):

$$\left(\frac{d}{2L}\right)_F = 0.8574(3.\dot{3} \times 10^{-5})^{1/2}(1 - 2q)^{1/2} = 4.9501 \times 10^{-3}(1 - 2q)^{1/2}$$

giving the extreme dimensions $(d/2L)_F \approx 3.796 \times 10^{-3}$ and 3.7614×10^{-3} for $q = 0.2091$ and 0.2113, respectively.
The transition q_c in the second region of the beam $0.2113 \leq q_c \leq 0.50$, where a shear design is displaced by one of bending, follows from the solution to equation (13.25b):

$$\left[q_c(1 - q_c) - \frac{1}{6}\right]^{1/3} = 0.14192(1 - 2q_c)^{1/2}$$

giving, by trial, $q_c = 0.2135$. Thus, for the region $0.2113 \leq q \leq 0.2135$, the normalised radius is supplied by the shear criterion (13.23) :

$$\left(\frac{d}{2L}\right)_F = 0.8574\left(3.\dot{3} \times 10^{-5}\right)^{1/2}(1 - 2q)^{1/2} = 4.9501 \times 10^{-3}(1 - 2q)^{1/2}$$

giving the extreme dimensions $(d/2L)_F = 3.7614 \times 10^{-3}$ and 3.747×10^{-3} for $q = 0.2113$ and 0.2135, respectively.
For the region $0.2135 \leq q \leq 0.50$, the normalised radius is supplied by the bending criterion (13.24b):

$$\left(\frac{d}{2L}\right)_M = 1.08385\,(3.\dot{3} \times 10^{-5})^{1/3}\left[q(1 - q) - \frac{1}{6}\right]^{1/3} = 0.03488\left[q(1 - q) - \frac{1}{6}\right]^{1/3}$$

giving the extreme dimensions $(d/2L)_M \approx 3.766 \times 10^{-3}$ and 0.01524 for $q = 0.2135$ and 0.50, respectively. For a 1 m length beam, the physical, rounded dimensions over the full length and depth, as derived from these values of q and $d/2L$, appear as continuous lines in Figure 13.8. The full shape of the beam, which reflects the half-length and half-depth profile into the second half of the length and into each lower depth, shows where the two constrictions must lie if weight is to be minimised.

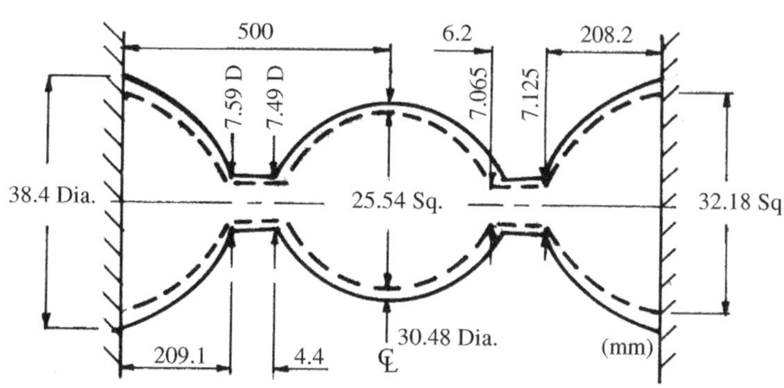

Figure 13.8 Length and depth profiles for circular and square sections

The removal of material required, say, from a uniform circular bar at the maximum safe diameter $d = 38.4$ mm (i.e. the non-optimum design) results in a considerable weight saving of more than 50%. To show this, note, firstly, that the weight of the uniform section bar with this diameter is

$$W = \rho \times \pi(d^2/4)L = 2710 \times \pi \times (0.0384^2/4) \times 1 = 3.1385 \text{ kg}$$

The weight of the optimum bar design follows from applying equations (13.27)–(13.30) to the two regions identified previously within the half-length as follows.
Taking $0 \le q \le 0.2113$ ($q_c = 0.2091$) first, the normalised weight contribution from bending $0 \le q \le 0.2091$ is

$$n_M = \pi(0.03488)^2 \int_0^{0.2091} \left[\frac{1}{6} - q(1-q)\right]^{2/3} dq = 1.3499 \times 10^{-4}$$

which was evaluated using Simpson's rule. The contribution to the weight from shear applies to the region $0.2091 \le q \le 0.2113$ as:

$$n_F = \pi(4.9501 \times 10^{-3})^2 \int_{0.2091}^{0.2113} (1-2q)dq = 7.698 \times 10^{-5}\, |q(1-q)|_{0.2091}^{0.2113} = 9.816 \times 10^{-8}$$

Hence the total, non-dimensional weight of this region is

$$n = n_M + n_F = 1.3509 \times 10^{-4}$$

Turning to $0.2113 \le q_c \le 0.50$ ($q_c = 0.2135$), within the intermediate region $0.2113 \le q \le 0.2135$ the weight contribution from shear is

$$n_F = \pi(4.9501 \times 10^{-3})^2 \int_{0.2113}^{0.2135} (1-2q)dq$$

$$= 7.698 \times 10^{-5}\, |q(1-q)|_{0.2113}^{0.2135} = 9.7414 \times 10^{-8}$$

and, within the remaining region $0.2135 \le q \le 0.50$, the contribution from bending is

$$n_M = \pi(0.03488)^2 \int_{0.2135}^{0.50} [q(1-q) - 1/6]^{2/3} dq = 1.1082 \times 10^{-4}$$

Hence the total weight of this region is

$$n = n_M + n_F = 1.1092 \times 10^{-4}$$

The full, normalised weight for the optimum design is therefore:

$$\sum n = 2[(1.3509 \times 10^{-4}) + (1.1092 \times 10^{-4})] = 4.9202 \times 10^{-4}$$

from which the total weight follows as

$$W_{\text{opt}} = \rho \times L^3 \times \sum n = 2710 \times 1^3 \times 4.9202 \times 10^{-4} = 1.333 \text{ kg}$$

and this reveals a 57.5% weight saving over the uniform circular section bar.

(b) Square Section
The transition q_c, in the region $0 \leq q_c \leq 0.2113$ where a bending design is displaced by one of shear, follows from the solution to equation (13.33a):

$$\left[\frac{1}{6} - q_c(1 - q_c)\right]^{1/3} = 0.15913(1 - 2q_c)^{1/2}$$

giving, by trial, $q_c = 0.2082$. Thus, for the region $0 \leq q \leq 0.2082$, the normalised half-depth is supplied by the bending criterion (13.32a):

$$\left(\frac{a}{2L}\right)_M = 0.9086(3.\dot{3} \times 10^{-5})^{1/3}\left[\frac{1}{6} - q(1 - q)\right]^{1/3} = 0.02924\left[\frac{1}{6} - q(1 - q)\right]^{1/3}$$

giving the extreme dimensions $(a/2L)_M = 0.01609$ and 3.5656×10^{-3} for $q = 0$ and 0.2082, respectively. For the region $0.2082 \leq q \leq 0.2113$, the half-depth is supplied by the shear criterion (13.31):

$$\left(\frac{a}{2L}\right)_F = 0.8059(3.\dot{3} \times 10^{-5})^{1/2}(1 - 2q)^{1/2} = 4.6530 \times 10^{-3}(1 - 2q)^{1/2}$$

giving the extreme dimensions $(a/2L)_\text{F} \approx 3.5546 \times 10^{-3}$ and 3.5357×10^{-3} for $q =$ 0.2082 and 0.2113, respectively.
The transition q_c, in the region $0.2113 \leq q_c \leq 0.50$, where a shear design is displaced by one of bending, follows from the solution to equation (13.33b):

$$\left[q_c(1 - q_c) - \frac{1}{6}\right]^{1/3} = 0.15913(1 - 2q_c)^{1/2}$$

giving, by trial, $q_c = 0.2144$. Thus, for the region $0.2113 \leq q \leq 0.2144$, the normalised half-depth is supplied by the shear criterion (13.31):

$$\left(\frac{a}{2L}\right)_F = 0.8059(3.\dot{3} \times 10^{-5})^{1/2}(1 - 2q)^{1/2} = 4.6530 \times 10^{-3}(1 - 2q)^{1/2}$$

giving the extreme dimensions$(a/2L)_F = 3.5357 \times 10^{-3}$ and 3.5167×10^{-3} for $q =$ 0.2113 and 0.2144, respectively. For the region $0.2144 \leq q \leq 0.50$, the half-depth is supplied by the bending criterion (13.32b):

$$\left(\frac{a}{2L}\right)_M = 0.9086(3.\dot{3} \times 10^{-5})^{1/3}[q(1 - q) - 1/6]^{1/3} = 0.02924[q(1 - q) - 1/6]^{1/3}$$

giving the extreme dimensions$(a/2L)_M \approx 3.5347 \times 10^{-3}$ and 0.01277 for $q = 0.2144$ and 0.50, respectively. Rounding errors within the trial solution for q_c are responsible for the small difference seen in $(a/2L)$ from the shear and bending designs at q_c, where they should agree. For a 1 m length beam, the physical dimensions, as derived from these q and $a/2L$ values, are superimposed as dashed lines upon the circular profile in Figure 13.8. The full shape of the beam, is found by reflecting this profile about the vertical and horizontal centre lines. The square-section profile again shows two constrictions, similar to the circular section, but with a smaller amount of material used.

In minimising the weight, the removal of material from a uniform square bar of maximum size, $a = 2 \times 16.09 = 32.18$ mm (i.e. the non-optimum safe design), results again in a weight saving of more than 50%. To show this, note, firstly, that the weight of the uniform square bar is

$$W = \rho \times a^2 L = 2710 \times 0.03218^2 \times 1 = 2.8064 \text{ kg}$$

The weight of the optimum bar design follows from applying equations (13.35)–(13.38) to the two sub-regions identified previously within the half-length as follows.

Taking $0 \leq q \leq 0.2113$ ($q_c = 0.2082$) first, for $0 \leq q \leq 0.2082$, the contribution to the normalised weight from bending is

$$n_M = 4(0.02924)^2 \int_0^{0.2082} [1/6 - q(1-q)]^{2/3} \mathrm{d}q = 1.18784 \times 10^{-4}$$

which was evaluated using Simpson's rule. For $0.2082 \leq q \leq 0.2113$, the contribution to the weight from shear is

$$n_F = 4(4.6530 \times 10^{-3})^2 \int_{0.2082}^{0.2113} (1 - 2q)\mathrm{d}q$$

$$= 8.6602 \times 10^{-5} \left| q(1-q) \right|_{0.2082}^{0.2113} = 1.5483 \times 10^{-7}$$

Hence the total weight of this region is

$$n = n_M + n_F = 1.1894 \times 10^{-4}$$

Turning to $0.2113 \leq q_c \leq 0.50$ ($q_c = 0.2144$), within the sub-region $0.2113 \leq q \leq 0.2144$ the contribution from shear is

$$n_F = 4(4.6530 \times 10^{-3})^2 \int_{0.2113}^{0.2144} (1 - 2q)\mathrm{d}q$$

$$= 8.6602 \times 10^{-5} \left| q(1-q) \right|_{0.2113}^{0.2144} = 1.5418 \times 10^{-8}$$

Within the remaining sub-region, $0.2144 \leq q \leq 0.50$, the contribution from bending is

$$n_M = 4(0.02924)^2 \int_{0.2144}^{0.50} [q(1-q) - 1/6]^{2/3} \mathrm{d}q = 1.0032 \times 10^{-4}$$

Hence, the total weight of this region is

$$n = n_M + n_F = 1.0034 \times 10^{-4}$$

The full normalised weight for the optimum design is therefore

$$\sum n = 2[(1.1894 \times 10^{-4}) + (1.0034 \times 10^{-4})] = 4.3855 \times 10^{-4}$$

from which the actual weight follows as

$$W_{\text{opt}} = \rho \times L^3 \times \sum n = 2710 \times 1^3 \times 4.3855 \times 10^{-4} = 1.1885 \text{ kg}$$

This reveals weight savings of 57.7% compared to the uniform square beam and 10.8% compared to the optimum circular-section beam bearing the same load.

13.4 Combined Loads

The third standard case refers to the combination of a central-concentrated load P and a distributed load w per unit length (see Figure 13.9a). With a discontinuity arising in the shear force diagram beneath the concentrated load at mid-span (see Figure 13.9b), it becomes necessary to work with ordinates $|F(z)|$ for each half of the beam with z taken from one end towards the centre. Ordinates $|M(z)|$ in the net moment diagram (see Figure 13.9c) are found by taking the difference between ordinates in the 'free' and

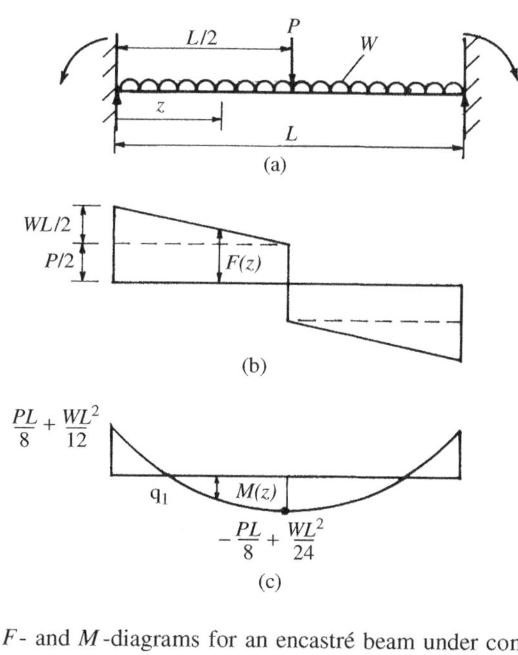

Figure 13.9 F- and M-diagrams for an encastré beam under combined loading

'fixing' moment diagrams, whose areas are equal. Alternatively, in the method adopted here we shall add the ordinates of the F- and M-diagrams in Figures 13.1b,d and 13.5b,d for when the loads act separately.

With reflective symmetry in each diagram at the beam centre, the total weight becomes twice that of one half. Within the half-length $0 \le z \le L/2$, the positive ordinate for the shear force diagram in Figure 13.9b may be expressed as the sum of the two ordinates previously considered in §§ 13.1 and 13.2. That is, we can write the sum of equations (13.1) and (13.21) as

$$\left|F\left(\frac{z}{L}\right)\right| = F_1 + F_2\left[1 - 2\left(\frac{z}{L}\right)\right], \quad \Rightarrow \quad F(q) = F_1 + F_2(1 - 2q) \tag{13.39}$$

where $q = z/L$. The reactions $F_1 = P/2$ and $F_2 = wL/2$ at each fixing sum to the maximum ordinate within the F-diagram (see Figure 13.9b). Where the difference between the fixing and free moment diagrams is positive, the ordinate in the net moment diagram (see Figure 13.9c) is expressed as the sum of equations (13.2a) and (13.22a),

$$\left|M\left(\frac{z}{L}\right)\right| = F_1 L\left(\frac{1}{4} - \frac{z}{L}\right) + F_2 L\left[\frac{1}{6} - \frac{z}{L}\left(1 - \frac{z}{L}\right)\right]$$

$$|M(q)| = F_1 L\left(\frac{1}{4} - q\right) + F_2 L\left[\frac{1}{6} - q(1 - q)\right] \tag{13.40a}$$

for $0 \le q \le q_1$, in which q_1 will follow from setting $M(q) = 0$ in equation (13.40a). Within the region $q_1 \le q \le 1/2$, where the net moment diagrams is negative, the magnitude of its ordinates becomes the sum of equations (13.2b) and (13.22b):

$$\left|M\left(\frac{z}{L}\right)\right| = F_1 L\left(\frac{z}{L} - \frac{1}{4}\right) + F_2 L\left[\frac{z}{L}\left(1 - \frac{z}{L}\right) - \frac{1}{6}\right]$$

$$|M(q)| = F_1 L\left(q - \frac{1}{4}\right) + F_2 L\left[q(1 - q) - \frac{1}{6}\right] \tag{13.40b}$$

13.4.1 Optimum Circular Section

For a circular cross-section, the optimum diameter d in bending is again found by combining each $|M(q)|$ expression (13.40a,b) with equation (3.3b). This gives the dependence of a normalised radius $d/(2L)$ upon the normalised length $q = z/L$ within two regions across the half length as

$$\left(\frac{d}{2L}\right)_M = \left(\frac{4}{\pi}\right)^{1/3}\left\{s_1\left(\frac{1}{4} - q\right) + s_2\left[\frac{1}{6} - q(1 - q)\right]\right\}^{1/3} \quad \text{for } 0 \le q \le q_1 \tag{13.41a}$$

$$\left(\frac{d}{2L}\right)_M = \left(\frac{4}{\pi}\right)^{1/3}\left\{s_1\left(q - \frac{1}{4}\right) + s_2\left[q(1 - q) - \frac{1}{6}\right]\right\}^{1/3} \quad \text{for } q_1 \le q \le \frac{1}{2} \tag{13.41b}$$

in which $s_1 = F_1/(\sigma_y L^2)$ and $s_2 = F_2/(\sigma_y L^2)$ are the normalised structural indices for each of the applied loads. We have seen previously how each s identifies short beams when large and long beams when small. When the design is based upon shear, the normalised radius is found by combining equations (5.15b) and (13.39). This gives the radius within the half-length

$$\left(\frac{d}{2L}\right)_F = \left(\frac{4}{\sqrt{3}\pi}\right)^{1/2} [s_1 + s_2(1 - 2q)]^{1/2} \quad \text{for } 0 \leq q \leq \frac{1}{2} \tag{13.42}$$

where $\sigma_y = \sqrt{3}\tau_y$ by the von Mises criterion. For given values of s_1 and s_2, equations (13.41a,b) and (13.42) show that the optimum diameter is variable within their respective lengths by both the bending and shear designs. The limiting length position $q = q_1$ that decides between equations (13.43a,b) is found by equating $M(q) = 0$ in equation (13.40a). This gives

$$\left(\frac{1}{4} - q_1\right) + \frac{s_2}{s_1}\left[\frac{1}{6} - q_1(1 - q_1)\right] = 0 \tag{13.43}$$

where $s_2/s_1 = F_2/F_1 = wL/P$ is the load (or index) ratio. For low-valued indices s_1 (to be determined), within a given ratio s_2/s_1, there will lie in the region $0 \leq q \leq q_1$ a critical position $q = q_c$ at which the diameters from both the bending and shear designs agree. It follows that q_c may be found by equating (13.41a) and (13.42),

$$1.2641\left\{\left(\frac{1}{4} - q_c\right) + \frac{s_2}{s_1}\left[\frac{1}{6} - q_c(1 - q_c)\right]\right\}^{1/3} = s_1^{1/6}\left[1 + \frac{s_2}{s_1}(1 - 2q_c)\right]^{1/2} \tag{13.44a}$$

and, similarly, within the region $q_1 \leq q \leq ½$, the critical position q_c follows from equating (13.41b) and (13.42),

$$1.2641\left\{\left(q_c - \frac{1}{4}\right) + \frac{s_2}{s_1}\left[q_c(1 - q_c) - \frac{1}{6}\right]\right\}^{1/3} = s_1^{1/6}\left[1 + \frac{s_2}{s_1}(1 - 2q_c)\right]^{1/2} \tag{13.44b}$$

We shall now proceed to solve the forgoing equations where the structural index (or load) ratio is $s_2/s_1 = 1$. The appropriate root to the quadratic equation (13.43) is $q_1 = 0.2362$ within the first half-length. The second root $q_1 = 1.7638$ applies to the second half-length which we need not consider further, given the loading symmetry.

For the initial sub-region $0 \leq q \leq 0.2362$, lying within the first half-length, equation (13.44a) becomes

$$[0.4167 - q_c(2 - q_c)]^{1/3} = 1.1188 s_1^{1/6}(1 - q_c)^{1/2} \tag{13.45a}$$

where q_c is the transition position within this region where a bending design switches to a shear design. At the limit of equation (13.45a) we set $q_c = 0$ to find the critical index s_{1MF} at which this bending plus shear design is displaced by an all-shear design. This gives

$$0.7469 = 1.1188\, s_{1MF}^{1/6}, \quad \Rightarrow \quad s_{1MF} = 0.0885 \tag{13.45b}$$

and hence the all-shear design applies when $s_1 \geq 0.0885$. Consequently, as the combined bending and shear design applies to $s_1 < 0.0885$, this restricts the 'MF' analysis to two s_1-values, 0.01 and 0.001, within the span of five decades to be taken for s_1 (i.e. from 0.001 to 100). For $s_1 = 0.01$, equation (13.45a) becomes

$$[0.4167 - q_c(2 - q_c)]^{1/3} = 0.5193(1 - q_c)^{1/2} \tag{13.46a}$$

The trial solution to equation (13.46a) gives $q_c = 0.1698$. For $s_1 = 0.001$, equation (13.45a) becomes

$$[0.4167 - q_c(2 - q_c)]^{1/3} = 0.3538(1 - q_c)^{1/2} \tag{13.46b}$$

The trial solution to equation (14.46b) gives $q_c = 0.216$.

For the second sub-region, $0.2362 \leq q \leq 0.5$, lying within the first half-length, equation (13.44b) becomes

$$[q_c(2 - q_c) - 0.4167]^{1/3} = 1.1188 s_1^{1/6}(1 - q_c)^{1/2} \tag{13.47a}$$

To find $s = s_{1FM}$, at which the transition from an all-shear design to a shear plus bending design applies, we set $q_c = 0.5$ in equation (13.47a). This leads to

$$0.6933 = 0.7911 s_{1FM}^{1/6}, \quad \Rightarrow \quad s_{1FM} = 0.4532 \tag{13.47b}$$

and hence the all-shear design will apply to $s_1 \geq 0.4532$. Consequently, as the combined design applies to $s_1 \leq 0.4532$, this restricts the 'FM' analysis to three s_1-values, 0.1, 0.01 and 0.001, within our span of five decades for the structural index s_1. For $s_1 = 0.1$, equation (13.47a) becomes

$$[q_c(2 - q_c) - 0.4167]^{1/3} = 0.7622(1 - q_c)^{1/2} \tag{13.48a}$$

The trial solution to equation (13.48a) gives $q_c = 0.391$. For $s_1 = 0.01$, equation (13.47a) becomes

$$[q_c(2 - q_c) - 0.4167]^{1/3} = 0.5193(1 - q_c)^{1/2} \tag{13.48b}$$

The trial solution to equation (13.48b) gives $q_c = 0.293$. Finally, for $s_1 = 0.001$, equation (13.47a) becomes

$$[q_c(2 - q_c) - 0.4167]^{1/3} = 0.3538(1 - q_c)^{1/2} \tag{13.48c}$$

Here the trial solution to equation (13.48c) gives $q_c = 0.256$.

13.4.2 Minimum Weight of Circular Section Beam ($s_2/s_1 = 1$)

We have seen for $s \geq s_{1MF} = 0.0885$ that the beam's diameter and weight within the region $0 \leq q \leq 0.2362$ follow solely from the all-shear design. For an elemental, normalised length δq lying within this region, the normalised weight $n_F = (1/\rho)(W_F/L^3)$

is found from equation (13.42):

$$n_F = \pi \int_0^{q_1} \left(\frac{d}{2L}\right)^2 \mathrm{d}q = \frac{4s_1}{\sqrt{3}} \int_0^{0.2362} \left[1 + \frac{s_2}{s_1}(1 - 2q_c)\right] \mathrm{d}q \tag{13.49a}$$

When $s_2/s_1 = 1$, the integration of equation (13.49a) leads to the beam weight within this region:

$$n_F = \frac{4s_1}{\sqrt{3}} \left| q(2 - q) \right|_0^{0.2362} = 0.9623 s_1 \tag{13.49b}$$

With $s < s_{MF}$, both bending and shear contribute to the minimum weight. The contribution from bending, which lies within $0 \le q \le q_c$, follows from equation (13.41a):

$$n_M = \pi \int_0^{q_c} \left(\frac{d}{2L}\right)^2 \mathrm{d}q = \pi \left(\frac{4s_1}{\pi}\right)^{2/3} \int_0^{q_c} \left\{\left(\frac{1}{4} - q\right) + \frac{s_2}{s_1}\left[\frac{1}{6} - q(1 - q)\right]\right\}^{2/3} \mathrm{d}q \tag{13.50a}$$

The contribution from shear, which lies within $q_c \le q \le q_1$ ($q_1 = 0.2362$), follows from equation (13.42):

$$\begin{aligned} n_F &= \pi \int_{q_c}^{q_1} \left(\frac{d}{2L}\right)^2 \mathrm{d}q = \frac{4s_1}{\sqrt{3}} \int_{q_c}^{0.2362} \left[1 + \frac{s_2}{s_1}(1 - 2q)\right] \mathrm{d}q \\ &= 2.309 s_1 \left| q\left[1 + \frac{s_2}{s_1}(1 - q)\right]\right|_{q_c}^{0.2362} \end{aligned} \tag{13.50b}$$

We may now substitute into equations (13.50a,b) the ratio s_2/s_1 and the value of q_c corresponding to s_1 within its chosen range. For example, with $s_2/s_1 = 1$ and $s_1 = 0.01$, where we had found previously that $q_c = 0.1698$, equation (13.50a) becomes

$$n_M = 0.1713 \int_0^{0.1698} [0.4167 - q(2 - q)]^{2/3} \mathrm{d}q = 0.01156 \tag{13.51a}$$

where the integral was evaluated using Simpson's rule. Equation (13.50b) provides the shear contribution directly as

$$n_F = 0.02309 \left| q(2 - q) \right|_{0.1698}^{0.2362} = 2.444 \times 10^{-3} \tag{13.51b}$$

Hence, by this design, the total weight contribution from this region is the sum of equations (13.51a,b): $n_M + n_F = 0.014$.

Within the region $0.2362 \le q \le 0.5$, we have seen for $s_1 \ge s_{1FM} = 0.4532$ that the beam's diameter and weight follow from an all-shear design. Thus, for an elemental, normalised length $\delta q = \delta(z/L)$ within this region, the minimum weight $n_F = (1/\rho)(W_F/L^3)$ from this design follows from equation (13.42):

$$n_F = \pi \int_{q_1}^{0.5} \left(\frac{d}{2L}\right)^2 dq = \frac{4s_1}{\sqrt{3}} \int_{0.2362}^{0.5} \left[1 + \frac{s_2}{s_1}(1 - 2q)\right] dq \qquad (13.52a)$$

When $s_2/s_1 = 1$, the integration of equation (13.52a) leads to the beam weight within this region:

$$n_F = \frac{4s_1}{\sqrt{3}} |q(2-q)|_{0.2362}^{0.5} = 0.7699 s_1 \qquad (13.52b)$$

With $s < s_{FM}$, shear and then bending contribute to the minimum weight within their respective regions: $0.2362 \le q \le q_c$ and $q_c \le q \le 0.5$. The contribution from shear is found from equation (13.42):

$$n_F = \pi \int_{q_1}^{q_c} \left(\frac{d}{2L}\right)^2 dq = \frac{4s_1}{\sqrt{3}} \int_{0.2362}^{q_c} \left[1 + \frac{s_2}{s_1}(1 - 2q)\right] dq$$

$$= 2.309 s_1 \left| q\left[1 + \frac{s_2}{s_1}(1 - q)\right]\right|_{0.2362}^{q_c} \qquad (13.53a)$$

The contribution from bending is found from equation (13.41b):

$$n_M = \pi \int_{q_c}^{0.5} \left(\frac{d}{2L}\right)^2 dq = \pi\left(\frac{4s_1}{\pi}\right)^{2/3} \int_{q_c}^{0.5} \left\{\left(q - \frac{1}{4}\right) + \frac{s_2}{s_1}\left[q(1-q) - \frac{1}{6}\right]\right\}^{2/3} dq \qquad (13.53b)$$

We may now substitute into equations (13.53a,b) the ratio s_2/s_1 and the value of q_c corresponding to s_1 within its chosen range. For example, with $s_2/s_1 = 1$ and $s_1 = 0.1$, we found previously that $q_c = 0.391$. Equation (13.53a) provides the shear contribution directly:

$$n_F = 0.2309 \left| q(2-q) \right|_{0.2362}^{0.391} = 0.04908 \qquad (13.54a)$$

Equation (13.53b) gives the bending contribution:

$$n_M = 0.7951 \int_{0.391}^{0.5} [q(2-q) - 0.4167]^{2/3} dq = 0.03626 \qquad (13.54b)$$

where Simpson's rule was used to evaluate the integral. Hence, by this design, the total weight contribution from this region is the sum of equations (13.54a,b): $n_M + n_F = 0.08534$.

The two weights calculated above appear in their appropriate positions in Table 13.10, where the further weights shown have been calculated in a similar manner for the remaining s_1-values. When bending plays the major role, i.e. for $s_1 = 0.001$, it is necessary to evaluate the bending contribution from the repeated application of Simpson's rule. The full weight of the beam appears within the final row, which is *twice* the sum of the entries in the preceding rows at each of the s_1-values given.

Figure 13.10 plots the final row in Table 13.10 against s_1. This provides the minimum beam weights across the full range of s_1, including the transitional values $s_{1MF} = 0.0885$ and $s_{1FM} = 0.4532$, between the shear and combined shear and bending designs, lying within each sub-region in Table 13.10. Figure 13.10 also examines the effect of the structural index upon the weight of encastré beams under three further load ratios, including $s_2/s_1 = 0$, for an isolated, central-concentrated load. The weights of this beam are taken from Figure 13.2 and applied, after dividing s by 2, to Figure 13.10 according to the respective definitions given in their captions. This shows that for $s_2/s_1 = 0$, an encastré beam in circular section has least weight when compared to the optimum weights of beams under positive load ratios $s_2/s_1 = 0.1$, 1 and 10.

The minimum weights of encastré beams bearing loads under ratios $s_2/s_1 = 10$ and 0.10 are summarised in Tables 13.11 and 13.12. These weight calculations were made in a similar manner to those outlined above for $s_2/s_1 = 1$. Figure 13.10 shows that the plots for the combined load ratios $s_2/s_1 = 0.1$ and 10 straddle the plot for $s_2/s_1 = 1$.

In Table 13.11, the transitional values of s_1, $s_{1MF} = 0.01126$ and $s_{1FM} = 4.788$, apply to their respective sub-regions $0 \le q \le 0.2171$ and $0.2171 \le q \le 0.5$. Thus, for the tabulated $s_1 \ge s_{1MF}$ a pure shear design applies to the first region. Similarly, a pure shear design applies to the second region where $s_1 \ge s_{1FM}$.

In Table 13.12 the transitional values $s_{1MF} = 0.2184$ and $s_{1FM} = 0.2715$ apply to their respective regions $0 \le q \le 0.2480$ and $0.2480 \le q \le 0.5$, so that where $s_1 \ge s_{1MF}$ a pure shear design applies to the first region and where $s_1 \ge s_{1FM}$ a pure shear design applies to the second region.

A consistent feature of Tables 13.10–13.12 is that as the beam's length increases, i.e. as s_1 decreases, the weight contribution from shear (n_F) diminishes within both sub-regions that comprise the half-length. Here the weight contribution from bending (n_M) dictates the optimum design.

Table 13.10 Minimum weights across five decades of index s_1 for load ratio $s_2/s_1 = 1$

s_1	0.001	0.01	0.10	1.0	10	100
$(0 \le q \le 0.2362)$						
$n_M + n_F$	2.8393×10^{-3}	0.014	0.0962	0.9621	9.621	96.21
$(0.2362 \le q \le 0.5)$						
$n_F + n_M$	2.9373×10^{-3}	0.01433	0.08534	0.7699	7.699	76.993
$\sum n$	11.553×10^{-3}	0.05667	0.3631	3.4642	34.641	346.41

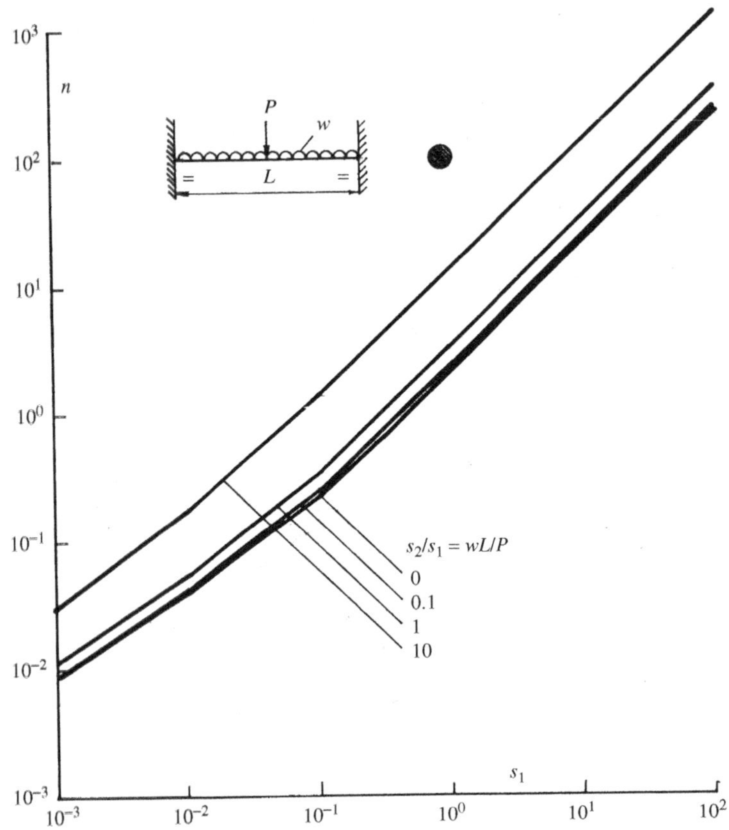

Figure 13.10 Minimum weight of encastré, circular-section beams under combined loading. (note: $s_1 = (1/\sigma_y)(F_1/L^2)$ where $F_1 = P/2$; $s_2 = (1/\sigma_y)(F_2/L^2)$ where $F_2 = wL/2$ and $n = (1/\rho)(W/L^3)$)

Table 13.11 Minimum weights for the load ratio $s_2/s_1 = 10$

s_1	0.001	0.01	0.10	1.0	10	100
$(0 \leq q \leq 0.2171)$						
$n_M + n_F$	7.15×10^{-3}	0.0444	0.4427	4.4266	44.266	442.66
$(0.2171 \leq q \leq ½)$						
$n_F + n_M$	7.90×10^{-3}	0.0415	0.2831	2.527	25.016	250.16
$\sum n$	0.0301	0.1717	1.4516	13.907	138.56	1385.84

13.4.3 Circular Section Profiles $(s_2/s_1 = 1)$

The effect of the bending and shear criteria upon this beam's normalised radius profile will now be examined for the load ratio $s_2/s_1 = 1$ at selected values of s_1: 1, 0.1 and 0.01. It will be necessary to consider design criteria within the two regions of the half-length separately.

Table 13.12 Minimum weights for the load ratio $s_2/s_1 = 0.10$

s_1	0.001	0.01	0.10	1.0	10	100
$(0 \le q \le 0.2480)$						
$n_M + n_F$	2.253×10^{-3}	0.01089	0.06465	0.6158	6.158	61.58
$(0.2480 \le q \le 1/2)$						
$n_F + n_M$	2.266×10^{-3}	0.01094	0.06408	0.5966	5.966	59.66
$\sum n$	9.037×10^{-3}	0.0437	0.2575	2.4248	24.248	242.48

(a) $s_1 = 1$
The calculations in Table 13.10 have revealed that an all-shear design applies to both sub-regions when the index $s_1 = 1$. To see how this influences the $d/2L$ profile within the first sub-region, $0 \le q \le 0.2362$, we have, from equations (13.41a) and (13.42),

$$\left(\frac{d}{2L}\right)_M = \left(\frac{4}{\pi}\right)^{1/3} 1^{1/3}\left[\frac{1}{4} - q + \frac{1}{6} - q(1-q)\right]^{1/3} = 1.0838[0.4167 - q(2-q)]^{1/3} \tag{13.55a}$$

$$\left(\frac{d}{2L}\right)_F = \left(\frac{4}{\sqrt{3}\pi}\right)^{1/2} \times 1^{1/2} \times [2(1-q]^{1/2} = 1.2125(1-q)^{1/2} \tag{13.55b}$$

where subscripts F and M refer to the force and moment design criteria, respectively. Within the second region, $0.2362 \le q \le 0.5$, the profiles follow from equations (13.42) and (13.41b) as

$$\left(\frac{d}{2L}\right)_F = 1.2125(1-q)^{1/2} \tag{13.55b}$$

$$\left(\frac{d}{2L}\right)_M = \left(\frac{4}{\pi}\right)^{1/3} 1^{1/3}\left[q - \frac{1}{4} + q(1-q) - \frac{1}{6}\right]^{1/3} = 1.0838[q(2-q) - 0.4167]^{1/3} \tag{13.55c}$$

Figure 13.11(a) shows the variations in the normalised radii over the half-length according to equations (13.55a–c) as continuous lines. Clearly, a larger diameter beam is required to support the shear loading and therefore the section should be designed according to the shear criterion (F) for its whole length. The maximum ordinates shown, for the ends and centre, are 1.2125 and 0.8574, respectively. These indicate a short, squat pin where, if the length is 1 cm, the end and central diameters become 2.425 cm and 1.7148 cm. Between these extremes equation (13.55b) provides a smooth variation. The bending criteria in equations (13.55a,c) provide the greater variations in diameter (M) but as these diameters are consistently lower than those found from shear (F) they do not feature within the optimum design.

(b) $s_1 = 0.1$
The calculations in Table 13.10 have revealed an all-shear design for the region $0 \le q \le 0.2362$ and a combined shear and bending design for the region $0.2362 \le q \le 0.5$.

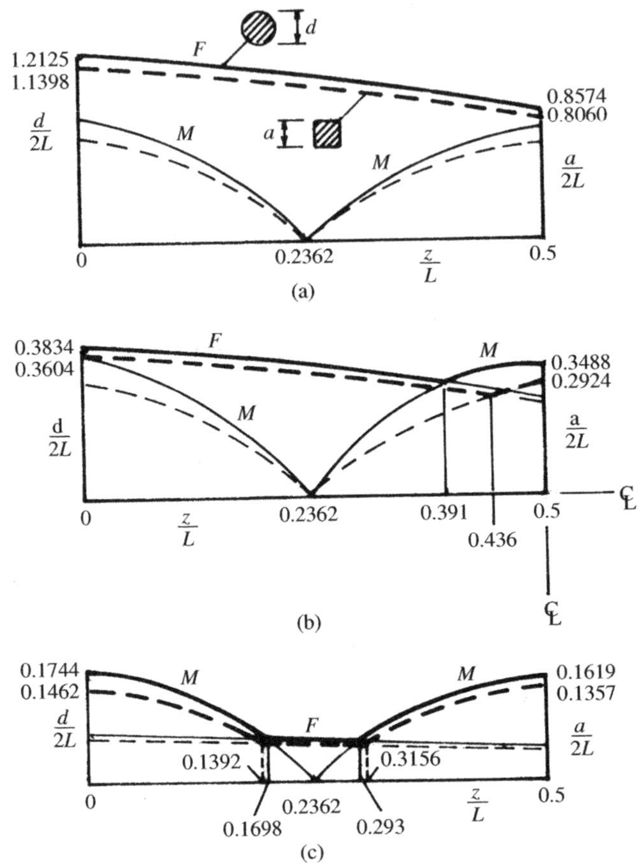

Figure 13.11 Normalised half-length and half-depth profiles for circular- and square-section encastré beams with $s_2/s_1 = 1$ for (a) $s_1 = 1$, (b) $s_1 = 0.1$ and (c) $s_1 = 0.01$

To establish the $d/2L$ profile within the region $0 \leq q \leq 0.2362$, we have, according to equations (13.41a) and (13.42):

$$\left(\frac{d}{2L}\right)_M = \left(\frac{4}{\pi}\right)^{1/3} 0.1^{1/3}\left[\frac{1}{4} - q + \frac{1}{6} - q(1-q)\right]^{1/3} = 0.5031[0.4167 - q(2-q)]^{1/3} \tag{13.56a}$$

$$\left(\frac{d}{2L}\right)_F = \left(\frac{4}{\sqrt{3}\pi}\right)^{1/2} \times 0.1^{1/2} \times [2(1-q]^{1/2} = 0.3834(1-q)^{1/2} \tag{13.56b}$$

and within the $0.2362 \leq q \leq 0.5$, the profiles follow from equations (13.42) and (13.41b), these being separated at $q_c = 0.391$:

$$\left(\frac{d}{2L}\right)_F = 0.3834(1-q)^{1/2} \tag{13.56b}$$

$$\left(\frac{d}{2L}\right)_M = \left(\frac{4}{\pi}\right)^{1/3} 0.1^{1/3}\left[q - \frac{1}{4} + q(1-q) - \frac{1}{6}\right]^{1/3} = 0.5031[q(2-q) - 0.4167]^{1/3} \tag{13.56c}$$

Figure 13.11(b) shows the variation in the normalised radii over the half-length according to equations (13.56a–c) . This shows that the beam ends should be designed according to a shear criterion (F) and the centre according to the bending criterion (M), so that the profile should be machined with an enlarged central region. The maximum ordinates by this combined design are 0.3834 and 0.3488 for the ends and centre, respectively. Between these extremes the diameter of the circular-section, encastré beam should vary in the manner shown in order to achieve the optimum design.

(c) $s_1 = 0.01$

The calculations in Table 13.10 have revealed a combined design (by shear and bending) in both regions. Correspondingly, the $d/2L$ profiles within the region $0 \le q \le 0.2362$ are again given by equations (13.41a) and (13.42), these being separated at $q_c = 0.1698$:

$$\left(\frac{d}{2L}\right)_M = \left(\frac{4}{\pi}\right)^{1/3} 0.01^{1/3}\left[\frac{1}{4} - q + \frac{1}{6} - q(1-q)\right]^{1/3} = 0.2335[0.4167 - q(2-q)]^{1/3} \tag{13.57a}$$

$$\left(\frac{d}{2L}\right)_F = \left(\frac{4}{\sqrt{3}\pi}\right)^{1/2} \times 0.01^{1/2} \times [2(1-q]^{1/2} = 0.1213(1-q)^{1/2} \tag{13.57b}$$

Within the region $0.2362 \le q \le 0.5$, the profiles defined by equations (13.42) and (13.41b) are separated at $q_c = 0.293$:

$$\left(\frac{d}{2L}\right)_F = 0.1213(1-q)^{1/2} \tag{13.57b}$$

$$\left(\frac{d}{2L}\right)_M = \left(\frac{4}{\pi}\right)^{1/3} 0.01^{1/3}\left[q - \frac{1}{4} + q(1-q) - \frac{1}{6}\right]^{1/3} = 0.2335[q(2-q) - 0.4167]^{1/3} \tag{13.57c}$$

Figure 13.11(c) shows the variation in the normalised radii over the half-length according to equations (13.57a–c) . The combined design applies to both regions. The largest radii, provided by bending, at the ends and at the centre, are 0.1744 and 0.1619, respectively. The enlarged sections at these positions are spanned by the smaller shear-based diameters as shown. The positions in the length, $q_c = 0.1698$ and 0.293, at which the bending–shear transitions occur were found previously by equating the governing $d/2L$ equations given above. We see, for this low s_1-value, that the section design is largely controlled by bending (M) but that shear (F) must take over where the bending moment falls to zero since the moment criterion deems that there would be no requirement for providing any section depth at all!

Where weight reduction is paramount, say in the design of vehicles where legislation limits fuel consumption and exhaust emissions, the uniform beam section may be abandoned for optimum, variable profiles derived here. Numerical machining is capable of reproducing the optimum profiles provided by this theory.

13.4.4 Optimum Square Section

The expressions (13.39) and (13.40a,b) for the dependence of the force and moment upon length again apply. Given a square-section beam, the dependence of its normalised semi-depth $a/2L$ upon normalised length z/L is required. When the design is to be based upon shear, the section's depth is found by combining equation (13.39) with equation (5.4c). This gives the shear-based geometry for the full half-length as

$$\left(\frac{a}{2L}\right)_F = \left(\frac{3\sqrt{3}}{8}\right)^{1/2}\left[s_1 + s_2\left(1 - 2\frac{z}{L}\right)\right]^{1/2} \quad \text{for } 0 \le \frac{z}{L} \le \frac{1}{2}$$

$$\left(\frac{a}{2L}\right)_F = 0.8059 s_1^{1/2}[1 + (s_2/s_1)(1 - 2q)]^{1/2}, \quad 0 \le q \le 0.5 \tag{13.58}$$

where $Q = \sqrt{3}$ by the von Mises criterion. The normalised structural indices for each of the applied loads are $s_1 = F_1/(\sigma_y L^2)$ and $s_2 = F_2/(\sigma_y L^2)$, where $F_1 = F/2$ and $F_2 = wL/2$. When the design is to be based upon bending, each $M(z/L)$ equation (13.40a,b) is combined with equation (3.6b). Thus, the section dimension within the two regions of the half-length becomes

$$\left(\frac{a}{2L}\right)_M = \left(\frac{3}{4}\right)^{1/3}\left\{s_1\left(\frac{1}{4} - \frac{z}{L}\right) + s_2\left[\frac{1}{6} - \frac{z}{L}\left(1 - \frac{z}{L}\right)\right]\right\}^{1/3} \quad \text{for } 0 \le \frac{z}{L} \le \frac{z_1}{L}$$

$$\left(\frac{a}{2L}\right)_M = 0.9086 s_1^{1/3}\left\{\left(\frac{1}{4} - q\right) + \frac{s_2}{s_1}\left[\frac{1}{6} - q(1 - q)\right]\right\}^{1/3}, \quad 0 \le q \le q_1 \tag{13.59a}$$

$$\left(\frac{a}{2L}\right)_M = \left(\frac{3}{4}\right)^{1/3}\left\{s_1\left(\frac{z}{L} - \frac{1}{4}\right) + s_2\left[\frac{z}{L}\left(1 - \frac{z}{L}\right) - \frac{1}{6}\right]\right\}^{1/3} \quad \text{for } \frac{z_1}{L} \le \frac{z}{L} \le \frac{1}{2}$$

$$\left(\frac{a}{2L}\right)_M = 0.9086 s_1^{1/3}\left\{\left(q - \frac{1}{4}\right) + \frac{s_2}{s_1}\left[q(1 - q) - \frac{1}{6}\right]\right\}^{1/3}, \quad q_1 \le q \le 1/2 \tag{13.59b}$$

where $s_2/s_1 = F_2/F_1 = wL/F$ is the load (index) ratio. Equation (13.40a) provides the position $q_1 = z_1/L$ at which $M(z_1/L) = 0$:

$$\left(\frac{1}{4} - q_1\right) + \frac{s_2}{s_1}\left[\frac{1}{6} - q_1(1 - q_1)\right] = 0 \tag{13.60}$$

We have seen previously how index s_1 identifies short beams when large and long beams when small. For given values of s_1 and s_2, equations (13.58) and (13.59a,b) show that the optimum dimension a is a variable within the respective lengths appropriate to the bending and shear designs. For low s_1-values (to be determined), within the ratio s_2/s_1, there will lie in the region $0 \le q \le q_1$ a critical position $q = q_c$ at which the section depths from both the bending and shear designs agree. Here q_c may be found by equating (13.58) and (13.59a):

$$1.1274\left\{\left(\frac{1}{4}-q_c\right)+\frac{s_2}{s_1}\left[\frac{1}{6}-q_c(1-q_c)\right]\right\}^{1/3}=s_1^{1/6}\left[1+\frac{s_2}{s_1}(1-2q_c)\right]^{1/2} \quad (13.61a)$$

Similarly, within the region $q_1 \le q \le 1/2$, the critical position q_c follows from equating (13.58) and (13.59b):

$$1.1274\left\{\left(q_c-\frac{1}{4}\right)+\frac{s_2}{s_1}\left[q_c(1-q_c)-\frac{1}{6}\right]\right\}^{1/3}=s_1^{1/6}\left[1+\frac{s_2}{s_1}(1-2q_c)\right]^{1/2} \quad (13.61b)$$

We shall now proceed to solve the foregoing equations for three values of the load (index) ratio: $s_2/s_1 = 1$, 0.1 and 10. These three ratios are sufficient to establish trends that enable interpolation for minimum weight at any intermediate ratio.

Load Ratio $s_2/s_1 = 1$

When $s_2/s_1 = 1$ the root of the quadratic equation (13.60) within the first half-length is $q_1 = 0.2362$. The second root $q_1 = 1.7638$ applies to the second half-length which we need not consider, given the loading symmetry.

For the first sub-region $0 \le q \le 0.2362$, within the half-length, equation (13.61a) becomes

$$[0.4167 - q_c(2-q_c)]^{1/3} = 1.2545 s_1^{1/6}(1-q_c)^{1/2} \quad (13.62a)$$

Firstly, it is required to find $s_1 = s_{1MF}$ from equation (13.62a), at which the transition from a combined bending and shear design to an all-shear design occurs. Setting $q_c = 0$ in equation (13.62a) leads to

$$0.7469 = 1.2545 s_{1MF}^{1/6}, \quad \Rightarrow \quad s_{1MF} = 0.0446 \quad (13.62b)$$

and hence the all-shear design will apply to $s_1 \ge 0.0446$. Consequently, as the combined design applies to $s_1 \le 0.0446$, this restricts our analysis to two s_1-values, 0.01 and 0.001, within the span of five decades (i.e. from 0.001 to 100) to be taken for s_1. For $s_1 = 0.01$, equation (13.62a) becomes

$$[0.4167 - q_c(2-q_c)]^{1/3} = 0.5823(1-q_c)^{1/2} \quad (13.63a)$$

A trial solution to equation (13.63a) gives $q_c = 0.1392$. For $s_1 = 0.001$, equation (13.62a) becomes

$$[0.4167 - q_c(2-q_c)]^{1/3} = 0.3967(1-q_c)^{1/2} \quad (13.63b)$$

A trial solution to equation (13.63b) gives $q_c = 0.2078$.

For the second sub-region, $0.2362 \le q \le 0.5$, lying within the half-length, equation (13.61b) becomes

$$[q_c(2-q_c) - 0.4167]^{1/3} = 1.2545 s_1^{1/6}(1-q_c)^{1/2} \quad (13.64a)$$

Here, to find $s = s_{1FM}$, at which the transition from an all-shear design to a combined shear and bending design applies, we set $q_c = 0.5$ in equation (13.64a). This leads to

$$0.6933 = 0.887 s_{1FM}^{1/6}, \quad \Rightarrow \quad s_{1FM} = 0.2281 \tag{13.64b}$$

and therefore the all-shear design will apply to $s_1 \geq 0.2281$. Consequently, as a combined design applies to $s_1 \leq 0.2281$, this restricts the analysis to three s_1 values, 0.1, 0.01 and 0.001, within our span of five decades for s_1. For $s_1 = 0.1$, equation (13.64a) becomes

$$[q_c(2 - q_c) - 0.4167]^{1/3} = 0.8547(1 - q_c)^{1/2} \tag{13.65a}$$

The solution to equation (13.65a) is $q_c = 0.4362$. For $s_1 = 0.01$, equation (13.64a) becomes

$$[q_c(2 - q_c) - 0.4167]^{1/3} = 0.5823(1 - q_c)^{1/2} \tag{13.65b}$$

The solution to equation (13.65b) is $q_c = 0.3156$. Finally, for $s_1 = 0.001$, equation (13.64a) becomes

$$[q_c(2 - q_c) - 0.4167]^{1/3} = 0.3967(1 - q_c)^{1/2} \tag{13.65c}$$

and here the trial solution to equation (13.65c) gives $q_c = 0.2627$.

13.4.5 Minimum Weight of Square Section Beam ($s_2/s_1 = 1$)

In the region $0 \leq q \leq 0.2362$, we have seen for $s_1 \geq s_{1MF} = 0.0446$ that the beam's depth and weight must follow from an all-shear design. Thus, for an element of normalised length $\delta q = \delta z/L$ within this region, the normalised weight $n_F = (1/\rho)(W_F/L^3)$ of the shear design follows from equation (13.58):

$$n_F = 4\int_0^{q_1} \left(\frac{a}{2L}\right)^2 dq = 2.598 s_1 \int_0^{0.2362} \left[1 + \frac{s_2}{s_1}(1 - 2q)\right] dq \tag{13.66a}$$

With $s_2/s_1 = 1$, the integration of equation (13.66a) leads to the beam weight within this region:

$$n_F = 2.598 s_1 \left| q(2 - q) \right|_0^{0.2362} = 1.0824 s_1 \tag{13.66b}$$

With $s < s_{MF}$, both bending and shear contribute to the minimum weight within this region. The contribution from bending is found from equation (13.59a),

$$n_M = 4\int_0^{q_c} \left(\frac{a}{2L}\right)^2 dq = 4\left(\frac{3s_1}{4}\right)^{2/3} \int_0^{q_c} \left\{\left(\frac{1}{4} - q\right) + \frac{s_2}{s_1}\left[\frac{1}{6} - q(1 - q)\right]\right\}^{2/3} dq \tag{13.67a}$$

and from equation (13.58), the shear weight contribution is

$$n_F = 4\int_{q_c}^{q_1}\left(\frac{a}{2L}\right)^2 \mathrm{d}q = \frac{3\sqrt{3}}{2}s_1\int_{0.2362}^{0.2362}\left[1+\frac{s_2}{s_1}(1-2q)\right]\mathrm{d}q$$

$$= 2.598s_1\left|q\left[1+\frac{s_2}{s_1}(1-q)\right]\right|_{q_c}^{q_c} \tag{13.67b}$$

We may now substitute into equations (13.67a,b) the ratio s_2/s_1 and the value of q_c corresponding to each s_1 within its chosen range. For example, with $s_2/s_1 = 1$ and $s_1 = 0.01$, where we had found previously that $q_c = 0.1392$, equation (13.67a) becomes

$$n_M = 0.1533\int_0^{0.1392}[0.4167 - q(2-q)]^{2/3}\mathrm{d}q = 9.1442\times10^{-3} \tag{13.68a}$$

in which the integral was evaluated numerically using Simpson's rule. Correspondingly, equation (13.67b) provides the shear contribution directly:

$$n_F = 0.02598\left|q(2-q)\right|_{0.1392}^{0.2362} = 4.0943\times10^{-3} \tag{13.68b}$$

Hence, by this combined bending and shear design, the total weight in this region is the sum of equations (13.68a,b): $n_M + n_F = 0.01324$.

Similarly, for $s_2/s_1 = 1$ and $s_1 = 0.001$, where previously $q_c = 0.2078$, the weight from bending follows from equation (13.67a):

$$n_M = 0.03302\int_0^{0.2078}[0.4167 - q(2-q)]^{2/3}\mathrm{d}q = 2.4418\times10^{-3} \tag{13.69a}$$

and from shear, equation (13.67b) gives

$$n_F = 0.002598\left|q(2-q)\right|_{0.2078}^{0.2362} = 1.1481\times10^{-4} \tag{13.69b}$$

Hence, the total weight in this region is the sum of equations (13.69a,b): $n_M + n_F = 2.5566\times10^{-3}$. These two summations appear as the first two entries within Table 13.13. The pure shear design depth, given by equation (13.66b), controls the section size for $s_1 > 0.0446$.

In the region $0.2362 \le q \le 0.5$, we have seen that for $s_{FM} \ge 0.2281$ the beam's square cross-section and weight follow from the all-shear design. Thus, for an element of its normalised length $\delta q = \delta(z/L)$, the minimum weight $n_F = (1/\rho)(W_F/L^3)$ for this region follows from the integration

$$n_F = 4\int_{q_1}^{0.5}\left(\frac{a}{2L}\right)^2\mathrm{d}q = \frac{3\sqrt{3}}{2}s_1\int_{0.2362}^{0.5}\left[1+\frac{s_2}{s_1}(1-2q_c)\right]\mathrm{d}q \tag{13.70a}$$

Table 13.13 Minimum beam weights for the load ratio $s_2/s_1 = 1$

s_1	0.001	0.01	0.10	1.0	10	100
$(0 \le q \le 0.2362)$						
$n_M + n_F$	2.5566×10^{-3}	0.01324	0.10824	1.0824	10.824	108.24
$(0.2362 \le q \le 0.5)$						
$n_F + n_M$	2.6459×10^{-3}	0.01336	0.08926	0.8662	8.662	86.62
$\sum n$	10.41×10^{-3}	0.0532	0.395	3.8972	38.972	389.72

in which equation (13.58) gives $a/2L$. When $s_2/s_1 = 1$, the integration of equation (13.70a) leads to the beam weight within this region:

$$n_F = 2.5981 s_1 \left| q(2-q) \right|_{0.2362}^{0.5} = 0.8662 s_1 \tag{13.70b}$$

With $s_1 < s_{1FM}$, shear and then bending contribute to the minimum normalised weight. The contribution from shear follows from equation (13.58) as

$$n_F = 4\int_{q_1}^{q_c} \left(\frac{a}{2L}\right)^2 dq = \frac{3\sqrt{3}}{2} s_1 \int_{0.2362}^{q_c} \left[1 + \frac{s_2}{s_1}(1-2q)\right] dq$$

$$= 2.5981 s_1 \left| q\left[1 + \frac{s_2}{s_1}(1-q)\right]\right|_{0.2362}^{q_c} \tag{13.71a}$$

The contribution from bending is found from equation (13.59b):

$$n_M = 4\int_{q_c}^{0.5} \left(\frac{a}{2L}\right)^2 dq = 4\left(\frac{3s_1}{4}\right)^{2/3} \int_{q_c}^{0.5} \left\{\left(q - \frac{1}{4}\right) + \frac{s_2}{s_1}\left[q(1-q) - \frac{1}{6}\right]\right\}^{2/3} dq \tag{13.71b}$$

We may now substitute into equations (13.71a,b) the ratio s_2/s_1 and the value of q_c corresponding to s_1 within its chosen range. For example, with $s_2/s_1 = 1$ and $s_1 = 0.1$, we found previously that $q_c = 0.4362$. Equation (13.71a) gives the shear contribution as

$$n_F = 0.2598 \left| q(2-q) \right|_{0.2362}^{0.4362} = 0.06898 \tag{13.72a}$$

and equation (13.71b) the bending contribution as

$$n_M = 0.7114 \int_{0.4362}^{0.5} [q(2-q) - 0.4167]^{2/3} dq = 0.02028 \tag{13.72b}$$

in which the integral was evaluated numerically using Simpson's rule. Hence, by this design, the total weight in this region is the sum of equations (13.72a,b): $n_M + n_F =$

0.08926. This partial beam weight appears in its appropriate positions in Table 13.13, where the further weights shown have been calculated in a similar manner for other s_1-values given. Here it has been necessary to evaluate the bending contribution from the repeated application of Simpson's rule when bending plays the major role, e.g. when $s_1 = 0.001$. The full weight of the beam appears within the final row, which is twice the sum of the entries in the preceding rows at each of the s_1-values given.

Figure 13.12 plots $\sum n$, from the final row in Table 13.13, against s_1 to reveal where the shear and combined shear and bending designs apply. The index s_1 crosses its transitional values $s_{1MF} = 0.0446$ and $s_{1FM} = 0.2281$ within the respective regions $0 \leq q \leq 0.2362$ and $0.2362 \leq q \leq 0.5$. As with a circular section (compare with Figure 13.10), Figure 13.12 shows again that the minimum weight of a beam under combined loadings ($s_2/s_1 = 0.1$, 1 and 10) is always greater than for a beam where a single load is concentrated at the centre ($s_2/s_1 = 0$). Minimum weights for the latter beam are taken from Figure 13.2, where s is halved before applying to Figure 13.12. It should be noted that in the absence of a central prop, the least weight that can be achieved in encastré beams

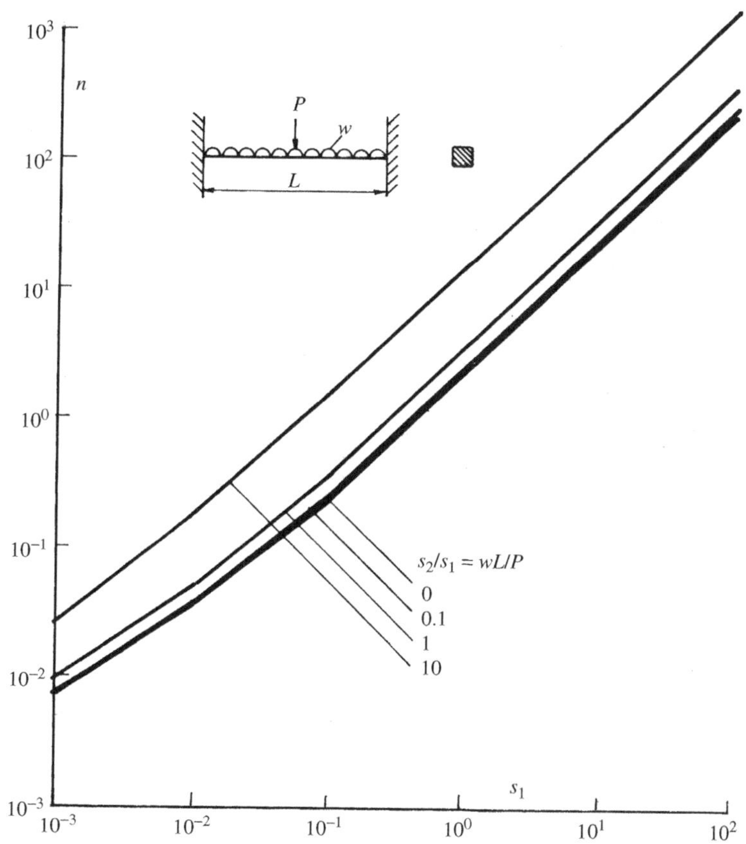

Figure 13.12 Minimum weight of a square-section encastré beam under combined loading (note: $s_1 = (1/\sigma_y)(F_1/L^2)$ where $F_1 = P/2$; $s_2 = (1/\sigma_y)(F_2/L^2)$ where $F_2 = wL/2$; and $n = (1/\rho)(W/L^3)$)

Table 13.14 Minimum weights for the load ratio $s_2/s_1 = 10$

s_1	0.001	0.01	0.10	1.0	10	100
$(0 \leq q \leq 0.2171)$						
$n_M + n_F$	6.686×10^{-3}	0.04981	0.4981	4.9805	49.805	498.05
$(0.2171 \leq q \leq 0.5)$						
$n_F + n_M$	7.2116×10^{-3}	0.04028	0.30155	2.8215	28.143	281.43
$\sum n$	0.0278	0.1802	1.5993	15.453	155.896	1558.96

in both these cross-sections arises when concentrated loading is replaced by uniformly distributed loading (see Figure 13.6). Figure 13.12 shows that plots of $\sum n$ against s_1 for the loading ratios $s_2/s_1 = 0.1$ and 10 straddle the case described above ($s_2/s_1 = 1$). The weights for each additional load ratio are summarised in Tables 13.14 and 13.15, whose calculations were done in a similar manner to those outlined above. Having established the trend between the four combined loading ratios 0, 0.1, 1 and 10, it follows that minimum weights may be interpolated reliably from Figure 13.12 at intermediate ratios.

In Table 13.14 the transitional values $s_{1MF} = 0.00567$ and $s_{1FM} = 2.4098$ apply to their respective regions, $0 \leq q \leq 0.2171$ and $0.2171 \leq q \leq 0.5$. Hence, where $s_1 \geq s_{1MF}$, a pure shear design applies to the first region and, where $s_1 \geq s_{1FM}$, a pure shear design will apply to the second region. Note that with $s_2/s_1 = 10$, the position $q_1 = 0.2171$ is found from equation (13.60). The bending moment is zero at this position, indicating where the moment changes sign.

In Table 13.15 the transitional s_1-values are $s_{1MF} = 0.1097$ and $s_{1FM} = 0.137$ appropriate to the respective regions, $0 \leq q \leq 0.2480$ and $0.2480 \leq q \leq 0.5$. Correspondingly, where $s_1 \geq s_{1MF}$ and $s_1 \geq s_{1FM}$, a pure shear design applies to each region. Here, $q_1 = 0.2480$ is found from equation (13.60), with $s_2/s_1 = 0.10$. Consistently across all these solutions, as the beam's length increases, i.e. as s_1 decreases, the contribution from shear diminishes within both sub-regions comprising the half-length.

13.4.6 Minimum Weight of Circular- versus Square-Section Beams

In the absence of specific numerical examples, it is not obvious from Figures 13.10 and 13.12 how the weights of circular and square sections compare. However, taking the final row of Tables 13.10–13.15 allows a weight comparison between the two sections for similar load ratios across the range of structural indices. Table 13.16 provides this

Table 13.15 Minimum weights for the load ratio $s_2/s_1 = 0.10$

s_1	0.001	0.01	0.10	1.0	10	100
$(0 \leq q \leq 0.2480)$						
$n_M + n_F$	2.0085×10^{-3}	0.010	0.06933	0.6928	6.928	69.28
$(0.2480 \leq q \leq 0.5)$						
$n_F + n_M$	2.027×10^{-3}	0.0101	0.06762	0.6712	6.712	67.12
$\sum n$	8.071×10^{-3}	0.0402	0.2739	2.7278	27.278	272.78

Table 13.16 Minimum weights of circular- and square-section encastré beams

s_2/s_1 ↓	Section ↓		$s_1 \to 0.001$	0.01	0.10	1.0	10	100
0.1	●	$\Sigma n \to$	9.037×10^{-3}	0.0437	0.2575	2.4248	24.248	242.48
0.1	■	$\Sigma n \to$	8.071×10^{-3}	0.0402	0.2739	2.7278	27.278	272.78
1.0	●	$\Sigma n \to$	11.553×10^{-3}	0.05667	0.3631	3.4642	34.641	346.41
1.0	■	$\Sigma n \to$	10.41×10^{-3}	0.0532	0.395	3.8972	38.972	389.72
10	●	$\Sigma n \to$	0.0301	0.1717	1.4516	13.907	138.56	1385.84
10	■	$\Sigma n \to$	0.0278	0.1802	1.5993	15.453	155.896	1558.96

comparison for both square and circular sections in ascending indices and load ratios. The trends under combined loading appear similar to those that would be found by superimposing Figures 13.2 and 13.6, namely that the circular section carries more weight in the region of bending (low s_1) but less weight in the region of shear (high s_1).

13.4.7 Square Section Profiles for $s_2/s_1 = 1$

The effect of bending and shear upon the beam's half-depth profile will now be examined at the load ratio $s_2/s_1 = 1$ for selected values of s_1. It is necessary to consider the profile design criteria within the two regions of the half-length separately. The normalised half-depth $a/2L$ for each beam profile is examined for three s_1-values 1, 0.1 and 0.01, as follows.

(a) $s_1 = 1$

Table 13.13 reveals that an all-shear design applies to both regions. To see how this influences the $a/2L$ profile within the first region, $0 \le q \le 0.2362$, we have, from equations (13.59a) and (13.58), the corresponding bending and shear designs:

$$\left(\frac{a}{2L}\right)_M = \left(\frac{3}{4}\right)^{1/3} 1^{1/3}\left[\frac{1}{4} - q + \frac{1}{6} - q(1-q)\right]^{1/3} = 0.9086[0.4167 - q(2-q)]^{1/3} \tag{13.73a}$$

$$\left(\frac{a}{2L}\right)_F = \left(\frac{3\sqrt{3}}{8}\right)^{1/2} 1^{1/2}[2(1-q)]^{1/2} = 1.1398(1-q)^{1/2} \tag{13.73b}$$

where subscripts F and M refer to the force and moment design criteria, respectively. Within the second region, $0.2362 \le q \le 0.5$, the respective profiles follow from equations (13.58) and (13.59b):

$$\left(\frac{a}{2L}\right)_F = \left(\frac{3\sqrt{3}}{8}\right)^{1/2} 1^{1/2}[2(1-q)]^{1/2} = 1.1398(1-q)^{1/2} \tag{13.73b}$$

$$\left(\frac{a}{2L}\right)_M = \left(\frac{3}{4}\right)^{1/3} 1^{1/3}\left[q - \frac{1}{4} + q(1-q) - \frac{1}{6}\right]^{1/3} = 0.9086[q(2-q) - 0.4167]^{1/3} \tag{13.73c}$$

Figure 13.11(a) shows the square section's variation over the beam's half-length (dashed lines). According to equations (13.73a–c) , a larger square section is required everywhere to support the shear loading and therefore the optimum section should be designed safely according to a shear criterion for its whole length. The maximum ordinates shown at the ends and centre are 1.1398 and 0.806, indicating a short, squat beam: for example, if the length is to be 10 mm, when acting as a pin or rivet, then the end and central depths become 22.796 mm and 16.12 mm, respectively. Between these positions equation (13.73b) provides a smooth variation. The bending criteria (13.73a) and (13.73c) provide the lesser depth variations shown and therefore these do not influence the optimum design. Note that Table 13.16 shows, for the given structural index $s_1 = 1$, that a shear-based normalised weight $\sum n$ for a square-section beam is greater than the weight of a circular-section beam.

(b) $s_1 = 0.1$
Table 13.13 reveals an all-shear design for the region $0 \le q \le 0.2362$ and a combined design for the region $0.2362 \le q \le 0.5$. To establish the $a/2L$ profile within the first region we have, according to equations (13.59a) and (13.58),

$$\left(\frac{a}{2L}\right)_M = \left(\frac{3}{4}\right)^{1/3} 0.1^{1/3}\left[\frac{1}{4} - q + \frac{1}{6} - q(1-q)\right]^{1/3} = 0.4217[0.4167 - q(2-q)]^{1/3} \tag{13.74a}$$

$$\left(\frac{a}{2L}\right)_F = \left(\frac{3\sqrt{3}}{8}\right)^{1/2} \times 0.1^{1/2} \times [2(1-q)]^{1/2} = 0.3604(1-q)^{1/2} \tag{13.74b}$$

and within the region $0.2362 \le q \le 0.5$, the profiles follow from equations (13.58) and (13.59b) as

$$\left(\frac{a}{2L}\right)_F = 0.3604(1-q)^{1/2} \tag{13.74b}$$

$$\left(\frac{a}{2L}\right)_M = \left(\frac{3}{4}\right)^{1/3} 0.1^{1/3}\left[q - \frac{1}{4} + q(1-q) - \frac{1}{6}\right]^{1/3} = 0.4217[q(2-q) - 0.4167]^{1/3} \tag{13.74c}$$

Figure 13.11(b) shows the variation in the normalised dimension over the half-length according to equations (13.74a–c). These confirm that the beam should be designed according to a shear criterion for all of the first region ($0 \le q \le 0.2362$) and for the initial part of the second region ($0.2363 \le q \le 0.436$). However, within the final part of the second region ($0.436 \le q \le 0.5$), the bending design dominates and therefore the optimum profile should be machined with an enlarged central portion. The maximum ordinates by this combined design are 0.3604 and 0.2924 for the ends and centre, respectively. Between these the depth of the square section should vary in the manner shown to achieve an optimum design of encastré beam. Here again Table 13.16 shows that the square section has slightly more weight than the circular section for $s_1 = 0.1$.

(c) $s_1 = 0.01$
Table 13.13 reveals a combined (shear and bending) design in both regions. Correspondingly, the $a/2L$ profiles within the region $0 \le q \le 0.2362$ are, according to equations (13.59a) and (13.58),

$$\left(\frac{a}{2L}\right)_M = \left(\frac{3}{4}\right)^{1/3} 0.01^{1/3}\left[\frac{1}{4} - q + \frac{1}{6} - q(1-q)\right]^{1/3} = 0.1957[0.4167 - q(2-q)]^{1/3} \quad (13.75a)$$

$$\left(\frac{a}{2L}\right)_F = \left(\frac{3\sqrt{3}}{8}\right)^{1/2} \times 0.01^{1/2} \times [2(1-q)]^{1/2} = 0.114(1-q)^{1/2} \quad (13.75b)$$

Within the second region, $0.2362 \le q \le 0.5$, the circular-section profile follows from equations (13.58) and (13.59b):

$$\left(\frac{a}{2L}\right)_F = 0.114(1-q)^{1/2} \quad (13.75b)$$

$$\left(\frac{a}{2L}\right)_M = \left(\frac{3}{4}\right)^{1/3} 0.01^{1/3}\left[q - \frac{1}{4} + q(1-q) - \frac{1}{6}\right]^{1/3} = 0.1957[q(2-q) - 0.4167]^{1/3} \quad (13.75c)$$

Figure 13.13c shows the variation in the normalised dimension over the half-length according to equations (13.75a–c) . Here the combined design applies to both regions within which two transitions occur, $0 \le q_c \le 0.2362$ and $0.2362 \le q_c \le 0.5$. The largest dimensions, provided by bending at the ends and the centre, are 0.1462 and 0.1357, respectively. These diminish into the length but are interrupted by shear-based design sizes as shown. The normalised positions in the length, $q_c = 0.1392$ and 0.3156, at which the transitions occur were found (previously) by equating the governing $a/2L$ equations given above. We see, for this low s_1-value, that the section design is largely controlled by bending but that shear must take over where the bending moment falls to zero. As a consequence, Table 13.16 shows that the square section has less weight than the circular section. Hence, low s_1-values provide a greater weight saving in square-section beams, which we shall now quantify in the following example.

Example 13.4 Design the optimum contours of encastré steel beams, having circular and square cross-sections, that are to support a central concentrated load of 5 kN, together with a uniformly distributed load of 10 kN/m within a length of 0.5 m. Compare the weights of each design. Take $\rho = 7850$ kg/m^3 and $\sigma_y = 300$ MPa.
Here we identify $P = 5$ kN and $w = 10$ kN/m to give in our theory the separated reactions:

$$F_1 = \frac{P}{2} = 2.5 \text{ kN} \quad \text{and} \quad F_2 = \frac{wL}{2} = \frac{10 \times 0.5}{2} = 2.5 \text{ kN}$$

Hence, the load (index) ratio follows as

$$\frac{s_2}{s_1} = \frac{wL}{P} = 1$$

At this load ratio we have found from equation (13.40a) that the bending moment changes sign at position $q_1 = z_1/L = 0.2362$. The structural index is

$$s_1 = \frac{1}{\sigma_y}\frac{F_1}{L^2} = \frac{1}{300 \times 10^6}\frac{2.5 \times 10^3}{0.5^2} = 3.\dot{3} \times 10^{-5}$$

for which bending and shear design will apply to sub-regions of the half-length. The following analyses apply to the design of the circular and square sections.

(a) Circular Section

(i) Region $0 \leq q \leq 0.2362$
Referring to equations (13.44a), the diameters from the bending and shear designs agree at a position q_c in this region when

$$[0.4167 - q_c(2 - q_c)]^{1/3} = 0.2007(1 - q_c)^{1/2}$$

from which a trial solution gives $q_c = 0.2327$.

(ii) *Region* $0.2362 \leq q \leq 0.50$
Referring to equations (13.44b), the diameters from a bending and shear designs agree at a position q_c in this region when

$$[q_c(2 - q_c) - 0.4167]^{1/3} = 0.2007(1 - q_c)^{1/2}$$

and here a trial solution gives $q_c = 0.2398$. From these q_c-values it is convenient to identify three sub-regions where the design alternates between bending and shear.

(iii) *Diameters and Contour Shape*
At each q_c-value given in (i) and (ii) above a transition occurs between diameters provided by the two designs within three sub-regions of the half-length. Within the first sub-region, $0 \leq q \leq 0.2327$, a bending design applies. Equation (13.41a) gives the normalised radius:

$$\left(\frac{d}{2L}\right)_M = 1.0839 s_1^{1/3}[0.4167 - q(2 - q)]^{1/3} = 0.03488[0.4167 - q(2 - q)]^{1/3}$$

which provides the following normalised radii at three selected positions:

$$q = 0, \qquad (d/2L)_M = 0.02606$$
$$q = 0.1164, (d/2L)_M = 0.02031$$
$$q = 0.2327, (d/2L)_M = 6.138 \times 10^{-3}$$

Within the second sub-region, $0.2327 \leq q \leq 0.2398$, a shear design applies, where from equation (13.42), the normalised radii are given as

$$\left(\frac{d}{2L}\right)_F = 1.2126 s_1^{1/2}(1-q)^{1/2} = 7.001 \times 10^{-3}(1-q)^{1/2}$$

which provides radii at three selected positions:

$$q = 0.2327,\ (d/2L)_F = 6.133 \times 10^{-3}$$
$$q = 0.2362,\ (d/2L)_F = 6.119 \times 10^{-3}$$
$$q = 0.2398,\ (d/2L)_F = 6.104 \times 10^{-3}$$

Finally, within the third sub-region, $0.2398 \leq q \leq 0.5$, a bending design applies for radii given by equation (13.41b) as:

$$\left(\frac{d}{2L}\right)_M = 0.03488[q(2-q) - 0.4167]^{1/3}$$

which provides radii at three selected positions:

$$q = 0.2398,\ (d/2L)_M = 6.118 \times 10^{-3}$$
$$q = 0.3699,\ (d/2L)_M = 0.0199$$
$$q = 0.50,\quad (d/2L)_M = 0.0242$$

Allowing for rounding errors, the radii agree at the common points within each sub-region. Figure 13.13a shows the optimised, circular-section profile according to these calculations. The shape revealed is mirrored into the second half to show where material can be removed from the beam without impairing its load carrying capacity. The largest diameter, which determines the sock size, is that for the ends. This is found from $d/2L = 0.02606$ as $d = 26.06$mm, given $L = 0.5$m. The smallest diameter is seen from the above calculations to be 6.11 mm in the region where the shear design applies.

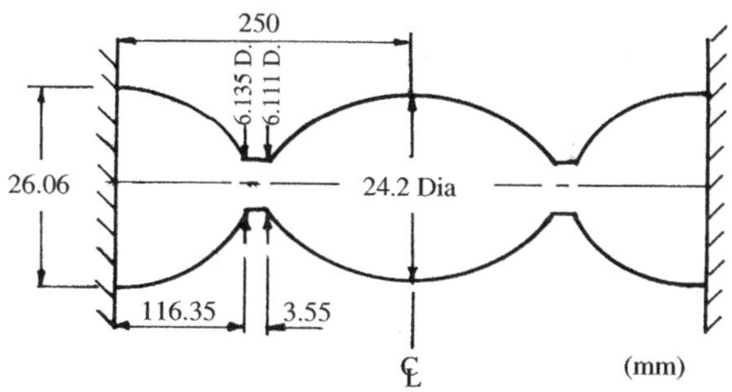

Figure 13.13a Circular section beam contour for load ratio $s_2/s_1 = 1$

(iv) *Weights*
The total, normalised weight for the optimised section is found by adding the contributions from the three sub-regions identified above. This gives:

$$\sum n = \left| n_M \right|_0^{0.2327} + \left| n_F \right|_{0.2327}^{0.2398} + \left| n_M \right|_{0.2398}^{0.5} \tag{i}$$

where n_M and n_F are integrals similar to equations (13.50a,b), (13.52a,b) and (13.53a,b), evaluated as follows:

$$n_M = 3.6905\left(3.\dot{3} \times 10^{-5}\right)^{2/3} \int_0^{0.2327} [0.4167 - q(2-q)]^{2/3} \mathrm{d}q = 2.6189 \times 10^{-4}$$

$$n_F = 2.3094\left(3.\dot{3} \times 10^{-5}\right) \int_{0.2327}^{0.2398} 2(1-q)\mathrm{d}q = 8.3486 \times 10^{-7}$$

$$n_M = 3.6905\left(3.\dot{3} \times 10^{-5}\right)^{2/3} \int_{0.2398}^{0.50} [q(2-q) - 0.4167]^{2/3} \mathrm{d}q = 2.5435 \times 10^{-4}$$

Hence, from equation (i), the normalised weight of the half beam is $\sum n = 5.1717 \times 10^{-4}$, from which the beam's full weight is found:

$$W = 2\rho L^3 \sum n = 2 \times 7850 \times 0.5^3 \times 5.1717 \times 10^{-4} = 1.015 \text{ kg}$$

Note that the weight of a uniform 26.06 mm diameter bar, 0.5 m long, is 2.094 kg. Hence the optimum design provides a 52% weight saving over a uniform section beam.

(b) Square Section

(i) *Region* $0 \le q \le 0.2362$
Referring to equation (13.61a), the square-section sizes from the bending and shear designs agree at a position q_c in this region when

$$[0.4167 - q_c(2 - q_c)]^{1/3} = 0.2251(1 - q_c)^{1/2}$$

from which a trial solution gives $q_c = 0.2312$.

(ii) *Region* $0.2362 \le q \le 0.50$
Referring to equation (13.61b), the section size from bending and shear designs are the same at position q_c in this region when

$$[q_c(2 - q_c) - 0.4167]^{1/3} = 0.2251(1 - q_c)^{1/2}$$

and here a trial solution gives $q_c = 0.2412$.

(iii) *Section Size and Contour Shape*
At each q_c-value given above a transition occurs between the section size provided by the two designs within three sub-regions. Within the first sub-region, $0 \leq q \leq 0.2312$, a bending design applies, where equation (13.59a) gives the normalised half-depth of the square section:

$$\left(\frac{a}{2L}\right)_M = 0.90856 s_1^{1/3}[0.4167 - q(2-q)]^{1/3} = 0.02924[0.4167 - q(2-q)]^{1/3}$$

which provides the following sizes at three selected positions:

$$q = 0, \qquad (a/2L)_M = 0.02184$$
$$q = 0.1156, (a/2L)_M = 0.01707$$
$$q = 0.2312, (a/2L)_M = 5.7874 \times 10^{-3}$$

Within the second sub-region, $0.2312 \leq q \leq 0.2412$, a shear design applies where, from equation (13.58), the normalised dimension is given as:

$$\left(\frac{a}{2L}\right)_F = 1.1398 s_1^{1/2}(1-q)^{1/2} = 6.5804 \times 10^{-3}(1-q)^{1/2}$$

which provides section sizes required at three selected positions:

$$q = 0.2312, (a/2L)_F = 5.7698 \times 10^{-3}$$
$$q = 0.2362, (a/2L)_F = 5.7510 \times 10^{-3}$$
$$q = 0.2412, (a/2L)_F = 5.7321 \times 10^{-3}$$

Within the third sub-section, $0.2412 \leq q \leq 0.5$, a bending design is given by equation (13.59b):

$$\left(\frac{a}{2L}\right)_M = 0.90856 s_1^{1/3}[q(2-q) - 0.4167]^{1/3} = 0.02924[q(2-q) - 0.4167]^{1/3}$$

which provides the section size required at the selected positions:

$$q = 0.2412, (a/2L)_M = 5.7293 \times 10^{-3}$$
$$q = 0.3706, (a/2L)_M = 0.01673$$
$$q = 0.50, \qquad (a/2L)_M = 0.02027$$

Figure 13.13b shows the optimised, square-section profile arising from these calculations. Allowing for rounding errors, $a/2L$ agrees at the common points between the two design criteria. The shape found is mirrored into the second half to reveal where material can be removed from the beam without impairing its load carrying capacity. The largest a, which determines the sock size, is that for the ends. This is calculated from $a/2L = 0.02184$ to be $a = 21.84$mm, given $L = 0.5$m. The smallest depth a is seen from the above

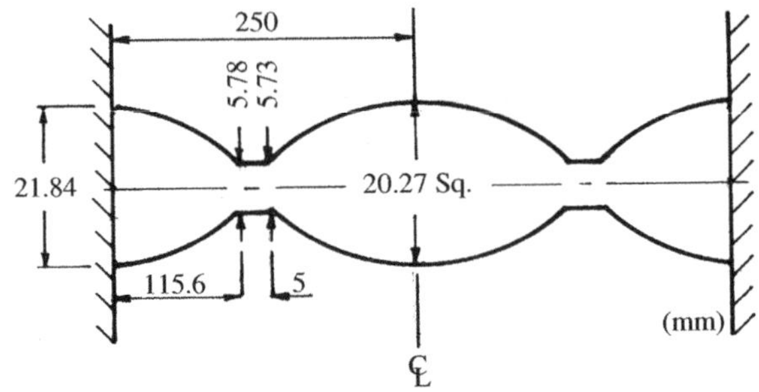

Figure 13.13b Square-section beam contour for load ratio $s_2/s_1 = 1$

calculations to be 5.73 mm in the region where the shear design applies. The half-depths, corresponding to the end and mid-range positions, are consistently less than the radii ($d/2L$) found for the circular section (see Figure 13.13a).

(iv) *Weights*
The total, normalised weight of the square-section beam is again found by adding the contributions from the three sub-regions identified above. This gives

$$\sum n = |n_M|_0^{0.2312} + |n_F|_{0.2312}^{0.2412} + |n_M|_{0.2412}^{0.5} \tag{ii}$$

where n_M and n_F are similar to the integrals in equations (13.69a,b)–(13.71a,b). These are evaluated for each sub-region as:

$$n_M = 3.3019\left(3.\dot{3} \times 10^{-5}\right)^{2/3} \int_0^{0.2312} [0.4167 - q(2-q)]^{2/3} \mathrm{d}q = 2.3604 \times 10^{-4}$$

$$n_F = 2.5981\left(3.\dot{3} \times 10^{-5}\right) \int_{0.2312}^{0.2412} 2(1-q) \mathrm{d}q = 1.323 \times 10^{-6}$$

$$n_M = 3.3019\left(3.\dot{3} \times 10^{-5}\right)^{2/3} \int_{0.2412}^{0.50} [q(2-q) - 0.4167]^{2/3} \mathrm{d}q = 2.2972 \times 10^{-4}$$

Hence, from equation (ii), the normalised weight of the half beam is $\sum n = 4.6708 \times 10^{-4}$, from which the beam's full weight is found:

$$W = 2\rho L^3 \sum n = 2 \times 7850 \times 0.5^3 \times 4.6708 \times 10^{-4} = 0.9167 \text{ kg}$$

Note that a 0.5 m length of uniform bar, with 21.84 mm square section, is 1.872 kg. Hence, the optimum design provides a weight saving of 51% compared to a uniform square bar. Moreover, the optimum square beam provides a further saving of 9.68% compared to the weight (1.015 kg) of an optimum circular section bearing similar loading.

Where weight reduction is paramount, say in the design of vehicles where legislation limits fuel consumption and exhaust emissions, the uniform beam section may be abandoned in favour of the optimum profiles derived here. Numerical machining is capable of reproducing the profiles provided by an optimum design.

References

[1] Bhatt, P. and Nelson, H. M. *Marshal and Nelson's Structures*, 3rd edition, Longman, 1990.
[2] Borresi, A. P. and Sidebottom, O. M. *Mechanics of Materials*, Wiley, 1985.
[3] Ford, H. and Alexander, J. M. *Advanced Mechanics of Materials*, Longman, 1963.
[4] Lancaster, P. R. and Michell, D. *Advanced Solid Mechanics*, Macmillan Press, 1980.
[5] Ugural, A. C and Fenster, S. K. *Advanced Strength and Applied Elasticity*, Elsevier, 1987.
[6] Polakowski, N. H. and Ripling, E. J. *Strength and Structure of Engineering Materials*, Prentice Hall, 1966.
[7] Popov, E. P. *Mechanics of Materials*, Prentice Hall, 1976.
[8] Rees, D. W. A. *Basic Solid Mechanics*, Macmillan, 1997.
[9] Rees, D. W. A. *Mechanics of Solids and Structures*, Imperial College Press, 2000.
[10] Rhodes, J. *Virtual Work and Energy Concepts*, Chatto and Windus, 1975.
[11] Richards, T. H. *Energy Methods in Stress Analysis*, Ellis-Horwood, 1977.
[12] Ryder, G. I. *Strength of Materials*, Macmillan, 1971.
[13] Seely, F. B. and Smith, J. O. *Advanced Mechanics of Materials*, Wiley, 1952.
[14] Stephens, R. C. *Strength of Materials*, Edward Arnold, 1971.
[15] Olsen, G. A. *Elements of Mechanics of Materials*, Allen & Unwin, 4th edition, 1951.
[16] Warnock, F. and Benham, P. *Strength of Materials and Theory of Structures*, Pitman, 1963.

Exercises

13.1 If the weight W and length L of an encastré steel beam are to be restricted to 1 kg and 250 mm respectively, what is the greatest central-concentrated load P a square-section beam is capable of supporting when: (i) the section is uniform; (ii) the section is allowed to vary optimally? Determine the beam's depth versus length profile for (ii). Take $\rho = 7850$ kg/m^3 and $\sigma_y = 300$ MPa.

13.2 Compare the optimum section profiles for 500 mm long, aluminium alloy beams with circular cross-section and fixed ends, required to support 50 kN when this load is: (i) applied at the centre; (ii) uniformly distributed over the length. Repeat the comparison for square-section aluminium alloy beams bearing the same loading. Take $\rho = 2700$ kg/m^3 and $\sigma_y = 100$ MPa.

13.3 Examine the changes necessary to the central-concentrated load, encastré beam analysis given in § 13.2, when the load position divides the length of a circular section beam into $A + B = L$, where A is the distance of the load from the left-hand fixed end.

13.4 Show, for an encastré, circular-section beam, loaded as in Figure 13.9, that the bending moment is zero for a normalised length position $q_1 = z/L = 0.2171$ from each end-fixing, when the load (index) ratio is $wL/P = 10$. Thus, for the first region of the half-length, $0 \leq q \leq 0.2171$, show that shear only controls the section diameter for a central load index $s_1 = (1/\sigma_y)(F_1/L^2) \geq 0.01126$. Similarly, show that for the second region of the half length, $0.2171 \leq q \leq 0.50$, shear only controls the section diameter for $s_1 \geq 4.7875$. Finally, determine the normalised weight of the beam's full length for structural indices $s_1 = 0.01$ and 10. [Answers: $\sum n = 0.1717$ and $\sum n = 138.56$; see Table 13.11]

13.5 Show for an encastré, circular-section beam, loaded as in Figure 13.9, with an index ratio $s_2/s_1 = 0.1$, that the bending moment is zero for a normalised length position $q_1 = z/L = 0.2480$ from each end-fixing. Thus, for the first region of the half-length, $0 \leq q \leq 0.2480$, show that both shear and bending control the section diameter when the central load index $s_1 = (1/\sigma_y)(F_1/L^2) < 0.2184$. Also show for the second region of the half length, $0.2480 \leq q \leq 0.50$, that shear only controls the section diameter when $s_1 \geq 0.2715$. Finally, determine the normalised weight of the beam's full length for structural indices $s_1 = 0.001$ and 1. [Answers: $\sum n = 9.037 \times 10^{-3}$ and $\sum n = 2.4248$; see Table 13.12]

13.6 Show for an encastré, square-section beam, loaded as in Figure 13.9, for a load ratio of 10, that the bending moment is zero for a normalised length position $q_1 = z/L = 0.2171$ from each end fixing. Thus, for the first region of the half-length, $0 \leq q \leq 0.2171$, show that shear only controls the section diameter for the central load parameter $s_1 = (1/\sigma_y)(F_1/L^2) \geq 0.00567$. Similarly, show for the second region of the half-length, $0.2171 \leq q \leq 0.50$, that shear only controls the section diameter for $s_1 \geq 2.4098$. Finally, determine the normalised weight of the beam's full length for $s_1 = 0.01$ and 10. [Answers: $\sum n = 0.1802$ and $\sum n = 155.896$; see Table 13.14]

13.7 Show for the encastré, square-section beam, loaded as in Figure 13.9, for an index ratio $s_2/s_1 = 0.1$, that the bending moment is zero for a normalised length position $q_1 = z/L = 0.2480$ from each end-fixing. Thus, for the first region of the half length, $0 \leq q \leq 0.2480$, show that both shear and bending control the section diameter when the central load index $s_1 = (1/\sigma_y)(F_1/L^2) < 0.1097$. Also show for the second region of the half length, $0.2480 \leq q \leq 0.50$, that a wholly shear design controls the section diameter when $s_1 \geq 0.137$. Finally, determine the normalised weight of the beam's full length for structural indices $s_1 = 0.001$ and 1. [Answer: $\sum n = 8.071 \times 10^{-3}$ and $\sum n = 2.7278$; see Table 13.15]

13.8 Using the results from Exercises 13.4 and 13.6, compare circular- and square-section beam profiles (i.e. $d/2L$ and $a/2L$ versus z/L) for an index ratio $s_2/s_1 = 10$ with two s_1 values, 0.01 and 10.

13.9 Using the results from Exercise 13.5 and 13.7, compare the circular- and square-section beam profiles (i.e. $d/2L$ and $a/2L$ versus z/L) for an index ratio $s_2/s_1 = 0.1$ at two s_1 values, 0.001 and 1.

13.10 Examine the changes necessary to the combined load theory given in § 13.4 when the concentrated load position divides the length of a square-section beam into $A + B = L$, where A is the distance of the load from the left-hand, built-in end.

13.11 Derive plots similar to those shown in Figure 13.2 for an encastré beam with a central-concentrated load for a rectangular cross-section in which the depth is a multiple of the breadth, i.e. $d = \alpha b$. Compare minimum weights within these plots for $\alpha = 1, 2, \ldots, 5$. Assume that lateral and torsional buckling of thin rectangular sections does not occur.

13.12 Derive plots similar to those shown in Figure 13.6 for an encastré beam with uniformly distributed loading for a rectangular cross-section in which the depth is a multiple of the breadth, i.e. $d = \alpha b$. Compare minimum weights within these plots for $\alpha = 1, 2, \ldots, 5$. Assume that lateral and torsional buckling of thin rectangular sections does not occur.

13.13 Determine the weight saving between the two beam loadings shown in Figures 13.2 and 13.6 when the total load carried by each beam is to be 10 kN, for circular, square and rectangular ($d = 2b$) cross-sections in a 1 m length. Take $\rho = 7850$ kg/m^3 and $\sigma_y = 300$ MPa.

14

Plastic Collapse of Beams and Frames

14.1 Introduction

Using a *load factor*, it is possible to design to the ultimate strength of a structure within the region of plasticity beyond the elastic limit. This makes more economical use of the material, given that the plastic strain range can exceed the elastic strain range by many orders of magnitude. For example, the ratio between the ultimate plastic strain and the strain at yield for many lightweight materials lies in the range 12–45. For a heavier construction steel, where the tensile strain to failure is typically 20%, $\varepsilon^P/\varepsilon^E$ may exceed 150! Figure 14.1 shows a typical stress–strain curve (dashed) for such materials in which the two strains are indicated. Clearly, the elastic strain at yield is small, this strain being given by the ratio between the yield stress σ_y and the modulus E:

$$\varepsilon^E = \sigma_y/E \tag{14.1}$$

Taking σ_y and E from the property table in Appendix A, the yield strain for aluminium alloy and mild steel are comparable at $\varepsilon^E = 0.15\%$. In fact, offset (plastic) strains of 0.1% and 0.2% are used to determine the yield stress value when, as is often the case for non-ferrous alloys, the yield point is not obvious within their very gradual transition from elasticity to plasticity. Elastic designs are safe but uneconomical, especially where a design criterion is to be limited only by the strength at failure. Here the dimensions, which apply to the ultimate strength of a material in its plastic condition, provide a lighter structure. Designing for plastic collapse is possible when the spread of plasticity is sufficient to reduce the structure to a mechanism at failure. All the intermediate elastic-plastic stages are ignored in consideration of this collapse condition. That a number of possible collapse mechanisms are possible allows for the selection of the one involving the least loading. Here, the required degree of safety is imposed upon the structure by dividing the least collapse load by a load factor to give the working load. Thus, the load factor is defined to be greater than unity:

$$\text{Load factor} = \frac{\text{Collapse load}}{\text{Working load}} \tag{14.2}$$

Mechanics of Optimal Structural Design: Minimum Weight Structures D. W. A. Rees

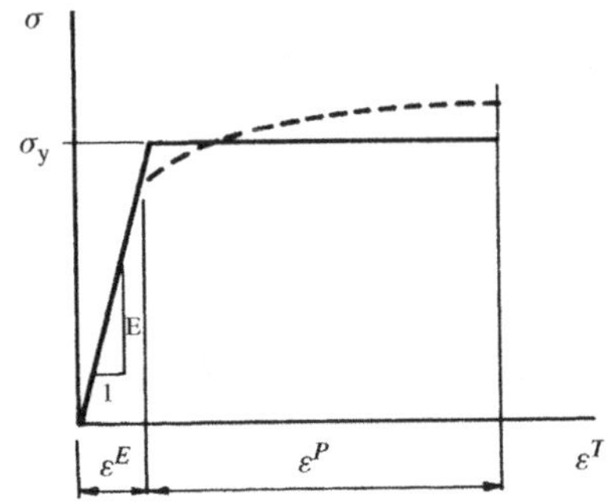

Figure 14.1 Stress–strain plot showing real (dashed line) and idealised plastic behaviour

Two points should be noted. Firstly, a check needs to be made that failure from an alternative buckling mode does not occur before the most likely mechanism of collapse is reached. Secondly, the collapse mechanism cannot limit displacement if this is to be the basis of a design. By implication, the strain and hence displacement must become infinite within every mechanism of collapse. The bases of elastic and plastic designs can be in stark contrast. The separated elastic and plastic stress versus strain plots shown in Figure 14.2 allow for two idealised material models. In each model plasticity is taken to occur without hardening, i.e. under the constant yield (flow) stress σ_y, that does not depend upon the strain. This strain component separation is adopted for an *elastic-perfectly plastic* analysis of structural members, the simplest of these being a tie bar in a plane frame. The idealised, *rigid-plastic* response removes elasticity and ignores the work hardening, as in Figure 14.2b.

14.2 Plane Frames

With the axial loading of a bar of section area A, the respective loads follow directly from the stress in Figure 14.2 as:

$$\text{Elastic Load}: P_E = \sigma \times A \ (\sigma < \sigma_y) \tag{14.3a}$$

$$\text{Plastic Load}: P_P = \sigma_y \times A \ (\sigma = \sigma_y) \tag{14.3b}$$

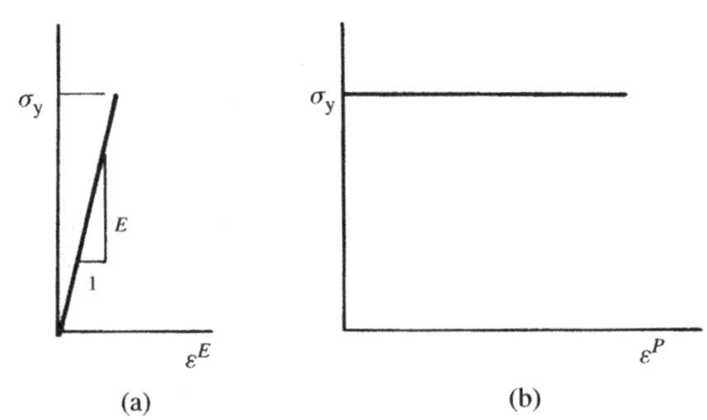

Figure 14.2 (a) Linear elastic and (b) idealised, perfectly plastic responses

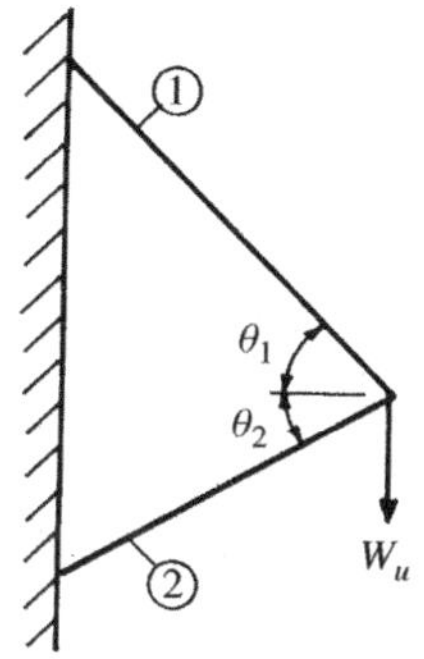

Figure 14.3 Two-bar cantilever frame

The ultimate load at collapse is the plastic load P_P. In this simple example the stresses σ and σ_y in equations (14.3a,b) are uniformly distributed within the cross-section, whether it be circular, square, tubular, etc. When we extend this to the statically determinate two-bar frame, shown in Figure 14.3, it is only necessary to apply joint equilibrium to determine its collapse load. Thus, assuming bar tension at the joint where the load is applied in Figure 14.3,

$$P_1 \sin\theta_1 - P_2 \sin\theta_2 - W_u = 0 \tag{14.4a}$$

$$-P_1 \cos\theta_1 - P_2 \cos\theta_2 = 0 \tag{14.4b}$$

where, from equations (14.3b),

$$P_1 = \sigma_y \times A_1 \text{ and } P_2 = \sigma_y \times A_2 \tag{14.5a,b}$$

Hence, from equations (14.4a) and (14.5a,b), the plastic collapse load for the structure becomes:

$$W_u = \sigma_y(A_1 \sin\theta_1 - A_2 \sin\theta_2) \tag{14.6a}$$

If both bars are to fail simultaneously, having reached their collapse loads at the same instant, 1 in tension and 2 in compression, equations (14.4b) and (14.5a,b) show further that their cross-sectional areas should be in the ratio

$$A_1/A_2 = \cos\theta_2/\cos\theta_1 \tag{14.6b}$$

Normally equation (14.6b) would not be satisfied so that as one bar fails first there would be a delay as the stress redistributes to attain the collapse condition within the second bar. Here, the compression member is assumed to be short such that failure occurs from plastic collapse and not by Euler buckling.

The collapse condition within a statically indeterminate frame also follows from joint equilibrium. The additional compatibility condition, normally required for the determination of elastic bar forces, is not required to establish this limiting plastic condition. For example, the three-bar plane frame in Figure 14.4 is said to have one degree of

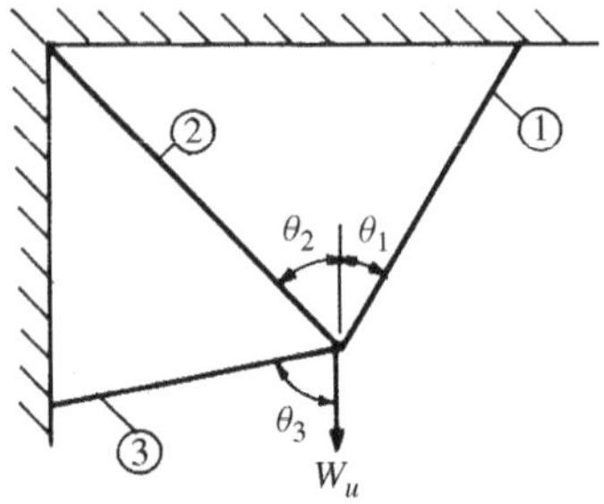

Figure 14.4 Three-bar corner frame

redundancy, as it has one bar more than is the essential number of bars required for the frame to support the load. When all members become fully plastic, the bar forces are known to be

$$P_i = \sigma_y \times A_i (i = 1, 2, 3) \tag{14.7}$$

As before, the collapse load follows directly from applying joint equilibrium equations:

$$P_1 \cos\theta_1 + P_2 \cos\theta_2 - P_3 \cos\theta_3 - W_u = 0 \tag{14.8a}$$

$$P_1 \sin\theta_1 - P_2 \cos\theta_2 - P_3 \sin\theta_3 = 0 \tag{14.8b}$$

Substituting equations (14.7) into equation (14.8a) gives

$$W_u = \sigma_y (A_1 \cos\theta_1 + A_2 \cos\theta_2 - A_3 \cos\theta_3) \tag{14.9a}$$

Substituting equations (14.7) into the horizontal force equilibrium equation (14.8b) connects areas to their inclinations:

$$A_1 \sin\theta_1 - A_2 \sin\theta_2 + A_3 \sin\theta_3 = 0 \tag{14.9b}$$

Again, if equation (14.9b) is not satisfied then as one bar fails first there would be a delay as stress redistributes in the remaining bars to attain the collapse condition at the ultimate level.

14.3 Beam Plasticity

In the case of beam bending, the stress redistribution within the section depth must be accounted for. The initially, linear elastic stress distribution is modified as zones of plasticity penetrate inwards from the top and bottom surfaces, as shown in Figure 14.5. For the fully elastic stress distribution (Figure 14.5a), the surface stress $\sigma_{max} < \sigma_y$. With the onset of yield (Figure 14.5b), σ_y is first reached at the outer surfaces to produce the 'fully elastic' section. The corresponding moment at first yield acts as a reference value for loads and moments in plastic analyses. Here, we are concerned only with the fully plastic collapse condition for the beam, shown in Figure 14.5d, and not the intermediate elastic-plastic (partially yielded) state, shown in Figure 14.5c.

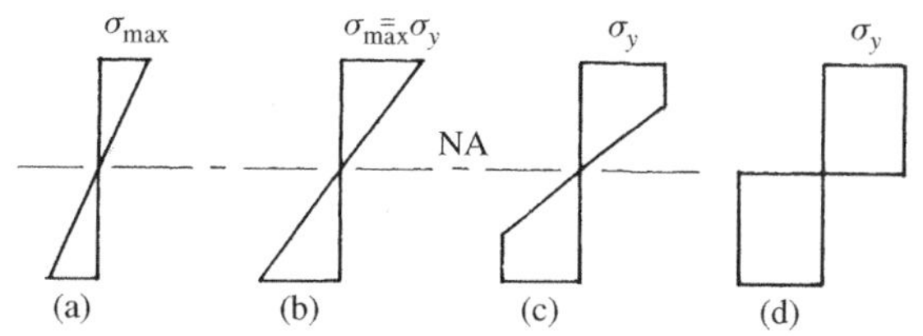

Figure 14.5 Plastic penetrations in an asymmetric cross-section of a beam

14.3.1 Plastic Moment

Of particular use when finding the collapse loading for a beam is the fully plastic moment M_P for its cross-section. This moment is found from the two rectangular, tensile and compressive stress distributions aside the neutral axis in Figure 14.5d. Since there can be no net axial force upon the beam, each stress distribution exerts an equal and opposite force to the other. With each distribution having a constant stress magnitude σ_y, this equilibrium requirement becomes

$$\sum F_x = 0 : \quad \Rightarrow \quad A_1\sigma_y - A_2\sigma_y = 0 \tag{14.10a}$$

from which $A_1 = A_2$, i.e. the position of the neutral axis divides the area of the section equally. Equation (14.10a) will always apply to the collapse condition, irrespective of the section shape. For axisymmetric sections, such as circular, square, rectangular and I-sections, the neutral axis (NA) bisects their areas. It follows that the neutral axis remains fixed in position in each part of Figure 14.5. This applies to all sections whose two halves are symmetrical about the neutral axis. However, for asymmetric sections, such as T-, U- and L-sections, once the initial yield point is reached, the penetration of further plasticity occurs with a shift in the neutral axis. The maximum shift of this axis lies between its initial position through the centroid of the section, and its final position at collapse, where it divides the section area equally. Here, each rectangular stress distribution exerts the same force $F = A_1\sigma_y = A_2\sigma_y$ with a separation h that is the distance between the individual centroids of each half-area. Thus, for the T-section shown in Figure 14.6a, the fully plastic moment M_P is identified with the couple taken from the fully plastic stress

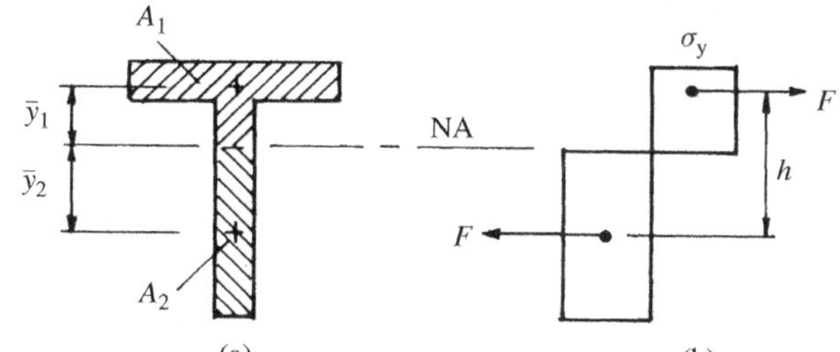

Figure 14.6 Fully plastic T-section

distribution diagram (see Figure 14.6b),

$$M_P = Fh = \frac{A\sigma_y}{2}(\overline{y}_1 + \overline{y}_2) \tag{14.10b}$$

where $\overline{y}_1$ and $\overline{y}_2$ locate the centroids of equal areas A_1 and A_2 ($= A/2$) lying on either side of the neutral axis as shown.

14.3.2 Shape Factor

This is a plastic property of the cross-section defined as

$$\text{Shape factor} = \frac{\text{Fully plastic moment}}{\text{Moment at first yield}}, \quad \Rightarrow \quad S = \frac{M_P}{M_y} \tag{14.11a,b}$$

where we have seen (Figure 14.6b) that M_P is the couple identified in equation (14.10b). Similarly, M_y may be found either as the couple within the fully elastic stress distribution (see Figure 14.5b), or, equivalently, from elastic bending theory. When the stress within one outer surface (asymmetric section) or both outer surfaces (symmetric section) has reached the yield value, this theory gives

$$M_y = \sigma_y I / y \tag{14.11c}$$

where I is the second moment of area for the fully elastic section about its centroidal axis and y is the distance from this axis to the outer, yielding surface. For example, in a rectangular section of breadth b and depth d, equations (14.10b) and (14.11c) become

$$M_P = \frac{bd\sigma_y}{2}\left(\frac{d}{2} + \frac{d}{2}\right) = \frac{bd^2\sigma_y}{4}, \quad M_y = \frac{\sigma_y(bd^3/12)}{d/2} = \frac{bd^2\sigma_y}{6} \tag{14.12a,b}$$

Substituting equations (14.12a,b) into equation (14.11b) gives the shape factor $S = 3/2$, i.e. the collapse moment required is 50% greater than that moment at which yielding begins. Here S is a numerical value though it may be an expression, as the following example shows:

Example 14.1 Find the shape factor for: (a) a solid circular section of radius R; (b) a tubular section with outer and inner radii R_o and R_i, respectively (see Figure 14.7).

The collapse moments require that the centroidal positions $\overline{y}$ are located within each half-section, as shown in Figure 14.7. The neutral axis, i.e. the x-axis, remains in position to divide the section area A equally throughout plastic penetration as is necessary to attain a fully plastic collapse condition.

(a) *Solid Circular Section*
In Figure 14.7a, the corresponding fully plastic and elastic moments follow from equations (14.10b) and (14.11c):

$$M_P = 2F\overline{y} = A\sigma_y\overline{y}, \quad M_y = \frac{\sigma_y(\pi R^4/4)}{R} = \frac{\pi R^3\sigma_y}{4} \tag{i, ii}$$

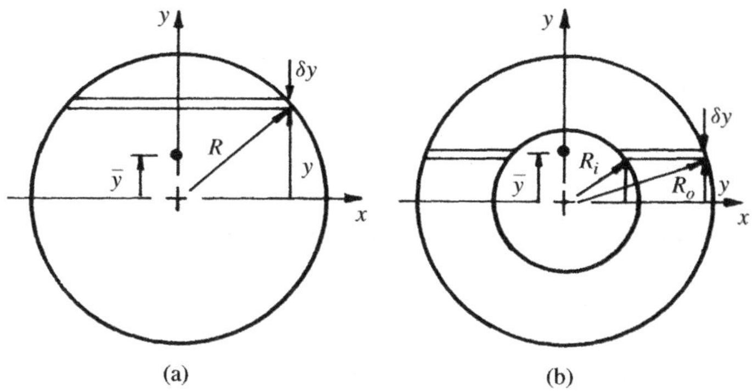

Figure 14.7 Solid circular and tubular beam sections

where A is the whole section area and $\bar{y}$ follows from the first moment of the half-section area about the x-axis. From Figure 14.7a, this is given by

$$\frac{A}{2} \times \bar{y} = \int_A y\,\mathrm{d}A \qquad \text{(iii)}$$

The area of the elemental strip is $\delta A = 2x\delta y$ and its first moment of area is $y\delta A = 2xy\delta y$. Now, as $x = \sqrt{(R^2 - y^2)}$, substituting into equation (iii) provides the following equation in $\bar{y}$:

$$\frac{\pi R^2}{2} \times \bar{y} = 2\int_y xy\,dy = 2\int_0^R y(R^2 - y^2)^{1/2}\,\mathrm{d}y$$

Carrying out the integration leads to

$$\frac{\pi R^2}{2} \times \bar{y} = -\frac{2}{3}\left|(R^2 - y^2)^{3/2}\right|_0^R = \frac{2R^3}{3}$$

from which

$$\bar{y} = \frac{4R}{3\pi} \qquad \text{(iv)}$$

Equations (i) and (iv) give the plastic moment:

$$M_P = \pi R^2 \sigma_y \times \frac{4R}{3\pi} = \frac{4R^3\sigma_y}{3} \qquad \text{(v)}$$

Dividing equations (v) and (ii) gives the shape factor for a solid circular section:

$$S = \frac{M_P}{M_y} = \frac{(4R^3\sigma_y/3)}{(\pi R^3\sigma_y/4)} = \frac{16}{3\pi} = 1.698 \qquad \text{(vi)}$$

(b) *Tubular Section*

For the tubular section shown in Figure 14.7b, the corresponding fully plastic and fully elastic moments follow from equations (14.10b) and (14.11c):

$$M_P = 2F\overline{y} = A\sigma_y\overline{y}, \quad M_y = \frac{\sigma_y\left[\pi\left(R_o^4 - R_i^4\right)/4\right]}{R_o} = \frac{\pi\left(R_o^4 - R_i^4\right)\sigma_y}{4R_o} \qquad \text{(vii, viii)}$$

where $\overline{y}$ follows from equation (iii). Here we need to subtract first moment of area for the inner semi-circle from that for the outer semi-circle, so that equation (iii) becomes

$$\frac{\pi(R_o^2 - R_i^2)}{2} \times \overline{y} = 2\left[\int_0^{R_o} y\left(R_o^2 - y^2\right)^{1/2} \, dy - \int_0^{R_i} y\left(R_i^2 - y^2\right)^{1/2} \, dy\right]$$

Upon integrating, we obtain

$$\frac{\pi}{2}(R_o^2 - R_i^2) \times \overline{y} = \frac{2}{3}(R_o^3 - R_i^3)$$

whence

$$\overline{y} = \frac{4(R_o^3 - R_i^3)}{3\pi(R_o^2 - R_i^2)} \qquad \text{(ix)}$$

Equations (vii) and (ix) give the plastic moment:

$$M_P = \pi(R_o^2 - R_i^2)\sigma_y \times \frac{4(R_o^3 - R_i^3)}{3\pi(R_o^2 - R_i^2)} = \frac{4\sigma_y(R_o^3 - R_i^3)}{3} \qquad \text{(x)}$$

Dividing equation (x) by equation (viii) gives the shape factor,

$$S = \frac{M_P}{M_y} = \frac{4\sigma_y(R_o^3 - R_i^3)/3}{\pi\sigma_y(R_o^4 - R_i^4)/(4R_o)} = \frac{16R_o(R_o^3 - R_i^3)}{3\pi(R_o^4 - R_i^4)} \qquad \text{(xi)}$$

in which equation (vi) is a special case of equation (xi) for $R_i = 0$.

14.3.3 Collapse Mechanism

Plastic collapse of a structure will occur from the formation of hinges at the positions where M_P has been reached. One, two or three hinges may be necessary to form a mechanism that results in the collapse of a part or the whole of a structure. A cantilever requires only a single hinge to form at its fixing, but a beam, with both ends built in, will require three hinges, one at each fixing and another within its span. In contrast, a simply supported beam will collapse from a single hinge within its span as the ends rotate on their supports. Generally, two hinges form with the collapse of components of a structure, e.g. the beam in a portal frame and the individual spans within a continuous beam or a propped cantilever. In all these cases the hinge rotations and load point deflection result in a triangular configuration of the collapsing elements of a structure. Collapse may be treated as a mechanism in which one of the following energy conservation principles is applied.

Work and Energy Interchange

A convenient method for finding the collapse loading of beams and frames is to equate the work done W by the applied forces to the internal energy E that arises within the collapse mechanism. Generally, W is the product of applied forces and their in-line displacements and E is the product of the plastic moments and hinge rotations. For N forces and n hinges,

$$W = E \quad \Rightarrow \quad \sum_{i=1}^{N} F_i \Delta_i = \sum_{i=1}^{n} M_i \theta_i \tag{14.13a,b}$$

For collapsing beams and other rigid structures, we shall see that the internal energy arises solely from rotations at hinges that form under the fully plastic collapse moment.

Principle of Virtual Work

An alternative, but equivalent, approach to the energy equation (14.13b) is to treat collapse as a rigid mechanism under an equilibrium system of N coplanar, non-concurrent forces $\mathbf{F}_i$ $(i = 1, 2, \ldots, N)$. The equilibrium conditions for the x, y components of $\mathbf{F}_i$ are

$$\sum_{i=1}^{N} (F_x)_i = 0, \quad \sum_{i=1}^{N} (F_y)_i = 0, \quad \sum_{i=1}^{N} (M_q)_i = 0 \tag{14.14a,b,c}$$

where the force components in the x- and y-directions sum to zero and the moments of these forces must also sum to zero at any point q. Since the in-line displacements and rotations that occur in a collapsing structure should not alter the equilibrium equations (14.14a,b,c), we say that they are *virtual* and write them as Δ^v and θ^v, respectively. Hence the equilibrium system of *real* forces and moments does zero *virtual* work. That is,

$$\sum_{i=1}^{N} F_i \Delta_i^v = 0 \quad \text{and} \quad \sum_{i=1}^{N} M_i \theta_i^v = 0 \tag{14.15a,b}$$

where F_i are the magnitudes of $\mathbf{F}_i$. Combining equations (14.15a,b) leads to a useful form for finding collapse loading:

$$\sum_{i=1}^{N} F_i \Delta_i^v = \sum_{i=1}^{N} M_i \theta_i^v = 0 \tag{14.16}$$

In equation (14.16), θ_i^v refers to the rotations that occur at each hinge under the collapse moment and Δ_i^v refers to the in-line deflections that occur beneath the applied forces. Figure 14.8 shows a simplified non-concurrent force system, with each force having undergone a virtual in-line displacement for a rigid translation. In the absence of rotation of the body equation (14.15a) applies, for $N = 3$:

$$\sum_{i=1}^{3} F_i \Delta_i^v = F_1 \Delta_1^v + F_2 \Delta_2^v + F_3 \Delta_3^v = 0$$

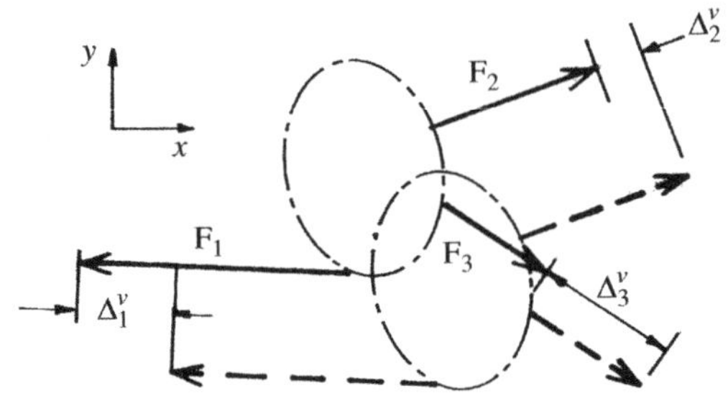

Figure 14.8 Virtual displacements in an equilibrium force system

Simultaneously, equations (14.14a,b) apply to the x, y components of F_i. To apply either equation (14.13b) or (14.16) to the collapse condition a geometry must be assumed. Where a number of collapse mechanisms are possible, the true collapse load is that which minimises the internal energy term $\sum M_i\theta_i$. Consideration of the following collapsing structures illustrates this.

14.4 Collapse of Simple Beams

14.4.1 Simple Cantilever

Figure 14.9 shows a cantilever with a concentrated end-load P applied at a distance L from the encastré fixing. The maximum moment in the cantilever, which occurs at the fixing, is $M = PL$. Hence, it follows that collapse will occur at the fixing from hinge rotation θ under the fully plastic moment M_P. Therefore, to avoid collapse, the beam section is chosen such that $M_P \geq PL$. The energy equation (14.13b) confirms this collapse condition as the beam hinges at its fixing under M_P, with a rotation θ and an end-displacement δ, as shown. This gives

$$P\delta = M_P\theta \tag{14.17}$$

and, since $\theta = \delta/L$, we see again that $M_P > PL$ if collapse is to be avoided. Example 14.1 showed that M_P will depend upon both the section dimensions and the yield stress of the section material. In practice, M_P must be chosen from a limited range of standard sections available in structural materials. The material becomes a crucial factor where the final choice of M_P rests with a minimum weight requirement.

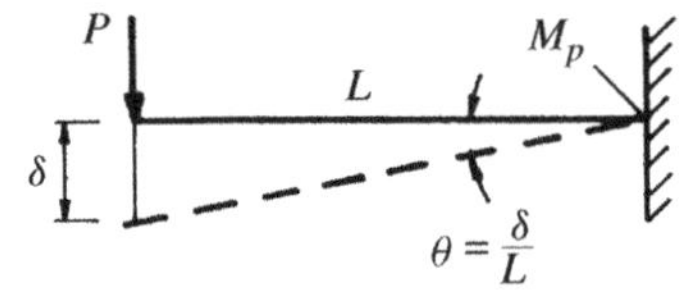

Figure 14.9 Collapse mechanism in a simple cantilever

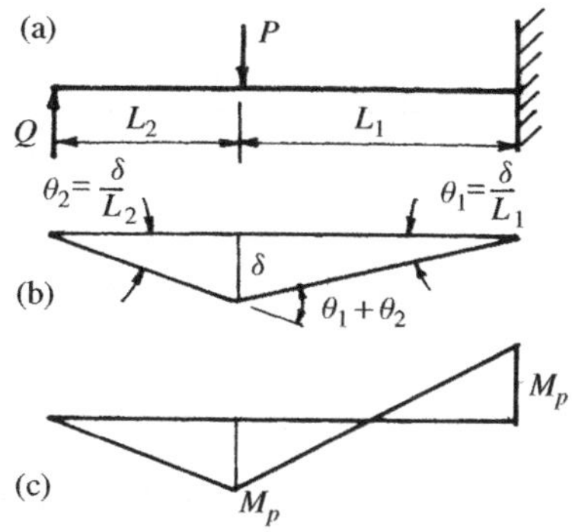

Figure 14.10 Collapse mechanism for an end-propped cantilever

14.4.2 Propped Cantilever, Concentrated Applied Load

In its elastic analysis, this problem (see Figure 14.10a) is statically indeterminate. This means that the prop force Q is required to satisfy force and moment equilibrium equations (14.14a–c) whilst remaining compatible with the specified displacement at the prop position. However, the solution to plastic collapse (see Figure 14.10b,c) requires only that the equilibrium condition is met. In particular, with the load point displacement δ and end rotations θ_1 and θ_2 (Figure14.10b), within the two hinges required for collapse under M_P (Figure 14.10c), equation (14.13b) becomes

$$P\delta = M_P\theta_1 + M_P(\theta_1 + \theta_2) \tag{14.18}$$

Initially, where displacements and rotations are small, the limbs of the collapse mechanism are assumed to be straight. Hence, the geometry in Figure 14.10b gives the rotations in radians:

$$\theta_1 = \delta/L_1 \text{ and } \theta_2 = \delta/L_2 \tag{14.19a,b}$$

Substituting equations (14.19a,b) into equation (14.18),

$$P\delta = \frac{M_P\delta}{L_1} + M_P\left(\frac{\delta}{L_1} + \frac{\delta}{L_2}\right) \tag{14.20a}$$

from which the plastic moment for the beam cross-section must satisfy

$$M_P \geq P\left(\frac{L_1L_2}{L_1 + 2L_2}\right) \tag{14.20b}$$

In the case of a rectangular section, say, then $M_P = bd^2\sigma_y/4$ in equation (14.20b). The collapse mechanism requires that two hinges form each under M_P. For this condition to be met, the moments redistribute as plasticity penetrates the depth at each hinge position, finally achieving the moment diagram at collapse shown in Figure 14.10c.

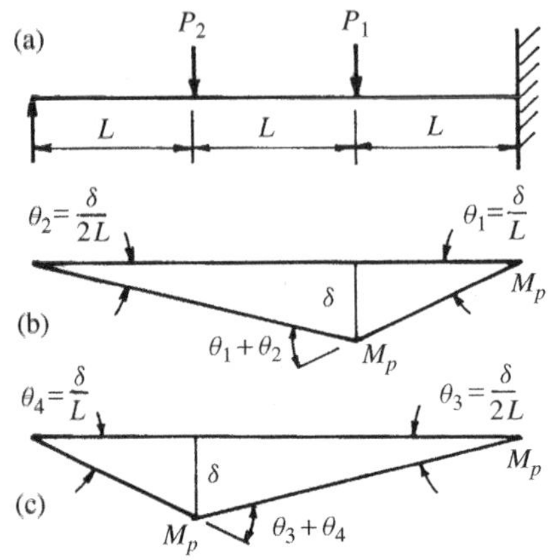

Figure 14.11 Possible collapse modes in a propped cantilever

14.4.3 Propped Cantilever, Concentrated Applied Loads

With more than one concentrated load, all possible collapse modes must be considered. Figure 14.11a shows a propped cantilever with two equally spaced loads. Figures 14.11b,c give two possible collapse modes I and II, each having formed a hinge at the fixing, with another hinge formed either beneath load P_1 (Figure 14.11b) or beneath P_2 (Figure 14.11c).

For mode I (Figure 14.11b), let the load point displacement be δ with end-rotations θ_1 and θ_2. Within the two hinges required for the collapse, equation (14.13b) becomes

$$P_1\delta + P_2\delta/2 = M_P\theta_1 + M_P(\theta_1 + \theta_2) \tag{14.21a}$$

Again, straight limbs are assumed initially within the collapse mechanism, where displacements and rotations are small. The geometry then gives the rotations in radians as

$$\theta_1 = \frac{\delta}{L} \quad \text{and} \quad \theta_2 = \frac{\delta}{2L} \tag{14.21b,c}$$

Substituting equations (14.21b,c) into equation (14.21a):

$$P_1\delta + P_2\frac{\delta}{2} = \frac{M_P\delta}{L} + M_P\left(\frac{\delta}{L} + \frac{\delta}{2L}\right) \tag{14.22a}$$

Cancelling δ, the plastic moment for the cross-section becomes

$$M_P = \frac{2}{5}\left(P_1 + \frac{P_2}{2}\right)L \tag{14.22b}$$

Within the two hinges required for the collapse mode II, (Figure 14.11c) equation (14.13b) becomes

$$P_1\delta/2 + P_2\delta = M_P\theta_3 + M_P(\theta_3 + \theta_4) \tag{14.23a}$$

in which

$$\theta_3 = \frac{\delta}{2L} \quad \text{and} \quad \theta_4 = \frac{\delta}{L} \tag{14.23b,c}$$

Substituting equations (14.23b,c) into equation (14.23a):

$$P_1\frac{\delta}{2} + P_2\delta = \frac{M_P\delta}{2L} + M_P\left(\frac{\delta}{2L} + \frac{\delta}{L}\right) \tag{14.24a}$$

Cancelling δ, the plastic moment for the cross-section gives

$$M_P = \frac{1}{2}\left(\frac{P_1}{2} + P_2\right)L \tag{14.24b}$$

The final choice of section is based upon that mode which provides the greater of the two plastic moments from equations (14.22b) and (14.24b).

14.4.4 Propped Cantilever, Uniformly Distributed Loading

Finding the uniformly distributed collapse loading w for a propped cantilever in Figure 14.12a is a made difficult with the uncertain location of the second hinge required. Clearly, this hinge must lie within the span length l when, as in Figure 14.12a, the cantilever is propped at its end. Therefore, we may assume the mechanism of collapse shown in Figure 14.12b. As with the previous cantilever beams, the fully plastic collapse moment is to be expressed in terms of the loading w and the beam length l.

Firstly, the position z_o of the second hinge is to be determined. For this, the principle of virtual work equation (14.16) is applied to the virtual displacement Δ^v and virtual rotations θ^v and ψ^v, shown in Figure 14.12b:

$$M_P\theta^v + M_P(\theta^v + \psi^v) = w\int_0^l v^v \,\mathrm{d}z \tag{14.25}$$

Note that the work term $F\delta^v$ in equation (14.16) has been replaced by the integral in which v^v is the virtual deflection at position z and $w \times \mathrm{d}z$ is the elemental force at this position. As the integrand in equation (14.25) is the strip area, the integral is the area

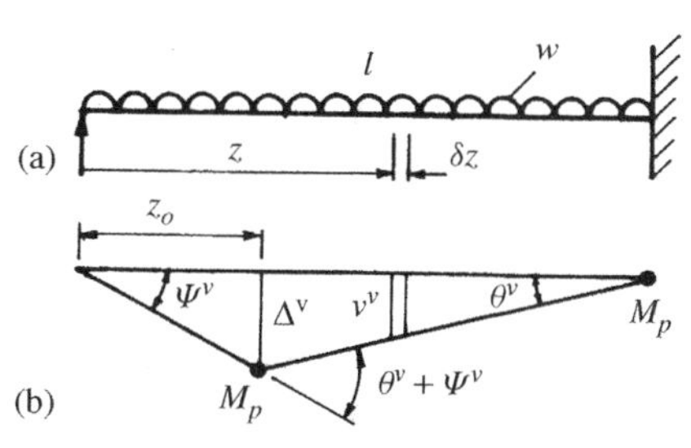

Figure 14.12 Collapse mechanism in a propped cantilever beam carrying distributed loading

enclosed by the mechanism:

$$\int_0^l v^v \, dz = \frac{l\Delta^v}{2} \tag{14.26a}$$

The rotation ψ^v is found from the geometrical relationship

$$\Delta^v = (l - z_o)\theta^v = z_o\psi^v \tag{14.26b}$$

from which

$$\psi^v = \frac{(l - z_o)\theta^v}{z_o} \tag{14.26c}$$

Substituting equations (14.26a–c) into equation (14.25), the virtual rotations cancel to give

$$M_P = \frac{wlz_o(l - z_o)}{2(l + z_o)} \tag{14.27a}$$

Differentiating equation (14.27a) to find the z_o-value which minimises M_P,

$$\frac{dM_P}{dz_o} = z_o^2 + 2z_ol - l^2 = 0 \tag{14.27b}$$

for which $z_o = 0.4141l$. Hence, to avoid collapse, it follows from equation (14.27a) that the plastic moment for the beam section must obey

$$M_P \geq 0.086\, wl^2 \tag{14.27c}$$

14.5 Encastré Beams

14.5.1 Single Concentrated Force

Figure 14.13a shows a concentrated load P offset from the centre of an encastré beam. Clearly, three hinges are necessary for collapse. That all hinges must form under M_P suggests an obvious collapse mechanism in Figure 14.13b. In the elastic case, the moments at the three hinge positions are unequal. Therefore, in loading beyond the elastic limit collapse will occur when the greatest moment at one section position attains the value of M_P. Thereafter, a stress redistribution occurs such that plastic hinges form at their remaining two positions.

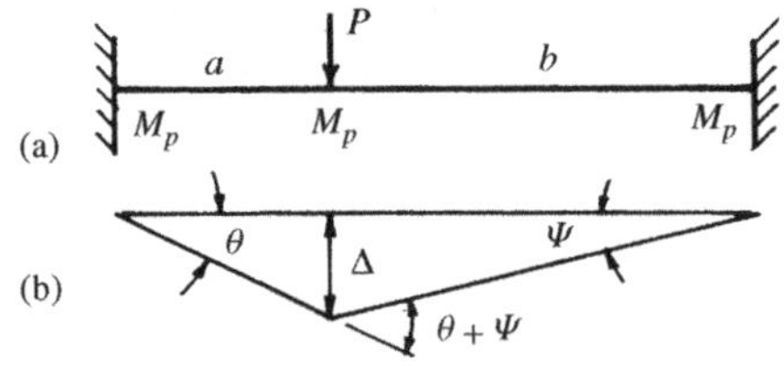

Figure 14.13 Collapse mechanism for an encastré beam supporting a concentrated load

Applying the work–energy equation (14.13b) to this mechanism,

$$M_P\theta + M_P(\theta + \psi) + M_P\psi = P\Delta \tag{14.28a}$$

Equation (14.28a) may be expressed in terms of one rotation (say, θ) by the relation

$$\Delta = a\theta = b\psi \tag{14.28b}$$

Combining equations (14.28a,b) gives

$$M_P\theta + M_P\theta(1 + a/b) + M_P\theta(a/b) = P(a \times \theta) \tag{14.29a}$$

The rotation cancels within equation (14.29a) to give an expression for the plastic moment M_P required for this beam section as

$$M_P \geq \frac{Pa}{2(1 + a/b)} \quad \Rightarrow \quad M_P \geq \frac{Pab}{2l} \tag{14.29b}$$

When the load is central $a = b = l/2$ in equation (14.29b), giving $M_P = Pl/8$.

14.5.2 Uniformly Distributed Loading

The beam in Figure 14.14a collapses when the plastic moment is attained in each of three hinges required for the collapse mechanism (Figure 14.14b). Applying the work–energy equation (14.13b) to this mechanism,

$$2M_P\theta + M_P(2\theta) = 2w\int_0^{l/2} v\,\mathrm{d}z = \frac{4w\Delta}{l}\int_0^{l/2} z\,\mathrm{d}z \tag{14.30a}$$

in which $v = 2z\Delta/l$. Integrating equation (14.30a) and then substituting $\Delta = l\theta/2$ gives

$$4M_P\theta = wl\Delta/2 = (wl/2)(l\theta/2) \tag{14.30b}$$

from which the moment M_P required for collapse of this beam is

$$M_P = wl^2/16 \tag{14.30c}$$

Hence, in knowing M_P for a given section, we can see what uniformly distributed loading will cause a given length of encastré beam to collapse.

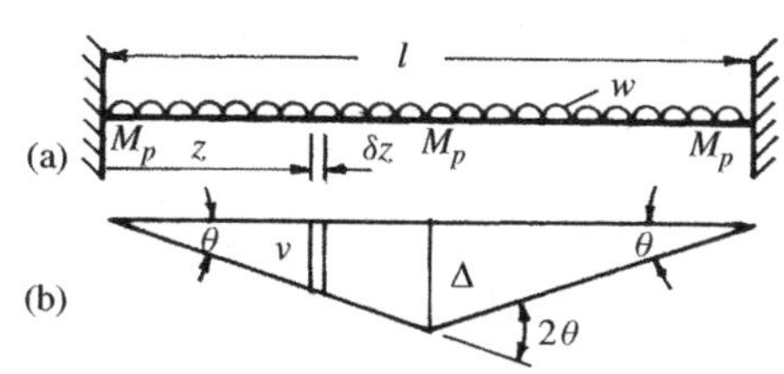

Figure 14.14 Collapse mechanism for an encastré beam carrying uniformly distributed loading

14.5.3 Concentrated and Distributed Loading

Figure 14.15a shows a combination of the previous two cases in which the third hinge is taken to form beneath the offset concentrated load P. Taking the collapse mechanism (Figure 14.15b) in two halves with their origins for z as shown, the work–energy equation (14.13b) becomes

$$M_P\theta + M_P(\theta + \psi) + M_P\psi = P\Delta + w\left(\int_0^a v\,\mathrm{d}z + \int_0^b v\,\mathrm{d}z\right) \tag{14.31a}$$

Substituting $v = \Delta \times z/a$ into the first integral, $v = \Delta \times z/b$ into the second integral and using the relationship $\Delta = a\theta = b\psi$ leads to

$$2M_P\theta\left(1 + \frac{a}{b}\right) = Pa\theta + \frac{w\theta}{2}\left[a^2 + b^2\left(\frac{a}{b}\right)\right] \tag{14.31b}$$

which gives

$$M_P \geq \frac{Pab}{2(a+b)} + \frac{wab}{4}, \quad \Rightarrow \quad M_P \geq \frac{ab}{2}\left(\frac{P}{l} + \frac{w}{2}\right) \tag{14.31c}$$

where $l = a + b$. When the concentrated loading is central $a = b = l/2$ and equation (14.31c) gives

$$M_P \geq \frac{Pl}{8} + \frac{wl^2}{16} \tag{14.32}$$

Equation (14.32b) restricts the load combination to a collapse condition in which equations (14.29b) and (14.30c) appear as a sum. Thus, if w is constant then the limiting P follows from equation (14.32) as

$$P \leq \frac{8M_P}{l} - \frac{wl}{2}$$

where, for a given material, M_P may be regarded as a property of the cross-section and *not* the loading. For example, with a rectangular section, $M_P = \sigma_y bd^2/4$.

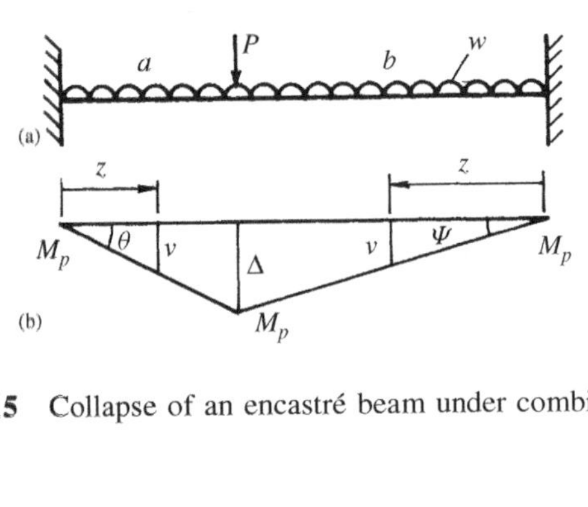

Figure 14.15 Collapse of an encastré beam under combined loading

14.6 Continuous Beams

Multi-span beams that rest upon three of more supports require multiple collapse mode solutions, as with the propped cantilever in Figure 14.11. In a continuous beam each span is assumed to have a constant cross-section but the section size may vary from span to span. The objective is to find the least collapse loading and thereby minimise weight from among the collapse moment M_P expressions across all modes. A number of different collapse modes will appear to be possible, with each depending upon a particular combination of plastic hinges that reduce the beam to a suitable mechanism for collapse. All such collapse mechanisms need to be considered as in the following example.

14.6.1 Two-Bays with Varying Section

Figure 14.16a shows a stepped, two-span beam in which the plastic collapse moment for span B is twice that of span A, i.e. $M_{P_B} = 2M_{P_A}$. The beam is fixed at one end of span B and rests upon two simple supports: one at the end of span A and the other between the

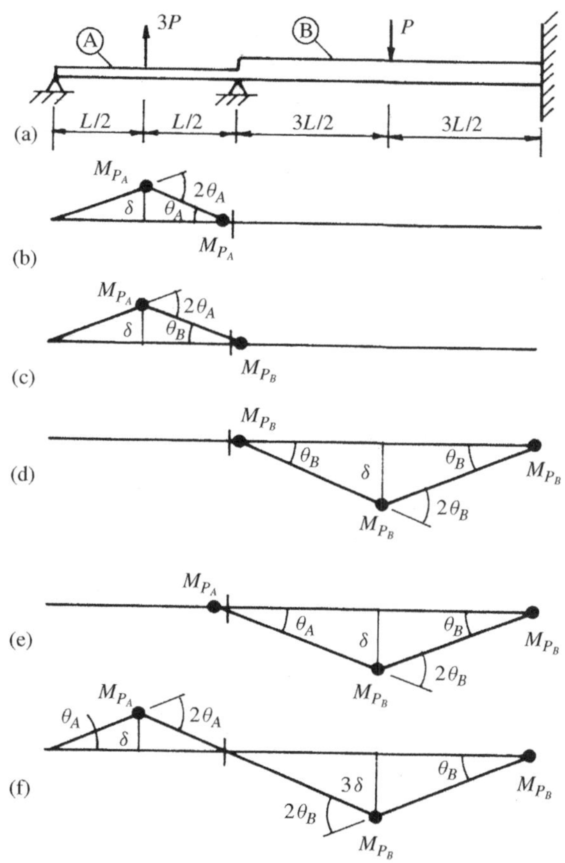

Figure 14.16 Continuous beam showing five possible collapse modes (b)–(f)

spans. If we wish to determine the collapse load P then five possible modes of failure, shown in Figure 14.16b–f, allow equation (14.13b) to be applied.

In Figure 14.16b, M_{P_A} is reached at the centre and at one end of span A, while span B is unaffected. Therefore,

$$M_{P_A}(2\theta_A) + M_{P_A}\theta_A = 3P\delta$$

and, with $\theta_A = 2\delta/L$,

$$\frac{M_{P_A}}{PL} \geq 0.5 \tag{14.33a}$$

Figure 14.16c is similar to Figure 14.16b in that M_{P_A} is reached within span A as before, but the second hinge now forms under M_{P_B} in span B. This gives

$$M_{P_A}(2\theta_A) + M_{P_B}\theta_B = 3P\delta$$

and, with $\theta_A = \theta_B = 2\delta/L$,

$$\frac{4}{3}\frac{M_{P_A}}{PL} + \frac{2}{3}\frac{M_{P_B}}{PL} \geq 1.0 \tag{14.33b}$$

In Figure 14.16d, the collapse moment M_{P_B} is reached at the centre and at both ends of span B, while span A is unaffected. This gives

$$4M_{P_B}\theta_B = P\delta$$

and, with $\theta_B = 2\delta/3L$,

$$\frac{M_{P_B}}{PL} \geq 0.375 \tag{14.33c}$$

Figure 14.16e is similar to Figure14.16d, except that one hinge M_{P_A} is reached within span A together with two further hinges in span B. This gives

$$M_{P_A}\theta_A + 3M_{P_B}\theta_B = P\delta$$

and, with $\theta_A = \theta_B = 2\delta/3L$,

$$\frac{2}{3}\frac{M_{P_A}}{PL} + 2\frac{M_{P_B}}{PL} \geq 1.0 \tag{14.33d}$$

Finally, in Figure 14.16f, one hinge forms within span A and two hinges form within span B: one at its centre and another at its fixing. No hinge forms at the central support, this acting as a pivot for the failure mechanism to occur. Therefore, by equation (14.13b),

$$2M_{P_A}\theta_A + 3M_{P_B}\theta_B = 3P\delta - P(-3\delta)$$

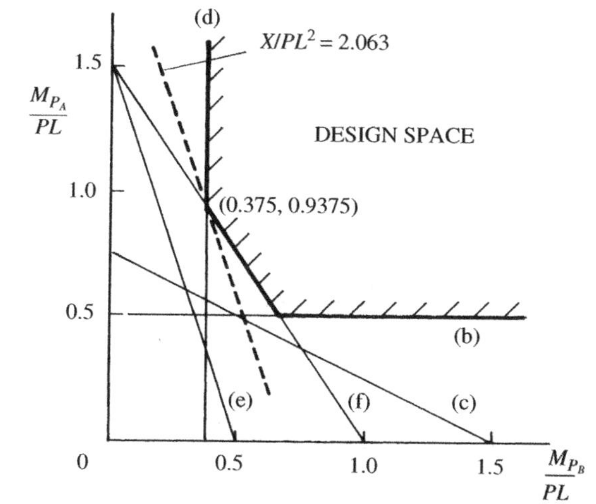

Figure 14.17 Design space derived from collapse modes (b)–(f)

and, with $\theta_A = \theta_B = 2\delta/L$,

$$\frac{2}{3}\frac{M_{P_A}}{PL} + \frac{M_{P_B}}{PL} \geq 1.0 \tag{14.33e}$$

The inequalities in equations (14.33a–e) provide conditions for the combination of moments in potential collapse modes to provide a safe design. It is seen that when failure modes (b)–(f) n equations (14.33a–e) are taken with equality, they appear as straight lines in the plot of M_{P_A}/PL versus M_{P_B}/PL shown in Figure 14.17. In particular, the modes corresponding to equations (14.33b,d,f) define a 'design space' for this beam. Within this space any combination of plastic moments for spans A and B will avoid collapse. More usually, this space is used to locate a point at which the beam's weight can be minimised. To do this the following weight function is employed.

14.6.2 Weight Function

It is possible to connect weight to M_P, theoretically, given that M_P depends upon the section's dimensions. For example, equation (14.12a) applies to a rectangular section. When the section's breadth and depth are increased proportionately as $b/d = \alpha$, the beam's weight per unit length is

$$w = \rho bd = \alpha\rho d^2 \tag{14.34a}$$

and the plastic moment becomes

$$M_P = \frac{bd^2\sigma_y}{4} = \frac{\alpha\sigma_y d^3}{4} \tag{14.34b}$$

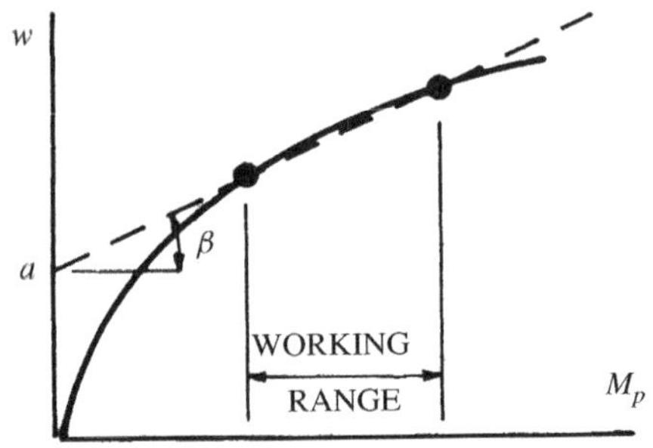

Figure 14.18 Parabolic dependence between w and M_P

Eliminating d between equations (14.34a,b) provides the relationship required:

$$w = \left(\frac{16\rho^3\alpha}{\sigma_y^2}\right)^{1/3} M_P^{2/3} = C_1 M_P^{2/3} \quad (14.34c)$$

Experiments by Neal [1] showed the index of M_P to be 0.6, in good agreement with equation (14.34c). However, much of the parabolic curve described by equation (14.34c) would not apply in a design practice where only a limited range of standard sections are available upon which to minimise weight, i.e. the 'working range', as shown in Figure 14.18. It is usually sufficiently accurate to employ a linear approximation to this working range:

$$w = a + bM_P \quad (14.35a)$$

where a is the intercept and $b = \tan\beta$ is the gradient, as shown. An error of 1% arose from applying equation (14.35a) to a restricted range of sections in which the largest section was twice the size of the smallest [1]. When equation (14.35a) is applied to a continuous beam consisting of a number of different span lengths, the total weight becomes

$$W = \sum(wL) = \sum(a + bM_P)L = a\sum L + b\sum M_P L \quad (14.35b)$$

Within equation (14.35b), $a\sum L$ is a constant and so W is a minimum when $\sum M_P L$ is a minimum. The summation term $\sum M_P L$ is known as the weight function X:

$$X = \sum M_P L \quad (14.36a)$$

Applying equation (14.36a) to our continuous beam example,

$$X = M_{P_A} L + M_{P_B}(3L) \quad (14.36b)$$

and dividing equation (14.36b) by PL^2,

$$\frac{M_{P_A}}{PL} + 3\frac{M_{P_B}}{PL} = \frac{X}{PL^2} \quad (14.37a)$$

When lines of increasing X/PL^2 are superimposed upon the design space, the required value of the weight function applies to that line which first makes contact with the design space boundary (see Figure 14.17). This occurs where the (dashed) line contacts the intersection point between modes (d) and (f) as shown. The coordinates of the intersection point are found from the simultaneous solution to equations (14.33c,e) . This gives

$$\left(\frac{M_{P_A}}{PL}\right)_{\text{opt}} = 0.9375, \quad \left(\frac{M_{P_B}}{PL}\right)_{\text{opt}} = 0.375 \qquad (14.37\text{b,c})$$

Substituting equations (14.37b,c) into equation (14.37a),

$$\left(\frac{X}{PL^2}\right)_{\text{opt}} = \left(\frac{M_{P_A}}{PL}\right)_{\text{opt}} + 3\left(\frac{M_{P_B}}{PL}\right)_{\text{opt}} = 2.063 \qquad (14.37\text{d})$$

The true minimum weight is achieved when the beam section satisfies equations (14.37b–d) but, again, the limited number of standard sections available may not render this solution possible. However, we may choose a section to approach the minimum weight most closely in the manner of Figure 14.19.

For example, should six sections I–VI be available in the chosen material, their M_P-values (normalised with PL) are plotted along both axes and extended over the design space as shown. Given that the minimum weight point within the design space has been established by following the analysis above, this construction reveals the closest proximity to that point with safe intersections between particular two-section combinations. Three safe contenders for minimum weight appear in Figure 14.18, one of which is a choice of section V for bay A and section II for bay B. Two further combinations alternate sections III and IV between each bay. Each combination would need to be investigated from the modifications to the values in equations (14.37b–d) arising from these section dimensions being slightly different from the ideal values. Note that the combination of sections I and V, lying just outside the design space, is likely to be at risk of failure under mode (f).

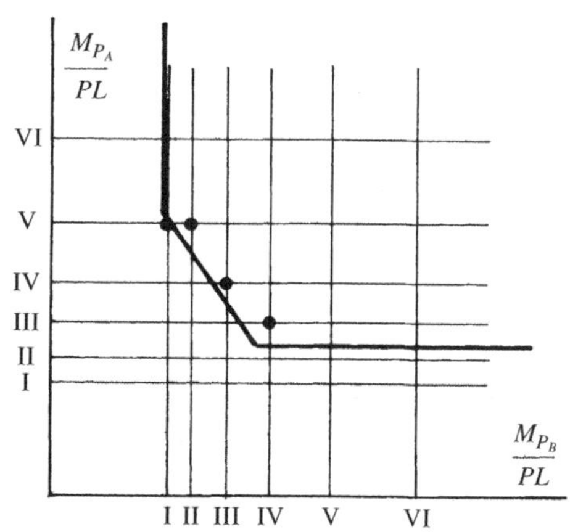

Figure 14.19 Minimum weight approximation within a restricted section range

14.7 Portal Frames

The analysis of collapse in a portal frame involves the application of equation (14.13b) to all possible failure modes [2]. Let the frame ABCDE in Figure 14.20a carry both vertical and horizontal loading. When the supports at A and E are hinged they carry no moments. Consequently, collapse can occur in the horizontal beam (Figure 14.20b) or the stanchions may sway (Figure 14.20c). In Figure 14.20d, collapse arises from a combination of collapse and sway.

Where different cross-sections exist between the beam and stanchions, the corner collapse will occur in the adjacent limb with the lower collapse moment. Assuming that the stanchions have a 50% greater moment capacity than the beam, all the hinges will form on the beam side of the corners under M_P, as shown. Failure from each mechanism occurs from the hinge formation at: B, C and D in Figure 14.20b; B and D in Figure 14.20c; and C and D in Figure 14.20d. Applying equation (14.13b) to each mode in Figure 14.20b–d, respectively:

$$M_P\theta + M_P(\theta + \psi) + M_P\psi = W\Delta \tag{14.38a}$$

$$2M_P\theta = 2W\delta \tag{14.38b}$$

$$M_P(\theta + \psi) + M_P(\theta + \psi) = W\Delta + 2W\delta \tag{14.38c}$$

Substituting $\psi = \theta/2, \delta = L\theta/2$ and $\Delta = L\theta/3$ into equations (14.38a,b,c) leads to the respective collapse loads: $9M_P/L, 2M_P/L$ and $9M_P/4L$. The least of these loads shows that a sway collapse (Figure 14.20c) would occur. Note that these results reveal that the collapse moment ($1.5M_P$) for each stanchion is not reached. Therefore, with a hinge at the base of each stanchion it would be more economical to select a constant section, with a single collapse moment M_P, for the frame. If this structure had the base of each

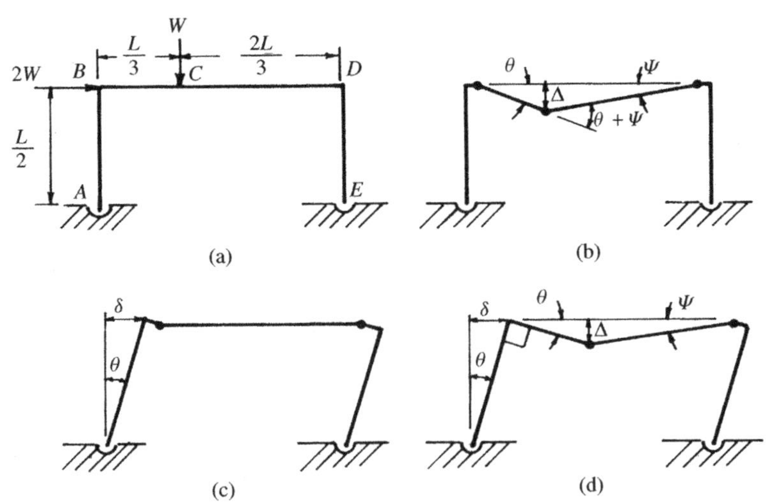

Figure 14.20 Collapse modes in a portal frame with hinged-base stanchions

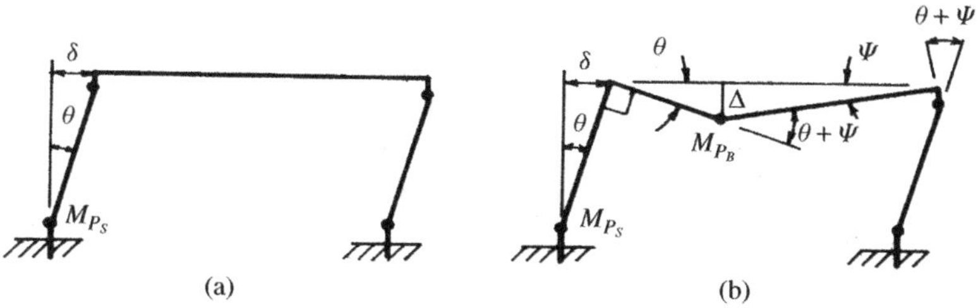

Figure 14.21 Collapse modes in a portal frame with fixed-base stanchions

stanchion built in, further hinges would form at A and E in Figure 14.20c,d, for which equations (14.38b,c) are modified:

$$2M_P\theta + 2(3M_P/2)\theta = 2W\delta \tag{14.39a}$$

$$M_P(\theta+\psi) + M_P(\theta+\psi) + 2(3M_P/2)\theta = W\Delta + 2W\delta \tag{14.39b}$$

The respective collapse loads are $W = 5M_P/L$ and $9M_P/2L$, the lesser indicating that a combined mode of failure will occur.

The encastré frame can be designed more economically when both the sway and the combined mode failures occur simultaneously. Let $M_{P_S}(< M_{P_B})$ be the modified collapse moment for the stanchions under similar loading (as in Figure 14.20a). The two collapse modes of interest are as shown in Figure 14.21. Equations (14.39a,b) are modified to

$$2M_{P_S}\theta + 2M_{P_S}\theta = 2W\delta \tag{14.40a}$$

$$2M_{P_S}\theta + M_{P_B}(\theta+\psi) + M_{P_S}(\theta+\psi) = W\Delta + 2W\delta \tag{14.40b}$$

Using $\psi = \theta/2$, $\delta = L\theta/2$ and $\Delta = L\theta/3$, equation (14.40a) gives $M_{P_S} = WL/4$. Substituting this into equation (14.40b) gives $M_{P_B} = 11WL/36$. Therefore $W = 36M_{P_B}/11L$ and $M_{P_S} = 9M_{P_B}/11$, which confirms that the stanchions in Figure 14.20a (i.e. with $M_{P_S} = 3M_{P_B}/2$) were over-designed. The following example illustrates how an optimum solution applies to a structure of uniform section and thereby will minimise its weight. A more general graphical interpretation of the optimum design of a collapsing structure then follows.

Example 14.2 The portal frame shown in Figure 14.22 has encastré stanchions and carries both vertical and horizontal loads W and P applied at the centre and right-hand end of the horizontal beam. Determine the most likely collapse mode when $H = 3L/4$ and $P = W/3$. Assume that the cross-section is uniform throughout.

Figure 14.22 shows all the possible collapse modes in which the uniform section implies a similar collapse moment for both stanchions and the beam. The work–energy interchange (14.13b) or the virtual work principle (14.16) apply to each mode as follows:

(a) For central load failure from three beam hinges (see Figure 14.22a),

$$M_P\theta + M_P(2\theta) + M_P\theta = W\Delta$$

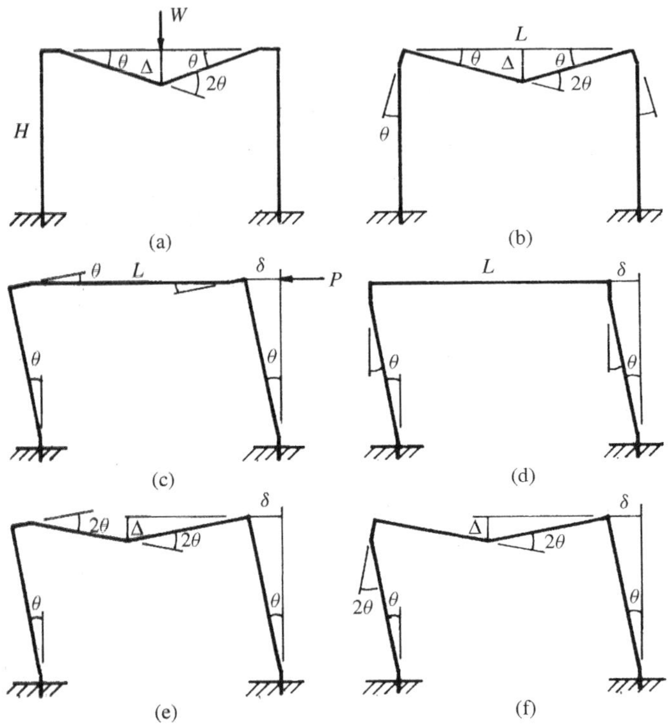

Figure 14.22 Collapse modes in a portal frame with encastré stanchions

Substituting $\Delta = \theta L/2$,

$$4M_P\theta = W\theta L/2$$

from which

$$W = 8M_P/L$$

(b) For central load failure from two stanchion hinges and one beam hinge (see Figure 14.22b),

$$2M_P\theta + M_P(2\theta) = W\Delta$$

The geometry reveals that θ is preserved at each hinge within the rigid collapse mechanism shown. Substituting $\Delta = \theta L/2$,

$$4M_P\theta = W\theta L/2$$

from which

$$W = 8M_P/L$$

We note that the vertical collapse load will differ between (a) and (b) only when M_P for the beam differs from that for the stanchion.

(c) For side-sway failure with two stanchion hinges and two beam hinges (see Figure 14.22c),

$$2M_P\theta + 2M_P\theta = P\delta$$

Substituting $P = W/3$ and $\delta = H\theta = (3L/4)\theta$,

$$4M_P\theta = (W/3)(3L/4)\theta$$

from which

$$W = 16M_P/L$$

(d) For side-sway failure with four stanchion hinges (see Figure 14.22d),

$$2M_P\theta + 2M_P\theta = P\delta$$

Substituting $P = W/3$ and $\delta = H\theta = (3L/4)\theta$,

$$4M_P\theta = (W/3)(3L/4)\theta$$

from which

$$W = 16M_P/L$$

Again, we note that the collapse load will differ between (c) and (d) only when M_P varies.

(e) For combined failure with side-sway arising from two stanchion hinges and two beam hinges (see Figure 14.22e) working clockwise from the lower left stanchion fixing,

$$M_P\theta + M_P(2\theta) + M_P(2\theta) + M_P\theta = P\delta + W\Delta$$

Substituting $P = W/3$, $\delta = (3L/4)\theta$ and $\Delta = \theta L/2$,

$$6M_P\theta = (W/3)(3L/4)\theta + W(\theta L/2)$$

from which

$$W = 8M_P/L$$

(f) For combined failure with side sway arising from three stanchion hinges and one beam hinge (see Figure 14.22f),

$$M_P\theta + M_P(2\theta) + M_P(2\theta) + M_P\theta = P\delta + W\Delta$$

Substituting $P = W/3$, $\delta = (3L/4)\theta$ and $\Delta = \theta L/2$,

$$6M_P\theta = (W/3)(3L/4)\theta + W(\theta L/2)$$

from which

$$W = 8M_P/L$$

Once again, we note that the collapse load would differ between (e) and (f) only when M_P for the beam differs from that of the stanchion. Of the possible six failure mechanisms (a)–(f), the lowest collapse load $W = 8M_P/L$ appears consistently between four of them, (a), (b), (e) and (f). Hence, as these four modes would occur simultaneously, we may conclude that the structure appears to be economically designed.

14.7.1 Economical Design

With different load combinations and length ratios it may be necessary to adjust the cross-section geometry to achieve a similar optimisation to that found in the previous example. Thus, in Figure 14.23a we write M_{P_S} for each stanchion, M_{P_B} for the beam and M'_P for the corners, with loads and lengths in the ratios $\alpha = H/L$ and $\beta = P/W$. The six collapse modes, given in Example 14.2, may now be reduced to the three shown in Figure 14.23b–d. For Figure 14.23b we have

$$M'_P\theta + M_{P_B}(2\theta) + M'_P\theta = W\Delta$$

Substituting $\Delta = \theta L/2$,

$$M'_P + M_{P_B} = WL/4 \tag{14.41a}$$

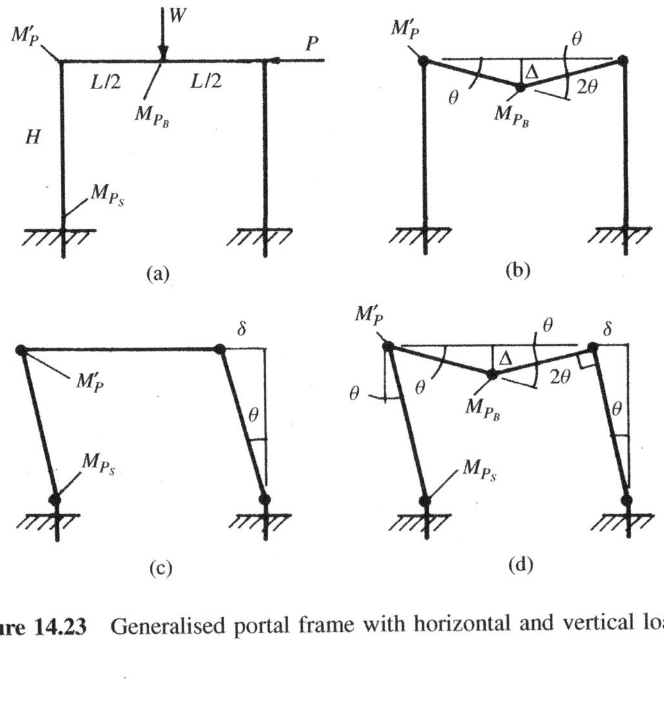

Figure 14.23 Generalised portal frame with horizontal and vertical loading

For Figure 14.23c we have

$$2M_{P_S}\theta + 2M'_P\theta = P\delta$$

Substituting $P = \beta W$ and $\delta = H\theta = (\alpha L)\theta$:

$$M'_P + M_{P_S} = \alpha\beta WL/2 \tag{14.41b}$$

Finally, for Figure 14.23d we have

$$M_{P_S}\theta + M'_P(2\theta) + M_{P_B}(2\theta) + M_{P_S}\theta = P\delta + W\Delta$$

Substituting $P = \beta W$ and $\delta = H\theta = (\alpha L)\theta$ and $\Delta = \theta L/2$

$$M'_P + M_{P_B} + M_{P_S} = WL/4 + \alpha\beta WL/2 \tag{14.41c}$$

Material is used most economically in the construction of the frame when equation (14.41a) is satisfied simultaneously with either equations (14.41b) or (14.41c). This requires the assumption that $M'_P = M_{P_S}$, as the following example shows. Normally, with the imposition of a load factor from equation (14.2), the collapse loading will never be reached. Safe section dimensions follow from the collapse moments within the shape factor definition given in equation (14.11a).

Example 14.3 The portal frame shown in Figure 14.23 has length and load ratios $\alpha = 5/8$ and $\beta = 3/8$, respectively. The beam is 8 m long and carries a central, vertical, working load of 100 kN. Determine the collapse moments required for the beam and stanchion sections when they are to operate under a load factor of 2. Find the diameters required when the frame is constructed from (a) solid bar, (b) tube with a diameter ratio $K = d_o/d_i = 3/2$, in both cases of mild steel having a yield stress $\sigma_y = 300$ MPa.

It follows from equation (14.2) that collapse occurs when the working loads are doubled. Hence, the right-hand sides of equations (14.41a,b) become

$$WL/4 = 2 \times 100 \times 8/4 = 400 \text{ kN m}$$

$$\alpha\beta WL/2 = \frac{5}{8} \times \frac{3}{8} \times 2 \times 100 \times \frac{8}{2} = 187.5 \text{ kN m}$$

Equations (14.41a,c) are satisfied when

$$M_{P_S} = \alpha\beta WL/2 = 187.5 \text{ kN m}, \quad M'_P = M_{P_S}, \quad M_{P_B} = 400 - 187.5 = 212.5 \text{ kN m}$$

Equations (14.41b,c) are satisfied when

$$M_{P_B} = WL/4 = 400 \text{ kN m}, \quad M'_P = M_{P_S}, \quad M_{P_S} = \alpha\beta WL/4 = 93.75 \text{ kN m}$$

The first solution provides near-matching sections for the beam and stanchions which are preferable for construction in bar and tube.

(a) Solid Bar

Diameters are found from the shape factor in which equation (14.11b) gives $M_y = M_P/S$, where M_y applies to elastic bending theory:

$$\frac{M_y}{\pi d^4/64} = \frac{M_P/S}{\pi d^4/64} = \frac{2\sigma_y}{d} \tag{i}$$

The bar diameter follows from equation (i):

$$d = \left(\frac{32M_P}{\pi \sigma_y S}\right)^{1/3} \tag{ii}$$

For a solid circular section $S = 1.7$, so that equation (ii) gives the stanchion diameter

$$d_S = \left(\frac{32M_{P_S}}{\pi \sigma_y S}\right)^{1/3} = \left(\frac{32 \times 187.5 \times 10^6}{\pi \times 300 \times 1.7}\right)^{1/3} = 155.3 \text{ mm}$$

and the beam diameter

$$d_B = \left(\frac{32M_{P_B}}{\pi \sigma_y S}\right)^{1/3} = \left(\frac{32 \times 212.5 \times 10^6}{\pi \times 300 \times 1.7}\right)^{1/3} = 161.9 \text{ mm}$$

Commercial solid sections will not conform exactly to these dimensions, but the analysis allows for the selection of the nearest stock sizes. The British Standard [3] accommodates a suitable selection as it provides hot rolled steel bars to 150 mm diameter, bright ferrous steel round bars to 200 mm, alloy steels, including stainless steel, to 280 mm.

(b) Circular Tube

The design in tube is complicated by its shape factor expression (see Example 14.1):

$$S = \frac{M_P}{M_y} = \frac{16d_o(d_o^3 - d_i^3)}{3\pi(d_o^4 - d_i^4)} = \frac{16K(K^3 - 1)}{3\pi(K^4 - 1)} \tag{iii}$$

Equation (iii) may be combined with the moment at yield expression:

$$\frac{M_y}{\pi(d_o^4 - d_i^4)/64} = \frac{2\sigma_y}{d_o} \tag{iv}$$

With the given diameter ratio $K = d_o/d_i = 3/2$, equation (iv) gives the inner diameter as

$$d_i = \left[\frac{32KM_P}{\pi \sigma_y S(K^4 - 1)}\right]^{1/3} = \left[\frac{6M_P}{\sigma_y(K^3 - 1)}\right]^{1/3} \tag{v}$$

When $M_P = M_{P_S}$, equation (v) gives the stanchion diameters

$$d_i = \left[\frac{6 \times 187.5 \times 10^6}{300(1.5^3 - 1)}\right]^{1/3} = 116.5 \text{ mm}, \quad d_o = 174.8 \text{ mm}$$

With $M_P = M_{P_B}$, the beam diameter is

$$d_i = \left[\frac{6 \times 212.5 \times 10^6}{300(1.5^3 - 1)}\right]^{1/3} = 121.4 \text{ mm}, \quad d_o = 182.1 \text{ mm}$$

The British Standard [3] is less accommodating to each d_o within its range of steel tubes. Nominal bore sizes are set at 100, 125 and 150 mm with corresponding outer diameters of 114, 140 and 165 mm. Alternative British Standards [4, 5] give a greater range of tubular sections in steel and aluminium quoting outer diameters and thicknesses. It is clear that a redesign would be necessary when selecting from the K-ratios available to standard sections. The reader may wish to investigate whether a better basis for selection is provided by the section modulus ($z = I/y$) as this property is listed in all standards (see Exercise 14.4).

14.7.2 Minimum Weight Solution

We may employ the virtual work principle to design a beam or frame to have minimum weight. To do this, even the most unlikely failure mode has to be accounted for. Consider a portal frame, bearing an inclined load, with the general dimensions shown in Figure 14.24a. The resolved force components and the ratio between them are

$$P = Q\sin\gamma, \quad W = Q\cos\gamma, \quad \alpha = \tan\gamma = P/W, \quad \text{for } 0 \le \gamma \le 90^\circ \quad (14.42a,b,c)$$

The following ratios apply to the lengths:

$$\beta = \frac{H}{L}, \quad \varepsilon = \frac{b}{a}, \quad a = \frac{L}{1+\varepsilon}, \quad b = a\varepsilon = \frac{L\varepsilon}{1+\varepsilon} \quad (14.43a\text{-}d)$$

There are six collapse modes, as shown in Figure 14.24b–g. With the deflections Δ and δ indicated, the rotation θ and collapse moments M_P are assigned appropriate subscripts B and S for the beam and stanchions. The application of equation (14.16) to each mode leads to a system of six non-dimensional moment equations. For Figure 14.24b,

$$M_{P_B}\theta_{1_B} + M_{P_B}(\theta_{1_B} + \theta_{2_B}) + M_{P_B}\theta_{2_B} = W\Delta \quad (14.44a)$$

Using the relationships

$$\Delta = a\theta_{1_B} = b\theta_{2_B} = \frac{L\theta_{1_B}}{1+\varepsilon}, \quad \theta_{2_B} = \frac{a\theta_{1_B}}{b} = \frac{\theta_{1_B}}{\varepsilon} \quad (14.44b,c)$$

leads to

$$\frac{M_{P_B}}{QL} = \frac{\varepsilon\cos\gamma}{2(1+\varepsilon)^2} \quad (14.44d)$$

For Figure 14.24c,

$$M_{P_S}\theta_{1_S} + M_{P_B}(\theta_{1_S} + \theta_{2S}) + M_{P_S}\theta_{2S} = (Q\cos\gamma)\Delta \quad (14.45a)$$

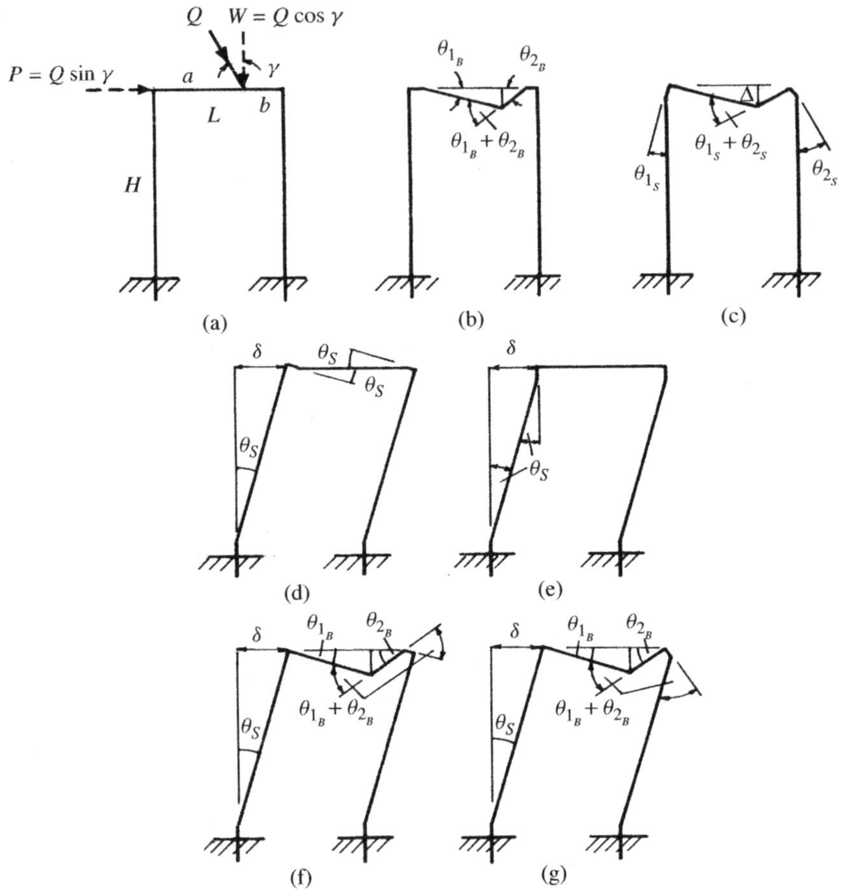

Figure 14.24 Possible failure modes in a portal frame with inclined load

and using the relationships

$$\Delta = a\theta_{1_S} = b\theta_{2_S} = \frac{L\theta_{1_S}}{1+\varepsilon}, \quad \theta_{2_S} = \frac{a\theta_{1_S}}{b} = \frac{\theta_{1_S}}{\varepsilon} \tag{14.45b,c}$$

leads to

$$\frac{M_{P_S}}{QL} + \frac{M_{P_B}}{QL} = \frac{\varepsilon \cos\gamma}{(1+\varepsilon)^2} \tag{14.45d}$$

For Figure 14.24d (with $\theta_S = \theta_B$),

$$2M_{P_S}\theta_S + 2M_{P_B}\theta_S = P\delta \tag{14.46a}$$

and using the relationships

$$P = Q\sin\gamma, \quad \delta = H\theta_S = \beta L\theta_S \tag{14.46b,c}$$

leads to

$$\frac{M_{P_S}}{QL} + \frac{M_{P_B}}{QL} = \frac{\beta}{2}\sin\gamma \tag{14.46d}$$

For Figure 14.24e,

$$P\delta = 4M_{P_S}\theta_S \tag{14.47a}$$

where, from equations (14.46b,c),

$$\frac{M_{P_S}}{QL} = \frac{\beta}{4}\sin\gamma \tag{14.47b}$$

For Figure 14.24f,

$$2M_{P_S}\theta_S + 2M_{P_B}(\theta_{1_B} + \theta_{2_B}) = P\delta + W\Delta \tag{14.48a}$$

and using the relationships

$$\theta_S = \theta_{1_B}, \quad \Delta = a\theta_{1_B} = b\theta_{2_B}, \quad \theta_{2_B} = \frac{a}{b}\theta_S = \frac{\theta_S}{\varepsilon} \tag{14.48b,c,d}$$

leads to

$$\frac{M_{P_S}}{QL} + \frac{1+\varepsilon}{\varepsilon}\frac{M_{P_B}}{QL} = \frac{1}{2}\left(\beta\sin\gamma + \frac{\cos\gamma}{1+\varepsilon}\right) \tag{14.48e}$$

For Figure 14.24g,

$$2M_{P_S}\theta_S + M_{P_B}(\theta_{1_B} + \theta_{2_B}) + M_{P_S}(\theta_{1_B} + \theta_{2_B}) = P\delta + W\Delta \tag{14.49a}$$

where, from equations (14.45b,c) and (14.48b–d),

$$\frac{1+3\varepsilon}{\varepsilon}\frac{M_{P_S}}{QL} + \frac{1+\varepsilon}{\varepsilon}\frac{M_{P_B}}{QL} = \beta\sin\gamma + \frac{\cos\gamma}{1+\varepsilon} \tag{14.49b}$$

For example, take the portal frame to be described by $\gamma = 30°$, $\beta = 1.25$ and $\varepsilon = 0.75$. Substituting these into equations (14.44d), (14.45d), (14.47b), (14.48e) and (14.49b) leads to a system of six straight-line equations within coordinate axes of $x = M_{P_S}/QL$ and $y = M_{P_B}/QL$:

$$y = 0.106 \tag{14.50a}$$

$$y = -x + 0.2121 \tag{14.50b}$$

$$y = -x + 0.3125 \tag{14.50c}$$

$$x = 0.1563 \tag{14.50d}$$

$$y = -0.429x + 0.24 \tag{14.50e}$$

$$y = -1.857x + 0.48 \tag{14.50f}$$

Figure 14.25 shows the six equations (14.50a–f) on the common axes x and y.

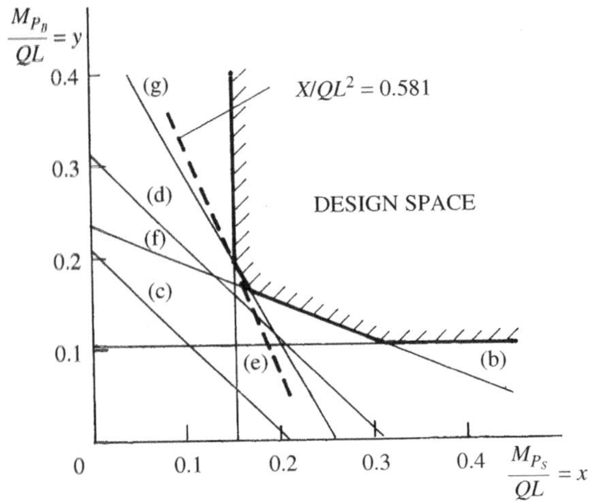

Figure 14.25 Design space for portal frame collapse under modes (b)–(g)

14.7.3 Weight Function

As with our continuous beam example considered earlier, the composite plot of straight-line equations (14.50a–f) reveals the working *design space* shown in Figure 14.25. If we now superimpose a weight function X upon Figure 14.25, one point on the design space boundary will provide a safe, minimum weight solution. We have seen previously that when the dependence of the total weight per unit length upon the collapse (ultimate) moment is taken to be linear we may interpret a weight function for a whole structure [1] as:

$$X = \sum M_P L \tag{14.51a}$$

For the portal frame in Figure 14.24a, equation (14.51a) becomes

$$X = (M_{P_B} \times L) + 2(M_{P_S} \times 1.25L) \tag{14.51b}$$

The corresponding non-dimensional form of equation (14.51b) is

$$\frac{X}{QL^2} = \frac{M_{P_B}}{QL} + 2.5\frac{M_{P_S}}{QL}, \quad \Rightarrow \quad y = -2.5x + X/QL^2 \tag{14.51c}$$

Equation (14.51c) describes a straight line of gradient -2.5 (dashed line in Figure 14.25). A minimum value of the intercept $X/(QL^2)$ with the y-axis is found when this line passes through the corner where the collapse modes (e) and (g) intersect. This requires a simultaneous solution to the corresponding equations (14.50d,f) , i.e. $x = 0.1563$, $y = 0.1898$, which gives the minimum weight conditions as

$$\frac{M_{P_S}}{QL} = 0.1563, \quad \frac{M_{P_B}}{QL} = 0.1898 \quad \text{and} \quad \frac{X}{QL^2} = 0.5805 \tag{14.52a,b,c}$$

Commercial sections will not conform exactly to the conditions in equations (14.52a–c) but the selection from within the design space is guided by the minimum weight point. This is the essence of the dynamic programming method developed in Chapter 15.

14.8 Effect of Axial Loading upon Collapse

When an axial load P is applied to a beam in bending, its moment carrying capacity M will be reduced. The manner in which a tensile (or a compressive) axial force will modify the stress distribution under a fully plastic moment depends upon the beam cross-section.

14.8.1 Rectangular Section $b \times d$ (see Figure 14.26a)

The fully plastic stress distribution in a rectangular cross-section under bending is shown in Figure 14.5d. An additional axial force applied in tension or in compression will modify this distribution. The effect of an axial tensile force P is shown in Figure 14.26b.

It is convenient to split Figure 14.26b into the sum of two components and use a multiplication factor α to locate the neutral axis (see Figure 14.26c,d). From these figures the moment and force carrying capacities become

$$M = \sigma_y b \times {}^1\!/\!_2(d - \alpha d) \times {}^1\!/\!_2(d + \alpha d) = (\sigma_y bd^2/4)(1 - \alpha^2) = M_P(1 - \alpha^2) \quad (14.53a)$$

$$P = \sigma_y b\alpha d = \alpha P_P \quad (14.53b)$$

where $M_P = \sigma_y bd^2/4$ (see equation (14.12a)) and $P_P = \sigma_y bd$ refer to collapse conditions under the separate actions of a bending moment and an axial force. Equation (14.53b) gives the shift in the neutral axis as

$$\alpha d/2 = (P/P_P)(d/2) \quad (14.53c)$$

Eliminating α between equations (14.53a,b) leads to an interaction equation

$$M/M_P = 1 - (P/P_P)^2 \quad (14.53d)$$

Equation (14.53d) provides the collapse moment M in the presence of an axial tensile or compressive force P.

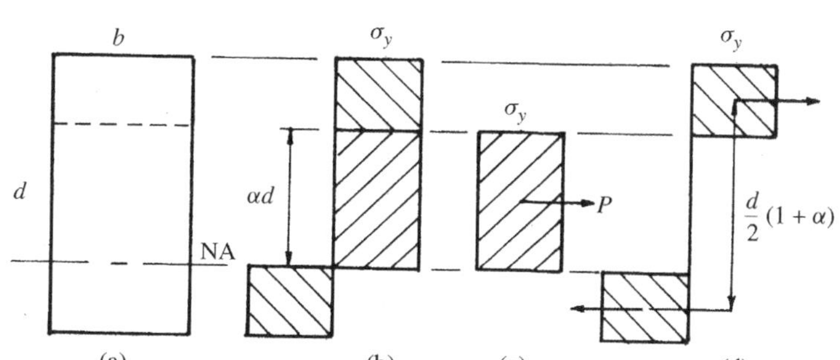

Figure 14.26 Stress distribution under combined tension and bending

Example 14.4 A rectangular section, of breadth $b = 75$ mm and depth $d = 50$ mm, is subjected to an axial tensile force $P = 150$ kN. By what percentage is the plastic moment carrying capacity of the section reduced? Take $\sigma_y = 250$ MPa.

Working in units of N and mm and noting that 1 MPa = 1 N/mm^2,

$$M_P = \sigma_y bd^2/4 = 250 \times 75 \times 50^2/4 = 11.719 \times 10^6 \text{ N mm} = 11.719 \text{ kN m}$$

$$P_P = \sigma_y bd = 250 \times 75 \times 50 = 937.5 \times 10^3 \text{ N} = 937.5 \text{ kN}$$

and from equation (14.53b),

$$\alpha = P/P_P = 150/937.5 = 0.16$$

Hence, from equation (14.53a), the moment carrying capacity becomes

$$M = 11.719(1 - 0.16^2) = 11.419 \text{ kN m}$$

i.e. a 2.56% reduction to M_P where, from equation (14.53c), the shift in the neutral axis is

$$\alpha d/2 = 0.16 \times 50/2 = 4 \text{ mm}$$

This shows that the axial force must be a greater fraction of P_P if neutral axis shift is to influence M_P significantly.

14.8.2 I-Section

For an I-section the neutral axis (NA) may lie in the web or in the flange, depending upon the magnitude of P. For the web position in Figure 14.27a, the moment capacity follows from the stress in the flanges and in part of the web (the left-shaded areas \\\ in Figure 14.27b). The moment capacity is the sum of the moments from the flange and the web:

$$M = \sigma_y[bt_f(d - t_f) + t_w(d/2 - t_f - \alpha d/2)(d/2 - t_f + \alpha d/2)] \tag{14.54a}$$

The force capacity applies to the right-shaded rectangular area /// in Figure 14.27b:

$$P = \alpha d t_w \sigma_y \tag{14.54b}$$

from which the neutral axis shift is $\alpha d/2 = P/(2t_w\sigma_y)$. When P is absent, $\alpha = 0$ and the fully plastic moment is

$$M_P = \sigma_y[bt_f(d - t_f) + (t_w/4)(d - 2t_f)^2] \tag{14.54c}$$

Combining equations (14.54a,b,c) gives the interaction equation

$$M = M_P - P^2/(4\sigma_y t_w) \tag{14.54d}$$

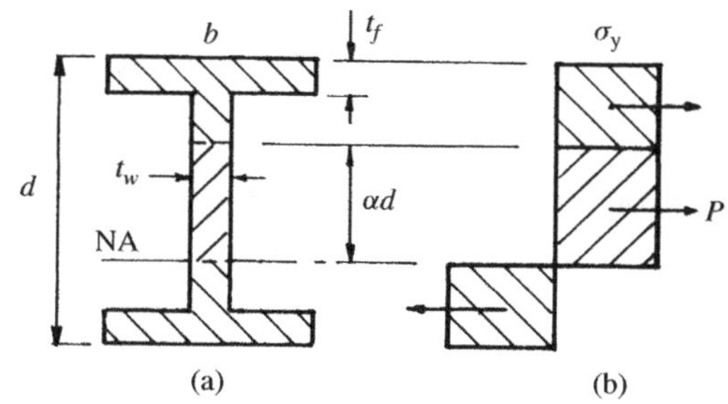

Figure 14.27 I-section with neutral axis (NA) in web

Now let the neutral axis lie in the flange, the shift being αt_f from the bottom surface, as shown in Figure 14.28a. We have, from Figure 14.28b,

$$M = b\alpha t_f \sigma_y (d - \alpha t_f) \tag{14.55a}$$

$$P = \sigma_y [2bt_f(1 - \alpha) + (d - 2t_f)t_w] \tag{14.55b}$$

The position of the neutral axis is, from equation (14.55b),

$$\alpha t_f = (P_P - P)/(2b\sigma_y) \tag{14.55c}$$

where $P_P = [2bt_f + (d - 2t_f)t_w]\sigma_y$ is the fully plastic tensile force. Combining equations (14.55a–c) leads to the interaction equation

$$M = (P_P - P)d/2 - (P_P - P)^2/(4b\sigma_y) \tag{14.55d}$$

Equation (14.55d) may be applied to determine M in the presence of P when the position of the neutral axis is such that $\alpha < 1$ in equation (14.55c). If it is found that $\alpha > 1$ then equation (14.54d) must be used. In larger commercial sections, as the following example will show, an axial load must be very high if it is to reduce M_P appreciably.

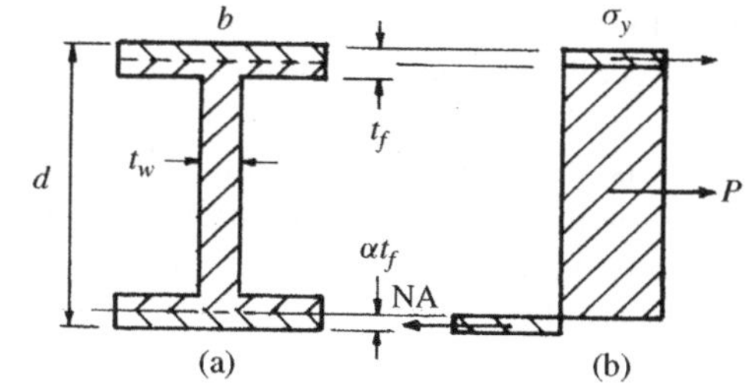

Figure 14.28 I-section with neutral axis (NA) in the flange

Example 14.5 The I-section shown in Figure 14.27a has dimensions $b = 300$ mm, $d = 150$ mm, $t_f = 15$ mm and $t_w = 10$ mm. If a beam of this section is to carry an axial load $P = 175$ kN in addition to a bending moment, what is the effect of this load upon its collapse moment? Take $\sigma_y = 280$ MPa.

The fully plastic tensile force is

$$P_P = [2bt_f + (d - 2t_f)t_w]\sigma_y$$

$$= [(2 \times 300 \times 15) + (150 - 30)10]280 = 2.856 \times 10^6 \text{ N} = 2.856 \text{ MN}$$

when, from equation (14.55c),

$$\alpha = (P_P - P)/(2bt_f\sigma_y)$$

$$= [(2.856 \times 10^6) - (175 \times 10^3)]/(2 \times 300 \times 15 \times 280) = 1.068$$

Since $\alpha > 1$, the neutral axis lies in the web and the interaction equation (14.54d) applies. Within this equation, the fully plastic moment is, from equation (14.54c),

$$M_P = \sigma_y[bt_f(d - t_f) + (t_w/4)(d - 2t_f)^2]$$

$$= 280[300 \times 15(150 - 15) + (10/4)(150 - 30)^2]$$

$$= 180.18 \times 10^6 \text{ N mm} = 180.18 \text{ kN m}$$

The interaction equation (14.54d) gives the reduced moment capacity

$$M = M_P - P^2/(4\sigma_y t_w)$$

$$= (180.18 \times 10^6) - (175 \times 10^3)^2/(4 \times 280 \times 10)$$

$$= 177.45 \times 10^6 \text{ N mm} = 177.45 \text{ kN m}$$

14.9 Effect of Shear Force upon Collapse

In general, a shear force will accompany a bending moment where the latter arises from transverse forces applied to a beam. The effect of shear force upon the collapse moment has not been considered so far. Earlier chapters in this book recognised that while bending moments dominate the design of long beams, shear force becomes important to consider in short beams. Recall that the maximum bending stress occurs at the outer surfaces while the neutral axis remains unstressed. Conversely, the maximum shear stress occurs at the neutral axis while the outer surfaces remain unstressed. Depending upon the relative magnitudes of the linear bending stress distribution and the parabolic shear stress distribution, each lying within the section's depth, two interactions between these stress distributions are possible.

14.9.1 Initial Yielding by Bending

It is assumed, firstly, that as the loading increases a plastic penetration occurs in a rectangular section to depth $\alpha_o d/2$ by bending just as the shear yield stress τ_y is reached at

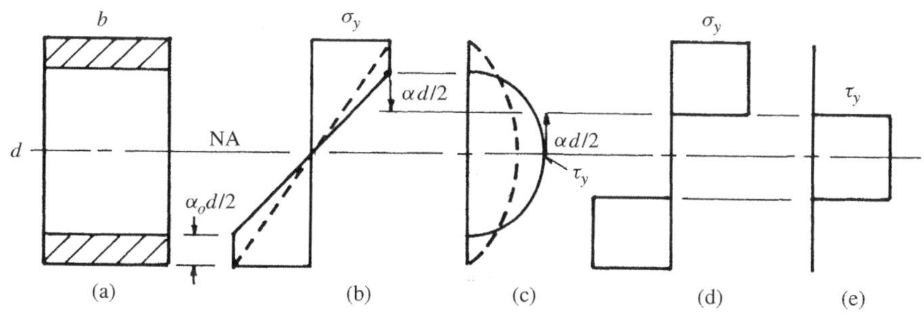

Figure 14.29 Plastic penetration by bending and then shear

the neutral axis (see Figure 14.29a–c). The initial, fully elastic, bending and shear stress distributions appear as dashed lines in Figure 14.29b,c respectively. To accommodate the initial plastic penetration without hardening, the low-valued, elastic shear stresses must be removed from the penetration zone (see Figure 14.29c). Thereafter, further penetrations occur by bending and shear at the same rate but in opposing directions. When these meet, the section becomes fully plastic under bending and shear, as shown in Figure 14.29d,e. Let these further simultaneous penetrations by shear and bending each be $\alpha d/2$, so that

$$2[(\alpha d/2) + (\alpha d/2) + (\alpha_o d/2)] = d$$

$$2\alpha + \alpha_o = 1 \tag{14.56a}$$

The plastic moment is, from Figure 14.29d,

$$\begin{aligned} M &= \sigma_y b(d/4)(1+\alpha_o)\left[(d/2)(1-\alpha_o) + (d/4)(1+\alpha_o)\right] \\ &= {}^1\!/\!_4(\sigma_y bd^2/4)(1+\alpha_o)(3-\alpha_o) \\ &= \frac{M_P}{4}(1+\alpha_o)(3-\alpha_o) \end{aligned} \tag{14.56b}$$

where $M_P = \sigma_y bd^2/4$ is the fully plastic moment for a rectangular section in the absence of shear. The plastic shear force is, from Figure 14.29e,

$$\begin{aligned} F &= \tau_y b(d/2)(1-\alpha_o) \\ &= \frac{F_P}{2}(1-\alpha_o) \end{aligned} \tag{14.56c}$$

From equation (14.56c),

$$\alpha_o = 1 - 2F/F_P \tag{14.56d}$$

and eliminating α_o between equations (14.56b,c) leads to the interaction equation

$$M/M_P = 1 - (F/F_P)^2 \tag{14.56e}$$

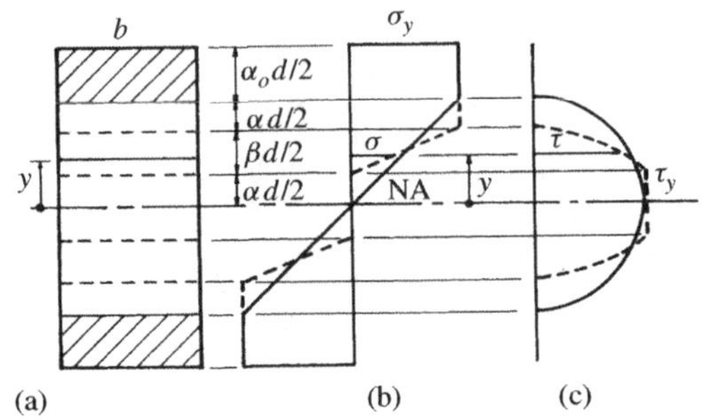

Figure 14.30 Elastic-plastic section under bending and shear

The spread of plasticity, by the simultaneous actions of bending and shear in Figure 14.29a–c, is confined to a central rectangular area of breadth b and depth $(1-\alpha_o)d$. This is shown to an enlarged scale in Figure 14.30a. Figures 14.30b,c reveal the partial spread of plasticity to intermediate depths $\alpha d/2$ by bending and shear within this rectangle.

Within the remaining non-penetrated, elastic region, of depth $\beta d/2$, there exist linear bending stress and a parabolic shear stress distributions (dashed lines in Figure 14.30b,c). Note that the beam depth is $(2\alpha+\beta+\alpha_o)d = d$ and therefore $\beta = 1-\alpha_o-2\alpha$. Hence, the elastic zone may be defined relative to the neutral axis in a non-dimensional form:

$$\alpha \le 2y/d \le (1-\alpha_o-\alpha) \tag{14.57}$$

At the extremes of the zone, defined by equation (14.57), the two elastic stresses must match their known boundary conditions:

(i) when $2y/d=\alpha, \sigma=0$ and $\tau=\tau_y$;
(ii) when $2y/d=(1-\alpha_o-\alpha)=(\alpha+\beta), \sigma=\sigma_y$ and $\tau=0$

Within the zone of equation (14.57), the elastic stress expressions are

$$\frac{\sigma}{\sigma_y} = \frac{2y/d-\alpha}{1-\alpha_o-2\alpha} \tag{14.58a}$$

$$\frac{\tau}{\tau_y} = \frac{(\alpha+\beta)^2-(2y/d)^2}{\beta(\beta+2\alpha)} = \frac{(1-\alpha_o-\alpha)^2-(2y/d)^2}{(1-\alpha_o-2\alpha)(1-\alpha_o)} \tag{14.58b}$$

The extent to which the elastic region remains unyielded may be examined from the von Mises effective stress definition

$$\left(\frac{\overline{\sigma}}{\sigma_y}\right)^2 = \left(\frac{\sigma}{\sigma_y}\right)^2 + \left(\frac{\tau}{\tau_y}\right)^2 \tag{14.59a}$$

where $\sigma_y^2 = 3\tau_y^2$. The material will have yielded under σ and τ when $\overline{\sigma}/\sigma_y = 1$ but it remains elastic when $\overline{\sigma}/\sigma_y < 1$. For example, if we take $\alpha_o = 0$, when initial yielding in

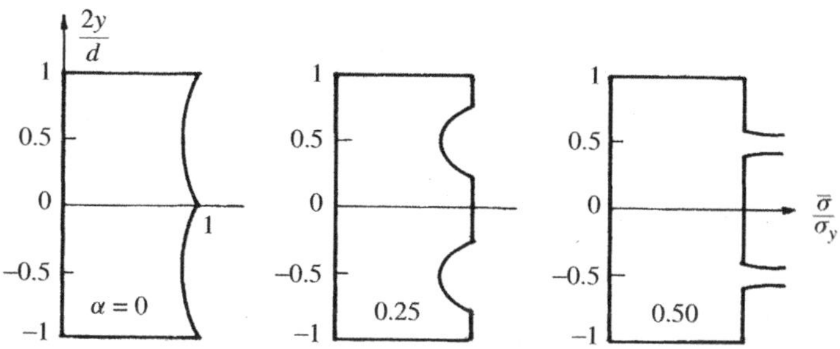

Figure 14.31 Effective stress with plastic penetration of a rectangular section

bending and shear occur simultaneously, then $0 \leq \alpha \leq 0.5$ for plastic penetration into each half-section. Substituting from equations (14.58a,b) into the right-hand side of equation (14.59a) provides an effective stress at a given depth $2y/d$ in the range $\alpha \leq 2y/d \leq 1 - \alpha$:

$$\left(\frac{\bar{\sigma}}{\sigma_y}\right)^2 = \left(\frac{2y/d - \alpha}{1 - 2\alpha}\right)^2 + \left[\frac{(1-\alpha)^2 - (2y/d)^2}{1 - 2\alpha}\right]^2 \qquad (14.59b)$$

Figure 14.31 shows the effective stress distributions corresponding to (a) the elastic limit, when $\alpha = 0$, (b) a partial penetration, where $\alpha = 0.25$ and (c) full plasticity, where $\alpha = 0.5$.

14.9.2 Initial Yielding by Shear

Plastic penetration occurs in shear outwards on both sides of the neutral axis to depths $\alpha_o d/2$, before the yield stress σ_y in bending is reached at the outer edges (see Figure 14.32a). The initial, fully elastic, bending and shear stress distributions appear as dashed lines in Figure 14.32b,c, respectively. To accommodate the shear penetration without hardening, the elastic bending stresses must be removed from within the shear penetration zone (see Figure 14.32c). With increased loading, further plastic penetrations occur by bending and shear at the same rate in opposing directions (see Figure 14.32b,c). When these meet the section becomes fully plastic, as shown in Figure 14.32d,e. Let the subsequent penetrations be $\alpha d/2$, so that $2\alpha + \alpha_o = 1$. The plastic moment is found from Figure 14.32e:

$$\begin{aligned} M &= \sigma_y b(d/4)(1 - \alpha_o)[(d/2)(1 + \alpha_o) + (d/4)(1 - \alpha_o)] \\ &= (M_P/4)(1 - \alpha_o)(3 + \alpha_o) \end{aligned} \qquad (14.60a)$$

The plastic shear force is found from Figure 14.32d:

$$\begin{aligned} F &= \tau_y b(d/2)(1 + \alpha_o) \\ &= \frac{F_P}{2}(1 + \alpha_o) \end{aligned} \qquad (14.60b)$$

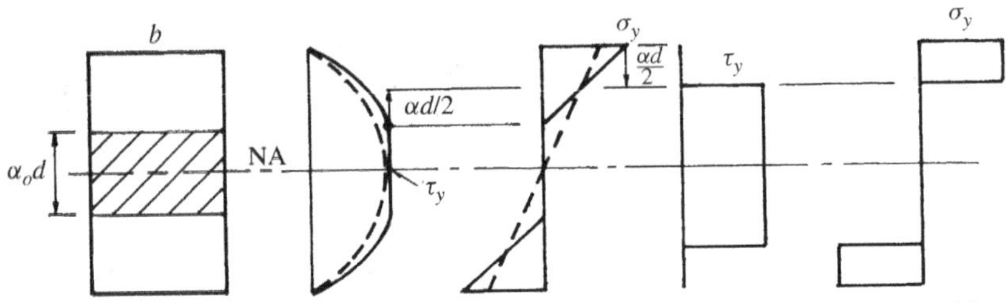

Figure 14.32 Plastic penetration by shear and then bending

From equation (14.60b),

$$\alpha_o = 2F/F_P - 1 \tag{14.60c}$$

and eliminating α_o between equations (14.60a,b) leads to an interaction equation similar to equation (14.56e):

$$M/M_P = 1 - (F/F_P)^2 \tag{14.60d}$$

Here the spread of plasticity, by the simultaneous actions of bending and shear in Figure 14.32a,b is confined to two outer rectangles, each of breadth b and depth $(1 - \alpha_o)d/2$. The enlarged scale of Figure 14.33a–c reveals the partial spread of plasticity to depths $\alpha d/2$ by shear and bending within each outer rectangle. Within the remaining non-penetrated, elastic regions of depth $\beta d/2$, there exist linear bending stress and a parabolic shear stress distributions (broken lines). Figures 14.33b,c show again that $\beta = 1 - \alpha_o - 2\alpha$ but the position of the elastic zone relative to the neutral axis is modified:

$$\alpha + \alpha_o \leq 2y/d \leq 1 - \alpha \tag{14.61}$$

At the extremes of the zone defined by equation (14.61), the two elastic stresses must match their known boundary conditions:

(i) when $2y/d = \alpha + \alpha_o$, $\sigma = 0$ and $\tau = \tau_y$;
(ii) when $2y/d = 1 - \alpha$, $\sigma = \sigma_y$ and $\tau = 0$.

Within the zone of equation (14.61), the elastic stress expressions are

$$\frac{\sigma}{\sigma_y} = \frac{2y/d - \alpha_o - \alpha}{1 - \alpha_o - 2\alpha} \tag{14.62a}$$

$$\frac{\tau}{\tau_y} = \frac{(1-\alpha)^2 - (2y/d)^2}{(1 - \alpha_o - 2\alpha)(1 + \alpha_o)} \tag{14.62b}$$

The fact that equations (14.62a,b) agree with equations (14.58a,b), when $\alpha_o = 0$ confirms the effective stress distributions for simultaneous yielding given in Figure 14.31.

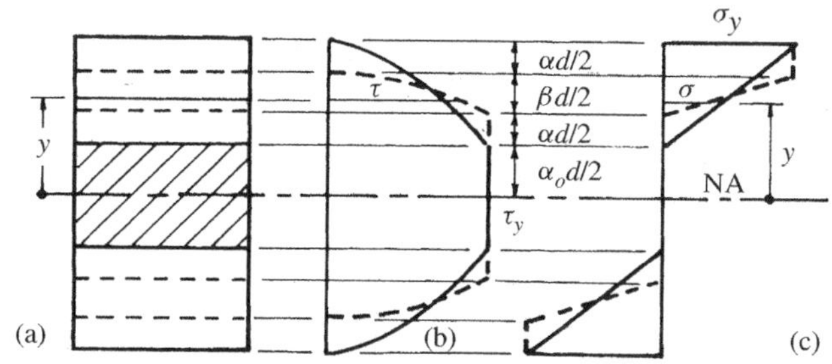

Figure 14.33 Elastic-plastic section under shear and bending

However, when $\alpha_o \neq 0$ in equations (14.62a,b) the effective stress within the elastic zones and the positions of these zones in the depth will depend upon the particular sequence of each plastic penetration mode.

14.10 Effect of Hardening upon Collapse

The elastic-perfectly plastic analyses above may be refined further when an allowance is made for work-hardening in the plastic region. Linear and parabolic schemes are shown in Figure 14.34. The simplest linear hardening material has a constant rate of hardening $n = \mathrm{d}\sigma/\mathrm{d}\varepsilon$, which has been used [6] to define the gradient of the plastic penetration zone within a rectangular beam section $b \times d$. It is assumed that plastic strain ε is proportional to the non-dimensional depth measure $2y/d$:

$$\varepsilon = C\left(\frac{2y}{d}\right) \tag{14.63a}$$

where C is a scaling factor. The hardening rate $n = \mathrm{d}\sigma/\mathrm{d}\varepsilon$ follows from equation (14.63a) as

$$Cn = \frac{\mathrm{d}\sigma}{\mathrm{d}(2y/d)} = \frac{d}{2}\left(\frac{\mathrm{d}\sigma}{\mathrm{d}y}\right) \tag{14.63b}$$

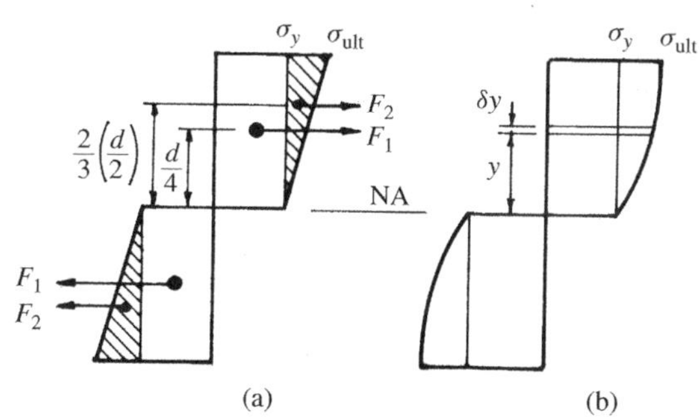

Figure 14.34 Effects of hardening on plastic penetration in a rectangular section beam

Integrating equation (14.63b) for the surface stress under the collapse condition in Figure 14.34a,

$$\int_{\sigma_y}^{\sigma} \mathrm{d}\sigma = \frac{2Cn}{d} \int_0^{d/2} \mathrm{d}y, \quad \Rightarrow \quad \sigma = \sigma_y + Cn \tag{14.63c}$$

Hence the yield stress has increased by an amount proportional to the hardening rate. This allows the outer surface stress in equation (14.63c) to be equated to the ultimate strength of the material. Therefore, C follows as

$$C = \frac{\sigma_{\text{ult}} - \sigma_y}{n} \tag{14.63d}$$

This account of linear hardening introduces an additional term to the fully plastic moment expression. For a rectangular section $b \times d$, the fully plastic moment in equation (14.12a) will be increased by the contribution from the shaded triangles in Figure 14.34a. The total plastic moment becomes the sum of the two couples $2 \times F_1 \times d/4$ and $2 \times F_2 \times (2/3)(d/2)$. This is

$$M'_P = \left(\frac{bd\sigma_y}{2} \times \frac{d}{2}\right) + \left(\frac{Cn}{2} \times \frac{bd}{2} \times 2 \times \frac{2}{3} \times \frac{d}{2}\right) = \frac{bd^2\sigma_y}{4} + \frac{Cnbd^2}{6} \tag{14.64a}$$

Substituting C from equation (14.63d) into equation (14.64a),

$$M'_P = \frac{bd^2}{12}(\sigma_y + 2\sigma_{\text{ult}}) \tag{14.64b}$$

Equation (14.64b) is most appropriate to the plastic penetration of half-hard metals which often harden linearly. Alternatively, since only the collapse condition is required, another estimate of the fully plastic moment is provided when the ultimate strength replaces the yield stress in the appropriate expression:

$$M'_P = \frac{bd^2\sigma_{\text{ult}}}{4} \tag{14.64c}$$

Comparing equation (14.64c) with equation (14.64b), the former is likely to be an over-estimate where the hardening rate is not linear but diminishes with increasing plastic strain. This is especially true for annealed metals where the Ludwik law [7] describes this hardening behaviour:

$$\frac{\sigma}{\sigma_y} = 1 + \left(\frac{\varepsilon^P}{\varepsilon_o}\right)^n \tag{14.65a}$$

The constant C in equation (14.63a) now becomes

$$C = \varepsilon_o(\sigma_{\text{ult}}/\sigma_y - 1)^{1/n} \tag{14.65b}$$

Correspondingly, it can be shown from Figure 14.34b that the modification to the second term in equation (14.64a) becomes the contribution to the fully plastic moment from parabolic hardening.

$$M'_P = \frac{bd^2\sigma_y}{4} + \frac{bd^2(\sigma_{ult} - \sigma_y)}{2(n+2)} \qquad (14.65c)$$

which returns to equation (14.64a) when $n = 1$. The derivation of this expression appears in Exercise 14.11 for the reader to confirm.

References

[1] Neal, B. G. *Plastic Methods of Structural Analysis*, Chapman & Hall, 1964.
[2] Trahiar, N. S. and Bradford, M. A. *The Behaviour and Design of Steel Structures*, Chapman & Hall, 1988.
[3] BS 4229 Part 2; 1969, Solid circular and hollow sections.
[4] BS 4848 Part 2; 1975, 1991, Hot rolled structural steel sections: specifications for hot finished, hollow sections.
[5] BS 4300; 1969, Wrought aluminium alloys for general engineering purposes; Part 10 Drawn tube; Part 12 Bars, extruded round tube and sections.
[6] Nadai, A. *Theory of Flow and Fracture of Solids*, McGraw-Hill, 1950.
[7] Ludwik, P. *Elemente der Technologischen Mechanik*, Springer, Berlin, 1909.

Exercises

14.1 Show that a shape factor $S = 2$ applies to a square section of side length l when the diagonal becomes the neutral axis of bending (see Figure 14.35a).

14.2 Show that the shape factor for the symmetrical I-section in Figure 14.35b is given by

$$S = \frac{3\left[1 - (bd^2)/(BD^2)\right]}{2\left[1 - (bd^3)/(BD^3)\right]}$$

14.3 Can the portal frame in Figure 14.22 be designed more economically if the load and length ratios P/W and H/L are altered? What should these ratios be?

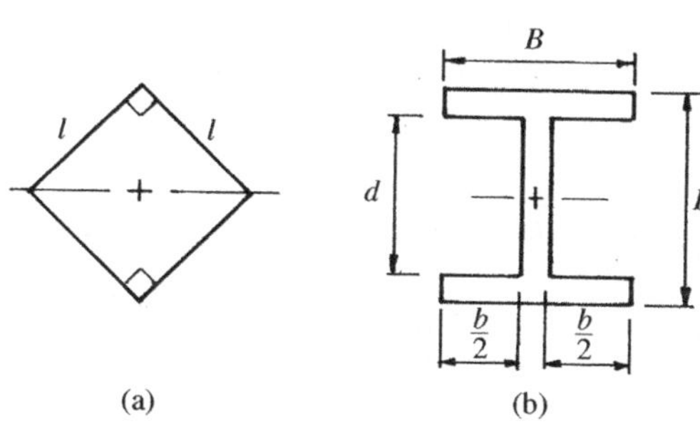

Figure 14.35 Solid beam cross-sections

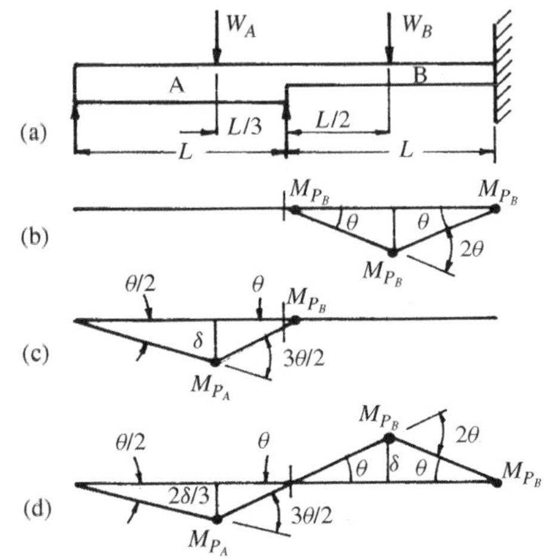

Figure 14.36 Propped cantilever

14.4 Imposing similar design constraints upon the portal frame given in Example 14.3, select suitable tubular sections for the beam and stanchions in Figure 14.23 using the tube's $K = r_2/r_1$ ratio and its section modulus $z = I/y$ as bases for selection. Refer to the standard sizes listed in BS 4848 [4].

14.5 The propped cantilever in Figure 14.36a carries the concentrated loads W_A and W_B within its respective spans A and B. The collapse moment for bay A, which carries the greater load, is M_{P_A} and that for bay B, carrying the lesser load, is M_{P_B}. Determine the collapse loads when: (i) they are applied in the ratio $W_A/W_B = 2$ for a beam with plastic moments $M_{P_A} = 2M_{P_B}$; (ii) they are applied in the ratio $W_A/W_B = 3$ for a beam with plastic moments $M_{P_A} = 3M_{P_B}$. Collapse loading should be found from a consideration of the possible modes of failure shown in Figure 14.36b–d. Hence determine suitable dimensions when each bay is constructed from rectangular steel section in which the depth is to be three times the breadth. Take $\sigma_y = 310$ MPa. [Answer: (i) $M_{P_B} = WL/6$]

14.6 The portal frame shown in Figure 14.37a has two vertical, encastré stanchions welded to a horizontal beam. The collapse moment for each stanchion is 20% greater than that for the beam (M_P). Their respective lengths are in the ratio 3 : 4 as shown. A vertical load W is offset to the right of centre by 1/8 of its length. A horizontal load $W/2$ is applied to the top of the left stanchion in the line of the beam axis. Determine the collapse load W for each of three possible failure modes: (i) beam collapse; (ii) stanchion sway; (iii) simultaneous stanchion sway and beam collapse. Hence determine the true failure mode. [Answers: (i) $8.53\,M_P/L$, (ii) $11.73\,M_P/L$, (iii) $7.86\,M_P/L$]

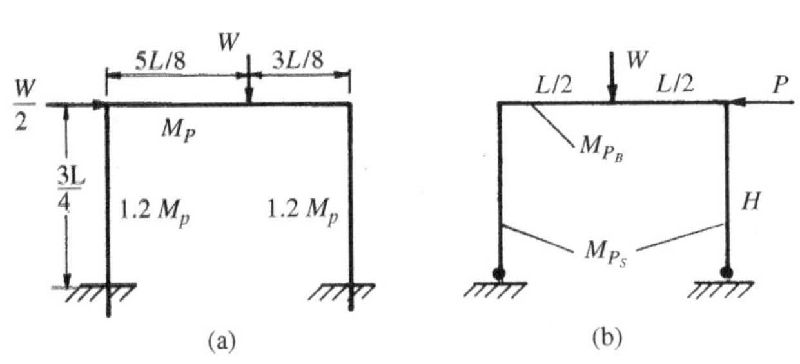

Figure 14.37 Portal frames

14.7 The portal frame shown in Figure 14.37b has two vertical stanchions welded to a horizontal beam. The collapse moment for each stanchion (M_{P_S}) is 70% of the collapse moment for the beam (M_{P_B}). Their respective lengths are in the ratio $L/H = 3/2$. Loads W and P are applied in the ratio $W/P = 3/1$. Determine the critical loading for: (i) beam collapse; (ii) stanchion sway; (iii) a combination of (i) and (ii), when the base of each stanchion is fixed. [Answers: (i) $W = 6.8\,M_{P_B}/L$, (ii) $W = 12.6\,M_{P_B}/L$, (iii) $W = 6.646\,M_{P_B}/L$]

14.8 The portal frame shown in Figure 14.37b has two vertical stanchions welded to a horizontal beam. The collapse moment for the beam (M_{P_B}) is 25% greater than the collapse moment (M_{P_S}) for each stanchion. Their respective lengths are in the ratio $L/H = 4/3$. Assuming that collapse will occur under the combined action of W and P, applied in the ratio $W/P = 2/1$, determine the collapse loading (W) when the base of each stanchion is: (i) hinged; (ii) fixed. If the stanchions are to be made from solid rectangular section $b \times d$ in a material with yield stress σ_y, explain how the section dimensions may be chosen to ensure a safe design. [Answers: (i) $36\,M_{P_S}/7L = 9bd^2\sigma_y/7L$, (ii) $52\,M_{P_S}/7L = 13bd^2\sigma_y/7L$]

14.9 Referring to Exercise 14.8, find the collapse load W (where $P = W/2$, $L/H = 4/3$, $M_{P_B} = 1.25M_{P_S}$) in Figure 14.37b when beam collapse and sway are taken separately. [Answers: Hinged base: beam collapse $14M_{P_S}/L$; sway $5.33M_{P_S}$/L. Fixed base: beam collapse $14M_{P_S}/L$; sway $10.67M_{P_S}$/L]

14.10 The portal frame shown in Figure 14.37b has two vertical stanchions welded to a horizontal beam. The collapse moment for the beam (M_{P_B}) is 30% greater than that for each stanchion (M_{P_S}). Their respective lengths are in the ratio $L/H = 3/2$. Assuming that collapse will occur under the combined action of W and P, applied in the ratio $W/P = 2/1$, determine the collapse loading when the base of each stanchion is: (i) hinged; (ii) fixed. If the stanchions are to be made from solid rectangular section $b \times d$ in a material with yield stress σ_y, explain how b and d may be chosen to ensure a safe design. [Answers: (i) $W = 5.52\,M_{P_S}$/L, (ii) $W = 7.92\,M_{P_S}$/L where $M_{P_S} = bd^2\sigma_y/4$]

14.11 Plot, in the manner of Figure 14.31a–c, the effective stress distributions in the case of penetrations by the simultaneous action of shear and bending ($\alpha_o = 0$) when $\alpha = 0.2$ and $\alpha = 0.4$. Confirm, from equation (14.59b), the effective elastic stress values at the depth positions $2y/d$ given below:

$\alpha = 0.2$

$2y/d$	0.2	0.3	0.4	0.5	0.6	0.7	0.8
$\overline{\sigma}/\sigma_y$	1	0.932	0.867	0.820	0.814	0.870	1.0

$\alpha = 0.4$

$2y/d$	0.4	0.45	0.5	0.55	0.6
$\overline{\sigma}/\sigma_y$	1	0.826	0.743	0.803	1.0

14.12 Show, from Figure 14.34b, that the fully plastic moment, when based upon Ludwik's parabolic hardening law (14.65a), is given by

$$M'_P = 2b\sigma_y \int_0^{d/2} \left[1 + \left(\frac{2yC}{\varepsilon_o d}\right)^n\right] y\,dy$$

Substituting C from equation (14.65b), confirm the second term in equation (14.65c) as the contribution to the fully plastic moment from parabolic hardening.

15

Dynamic Programming

15.1 Introduction

The previous chapter showed how the two-dimensional design space provided the optimum plastic moments for the spans within continuous beams and portal frames. These moments were selected on the basis of a minimum weight function corresponding to a point within the design space. This approach is limited to three unknowns, two plastic moments and the minimum weight, that arose in two-span beams and in the simplest of portal frames. Geometrically, this minimum weight solution lies at the intersection between two plane diagrams when overlaid, one for the design space and the other for the minimum weight function. In the case of a greater number of unknown plastic moments appearing in multiple-span beams, or two-tier portal frames, the use of further design spaces becomes impractical and the dynamic programming (DP) method is used. This technique makes its decisions on the optimum plastic moment required for each span in sequence based upon a similar optimisation for the preceding spans. The DP method will be illustrated using the example of a continuous beam having a variable cross-section, but where the section remains uniform within each span. The objective is again to achieve that combination of plastic moments M_P which achieve minimum weight W. Given that there are a limited number of standard sections available, the nearest approach to an optimum condition is sought here. For example, Table 15.1 shows eight typical rectangular steel sections A–H with plastic moments $M_P = \sigma_y bd^2/4$ (kN m) and weights per unit length $W/L = \rho bd$ (N/m) as given. The section sizes, given in mm, refer to the outer dimensions $b \times d$ for each rectangular section, as taken from [1].

15.2 Single-Span Beam

As a prelude to the analysis of multiple-span beams, a statement of the collapse condition for the propped cantilever in Figure 15.1a will be given. This provides a useful illustration of how its two greatest moments, at the fixing and beneath the load, are adjusted at collapse to provide the optimum cross-section. The lightest beam is one where the bending moment distribution has first been arranged to equalise the maximum moments at collapse. This

Mechanics of Optimal Structural Design: Minimum Weight Structures D. W. A. Rees

Table 15.1 Standard rectangular sections for beams

Section	$(b \times d)$, mm	M_P, kN m	W/L, N/m
A	203 × 133	65	245
B	203 × 133	78	294
C	254 × 146	99	304
D	305 × 127	135	363
E	305 × 165	156	392
F	356 × 172	193	441
G	356 × 172	223	500
H	381 × 152	240	510

follows where there is an equality between the maximum moments in the net moment diagram (shaded in Figure 15.1b). The free moment at B is 240 kN m. Since the net moment is the difference between the free and fixing moments, the ordinates M_B and M_C in Figure 15.1b are each made equal to M_P when

$$240 - M_P/2 = M_P$$

Thus in Figure 15.1b the arrangement $M_B = M_C = 160$ kN m applies to collapse. Hence, the selection of the beam section requires only that $M_P \geq 160$ kN m subject to the minimum weight requirement.

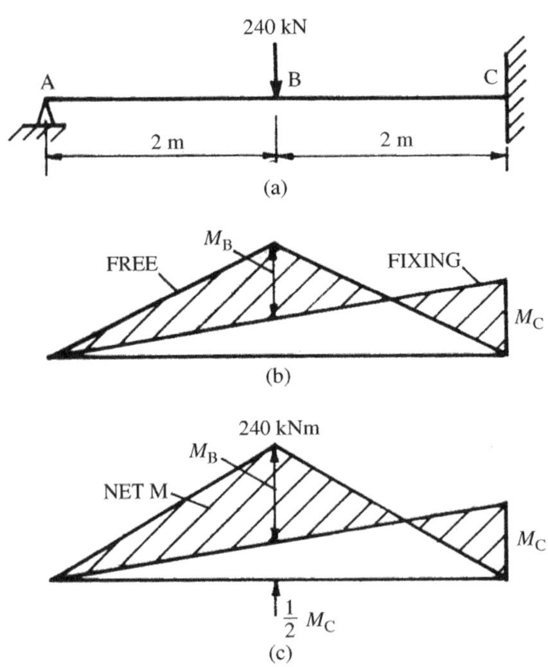

Figure 15.1 Single span beam with maximum moments at B and C

In the more general case, AC would be one span within a multi-span beam, making it necessary to account for the interactive effects between the moments from adjacent spans. Let the degree of constraint at C vary between the simply supported and clamped conditions, for when ABC lies between or at the end of additional spans. Here M_C will not then equal M_B, as shown in Figure 15.1c. Hence, the collapse condition is based upon the greater of the two moments, M_C and $M_B = 240 - ½M_C$ attaining M_P. The optimum design condition is

$$M_P \geq M_{\max}, \quad \Rightarrow \quad \begin{matrix} \text{either} \\ \text{or} \end{matrix} \begin{cases} M_P \geq 240 - ½M_C \\ M_P \geq M_C \end{cases} \qquad (15.1\text{a,b,c})$$

where M_P is taken from Table 15.1, i.e. within the range of beam sections. When M_C is allowed to vary from zero across the range of permitted M_P-values in Table 15.1, it becomes necessary to investigate which of the two moments M_B or M_C provides the value of the maximum moment ($M_{\max}$) in equation (15.1a). In the first and second columns of Table 15.2 the available M_C values are listed to enable the calculation of $M_B = 240 - ½M_C$. The third column lists the greater of these two moments as required by equations (15.1b,c). The lowest M_P is then taken from Table 15.1 to accommodate $M_{\max}$ as shown in Table 15.2. Finally, the full weight of the beam is found by multiplying the appropriate figure in final column of Table 15.1 by the 4 m length. An inspection of the weights produced in this manner from all the available sections allows that section(s) with the minimum weight to be selected. In this case, the weight calculation given in the final column of Table 5.2 show that the lightest design may be selected from among four equals (underlined).

15.3 Two-Span Beam

Consider the two-span beam shown in Figure 15.2a. The DP method is able to find those cross-sections for each span AC and CE that will provide the lightest overall design.

Again, the ideal situation would be if the peak net moments, M_B, M_C, M_D and M_E, lying in the shaded regions between the free- and fixing-moment diagrams, were equal, to allow the selection of an appropriate section directly. However, the interaction between

Table 15.2 Minimum weight design of a single span ABC

M_C	M_B	$M_{\max}$		M_P (kNm)	Section	W (kN)	
←	(kNm)	→			(from Table15.1)		
0	240	240		240	H	2.04	
65	208	208		223	G	2.0	
78	201	201	@ B	223	G	2.0	
99	191	191		193	F	<u>1.764</u> ←	$W_{\max}$
135	173	173		193	F	<u>1.764</u> ←	$W_{\max}$
156	162	162		193	F	<u>1.764</u> ←	$W_{\max}$
193	144	193		193	F	<u>1.764</u> ←	$W_{\max}$
223	129	223	@ C	223	G	2.0	
240	120	240		240	H	2.040	

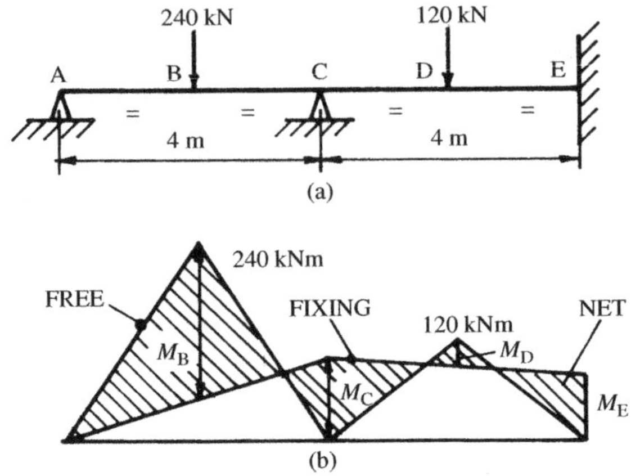

Figure 15.2 Two-span beam showing net peak moments at B, C, D and E

spans prevents a direct selection from being made. For example, if M_B is made equal to M_C in Figure 15.2b, when each is used to calculate M_D and M_E, it would be found that $M_D \neq M_B$ and $M_E \neq M_B$, in general. Moreover, as we have seen, the range of sections available does not allow an exact match between $M_{\max}$ and the available M_P values. This mismatch will also have an interactive effect upon the minimum weight of the two spans. Hence, the choice of M_P within each span AC and CE must rest upon the maximum moment being taken from the following two selections between their peak moments (from left to right in Figure 15.2a):

$$\text{Span ABC: } M_{P_{AC}} \geq \text{the greatest of: } \begin{cases} [M_A] = 0 \\ [M_B] = [240 - \tfrac{1}{2}M_C] \\ [M_C] \end{cases} \tag{15.2a}$$

$$\text{Span CDE: } M_{P_{CE}} \geq \text{the greatest of: } \begin{cases} [M_C] \\ [M_D] = [120 - \tfrac{1}{2}(M_C + M_E)] \\ [M_E] \end{cases} \tag{15.2b}$$

The basis of this search is to achieve a minimum total weight, i.e. $(W_{AC} + W_{CE})_{\min}$. To do this the weight is taken to be linear in M_P (see Figure 14.18) for the restricted range of sections available in Table 15.1. The reader will recall that a similar assumption was made in the construction of the design space for a continuous beam in the previous chapter. It follows from equation (14.35b) that the minimum total weight is given as:

$$(W_{AC} + W_{CE})_{\min} = (M_{P_{AC}} L_{AC} + M_{P_{CE}} L_{CE})_{\min} + \text{constant} \tag{15.3}$$

In this case, as $L_{AC} = L_{CE}$, equation (15.3) shows that the minimum, total weight corresponds to $(M_{P_{AC}} + M_{P_{CE}})_{\min}$, where the two moments follow from equation (15.2a,b). Table 15.3 shows the manner in which the search for $M_{P_{AC}}$ and $M_{P_{CE}}$ within equations (15.2a,b) is made by varying the values of M_C and M_E over the permitted range of

Table 15.3 Search for moments that provide a total, minimum weight

$M_C \downarrow$ / $M_E \rightarrow$	0	-	-	-	-	-	223
0	Total weight for the two spans.						
-							
-	The choice of optimum M_C and M_E						
-							
-	will give the total, minimum weight						
223							

sections. Note that it is not necessary to include $M_P = 240$ kN m in Table 15.3 because this value appears directly from equation (15.2a) at $M_C = 0$. Table 15.4 illustrates the calculations required when $M_E = 0$. The entries for AC apply to that span when taken alone (see Table 15.2).

Table 15.4 Search for minimum total weight $\sum W$ when $M_E = 0$ (Units: W in N, M in kNm)

	SPAN AC (Table 15.2)				SPAN CE				
M_C	M_{max}		$(M_P)_{min}$	W_{AC}	M_{max}		$(M_P)_{min}$	W_{CE}	ΣW
0	240		240	2040	120		135	1452	3492
65	208		223	2000	88	@ D	99	1216	3216
78	201	@ B	223	2000	81		99	1216	3216
99	191		193	1764	99		99	1216	2980 ← MIN
135	173		193	1764	135		135	1452	3216
156	162		193	1764	156	@ C	156	1568	3332
193	193	@ C	193	1764	193		193	1764	3528
223	223		223	2000	223		223	2000	4000

Next, the process is repeated for the full range of M_E-values. The results of these repetitive analyses, which are well suited to computer programming, are given in Table 15.5. The minimum weight design for each value of M_E is underlined in Table 15.5. Of the four contenders, giving $W = 2.98$ kN, Table 15.5 shows that an overall minimum weight design is achieved by varying the degree of fixity at E from 0 to 99 kN m. If $M_E > 99$ kN m, where the support at E clamps the beam with greater fixity, a weight penalty is incurred.

15.4 Three-Span Beam

In Figure 15.3a, an additional, third span EFG, 6 m long and carrying a central-concentrated load of 160 kN, has extended the two spans previously examined. In addition to satisfying equations (15.2a,b) within Figure 15.3b, we now have a further

Table 15.5 Search for minimum total weight across full range of M_E

M_C \ M_E	0	65	78	99	135	156	193	223
0	3492	3257	3218	3257	3492	3610	3806	4042
65	3216	2980	3178	3218	3453	3571	3767	4002
78	3216	3178	3178	3218	3453	3571	3767	4002
99	2980	2980	2980	2980	3218	3335	3532	3767
135	3216	3218	3218	3218	3218	3355	3532	3767
156	3332	3335	3335	3335	3335	3335	3532	3767
193	3528	3532	3532	3532	3532	3532	3532	3767
223	4000	4002	4002	4002	4002	4002	4002	4002

requirement for the peak net moment within the additional span EFG:

$$\text{Span EFG: } M_{P_{EG}} \geq \text{the greatest of: } \begin{cases} [M_E] \\ [M_F] = [240 - ½(M_E + M_G)] \\ [M_G] \end{cases} \tag{15.4}$$

An example of the calculations required from equation (15.4) is given in Table 15.6 for $M_G = 0$. Note that this search for the total minimum weight employs the optimum design already established for the double span ACE in column 5. A similar search process is repeated across the full range of M_G values to give the results within Table 15.7. The minimum values, found for M_E at each M_G value, are underlined. From these it becomes apparent that the overall minimum weight for the third span applies to $M_G = 135$ kNm when $M_E = 78$ or 99 kN m. Correspondingly, when seeking the overall minimum weight design for the first two spans (see Table 15.5) we may adjust the fixity of M_E to lie within this narrower range. Next, we compare this DP solution with that from the beam's design space.

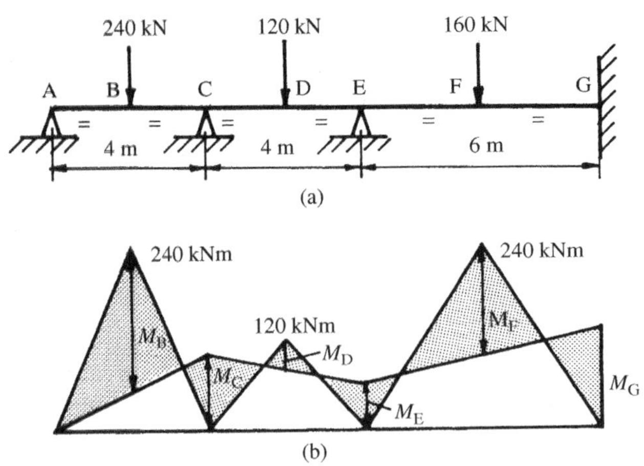

Figure 15.3 Three-span beam with minimum moments at B and C

Table 15.6 Search for minimum total weight when $M_G = 0$

M_E	$M_{\max}$	$(M_P)_{\min}$	W_{EG}	$(W_{ACE})_{min}$	$\sum W$
0	240	240	3061	2980	6041
65	208	223	3004	2980	5984
78	201	223	3004	2980	5984
99	191	193	2649	2980	5629
135	173	193	2649	3218	5867
156	162	193	2649	3335	5984
193	193	193	2649	3532	6181
223	223	223	3002	3767	6769

Table 15.7 Search for minimum total weight across full range of M_E (kNm)

M_E \ M_G	0	65	78	99	135	156	193	223
0	6041	5984	5984	5629	5629	5629	5629	5984
65	5984	5629	5629	5629	5337	5337	5629	5984
78	5984	5629	5629	5337	5160	5337	5629	5984
99	5629	5629	5337	5337	5160	5337	5629	5984
135	5867	5572	5396	5396	5395	5572	5827	6220
156	5984	5690	5690	5690	5690	5690	5984	6337
193	6181	6181	6181	6181	6181	6181	6181	6533
223	6769	6769	6769	6769	6769	6769	6769	6769

15.5 Design Space

In the two-span beam (Figure 15.2), when the fixing moment at E is zero, a minimum total weight is achieved between the spans at AC and CE (see Tables 15.4 and 15.5). This optimisation identifies the continuous beam resting upon three simple supports, in Figure 15.4a. The five possible plastic collapse modes for this beam are shown in Figure 15.4b–f. The energy exchange equation (14.13b) is applied to each mode, with moments in kN m, as follows:

$$\text{Figure 15.4b,}\quad 240\delta = 3M_{P_1}\theta_1 \quad\Rightarrow M_{P_1} \geq 160 \tag{15.5a}$$

$$\text{Figure 15.4c,}\quad 240\delta = 2M_{P_1}\theta_1 + M_{P_2}\theta_2 \quad\Rightarrow M_{P_1} + {}^1\!/_2 M_{P_2} \geq 240 \tag{15.5b}$$

$$\text{Figure 15.4d,}\quad 120\delta = 3M_{P_2}\theta_2 \quad\Rightarrow M_{P_2} \geq 80 \tag{15.5c}$$

$$\text{Figure 15.4e,}\quad 120\delta = M_{P_1}\theta_1 + 2M_{P_2}\theta_2 \quad\Rightarrow {}^1\!/_2 M_{P_1} + M_{P_2} \geq 120 \tag{15.5d}$$

$$\text{Figure 15.4f,}\quad 240\delta - 120\delta = 2M_{P_1}\theta_1 + 2M_{P_2}\theta_2 \Rightarrow M_{P_1} + M_{P_2} \geq 120 \tag{15.5e}$$

Using a linear approximation to the distributed weight (see Figure 14.18) and with each span being 4 m long, the weight function (14.36a) becomes

$$X = \sum(M_P L) = 4(M_{P_1} + M_{P_2}) \tag{15.6}$$

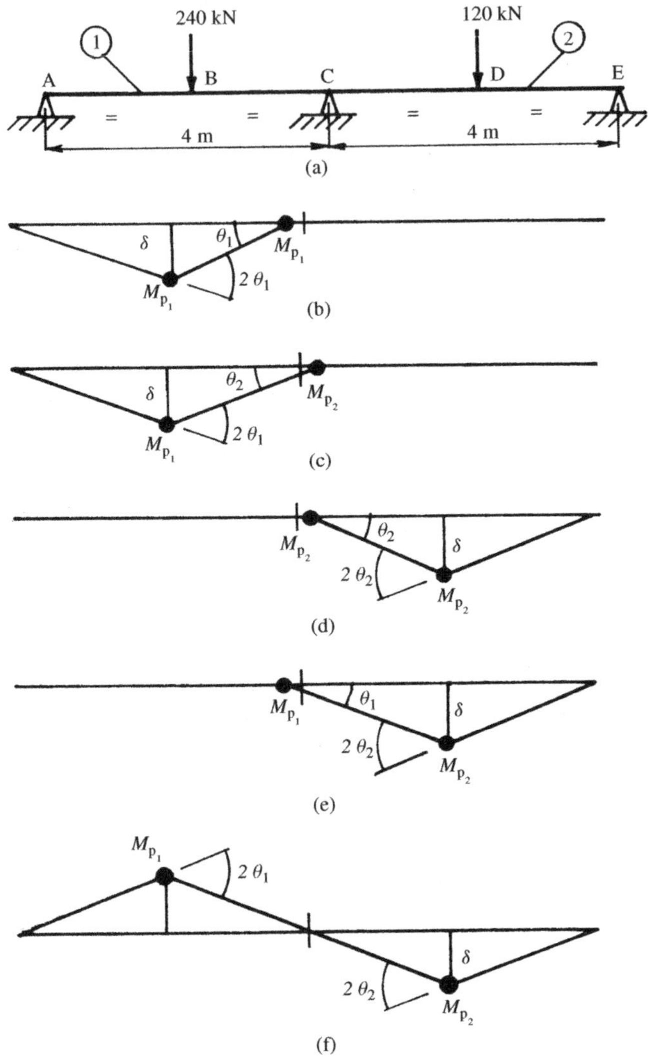

Figure 15.4 Two-span continuous beam showing 5 collapse modes

The equalities in equations (15.5a–e), found from the plastic collapse modes Figure 15.4b–f, appear as straight lines in a design space within axes M_{P_1}, M_{P_2} (see Figure 15.5). When the linear contours (dashed) from the weight function (15.6) are superimposed upon Figure 15.5, for values of X between 800 and 1600 kN m^2, it is seen that a line with $X = 1120$ kN m^2 contacts that corner of the design space where modes (c) and (d) intersect. The intersection point corresponds to $M_{P_1} = 200$ kN m and $M_{P_2} = 80$ kN m. To establish the combination of plastic moments that lies closest to these intersection coordinates, the plastic moments from the range of available sections (see Table 15.1) are superimposed along both axes of the design space, as shown in Figure 15.6. The

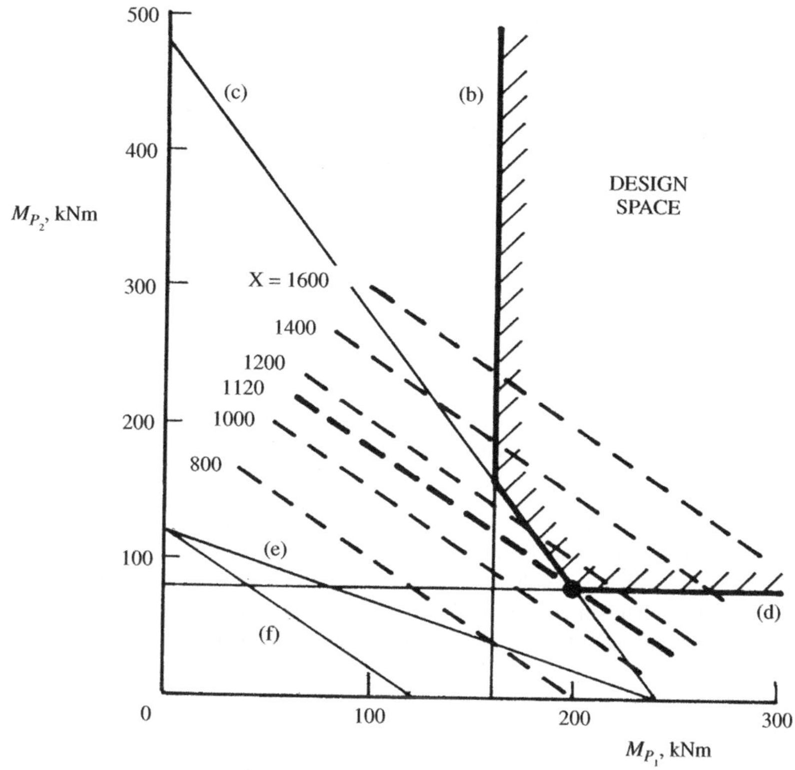

Figure 15.5 Design space for the two-bay, continuous beam shown in Figure 15.4a

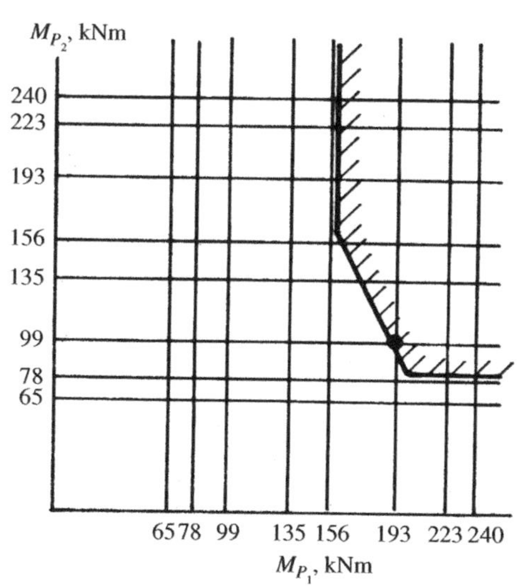

Figure 15.6 Design space with superimposed grid of available plastic moments

grid reveals that sections with plastic moments $M_{P_1} = 193$ kN m and $M_{P_2} = 99$ kN m lie closest to the intersection point. We see that these confirm the earlier result found from dynamic programming (see Tables 15.2 and 15.4). The 'weight' obtained with these available plastic moments becomes $X = 4(193 + 99) = 1168$ kN m^2, compared to the theoretical minimum 1120 kN m^2 shown in Figure 15.5.

Reference

[1] BCSA 28, 1965, Plastic design, British Construction and Steelwork Association.

Exercises

15.1 Conduct searches for the minimum total weight of the beam in Figure 15.2 when the moment at E is preset at (i) 135 kN m and (ii) 223 kN m. Hence confirm the entries for $\sum W$, given in Table 15.5, as 3218 and 3767 N.

15.2 Use the dynamic programming method to establish the right-hand fixing moment when minimising the weight of the continuous beam to support the loading in Figure 14.16a for $L = 1$ m and $P = 200$ kN. Compare with the weight found from the previous solution, given in Chapter 14, in which a plastic hinge mechanism applies to the right-hand end when it is to remains fixed. Take the plastic moments from the range of available sections given in Table 15.1.

Appendix A

Mechanical Properties

A.1 Non-Metals

A few non-metals are listed in Table A.1. Collectively the majority of these have limited load bearing capacity where they show (i) low strength, (ii) brittle failure and (ii) a tendency to creep. Compared to metallic materials (see Table A.2) the properties of non-metals may appear undefined on account of the absence of a linear-elastic region for many polymers and the directional behaviour of woods and composites. The latter have by far the greatest application to weight/strength applications [1–4] where they can be used to maximum effect by aligning the grain/fibre direction with the direction of the major principal tensile stress arising from the loading applied.

Table A.1 Weight, stiffness and strength properties of polymers and composites

Non-Metal[1–4]	Density ρ, kg/m^3	Modulus[a] E, GPa	Strength[c] MPa	% El[d]	Comments
ABS	1045	1.4–3.1	17–58	10–140	
Acetate	1220–1340	1.0–2.0	25–65	5–55	Cellulose
Acrylic	1185	2.7–3.5	50–70	5–8	
Cartilage		0.024			
Carbon	3300	1200		–	Diamond
Carbon	1900–2300	700	20	–	Graphite fibre
Ceramics	3600	276	2480/220	–	90% alumina, C/T
Carbides	1500	400	5170–5930	–	94%WC, 6% Co
CFR epoxy	1700	165	1500	2	Along fibre
Concrete	2400	14	4	2	Compression, 70% water
Douglas fir	500–520	12.5	50/138	5	Along grain (C/T)
Epoxy	1150	3.5–8	50–100	–	Cast, thermoset
GFR polyester	1800	20	350	4	50/50 along fibre

(continued overleaf)

Mechanics of Optimal Structural Design: Minimum Weight Structures D. W. A. Rees

Table A.1 *(continued)*

Non-Metal[1-4]	Density ρ, kg/m^3	Modulus[a] E, GPa	Strength[c] MPa	% El[d]	Comments
Glasses	2600–4100	70	50	–	Al, B, soda silicates
HDPE	935–970	0.4–1.3	22–38	50–300	
LDPE	910–970	0.1–0.3	8–16	300–800	
Nylon 66	1150	2.8–3.3	60–80	60–300	Polyamide
Oak	650	12	100	5	Along grain
Perspex	1190	3	50	2–7	
Polyester	1100–1350	2.4	40–55	650	Thermoset
PTFE	2200	0.34	14–35	200–600	
PC	1200	2.1–2.4	55–65	60–100	
PP	900	1.1–1.6	30–40	50–600	
PS	1060	2.5–4.1	35–60	2–40	
PU	1100–1250	0.01	40	650	
PVC	1400–1700	2.4–4.1	48–58	2–40	Rigid unplasticised
Rubber	910	0.007	17	500	Polyisoprene
Tendon		0.6			

Note: C/T = compression/tension.
For footnotes, see Table A.2.

A.2 Metals and Alloys

The properties listed in Table A.2 apply to both metals and alloys. Particular applications of these materials in engineering structures are normally confined to those alloys that

Table A.2 Weight, stiffness and strength properties of metals and alloys

Metal/Alloy[5-16]	Density ρ, kg/m^3	Elastic moduli[a] E, GPa	G, GPa	0.1% PS[b] σ_y, MPa	UTS[c] σ_{ult}, MPa	% El[d]	Comments
Aluminium[10,11]	2710	70–71	27	50	80	30	99.5% pure
Al-Cu alloy[11,12]	2800	75	28.5	290	425	20	4.4% Cu (2014 -T4)
Al-Mg alloy[11,12]	2725	71	26.5	270	330	8	3.5% Mg (5154- H38)
Alloy steel[5,6]	7900	210	83	750	1000	15	2.5% Ni, Cr, Mo-En25
Antimony	6680	78	29.3				
Bismuth	9800	32	12				
Brass[13]	8500	104	39	140	440	8	Free cutting
Bronze[13]	8800	117	45	100	190	10	+ phosphorous
Cast iron	7150	97	41	310	500(C)	7	Nodular type (BS 2789)
Cobalt	8900	206	79		500		
Constantan	8880	170	63.9				

Table A.2 *(continued)*

Metal/Alloy[5–16]	Density ρ, kg/m^3	Elastic moduli[a] E, GPa	G, GPa	0.1% PS[b] σ_y, MPa	UTS[c] σ_{ult}, MPa	% El[d]	Comments
Copper	8950	96–117	38	65	175	45	
Gold	19300	79	27	80	120	40	
Invar[15]	8000	145	56	280	480	40	64% Fe, 36% Ni
Iron	7850	200–206	82	165	300	45	Pure
Lead	11370	17–18	6	12	15	50	
Magnesium	1740	44	17.1	95	190	5	
Manganese	8500	120	45.1				
Mild-steel[6,7]	7860	207	81	300	510	35	0.4% C
Monel	8800	206	79	240	420–520	40	70% Ni, 30% Cu
Nickel	8900	198	80	60	300	30	
Ni-Alloy[16]	9000	200	79	800	1000	10	76% Ni, 20% Cr
P-Bronze[13]	8800	120	43.5	420	560	30	C510 wrought
Platinum	21040	164	51	250	350		
Silver	10530	78	29	150	180	45	
Stainless steel[7,8]	7930	200	77	210	510	60	18% Cr, 8% Ni (austenitic)
Tin	7300	40	14.7		30		
Titanium	4540	118	45	480	620	20	
Ti-Al alloy[9]	4430	110	42	800	860	15	5% Al (α-alloy)
Ti-Cu alloy[9]	4700	115	44	700	750	15	2.5% Cu
Ti-Sn alloy[9]	4600	105	40	1000	1300	12	11% Sn
Tungsten	19300	410	157				
Zinc	6860	86	38	100	150	50	
Zn-Alloy[14]	6800	80	31	250	330	50	4% Al (β-alloy)

[1–16]These refer to the list of works at the end of this appendix.
[a]E is the modulus of elasticity (Young's modulus), as found from the gradient to the linear-elastic, stress–strain response in a tensile test. Where non-linear behaviour is observed, in polymers and rubbers, E expresses the gradient to the stress–strain curve at its origin. G is the modulus of rigidity (shear modulus) most usually found from the gradient to the linear-elastic, shear stress versus shear strain response in a torsion test. Fewer values of G are available for non-metals than metals though given E for a non-metal, a value for G may be approximated by $G = E/3$.
[b]0.1% proof stress (PS) refers to the stress corresponding to an offset strain of 0.1%. This is a common measure of the yield strength for a metallic material that does not show a sharply defined yield point. The offset is the amount of permanent set (plastic strain) that would arise following an unloading from the quoted proof stress level. Offsetting by lesser or greater amounts of strain (typically 0.05% or 0.2%) is permitted depending upon the amount of permanent set that is tolerable. This yield strength measure does not appear as a listed property for non-metals that respond to stress either in a non-linear or in a brittle manner.
[c]Ultimate tensile strength (UTS) is the greatest stress a material sustains at failure. It would normally be calculated by dividing the maximum load observed in a tensile test by the original cross-sectional area, i.e. without correction for the area reduction under maximum load.
[d]Percentage elongation (% El) refers to the percentage amount by which the initial parallel length of a tensile testpiece has been stretched to failure. It is also the limiting percentage engineering strain for a material since this strain is also defined identically as (extension $\div$ original length) $\times$ 100.

can offer the required strength and safety imposed by the loading. The most common load bearing alloys of steel, aluminium and titanium, as they appear in various standards [5–16], are too numerous to list here. Table A.2 is intended to provide a comparison between selected mechanical properties of these materials and similar properties of other metals and alloys that are required occasionally bear load.

References

[1] Bolton, W. *Engineering Materials Pocket Book*, 2nd edition, Newnes, 1996.
[2] Matthews, C. *Engineers' Data Book*, PEP Ltd., Institute of Mechanical Engineers, 1998.
[3] Tennent, R. M. *Science Data Book*, Oliver and Boyd, 1971.
[4] Ashby, M. F. *Materials Selection in Mechanical Design*: *Materials and Process Selection Charts* (Appendix C), Pergamon Press, 1985.
[5] BS 970: 1955, En steel specifications.
[6] SAE 1000-1095, 1958, Plain carbon and alloy steels.
[7] BS 970 Part 1: 1986, Requirements for carbon, carbon manganese, alloy and stainless steels.
[8] AISI 201-502: 1955, Stainless and heat resisting steels.
[9] BS En 2858-1: 1994, Titanium and titanium alloys.
[10] BS 1490: 1988, Specification for aluminium and aluminium alloy ingots and castings.
[11] BS 1470–1477: 1972, Wrought aluminium and aluminium alloys (sheet, strip, plate, tube and wire).
[12] ADA, Inf. Bull. No.2, *The Properties of Aluminium and its Alloys*, Aluminium Development Association, 1995.
[13] BS 2870: 1980, Rolled copper and copper alloys.
[14] BS1004: 1985, Specification of zinc alloy die castings.
[15] BS 3072: 1996, Specification for nickel and nickel alloys – sheet and plate.
[16] ASTM B574: 1994, Specification for low carbon, nickel-chromium and other alloys.

Appendix B

Plate Buckling Under Uniaxial Compression

B.1 Wide and Slender Struts

Consider the axial compression of a wide, thin plate, of thickness t, with similar length and width dimensions, a and b respectively (see Figure. B.1a). Having simply supported (pinned) ends and unsupported sides, the strut deflects laterally with uniform curvature and surface strains in opposite sense. The cross-section $b \times t$ remains rectangular during bending when biaxial stresses σ_1 and σ_2 exist in the plane of the plate as shown.

The lateral stress σ_2 arises from there being no strain in the 2-direction:

$$\varepsilon_2 = 0 = (1/E)(\sigma_2 - \nu\sigma_1), \Rightarrow \sigma_2 = \nu\sigma_1$$

$$\varepsilon_1 = (1/E)(\sigma_1 - \nu\sigma_2) = (1 - \nu^2)(\pm\sigma_1)/E \tag{B.1}$$

In a slender strut, where b is small compared to a, the presence of lateral strains ε_2 results in *anticlastic curvature*, i.e. the cross-section does not remain rectangular. The distortion arises from opposing ε_2 within the tensile and compressive surfaces (the Poisson effect). Here, as $\sigma_2 = 0$, the axial and lateral surface strains are $\varepsilon_1 = (\pm\sigma_1)/E$ and $\varepsilon_2 = -\nu(\pm\sigma_1)/E$. Comparing ε_1 with equation (B.1) reveals a difference in ε_1 between slender and wide struts. Consequently, it becomes necessary to modify the flexure equation for a wide strut by the factor $(1 - \nu^2)$. From Figure B.1b, the curvature is expressed in reciprocal form:

$$\mathrm{d}^2w/\mathrm{d}x^2 = 1/R = \varepsilon_1/z$$

Substituting from equation (B.1), with $M/I = \sigma_1/z$ (the sign of σ_1 depends upon sign of z),

$$\frac{\mathrm{d}^2w}{\mathrm{d}x^2} = \frac{1-\nu^2}{E}\frac{\sigma_1}{z} = \frac{(1-\nu^2)M}{EI}$$

Mechanics of Optimal Structural Design: Minimum Weight Structures D. W. A. Rees

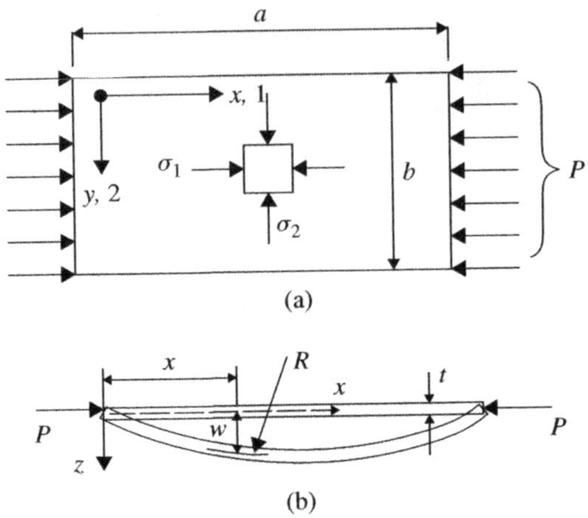

Figure B.1 Buckling of a wide plate strut

With a bending moment $M = -Pw$, the solution to the critical buckling load for a pinned-end plate is

$$P_{\mathrm{cr}} = \frac{\pi^2 EI}{(1-\nu^2)a^2}$$

Substituting $P_{cr} = \sigma_{cr} bt$ and $I = Ak^2$ leads to the critical buckling stress in the section

$$\sigma_{\mathrm{cr}} = \frac{\pi^2 E}{(1-\nu^2)(a/k)^2} \tag{B.2a}$$

Other edge fixings may be accounted for by rewriting equation (B.2a) in terms of an equivalent length L_e as

$$\sigma_{\mathrm{cr}} = \frac{\pi^2 E}{(L_e/k)^2} \tag{B.2b}$$

in which the equivalent lengths are:

$$L_e = a(1-\nu^2)^{1/2} \quad \text{for pinned ends} \tag{B.3a}$$

$$L_e = (a/2)(1-\nu^2)^{1/2} \quad \text{for fixed ends} \tag{B.3b}$$

$$L_e = (a/\sqrt{2})(1-\nu^2)^{1/2} \quad \text{for pinned-fixed ends} \tag{B.3c}$$

$$L_e = 2a(1-\nu^2)^{1/2} \quad \text{for fixed-free ends} \tag{B.3d}$$

Moreover, an inelastic buckling stress for wide-plate struts may be estimated from equation (B.2b) from these equivalent lengths when E in equation (B.2b) is replaced by the tangent modulus E_T. In this book we write the plate buckling stress formulae in an alternative form. Setting $k^2 = t^2/12$ in equation (B.2a) reveals the form adopted:

$$\sigma_{cr} = \frac{\pi^2}{12}\left(\frac{b}{a}\right)^2\left(\frac{E}{1-\nu^2}\right)\left(\frac{t}{b}\right)^2 = KE\left(\frac{t}{b}\right)^2 \tag{B.4}$$

where the buckling coefficient K depends upon the plate geometry, the elastic constant ν and the manner in which its ends are supported (see equations (B.3a–d)).

B.2 Plates with Supported Sides

More generally, with the unidirectional compression of thin, wide plates the sides may be free, simply supported or clamped. In the case of free, or unsupported, sides we return to the wide-strut problem considered above. Compressive loading of thin, wide plates, as opposed to slender struts, refers to plates whose side lengths a and b (see Figure B.1) do not differ by more than an order of magnitude. The theory outlined in Appendix C accounts for buckling of thin plates with similar geometry under biaxial compression when their four sides have various types of edge support. We may derive from this theory the following reduction to axial compression as required here.

B.2.1 Simple Supports

Under a unidirectional compression, the critical buckling stress σ_1, for a plate with all four sides simply supported, follows from setting $\sigma_2 = 0\,(\beta = 0)$ in equation (C.1). This gives [1]:

$$\sigma_{cr} = \frac{D\pi^2}{t}\left[\left(\frac{m}{a}\right)^2 + \left(\frac{n}{b}\right)^2\right]^2\left(\frac{a}{m}\right)^2 = \frac{D\pi^2}{tb^2}\left(\frac{mb}{a} + \frac{n^2 a}{mb}\right)^2 \tag{B.5a}$$

where $D = Et^3/[12(1-\nu^2)]$ is the flexural stiffness, and t, a and b are the plate thickness, length and breadth, respectively. The number of half-waves of buckling in the 1- and 2-directions are denoted by m and n, respectively. For example, the simply supported plate in Figure B.2 has buckled with a single half-wave ($n = 1$) in the 2-direction and three half-waves ($m = 3$) in the 1-direction.

When $n = 1$ in equation (B.5a), m will correspond to a minimum in the buckling stress expression:

$$\sigma_{cr} = \frac{D\pi^2}{tb^2}\left(\frac{m}{r} + \frac{r}{m}\right)^2 \tag{B.5b}$$

where $r = a/b$. Differentiating for the m-values that minimise equation (B.5b) with integral r:

$$d\sigma_{cr}/dm = 2(m/r + r/m)(1/r - r/m^2) = 0$$

$$m/r^2 + 1/m - 1/m - r^2/m^3 = 0$$

$$m^4 - r^4 = 0$$

$$(m - r)(m + r)(m^2 + r^2) = 0$$

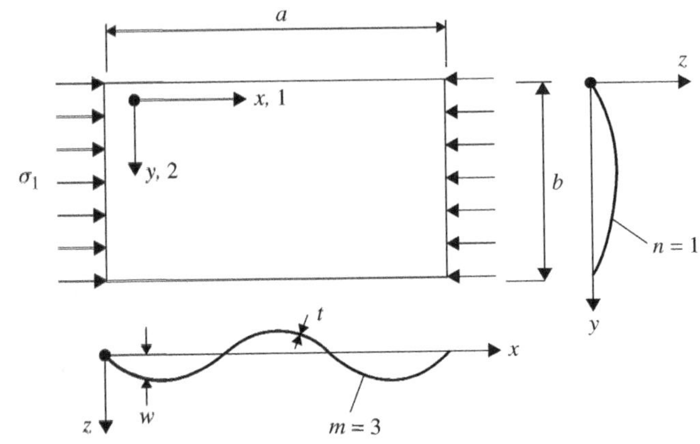

Figure B.2 Buckling of a thin, simply supported plate under uniaxial compression

The condition $m = r$ implies that the plate will buckle into an integral number of square cells $b \times b$ each under the same stress. That is, from equation (B.5b),

$$\sigma_{cr} = \frac{4D\pi^2}{tb^2} = \frac{\pi^2 E}{3(1-\nu^2)}\left(\frac{t}{b}\right)^2 \quad \text{(B.6a,b)}$$

which has a similar form to the wide strut buckling stress in equation (B.4). However, K will now take a different value depending upon r and m. Figure B.3 provides the buckling stress when both integer and non-integer values of r are taken with values of $m = 1$, 2, 3 and 4 in equation (B.5b). The trough in each curve corresponds to equation (B.6a). At the intersections of these curves, where m has been increased by one successively:

$$[m/r + r/m] = [(m+1)/r + r/(m+1)]$$

$$\therefore r = \sqrt{m(m+1)}$$

from which $r = \sqrt{2}$ for $m = 1$, $r = \sqrt{6}$ for $m = 2$, $r = \sqrt{12}$ for $m = 3$, etc., as shown.

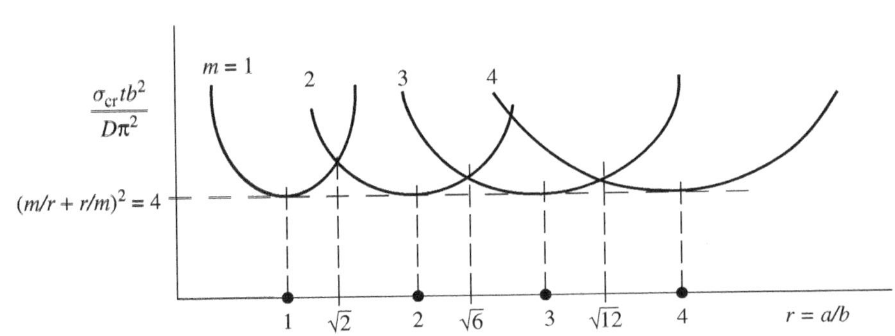

Figure B.3 Effect of r and m on uniaxial buckling stress

B.2.2 Other Edge Fixings

Various methods account for the effect of constraining the plate's edges. The simplest of these [2] generalise equation (B.6a) by introducing elastic constraint coefficients K as follows:

$$\sigma_{\text{cr}} = \frac{K_r D\pi^2}{tb^2} = K_e E\left(\frac{t}{b}\right)^2 \qquad \text{(B.7a,b)}$$

where equation (B.7a) applies to any edge fixings when

$$K_r = (m/r)^2 + p + q(r/m)^2 \qquad \text{(B.7c)}$$

Clearly, with edge rotational restraint factors $p = 2$ and $q = 1$, equation (B.7c) contains equation (B.5b) in the special case of simple supports. The dependence of K_r upon the edge restraint factors (p and q), the plate aspect ratio ($r = a/b$) and the buckling mode m, has been established experimentally in certain cases. For example, Table B.1 applies to a plate with fixed sides. As r increases, the effect of edge restraint lessens and K_r approaches the minimum value of 4 as found from equation (B.6a) for a plate with simply supported edges. A more convenient graphical approach employs equation (B.7b) within design curves that provide the restraint coefficient $K_e = K_r \times \pi^2/[12(1 - \nu^2)]$ for both uniaxial and biaxial loadings [3]. These curves present the ratio between the critical elastic buckling stress $(\sigma_{cr})_e$ for a plate with a particular edge fixings to the buckling stress for a simply supported plate (i.e. equation (B.6b)) with a similar aspect ratio $r = a/b$. Figure B.4 shows how this stress ratio varies with r for clamped and various mixed-edge fixings for plates under uniaxial compression.

To apply Figure B.4, for example, let a 5 mm thick aluminium alloy plate with $a = 400$ mm and $b = 200$ have compression applied through a rigid clamp to its shorter sides. The buckling stress is required when the longer side is simply supported. Firstly, the reference stress, used to normalise the ordinate in Figure B.4, is given by equation (B.6b):

$$\sigma_{\text{cr}} = \frac{\pi^2 E}{3(1 - \nu^2)}\left(\frac{t}{b}\right)^2 = \frac{\pi^2 \times 70000}{3(1 - 0.33^2)}\left(\frac{5}{200}\right)^2 = 25.9 \text{ MPa}$$

in which the elastic constants are $E = 70$ GPa and $\nu = 0.33$. Reading from the appropriate graph in Figure B.4 gives the elastic buckling coefficient for $a/b = 2$:

$$\eta = (\sigma_{\text{cr}})_e/\sigma_{\text{cr}} = 1.22$$

Table B.1 Restraint coefficients for a plate with fixed edges

$r = a/b$	0.75	1.0	1.5	2.0	2.5	3.0
K_r	11.69	10.07	8.33	7.88	7.57	7.37

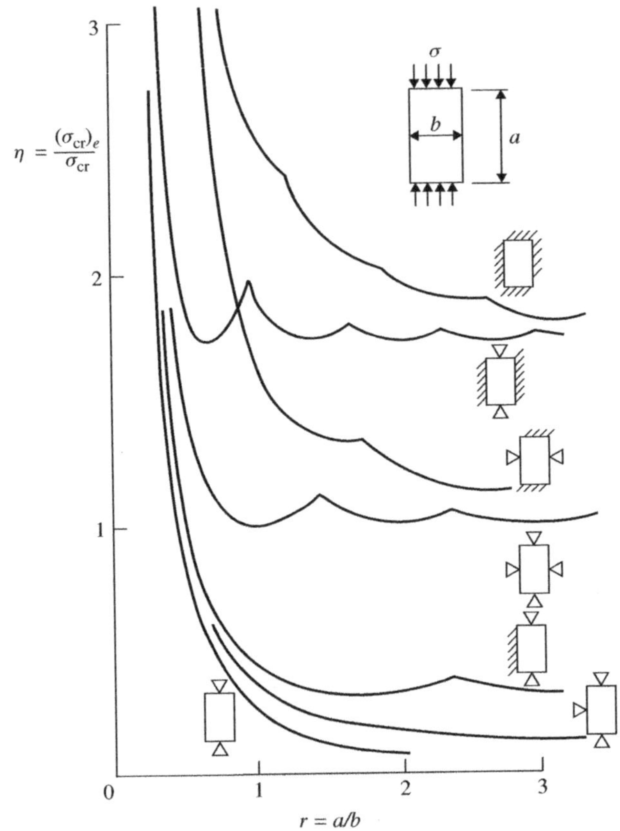

Figure B.4 Plate buckling under compression

Hence, the critical, elastic buckling stress follows as

$$(\sigma_{cr})_e = \eta\,\sigma_{cr} = 1.22 \times 25.9 = 31.6 \text{ MPa}$$

B.3 Inelastic Buckling

In thicker plates the buckling stress $(\sigma_{cr})_e$, calculated from Figure B.4 and equation (B.6b), in the manner of the previous example, may exceed the yield stress σ_y of the plate material. The solution will be invalid because linear elasticity cannot then be assumed at the critical stress level. Elastic buckling analyses employ the elastic compressive modulus which is constant up to the proportional limit. A correction for plastic buckling accounts for the decreasing gradient to the flow curve as the stress increases within the plastic range (see Figure B.5a).

In a deflected, slender strut a stress gradient arises where the sum of direct and bending stresses amounts to a greater compression on its concave side. Where the elastic limit is exceeded a tangent modulus E_T (see Figure B.5b) is often employed with Euler's strut

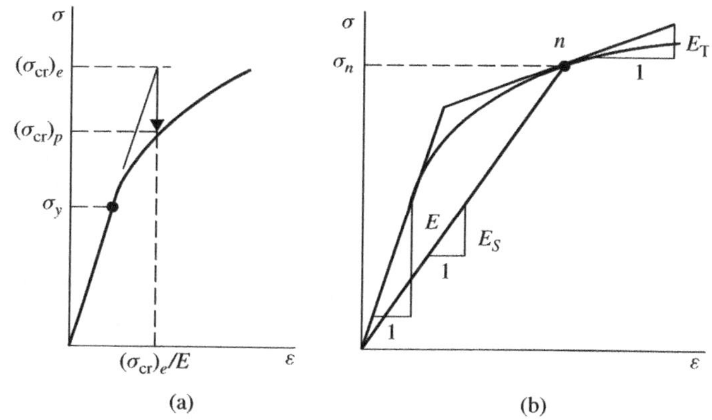

Figure B.5 Tangent and secant moduli

theory to provide an average, compressive stress at buckling. Equally, E_T may replace E for an account of plastic buckling of thin plates, particularly where through-thickness stress gradients are less severe. Figure B.5a shows that where plasticity has occurred it reduces the buckling stress to a lower level $(\sigma_{cr})_p$. Figure B.5b shows how both the tangent and secant moduli, E_T and E_S respectively, are used to account for a plastic stress level. The secant modulus gives the total strain at a reference point n as

$$\varepsilon_n = \sigma_n / E_S \tag{B.8a}$$

The Ramberg–Osgood description [4] to a stress–strain curve gives the total strain under a plastic stress level as

$$\varepsilon = \sigma/E + \alpha(\sigma/E)^m \tag{B.8b}$$

Combining equations (B.8a,b) at the reference stress σ_n gives

$$E/E_S - 1 = \alpha(\sigma_n/E)^{m-1} \tag{B.9a}$$

Differentiating equation (B.8b) gives the gradient $E_T = \mathrm{d}\sigma/\mathrm{d}\varepsilon$ at point n:

$$(1/m)(E/E_T - 1) = \alpha(\sigma/E)^{m-1} \tag{B.9b}$$

Now let σ_n be the stress level at which $E_T = E/2$. Equation (B.9b) gives

$$\alpha = (1/m)(\sigma_n/E)^{1-m} \tag{B.9c}$$

Substituting equation (B.9c) into equation (B.8b) and multiplying through by E/σ_n leads to a normalised stress–total strain relationship

$$\varepsilon = \frac{\sigma_n}{E}\left[\frac{\sigma}{\sigma_n} + \frac{1}{m}\left(\frac{\sigma}{\sigma_n}\right)^m\right] \tag{B.10}$$

Table B.2 m-values for aluminium (L and DTD - grades) and steel (S - grades)

Material	L71	L73	DTD687	L65	S520	S514	S515	S134	S96
m	16	17	15	29	4.9	17	8.4	7	105
Form	S	S	P	B	S	S	S	B	B

where σ_n and m are material properties found from fitting equation (B.10) to a stress–strain curve. Table B.2 gives typical m-values that lie in the range 16–29 for aluminium alloys and 5–105 for steels [5]. The m-value depends upon whether the material is in sheet (S), plate (P) or bar (B) form.

Figure B.5a shows that the elastic strain under $(\sigma_{cr})_e$ is $\varepsilon = (\sigma_{cr})_e/E$. Substituting this value for ε in equation (B.10) allows $\sigma = (\sigma_{cr})_p$ to be found. Design data [6] adopt a graphical solution to $(\sigma_{cr})_p$, employing a *plasticity reduction factor* $\mu < 1$ in

$$\mu = (\sigma_{\text{cr}})_p/(\sigma_{\text{cr}})_e \tag{B.11}$$

Both m and σ_n influence μ in the manner shown in Figure B.6. When applying Figure B.6, firstly, we find from Figure B.4, the critical elastic buckling stress $(\sigma_{\text{cr}})_e$. This determines the ratio $(\sigma_{\text{cr}})_e/\sigma_n$ from which a μ-value is found from Figure B.6. Equation (B.11) is then employed to find $(\sigma_{\text{cr}})_p$.

Note that equation (B.6) may be used to find $(\sigma_{\text{cr}})_e$ for a simply supported plate subjected to uniaxial compression across its shorter sides. For example, consider the buckling of a 5 mm thick S520 steel plate with sides $a = 600$ mm and $b = 300$ mm under this condition. Given this material's properties ($E = 210$ GPa, $v = 0.27$, $m = 4.9$, $\sigma_y = 300$ MPa and $\sigma_n = 450$ MPa), we find from equation (B.6b) the critical elastic buckling stress $\sigma_{\text{cr}} = (\sigma_{\text{cr}})_e$:

$$(\sigma_{\text{cr}})_e = \frac{\pi^2 E}{3(1-v^2)}\left(\frac{t}{b}\right)^2 = \frac{\pi^2(210\times 10^3)}{3(1-0.27^2)}\left(\frac{5}{200}\right)^2 = 465.75 \text{ MPa}$$

As this stress exceeds $\sigma_y = 300$ MPa, it requires a correction for plasticity. The abscissa in Figure B.6 is $465.75/450 = 1.035$, from which its ordinate $\mu = 0.83$ applies to $m = 4.9$.

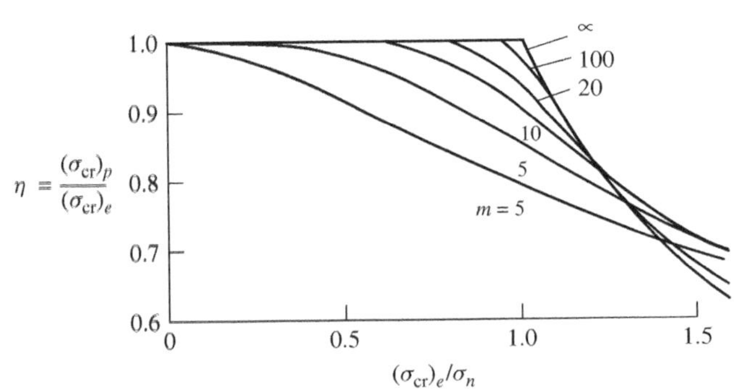

Figure B.6 Plasticity reduction factor for a simply supported plate

Hence, from equation (B.11), the critical plastic buckling stress becomes

$$(\sigma_{cr})_p = \mu(\sigma_{cr})_e = 0.83 \times 465.75 = 386.57 \text{ MPa}$$

We should note here that Figure B.6 is constructed for a plate having all four sides simply supported in a material with Poisson's ratios 0.3 and 0.5 within the elastic and plastic ranges, respectively. Fortunately, the greatest possible deviations in ν (from 0.25 to 0.35) can have only a negligible influence upon the reduction factor and therefore this effect may be ignored. The elastic Poisson's ratio effect in Figure B.4 should, however, be accounted for. The effect of the plate supports upon the reduction factor μ has been discussed further in [7, 8]. This amounts to a vertical shift in the curves in Figure B.6 when the sides are supported by other means. For example, if this plate has clamped edges at both ends and along both sides, we read from the upper curve in Figure B.4 the following stress ratio, corresponding to the integral number of buckled panels (i.e. $r = a/b = 2$) in the sheet:

$$\eta = (\sigma_{cr})_e/\sigma_{cr} = 2$$

As the denominator has been identified with 465.75 MPa for a simply supported plate, the required elastic buckling stress for the clamped plate follows as $(\sigma_{cr})_e = 2 \times 465.75 = 931.5$ MPa. Strictly speaking, three further corrections should be applied to this stress magnitude:

(i) $(\sigma_{cr})_e$, as calculated above, should be multiplied by the factor $(1 - 0.3^2)/(1 - \nu^2)$ for a Poisson's ratio different from 0.3 (the basis of Figure B.4). This gives

$$(\sigma_{cr})_e = \frac{1 - 0.30^2}{1 - 0.27^2} \times 931.5 = 914.3 \text{ MPa}$$

(ii) $(\sigma_{cr})_e$ in (i) above should then be corrected for plasticity using Figure B.6 with an abscissa value of $914.3/450 = 2.03$. This gives $\mu \approx 0.62$, at $m = 4.9$, for a plate with simply supported edges.

(iii) With clamped plate edges, a downward shift in the respective curve in Figure B.6 has shown [6] that $\mu \approx 0.58$, i.e. a multiplying factor of 0.935 applies to the plasticity reduction factor found in (ii). Hence, the critical plastic buckling stress is estimated as

$$(\sigma_{cr})_p = \mu(\sigma_{cr})_e = 0.58 \times 914.3 = 530.3 \text{ MPa}$$

In the event of a near-ultimate strength value being found from this procedure the implication is that, in the manner of its fixing, this plate cannot buckle without a significant amount of accompanying plasticity. Elastic-plastic buckling at a lesser critical stress would be more likely to arise in a plate of reduced thickness.

B.4 Post-Buckling

When a plate has buckled the load may be increased further as the axial compressive stress increases in the material adjacent to the side supports (see Figure B.7). Only a

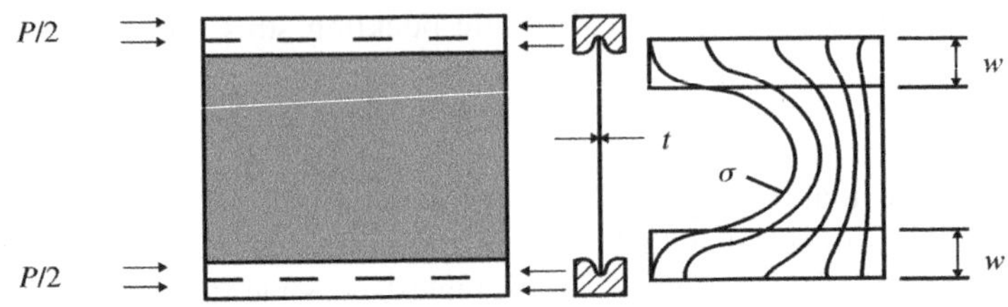

Figure B.7 Stress distribution in a buckled plate

slight increase in axial stress occurs in the central buckled material. Consequently, if we assume that the whole load P is carried by two edge strips of effective width $2w$, over which σ is assumed constant, the load supported becomes

$$P = 2wt\sigma \quad \text{(B.12a)}$$

With the edges of our equivalent elastic plate all simply supported, the critical buckling stress becomes [2]

$$\sigma_{cr} = \frac{\pi^2 E}{3(1-\nu^2)}\left(\frac{t}{2w}\right)^2 \quad \text{(B.12b)}$$

from which w may be found once σ_{cr} attains the yield stress σ_y:

$$w = \frac{\pi t}{2}\sqrt{\frac{E}{3\sigma_y(1-\nu^2)}} \quad \text{(B.12c)}$$

Taking $\nu = 0.3$ in equation (B.12c) gives

$$w = 0.95t\sqrt{E/\sigma_y} \quad \text{(B.13a)}$$

Hence when $b = 2w$ the full plate width is used most effectively in post-buckling. With alternative support along the longer side, the asymptotic values of the critical elastic buckling stress ratio apply. These appear in Figure B.4, in which equation (B.12b) defines the denominator for the ordinate $\eta = (\sigma_{cr})_e/\sigma_{cr}$. For example, when one long side is simply supported and the other parallel side is free, the second lowest plot in Figure B.6 shows that the critical stress ratio is 0.106. This modifies the semi-effective width for the plate:

$$w = 0.95t\sqrt{0.106E/\sigma_y} = 0.31t\sqrt{E/\sigma_y} \quad \text{(B.13b)}$$

Experiment [9] has shown that the coefficient in equation (B.13a) is nearer 0.85. However, when the 'plate' forms the compressive surface of a box section in bending the coefficient is 1.14.

References

[1] Vinson, J. R. *Structural Mechanics*, Wiley, 1974.
[2] Timoshenko, S. P. and Gere, J. *Theory of Elastic Stability*, 2nd edition, McGraw-Hill, 1961.

[3] ESDU 72019, Buckling of flat isotropic plates under uniaxial and biaxial loading, August 1972.
[4] Ramberg, W. and Osgood, W. R. Description of stress-strain curves by three parameters, NACA Tech Note 902, April 1943.
[5] ESDU 76016, Generalisation of smooth, continuous stress-strain curves for metallic materials, May 1985.
[6] ESDU 83044, Plasticity correction factors for plate buckling, December 1983.
[7] Gerard, G. and Becker, H. *Handbook of Structural Stability*, Part I - Buckling of flat plates, NACA Tech Note 3781, October 1954.
[8] Gerard, G. *Introduction to Structural Stability Theory*, McGraw-Hill, 1962.
[9] Yu, W., *Cold Formed Steel Structures*, McGraw-Hill, 1973.

Appendix C

Plate Buckling Under Biaxial Compression and Shear

C.1 Biaxial Compression

Shear buckling and uniaxial compressive buckling of wide plates are special cases of plate buckling under biaxial compression. An outline of the latter theory is given here. The appropriate reduction to shear is made in § C.2 and to uniaxial compression in Appendix B.

Consider a simply supported, thin rectangular plate $a \times b$ with thickness t. Let uniform compressive stresses σ_1 act normal to $b \times t$ and σ_2 act normal to $a \times t$, as shown in Figure C.1. When the stresses increase proportionately in the ratio $\beta = \sigma_2/\sigma_1$, the actual number of half-waves of buckling is that which minimises σ_1. In general, this stress is expressed as [1]

$$(\sigma_1)_{cr} = \frac{D\pi^2[(m/a)^2 + (n/b)^2]^2}{t[(m/a)^2 + \beta(n/b)^2]} \tag{C.1}$$

where $D = Et^3/[12(1 - \nu^2)]$ is the flexural stiffness, and m and n are the respective number of half-waves of buckling in the x- and y-directions. For example, in Figure C.1 a buckled plate is shown where $m = 3$ and $n = 1$. For a square plate, where $a = b$, buckling occurs with one half-wave in each direction. That is, $m = n = 1$ and equation (C.1) reduces to

$$(\sigma_1)_{cr} = \frac{\pi^2 E}{3(1 - \nu^2)(1 + \beta)}\left(\frac{t}{a}\right)^2 \tag{C.2a}$$

In the case of equi-biaxial compression, where $\beta = 1$, equation (C.2a) gives

$$(\sigma_1)_{cr} = \frac{\pi^2 E}{6(1 - \nu^2)}\left(\frac{t}{a}\right)^2 \tag{C.2b}$$

Mechanics of Optimal Structural Design: Minimum Weight Structures D. W. A. Rees

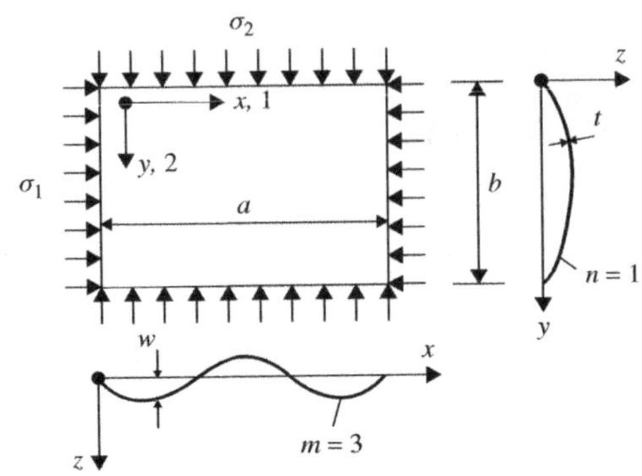

Figure C.1 Buckling of a thin plate under biaxial compression

A useful buckling stress formula $(\sigma_1)_{cr}$ applies to a slender plate with aspect ratio $r = a/b$ in which $b \ll a$. Firstly, we must find the number of half- wavelengths for the buckled shape for the x-direction. Taking $n = 1$ for the much smaller b-dimension, equation (C.1) becomes

$$(\sigma_1)_{cr} = \frac{D\pi^2(m^2 + r^2)^2}{a^2t(m^2 + \beta r^2)} = \frac{D\pi^2(m/r + r/m)^2}{b^2t[1 + \beta(r/m)^2]} \tag{C.3}$$

and m is found from the condition that $(\sigma_1)_{cr}$ in equation (C.3) is a minimum. This gives

$$\begin{aligned} \mathrm{d}(\sigma_1)_{cr}/\mathrm{d}m &= [1 + \beta(r/m)^2]2(m/r + r/m)(1/r - r/m^2) \\ &\quad - (m/r + r/m)^2 \times 2\beta(r/m)(-r/m^2) = 0 \\ [1 - (r/m)^2][1 &+ \beta(r/m)^2] + \beta(r/m)^2[1 + (r/m)^2] = 0 \\ \therefore (r/m) &= 1/(1 - 2\beta)^{1/2} \end{aligned} \tag{C.4a}$$

whereupon the number of half- wavelengths in the 1-direction becomes

$$m = r(1 - 2\beta)^{1/2} \tag{C.4b}$$

The corresponding half- wavelength is

$$a/m = b(r/m) = b/(1 - 2\beta)^{1/2} \tag{C.4c}$$

Substituting equation (C.4a) into equation (C.3),

$$(\sigma_1)_{cr} = \frac{D\pi^2[(1 - 2\beta) + 2 + 1/(1 - 2\beta)]}{bt^2[1 + \beta/(1 - 2\beta)]} = \frac{4D\pi^2}{b^2t}(1 - \beta) \tag{C.5a}$$

It is apparent from equation (C.4c) that as $\beta \to 1/2$ the half-wavelength approaches infinity. The corresponding buckling stress is, from equation (C.5a),

$$(\sigma_1)_{cr} \to (2D\pi^2)/(b^2t) \tag{C.5b}$$

For each solution to $(\sigma_1)_{cr}$ the critical lateral stress follows as $(\sigma_2)_{cr} = \beta(\sigma_1)_{cr}$. The two critical stresses remain elastic provided they satisfy a von Mises yield criterion

$$\sigma_1^2 - \sigma_1\sigma_2 + \sigma_2^2 \leq \sigma_y^2 \tag{C.6a}$$

where σ_y is the uniaxial yield stress. For a given β, equation (C.6a) limits σ_1 to

$$\sigma_{1y} \leq \sigma_y/(1 - \beta + \beta^2)^{1/2} \tag{C.6b}$$

If $(\sigma_1)_{cr}$ from equation (C.5a) exceeds the limiting elastic stress value in equation (C.6b) it must be corrected for plasticity. For this correction we can use the uniaxial plasticity reduction factor in Figure B.6. However, when applying this figure to biaxial compression it becomes necessary to redefine the abscissa as $(\sigma_1)_{cr}/\sigma_n'$. Here σ_n' follows from equation (C.6b) as $\sigma_n' = \sigma_n/(1 - \beta - \beta^2)^{1/2}$ where σ_n refers to the plastic stress value corresponding to a secant modulus $E_S = E/2$ under uniaxial stressing.

C.2 Pure Shear

When the sides of a thin plate $a \times b \times t$ are subjected to shear stress τ, the principal stress state (σ_1, σ_2) within the plate becomes one of diagonal tension and compression. The principal stresses are equal in magnitude but have opposite sense in compression and tension: $\sigma_1 = -\tau$ and $\sigma_2 = \tau$ respectively, as shown in Figure C.2a. Hence, we have a principal, biaxial stress ratio $\beta = -1$ under pure shear for which a buckling condition will arise. Shear buckling of flat plates occurs with parallel wrinkles lying perpendicular to the compressive stress σ_1 (see Figure C.2b).

When buckled, the plate cannot sustain a further increase in diagonal compression, though an increase in diagonal tension is possible. To achieve the pure shear state in Figure C.1 the direction of σ_2 is reversed to tension and its magnitude is made equal to that of σ_1. On setting $\beta = -1, n = 1$ and $r = a/b$ in equation (C.1), the critical, elastic compressive stress σ_1 for a plate in shear becomes

$$(\sigma_1)_{cr} = \frac{D\pi^2}{a^2 t}\frac{(m^2 + r^2)^2}{m^2 - r^2} \tag{C.7a}$$

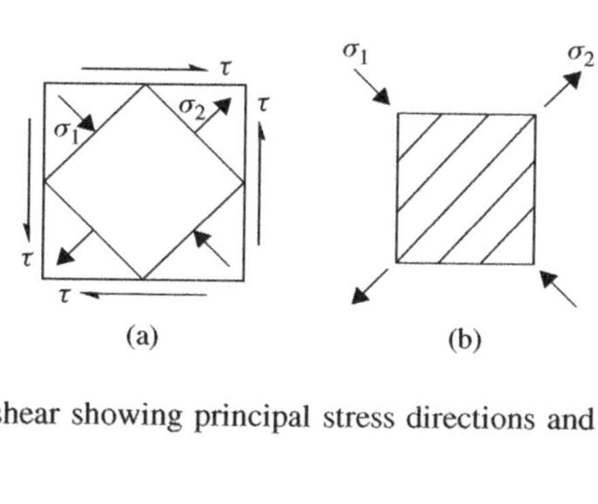

Figure C.2 Plate in shear showing principal stress directions and wrinkling normal to σ_1

where m, the number of half-waves in the 1-direction, is found when $(\sigma_1)_{cr}$ in equation (C.7a) is a minimum. That is,

$$\frac{d(\sigma_1)_{cr}}{dm} = (m^2 - r^2) \times 2(m^2 + r^2) \times 2m - (m^2 + r^2)^2 \times 2m = 0 \qquad \text{(C.7b)}$$

which gives $m = \sqrt{3}r$. Taking integral values of $m = 1, 2, 3, \ldots$ within the length a, the corresponding values of the ratio $r = 1/\sqrt{3}, 2/\sqrt{3}, 3/\sqrt{3}, \ldots$ apply to these minima. From equation (C.7a), each $(m,\ r)$ combination gives the common buckling stress expression:

$$(\sigma_1)_{cr} = \frac{8\pi^2 D}{tb^2} = \frac{2\pi^2 E}{3(1-\nu^2)}\left(\frac{t}{b}\right)^2 \qquad \text{(C.8a)}$$

Given that $\tau_{cr} = (\sigma_1)_{cr}$, the critical elastic shear stress for any (non minimal) r-value takes a similar form to equation (C.8a):

$$\tau_{cr} = K_e E\left(\frac{t}{b}\right)^2 \qquad \text{(C.8b)}$$

where b is the shorter side and K_e depends upon the manner in which the sides are supported. Figure C.3 provides the elastic shear buckling coefficient K_e for various edge fixings when $\nu = 0.3$ [2]. If $\nu \neq 0.3$ the K_e value should be multiplied by $(1 - 0.3^2)/(1 - \nu^2)$.

Pure shear also arises when thin strips and plates are subjected to axial torsion. Thus, for the thin strip shown in Figure 4.6, the length L replaces a so that we may find the elastic buckling coefficient from Figure 4.7. This is identical to Figure C.3 in all but the labelling of the abscissa, where b/L has replaced the plate aspect ratio $b/a \leq 1$. Figures C.3 and 4.7 are used extensively to provide optimum designs arising from the coincidence of critical shear stress levels within appropriate failure criteria including local plate buckling.

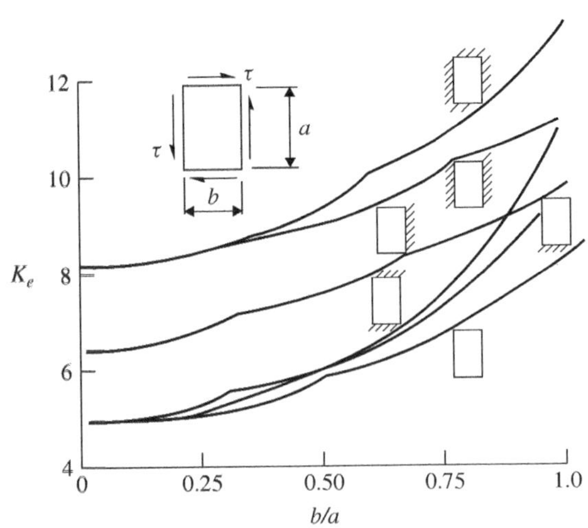

Figure C.3 Dependence of shear buckling coefficient upon plate geometry and edge-fixing ($\nu = 0.3$)

C.3 Inelastic Shear Buckling

Equation (C.8b) supplies a critical *elastic* buckling stress $\tau_{cr} = (\tau_{cr})_e$, but it will need to be corrected if this stress is found to exceed the shear yield stress τ_y of the plate material [3]. The correction is similar to that described in Appendix B for the effect of plasticity upon the compressive buckling stress. Again, the Ramberg–Osgood empirical description [4, 5] of the material's uniaxial flow curve is used (see equation (B.10)) to derive a plasticity reduction factor μ for shear. This factor [2, 3] is provided graphically in Figure 4.8 for the calculation of a critical *plastic* buckling stress for plates either in shear or in torsion:

$$(\tau_{cr})_p = \mu(\tau_{cr})_e \tag{C.9}$$

As an example of the application of equation (C.9) consider a 4.5 mm thick DTD 687 aluminium alloy plate with $b/a = 200$ mm/267 mm $= 0.75$. Figure C.3 gives the elastic shear buckling coefficient $K_e = 10$ for a plate, say, with both its longer sides clamped and its shorter sides simply supported. Given $E = 73$ GPa for this material, the critical, elastic buckling stress follows from equation (C.8b) as:

$$\tau_{cr} = 10 \times 73 \times 10^3 (4.5/200)^2 = 369.6 \text{ MPa}$$

From Table B.2, $m = 15$ applies to the plate material. Taking $\sigma_n = 340$ MPa, the abscissa in Figure 4.8 becomes

$$\frac{(\tau_{cr})_e}{\sigma_n} = \frac{\tau_e}{\sigma_n} = \frac{369.6}{340} = 1.087$$

Correspondingly, Figure 4.8 gives $\mu = 0.56$, and from equation (C.9) the plastic buckling stress is

$$(\tau_{cr})_p = 0.56 \times 369.6 = 207 \text{ MPa}$$

Note that Figure 4.8 shows $\mu = 1$ when $\tau_e/\sigma_n \leq 0.4$ and, therefore, no correction is required.

References

[1] Vinson, J. R. *Structural Mechanics*, Wiley, 1974.
[2] ESDU 71005, Buckling of flat plates in shear, February 1971.
[3] Gerard, G. Critical shear stress of plates above the proportional limit, *Journal of Applied Mechanics*, **15**(1), 7–12, March 1948.
[4] Ramberg, W. and Osgood, W. R. Description of stress-strain curves by three parameters, NACA Tech Note, 902, April 1943.
[5] ESDU 76016, Generalised equations for stress-strain curves, August 1976.

Appendix D

Secondary Buckling

D.1 Buckling Modes

Primary buckling refers to the lateral shift in a slender strut's transverse sections away from its longitudinal axis [1]. A similar shift can occur in a wide strut, a shear web, a beam or a torsion bar. All are susceptible to the primary buckling mode where flexure promotes a combined translation and rotation of every cross-section to a varying degree, depending upon its position in the length, as in the buckling of a strut or the bending of a beam. The primary compressive buckling of sheet between rivets along the rivet line is referred to as inter-rivet buckling [2, 3]. Where this occurs with a wavelength that differs from the rivet pitch it has been influenced by the nearness of webs and flanges. When the wavelength is less than the rivet pitch the undulations resemble the wrinkling that occurs in thin unsupported sheet under compression.

Secondary buckling refers to any other mode of buckling that is not primarily flexural. The straight, thin walls of an open section may distort with translation and rotation in its plane. Localised stress concentrations at corners may exceed the yield stress and cause either a local crippling failure or a reduction in the resistance to buckling by other modes. In particular, local buckling is confined to a section's individual 'flats' that form the web, flange and stiffener. Thus, the local buckling mode considered here refers to the accompanying distortion within the cross-section itself as the flats bow and rotate under the action of the compression and shear they are expected to support [1]. Both open and closed, thin-walled, slender strut sections are prone to local compressive buckling. For example, local buckling modes are shown within the walls of a box section in Figure 1.6b and within both flanges and the web of an I-section in Figure 1.8b. A similar bucking mode can occur within the thin walls of wide strut sections. Examples are shown for Z- and integrally stiffened stringers in Figures 2.15 and 2.18, respectively. Where the stiffener is rivetted or spot-welded to a wide plate strut or shear web the local mode prevails when the individual flats, lying between the attachment points, are placed either under compression or in shear.

In these examples the junctions between plate elements remain straight while the centres or the free edges bow out. Local buckling refers to a half-wavelength of bowing that is generally of the same order as the flat width. The local buckling stress is independent of

Mechanics of Optimal Structural Design: Minimum Weight Structures D. W. A. Rees

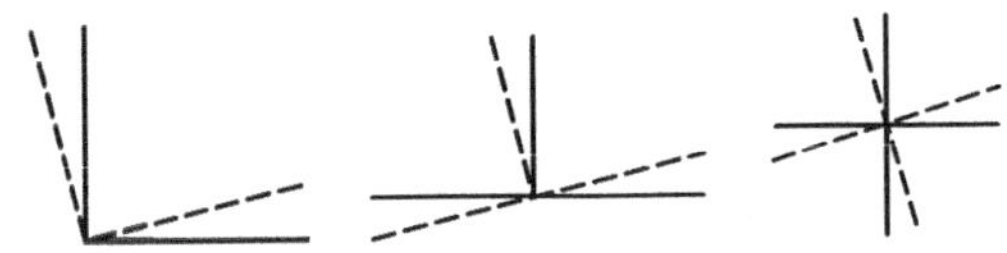

Figure D.1 Local buckling in sections with equal flats

axial length when this is at least three times the width of the largest flat. Stresses at the junctions grow more rapidly that the average section stress and can become sufficiently high to cause a local collapse. In the analysis of a local buckling stress due consideration must be given to the manner of the flat support. At the junction between two flats a rotational restraint is exerted by one flat upon the other. Thus a flange, for example, has one edge free and the other supported. The plates within a box section as in Figure 1.6b have both edges supported with a degree of restraint lying between a simple support and the fully clamped condition. Sections with equal-width flats (see Figure D.1) may be taken as having their edges simply supported along the join since neither flat will exert a restraint upon the other. Clearly, in cross-sections having equal flats we should expect all flats to buckle simultaneously so that the critical stress for the section becomes that for a single flat. A similar, coincident buckling behaviour can be arranged in sections with unequal flats by adjusting their thicknesses. It follows that failure criteria, used in analyses of minimum weight in thin-walled sections, must apply to the geometry of the section's individual flats.

D.2 Local Compressive Buckling

D.2.1 Equal Flats

Here the uniaxial plate buckling formula (B.7b) is applied to each limb where K_e provides for the appropriate restraint exerted upon it by neighbouring limbs. We may apply Figure B.4 to each limb when it is taken to be a plate under uniform compression with specified edge supports. A similar approach is taken to find the buckling coefficient for the horizontal plates in thin-walled sections placed under bending. Note here that vertical plates under bending experience a stress gradient, from tension to compression, which renders them with a greater resistance to local buckling. More usually, Figure B.4 is used to find the critical, elastic compressive stress for local buckling to occur within each flat of a thin-walled section under uniform compression. This figure supplies the scaling factor η, for the plate's length/width ratio (a/b) given its edge-support conditions. The critical compressive, buckling stress is

$$(\sigma_{cr})_e = \eta(\sigma_{cr}) = \frac{\eta\pi^2 E}{3(1-\nu^2)}\left(\frac{t}{b}\right)^2 = K_e E\left(\frac{t}{b}\right)^2 \tag{D.1a}$$

in which equation (B.6b) provides the reference stress σ_{cr} for a simply supported plate with identical loading and geometry a/b and t. The elastic restraint (buckling) coefficient for other edge supports follows from equation (D.1a) and Figure B.4 as

$$K_e = \frac{\eta \pi^2}{3(1 - \nu^2)} \tag{D.1b}$$

The buckling stress, calculated for the actual support condition, may, in turn, need scaling down should plasticity arise, i.e. where $(\sigma_{cr})_e$ exceeds the stress at the material's proportional limit. Figure B.6 supplies the 'plasticity reduction factor' μ, based upon a Ramberg–Osgood hardening law (see § B.3). The reduction factor is applied to equation (D.1a) as follows:

$$(\sigma_{cr})_p = \mu(\sigma_{cr})_e = \mu \eta \sigma_{cr} \tag{D.2a}$$

Substituting equation (D.1a) into equation (D.2a),

$$(\sigma_{cr})_p = \frac{\mu \eta \pi^2 E}{3(1 - \nu^2)} \left(\frac{t}{b}\right)^2 = \mu K_e E \left(\frac{t}{b}\right)^2 \tag{D.2b}$$

An example of the application of equation (D.2b) is given in Appendix B.

D.2.2 Unequal Flats

Open sections, consisting of straight, unequal sides (such as I, T, Z, L and U), as well as thin-walled closed tubes, are prone to local buckling at their junctions. When the strut length is at least four times the section depth h, each limb (flat) may be treated as a plate with simple side support(s) from its neighbouring limb(s). An approximate buckling load for the whole section can be obtained by adding the buckling loads for individual flats, assumed to be simply supported at their junctions. A more exact analysis accounts for the degree to which a supported side restrains an unsupported side. In providing a local elastic compressive buckling stress, equation (D.2b) takes the common form

$$(\sigma_{cr})_e = K_e E \left(\frac{t}{h}\right)^2 \tag{D.3}$$

Referring to Figure D.2, it is seen how the buckling coefficient K_e depends upon the d/h ratio and the neighbouring restraints for the four sections shown [4, 5]. The elastic coefficient K_e applies to wall thickness $t < d/5$ and Poisson's ratio $\nu = 0.3$. When $\nu \neq 0.3$, K_e should be corrected by a multiplying factor $(1 - 0.3^2)/(1 - \nu^2)$. If $(\sigma_{cr})_e$, from equation (D.3), is found to exceed the yield stress of the strut material it becomes necessary to employ the plasticity reduction factor μ in equation (D.2a), as described above.

D.3 Global Buckling

In medium length struts, local instability results in a loss of stiffness but without complete failure. However, the influence of local instability upon global buckling, arising from

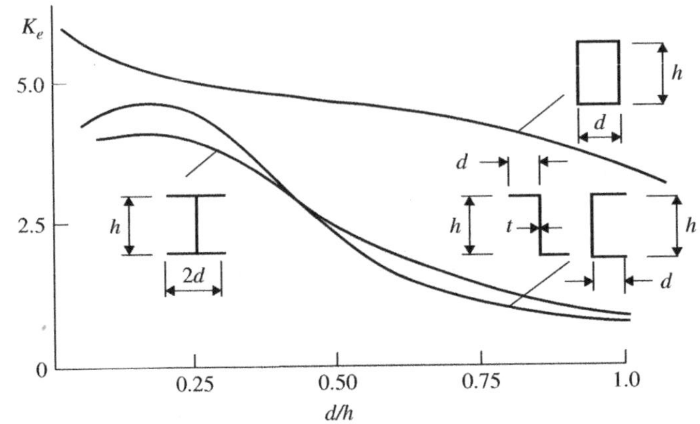

Figure D.2 Local buckling coefficient for uniform thin sections

flexure and torsion, is important. If the strut is very long the global buckling stresses are attained well before the onset of local buckling. On the other hand, a short strut may carry more compression than that which produces local buckling. Final failure is estimated to occur under a crippling stress σ_c, found from [4]:

$$\sigma_c = (\sigma_{cr}\sigma_{0.1})^{1/2} \tag{D.4}$$

where σ_{cr} is taken to be the local buckling stress, either elastic or plastic, estimated from equations (D.1)–(D.3) and $\sigma_{0.1}$ refers to the 0.1% compressive proof stress for the strut material.

Test data shows that equation (D.4) is accurate to within 10%. For example, let us find the local buckling and crippling stresses for a short aluminium strut with the Z-section, shown in Figure D.2. Take dimensions $h = 125$ mm, $d = 25$ mm and $t = 4$ mm, for an aluminium section with $E = 74$ GPa, $v = 0.3$ and $m = 10$. The 0.1% proof stress, $\sigma_{0.1} = 280$ MPa and $\sigma_n = 325$ MPa corresponds to a tangent modulus $E_T = E/2$ (Figure B.5b). From Figure D.2, we find $K_e = 4$ for $d/h = 0.2$. Hence, equation (D.3) gives the local elastic compressive buckling stress:

$$(\sigma_{cr})_e = 4 \times 74 \times 10^3 (4/125)^2 = 303.1 \text{ MPa}$$

This exceeds the proof stress $\sigma_{0.1} = 280$ MPa and so the ratio $(\sigma_{cr})_e/\sigma_n = 303.1/325 = 0.933$ is used with Figure B.6 to give a plasticity reduction factor $\mu = 0.877$. Equation (D.2) provides the corrected, compressive, plastic buckling stress as

$$(\sigma_{cr})_p = \mu(\sigma_{cr})_e = 0.877 \times 303.1 = 265.82 \text{ MPa}$$

and, using equation (D.4), a crippling stress is estimated as

$$\sigma_c = (265.82 \times 280)^{1/2} = 272.8 \text{ MPa}$$

D.4 Local Shear Buckling

A local buckling is also possible in the plates of stiffened and corrugated sections when these act as shear webs (see Chapter 8). Here the plate, shear buckling formula (see Appendix C) is applied to the ratio between each individual flat length and the web depth. Given that the plate's sides are simply supported along the folds of the corrugation, we may read from Figure C.3 a local buckling coefficient for each flat when it is taken to be a plate under uniform shear. Figure C.3 also supplies the critical shear stress for local buckling of flat plate elements within a section (i.e. a flange, web or thin wall) when placed under shear or torsion. Here the particular edge condition governs the critical shear stress at buckling,

$$\tau_{cr} = K_e E \left(\frac{t}{b}\right)^2 \tag{D.5}$$

where the local, elastic, shear buckling coefficient, K_e is read directly from the figure for the flat's width/length ratio (b/a). When the widths b of adjoining flats are different, equation (D.5) shows that the flat with greater b buckles first. Thus, the critical stress for the whole section is based upon the weaker flat but the stress does depend upon the restraint exerted by the other flat(s). For example, where $a \gg b$, giving $b/a \approx 0$, Figure C.3 shows $K_e \approx 4.84$ for a plate with its four sides simply supported. Further design data are given in Appendix C to account for plasticity effects where τ_{cr}, from equation (D.5), is found to exceed the proportional limit.

References

[1] ESDU 78021, Guide to items on the strength and stability of struts, October 1978
[2] ESDU 02.01.08, Buckling in compression of sheet between rivets, October 1962.
[3] ESDU 02.01.09, Inter-rivet buckling curves for specific materials, October 1962.
[4] ESDU 78020, Local buckling and crippling of I, Z and channel section struts, July 1978.
[5] ESDU 01 01 09, Local buckling and crippling of rectangular tube section struts, November 1978.

Bibliography

A limited number of existing books [1–9] share certain topics in common with the present book. Cox [1], Farkas [2], Zenman and Irvine [3] and Shanley [4] are the closest, overlapping with material given here in earlier chapters despite their differing analyses. Other authors [5–9] rely upon numerical procedures, whereas here the tendency has been to adopt alternative analytical techniques based upon classical mechanics. Cox [1] offers the closest approach to that adopted for beams and struts in Chapters 1–3 and for webs and frames in Chapters 8 and 9. Other, more dedicated texts [10–13] overlap with the plasticity and framework analyses given in Chapters 14 and 15. Farkas [2], Zenman and Irvine [3] and Kirsch [8] also consider plastic collapse of structures as presented in these chapters, but in greater detail. Most books are rather specialised and are not pitched, as with the present work, at the introductory level. None of the remaining least weight topics given here appears in other books. It would be true to say that none of these earlier structural optimisation texts matches the full content of the present work, where many original minimum weight analyses can be found. Structural mechanics texts [14–20] published more recently cover various topics such as bending, torsion, buckling and fracture in a classical manner with no recourse to weight minimisation. The techniques adopted for their analyses fall within the areas of elasticity, plasticity and introductory finite elements. The final text [21] provides a state-of-the-art collection of papers on the subject of optimum structures. The breadth of the subject of 'optimum structures' becomes clear from this review, covering shape, topology, acoustics, robustness, life cycle and cost. However, a least weight to strength optimisation does not appear among them.

[1] Cox, H. L. *The Design of Structures of Least Weight*, Pergamon, 1965.
[2] Farkas, J. *Optimum Design of Metal Structures*, Ellis-Horwood, 1984.
[3] Zenman, P. and Irvine, H. M. *Plastic Design – an Imposed Hinge Rotation Approach*, Allen & Unwin, 1986.
[4] Shanley, P. *Weight-Strength Analysis of Aircraft Structures*, McGraw-Hill, 1952.
[5] Spunt, L. *Optimum Structural Design*, Prentice Hall, 1971.
[6] Hemp, W. S. *Optimum Structures*, Clarendon Press, 1973.
[7] Majid, K. I. (ed.). *Optimum Design of Structures*, Newnes-Butterworth, 1974.
[8] Kirsch, U. *Optimum Structural Design*, McGraw-Hill, 1981.

Mechanics of Optimal Structural Design: Minimum Weight Structures D. W. A. Rees

[9] Gallagher, R. H. and Zienkiewicz, O. (eds). *Optimum Structural Design*, Wiley, 1973.
[10] Baker, J. and Heyman, J. *Plastic Design of Frames*, Cambridge University Press, 1975.
[11] Neal, B. G. *Plastic Methods of Structural Analysis*, Chapman & Hall, 1965.
[12] Hodge, P. G. Plastic Analysis of Structures, McGraw-Hill, 1959.
[13] Save, M and Prager, W. *Structural Optimisation*, Plenum Press, 1985.
[14] West, H. H. *Analysis of Structures*, Wiley, 1989.
[15] Carpinteri, A. *Structural Mechanics – A Unified Approach*, E & FN Spon, 1997.
[16] Williams, M. S. and Todd, J. D. *Structures, Theory and Analysis*, Macmillan, 2000.
[17] Al Nagem, H. *Structural Mechanics*, Prentice Hall, 2003.
[18] Jennings, A. *Structures from Theory to Practice*, Taylor and Francis, 2004.
[19] Chanakya, A. *Design of Structural Elements*, 2004
[20] MacLeod, I. A. *Modern Structural Analyses*, Thomas Telford, 2005.
[21] Arora, J. S. (Ed.) *Optimisation of Structural and Mechanical Systems*, World Scientific, 2007.

The following references, most of which are cited in consideration of buckling behaviour throughout the text, are compiled here for convenience. They fall within the following four themes for which the many ESDU (Engineering Science Data Unit) items can now be found on-line.

I Buckling of Plates

ESDU, Items on buckling of flat rectangular plates and panels in shear: 02.03.02 (1983), 71005 (1976), 74022 (1976), 77014 (1977).
ESDU, Items on buckling of curved plates and panels in shear: 02.03.18 (1977), 02.03.19 (1979), 77018 (1977).
ESDU 02.03.24, Strength of castellations in shear, 1949.
ESDU 01.01.08, Local instability of struts with flat sides, 1990.
ESDU 72019, Buckling of flat isotropic plates under uniaxial and biaxial loading, 1972.
ESDU, Information on the use of data sheets of the 00.07 series, April 1947.
ESDU 02.01.08, Buckling in compression of sheet between rivets, October 1962.
ESDU 02.01.09, Inter-rivet buckling curves for specific materials, October 1962.
Roark, R. J. and Young, A. C. *Formulas for Stress and Strain*, McGraw-Hill, 1975.
Timoshenko, S. J. and Gere, J. M. *Theory of Elastic Stability*, McGraw-Hill, 1961.

II Buckling of Open Sections

Catchpole, E. J. The optimum design of compression surfaces having unflanged integral stiffeners, *Journal of the Royal Aeronautical Society*, November 1954.
ESDU 76023, Buckling of struts, lipped and unlipped channel sections, 1976.
ESDU 77030, Buckling of struts, lipped and unlipped Z-sections, 1977.
ESDU 78020, Local buckling and crippling of I, Z and channel section struts, 1978.
ESDU 78021, Guide to items on the strength and stability of struts, 1990.
ESDU 89007, Flexural and torsional buckling of thin-walled open-section struts, 1989.
ESDU 01.01.10, Torsional instability of stringers and struts of angle section, 1978.

ESDU 77023, Shear centre and primary warping constant for lipped and unlipped channel and Z-sections, Sept 1977.

Farrar, D. J. The design of compression structures for minimum weight. *Journal of the Royal Aeronautical Society*, 53 (1949), 1041–1052.

Goodier, J. N. Torsional and flexural buckling of bars of thin-walled open section under compressive and bending loads, *Journal of Applied Mechanics*, 9, 1942, A103–A107.

Peterson, J. P. and Card, M. F. Buckling strength of corrugated webs in shear, NACA Tech. Note D-424, June 1960.

III Buckling of Tubes

Batdorf, S. B. A simplified method of elastic stability analysis for thin cylindrical shells, I – Donnell's equations, NACA Note No 1341, 1947.

Batdorf, S. B., Stein, M. and Schildercrout, M. Critical stress of thin-walled cylinders in torsion, NACA Tech Note 1345, 1947.

BS 4848: Part 2: 1975, Hollow sections (circular and square).

Donnell, L. H. Stability of thin-walled tubes under torsion, NACA Rpt 479, 1933.

ESDU 01.01.19, Local buckling and crippling of rectangular tube section struts, 1976.

Gerard G. Compressive and torsional buckling of thin-walled cylinders in yield region, NACA Tech Note, 3726, August 1956.

Gerard, G. *Introduction to Structural Stability Theory*, McGraw-Hill, 1962.

Gerard, G. and Becker, H. Handbook of structural stability, Part III: Buckling of curved plates and shells, NACA Tech Note 3783, 1957.

IV Plastic Buckling

BCSA 28, 1965, Plastic design, British Construction and Steelwork Association.

ESDU 76016, Generalisation of smooth continuous stress-strain curves for metallic materials, 1976.

ESDU 83044, Plasticity correction factors for plate buckling, December 1983.

Gerard, G. Critical shear stress of plates above the proportional limit, *Journal of Applied Mechanics*, 15(1), March 1948, 7–12.

Ramberg, W. and Osgood, W. R. Description of stress-strain curves by three parameters, NACA Tech Note 902, April 1943.

Rees, D. W. A. *Mechanics of Solids and Structures*, Imperial College Press, 2000.

Rees, D. W. A. *Basic Engineering Plasticity*, Elsevier, 2006.

Index

Printed and bound by CPI Group (UK) Ltd, Croydon, CR0 4YY
18/09/2026
14964993-0001